This is a volume in
COMPUTER SCIENCE AND APPLIED MATHEMATICS
A Series of Monographs and Textbooks

This series has been renamed
COMPUTER SCIENCE AND SCIENTIFIC COMPUTING

Editor: WERNER RHEINBOLT

A complete list of titles in this series is available from the publisher upon request.

The Theory of Matrices
Second Edition

with Applications

Peter Lancaster
Department of Mathematics
University of Calgary
Calgary, Alberta, Canada

Miron Tismenetsky
IBM Scientific Center
Technion City
Haifa, Israel

ACADEMIC PRESS, INC.
Boston San Diego New York
London Sydney Tokyo Toronto

This book is printed on acid-free paper. ♾

ACADEMIC PRESS, INC.
A Division of Harcourt Brace & Company
525 B Street, Suite 1900, San Diego, CA 92101-4495

United Kingdom Edition published by
ACADEMIC PRESS LIMITED
24-28 Oval Road, London NW1 7DX

Library of Congress Cataloging in Publication Data

Lancaster, Peter. Date.
The theory of matrices.
(Computer science and applied mathematics)
Includes index.
I. Matrices. I. Tismenetsky, M. (Miron) II. Title.
III. Series.
QA188.L36 1984 512.9'434 83-15775
ISBN 0-12-435560-9 (alk. paper).

Printed in the United States of America
94 95 96 QW 9 8 7 6 5

To our wives,
Edna and Fania

Contents

Preface xiii

1 Matrix Algebra

1.1 Special Types of Matrices 2
1.2 The Operations of Addition and Scalar Multiplication 4
1.3 Matrix Multiplication 7
1.4 Special Kinds of Matrices Related to Multiplication 10
1.5 Transpose and Conjugate Transpose 13
1.6 Submatrices and Partitions of a Matrix 16
1.7 Polynomials in a Matrix 19
1.8 Miscellaneous Exercises 21

2 Determinants, Inverse Matrices, and Rank

2.1 Definition of the Determinant 23
2.2 Properties of Determinants 27
2.3 Cofactor Expansions 32
2.4 Laplace's Theorem 36
2.5 The Binet–Cauchy Formula 39
2.6 Adjoint and Inverse Matrices 42
2.7 Elementary Operations on Matrices 47
2.8 Rank of a Matrix 53
2.9 Systems of Linear Equations and Matrices 56
2.10 The *LU* Decomposition 61
2.11 Miscellaneous Exercises 63

3 Linear, Euclidean, and Unitary Spaces

3.1 Definition of a Linear Space 71
3.2 Subspaces 75

3.3 Linear Combinations 78
3.4 Linear Dependence and Independence 80
3.5 The Notion of a Basis 83
3.6 Sum and Direct Sum of Subspaces 87
3.7 Matrix Representation and Rank 91
3.8 Some Properties of Matrices Related to Rank 95
3.9 Change of Basis and Transition Matrices 98
3.10 Solution of Equations 100
3.11 Unitary and Euclidean Spaces 104
3.12 Orthogonal Systems 107
3.13 Orthogonal Subspaces 111
3.14 Miscellaneous Exercises 113

4 Linear Transformations and Matrices

4.1 Linear Transformations 117
4.2 Matrix Representation of Linear Transformations 122
4.3 Matrix Representations, Equivalence, and Similarity 127
4.4 Some Properties of Similar Matrices 131
4.5 Image and Kernel of a Linear Transformation 133
4.6 Invertible Transformations 138
4.7 Restrictions, Invariant Subspaces, and Direct Sums of Transformations 142
4.8 Direct Sums and Matrices 145
4.9 Eigenvalues and Eigenvectors of a Transformation 147
4.10 Eigenvalues and Eigenvectors of a Matrix 152
4.11 The Characteristic Polynomial 155
4.12 The Multiplicities of an Eigenvalue 159
4.13 First Applications to Differential Equations 161
4.14 Miscellaneous Exercises 164

5 Linear Transformations in Unitary Spaces and Simple Matrices

5.1 Adjoint Transformations 168
5.2 Normal Transformations and Matrices 174
5.3 Hermitian, Skew-Hermitian, and Definite Matrices 178
5.4 Square Root of a Definite Matrix and Singular Values 180
5.5 Congruence and the Inertia of a Matrix 184
5.6 Unitary Matrices 188
5.7 Polar and Singular-Value Decompositions 190
5.8 Idempotent Matrices (Projectors) 194
5.9 Matrices over the Field of Real Numbers 200
5.10 Bilinear, Quadratic, and Hermitian Forms 202
5.11 Finding the Canonical Forms 205
5.12 The Theory of Small Oscillations 208
5.13 Admissible Pairs of Matrices 212
5.14 Miscellaneous Exercises 217

6 The Jordan Canonical Form: A Geometric Approach

6.1 Annihilating Polynomials 221
6.2 Minimal Polynomials 224
6.3 Generalized Eigenspaces 229
6.4 The Structure of Generalized Eigenspaces 232
6.5 The Jordan Theorem 236
6.6 Parameters of a Jordan Matrix 239
6.7 The Real Jordan Form 242
6.8 Miscellaneous Exercises 244

7 Matrix Polynomials and Normal Forms

7.1 The Notion of a Matrix Polynomial 246
7.2 Division of Matrix Polynomials 248
7.3 Elementary Operations and Equivalence 253
7.4 A Canonical Form for a Matrix Polynomial 256
7.5 Invariant Polynomials and the Smith Canonical Form 259
7.6 Similarity and the First Normal Form 262
7.7 Elementary Divisors 265
7.8 The Second Normal Form and the Jordan Normal Form 269
7.9 The Characteristic and Minimal Polynomials 271
7.10 The Smith Form: Differential and Difference Equations 274
7.11 Miscellaneous Exercises 278

8 The Variational Method

8.1 Field of Values. Extremal Eigenvalues of a Hermitian Matrix 283
8.2 Courant–Fischer Theory and the Rayleigh Quotient 286
8.3 The Stationary Property of the Rayleigh Quotient 289
8.4 Problems with Constraints 290
8.5 The Rayleigh Theorem and Definite Matrices 294
8.6 The Jacobi–Gundelfinger–Frobenius Method 296
8.7 An Application of the Courant–Fischer Theory 300
8.8 Applications to the Theory of Small Vibrations 302

9 Functions of Matrices

9.1 Functions Defined on the Spectrum of a Matrix 305
9.2 Interpolatory Polynomials 306
9.3 Definition of a Function of a Matrix 308
9.4 Properties of Functions of Matrices 310
9.5 Spectral Resolution of $f(A)$ 314
9.6 Component Matrices and Invariant Subspaces 320
9.7 Further Properties of Functions of Matrices 322

9.8 Sequences and Series of Matrices 325
9.9 The Resolvent and the Cauchy Theorem for Matrices 329
9.10 Applications to Differential Equations 334
9.11 Observable and Controllable Systems 340
9.12 Miscellaneous Exercises 345

10 Norms and Bounds for Eigenvalues

10.1 The Notion of a Norm 350
10.2 A Vector Norm as a Metric: Convergence 354
10.3 Matrix Norms 358
10.4 Induced Matrix Norms 362
10.5 Absolute Vector Norms and Lower Bounds of a Matrix 367
10.6 The Geršgorin Theorem 371
10.7 Geršgorin Disks and Irreducible Matrices 374
10.8 The Schur Theorem 377
10.9 Miscellaneous Exercises 380

11 Perturbation Theory

11.1 Perturbations in the Solution of Linear Equations 383
11.2 Perturbations of the Eigenvalues of a Simple Matrix 387
11.3 Analytic Perturbations 391
11.4 Perturbation of the Component Matrices 393
11.5 Perturbation of an Unrepeated Eigenvalue 395
11.6 Evaluation of the Perturbation Coefficients 397
11.7 Perturbation of a Multiple Eigenvalue 399

12 Linear Matrix Equations and Generalized Inverses

12.1 The Notion of a Kronecker Product 406
12.2 Eigenvalues of Kronecker Products and Composite Matrices 411
12.3 Applications of the Kronecker Product to Matrix Equations 413
12.4 Commuting Matrices 416
12.5 Solutions of $AX + XB = C$ 421
12.6 One-Sided Inverses 424
12.7 Generalized Inverses 428
12.8 The Moore–Penrose Inverse 432
12.9 The Best Approximate Solution of the Equation $A\mathbf{x} = \mathbf{b}$ 435
12.10 Miscellaneous Exercises 438

13 Stability Problems

13.1 The Lyapunov Stability Theory and Its Extensions 441
13.2 Stability with Respect to the Unit Circle 451

13.3 The Bezoutian and the Resultant 454
13.4 The Hermite and the Routh–Hurwitz Theorems 461
13.5 The Schur–Cohn Theorem 466
13.6 Perturbations of a Real Polynomial 468
13.7 The Liénard–Chipart Criterion 470
13.8 The Markov Criterion 474
13.9 A Determinantal Version of the Routh–Hurwitz Theorem 478
13.10 The Cauchy Index and Its Applications 482

14 Matrix Polynomials

14.1 Linearization of a Matrix Polynomial 490
14.2 Standard Triples and Pairs 493
14.3 The Structure of Jordan Triples 500
14.4 Applications to Differential Equations 506
14.5 General Solutions of Differential Equations 509
14.6 Difference Equations 512
14.7 A Representation Theorem 516
14.8 Multiples and Divisors 518
14.9 Solvents of Monic Matrix Polynomials 520

15 Nonnegative Matrices

15.1 Irreducible Matrices 528
15.2 Nonnegative Matrices and Nonnegative Inverses 530
15.3 The Perron–Frobenius Theorem (I) 532
15.4 The Perron–Frobenius Theorem (II) 538
15.5 Reducible Matrices 543
15.6 Primitive and Imprimitive Matrices 544
15.7 Stochastic Matrices 547
15.8 Markov Chains 550

Appendix 1: A Survey of Scalar Polynomials 553
Appendix 2: Some Theorems and Notions from Analysis 557
Appendix 3: Suggestions for Further Reading 560

Index 563

Preface

In this book the authors try to bridge the gap between the treatments of matrix theory and linear algebra to be found in current textbooks and the mastery of these topics required to use and apply our subject matter in several important areas of application, as well as in mathematics itself. At the same time we present a treatment that is as self-contained as is reasonably possible, beginning with the most fundamental ideas and definitions. In order to accomplish this double purpose, the first few chapters include a complete treatment of material to be found in standard courses on matrices and linear algebra. This part includes development of a computational algebraic development (in the spirit of the first edition) and also development of the abstract methods of finite-dimensional linear spaces. Indeed, a balance is maintained through the book between the two powerful techniques of matrix algebra and the theory of linear spaces and transformations.

The later chapters of the book are devoted to the development of material that is widely useful in a great variety of applications. Much of this has become a part of the language and methodology commonly used in modern science and engineering. This material includes variational methods, perturbation theory, generalized inverses, stability theory, and so on, and has innumerable applications in engineering, physics, economics, and statistics, to mention a few.

Beginning in Chapter 4 a few areas of application are developed in some detail. First and foremost we refer to the solution of constant-coefficient systems of differential and difference equations. There are also careful developments of the first steps in the theory of vibrating systems, Markov processes, and systems theory, for example.

The book will be useful for readers in two broad categories. One consists of those interested in a thorough reference work on matrices and linear algebra for use in their scientific work, whether in diverse applications or in mathematics itself. The other category consists of undergraduate or graduate students in a variety of possible programs where this subject matter is required. For example, foundations for courses offered in mathematics, computer science, or engineering programs may be found here. We address the latter audience in more detail.

The first seven chapters are essentially self-contained and require no formal prerequisites beyond college algebra. However, experience suggests that this material is most appropriately used as a second course in matrices or linear algebra at the sophomore or a more senior level.

There are possibilities for several different courses depending on specific needs and specializations. In general, it would not be necessary to work systematically through the first two chapters. They serve to establish notation, terminology, and elementary results, as well as some deeper results concerning determinants, which can be developed or quoted when required. Indeed, the first two chapters are written almost as compendia of primitive definitions, results, and exercises. Material for a traditional course in linear algebra, but with more emphasis on matrices, is then contained in Chapters 3–6, with the possibility of replacing Chapter 6 by Chapter 7 for a more algebraic development of the Jordan normal form including the theory of elementary divisors.

More advanced courses can be based on selected material from subsequent chapters. The logical connections between these chapters are indicated below to assist in the process of course design. It is assumed that in order to absorb any of these chapters the reader has a reasonable grasp of the first seven, as well as some knowledge of calculus. In this sketch the stronger connections are denoted by heavier lines.

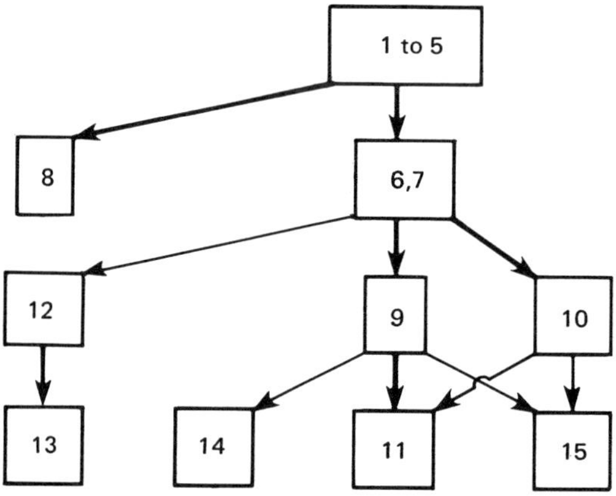

Prerequisite structure by chapters.

There are many exercises and examples throughout the book. These range from computational exercises to assist the reader in fixing ideas, to extensions of the theory not developed in the text. In some cases complete solutions are given, and in others hints for solution are provided. These are seen as an integral part of the book and the serious reader is expected to absorb the information in them as well as that in the text.

In comparison with the 1969 edition of "The Theory of Matrices" by the first author, this volume is more comprehensive. First, the treatment of material in the first seven chapters (four chapters in the 1969 edition) is completely rewritten and includes a more thorough development of the theory of linear spaces and transformations, as well as the theory of determinants.

Chapters 8–11 and 15 (on variational methods, functions of matrices, norms, perturbation theory, and nonnegative matrices) retain the character and form of chapters of the first edition, with improvements in exposition and some additional material. Chapters 12–14 are essentially extra material and include some quite recent ideas and developments in the theory of matrices. A treatment of linear equations in matrices and generalized inverses that is sufficiently detailed for most applications is the subject of Chapter 12. It includes a complete description of commuting matrices. Chapter 13 is a thorough treatment of stability questions for matrices and scalar polynomials. The classical polynomial criteria of the nineteenth century are developed in a systematic and self-contained way from the more recent inertia theory of matrices. Chapter 14 contains an introduction to the recently developed spectral theory of matrix polynomials in sufficient depth for many applications, as well as providing access to the more general theory of matrix polynomials.

The greater part of this book was written while the second author was a Research Fellow in the Department of Mathematics and Statistics at the University of Calgary. Both authors are pleased to acknowledge support during this period from the University of Calgary. Many useful comments on the first edition are embodied in the second, and we are grateful to many colleagues and readers for providing them. Much of our work has been influenced by the enthusiasms of co-workers I. Gohberg, L. Rodman, and L. Lerer, and it is a pleasure to acknowledge our continued indebtedness to them. We would like to thank H. K. Wimmer for several constructive suggestions on an early draft of the second edition, as well as other colleagues, too numerous to mention by name, who made helpful comments.

The secretarial staff of the Department of Mathematics and Statistics at the University of Calgary has been consistently helpful and skillful in preparing the typescript for this second edition. However, Pat Dalgetty bore the brunt of this work, and we are especially grateful to her. During the period of production we have also benefitted from the skills and patience demonstrated by the staff of Academic Press. It has been a pleasure to work with them in this enterprise.

P. Lancaster M. Tismenetsky
Calgary Haifa

CHAPTER 1

Matrix Algebra

An ordered array of mn elements a_{ij} $(i = 1, 2, \ldots, m;\ j = 1, 2, \ldots, n)$ written in the form

$$A = \begin{bmatrix} a_{11} & a_{12} & \cdots & a_{1n} \\ a_{21} & a_{22} & \cdots & a_{2n} \\ \vdots & \vdots & & \vdots \\ a_{m1} & a_{m2} & \cdots & a_{mn} \end{bmatrix}$$

is said to be a *rectangular* $m \times n$ *matrix*. These elements can be taken from an arbitrary field $\mathscr{F}$. However, for the purposes of this book, $\mathscr{F}$ will always be the set of all real or all complex numbers, denoted by $\mathbb{R}$ and $\mathbb{C}$, respectively.

A matrix A may be written more briefly in terms of its elements as

$$A = [a_{ij}]_{i,j=1}^{m,n}, \qquad \text{or} \qquad A = [a_{ij}],$$

where a_{ij} $(1 \le i \le m,\ 1 \le j \le n)$ denotes the element of the matrix lying on the intersection of the ith row and the jth column of A.

Two matrices having the same number of rows (m) and columns (n) are matrices of the *same size*. Matrices of the same size

$$A = [a_{ij}]_{i,j=1}^{m,n} \qquad \text{and} \qquad B = [b_{ij}]_{i,j=1}^{m,n}$$

are equal if and only if all the corresponding elements are identical, that is, $a_{ij} = b_{ij}$ for $1 \le i \le m$ and $1 \le j \le n$.

The set of all $m \times n$ matrices with real elements will be denoted by $\mathbb{R}^{m \times n}$. Similarly, $\mathbb{C}^{m \times n}$ is the set of all $m \times n$ matrices with complex elements.

1.1 Special Types of Matrices

If the number of rows of a matrix is equal to the number of columns, that is, $m = n$, then the matrix is *square* or *of order n*:

$$A = [a_{ij}]_{i,j=1}^{n} = \begin{bmatrix} a_{11} & a_{12} & \cdots & a_{1n} \\ a_{21} & a_{22} & \cdots & a_{2n} \\ \vdots & \vdots & & \vdots \\ a_{n1} & a_{n2} & \cdots & a_{nn} \end{bmatrix}.$$

The elements $a_{11}, a_{22}, \ldots, a_{nn}$ of a square matrix form its *main diagonal*, whereas the elements $a_{1n}, a_{2,n-1}, \ldots, a_{n1}$ generate the *secondary diagonal* of the matrix A.

Square matrices whose elements above (respectively, below) the main diagonal are zeros,

$$A_1 = \begin{bmatrix} a_{11} & 0 & \cdots & 0 \\ a_{21} & a_{22} & \ddots & \vdots \\ \vdots & \ddots & \ddots & 0 \\ a_{n1} & \cdots & a_{n,n-1} & a_{nn} \end{bmatrix}, \qquad A_2 = \begin{bmatrix} a_{11} & a_{12} & \cdots & a_{1n} \\ 0 & a_{22} & \ddots & \vdots \\ \vdots & \ddots & \ddots & a_{n-1,n} \\ 0 & \cdots & 0 & a_{nn} \end{bmatrix},$$

are called *lower-* (respectively, *upper-*) *triangular matrices.*

Diagonal matrices are a particular case of triangular matrices, for which all the elements lying outside the main diagonal are equal to zero:

$$A = \begin{bmatrix} a_{11} & 0 & \cdots & 0 \\ 0 & a_{22} & & \vdots \\ \vdots & & \ddots & 0 \\ 0 & \cdots & 0 & a_{nn} \end{bmatrix} = \operatorname{diag}[a_{11}, a_{22}, \ldots, a_{nn}].$$

If $a_{11} = a_{22} = \cdots = a_{nn} = a$, then the diagonal matrix A is called a *scalar matrix*;

$$A = \begin{bmatrix} a & 0 & \cdots & 0 \\ 0 & a & \ddots & \vdots \\ \vdots & \ddots & \ddots & 0 \\ 0 & \cdots & 0 & a \end{bmatrix} = \operatorname{diag}[a, a, \ldots, a].$$

In particular, if $a = 1$, the matrix A becomes the *unit*, or *identity matrix*

$$I = \begin{bmatrix} 1 & 0 & \cdots & 0 \\ 0 & 1 & \ddots & \vdots \\ \vdots & \ddots & \ddots & 0 \\ 0 & \cdots & 0 & 1 \end{bmatrix},$$

and in the case $a = 0$, a square *zero-matrix*

$$0 = \begin{bmatrix} 0 & 0 & \cdots & 0 \\ 0 & & \ddots & \vdots \\ \vdots & \ddots & & 0 \\ 0 & \cdots & 0 & 0 \end{bmatrix}$$

is obtained. A rectangular matrix with all its elements zero is also referred to as a zero-matrix.

A square matrix A is said to be a *Hermitian* (or *self-adjoint*) matrix if the elements on the main diagonal are real and whenever two elements are positioned symmetrically with respect to the main diagonal, they are mutually complex conjugate. In other words, Hermitian matrices are of the form

$$\begin{bmatrix} a_{11} & a_{12} & \cdots & a_{1n} \\ \bar{a}_{12} & a_{22} & \cdots & a_{2n} \\ \vdots & \vdots & \ddots & \vdots \\ \bar{a}_{1n} & \bar{a}_{2n} & \cdots & a_{nn} \end{bmatrix},$$

so that $a_{ji} = \bar{a}_{ij}$ for $i = 1, 2, \ldots, n$, $j = 1, 2, \ldots, n$, and $\bar{a}_{ij}$ denotes the complex conjugate of the number a_{ij}.

If all the elements located symmetrically with respect to the main diagonal are *equal*, then a square matrix is said to be *symmetric*:

$$\begin{bmatrix} a_{11} & a_{12} & \cdots & a_{1n} \\ a_{12} & a_{22} & \cdots & a_{2n} \\ \vdots & \vdots & \ddots & \vdots \\ a_{1n} & a_{2n} & \cdots & a_{nn} \end{bmatrix}.$$

It is clear that, in the case of a *real matrix* (i.e., consisting of real numbers), the notions of Hermitian and symmetric matrices coincide.

Returning to rectangular matrices, note particularly those matrices having only one column (*column-matrix*) or one row (*row-matrix*) of length, or size, n:

$$\boldsymbol{b} = \begin{bmatrix} b_1 \\ b_2 \\ \vdots \\ b_n \end{bmatrix}, \qquad \boldsymbol{c}^{\mathrm{T}} = [c_1 \quad c_2 \quad \cdots \quad c_n].$$

The reason for the T symbol, denoting a row-matrix, will be made clear in Section 1.5.

Such $n \times 1$ and $1 \times n$ matrices are also referred to as *vectors* or *ordered n-tuples*, and in the cases $n = 1, 2, 3$ they have an obvious geometrical

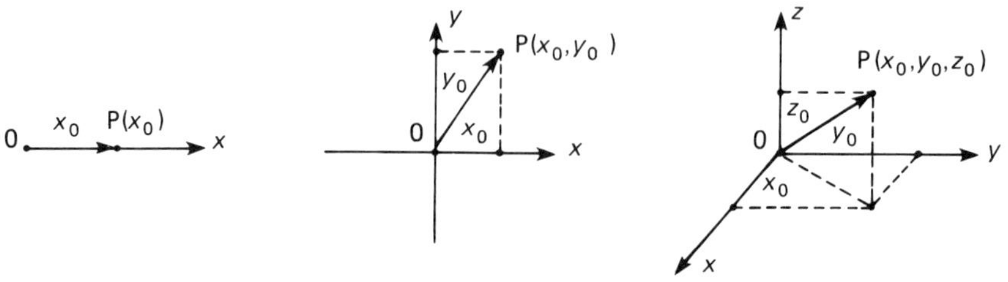

Fig. 1.1 Coordinates and position vectors.

meaning as the coordinates of a point P (or as components of the vector $\overrightarrow{OP}$) in one-, two-, or three-dimensional space with respect to the coordinate axes (Fig. 1.1).

For example, a point P in the three-dimensional Euclidean space, having Cartesian coordinates (x_0, y_0, z_0), and the vector $\overrightarrow{OP}$, are associated with the 1×3 row-matrix $[x_0 \quad y_0 \quad z_0]$. The location of the point P, as well as of the vector $\overrightarrow{OP}$, is described completely by this (*position*) vector.

Borrowing some geometrical language, the *length* of a vector (or position vector) is defined by the natural generalization of Euclidean geometry: for a vector $\boldsymbol{b}$ with elements $b_1, b_2, \ldots, b_n$ the length is

$$|\boldsymbol{b}| \triangleq (|b_1|^2 + |b_2|^2 + \cdots + |b_n|^2)^{1/2}.$$

Note that, throughout this book, the symbol $\triangleq$ is employed when a relation is used as a definition.

1.2 The Operations of Addition and Scalar Multiplication

Since vectors are special cases of matrices, the operations on matrices will be defined in such a way that, in the particular cases of column matrices and of row matrices, they correspond to the familiar operations on position vectors. Recall that, in three-dimensional Euclidean space, the sum of two position vectors is introduced as

$$[x_1 \quad y_1 \quad z_1] + [x_2 \quad y_2 \quad z_2] \triangleq [x_1 + x_2 \quad y_1 + y_2 \quad z_1 + z_2].$$

This definition yields the parallelogram law of vector addition, illustrated in Fig. 1.2.

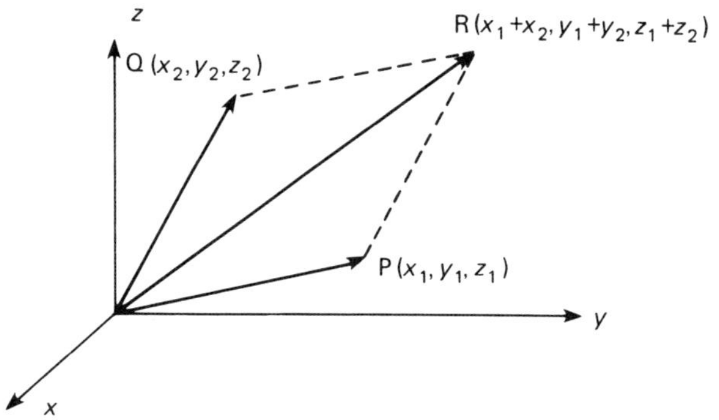

Fig. 1.2 The parallelogram law.

For ordered n-tuples written in the form of row- or column-matrices, this operation is naturally extended to

$$\begin{aligned}&[x_1 \quad x_2 \quad \cdots \quad x_n] + [y_1 \quad y_2 \quad \cdots \quad y_n] \\ &\qquad \triangleq [x_1 + y_1 \quad x_2 + y_2 \quad \cdots \quad x_n + y_n]\end{aligned}$$

or

$$\begin{bmatrix} x_1 \\ x_2 \\ \vdots \\ x_n \end{bmatrix} + \begin{bmatrix} y_1 \\ y_2 \\ \vdots \\ y_n \end{bmatrix} \triangleq \begin{bmatrix} x_1 + y_1 \\ x_2 + y_2 \\ \vdots \\ x_n + y_n \end{bmatrix}.$$

That is, the *elements* (or *components*, or *coordinates*) of the resulting vector are merely the sums of the corresponding elements of the vectors. Note that only vectors of the same size may be added.

Now the following definition of the sum of two matrices $A = [a_{ij}]_{i,j=1}^{m,n}$ and $B = [b_{ij}]_{i,j=1}^{m,n}$ of the same size is natural:

$$A + B \triangleq [a_{ij} + b_{ij}]_{i,j=1}^{m,n}.$$

The properties of the real and complex numbers (which we refer to as *scalars*) lead obviously to the *commutative* and *associative laws* of matrix addition.

Exercise 1. Show that, for matrices of the same size,

$$\begin{aligned} A + B &= B + A, \\ (A + B) + C &= A + (B + C). \quad \square \end{aligned}$$

These rules allow easy definition and computation of the sum of several matrices of the same size. In particular, it is clear that the sum of any number of $n \times n$ upper- (respectively, lower-) triangular matrices is an upper- (respectively, lower-) triangular matrix. Note also that the sum of several diagonal matrices of the same order is a diagonal matrix.

The operation of subtraction on matrices is defined as for numbers. Namely, the *difference* of two matrices A and B of the same size, written $A - B$, is a matrix X that satisfies

$$X + B = A.$$

Obviously,

$$A - B = [a_{ij} - b_{ij}]_{i,j=1}^{m,n},$$

where

$$A = [a_{ij}]_{i,j=1}^{m,n}, \qquad B = [b_{ij}]_{i,j=1}^{m,n}.$$

It is clear that the zero-matrix plays the role of the zero in numbers: a matrix does not change if the zero-matrix is added to it or subtracted from it.

Before introducing the operation of multiplication of a matrix by a scalar, recall the corresponding definition for (position) vectors in three-dimensional Euclidean space: If $\boldsymbol{a}^{\mathrm{T}} = [a_1 \quad a_2 \quad a_3]$ and α denotes a real number, then the vector $\alpha\boldsymbol{a}^{\mathrm{T}}$ is defined by

$$\alpha\boldsymbol{a}^{\mathrm{T}} \triangleq [\alpha a_1 \quad \alpha a_2 \quad \alpha a_3].$$

Thus, in the product of a vector with a scalar, each element of the vector is multiplied by this scalar.

This operation has a simple geometrical meaning for real vectors and scalars (see Fig. 1.3). That is, the length of the vector $\alpha\boldsymbol{a}^{\mathrm{T}}$ is $|\alpha|$ times the length of the vector $\boldsymbol{a}^{\mathrm{T}}$, and its orientation does not change if $\alpha > 0$ and it reverses if $\alpha < 0$.

Passing from (position) vectors to the general case, the product of the matrix $A = [a_{ij}]$ with a scalar α is the matrix C with elements $c_{ij} = \alpha a_{ij}$, that is, $C \triangleq [\alpha a_{ij}]$. We also write $C = \alpha A$. The following properties of scalar

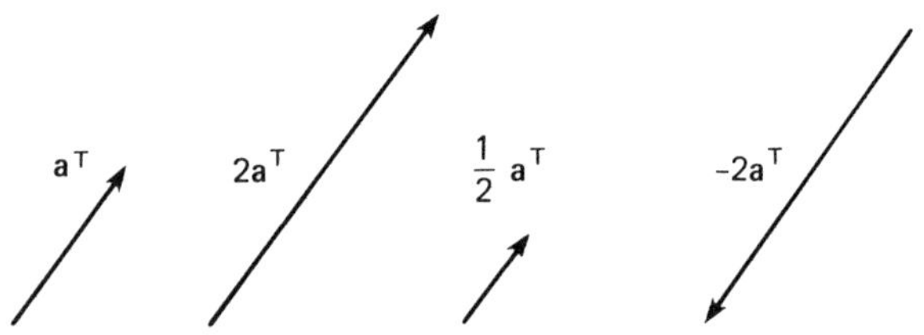

Fig. 1.3 Scalar multiplication.

multiplication and addition of matrices follow immediately from the definitions and from the corresponding properties of numbers.

Exercise 2. Check that for any two matrices of the same size and any scalars α, β

$$\begin{aligned} 0A &= 0, \\ (\alpha + \beta)A &= \alpha A + \beta A, \\ \alpha(A + B) &= \alpha A + \alpha B, \\ \alpha(\beta A) &= (\alpha\beta)A. \quad \square \end{aligned}$$

Note that by writing $(-1)B \triangleq -B$ we may, alternatively, define the difference $A - B$ as the sum of the matrices A and $-B$.

1.3 Matrix Multiplication

Like other operations on matrices, the notion of the product of two matrices can be motivated by some concepts of vector algebra. Recall that in three-dimensional Euclidean space, the *scalar* (or *dot*, or *inner*) *product* of the (position) vectors $\boldsymbol{a}$ and $\boldsymbol{b}$ is an important and useful concept. It is defined by

$$\boldsymbol{a} \cdot \boldsymbol{b} \triangleq |\boldsymbol{a}||\boldsymbol{b}| \cos \alpha, \tag{1}$$

where α denotes the angle between the given vectors. Clearly, for nonzero vectors $\boldsymbol{a}$ and $\boldsymbol{b}$, the scalar product is equal to zero if and only if $\boldsymbol{a}$ and $\boldsymbol{b}$ are orthogonal.

If the coordinates a_1, a_2, a_3 and b_1, b_2, b_3 of the vectors $\boldsymbol{a}$ and $\boldsymbol{b}$, respectively, are known, then the computation of the scalar product is simple;

$$\boldsymbol{a} \cdot \boldsymbol{b} = a_1 b_1 + a_2 b_2 + a_3 b_3. \tag{2}$$

The crux of the definition of matrix multiplication lies in the convention of *defining* the product of *row vector* $\mathbf{a}^{\mathrm{T}}$ with *column vector* $\boldsymbol{b}$ (in this order) to be the sum on the right of Eq. (2). Thus,

$$\boldsymbol{a}^{\mathrm{T}}\boldsymbol{b} = [a_1 \quad a_2 \quad a_3] \begin{bmatrix} b_1 \\ b_2 \\ b_3 \end{bmatrix} \triangleq a_1 b_1 + a_2 b_2 + a_3 b_3, \tag{3}$$

and this definition will now supersede the "dot" notation of Eqs. (1) or (2). Not unnaturally, it is described as "row-into-column" multiplication.

In contrast to Eq. (1), the definition in (3) lends itself immediately to broad generalizations that will be very useful indeed. First, the elements of $\boldsymbol{a}$ and $\boldsymbol{b}$ may be from any field (i.e., they are not necessarily real numbers), and second, the vectors may equally well have n elements each, for any positive integer n.

Thus, if $\boldsymbol{a}$, $\boldsymbol{b}$ are each vectors with n elements (the elements can be either real or complex numbers), define

$$\boldsymbol{a}^{\mathrm{T}}\boldsymbol{b} = [a_1 \quad a_2 \quad \cdots \quad a_n]\begin{bmatrix} b_1 \\ b_2 \\ \vdots \\ b_n \end{bmatrix} \triangleq \sum_{r=1}^{n} a_r b_r. \tag{4}$$

This idea will now be exploited further in the full definition of matrix multiplication.

Let A be any matrix consisting of m rows $\boldsymbol{a}_i^{\mathrm{T}}$ $(1 \le i \le m)$ of length l (so A is $m \times l$) and let B denote a matrix containing n columns $\boldsymbol{b}_j$ $(1 \le j \le n)$ of the same length l (so that B is $l \times n$). Then the *product* of A and B is defined to be the $m \times n$ matrix $C = [c_{ij}]$, where

$$c_{ij} = \boldsymbol{a}_i^{\mathrm{T}}\boldsymbol{b}_j \qquad (1 \le i \le m, \quad 1 \le j \le n).$$

Thus

$$AB = \begin{bmatrix} \boldsymbol{a}_1^{\mathrm{T}} \\ \boldsymbol{a}_2^{\mathrm{T}} \\ \vdots \\ \boldsymbol{a}_m^{\mathrm{T}} \end{bmatrix} [\boldsymbol{b}_1 \quad \boldsymbol{b}_2 \quad \cdots \quad \boldsymbol{b}_n]$$

$$\triangleq \begin{bmatrix} \boldsymbol{a}_1^{\mathrm{T}}\boldsymbol{b}_1 & \boldsymbol{a}_1^{\mathrm{T}}\boldsymbol{b}_2 & \cdots & \boldsymbol{a}_1^{\mathrm{T}}\boldsymbol{b}_n \\ \boldsymbol{a}_2^{\mathrm{T}}\boldsymbol{b}_1 & \boldsymbol{a}_2^{\mathrm{T}}\boldsymbol{b}_2 & \cdots & \boldsymbol{a}_2^{\mathrm{T}}\boldsymbol{b}_n \\ \vdots & \vdots & & \vdots \\ \boldsymbol{a}_m^{\mathrm{T}}\boldsymbol{b}_1 & \boldsymbol{a}_m^{\mathrm{T}}\boldsymbol{b}_2 & \cdots & \boldsymbol{a}_m^{\mathrm{T}}\boldsymbol{b}_n \end{bmatrix} \equiv [\boldsymbol{a}_i^{\mathrm{T}}\boldsymbol{b}_j]_{i,j=1}^{m,n}.$$

Using (4) we get the multiplication formula

$$AB = \left[\sum_{k=1}^{l} a_{ik} b_{kj}\right]_{i,j=1}^{m,n}. \tag{5}$$

Exercise 1. Show that

$$\begin{bmatrix} 2 & -1 & 0 \\ 0 & 3 & 1 \end{bmatrix}\begin{bmatrix} 3 & 0 \\ 2 & 1 \\ 1 & 0 \end{bmatrix} = \begin{bmatrix} 4 & -1 \\ 7 & 3 \end{bmatrix}. \quad \square$$

It is important to note that the product of two matrices is defined if and only if the number of columns of the first factor is equal to the number of rows of the second. When this is the case, A and B are said to be *conformable* (with respect to matrix multiplication).

Exercise 2. If $\boldsymbol{a}$ is a column matrix of size m and $\boldsymbol{b}^{\mathrm{T}}$ is a row matrix of size n show that $\boldsymbol{a}\boldsymbol{b}^{\mathrm{T}}$ is an $m \times n$ matrix. (This matrix is sometimes referred to as an *outer product* of $\boldsymbol{a}$ and $\boldsymbol{b}$.) ☐

The above definition of matrix multiplication is standard; it focusses on the *rows* of the first factor and the *columns* of the second. There is another representation of the product AB that is frequently useful and, in contrast, is written in terms of the columns of the first factor and the rows of the second.

First note carefully the conclusion of Exercise 2. Then let A be $m \times l$ and B be $l \times n$ (so that A and B are conformable for matrix multiplication). Suppose that the columns of A are $\boldsymbol{a}_1, \boldsymbol{a}_2, \ldots, \boldsymbol{a}_l$ and the rows of B are $\boldsymbol{b}_1^{\mathrm{T}}, \boldsymbol{b}_2^{\mathrm{T}}, \ldots, \boldsymbol{b}_l^{\mathrm{T}}$. Then it is claimed that

$$AB = \sum_{k=1}^{l} \boldsymbol{a}_k \boldsymbol{b}_k^{\mathrm{T}}, \tag{6}$$

and of course each product $\boldsymbol{a}_k \boldsymbol{b}_k^{\mathrm{T}}$ is an $m \times n$ matrix.

Exercise 3. Establish the formula in (6). ☐

Matrix multiplication preserves *some* properties of multiplication of scalars.

Exercise 4. Check that for any matrices of appropriate orders

$$AI = IA = A,$$

$$A(B + C) = AB + AC, \qquad (B + C)D = BD + CD \qquad \text{(distributive laws)},$$

$$A(BC) = (AB)C \qquad \text{(associative law)}. \quad \square$$

However, the matrix product does *not* retain some important properties enjoyed by multiplication of scalars, as the next exercises demonstrate.

Exercise 5. Show, setting

$$A = \begin{bmatrix} 2 & 1 \\ 0 & 0 \\ 1 & 0 \end{bmatrix} \quad \text{and} \quad B = \begin{bmatrix} 0 & 1 & 2 \\ -1 & 0 & 0 \end{bmatrix}$$

or

$$A = \begin{bmatrix} 2 & 1 \\ 0 & 0 \end{bmatrix} \quad \text{and} \quad B = \begin{bmatrix} 1 & 0 \\ -1 & 0 \end{bmatrix},$$

that $AB \neq BA$. ☐

Thus, matrix multiplication is, in general, *not commutative*.

Exercise 6. Show that $AB = 0$ does not imply that either A or B is a zero-matrix. (In other words, construct matrices $A \neq 0$, $B \neq 0$ for which $AB = 0$).

Exercise 7. Check that if $AB = AC$, $A \neq 0$, it does not always follow that $B = C$. □

A particular case of matrix multiplication is that of multiplication of a matrix by itself. Obviously, such multiplication is possible if and only if the matrix is square. So let A denote a square matrix and let p be any positive integer. The matrix

$$A^p \triangleq \underbrace{AA \cdots A}_{p\,\text{times}}$$

is said to be the pth *power* of the matrix A, and A^0 is defined to be the identity matrix with the size of A. Note also that $A^1 \triangleq A$.

Exercise 8. Verify that

$$\begin{bmatrix} 2 & -1 \\ 0 & 2 \end{bmatrix}^3 = \begin{bmatrix} 8 & -12 \\ 0 & 8 \end{bmatrix}.$$

Exercise 9. Prove the *exponent laws*;

$$A^pA^q = A^{p+q},$$
$$(A^p)^q = A^{pq},$$

where p, q are any nonnegative integers. □

1.4 Special Kinds of Matrices Related to Multiplication

Some types of matrices having special properties with regard to matrix multiplication are presented in this section. As indicated in Exercise 1.3.5, two matrices do not generally commute: $AB \neq BA$. But when equality does hold, we say that the matrix A *commutes* with B; a situation to be studied in some depth in section 12.4.

Exercise 1. Find all matrices that commute with the matrices

$$\text{(a)} \quad \begin{bmatrix} 1 & 0 \\ -1 & 0 \end{bmatrix}, \qquad \text{(b)} \quad \begin{bmatrix} 0 & 1 & 0 \\ 0 & 0 & 1 \\ 0 & 0 & 0 \end{bmatrix}.$$

Answers.

$$\text{(a)}\quad \begin{bmatrix} d-c & 0 \\ c & d \end{bmatrix}, \qquad \text{(b)}\quad \begin{bmatrix} a & b & c \\ 0 & a & b \\ 0 & 0 & a \end{bmatrix},$$

where a, b, c, d denote arbitrary scalars. □

It follows from Exercise 1.3.4 that the identity matrix commutes with any square matrix. It turns out that there exist other matrices having this property. A general description of such matrices is given in the next exercise. Following that, other special cases of commuting matrices are illustrated.

Exercise 2. Show that a matrix commutes with any square conformable matrix if and only if it is a scalar matrix.

Exercise 3. Prove that diagonal matrices of the same size commute.

Exercise 4. Show that (nonnegative integer) powers of the same square matrix commute.

Exercise 5. Prove that if a matrix A commutes with a diagonal matrix $\operatorname{diag}[a_1, a_2, \ldots, a_n]$, where $a_i \neq a_j$, $i \neq j$, then A is diagonal. □

Another important class of matrices is formed by those that satisfy the equation

$$A^2 = A. \tag{1}$$

Such a matrix A is said to be *idempotent*.

Exercise 6. Check that the matrices

$$A_1 = \begin{bmatrix} 1 & 0 \\ 0 & 0 \end{bmatrix} \quad \text{and} \quad A_2 = \begin{bmatrix} 1 & 2 & 2 \\ 0 & 0 & -1 \\ 0 & 0 & 1 \end{bmatrix}$$

are idempotent.

Exercise 7. Show that if A is idempotent, and if p is a positive integer, then $A^p = A$.

Exercise 8. Describe the class of 2×2 idempotent matrices. □

A square matrix A is called *nilpotent* if there is a positive integer p such that

$$A^p = 0. \tag{2}$$

Obviously, it follows from (2) that any integer kth power of A with $k \geq p$ is also a zero matrix. Hence the notion of the *least* integer p_0 satisfying (2) is reasonable and is referred to as the *degree of nilpotency* of the matrix A.

Exercise 9. Check that the matrix

$$A = \begin{bmatrix} 0 & 1 & 0 \\ 0 & 0 & 1 \\ 0 & 0 & 0 \end{bmatrix}$$

is nilpotent with degree of nilpotency equal to 3.

Exercise 10. Prove that every upper-triangular matrix having zeros on the main diagonal is nilpotent.

Exercise 11. Find all nilpotent 2×2 matrices having degree of nilpotency equal to two.

Answer.

$$\begin{bmatrix} a & b \\ c & -a \end{bmatrix},$$

where $a^2 + bc = 0$ and not all of a, b, c are zero. □

The last type of matrix to be considered in this section is the class of *involutory matrices.* They are those square matrices satisfying the condition

$$A^2 = I. \tag{3}$$

Exercise 12. Check that for any angle θ,

$$A = \begin{bmatrix} \cos\theta & \sin\theta \\ \sin\theta & -\cos\theta \end{bmatrix}$$

is an involutory matrix.

Exercise 13. Show that any matrix having 1 or -1 in all positions on the secondary diagonal and zeros elsewhere is involutory.

Exercise 14. Show that a matrix A is involutory if and only if

$$(I - A)(I + A) = O.$$

Exercise 15. Find all 2×2 involutory matrices.

Answer.

$$\begin{bmatrix} a & b \\ c & -a \end{bmatrix},$$

with $a^2 + bc = 1$, or

$$\begin{bmatrix} \varepsilon & 0 \\ 0 & \varepsilon \end{bmatrix}$$

with $\varepsilon = \pm 1$.

Exercise 16. Show that if A is an involutory matrix, then the matrix $B = \frac{1}{2}(I + A)$ is idempotent. □

1.5 Transpose and Conjugate Transpose

In this section some possibilities for forming new matrices from a given one are discussed.

Given an $m \times n$ matrix $A = [a_{ij}]$, the $n \times m$ matrix obtained by interchanging the rows and columns of A is called the *transpose* of the matrix A and is denoted by A^T. In more detail, if

$$A = \begin{bmatrix} a_{11} & a_{12} & \cdots & a_{1n} \\ a_{21} & a_{22} & \cdots & a_{2n} \\ \vdots & \vdots & & \vdots \\ a_{m1} & a_{m2} & \cdots & a_{mn} \end{bmatrix}, \quad \text{then} \quad A^T \triangleq \begin{bmatrix} a_{11} & a_{21} & \cdots & a_{m1} \\ a_{12} & a_{22} & \cdots & a_{m2} \\ \vdots & \vdots & & \vdots \\ a_{1n} & a_{2n} & \cdots & a_{mn} \end{bmatrix}.$$

Note that the rows of A become the columns of A^T and the columns of A are the rows of its transpose. Furthermore, in view of this definition our notation $\mathbf{a}^T$ for a row matrix in the previous sections just means that it is the transpose of the corresponding column matrix $\mathbf{a}$.

Exercise 1. The matrices

$$A = \begin{bmatrix} 2 & 1 & 0 \\ -1 & 2 & 1 \end{bmatrix} \quad \text{and} \quad B = \begin{bmatrix} 2 & -1 \\ 1 & 2 \\ 0 & 1 \end{bmatrix}$$

are mutually transposed matrices: $A = B^T$ and $B = A^T$. □

The following properties of the transpose are simple consequences of the definition.

Exercise 2. Let A and B be any matrices of an appropriate order and let $\alpha \in \mathscr{F}$. Prove that

$$(A^T)^T = A,$$

$$(\alpha A)^T = \alpha A^T,$$

$$(A + B)^T = A^T + B^T,$$

$$(AB)^T = B^T A^T. \quad \square$$

Using the notion of transpose, a brief definition of a symmetric matrix (equivalent to that in Section 1.1) can be formulated.

Exercise 3. Show that a square matrix A is a symmetric matrix if and only if

$$A^{\mathrm{T}} = A. \quad \square \tag{1}$$

Some additional properties of symmetric matrices follow easily from (1).

Exercise 4. Prove that if A and B are symmetric matrices, then so is $A + B$.

Exercise 5. Show that for any square matrix A the matrices AA^{T}, $A^{\mathrm{T}}A$, and $A^{\mathrm{T}} + A$ are symmetric. □

A matrix A is called *skew-symmetric* if

$$A^{\mathrm{T}} = -A. \tag{2}$$

Exercise 6. Check that the matrix

$$A = \begin{bmatrix} 0 & -2 & -3 \\ 2 & 0 & -2 \\ 3 & 2 & 0 \end{bmatrix}$$

is skew-symmetric. □

Obviously the elements of an $n \times n$ skew-symmetric matrix $A = [a_{ij}]$ satisfy the conditions

$$a_{ij} = -a_{ji}, \qquad 1 \leq i, j \leq n,$$

and in particular, the elements of A on the main diagonal are zeros.

Exercise 7. Check that for any square matrix A the matrix $A - A^{\mathrm{T}}$ is skew-symmetric. □

The assertion of Exercise 7 together with Exercise 5 admits the following decomposition:

Exercise 8. Show that any square matrix A can be represented in the form

$$A = A_1 + A_2,$$

where A_1 is a symmetric matrix and A_2 a skew-symmetric matrix. □

Consider a matrix $A = [a_{ij}]_{i,j=1}^{m,n}$ with elements from $\mathbb{C}$. The *complex conjugate* $\bar{A}$ of A is defined as the matrix obtained from A by changing all its entries to their complex conjugates.† In other words, $\bar{A} \triangleq [\bar{a}_{ij}]_{i,j=1}^{m,n}$. It is clear that a matrix A is real if and only if $\bar{A} = A$.

† Recall that if $a = c + di$, $c, d \in \mathbb{R}$, then the complex conjugate is $\bar{a} \triangleq c - di$.

The operations of transposition and conjugation are combined in the important concept of the *conjugate transpose* A^* of a matrix A. It is defined by $A^* \triangleq \bar{A}^{\mathrm{T}}$. In more detail, for $A = [a_{ij}]_{i,j=1}^{m,n}$,

$$A^* \triangleq \begin{bmatrix} \bar{a}_{11} & \bar{a}_{21} & \cdots & \bar{a}_{m1} \\ \bar{a}_{12} & \bar{a}_{22} & \cdots & \bar{a}_{m2} \\ \vdots & \vdots & & \vdots \\ \bar{a}_{1n} & \bar{a}_{2n} & \cdots & \bar{a}_{mn} \end{bmatrix}.$$

It is clear that for a real matrix A the notions of transpose and conjugate transpose are equivalent.

Exercise 9. Check that the conjugate transposes of the matrices

$$A = \begin{bmatrix} 2 & -3 \\ 0 & 1 \\ 1 & 0 \end{bmatrix} \quad \text{and} \quad B = \begin{bmatrix} i & -3 \\ 0 & 2i + 1 \end{bmatrix}$$

are the matrices

$$A^* = \begin{bmatrix} 2 & 0 & 1 \\ -3 & 1 & 0 \end{bmatrix} = A^{\mathrm{T}} \quad \text{and} \quad B^* = \begin{bmatrix} -i & 0 \\ -3 & -2i + 1 \end{bmatrix},$$

respectively.

Exercise 10. (Compare with Exercise 2.) Show that $(A^*)^* = A$, $(\alpha A)^* = \bar{\alpha}A^*$, $(A + B)^* = A^* + B^*$, $(AB)^* = B^*A^*$. □

Applying the notion of the conjugate transpose, a nice characterization of a Hermitian matrix can be formulated.

Exercise 11. Prove that a square matrix A is Hermitian (as defined in Section 1.1) if and only if $A = A^*$.

Exercise 12. Show that for any matrix A, the matrices AA^* and A^*A are Hermitian.

Exercise 13. Check that, for any square matrix A, the matrices $A + A^*$ and $i(A - A^*)$ are Hermitian. □

Observe that the 1×1 complex matrices (i.e., those with only one scalar element) correspond precisely to the set of all complex numbers. Furthermore, in this correspondence the 1×1 Hermitian matrices correspond to the real numbers. More generally, it is useful to consider the embedding of the Hermitian matrices in $\mathbb{C}^{n \times n}$ as a generalization of the natural embedding of $\mathbb{R}$ in $\mathbb{C}$. In particular, note the analogy between the following representation of a complex square matrix and the familiar Cartesian decomposition of complex numbers.

Exercise 14. Show that every square matrix A can be written uniquely in the form $A = A_1 + iA_2$, where A_1 and A_2 are Hermitian matrices. □

A *skew-Hermitian matrix* A is defined in terms of the conjugate transpose by $A^* = -A$.

Exercise 15. Check that the matrix

$$A = \begin{bmatrix} i & i \\ i & i \end{bmatrix}$$

is skew-Hermitian.

Exercise 16. Show that the diagonal elements of a skew-Hermitian matrix are pure imaginary. □

1.6 Submatrices and Partitions of a Matrix

Given a matrix $A = [a_{ij}]_{i,j=1}^{m,n}$, if a number of complete rows or columns of A are deleted, or if some complete rows *and* complete columns are deleted, the new matrix that is obtained is called a *submatrix* of A.

Example 1. If

$$A = \begin{bmatrix} 2 & 3 & -2 & 4 \\ 0 & 1 & 1 & 0 \\ 2 & -1 & 0 & 0 \end{bmatrix},$$

then the matrices

$$\begin{bmatrix} 2 & 3 \\ 0 & 1 \end{bmatrix}, \qquad [1], \qquad [2 \quad 3 \quad -2 \quad 4], \qquad \begin{bmatrix} 2 & 4 \\ 2 & 0 \end{bmatrix}$$

are some of the submatrices of A. □

There is a special way of dividing a matrix into submatrices by inserting dividing lines between specified rows and between specified columns (each dividing line running the full width or height of the matrix array). Such a division is referred to as a *partition* of the matrix.

Example 2. The matrix

$$A = \begin{bmatrix} 2 & 3 & -2 & 4 \\ 0 & 1 & 1 & 0 \\ 2 & -1 & 0 & 0 \end{bmatrix}$$

can be partitioned as

$$A = \left[\begin{array}{cc|cc} 2 & 3 & -2 & 4 \\ 0 & 1 & 1 & 0 \\ \hline 2 & -1 & 0 & 0 \end{array}\right] \quad \text{or} \quad A = \left[\begin{array}{c|cc|c} 2 & 3 & -2 & 4 \\ \hline 0 & 1 & 1 & 0 \\ 2 & -1 & 0 & 0 \end{array}\right]$$

and in several other ways. In the first case, the matrix is partitioned into submatrices (or *blocks*) labeled

$$A = \begin{bmatrix} A_{11} & A_{12} \\ A_{21} & A_{22} \end{bmatrix}, \tag{1}$$

where

$$A_{11} = \begin{bmatrix} 2 & 3 \\ 0 & 1 \end{bmatrix}, \quad A_{12} = \begin{bmatrix} -2 & 4 \\ 1 & 0 \end{bmatrix}, \quad A_{21} = [2 \ \ -1], \quad A_{22} = [0 \ \ 0].$$

In the second partition A can be written in the form

$$A = \begin{bmatrix} A_{11} & A_{12} & A_{13} \\ A_{21} & A_{22} & A_{23} \end{bmatrix},$$

where

$$A_{11} = [2], \quad A_{12} = [3 \ \ -2], \quad A_{13} = [4],$$

$$A_{12} = \begin{bmatrix} 0 \\ 2 \end{bmatrix}, \quad A_{22} = \begin{bmatrix} 1 & 1 \\ -1 & 0 \end{bmatrix}, \quad A_{23} = \begin{bmatrix} 0 \\ 0 \end{bmatrix}.$$

Note that some blocks may be 1×1 matrices, that is, scalars. □

The submatrices of a partitioned matrix are referred to as the elements of a *block-matrix*. For example, the submatrices $A_{11}, A_{12}, A_{21}, A_{22}$ of (1) may be considered as the elements of the 2×2 block-matrix A.

The notions of a *square block-matrix*, *a diagonal block-matrix*, and so on, are defined as for matrices with scalar elements (see Section 1.1).

Example 3. The matrices

$$A = \begin{bmatrix} A_{11} & A_{12} \\ 0 & A_{22} \end{bmatrix}, \qquad B = \begin{bmatrix} B_{11} & 0 \\ 0 & B_{22} \end{bmatrix}$$

are upper-triangular and diagonal block-matrices, respectively. □

Square matrices admit some special partitions. If the diagonal blocks A_{ii} of a square matrix A are square matrices for each i, then such a partition of A is said to be *symmetric*.

Example 4. The matrix

$$A = \left[\begin{array}{rr:r} 1 & 2 & 2 \\ -1 & 0 & 1 \\ \hdashline 0 & 0 & 3 \end{array}\right]$$

is a symmetrically partitioned matrix.

Example 5. Let the matrices A and B be partitioned as follows,

$$A = \begin{bmatrix} A_{11} & A_{12} \\ A_{21} & A_{22} \end{bmatrix}, \qquad B = \begin{bmatrix} B_{11} & B_{12} \\ B_{21} & B_{22} \end{bmatrix},$$

and be of the same size. Recalling the definition of matrix addition, it is easily seen that if each pair of blocks A_{ij} and B_{ij} $(1 \leq i, j \leq 2)$ is of the same size, then

$$A \pm B = \begin{bmatrix} A_{11} \pm B_{11} & A_{12} \pm B_{12} \\ A_{21} \pm B_{21} & A_{22} \pm B_{22} \end{bmatrix}.$$

Clearly, if the partition of B differs from that of A, say,

$$B = \begin{bmatrix} B_{11} & B_{12} & B_{13} \\ B_{21} & B_{22} & B_{23} \end{bmatrix},$$

then A and B cannot be added block by block. □

Block-matrix multiplication can also be performed in some cases.

Exercise 6. Let the conformable matrices A and B be partitioned in the following way:

$$A = \left[\begin{array}{r:rr} 2 & 1 & 3 \\ 0 & -1 & 2 \end{array}\right] \triangleq [A_{11} \quad A_{12}],$$

$$B = \left[\begin{array}{rrr} 3 & -7 & -7 \\ \hdashline -2 & 1 & 4 \\ 0 & 2 & 4 \end{array}\right] \triangleq \begin{bmatrix} B_{11} \\ B_{21} \end{bmatrix}.$$

Check by a direct computation that

$$AB = [A_{11} \quad A_{12}]\begin{bmatrix} B_{11} \\ B_{21} \end{bmatrix} = A_{11}B_{11} + A_{12}B_{21}. \quad \square$$

The above examples show that in certain cases, the rules of matrix addition and multiplication carry over to block-matrices. The next exercises give the general results.

Exercise 7. Prove that if each pair of blocks A_{ij} and B_{ij} of the matrices A and B, respectively, are of the same size, then

$$A + B = [A_{ij} + B_{ij}]_{i,j=1}^{m,n}, \qquad A - B = [A_{ij} - B_{ij}]_{i,j=1}^{m,n}.$$

Exercise 8. Prove that if the $m \times l$ matrix A has blocks A_{ik} of sizes $m_i \times l_k$ where $1 \le i \le r$ and $1 \le k \le \alpha$, and the $l \times n$ matrix B consists of blocks B_{kj} of sizes $l_k \times n_j$, where $1 \le j \le s$, then [see Eq. (1.3.5)]

$$AB = \left[\sum_{k=1}^{\alpha} A_{ik} B_{kj}\right]_{i,j=1}^{r,s}. \quad \square$$

Note that two "conformable" conditions are required for performing block-matrix multiplication. The number of blocks in a row of A must be the same as the number of blocks in a column of B. Also, the number of columns in A_{ik} must be the same as the number of rows in B_{kj} for $1 \le k \le \alpha$ and all possible i and j.

1.7 Polynomials in a Matrix

Let A denote a square matrix and let

$$p(\lambda) = a_0 + a_1\lambda + \cdots + a_l\lambda^l \qquad (a_l \ne 0)$$

be a polynomial of degree l with scalar coefficients. The operations on matrices introduced in Sections 1.2 and 1.3 can be combined to define a matrix

$$p(A) \triangleq a_0 I + a_1 A + \cdots + a_l A^l$$

This is said to be a *polynomial in A*.

Exercise 1. Check that if $p(\lambda) = 2 - \lambda + \lambda^2$ and

$$A = \begin{bmatrix} 0 & 1 & 0 \\ 0 & 1 & 0 \\ 1 & 0 & 1 \end{bmatrix},$$

then

$$p(A) = 2I - A + A^2 = \begin{bmatrix} 2 & 0 & 0 \\ 0 & 2 & 0 \\ 0 & 1 & 2 \end{bmatrix}. \quad \square$$

Since every square matrix commutes with itself, all polynomials in a given matrix commute with one another. This fact is required for generalizing the following familiar properties of scalar polynomials to polynomials in a matrix.

Exercise 2. Show that if p, q are scalar polynomials and

$$p(\lambda) + q(\lambda) = h(\lambda), \qquad p(\lambda)q(\lambda) = t(\lambda),$$

then, for any square matrix A;

$$p(A) + q(A) = h(A), \qquad p(A)q(A) = t(A).$$

Verify also that if

$$p(\lambda) = q(\lambda)\, d(\lambda) + r(\lambda),$$

then, for any square matrix A;

$$p(A) = q(A)\, d(A) + r(A). \quad \square$$

In the expression $p(\lambda) = q(\lambda)\, d(\lambda) + r(\lambda)$, if $r(\lambda)$ is the zero polynomial or has degree less than the degree of $d(\lambda)$, then $q(\lambda)$ and $r(\lambda)$ are, respectively, the *quotient* and *remainder* obtained on division of $p(\lambda)$ by $d(\lambda)$. In the case $r(\lambda) = 0$, $d(\lambda)$ is a *divisor* of $p(\lambda)$. Exercise 2 shows that such a representation carries over to polynomials in a matrix and, in particular, implies that if $d(\lambda)$ is a divisor of $p(\lambda)$, then $d(A)$ is a divisor of $p(A)$ in the sense that $p(A) = Q\, d(A)$ for some square matrix Q of the same order as A.

However, the well-known and important fact that a scalar polynomial of degree l has exactly l complex zeros does not generalize nicely for a polynomial in a matrix. The next example illustrates this.

Example 3. Let

$$p(\lambda) = (\lambda - 1)(\lambda + 2)$$

and let A be an arbitrary 2×2 matrix. Then in view of Exercise 2,

$$p(A) = (A - I)(A + 2I) = A^2 + A - 2I. \tag{1}$$

In contrast to the scalar polynomial $p(\lambda)$, the polynomial $p(A)$ has more than two *zeros*, that is, matrices A for which $p(A) = 0$. Indeed, since the product of two matrices $A - I$ and $A + 2I$ may be the zero matrix even in the case of nonzero factors (see Exercise 1.3.6), the polynomial (1), in addition to the zeros I and $-2I$,[†] may have many others. In fact, it is not difficult to check that the matrix

$$A = \begin{bmatrix} -2 & a \\ 0 & 1 \end{bmatrix}$$

is also a zero of $p(A)$ for any number a.

[†] These zeros of $p(A)$ are scalar matrices. Obviously $p(A)$ has no other scalar matrices as zeros.

Exercise 4. Let $p(\lambda) = \lambda^3 + \lambda^2 + 2\lambda + 2$. Show that if

$$A = \begin{bmatrix} \lambda & 0 & 0 & 0 & 0 & 0 \\ 0 & \mu & 1 & 0 & 0 & 0 \\ 0 & 0 & \mu & 0 & 0 & 0 \\ 0 & 0 & 0 & v & 1 & 0 \\ 0 & 0 & 0 & 0 & v & 1 \\ 0 & 0 & 0 & 0 & 0 & v \end{bmatrix},$$

then

$$p(A) = \begin{bmatrix} p(\lambda) & 0 & 0 & 0 & 0 & 0 \\ 0 & p(\mu) & p'(\mu) & 0 & 0 & 0 \\ 0 & 0 & p(\mu) & 0 & 0 & 0 \\ 0 & 0 & 0 & p(v) & p'(v) & \frac{1}{2}p''(v) \\ 0 & 0 & 0 & 0 & p(v) & p'(v) \\ 0 & 0 & 0 & 0 & 0 & p(v) \end{bmatrix},$$

where p' and p'' denote the first and second derivatives of p, respectively.

Exercise 5. Let A be a symmetrically partitioned matrix of the form

$$A = \begin{bmatrix} A_{11} & A_{12} \\ O & I \end{bmatrix}.$$

Prove that for any positive integer n,

$$A^n = \begin{bmatrix} A_{11}^n & p_n(A_{11})A_{12} \\ O & I \end{bmatrix},$$

where $p_n(\lambda) = (\lambda^n - 1)/(\lambda - 1)$. □

1.8 Miscellaneous Exercises

1. Let a square matrix $A \in \mathbb{R}^{n \times n}$ with nonnegative elements be such that the sum of all elements of each of its rows (each *row sum*) is equal to one. Such a matrix is referred to as a *stochastic matrix*. Prove that the product of two stochastic matrices is a stochastic matrix.

2. A matrix $A \in \mathbb{R}^{n \times n}$ is said to be *doubly stochastic* if both A and A^T are stochastic matrices. Show that the product of two doubly stochastic matrices is a doubly stochastic matrix.

3. The *trace* of a matrix $A \in \mathbb{C}^{n \times n}$, written tr A, is defined as the sum of all the elements lying on the main diagonal of A. Prove the following.

(a) $\operatorname{tr}(\alpha A + \beta B) = \alpha \operatorname{tr} A + \beta \operatorname{tr} B$ for any $A, B \in \mathbb{C}^{n \times n}$; $\alpha, \beta \in \mathbb{C}$.
(b) If $A \in \mathbb{C}^{n \times m}$ and $B \in \mathbb{C}^{m \times n}$ then $\operatorname{tr} AB = \operatorname{tr} BA$.
(c) If $A = [a_{ij}]_{i,j=1}^{n} \in \mathbb{C}^{n \times n}$, then $\operatorname{tr} AA^* = \operatorname{tr} A^*A = \sum_{i,j=1}^{n} |a_{ij}|^2$.
(d) If $A, B \in \mathbb{C}^{n \times n}$ and A is idempotent, then $\operatorname{tr}(AB) = \operatorname{tr}(ABA)$.

4. Prove that if $n \geq 2$, there are no $n \times n$ matrices A and B such that

$$AB - BA = I_n.$$

Hint. Use Exercise 3.

5. Use induction to prove that if $X, Y \in \mathcal{F}^{n \times n}$, then for $r = 1, 2, \ldots,$

$$X^r - Y^r = \sum_{j=0}^{r-1} Y^j(X - Y)X^{r-1-j}.$$

CHAPTER 2

Determinants, Inverse Matrices, and Rank

The important concepts of the determinant of a matrix and the rank of a matrix are introduced and examined in this chapter. Some applications of these notions to the solution of systems of linear equations are also considered.

2.1 Definition of the Determinant

In this section we define the most important scalar-valued function associated with the elements of a square matrix—the determinant. Recall that the notion of the determinant of 2×2 and 3×3 matrices has its origin in solving linear systems like

$$\begin{aligned} a_{11}x + a_{12}y &= b_1, \\ a_{21}x + a_{22}y &= b_2 \end{aligned} \tag{1}$$

for x, y and

$$\begin{aligned} a_{11}x + a_{12}y + a_{13}z &= b_1, \\ a_{21}x + a_{22}y + a_{23}z &= b_2, \\ a_{31}x + a_{32}y + a_{33}z &= b_3 \end{aligned} \tag{2}$$

for x, y, z.

It is easily seen that in the case $a_{11}a_{22} - a_{12}a_{21} \neq 0$ the solution (x, y) of (1) can be written in the form

$$x = \frac{b_1 a_{22} - b_2 a_{12}}{a_{11}a_{22} - a_{12}a_{21}}, \qquad y = \frac{b_2 a_{11} - b_1 a_{21}}{a_{11}a_{22} - a_{12}a_{21}}, \tag{3}$$

where the denominator in the above expressions is said to be the *determinant* of the *coefficient matrix*

$$A = \begin{bmatrix} a_{11} & a_{12} \\ a_{21} & a_{22} \end{bmatrix}$$

of system (1). Thus, by definition

$$\det \begin{bmatrix} a_{11} & a_{12} \\ a_{21} & a_{22} \end{bmatrix} \triangleq a_{11}a_{22} - a_{12}a_{21}.$$

Note also that the numerators in (3) can be written as determinants of the matrices

$$A_x = \begin{bmatrix} b_1 & a_{12} \\ b_2 & a_{22} \end{bmatrix} \quad \text{and} \quad A_y = \begin{bmatrix} a_{11} & b_1 \\ a_{21} & b_2 \end{bmatrix},$$

respectively.

Hence, if $\det A \neq 0$, the solution of (1) is

$$x = \frac{\det A_x}{\det A}, \qquad y = \frac{\det A_y}{\det A}, \tag{4}$$

where the matrices A_x and A_y are obtained from the coefficient matrix A of the system by putting the column $[b_1 \quad b_2]^T$ in the place of the appropriate column in A. The formulas in (4) are known as *Cramer's rule* for finding the solution of the linear system (1).

Similarly, the solution of (2) can be found by Cramer's rule extended to the three-variable case:

$$x = \frac{\det A_x}{\det A}, \qquad y = \frac{\det A_y}{\det A}, \qquad z = \frac{\det A_z}{\det A}, \tag{5}$$

where the determinant of A is defined by

$$\det \begin{bmatrix} a_{11} & a_{12} & a_{13} \\ a_{21} & a_{22} & a_{23} \\ a_{31} & a_{32} & a_{33} \end{bmatrix} \triangleq a_{11}a_{22}a_{33} + a_{12}a_{23}a_{31} + a_{13}a_{21}a_{32} - a_{13}a_{22}a_{31} - a_{12}a_{21}a_{33} - a_{11}a_{23}a_{32} \tag{6}$$

and is assumed to be a nonzero number. The matrices A_x, A_y, A_z in (5) are obtained by interchanging the appropriate column of A with the column $[b_1 \quad b_2 \quad b_3]^T$ of right-hand terms from (2).

To define the notion of determinant for an arbitrary $n \times n$ square matrix so that Cramer's rule remains true, observe that for a 3×3 matrix the determinant consists of summands of the form $a_{1j_1}a_{2j_2}a_{3j_3}$, where j_1, j_2, j_3 are the numbers 1, 2, 3 written in any possible order, that is, (j_1, j_2, j_3) is a permutation of the numbers 1, 2, 3. In general, if $j_1, j_2, \ldots, j_n$ are the numbers $1, 2, \ldots, n$ written in any order, then $(j_1, j_2, \ldots, j_n)$ is said to be a *permutation* of $1, 2, \ldots, n$.

Let $A = [a_{ij}]_{i,j=1}^n$ denote an arbitrary square matrix. Keeping in mind the structure of determinants of small-sized matrices, we define a *diagonal* of A as a sequence of n elements of the matrix containing one and only one element from each row of A and one and only one element from each column of A. A diagonal of A is always assumed to be ordered according to the row indices; therefore it can be written in the form

$$a_{1j_1}, a_{2j_2}, \ldots, a_{nj_n}, \tag{7}$$

where $(j_1, j_2, \ldots, j_n)$ is a permutation of the numbers $1, 2, \ldots, n$. In particular, if $(j_1, j_2, \ldots, j_n) = (1, 2, \ldots, n)$, we obtain the *main* diagonal of A. In the case $(j_1, j_2, \ldots, j_n) = (n, n-1, \ldots, 1)$, the *secondary* diagonal of A is obtained (see Section 1.1). Clearly, an $n \times n$ matrix has exactly $n!$ distinct diagonals.

Exercise 1. Check that the determinants of 2×2 and 3×3 matrices are algebraic sums of products of the elements from each diagonal. □

For defining the signs of the products [in (6), for example], we need some facts about permutations. We say that a pair of numbers j_k and j_p in the permutation $(j_1, j_2, \ldots, j_n)$ form an *inversion* if $j_k > j_p$ while $k < p$, that is, if a larger number in the permutation precedes a smaller one. Each permutation $j = (j_1, j_2, \ldots, j_n)$ has a certain number of inversions associated with it, denoted briefly by $t(j)$. This number is uniquely defined if the inversions are counted successively, in the way illustrated in the next example. The permutation is called *odd* or *even* according to whether the number $t(j)$ is odd or even. This property is known as the *parity* of the permutation.

Exercise 2. Find $t(j)$, where $j = (2, 4, 3, 1, 5)$.

Solution. The number 2 occurs before 1 and therefore forms an inversion. The number 4 is in inversion with both 3 and 1, while 3 forms an inversion with 1. Thus, the total number of inversions $t(j)$ in the given permutation is 4, and j is an even permutation.

Exercise 3. Show that interchanging two neighbouring elements in an odd (respectively, even) permutation produces an even (respectively, odd) one.

Exercise 4. Prove that interchanging *any* two elements in a permutation changes it from odd to even or vice versa.

Hint. Show that if there are k numbers between the given elements a and b, then $2k + 1$ interchanges of neighbouring elements are required to interchange a and b.

Exercise 5. Show that the parity of a permutation j can be decided by counting the number of interchanges of pairs of elements required to transform j to the sequence j_0 in natural order, that is, $j_0 = (1, 2, \ldots, n)$. (In general, this transformation can be completed in many ways, but the number of interchanges required will always be even or odd as $t(j)$ is even or odd.) □

Now we are in position to give a definition of the determinant of an $n \times n$ matrix. Let A be a square matrix of order n. The *determinant* of A, denoted det A, is defined by

$$\det A \triangleq \sum_j (-1)^{t(j)} a_{1j_1} a_{2j_2} \cdots a_{nj_n}, \tag{8}$$

where $t(j)$ is the number of inversions in the permutation $j = (j_1, j_2, \ldots, j_n)$ and j varies over all $n!$ permutations of $1, 2, \ldots, n$.

In other words, det A is a sum of $n!$ products. Each product involves n elements of A belonging to the same diagonal. The product is multiplied by $+1$ or -1 according to whether the permutation $(j_1, j_2, \ldots, j_n)$ that defines the diagonal is even or odd, respectively.

Exercise 6. Verify that the above definition of determinant coincides with those for 2×2 and 3×3 matrices.

Exercise 7. Evaluate det A, where

$$A = \begin{bmatrix} 0 & a_{12} & 0 & 0 & 0 \\ 0 & 0 & 0 & a_{24} & 0 \\ 0 & 0 & a_{33} & a_{34} & 0 \\ a_{41} & a_{42} & 0 & 0 & 0 \\ 0 & 0 & 0 & a_{54} & a_{55} \end{bmatrix}.$$

SOLUTION. Obviously, if a diagonal of a matrix contains a zero, then the corresponding term for the determinant (8) is zero. Since the given matrix has only one diagonal of possibly nonzero elements, namely,

$$a_{12}, a_{24}, a_{33}, a_{41}, a_{55},$$

it thus follows that

$$|\det A| = a_{12}a_{24}a_{33}a_{41}a_{55}.$$

Consider the corresponding permutation $j = (2, 4, 3, 1, 5)$. The number $t(j)$ is equal to 4 (see Exercise 2) and hence the permutation is even. Thus,

$$\det A = a_{12}a_{24}a_{33}a_{41}a_{55}.$$

Exercise 8. Check that

$$\det \operatorname{diag}[a_{11}, a_{22}, \ldots, a_{nn}] = a_{11}a_{22} \cdots a_{nn}.$$

Exercise 9. Verify that the determinant of a square matrix containing a zero row (or column) is equal to zero.

Exercise 10. Check that the determinant of a triangular matrix is equal to the product of its elements on the main diagonal.

Exercise 11. Let the matrix A of order n have only zero elements above (or below) the secondary diagonal. Show that

$$\det A = (-1)^{n(n-1)/2} a_{1n}a_{2,n-1} \cdots a_{n1}. \quad \square$$

Some general methods for computing determinants will be suggested later after a study of the properties of determinants in the next section.

2.2 Properties of Determinants

In what follows A denotes an arbitrary $n \times n$ matrix. First we note some properties of determinants that are immediate consequences of the definition.

Exercise 1. Prove that if $\tilde{A}$ denotes a matrix obtained from a square matrix A by multiplying one of its rows (or columns) by a scalar k, then

$$\det \tilde{A} = k \det A. \quad \square$$

Recall that a function f is *homogeneous* over a field $\mathcal{F}$ if $f(kx) = kf(x)$ for every $k \in \mathcal{F}$. The assertion of Exercise 1 thus shows that the determinant is a homogeneous function of the ith row (or column) of the matrix A. Applying this property n times, it is easily seen that for any scalar k and $n \times n$ matrix A

$$\det(kA) = k^n \det A. \tag{1}$$

A function f is *additive* if $f(a + b) = f(a) + f(b)$ for every a, b in the domain of f. The following proposition shows that $\det A$ is an additive function of a row (or column) of the matrix A.

Exercise 2. Prove that if each element of the ith row of an $n \times n$ matrix A can be represented in the form

$$a_{ij} = b_j + c_j, \qquad j = 1, 2, \ldots, n,$$

then

$$\det A = \det B + \det C,$$

where all rows of B and C except the ith one are identical to those of A, and the ith rows are given by

$$B_i = [b_1 \quad b_2 \quad \cdots \quad b_n], \qquad C_i = [c_1 \quad c_2 \quad \cdots \quad c_n].$$

Note that a similar result holds concerning the columns of A. □

Combining the results of Exercises 1 and 2 yields the following.

Property 1. *The determinant of a square matrix is an additive and homogeneous function of the ith row (or column) of* A, $1 \leq i \leq n$.

We now continue to establish several other important properties of the determinant function.

Property 2. *The determinant of A is equal to the determinant of its transpose;*

$$\det A = \det A^{\mathrm{T}}.$$

PROOF. Recalling the definition of the transpose, it is not difficult to see that a diagonal

$$a_{1j_1}, a_{2j_2}, \ldots, a_{nj_n} \tag{2}$$

of A ordered, as usual, according to the row indices, and corresponding to the permutation

$$j = (j_1, j_2, \ldots, j_n) \tag{3}$$

of $1, 2, \ldots, n$, is also a diagonal of A^{T}. Since the element a_{ij} $(1 \leq i, j \leq n)$ of A is in the (j, i)th position in A^{T}, the *second* indices in (2) determine the row of A^{T} in which each element lies. To arrive at a term of $\det A^{\mathrm{T}}$, the elements must be permuted so that the second indices are in the natural order. Suppose that this reordering of the elements in (2) produces the diagonal

$$a_{k_1 1}, a_{k_2 2}, \ldots, a_{k_n n}. \tag{4}$$

Then the sign to be attached to the term (4) in det A^{T} is $(-1)^{t(k)}$, where k denotes the permutation

$$k = (k_1, k_2, \ldots, k_n). \tag{5}$$

To establish Property 2 it remains to show only that permutations (3) and (5) are both odd or both even. Consider pairs of indices

$$(1, j_1), \quad (2, j_2), \quad \ldots, \quad (n, j_n) \tag{6}$$

corresponding to a term of det A and pairs

$$(k_1, 1), \quad (k_2, 2), \quad \ldots, \quad (k_n, n) \tag{7}$$

obtained from (6) by a reordering according to the second entries. Observe that each interchange of pairs in (6) yields a simultaneous interchange of numbers in the permutations

$$j_0 = (1, 2, \ldots, n) \qquad \text{and} \qquad j = (j_1, j_2, \ldots, j_n).$$

Thus, a sequence of interchanges applied to j to bring it to the natural order j_0 (refer to Exercise 2.1.5) will simultaneously transform j_0 to k. But then the same sequence of interchanges applied in the reverse order will transform k to j. Thus, using Exercise 2.1.5, it is found that $t(j) = t(k)$. This completes the proof. ■

Note that this result establishes an equivalence between the rows and columns concerning properties of determinants. This allows the derivation of properties of determinants involving columns from those involving rows of the transposed matrix and vice versa. Taking advantage of this, subsequent proofs will be only of properties concerning the rows of a matrix.

Property 3. *If the $n \times n$ matrix B is obtained by interchanging two rows (or columns) of A, then det $B = -$det A.*

PROOF. Observe that the terms of det A and det B consist of the same factors, taking one and only one from each row and each column. It suffices, therefore, to show that the signs of each term are changed.

We first prove this result when B is obtained by interchanging the first two rows of A. We have

$$b_{1j_1} = a_{2j_1}, \qquad b_{2j_2} = a_{1j_2}, \qquad b_{kj_k} = a_{kj_k} \qquad (k = 3, 4, \ldots, n),$$

and so the expansion for det B becomes

$$\begin{aligned} \det B &= \sum_j (-1)^{t(j)} b_{1j_1} b_{2j_2} \cdots b_{nj_n} \\ &= \sum_j (-1)^{t(j)} a_{1j_2} a_{2j_1} a_{3j_3} \cdots a_{nj_n}, \end{aligned}$$

Let $\tilde{j}$ be the permutation $(j_2, j_1, j_3, \ldots, j_n)$ and j denote the permutation $(j_1, j_2, \ldots, j_n)$. By Exercise 2.1.4, $t(\tilde{j}) = t(j) \pm 1$ and therefore

$$\det B = -\sum_{j} (-1)^{t(\tilde{j})} a_{1j_2} a_{2j_1} a_{3j_3} \cdots a_{nj_n}.$$

But summing over all permutations $\tilde{j}$ gives the same $n!$ terms as summing over all permutations j. Thus $\det B = -\det A$, as required.

There is nothing new in principle in proving this result when any two neighbouring rows are interchanged: the indices 1 and 2 above are merely replaced by the numbers of the rows.

Now let the rows be in general position, with row numbers r, s, and $r < s$, say. Then with $s - r$ interchanges of neighbouring rows, rows $r, r + 1, \ldots, s - 1, s$ are brought into positions $r + 1, r + 2, \ldots, s, r$. A further $s - r - 1$ interchanges of neighbouring rows produces the required order $s, r + 1, r + 2, \ldots, s - 1, r$. Thus, a total of $2(s - r) - 1$ interchanges of neighbouring rows has the net effect of interchanging rows r and s. Since the number of such interchanges is always odd, the effect on the determinant is, once more, simply to change the sign. ■

The next property follows immediately from Property 3.

Property 4. *If the $n \times n$ matrix A has two rows (or columns) alike, then*

$$\det A = 0.$$

PROOF. The interchange of identical rows (or columns) obviously does not change the matrix. On the other hand, Property 3 implies that its determinant must change sign. Thus, $\det A = -\det A$, hence $\det A = 0$.

Exercise 3. Prove that if a row (or column) of A is a multiple of another row (or column) of the matrix, then $\det A = 0$.

Hint. Use Exercise 1 and Property 4. □

Combining the result in Exercise 3 with that of Exercise 2, we obtain the following.

Property 5. *Let B be the matrix obtained from A by adding the elements of its ith row (or column) to the corresponding elements of its jth row (or column) multiplied by a scalar α $(j \neq i)$. Then*

$$\det B = \det A.$$

When the operation described in Property 5 is applied several times, the evaluation of the determinant can be reduced to that of a triangular matrix. This is the essence of *Gauss's method* for computing determinants. It is illustrated in the next example.

Example 4. To compute the determinant of the matrix

$$A = \begin{bmatrix} 4 & -5 & 10 & 3 \\ 1 & -1 & 3 & 1 \\ 2 & -4 & 5 & 2 \\ -3 & 2 & -7 & -1 \end{bmatrix},$$

interchange the first two rows. Next add -4 (respectively, -2 and 3) times the first row of the new matrix to the second (respectively, third and fourth) rows to derive, using Properties 3 and 5,

$$\det A = -\det \begin{bmatrix} 1 & -1 & 3 & 1 \\ 0 & -1 & -2 & -1 \\ 0 & -2 & -1 & 0 \\ 0 & -1 & 2 & 2 \end{bmatrix}.$$

Now add -2 (respectively, -1) times the second row to the third (respectively, fourth) to obtain

$$\det A = -\det \begin{bmatrix} 1 & -1 & 3 & 1 \\ 0 & -1 & -2 & -1 \\ 0 & 0 & 3 & 2 \\ 0 & 0 & 4 & 3 \end{bmatrix}.$$

It remains now to add the third row multiplied by $-\frac{4}{3}$ to the last row to get the desired (upper-) triangular form:

$$\det A = -\det \begin{bmatrix} 1 & -1 & 3 & 1 \\ 0 & -1 & -2 & -1 \\ 0 & 0 & 3 & 2 \\ 0 & 0 & 0 & \frac{1}{3} \end{bmatrix}.$$

Thus, by Exercise 2.1.10, $\det A = 1$.

Exercise 5. If A is an $n \times n$ matrix and $\bar{A}$ denotes its complex conjugate, prove that $\det \bar{A} = \overline{\det A}$.

Exercise 6. Prove that the determinant of any Hermitian matrix is a real number.

Exercise 7. Show that for any matrix A, $\det(A^*A)$ is a nonnegative number.

Exercise 8. Show that the determinant of any skew-symmetric matrix of odd order is equal to zero.

Exercise 9. Show that the determinant of the $n \times n$ *companion matrix*

$$C_a = \begin{bmatrix} 0 & 1 & 0 & \cdots & 0 \\ \vdots & & 1 & & \vdots \\ \vdots & & \ddots & \ddots & 0 \\ 0 & \cdots & & 0 & 1 \\ -a_0 & -a_1 & -a_2 & \cdots & -a_{n-1} \end{bmatrix},$$

where $a_0, a_1, \ldots, a_{n-1} \in \mathbb{C}$, is equal to $(-1)^n a_0$.

Hint. Multiply the ith row ($i = 1, 2, \ldots, n - 1$) by a_i and add it to the last one.

Exercise 10. If $A = [a_{ij}]_{i,j=1}^{m,n}$, $m \le n$, and $B = [b_{ij}]_{i,j=1}^{n,m}$, check that

$$\det\begin{bmatrix} A & 0 \\ -I_n & B \end{bmatrix} = \det\begin{bmatrix} A & AB \\ -I_n & 0 \end{bmatrix},$$

where I_n stands for the $n \times n$ identity matrix.

Hint. Use Property 5 and add b_{jk} times the jth column ($1 \le j \le n$) of the matrix on the left to the $(n + k)$th column, $k = 1, 2, \ldots, m$. □

2.3 Cofactor Expansions

Another method for computation of determinants is based on their reduction to determinants of matrices of smaller sizes. This is in contrast to the method of Gauss (see Example 2.2.4). The following notion is important for this approach.

Let A be an $n \times n$ matrix. A *minor* of order $n - 1$ of A is defined to be the determinant of a submatrix of A obtained by striking out one row and one column from A. The minor obtained by striking out the ith row and jth column is written M_{ij} $(1 \le i, j \le n)$.

Exercise 1. Concerning the companion matrix of order 4 in Exercise 2.2.9,

$$M_{11} = \det\begin{bmatrix} 0 & 1 & 0 \\ 0 & 0 & 1 \\ -a_1 & -a_2 & -a_3 \end{bmatrix} = -a_1,$$

$$M_{41} = \det\begin{bmatrix} 1 & 0 & 0 \\ 0 & 1 & 0 \\ 0 & 0 & 1 \end{bmatrix} = 1. \quad \square$$

Minors can play an important role in computing the determinant, in view of the following result.

Theorem 1 (Cofactor expansion). *Let A be an arbitrary $n \times n$ matrix. Then for any i, j $(1 \le i, j \le n)$;*

$$\det A = a_{i1}A_{i1} + a_{i2}A_{i2} + \cdots + a_{in}A_{in} \tag{1}$$

or, similarly,

$$\det A = a_{1j}A_{1j} + a_{2j}A_{2j} + \cdots + a_{nj}A_{nj}, \tag{2}$$

where $A_{pq} = (-1)^{p+q}M_{pq}$.

Before proceeding with the details of the proof, note that the numbers A_{pq} $(1 \le p, q \le n)$ are called the *cofactors* of the elements a_{pq} and, therefore, formulas (1) and (2) are referred to as, respectively, *row* and *column cofactor expansions* of det A. Thus, the cofactor A_{pq} is just the minor M_{pq} multiplied by $+1$ or -1, according to whether the sum of the subscripts is even or odd.

PROOF. The proof begins with the observation that every term in Eq. (2.1.8) for det A contains an element of the ith row. Therefore, collecting together the terms containing a_{ij} $(j = 1, 2, \ldots, n)$, we obtain

$$\det A = a_{i1}\tilde{A}_{i1} + a_{i2}\tilde{A}_{i2} + \cdots + a_{in}\tilde{A}_{in} \qquad (1 \le i \le n). \tag{3}$$

Hence, to establish (1), it suffices to show that

$$\tilde{A}_{ij} = A_{ij} \tag{4}$$

for all i, j.

Consider first the case $i = j = 1$; we show that $\tilde{A}_{11} = A_{11}$. Indeed, from (2.1.8) and the construction of $\tilde{A}_{11}$ in (3), it follows that

$$\tilde{A}_{11} = \sum_{j} (-1)^{t(j)} a_{2j_2} a_{3j_3} \cdots a_{nj_n}, \tag{5}$$

where $t(j)$ denotes the number of inversions of the permutation $(1, j_2, j_3, \ldots j_n)$ or, equivalently, of permutation $(j_2, j_3, \ldots, j_n)$ of the numbers 2, 3, ..., n.

Using the definition of the determinant for the $(n-1) \times (n-1)$ matrix obtained by deleting the first row and the first column of A, it is easily concluded that the expression in (5) is that determinant, that is, M_{11}. Thus $\tilde{A}_{11} = M_{11} = (-1)^{1+1}M_{11} = A_{11}$.

To prove (4) in the general case, we first shift a_{ij} to the (1, 1) position by means of $i - 1$ successive interchanges of adjacent rows followed by $j - 1$ interchanges of adjacent columns. Call the rearranged matrix B. The minor associated with a_{ij} is the same in both A and B because the relative positions of rows and columns of the submatrix corresponding to this minor M_{ij} are unchanged. Hence, using the special case already proved,

$$\det B = a_{ij}M_{ij} + (\text{terms not involving } a_{ij}).$$

But Property 3 of determinants implies that

$$\det B = (-1)^{i-1}(-1)^{j-1} \det A = (-1)^{i+j} \det A.$$

Hence

$$\det A = (-1)^{i+j}a_{ij}M_{ij} + (\text{terms not involving } a_{ij}).$$

The result now follows on comparison with (3). Equation (2) can now be obtained using Eq. (1) and the fact that $\det A = \det A^{\mathrm{T}}$. ■

Exercise 2. The determinant of the matrix

$$A = \begin{bmatrix} 3 & 12 & -1 & 0 \\ -1 & 0 & 5 & 6 \\ 0 & 3 & 0 & 0 \\ 0 & 4 & -1 & 2 \end{bmatrix}$$

can be easily evaluated by use of the cofactor expansion along the third row. Indeed, the third row contains only one nonzero element and, therefore, in view of (1);

$$\det A = 3A_{32} = 3(-1)^{3+2} \det \begin{bmatrix} 3 & -1 & 0 \\ -1 & 5 & 6 \\ 0 & -1 & 2 \end{bmatrix} = (-3)46 = -138. \quad \square$$

The next exercise shows that if in expression (1) the cofactors A_{ij} $(1 \leq i \leq n, j = 1, 2, \ldots, n)$ are replaced by those corresponding to elements of another row, then the expression obtained is equal to zero. A similar result holds for columns.

Exercise 3. Show that if $i \neq r$, then

$$a_{i1}A_{r1} + a_{i2}A_{r2} + \cdots + a_{in}A_{rn} = 0, \tag{6}$$

and if $j \neq s$, then

$$a_{1j}A_{1s} + a_{2j}A_{2s} + \cdots + a_{nj}A_{ns} = 0. \tag{7}$$

Hint. Consider the matrix obtained from A by replacing row i by row r $(i \neq r)$ and use Property 4.

Exercise 4. Check that det A, where A is the $n \times n$ *Fibonacci matrix*

$$A = \begin{bmatrix} 1 & 1 & 0 & \cdots & 0 \\ -1 & 1 & 1 & \ddots & \vdots \\ 0 & -1 & \ddots & \ddots & 0 \\ \vdots & \ddots & \ddots & \ddots & 1 \\ 0 & \cdots & 0 & -1 & 1 \end{bmatrix},$$

is exactly the nth term in the Fibonacci sequence

$$1, 2, 3, 5, 8, 13, \ldots = \{a_n\}_{n=1}^{\infty},$$

where $a_n = a_{n-1} + a_{n-2}$ $(n \geq 3)$.

Exercise 5. (The Fibonacci matrix is an example of a "tridiagonal" matrix. This exercise concerns general tridiagonal matrices.) Let J_n denote the *Jacobi matrix* of order n;

$$J_n = \begin{bmatrix} a_1 & b_1 & 0 & & \cdots & 0 \\ c_1 & a_2 & b_2 & & & \vdots \\ 0 & c_2 & a_3 & b_3 & \cdots & 0 \\ \vdots & & \ddots & \ddots & \ddots & b_{n-1} \\ 0 & & \cdots & & c_{n-1} & a_n \end{bmatrix}.$$

Show that

$$|J_n| = a_n |J_{n-1}| - b_{n-1} c_{n-1} |J_{n-2}| \qquad (n \geq 3),$$

where the abbreviation $|A|$, as well as det A, denotes the determinant of A.

Exercise 6. Confirm that

$$\det \begin{bmatrix} 1 & 1 & \cdots & 1 \\ x_1 & x_2 & & x_n \\ x_1^2 & x_2^2 & & x_n^2 \\ \vdots & \vdots & & \vdots \\ x_1^{n-1} & x_2^{n-1} & \cdots & x_n^{n-1} \end{bmatrix} = \prod_{1 \leq j < i \leq n} (x_i - x_j). \tag{8}$$

The matrix in (8), written V_n, is called the *Vandermonde matrix* of order n. The relation in (8) states that its determinant is equal to the product of all differences of the form $x_i - x_j$ $(1 \leq j < i \leq n)$.

Hint. Use Property 5 of Section 2.2 and add the ith row multiplied by $-x_1$ to the row $i + 1$ $(i = n - 1, n - 2, \ldots, 1)$. Apply a column cofactor expansion formula for the reduction of det V_n to det V_{n-1}.

***Exercise* 7.** Show that if C_a denotes the companion matrix (see Exercise 2.2.9) associated with the (monic) polynomial

$$a(\lambda) = a_0 + a_1\lambda + \cdots + a_{n-1}\lambda^{n-1} + \lambda_n,$$

then

$$\det(\lambda I - C_a) = a(\lambda).$$

Deduce that $\det(\lambda_0 I - C_a) = 0$ if and only if λ_0 is a zero of $a(\lambda)$. □

2.4 Laplace's Theorem

The row and column cofactor expansions of a determinant can be generalized in a natural way, using the idea of a minor of order p $(1 \le p \le n)$. Recall that we have already defined minors of order $n - 1$ of an $n \times n$ determinant.† More generally, if A is an $m \times n$ matrix, then the determinant of a $p \times p$ submatrix of A $(1 \le p \le \min(m, n))$, obtained from A by striking out $m - p$ rows and $n - p$ columns, is called a *minor of order* p of A.

In more detail, if the rows and columns *retained* are given by subscripts

$$1 \le i_1 < i_2 < \cdots < i_p \le m, \qquad 1 \le j_1 < j_2 < \cdots < j_p \le n, \tag{1}$$

respectively, then the corresponding $p \times p$ minor is denoted by

$$A\begin{pmatrix} i_1 & i_2 & \cdots & i_p \\ j_1 & j_2 & \cdots & j_p \end{pmatrix} \triangleq \det[a_{i_k j_k}]_{k=1}^{p}. \tag{2}$$

For example, the minor M_{1n} of order $n - 1$ of an $n \times n$ matrix A is, in this notation,

$$M_{1n} = A\begin{pmatrix} 2 & 3 & \cdots & n \\ 1 & 2 & \cdots & n-1 \end{pmatrix},$$

and the minors of order one are merely the elements of the matrix.

The minors for which $i_k = j_k$ $(k = 1, 2, \ldots, p)$ are called the *principal minors* of A of order p, and the minors with $i_k = j_k = k$ $(k = 1, 2, \ldots, p)$ are referred to as the *leading principal minors* of A.

Before proceeding to a generalization of Eqs. (2.3.1) and (2.3.2), we also need the notion of *complementary minors*, that is, the determinant of the sub-

† That is, the determinant of an $n \times n$ matrix.

matrix of a *square* matrix A resulting from the *deletion* of the rows and columns listed in (1). We denote it by

$$A\begin{pmatrix} i_1 & i_2 & \cdots & i_p \\ j_1 & j_2 & \cdots & j_p \end{pmatrix}^{\mathrm{c}}.$$

The *complementary cofactor* to (2) is then defined by

$$A^{\mathrm{c}}\begin{pmatrix} i_1 & i_2 & \cdots & i_p \\ j_1 & j_2 & \cdots & j_p \end{pmatrix} \triangleq (-1)^s A\begin{pmatrix} i_1 & i_2 & \cdots & i_p \\ j_1 & j_2 & \cdots & j_p \end{pmatrix}^{\mathrm{c}},$$

where $s = (i_1 + i_2 + \cdots + i_p) + (j_1 + j_2 + \cdots + j_p)$. Obviously, in the case $p = 1$, this notion coincides with the previously defined cofactor of an element. Note that only the position of the index "c" indicates whether we are considering a complementary minor (on the right) or complementary cofactor (on the left).

Example 1. Consider a 5×5 matrix $A = [a_{ij}]_{i,j=1}^{5}$. Then, for instance,

$$A\begin{pmatrix} 2 & 3 & 5 \\ 1 & 2 & 4 \end{pmatrix} = \begin{vmatrix} a_{21} & a_{22} & a_{24} \\ a_{31} & a_{32} & a_{34} \\ a_{51} & a_{52} & a_{54} \end{vmatrix},$$

$$A\begin{pmatrix} 2 & 3 & 5 \\ 1 & 2 & 4 \end{pmatrix}^{\mathrm{c}} = A\begin{pmatrix} 1 & 4 \\ 3 & 5 \end{pmatrix} = \begin{vmatrix} a_{13} & a_{15} \\ a_{43} & a_{45} \end{vmatrix},$$

$$A^{\mathrm{c}}\begin{pmatrix} 2 & 3 & 5 \\ 1 & 2 & 4 \end{pmatrix} = (-1)^s \begin{vmatrix} a_{13} & a_{15} \\ a_{43} & a_{45} \end{vmatrix} = -\begin{vmatrix} a_{13} & a_{15} \\ a_{43} & a_{45} \end{vmatrix},$$

where $s = (2 + 3 + 5) + (1 + 2 + 4) = 17$. □

Theorem 1 (Laplace's theorem). *Let A denote an arbitrary $n \times n$ matrix and let any p rows (or columns) of A be chosen. Then* det A *is equal to the sum of the products of all $C_{n,p}$ minors*† *lying in these rows with their corresponding complementary cofactors:*

$$\det A = \sum_j A\begin{pmatrix} i_1 & i_2 & \cdots & i_p \\ j_1 & j_2 & \cdots & j_p \end{pmatrix} A^{\mathrm{c}}\begin{pmatrix} i_1 & i_2 & \cdots & i_p \\ j_1 & j_2 & \cdots & j_p \end{pmatrix}, \tag{3}$$

where the summation extends over all $C_{n,p}$ distinct sets of (column) indices $j_1, j_2, \ldots, j_p$ $(1 \le j_1 < j_2 < \cdots < j_p \le n)$. Or, equivalently, using columns,

$$\det A = \sum_i A\begin{pmatrix} i_1 & i_2 & \cdots & i_p \\ j_1 & j_2 & \cdots & j_p \end{pmatrix} A^{\mathrm{c}}\begin{pmatrix} i_1 & i_2 & \cdots & i_p \\ j_1 & j_2 & \cdots & j_p \end{pmatrix}. \tag{4}$$

where $1 \le i_1 < i_2 < \cdots < i_p \le n$.

† The symbol $C_{n,p}$ denotes the binomial coefficient $n!/(p!)(n-p)!$.

PROOF. The proof of this theorem can be arranged similarly to that of the cofactor expansion formula. However, we prefer here the following slightly different approach. First we prove that the product of the $p \times p$ leading principal minor with its complementary minor provides $p!(n-p)!$ elements of det A. Indeed, if

$$(-1)^{t(j')}a_{1j_1}a_{2j_2}\cdots a_{pj_p} \qquad \text{and} \qquad (-1)^{t(j'')}a_{p+1,j_{p+1}}a_{p+2,j_{p+2}}\cdots a_{nj_n}$$

are any terms of the given minor and its cofactor, then the product

$$(-1)^{t(j')+t(j'')}a_{1j_1}a_{2j_2}\cdots a_{nj_n}$$

is a term of det A, because j_k $(1 \le k \le p)$ is less than j_r $(p+1 \le r \le n)$. Therefore the number $t(j') + t(j'')$ is the number of inversions in the permutation $j = (j_1, j_2, \ldots, j_n)$, so it is equal to $t(j)$.

Since a $p \times p$ minor has $p!$ terms and the $(n-p) \times (n-p)$ corresponding complementary cofactor has $(n-p)!$ terms, their product provides $p!(n-p)!$ elements of det A.

Now consider an arbitrary minor

$$A\begin{pmatrix} i_1 & i_2 & \cdots & i_p \\ j_1 & j_2 & \cdots & j_p \end{pmatrix} \tag{5}$$

lying in the chosen rows of A. Starting with its i_1th row, we successively interchange the rows of A with the preceding ones so that eventually the minor in (5) lies in the first p rows of A. Note that $\sum_{k=1}^{p} i_k - \sum_{k=1}^{p} k$ interchanges of rows are performed. Similarly, the shifting of (5) to the position of the $p \times p$ leading submatrix of A requires $\sum_{k=1}^{p} j_k - \sum_{k=1}^{p} k$ interchanges of columns. Observe that the complementary minor to (5) did not change and that, using the part of the theorem already proved and Property 3 of determinants, its product with the minor provides $p!(n-p)!$ terms of $(-1)^s \det A$, where $s = \sum_{k=1}^{p} i_k + \sum_{k=1}^{p} j_k$.

It suffices now to replace the complementary minor by the corresponding complementary cofactor. Thus, each of $C_{n,p}$ distinct products of minors lying in the chosen p rows with their complementary cofactors provides $p!(n-p)!$ distinct element of det A. Consequently, the sum of these products gives exactly

$$C_{n,p}\,p!(n-p)! = n!$$

distinct terms of det A.

The proof of (3) is completed. The formula in (4) follows from (3) by use of Property 2 of determinants. ■

Exercise 2. If A_1 and A_4 are square matrices, show that

$$\det\begin{bmatrix} A_1 & 0 \\ A_3 & A_4 \end{bmatrix} = (\det A_1)(\det A_4).$$

Exercise 3. Show that if a square matrix A is of a block-triangular form, with square blocks A_i ($i = 1, 2, \ldots, k$) on the main (block) diagonal, then

$$\det A = \prod_{i=1}^{k} \det A_i.$$

Exercise 4. Prove that if the $n \times n$ matrix A is of the form

$$A = \begin{bmatrix} A_1 & A_2 \\ A_3 & 0 \end{bmatrix},$$

where A_2 and A_3 are $p \times p$ and $(n - p) \times (n - p)$ matrices, respectively, $p < n$, then

$$\det A = (-1)^{(n+1)p}(\det A_2)(\det A_3).$$

Exercise 5. Compute the determinant of the $n \times n$ matrix $[-D \quad B]$, where B is an $n \times m$ matrix ($m < n$) and D is obtained from the $n \times n$ identity matrix I_n by striking out the columns

$$j_1, j_2, \ldots, j_m \quad (1 \le j_1 < j_2 < \cdots < j_m \le n).$$

Answer.

$$\det[-D \quad B] = (-1)^t B\begin{pmatrix} j_1 & j_2 & \cdots & j_m \\ 1 & 2 & \cdots & m \end{pmatrix},$$

where

$$t = (n - m) + \left(\sum_{k=1}^{n} k + \sum_{k=1}^{m} j_k\right) - \sum_{k=1}^{n-m} k. \tag{6}$$

Hint. Observe that the submatrix $-D$ possesses only one nonzero minor of order $n - m$, and use (4).

Exercise 6. Show that if all minors of order k ($1 \le k < \min(m, n)$) of an $m \times n$ matrix A are equal to zero, then any minor of A of order greater than k is also zero.

Hint. Use Laplace's theorem for finding a larger minor by expansion along any k rows (or columns). □

2.5 The Binet–Cauchy Formula

As an important application of Laplace's theorem, we prove the following result concerning the determinant of a product of two matrices.

Theorem 1 (Binet–Cauchy formula). *Let A and B be $m \times n$ and $n \times m$ matrices, respectively. If $m \le n$ and $C = AB$, then*

$$\det C = \sum_{1 \le j_1 < j_2 < \cdots < j_m \le n} A\begin{pmatrix} 1 & 2 & \cdots & m \\ j_1 & j_2 & \cdots & j_m \end{pmatrix} B\begin{pmatrix} j_1 & j_2 & \cdots & j_m \\ 1 & 2 & \cdots & m \end{pmatrix}. \tag{1}$$

That is, the determinant of the product AB is equal to the sum of the products of all possible minors of (the maximal) order m of A with corresponding minors of B of the same order.[†]

PROOF. To prove the theorem, we first construct a block-triangular $(n + m) \times (n + m)$ matrix

$$\tilde{C} = \begin{bmatrix} A & 0 \\ -I_n & B \end{bmatrix},$$

where I_n denotes the $n \times n$ identity matrix. Now use Exercise 2.2.10 to deduce

$$\det \tilde{C} = \det\begin{bmatrix} A & AB \\ -I_n & 0 \end{bmatrix} = \det\begin{bmatrix} A & C \\ -I_n & 0 \end{bmatrix}$$

and hence, by Exercise 2.4.4,

$$\det \tilde{C} = (-1)^{(m+n+1)m}(\det C)(-1)^n = (-1)^{(m+n)(m+1)} \det C. \tag{2}$$

Now we evaluate $\det \tilde{C}$ in a different way. Since

$$\tilde{C}\begin{pmatrix} 1 & 2 & \cdots & m \\ j_1 & j_2 & \cdots & j_m \end{pmatrix} = 0$$

for at least one $j_r > n$ $(1 \le r \le m)$, it follows, using Laplace's theorem, that

$$\det \tilde{C} = \sum_{1 \le j_1 < j_2 \cdots < j_m \le n} A\begin{pmatrix} 1 & 2 & \cdots & m \\ j_1 & j_2 & \cdots & j_m \end{pmatrix} \tilde{C}^c\begin{pmatrix} 1 & 2 & \cdots & m \\ j_1 & j_2 & \cdots & j_m \end{pmatrix}. \tag{3}$$

Recalling the definition of $\tilde{C}$ and the result of Exercise 2.4.5, it is not difficult to see that

$$\tilde{C}^c\begin{pmatrix} 1 & 2 & \cdots & m \\ j_1 & j_2 & \cdots & j_m \end{pmatrix} = (-1)^s(-1)^t B\begin{pmatrix} j_1 & j_2 & \cdots & j_m \\ 1 & 2 & \cdots & m \end{pmatrix}, \tag{4}$$

where $s = \sum_{k=1}^m k + \sum_{k=1}^m j_k$ and t is given by Eq. (2.4.6).[‡] A little calculation gives

$$s + t = n(m + 1) + 2 \sum_{k=1}^{m} j_k,$$

and so

$$(s + t) - (m + n)(m + 1) = 2 \sum_{k=1}^{m} j_k - m(m + 1),$$

which is always even.

[†] For $m > n$ see Proposition 3.8.4.

[‡] Note that in the case $n = m$, we have $t = 0$.

Substituting (4) in (3) and comparing with (2), we complete the proof of the theorem. ■

By putting $m = n$ in (1), the following useful fact is obtained:

Corollary 1. *The determinant of the product of two square matrices A and B is the product of the determinants of the factors*:

$$\det AB = (\det A)(\det B). \tag{5}$$

Example 1. Let

$$A = \begin{bmatrix} 1 & 0 & 2 \\ -1 & 1 & 1 \end{bmatrix} \quad \text{and} \quad B = \begin{bmatrix} -1 & -1 \\ -2 & 0 \\ 1 & 1 \end{bmatrix}.$$

Then obviously

$$C = AB = \begin{bmatrix} 1 & 1 \\ 0 & 2 \end{bmatrix}$$

and $\det C = 2$. The same result is obtained by using the Binet–Cauchy formula:

$$\begin{aligned} \det C &= \sum_{1 \le j_1 < j_2 \le 3} A\begin{pmatrix} 1 & 2 \\ j_1 & j_2 \end{pmatrix} B\begin{pmatrix} j_1 & j_2 \\ 1 & 2 \end{pmatrix} \\ &= A\begin{pmatrix} 1 & 2 \\ 1 & 2 \end{pmatrix} B\begin{pmatrix} 1 & 2 \\ 1 & 2 \end{pmatrix} + A\begin{pmatrix} 1 & 2 \\ 1 & 3 \end{pmatrix} B\begin{pmatrix} 1 & 3 \\ 1 & 2 \end{pmatrix} + A\begin{pmatrix} 1 & 2 \\ 2 & 3 \end{pmatrix} B\begin{pmatrix} 2 & 3 \\ 1 & 2 \end{pmatrix} \\ &= \begin{vmatrix} 1 & 0 \\ -1 & 1 \end{vmatrix} \begin{vmatrix} -1 & -1 \\ -2 & 0 \end{vmatrix} + \begin{vmatrix} 1 & 2 \\ -1 & 1 \end{vmatrix} \begin{vmatrix} -1 & -1 \\ 1 & 1 \end{vmatrix} + \begin{vmatrix} 0 & 2 \\ 1 & 1 \end{vmatrix} \begin{vmatrix} -2 & 0 \\ 1 & 1 \end{vmatrix} \\ &= 2. \end{aligned}$$

Exercise 2. Check that for any $k \times n$ matrix A, $k \le n$,

$$\det AA^* = \sum_{1 \le j_1 < j_2 < \cdots < j_k \le n} \left| A\begin{pmatrix} 1 & 2 & \cdots & k \\ j_1 & j_2 & \cdots & j_k \end{pmatrix} \right|^2 \ge 0.$$

Exercise 3. Applying the Binet–Cauchy formula, prove that

$$\begin{vmatrix} \sum_{i=1}^{n} a_i c_i & \sum_{i=1}^{n} a_i d_i \\ \sum_{i=1}^{n} b_i c_i & \sum_{i=1}^{n} b_i d_i \end{vmatrix} = \sum_{1 \le j < k \le n} \begin{vmatrix} a_j & a_k \\ b_j & b_k \end{vmatrix} \begin{vmatrix} c_j & c_k \\ d_j & d_k \end{vmatrix}$$

and deduce the *Cauchy identity*:

$$\left(\sum_{i=1}^{n} a_i c_i\right)\left(\sum_{i=1}^{n} b_i d_i\right) - \left(\sum_{i=1}^{n} a_i d_i\right)\left(\sum_{i=1}^{n} b_i c_i\right)$$
$$= \sum_{1 \le j < k \le n} (a_j b_k - a_k b_j)(c_j d_k - c_k d_j).$$

Hint. Use the matrices

$$A = \begin{bmatrix} a_1 & a_2 & \cdots & a_n \\ b_1 & b_2 & \cdots & b_n \end{bmatrix}, \qquad B = \begin{bmatrix} c_1 & c_2 & \cdots & c_n \\ d_1 & d_2 & \cdots & d_n \end{bmatrix}^{\mathrm{T}}.$$

Exercise 4. Prove the Cauchy–Schwarz inequality,

$$\left(\sum_{i=1}^{n} a_i b_i\right)^2 \le \left(\sum_{i=1}^{n} a_i^2\right)\left(\sum_{i=1}^{n} b_i^2\right),$$

in real numbers.†

Hint. Use Exercise 3.

Exercise 5. Check that the determinant of the product of any (finite) number of square matrices of the same order is equal to the product of the determinants of the factors. □

2.6 Adjoint and Inverse Matrices

Let $A = [a_{ij}]_{i,j=1}^{n}$ be any $n \times n$ matrix and, as defined in Section 2.4, let $A_{ij} = A^{c}\begin{pmatrix} i \\ j \end{pmatrix}$ be the cofactor of a_{ij} $(1 \le i, j \le n)$. The *adjoint* of A, written adj A, is defined to be the transposed matrix of cofactors of A. Thus

$$\operatorname{adj} A \triangleq ([A_{ij}]_{i,j=1}^{n})^{\mathrm{T}}.$$

Exercise 1. Check that if

$$A = \begin{bmatrix} 2 & 1 & -1 \\ 0 & 2 & 0 \\ 1 & -1 & 3 \end{bmatrix},$$

† For a straightforward proof of this inequality (in a more general context), see Theorem 3.11.1.

then

$$\operatorname{adj} A = \begin{bmatrix} \begin{vmatrix} 2 & 0 \\ -1 & 3 \end{vmatrix} & -\begin{vmatrix} 0 & 0 \\ 1 & 3 \end{vmatrix} & \begin{vmatrix} 0 & 2 \\ 1 & -1 \end{vmatrix} \\ -\begin{vmatrix} 1 & -1 \\ -1 & 3 \end{vmatrix} & \begin{vmatrix} 2 & -1 \\ 1 & 3 \end{vmatrix} & -\begin{vmatrix} 2 & 1 \\ 1 & -1 \end{vmatrix} \\ \begin{vmatrix} 1 & -1 \\ 2 & 0 \end{vmatrix} & -\begin{vmatrix} 2 & -1 \\ 0 & 0 \end{vmatrix} & \begin{vmatrix} 2 & 1 \\ 0 & 2 \end{vmatrix} \end{bmatrix}^{\mathrm{T}} = \begin{bmatrix} 6 & -2 & 2 \\ 0 & 7 & 0 \\ -2 & 3 & 4 \end{bmatrix}. \quad \square$$

Some properties of adjoint matrices follow immediately from the definition.

Exercise 2. Check that for any $n \times n$ matrix A and any $k \in \mathscr{F}$,

$$\operatorname{adj}(A^{\mathrm{T}}) = (\operatorname{adj} A)^{\mathrm{T}},$$

$$\operatorname{adj}(A^*) = (\operatorname{adj} A)^*,$$

$$\operatorname{adj} I = I,$$

$$\operatorname{adj}(kA) = k^{n-1} \operatorname{adj} A. \quad \square$$

The following striking result, and its consequences, are the main reason for interest in the adjoint of a matrix.

Theorem 1. *For any $n \times n$ matrix A,*

$$A(\operatorname{adj} A) = (\det A)I. \tag{1}$$

PROOF. Computing the product on the left by use of the row expansion formula (2.3.1) and the result of Exercise 2.3.3, we obtain a matrix with the number $\det A$ in each position on its main diagonal and zeros elsewhere, that is, $(\det A)I$. ■

Exercise 3. Prove that

$$(\operatorname{adj} A)A = (\det A)I. \quad \square \tag{2}$$

Theorem 1 has some important consequences. For example, if A is $n \times n$ and $\det A \neq 0$, then

$$\det(\operatorname{adj} A) = (\det A)^{n-1}. \tag{3}$$

In fact, using the property $\det AB = (\det A)(\det B)$ with Eq. (1) and the known determinant of the diagonal matrix, the result follows.

The vanishing or nonvanishing of $\det A$ turns out to be a vitally important property of a square matrix and warrants a special definition: A square matrix A is said to be *singular* or *nonsingular* according to whether $\det A$ is zero or nonzero.

Exercise 4. Confirm that the matrices

$$A = \begin{bmatrix} 2 & 2 \\ 1 & 1 \end{bmatrix} \quad \text{and} \quad B = \begin{bmatrix} 1 & 1 & 2 \\ 0 & 1 & 1 \\ 0 & 0 & -4 \end{bmatrix}$$

are, respectively, singular and nonsingular. □

An $n \times n$ matrix B is referred to as an *inverse* of the $n \times n$ matrix A if

$$AB = BA = I, \tag{4}$$

and when this is the case, we write $B = A^{-1}$. Also, when A has an inverse, A is said to be *invertible*. We shall see that a square matrix is invertible if and only if it is nonsingular (Theorem 2 of this section).

Exercise 5. Check that the matrices

$$A = \begin{bmatrix} 1 & 0 & 0 \\ \frac{1}{2} & -\frac{1}{2} & 0 \\ 0 & -\frac{1}{2} & \frac{1}{2} \end{bmatrix} \quad \text{and} \quad B = \begin{bmatrix} 1 & 0 & 0 \\ 1 & -2 & 0 \\ 1 & -2 & 2 \end{bmatrix}$$

are mutually inverse matrices.

Exercise 6. Show that a square upper- (lower-) triangular matrix is invertible if and only if there are no zeros on the main diagonal. If this is the case, verify that the inverse is a matrix of the same kind.

Exercise 7. Check that the involutory matrices, and only they, are their own inverses.

Exercise 8. Check that if A^{-1} exists, then

$$(A^{-1})^{\mathrm{T}} = (A^{\mathrm{T}})^{-1}. \quad \square$$

The notion of an inverse matrix has its origin in the familiar field of the real (or complex) numbers. Indeed, if $a \neq 0$, we may consider the quotient $b = 1/a$ as the inverse (or reciprocal) of the number a, and b satisfies $ab = ba = 1$, as in Eq. (4). We also write $b = a^{-1}$, of course. However, the noncommutativity of matrix multiplication and, especially, the lack of some of the properties of numbers, as mentioned in Section 1.3, yield substantial differences between the algebraic properties of inverse matrices and reciprocals of numbers.

For instance, as the next exercise shows, not all nonzero matrices have an inverse.

Exercise 9. Check that if

$$A = \begin{bmatrix} 0 & 0 & 1 \\ 0 & 0 & 0 \\ 0 & 0 & 0 \end{bmatrix},$$

then there is no 3×3 matrix B such that

$$AB = I. \quad \square$$

It should be emphasized that, in view of the lack of commutativity of matrix multiplication, if B is an inverse of A the definition requires that it must simultaneously be a *left* inverse ($BA = I$) and a *right* inverse ($AB = I$). However, it will be seen shortly that a one-sided inverse for a *square* matrix is necessarily two-sided.

Furthermore, it should be noted that the definition (4) provides trivially a necessary condition for a matrix to be invertible. Namely, if a square matrix A has even a one-sided inverse, then A is nonsingular, that is, $\det A \neq 0$. Indeed, if for instance $AB = I$, then

$$\det AB = (\det A)(\det B) = 1$$

and, obviously, $\det A \neq 0$. It turns out that the condition $\det A \neq 0$ is also sufficient for A to have an inverse.

Theorem 2. *A square matrix is invertible if and only if it is nonsingular.*

PROOF. If $\det A \neq 0$, then formulas (1) and (2) show that the matrix

$$B = \frac{1}{\det A} \operatorname{adj} A \tag{5}$$

satisfies conditions (4) and, therefore, is an inverse of A. ■

The existence of an inverse in the case $\det A \neq 0$ easily implies its uniqueness. To see this, first check that the right inverse is unique, that is, if $AB_1 = I$ and $AB_2 = I$, then $B_1 = B_2$. Indeed, $AB_1 = AB_2$ implies $A(B_1 - B_2) = 0$, and since A has a right inverse, then $\det A \neq 0$. Applying Theorem 2, we deduce the existence of A^{-1}. Multiplying by A^{-1} from the left we have

$$A^{-1}A(B_1 - B_2) = 0,$$

and the required equality $B_1 = B_2$ is obtained.

Similarly, the uniqueness of the left inverse of a nonsingular matrix is established. Hence if $\det A \neq 0$, then both one-sided inverses are unique. Furthermore, since they are both given by Eq. (5), they are equal.

Exercise 10. Check, using Eq. (5), that

$$\begin{bmatrix} 2 & 1 & -1 \\ 0 & 2 & 0 \\ 1 & -1 & 3 \end{bmatrix}^{-1} = \tfrac{1}{14}\begin{bmatrix} 6 & -2 & 2 \\ 0 & 7 & 0 \\ -2 & 3 & 4 \end{bmatrix}. \quad \square$$

Numerical calculations with Eq. (5) for large matrices are, in general, notoriously expensive in computing time. Other methods for calculating the inverse will be presented in Exercise 2.7.4 and provide a better basis for general-purpose algorithms.

Exercise 11. Check that if A is nonsingular, then

$$\det(A^{-1}) = (\det A)^{-1}.$$

Exercise 12. Confirm that the product of any number of square matrices of the same order is invertible if and only if each factor is invertible.

Exercise 13. Prove that

$$(AB)^{-1} = B^{-1}A^{-1},$$

provided A and B are invertible matrices of the same size.

Exercise 14. Show that if A and B are nonsingular matrices of the same size and $\det A = \det B$, then there exists a matrix C with determinant 1 such that $A = BC$.

Exercise 15. Check that if A is a nonsingular matrix and D is square (A and D may be of different sizes), then

$$\det\begin{bmatrix} A & B \\ C & D \end{bmatrix} = \det(D - CA^{-1}B)\det A. \tag{6}$$

Hint. Use the relation

$$\begin{bmatrix} I & 0 \\ CA^{-1} & I \end{bmatrix}\begin{bmatrix} A & 0 \\ 0 & D - CA^{-1}B \end{bmatrix}\begin{bmatrix} I & A^{-1}B \\ 0 & I \end{bmatrix} = \begin{bmatrix} A & B \\ C & D \end{bmatrix}, \tag{7}$$

as well as Exercises 2.4.2 and 2.5.5. (It is interesting to observe the analogy between Eq. (6) and the formula for the determinant of a 2×2 matrix. Note also that formulas similar to Eq. (7) in the cases when B, C, or D are invertible can be easily derived.) The matrix $D - CA^{-1}B$ is known as the *Schur complement* of A in $\begin{bmatrix} A & B \\ C & D \end{bmatrix}$.

Exercise 16. A matrix $A \in \mathbb{R}^{n \times n}$ is said to be *orthogonal* if $A^{T}A = I$. Show that:

(a) An orthogonal matrix A is invertible, $\det A = \pm 1$, and $A^{-1} = A^{T}$.

(b) If $n = 2$, all orthogonal matrices have one of the two forms,

$$\begin{bmatrix} \cos\theta & -\sin\theta \\ \sin\theta & \cos\theta \end{bmatrix}, \qquad \begin{bmatrix} \cos\theta & \sin\theta \\ \sin\theta & -\cos\theta \end{bmatrix},$$

for some angle θ.

Exercise 17. A matrix $U \in \mathbb{C}^{n \times n}$ is said to be *unitary* if $U^*U = I$. Show that, in this case, U is invertible, $|\det U| = 1$ and $U^{-1} = U^*$. □

2.7 Elementary Operations on Matrices

Properties 1 through 5 of determinants indicate a close relationship between the determinant of a square matrix A and matrices obtained from A by the following operations:

(1) interchanging two rows (or columns) in A;
(2) multiplying all elements of a row (or column) of A by some nonzero number;
(3) adding to any row (or column) of A any other row (or column) of A multiplied by a nonzero number.

The operations 1 through 3 listed above are referred to as *elementary row* (respectively, *column*) *operations of types* 1, 2, and 3 respectively.

Now observe that these manipulations of the rows and columns of a matrix A can be achieved by pre- and postmultiplication (respectively) of A by appropriate matrices. In particular, the interchange of rows i_1 and i_2 $(i_1 < i_2)$ of A can be performed by multiplying A from the left by the $n \times n$ matrix

$$E^{(1)} = \begin{array}{c} \\ \\ \\ (i_1) \\ \\ \\ \\ (i_2) \\ \\ \\ \\ \end{array} \begin{bmatrix} 1 & & & & & & & & & & \\ & \ddots & & & & & & & & & \\ & & 1 & & & & & & & & \\ & & & 0 & & \cdots & & 1 & & & \\ & & & & 1 & & & & & & \\ & & & \vdots & & \ddots & & \vdots & & & \\ & & & & & & 1 & & & & \\ & & & 1 & & \cdots & & 0 & & & \\ & & & & & & & & 1 & & \\ & & & & & & & & & \ddots & \\ & & & & & & & & & & 1 \end{bmatrix}, \tag{1}$$

obtained by interchanging rows i_1 and i_2 of the identity matrix. Thus, the matrix $E^{(1)}A$ has the same rows as A except that rows i_1 and i_2 are interchanged.

Furthermore, the effect of multiplying the ith row of A by a nonzero number k can be achieved by forming the product $E^{(2)}A$, where

$$E^{(2)} = \begin{matrix} \\ \\ \\ (i) \\ \\ \\ \\ \end{matrix}\begin{bmatrix} 1 & & & & & \\ & \ddots & & & & \\ & & 1 & & & \\ & & & k & & \\ & & & & 1 & \\ & & & & & \ddots & \\ & & & & & & 1 \end{bmatrix}. \tag{2}$$

Finally, adding k times row i_2 to row i_1 of A is equivalent to multiplication of A from the left by the matrix

$$E^{(3)} = \begin{matrix} \\ \\ (i_1) \\ \\ \\ (i_2) \\ \\ \\ \end{matrix}\begin{bmatrix} 1 & & & & & \\ & \ddots & & & & \\ & & 1 & \cdots & k & \\ & & & 1 & \vdots & \\ & & & & \ddots & \\ & & & & 1 & \\ & & & & & 1 & \\ & & & & & & \ddots \\ & & & & & & & 1 \end{bmatrix} \quad \text{or} \quad E^{(3)} = \begin{matrix} \\ \\ (i_2) \\ \\ (i_1) \\ \\ \\ \end{matrix}\begin{bmatrix} 1 & & & & \\ & \ddots & & & \\ & & 1 & & \\ & & \vdots & \ddots & \\ & & k & \cdots & 1 \\ & & & & & \ddots \\ & & & & & & 1 \end{bmatrix}, \tag{3}$$

depending on whether $i_1 < i_2$ or $i_1 > i_2$. Again note that the operation to be performed on A is applied to I in order to produce $E^{(3)}$.

Similarly, the multiplication of A on the *right* by the appropriate matrices $E^{(1)}$, $E^{(2)}$, or $E^{(3)}$ leads to analogous changes in columns.

The matrices $E^{(1)}$, $E^{(2)}$, and $E^{(3)}$ are called *elementary matrices* of *types* 1, 2, and 3, respectively.

Properties 1, 3, and 5 of determinants may now be rewritten as follows:

$$\begin{aligned} \det E^{(1)}A &= \det AE^{(1)} = -\det A, \\ \det E^{(2)}A &= \det AE^{(2)} = k \det A, \\ \det E^{(3)}A &= \det AE^{(3)} = \det A. \end{aligned} \tag{4}$$

Exercise 1. Check that

$$\det E^{(1)} = -1, \qquad \det E^{(2)} = k, \qquad \det E^{(3)} = 1.$$

Exercise 2. Check that the inverses of the elementary matrices are also elementary of the same type. ☐

The application of elementary operations admits the reduction of a general $m \times n$ matrix to a simpler one, which contains, in particular, more zeros than the original matrix. This approach is known as the *Gauss reduction* (or *elimination*) *process* and is illustrated below. In fact, it was used in Example 2.2.4 for computing a determinant.

Example 3. Consider the matrix

$$A = \begin{bmatrix} 0 & 1 & 2 & -1 \\ -2 & 0 & 0 & 6 \\ 4 & -2 & -4 & -10 \end{bmatrix}.$$

Multiplying the second row by $-\frac{1}{2}$, we obtain

$$A_1 = E_1^{(2)} A = \begin{bmatrix} 0 & 1 & 2 & -1 \\ 1 & 0 & 0 & -3 \\ 4 & -2 & -4 & -10 \end{bmatrix},$$

where $E_1^{(2)}$ is a 3×3 matrix of the type 2 with $k = -\frac{1}{2}$, $i = 2$.

Multiplying the second row by -4 and adding it to the third, we deduce

$$A_2 = E_2^{(3)} A_1 = \begin{bmatrix} 0 & 1 & 2 & -1 \\ 1 & 0 & 0 & -3 \\ 0 & -2 & -4 & 2 \end{bmatrix},$$

where $E_2^{(3)}$ is a 3×3 elementary matrix of the third type with $k = -4$, $i_1 = 3$, $i_2 = 2$.

The interchange of the first two rows provides

$$A_3 = E_3^{(1)} A_2 = \begin{bmatrix} 1 & 0 & 0 & -3 \\ 0 & 1 & 2 & -1 \\ 0 & -2 & -4 & 2 \end{bmatrix},$$

and the first column of the resulting matrix consists only of zeros except one element equal to 1.

Employing a 3×3 matrix of type 3, where $k = 2$, $i_1 = 3$, $i_2 = 2$, we obtain

$$A_4 = E_4^{(3)} A_3 = \begin{bmatrix} 1 & 0 & 0 & -3 \\ 0 & 1 & 2 & -1 \\ 0 & 0 & 0 & 0 \end{bmatrix} \tag{5}$$

and it is clear that A_4 is the simplest form of the given matrix A that can be achieved by row elementary operations. This is the "reduced row-echelon form" for A, which is widely used in solving linear systems (see Section 2.9).

Performing appropriate elementary column operations, it is not difficult to see that the matrix A can be further reduced to the *diagonal* form:

$$A_5 = A_4 E_5^{(3)} E_6^{(3)} E_7^{(3)} = \begin{bmatrix} 1 & 0 & 0 & 0 \\ 0 & 1 & 0 & 0 \\ 0 & 0 & 0 & 0 \end{bmatrix},$$

where $E_5^{(3)}$, $E_6^{(3)}$, $E_7^{(3)}$ denote the following matrices of type (3):

$$E_5^{(3)} = \begin{bmatrix} 1 & 0 & 0 & 3 \\ 0 & 1 & 0 & 0 \\ 0 & 0 & 1 & 0 \\ 0 & 0 & 0 & 1 \end{bmatrix}, \quad E_6^{(3)} = \begin{bmatrix} 1 & 0 & 0 & 0 \\ 0 & 1 & -2 & 0 \\ 0 & 0 & 1 & 0 \\ 0 & 0 & 0 & 1 \end{bmatrix}, \quad E_7^{(3)} = \begin{bmatrix} 1 & 0 & 0 & 0 \\ 0 & 1 & 0 & 1 \\ 0 & 0 & 1 & 0 \\ 0 & 0 & 0 & 1 \end{bmatrix}.$$

Summarizing all of these reductions, we have

$$(E_4^{(3)} E_3^{(1)} E_2^{(3)} E_1^{(2)}) A (E_5^{(3)} E_6^{(3)} E_7^{(3)}) = \begin{bmatrix} 1 & 0 & 0 & 0 \\ 0 & 1 & 0 & 0 \\ 0 & 0 & 0 & 0 \end{bmatrix} \tag{6}$$

or

$$PAQ = \begin{bmatrix} 1 & 0 & 0 & 0 \\ 0 & 1 & 0 & 0 \\ 0 & 0 & 0 & 0 \end{bmatrix}, \tag{7}$$

where the matrices P and Q are products of elementary matrices and are, therefore, nonsingular. □

An application of the elimination procedure just used to the case of a generic rectangular $m \times n$ matrix leads to the following important result, which is presented without a formal proof.

Theorem 1. *Any nonzero $m \times n$ matrix can be reduced by elementary row operations to reduced row echelon form, which has the following defining properties:*

(1) *All zero rows are in the bottom position(s).*

(2) *The first nonzero element (reading from the left) in a nonzero row is a one (called a "leading one").*

(3) *For $j = 2, 3, \ldots, m$ the leading one in row j (if any) appears to the right of the leading one in row $j - 1$.*

(4) *Any column containing a leading one has all other elements equal to zero.*

The following matrices are all in reduced row echelon form:

$$\begin{bmatrix} 1 & 0 & -3 \\ 0 & 1 & 2 \end{bmatrix}, \quad \begin{bmatrix} 1 & 0 \\ 0 & 1 \\ 0 & 0 \end{bmatrix}, \quad \begin{bmatrix} 1 & 2 & 1 \\ 0 & 0 & 0 \end{bmatrix}, \quad \begin{bmatrix} 1 & -2 & 0 \\ 0 & 0 & 1 \end{bmatrix}.$$

Observe that these matrices cannot be reduced to simpler forms by the application of elementary row operations. If, however, elementary column operations are also permitted, then further reductions may be possible, as the next theorem shows. Again, a detailed proof is omitted.

Theorem 2. *Any nonzero $m \times n$ matrix can be reduced by the application of elementary row and column operations to an $m \times n$ matrix in one of the following forms*:

$$[I_m \quad 0], \quad \begin{bmatrix} I_n \\ 0 \end{bmatrix}, \quad \begin{bmatrix} I_r & 0 \\ 0 & 0 \end{bmatrix}, \quad I_n \tag{8}$$

Recalling the connection of elementary operations with elementary matrices, we derive the next theorem from Theorem 2 (see also Eqs. (6) and (7)).

Theorem 3. *Let A be an $m \times n$ matrix. There exists a finite sequence of elementary matrices $E_1, E_2, \ldots, E_{k+s}$ such that*

$$E_k E_{k-1} \cdots E_1 A E_{k+1} E_{k+2} \cdots E_{k+s} \tag{9}$$

is one of the matrices in (8).

Since elementary matrices are nonsingular (see Exercise 1), their product is also and, therefore, the matrix in Eq. (9) can be written in the form PAQ, where P and Q are nonsingular. This point of view is important enough to justify a formal statement.

Corollary 1. *For any $m \times n$ matrix A there exist a nonsingular $m \times m$ matrix P and a nonsingular $n \times n$ matrix Q such that PAQ is one of the matrices in* (8).

Consider now the set of all nonzero $m \times n$ matrices with fixed m and n. Since the number of matrices of the types in (8) is strictly limited, there must exist many different matrices having the same matrix from (8) as its reduced form.

Consider two such matrices A and B. In view of Theorem 3, there are two sequences of elementary matrices reducing A and B to the same matrix. Hence

$$E_k E_{k-1} \cdots E_1 A E_{k+1} \cdots E_{k+s} = \tilde{E}_l \tilde{E}_{l-1} \cdots \tilde{E}_1 B \tilde{E}_{l+1} \cdots \tilde{E}_{l+t},$$

and since the inverse of an elementary matrix is an elementary matrix of the same kind (Exercise 2), we can conclude that

$$A = (E_1^{-1}E_2^{-1} \cdots E_k^{-1}\tilde{E}_l \cdots \tilde{E}_1)B(\tilde{E}_{l+1} \cdots \tilde{E}_{l+t}E_{k+s}^{-1} \cdots E_{k+1}^{-1}).$$

Hence A can be obtained from B by a number of elementary operations.

In the language of Corollary 1, we have shown that

$$A = PBQ \tag{10}$$

for some nonsingular matrices P and Q of appropriate sizes. The next theorem summarizes these results.

Theorem 4. *If A and B are matrices of the same size, the following statements are equivalent:*

(1) *A and B have the same reduced form* (8).

(2) *Either of these matrices can be obtained from the other by elementary row and column operations.*

(3) *There exist nonsingular matrices P and Q such that $A = PBQ$.*

Each of the above statements can be used as a definition of the following notion of equivalence of two matrices. For instance, the matrices A and B are said to be *equivalent*, written $A \sim B$, if statement 3 applies.

Obviously, the relation $A \sim B$ is *an equivalence relation* in the sense that

(1) $A \sim A$ (a matrix is equivalent to itself). This is the property of *reflexivity*.

(2) If $A \sim B$, then $B \sim A$ (*symmetry*).

(3) If $A \sim B$ and $B \sim C$, then $A \sim C$ (*transitivity*).

Thus, the set of all $m \times n$ matrices with fixed m and n is split into nonintersecting classes of mutually equivalent matrices (*equivalence classes*). Each $m \times n$ matrix A belongs to one and only one class, which can be described as the set of all matrices PAQ where P and Q vary over all nonsingular $m \times m$ and $n \times n$ matrices, respectively.

Each equivalence class is represented and determined uniquely by a matrix of the form (8), called the *canonical* (*simplest*) *form* of the matrices of this class. For example, if $r < m < n$, then a matrix

$$\begin{bmatrix} I_r & 0 \\ 0 & 0 \end{bmatrix} \tag{11}$$

is a representative of a class; it is the canonical form of the matrices of its equivalence class. Obviously, an equivalence class is identified if the number r in (11) is known. This fact justifies a detailed investigation of the number r, as carried out in the next section.

We conclude this section with a few supplementary exercises. The first provides a method for finding the inverse of an invertible matrix via elementary row operations.

Exercise 4. Show that if A is an $n \times n$ matrix, then the matrices

$$[A \quad I_n] \qquad \text{and} \qquad [I_n \quad B] \tag{12}$$

of the same size are row equivalent (that is, each can be reduced to the other by elementary row operations) if and only if A is invertible and $B = A^{-1}$.

Hint. If the matrices in (12) are row equivalent, $[A \quad I_n] = P[I_n \quad B] = [P \quad PB]$. Thus, if the Gauss reduction process applied to $[A \quad I_n]$ gives the matrix $[I_n \quad B]$, then necessarily $B = A^{-1}$. If such a reduction is impossible then A fails to be invertible.

Exercise 5. Show that if two square matrices are equivalent, then they are either both nonsingular or both singular.

Exercise 6. Check that an $n \times n$ matrix is nonsingular if and only if it is row (or column) equivalent to the identity matrix I_n. Thus, the class of mutually equivalent matrices containing an identity matrix consists of all nonsingular matrices of the appropriate size.

Exercise 7. Show that any nonsingular matrix can be decomposed into a product of elementary matrices.

Exercise 8. Show that the reduced row-echelon form of a given matrix is unique. □

2.8 Rank of a Matrix

Let A be an $m \times n$ matrix. The size r $(1 \le r \le \min(m, n))$ of the identity matrix in the reduced (canonical) form for A of Theorem 2.7.2 is referred to as the *rank* of A, written $r = \operatorname{rank} A$. Recalling the possible canonical forms of A given by (2.7.8), we have

$$\operatorname{rank}[I_m \quad 0] = m, \qquad \operatorname{rank}\begin{bmatrix} I_n \\ 0 \end{bmatrix} = n,$$

$$\operatorname{rank}\begin{bmatrix} I_r & 0 \\ 0 & 0 \end{bmatrix} = r, \qquad \operatorname{rank} I_n = n.$$

Clearly, all matrices belonging to the same class of equivalent matrices have the same rank. In other words, mutually equivalent matrices are of the same rank.

Exercise 1. Check that the matrix in Exercise 2.7.3 has rank 2.

Exercise 2. Show that the rank of an $n \times n$ matrix A is equal to n if and only if A is nonsingular. □

The last exercise shows that in the case $\det A \neq 0$, the problem of finding the rank is solved: it is equal to the size of the (square) matrix.

In the general case of a rectangular or singular square matrix, we need a method of finding the rank that is more convenient than its reduction to canonical form. Note that the problem is important, since the rank determines uniquely the whole class of mutually equivalent matrices of the given size.

We start to develop a method for the direct evaluation of rank by examining the matrix A itself with an obvious but suggestive remark. Consider the representative (canonical form)

$$\begin{bmatrix} I_r & 0 \\ 0 & 0 \end{bmatrix} \tag{1}$$

of a class of equivalent matrices of rank r and observe that its rank r is equal to the order of the largest nonvanishing minor. The same fact clearly holds for other types of canonical forms in (2.7.8). It turns out that this property of the rank carries over to an arbitrary matrix.

Theorem 1. *The rank of an $m \times n$ matrix A is equal to the order of its largest nonzero minor.*

PROOF. Theorem 2.7.3 states, in fact, that the matrix A can be obtained from its canonical form (2.7.8) by multiplication by elementary matrices (inverses to those in (2.7.9)). Then, in view of the remark preceding Theorem 1, it suffices to show that multiplying a matrix by an elementary matrix does not change the order of its largest nonzero minor.

Indeed, the effect of multiplication by an elementary matrix is expressed in Eqs. 2.7.4: the elementary operations only transform nonzero minors to nonzero minors and zero minors to zero minors. Obviously, the same situation holds if repeated multiplication by elementary matrices is performed. ■

Example 3. The rank of the matrix

$$A = \begin{bmatrix} 3 & 2 & 1 & 1 \\ 6 & 4 & 2 & 2 \\ -3 & -2 & -1 & -1 \end{bmatrix}$$

is equal to 1, because there is a nonzero minor of order 1 (i.e., an element of the matrix) and each minor of order 2 or 3 is equal to zero. □

Some properties of the rank are indicated in the following exercises:

Exercise 4. Show that for any rectangular matrix A and any nonsingular matrices P, Q of appropriate sizes,

$$\operatorname{rank}(PAQ) = \operatorname{rank} A.$$

Exercise 5. Prove that for any rectangular matrix A,

$$\operatorname{rank} A = \operatorname{rank} A^{\mathrm{T}} = \operatorname{rank} A^*.$$

Exercise 6. Prove that for a block-diagonal matrix,

$$\operatorname{rank} \operatorname{diag}[A_{11}, A_{22}, \ldots, A_{kk}] = \sum_{i=1}^{k} \operatorname{rank} A_{ii}. \quad \square$$

For block-triangular matrices we can claim only the following:

Exercise 7. Confirm that

$$\operatorname{rank}\begin{bmatrix} A_{11} & A_{12} & \cdots & A_{1k} \\ 0 & A_{22} & \ddots & \vdots \\ \vdots & \ddots & \ddots & A_{k-1,k} \\ 0 & \cdots & 0 & A_{kk} \end{bmatrix} \geq \sum_{i=1}^{k} \operatorname{rank} A_{ii}. \quad \square$$

A case in which inequality occurs in Exercise 7 is illustrated in the next exercise.

Exercise 8. The rank of the matrix

$$\begin{bmatrix} A_{11} & A_{12} \\ 0 & A_{22} \end{bmatrix} = \left[\begin{array}{cc:cc} 1 & 0 & 0 & 0 \\ 0 & 0 & 0 & 1 \\ \hdashline 0 & 0 & 0 & 0 \\ 0 & 0 & 1 & 0 \end{array}\right]$$

is 3, although

$$\operatorname{rank}\begin{bmatrix} 1 & 0 \\ 0 & 0 \end{bmatrix} + \operatorname{rank}\begin{bmatrix} 0 & 0 \\ 1 & 0 \end{bmatrix} = 2. \quad \square$$

The rank of a matrix has now been investigated from two points of view: that of reduction by elementary operations and that of determinants. The concept of rank will be studied once more in Section 3.7, from a third point of view.

Exercise 9. Let A be an invertible $n \times n$ matrix and D be square. Confirm that

$$\operatorname{rank}\begin{bmatrix} A & B \\ C & D \end{bmatrix} = n$$

if and only if $D = CA^{-1}B$; that is, if and only if the Schur complement of A is the zero matrix.

Hint. See Eq. (2.6.7). □

2.9 Systems of Linear Equations and Matrices

Consider the set of m *linear algebraic equations* (*a linear system*)

$$\begin{aligned} a_{11}x_1 + a_{12}x_2 + \cdots + a_{1n}x_n &= b_1, \\ a_{21}x_1 + a_{22}x_2 + \cdots + a_{2n}x_n &= b_2, \\ &\vdots \\ a_{m1}x_1 + a_{m2}x_2 + \cdots + a_{mn}x_n &= b_m. \end{aligned} \tag{1}$$

in n *unknowns* $x_1, x_2, \ldots, x_n$.

An n-tuple $(x_1^0, x_2^0, \ldots, x_n^0)$ is said to be a *solution* of (1) if, upon substituting x_i^0 instead of x_i $(i = 1, 2, \ldots, n)$ in (1), equalities are obtained. We also say in this case that x_i^0 $(i = 1, 2, \ldots, n)$ *satisfy* each of the equations in (1). If the system (1) has no solutions, then it is *inconsistent*, while the system is *consistent* if it possesses at least one solution. It is necessary in many applications to *solve* such a system, that is, to find all its solutions or to show that it is inconsistent. In this section we first present the Gauss reduction method for solving linear systems and then some other methods, using matrix techniques developed earlier. A further investigation of linear systems and their solutions appears in Section 3.10.

Returning to the system (1), observe that the coefficients on the left-hand side form an $m \times n$ matrix

$$A = [a_{ij}]_{i,j=1}^{m,n}, \tag{2}$$

called the *coefficient matrix* of (1). Furthermore, the n unknowns and the m numbers on the right can be written as (column) vectors:

$$\mathbf{x} = \begin{bmatrix} x_1 \\ x_2 \\ \vdots \\ x_n \end{bmatrix}, \qquad \mathbf{b} = \begin{bmatrix} b_1 \\ b_2 \\ \vdots \\ b_m \end{bmatrix}, \tag{3}$$

and using the definition of the matrix product, we can rewrite (1) in the abbreviated (matrix) form

$$Ax = \boldsymbol{b}. \tag{4}$$

Obviously, solving (1) is equivalent to finding all vectors (if any) satisfying (4). Such vectors are referred to as the *solution vectors* (or just *solutions*) of the (matrix) equation (4), and together they form the *solution set*.

Note also that the system (1) is determined uniquely by its so-called *augmented matrix*

$$[A \quad \boldsymbol{b}]$$

of order $m \times (n + 1)$.

Recall now that the basic elimination (Gauss) method of solving linear systems relies on the replacement of the given system by a simpler one having the same set of solutions, that is, on the transition to an *equivalent system*. It is easily seen that the following operations on (1) transform the system to an equivalent one:

(1) Interchanging equations in the system;
(2) Multiplying an equation by a nonzero constant;
(3) Adding one equation, multiplied by a number, to another.

Recalling the effect of multiplication of a matrix from the left by elementary matrices, we see that the operations 1 through 3 correspond to left multiplication of the augmented matrix $[A \quad \boldsymbol{b}]$ by elementary matrices $E^{(1)}$, $E^{(2)}$, $E^{(3)}$, respectively, in fact, by multiplication of both sides of (4) by such matrices on the left. In other words, for any elementary matrix E_m of order m, the systems (4) and

$$E_m Ax = E_m \boldsymbol{b}$$

are equivalent. Moreover, since any nonsingular matrix E can be decomposed into a product of elementary matrices (see Exercise 2.7.8), the original system (4) is equivalent to the system $EAx = E\boldsymbol{b}$ for *any* nonsingular $m \times m$ matrix E.

Referring to Theorem 2.7.1, any matrix can be reduced by elementary row operations to reduced row-echelon form. Therefore the system (4) can be transformed by multiplication on the left by a nonsingular matrix $\tilde{E}$ to a simpler equivalent form, say,

$$\tilde{E}Ax = \tilde{E}\boldsymbol{b} \tag{5}$$

where $[\tilde{E}A \quad \tilde{E}\boldsymbol{b}]$ has reduced row-echelon form.

This is the *reduced row-echelon form of the system*, which admits calculation of the solutions very easily. The argument above establishes the following theorem.

Theorem 1. *Any system of m linear equations in n unknowns has an equivalent system in which the augmented matrix has reduced row-echelon form.*

In practice, the reduction of the system to its reduced row-echelon form is carried out by use of the Gauss elimination process illustrated in Example 2.7.3.

Exercise 1. Let us solve the system

$$\begin{aligned} x_2 + 2x_3 &= -1, \\ -2x_1 \qquad\qquad &= 6, \\ 4x_1 - 2x_2 - 4x_3 &= -10, \end{aligned}$$

having the augmented matrix

$$\begin{bmatrix} 0 & 1 & 2 & -1 \\ -2 & 0 & 0 & 6 \\ 4 & -2 & -4 & -10 \end{bmatrix}$$

considered in Example 2.7.3. It was shown there that its reduced row-echelon form is

$$\begin{bmatrix} 1 & 0 & 0 & -3 \\ 0 & 1 & 2 & -1 \\ 0 & 0 & 0 & 0 \end{bmatrix},$$

(see Eq. 2.7.5) which means that the system

$$\begin{aligned} 1 \cdot x_1 + 0 \cdot x_2 + 0 \cdot x_3 &= -3, \\ 0 \cdot x_1 + 1 \cdot x_2 + 2x_3 &= -1 \end{aligned}$$

is equivalent to the original one. Hence the relations

$$x_1 = -3, \qquad x_2 = -1 - 2t, \qquad x_3 = t,$$

where t may take any value, describe all of the (infinitely many) solutions of the given system. □

Proceeding to the case of a system of n linear equations with n unknowns, we recall that if the coefficient matrix A of the system is nonsingular, then A is row equivalent to the $n \times n$ identity matrix (Exercise 2.7.6). Hence there is a nonsingular matrix B such that $BA = I_n$, and (4) can be rewritten in an equivalent form,

$$\mathbf{x} = BA\mathbf{x} = B\mathbf{b}.$$

Obviously, $B = A^{-1}$ and then the solution of (4) is given by $\mathbf{x} = A^{-1}\mathbf{b}$ and is unique, since the inverse is unique.

Theorem 2. *If the coefficient matrix A of a linear system $A\boldsymbol{x} = \boldsymbol{b}$ is nonsingular, then the system has a unique solution given by $\boldsymbol{x} = A^{-1}\boldsymbol{b}$.*

It should be noted that in spite of the obvious importance of Theorem 2, it is rarely used in computational practice since calculation of the inverse is usually more difficult than direct solution of the system.

Example 2. Consider the system

$$\begin{aligned} 2x_1 + x_2 - x_3 &= 1, \\ 2x_2 &= -4, \\ x_1 - x_2 + 3x_3 &= 0. \end{aligned}$$

The matrix

$$A = \begin{bmatrix} 2 & 1 & -1 \\ 0 & 2 & 0 \\ 1 & -1 & 3 \end{bmatrix}$$

is invertible with the inverse (see Exercise 2.6.10)

$$A^{-1} = \tfrac{1}{14}\begin{bmatrix} 6 & -2 & 2 \\ 0 & 7 & 0 \\ -2 & 3 & 4 \end{bmatrix}.$$

Then the solution of the given system is

$$\boldsymbol{x} = \begin{bmatrix} x_1 \\ x_2 \\ x_3 \end{bmatrix} = A^{-1}\begin{bmatrix} 1 \\ -4 \\ 0 \end{bmatrix} = \begin{bmatrix} 1 \\ -2 \\ -1 \end{bmatrix}$$

and

$$x_1 = 1, \qquad x_2 = -2, \qquad x_3 = -1. \quad \square$$

Another approach to solving the linear system $A\boldsymbol{x} = \boldsymbol{b}$ with nonsingular A relies on the formula (2.6.5) for finding the inverse A^{-1}. Thus

$$\boldsymbol{x} = A^{-1}\boldsymbol{b} = \frac{1}{\det A}(\operatorname{adj} A)\boldsymbol{b}.$$

In more detail, if A_{ij} denotes the cofactor of a_{ij} in A, then

$$\begin{bmatrix} x_1 \\ x_2 \\ \vdots \\ x_n \end{bmatrix} = \frac{1}{\det A}\begin{bmatrix} A_{11} & A_{21} & \cdots & A_{n1} \\ A_{12} & A_{22} & \cdots & A_{n2} \\ \vdots & \vdots & \ddots & \vdots \\ A_{1n} & A_{2n} & \cdots & A_{nn} \end{bmatrix}\begin{bmatrix} b_1 \\ b_2 \\ \vdots \\ b_n \end{bmatrix}$$

or, what is equivalent,

$$x_i = \frac{1}{\det A} \sum_{j=1}^{n} b_j A_{ji} \quad (i = 1, 2, \ldots, n).$$

Observe now that this summation is just the cofactor expansion along the ith column of the determinant of the matrix $A^{(i)}$, where $A^{(i)}$ is obtained from A by replacing its ith column by $\boldsymbol{b} = [b_1 \quad b_2 \quad \cdots \quad b_n]^{\mathrm{T}}$. Thus we have the formulas

$$x_i = \frac{\det A^{(i)}}{\det A} \quad (i = 1, 2, \ldots, n). \tag{6}$$

This is *Cramer's rule* for solving linear systems of n equations with n unknowns in the case of a nonsingular coefficient matrix. Note that formulas (6) for $n = 2, 3$ were indicated in Section 2.1.

Example 3. Let us use Cramer's rule to solve the linear system

$$\begin{aligned} 2x_1 - x_2 - 2x_3 &= 5, \\ 4x_1 + x_2 + 2x_3 &= 1, \\ 8x_1 - x_2 + x_3 &= 5. \end{aligned}$$

We have

$$\det \begin{bmatrix} 2 & -1 & -2 \\ 4 & 1 & 2 \\ 8 & -1 & 1 \end{bmatrix} = 18 \neq 0,$$

and therefore Cramer's rule can be applied. Successively replacing the columns of the matrix by the column $[5 \quad 1 \quad 5]^{\mathrm{T}}$ and computing the determinants, we deduce

$$\det \begin{bmatrix} 5 & -1 & -2 \\ 1 & 1 & 2 \\ 5 & -1 & 1 \end{bmatrix} = 18, \quad \det \begin{bmatrix} 2 & 5 & -2 \\ 4 & 1 & 2 \\ 8 & 5 & 1 \end{bmatrix} = 18, \quad \det \begin{bmatrix} 2 & -1 & 5 \\ 4 & 1 & 1 \\ 8 & -1 & 5 \end{bmatrix} = -36.$$

Thus

$$x_1 = \tfrac{18}{18} = 1, \qquad x_2 = \tfrac{18}{18} = 1, \qquad x_3 = -\tfrac{36}{18} = -2$$

is the required solution of the given system. □

It should be mentioned that although Cramer's rule is an important theoretical result and a useful one when the size of A is not too big, it is not recommended as a general-purpose algorithm for the numerical solution of large sets of equations. Algorithms based on reduction by elementary operations are generally more efficient.

It will be seen in Section 3.10 that a complete description of the solution set of $A\mathbf{x} = \mathbf{b}$ for a general $m \times n$ matrix A can be made to depend on the solution set of the *homogeneous* equation $A\mathbf{x} = \mathbf{0}$ (which will always include the trivial solution $\mathbf{x} = \mathbf{0}$). We now conclude with two exercises concerning homogeneous linear systems.

Exercise 4. Show that if A is nonsingular, then the homogeneous system $A\mathbf{x} = \mathbf{0}$ has *only* the trivial solution. □

Note also the following important result concerning homogeneous systems with $m < n$.

Exercise 5. Check that any homogeneous linear system having more unknowns than equations always has infinitely many solutions.

Hint. Consider the reduced row-echelon form of the system. □

2.10 The *LU* Decomposition

Consider Theorem 2.7.3 once more, and suppose that the matrix A is square of size $n \times n$ and, in addition, is such that the reduction to canonical form can be completed *without* elementary operations of type 1. Thus, no row or column interchanges are required. In this case, the canonical form is

$$D_0 = \begin{bmatrix} I_r & 0 \\ 0 & 0 \end{bmatrix}$$

if $r < n$, and we interpret D_0 as I_n if $r = n$. Thus, Theorem 2.7.3 gives

$$(E_k E_{k-1} \cdots E_1) A (E_{k+1} E_{k+2} \cdots E_{k+s}) = D_0$$

and $E_1, E_2, \ldots, E_k$ are lower-triangular elementary matrices and $E_{k+1}, E_{k+2}, \ldots, E_{k+s}$ are upper-triangular elementary matrices. Since the inverses of elementary matrices are elementary matrices of the same type, it follows that we may write

$$A = LU, \tag{1}$$

where $L = E_1^{-1} E_2^{-1} \cdots E_k^{-1}$ and is a nonsingular lower-triangular matrix. Also, $U = D_0 E_{k+s}^{-1} \cdots E_{k+2}^{-1} E_{k+1}^{-1}$, an upper-triangular matrix, which is nonsingular if D_0, and hence A, is nonsingular. A factorization of A into triangular factors, as in Eq. (1), is known as an *LU decomposition* of A.

When such a factorization is known, solving the linear system $A\mathbf{x} = \mathbf{b}$ is relatively quick and simple. The system $L\mathbf{y} = \mathbf{b}$ is solved first. Recall that L is nonsingular so, by Theorem 2.9.2, there is a unique solution vector $\mathbf{y}$.

Also, because L is lower triangular, $\boldsymbol{y}$ is easily calculated if the components $y_1, y_2, \ldots, y_n$ are found successively.

Then the equation $U\boldsymbol{x} = \boldsymbol{y}$ is solved to obtain the solution(s) of $A\boldsymbol{x} = \boldsymbol{b}$. For $U\boldsymbol{x} = \boldsymbol{y}$ and $L\boldsymbol{y} = \boldsymbol{b}$ clearly imply $LU\boldsymbol{x} = A\boldsymbol{x} = \boldsymbol{b}$. But once more, the triangular form of U makes the derivation of $\boldsymbol{x}$ from $\boldsymbol{y}$ particularly easy. In this case, the components of $\boldsymbol{x}$ are found in the order $x_n, x_{n-1}, \ldots, x_2, x_1$.

The practical implementation and analysis of elimination algorithms for linear systems depend heavily on the possibility of triangular factorizations, and our next result is the fundamental theorem of this kind. The discussion to this point has required the unsatisfactory hypothesis that row and column interchanges would not be required in the reduction of A to canonical form. The theorem replaces this hypothesis by a condition on the leading principal minors of A (see Section 2.4 for the definition).

Theorem 1. *Let $A \in \mathbb{C}^{n \times n}$ and assume that the leading principal minors*

$$A\begin{pmatrix} 1 \\ 1 \end{pmatrix}, \quad A\begin{pmatrix} 1 & 2 \\ 1 & 2 \end{pmatrix}, \quad \ldots, \quad A\begin{pmatrix} 1 & 2 & \cdots & n-1 \\ 1 & 2 & \cdots & n-1 \end{pmatrix} \tag{2}$$

are all nonzero. Then there is a unique lower triangular L with diagonal elements all equal to one and a unique upper triangular matrix U such that $A = LU$.

Furthermore, $\det A = \det U = u_{11}u_{22}\cdots u_{nn}$.

PROOF. The proof is by induction on n; the size of matrix A. Clearly, if $n = 1$, then $a_{11} = 1 \cdot a_{11}$ provides the unique factorization of the theorem.

Now assume the theorem true for square matrices of size $(n-1) \times (n-1)$, and partition A as

$$A = \begin{bmatrix} A_{n-1} & \boldsymbol{a}_1 \\ \boldsymbol{a}_2^{\mathrm{T}} & a_{nn} \end{bmatrix}, \tag{3}$$

where $\boldsymbol{a}_1, \boldsymbol{a}_2 \in \mathbb{C}^{n-1}$. Then the theorem can be applied to A_{n-1} yielding the unique triangular factorization

$$A_{n-1} = L_{n-1}U_{n-1},$$

where L_{n-1} has diagonal elements all equal to one. Since

$$\det A_{n-1} = A\begin{pmatrix} 1 & 2 & \cdots & n-1 \\ 1 & 2 & \cdots & n-1 \end{pmatrix} \neq 0,$$

L_{n-1} and U_{n-1} are nonsingular.

Now consider $n \times n$ partitioned triangular matrices

$$L = \begin{bmatrix} L_{n-1} & \boldsymbol{0} \\ \boldsymbol{d}^{\mathrm{T}} & 1 \end{bmatrix} \quad \text{and} \quad U = \begin{bmatrix} U_{n-1} & \boldsymbol{c} \\ \boldsymbol{0}^{\mathrm{T}} & a_{nn} - \boldsymbol{d}^{\mathrm{T}}\boldsymbol{c} \end{bmatrix}, \tag{4}$$

where $\boldsymbol{c}$ and $\boldsymbol{d}$ are vectors in $\mathbb{C}^{n-1}$ to be determined. Form the product

$$LU = \begin{bmatrix} L_{n-1}U_{n-1} & L_{n-1}\boldsymbol{c} \\ \boldsymbol{d}^{\mathrm{T}}U_{n-1} & a_{nn} \end{bmatrix},$$

and compare the result with Eq. (3). It is clear that if $\boldsymbol{c}$ and $\boldsymbol{d}$ are chosen so that

$$L_{n-1}\boldsymbol{c} = \boldsymbol{a}_1, \qquad \boldsymbol{d}^{\mathrm{T}}U_{n-1} = \boldsymbol{a}_2^{\mathrm{T}}, \tag{5}$$

then the matrices of (4) will give the required factorization. But Eqs. (5) are uniquely solvable for $\boldsymbol{c}$ and $\boldsymbol{d}$ because L_{n-1} and U_{n-1} are nonsingular. Thus,

$$\boldsymbol{c} = L_{n-1}^{-1}\boldsymbol{a}_1 \qquad \text{and} \qquad \boldsymbol{d}^{\mathrm{T}} = \boldsymbol{a}_2^{\mathrm{T}}U_{n-1}^{-1},$$

and subsitution in (4) determines the unique triangular factors of A required by the theorem.

For the determinantal property, recall Exercise 2.1.10 and Corollary 2.5.1 to deduce that $\det L = 1$ and $\det A = (\det L)(\det U)$. Hence $\det A = \det U$. ■

Exercise 1. Under the hypotheses of Theorem 1, show that there are unique lower- and upper-triangular matrices L and U with all diagonal elements equal to one (for *both* matrices) and a unique diagonal matrix D such that $A = LDU$.

Exercise 2. In addition to the hypotheses of Theorem 1, assume that A is Hermitian. Show that there is a unique L (as in the theorem) and a unique real diagonal matrix D such that $A = LDL^*$.

Hint. Use Exercise 1 and especially the uniqueness property. (The factorization of this exercise is often known as the *Cholesky-factorization*.) □

2.11 Miscellaneous Exercises

1. Check that if $\alpha = \sqrt{\frac{2}{7}}$, $\beta = \sqrt{\frac{27}{28}}$, then

$$\det\begin{bmatrix} \lambda^2 + 2 & \alpha\lambda & \beta\lambda \\ \alpha\lambda & \lambda^2 + 2 & 0 \\ \beta\lambda & 0 & \lambda^2 + \frac{1}{4} \end{bmatrix} = (\lambda^2 + 1)^3.$$

2. If A is a square matrix and $A^2 + 2A + I = 0$, show that A is nonsingular and that $A^{-1} = -(A + 2I)$.

3. Given that D is a diagonal matrix and nonsingular, prove that if $D = (I + A)^{-1}A$, then A is diagonal.

Hint. Establish first that $D = A(I - D)$.

4. Let matrices I, A, B be $n \times n$. Check that if $I + AB$ is invertible, then $I + BA$ is invertible and

$$(I + BA)^{-1} = I - B(I + AB)^{-1}A.$$

5. Verify the *Sherman–Morrison formula*:

$$(A + \boldsymbol{u}\boldsymbol{v}^{\mathrm{T}})^{-1} = A^{-1} - \frac{(A^{-1}\boldsymbol{u})(\boldsymbol{v}^{\mathrm{T}}A^{-1})}{1 + \boldsymbol{v}^{\mathrm{T}}A^{-1}\boldsymbol{u}},$$

provided the matrices involved exist.

6. Let

$$P_n = \begin{bmatrix} 0 & & \cdots 0 & 1 \\ \vdots & & 1 & 0 \\ 0 & 1 \cdot^{\cdot^{\cdot}} & & \vdots \\ 1 & 0 & \cdots & 0 \end{bmatrix}$$

be an $n \times n$ matrix. Show that

$$d_n(\lambda) \triangleq \det(\lambda I_n - P_n) = \begin{cases} (\lambda^2 - 1)^p(\lambda - 1) & \text{if} \quad n = 2p + 1 \\ (\lambda^2 - 1)^p & \text{if} \quad n = 2p \end{cases}$$

Hint. Show first that $d_n(\lambda) = (\lambda^2 - 1)d_{n-2}$ $(n \geq 3)$.

7. Consider the $n \times n$ matrix

$$A = \begin{bmatrix} x + \lambda & x & \cdots & & x \\ x & x + \lambda & & & \vdots \\ \vdots & & & \ddots & x \\ x & x & \cdots & x & x + \lambda \end{bmatrix}.$$

(a) Prove that $\det A = \lambda^{n-1}(nx + \lambda)$.
(b) Prove that if A^{-1} exists, it is of the same form as A.

Hint. For part (a), add all columns to the first and then subtract the first row from each of the others. For part (b), observe that a matrix A is of the above form if and only if $PAP^{\mathrm{T}} = A$ for any $n \times n$ permutation matrix. (An $n \times n$ matrix P is a *permutation matrix* if it is obtained from I_n by interchanges of rows or columns.)

8. Let the elements of the $n \times n$ matrix $A = [\boldsymbol{a}_{ij}]$ be differentiable functions of x. If $A = [\boldsymbol{a}_1 \quad \boldsymbol{a}_2 \quad \cdots \quad \boldsymbol{a}_n]$, prove that

$$\frac{d}{dx}\det A = \det\begin{bmatrix}\frac{d}{dx}\boldsymbol{a}_1 & \boldsymbol{a}_2 & \cdots & \boldsymbol{a}_n\end{bmatrix} + \det\begin{bmatrix}\boldsymbol{a}_1 & \frac{d}{dx}\boldsymbol{a}_2 & \boldsymbol{a}_3 & \cdots & \boldsymbol{a}_n\end{bmatrix} + \cdots + \det\begin{bmatrix}\boldsymbol{a}_1 & \boldsymbol{a}_2 & \cdots & \frac{d}{dx}\boldsymbol{a}_n\end{bmatrix},$$

where the derivative of a vector is the vector of the derivatives of its elements.

Hint. Use the definition of det A and Property 1.

9. Let A be an $n \times n$ matrix and $\boldsymbol{x}$, $\boldsymbol{y}$ be vectors of order n. Confirm the relation

$$\det\begin{bmatrix} A & \boldsymbol{x} \\ \boldsymbol{y}^{\mathrm{T}} & \alpha \end{bmatrix} = \alpha \det A - \boldsymbol{y}^{\mathrm{T}}(\operatorname{adj} A)\boldsymbol{x}$$

for any complex number α. Then deduce

$$\det\begin{bmatrix} A & A\xi \\ \xi^{\mathrm{T}}A & \alpha \end{bmatrix} = (\det A)(\alpha - \xi^{\mathrm{T}}A\xi)$$

for any vector ξ of order n.

Hint. Expand by the last column and then expand the cofactors by the last row.

10. Let an $n \times n$ matrix $A = [a_{ij}]$ be represented in the form $A = [A_1 \quad A_2]$, where A_1 and A_2 are $n \times k$ and $n \times (n - k)$ matrices $(1 \le k \le n - 1)$. Prove that

$$|\det A|^2 \le (\det A_1^*A_1)(\det A_2^*A_2).$$

Hint. Use Laplace's expansion formula along the first k columns, the Cauchy–Schwartz inequality (Exercise 2.5.4), and Exercise 2.5.2.

11. Establish *Hadamard's inequality* for any square matrix $A = [a_{ij}]_{i,j=1}^n$:

$$|\det A|^2 \le \prod_{j=1}^{n} \sum_{i=1}^{n} |a_{ij}|^2.$$

Hint. Apply the result of Exercise 10, where A_1 is a column of A.

12. Check, using Hadamard's inequality, that

$$|\det A| \le M^n n^{n/2}$$

provided $A = [a_{ij}]_{i,j=1}^n$ and $|a_{ij}| \le M$ $(1 \le i, j \le n)$.

13. Evaluate det H_n, where H_n denotes the following *Hankel matrix*:

$$H_n = \begin{bmatrix} s_0 & s_1 & s_2 & \cdots & s_{n-1} \\ s_1 & s_2 & & \cdots & s_n \\ s_2 & & & & \\ \vdots & \vdots & & & \vdots \\ s_{n-1} & s_n & & \cdots & s_{2n-2} \end{bmatrix} \equiv [s_{i+j-2}]_{i,j=1}^{n},$$

and where the elements have the form

$$s_k = \sum_{i=1}^{n} x_i^k \qquad (0 \le k \le 2n-2)$$

for some numbers $x_1, x_2, \ldots, x_n \in \mathscr{F}$. (The numbers $s_0, s_1, \ldots, s_{2n-2}$ are called *Newton sums* of $x_1, x_2, \ldots, x_n$.)

Answer. $\det H_n = \prod_{1 \le j < i \le n} (x_i - x_j)^2$.

Hint. $H_n = V_n V_n^{\mathrm{T}}$, where V_n is the Vandermonde matrix (see Exercise 2.3.6).

14. Compute the determinant of the *circulant matrix*

$$C = \begin{bmatrix} c_0 & c_1 & & \cdots & c_{n-1} \\ c_{n-1} & c_0 & c_1 & \cdots & c_{n-2} \\ & & & \ddots & \vdots \\ \vdots & \ddots & \ddots & & c_1 \\ c_1 & \cdots & & c_{n-1} & c_0 \end{bmatrix}$$

Answer. $\det C = \prod_{i=1}^{n} f(\varepsilon_i)$, where $f(x) = \sum_{i=0}^{n-1} c_i x^i$ and $\varepsilon_1, \varepsilon_2, \ldots, \varepsilon_n$ denote the distinct nth roots of unity.

Hint. $CV_n = V_n \operatorname{diag}[f(\varepsilon_1), f(\varepsilon_2), \ldots, f(\varepsilon_n)]$, where V_n is the Vandermonde matrix constructed from the ε_i $(i = 1, 2, \ldots, n)$.

15. Prove that the determinant of a *generalized circulant matrix* $\tilde{C}$,

$$\tilde{C} = \begin{bmatrix} c_0 & c_1 & & \cdots & c_{n-1} \\ \alpha c_{n-1} & c_0 & c_1 & \cdots & c_{n-2} \\ \alpha c_{n-2} & \alpha c_{n-1} & c_0 & \ddots & \vdots \\ \vdots & \vdots & \ddots & \ddots & c_1 \\ \alpha c_1 & \alpha c_2 & \cdots & \alpha c_{n-1} & c_0 \end{bmatrix},$$

is equal to $\prod_{i=1}^{n} f(\varepsilon_i)$, where $f(x) = \sum_{i=0}^{n-1} c_i x^i$ and $\varepsilon_1, \varepsilon_2, \ldots, \varepsilon_n$ are all the nth roots of α.

16. Show that the inverses of the $n \times n$ triangular matrices

$$A_1 = \begin{bmatrix} 1 & -1 & 0 & \cdots & 0 \\ 0 & 1 & -1 & & \vdots \\ \vdots & & \ddots & \ddots & \\ & & & 1 & -1 \\ 0 & \cdots & & 0 & 1 \end{bmatrix} \quad \text{and} \quad A_2 = \begin{bmatrix} 1 & 0 & & \cdots & 0 \\ -1 & 1 & & & \vdots \\ 0 & -1 & \ddots & & \\ \vdots & & \ddots & 1 & 0 \\ 0 & \cdots & 0 & -1 & 1 \end{bmatrix}$$

are matrices with zeros below (respectively, above) the main diagonal and ones elsewhere.

17. Check that the inverses of the $n \times n$ matrices

$$A = \begin{bmatrix} 1 & 0 & & \cdots & 0 \\ -2 & 1 & & & \vdots \\ 1 & -2 & \ddots & & \\ \vdots & \ddots & \ddots & \ddots & 0 \\ 0 & \cdots & 1 & -2 & 1 \end{bmatrix} \quad \text{and} \quad B = \begin{bmatrix} 2 & -1 & \cdots & 0 \\ -1 & 2 & -1 & \vdots \\ \vdots & \ddots & \ddots & -1 \\ 0 & & -1 & 1 \end{bmatrix}$$

are the matrices

$$A^{-1} = \begin{bmatrix} 1 & 0 & \cdots & & 0 \\ 2 & \ddots & \ddots & & \vdots \\ 3 & & \ddots & & \\ \vdots & \ddots & \ddots & & \\ n-1 & & & 1 & 0 \\ n & n-1 & \cdots & 2 & 1 \end{bmatrix}$$

and

$$B^{-1} = \begin{bmatrix} 1 & 1 & & \cdots & 1 \\ 1 & 2 & & \cdots & 2 \\ \vdots & \vdots & \ddots & & \vdots \\ & & & n-1 & n-1 \\ 1 & 2 & \cdots & n-1 & n \end{bmatrix},$$

respectively, where row j of B^{-1} is $[1 \quad 2 \quad \cdots \quad j \quad \cdots \quad j]$.

Hint. Use $A = A_2^2, B = A_1 A_2$, where A_1, A_2 are defined in Exercise 16.

18. An $n \times n$ matrix $A = [a_{ij}]$, where

$$a_{ij} = a_{i-j}\,(0 \le i \le n-1, 0 \le j \le n-1),$$

that is,

$$A = \begin{bmatrix} a_0 & a_{-1} & \cdots & a_{-n+1} \\ a_1 & a_0 & \ddots & \vdots \\ \vdots & \ddots & \ddots & a_{-1} \\ a_{n-1} & \cdots & a_1 & a_0 \end{bmatrix},$$

is said to be a *Toeplitz matrix*, and it is said to be generated by the *quasi-polynomial*

$$p(\lambda) = a_{n-1}\lambda^{n-1} + \cdots + a_1\lambda + a_0 + a_{-1}\lambda^{-1} + \cdots + a_{-n+1}\lambda^{-n+1}.$$

(For example, matrices A and B of Exercise 17 are Toeplitz matrices.) Show that an invertible matrix $A = [a_{ij}]_{i,j=1}^n$ is Toeplitz if and only if there are companion matrices of the forms

$$C_p^{(1)} = \begin{bmatrix} -p_{n-1} & -p_{n-2} & \cdots & -p_0 \\ 1 & 0 & \cdots & 0 \\ \vdots & \ddots & \ddots & \vdots \\ 0 & \cdots & 1 & 0 \end{bmatrix}$$

and

$$C_p^{(2)} = \begin{bmatrix} 0 & \cdots & 0 & -p_0 \\ 1 & \ddots & \vdots & \vdots \\ \vdots & \ddots & 0 & -p_{n-2} \\ 0 & \cdots & 1 & -p_{n-1} \end{bmatrix}$$

such that

$$C_p^{(1)}A = AC_p^{(2)}.$$

Hint. $-[p_{n-1} \quad p_{n-2} \quad \cdots \quad p_0] = [a_{-1} \quad a_{-2} \quad \cdots \quad a_{-n}]A^{-1}$ for any a_{-n}.

19. Let the polynomial $p(\lambda) = 1 + a\lambda + b\lambda^2$ $(b \neq 0)$ with real coefficients have a pair of complex conjugate roots $re^{\pm i\varphi}$. Confirm the decomposition

$$\begin{bmatrix} 1 & & \cdots & & 0 \\ a & 1 & & & \\ b & a & 1 & & \vdots \\ \vdots & \ddots & \ddots & \ddots & \\ 0 & & b & a & 1 \end{bmatrix} = \begin{bmatrix} 1 & & \cdots & 0 \\ c_2 & 1 & & \vdots \\ \vdots & c_3 & \ddots & \\ 0 & \cdots & c_n & 1 \end{bmatrix} \begin{bmatrix} 1 & & \cdots & 0 \\ d_2 & 1 & & \vdots \\ \vdots & d_3 & \ddots & \\ 0 & \cdots & d_n & 1 \end{bmatrix},$$

where

$$c_k = -\frac{\sin(k-2)\varphi}{r\sin(k-1)\varphi}, \qquad d_k = -\frac{\sin k\varphi}{r\sin(k-1)\varphi} \qquad (k = 2, 3, \ldots, n).$$

20. Consider the $n \times n$ triangular Toeplitz matrix

$$A = \begin{bmatrix} 1 & 0 & & \cdots & & 0 \\ a_1 & & & & & \\ \vdots & & & & & \vdots \\ a_l & & \ddots & \ddots & & \\ 0 & \ddots & & & & \\ \vdots & & \ddots & & & 0 \\ 0 \cdots 0 & a_l & & \cdots & a_1 & 1 \end{bmatrix} \qquad (a_l \neq 0, l < n)$$

generated by the polynomial $p(\lambda) = 1 + \sum_{i=1}^{l} a_i \lambda^i$. Show that if

$$p(\lambda) = \prod_{j=1}^{k} (1 - \lambda r_j) \prod_{s=1}^{t} (1 + a_s \lambda + b_s \lambda^2) \qquad (k + 2t = l),^{\dagger}$$

then A can be decomposed as

$$A = \prod_{j=1}^{k} \begin{bmatrix} 1 & & \cdots & 0 \\ r_j & \ddots & & \\ \vdots & \ddots & \ddots & \vdots \\ 0 & \cdots & r_j & 1 \end{bmatrix} \prod_{s=1}^{t} \begin{bmatrix} 1 & & & \cdots & 0 \\ a_s & \ddots & & & \\ b_s & \ddots & \ddots & & \vdots \\ \vdots & \ddots & \ddots & \ddots & \\ 0 & \cdots & b_s & a_s & 1 \end{bmatrix}.$$

21. Prove that the companion matrix C_α associated with the monic polynomial $a(\lambda) = \sum_{i=0}^{n} a_i \lambda^i$ (where $a(\lambda)$ has distinct zeros $x_1, x_2, \ldots, x_n$) can be represented in the form

$$C_a = V_n \operatorname{diag}[x_1, x_2, \ldots, x_n] V_n^{-1},$$

where V_n denotes the Vandermonde matrix of $x_1, x_2, \ldots, x_n$.

22. Prove that if x_i $(1 \leq i \leq s)$ is a zero of $a(\lambda)$ with multiplicity k_i

$$\left(\sum_{i=1}^{s} k_i = n \right),$$

then

$$C_a = \tilde{V}_n \operatorname{diag}\left[\underbrace{\begin{bmatrix} x_1 & 1 & & \\ & \ddots & \ddots & \\ & & \ddots & 1 \\ & & & x_1 \end{bmatrix}}_{k_1}, \underbrace{\begin{bmatrix} x_2 & 1 & & \\ & \ddots & \ddots & \\ & & \ddots & 1 \\ & & & x_2 \end{bmatrix}}_{k_2}, \ldots, \underbrace{\begin{bmatrix} x_s & 1 & & \\ & \ddots & \ddots & \\ & & \ddots & 1 \\ & & & x_s \end{bmatrix}}_{k_s} \right] \tilde{V}_n^{-1}$$

† Observe that this is the factorization of the polynomial $q(\lambda) = \lambda^l p(1/\lambda)$ into linear and quadratic factors irreducible in $\mathbb{R}$.

where $\tilde{V}_n$ denotes the *generalized Vandermonde matrix*

$$\tilde{V}_n = [W_1 \quad W_2 \quad \cdots \quad W_s].$$

Here $W_i = [X_i^{(0)} \quad X_i^{(1)} \quad \cdots \quad X_i^{(k_i)}]$ is an $n \times k_i$ matrix $(1 \le i \le s)$, and

$$X_i^{(m)} = \frac{1}{m!}\frac{d^m}{dx_i^m}[1 \quad x_i \quad x_i^2 \quad \cdots \quad x_i^n]^{\mathrm{T}} \qquad (1 \le m \le k_i).$$

23. Let complex matrices A, B, C, D be of sizes $n \times n$, $n \times l$, $m \times n$, and $m \times l$, respectively. The system $\mathscr{S}$:

$$\left.\begin{aligned} \lambda \boldsymbol{x} &= A\boldsymbol{x} + B\boldsymbol{u} \\ \boldsymbol{y} &= C\boldsymbol{x} + D\boldsymbol{u} \end{aligned}\right\}, \qquad \boldsymbol{x} \in \mathbb{C}^n, \quad \boldsymbol{u} \in \mathbb{C}^l, \quad \mathbb{C} \in \mathbb{C}^m,$$

where λ is a complex parameter, plays a fundamental role in "systems theory."

(a) Show that

$$\boldsymbol{y} = [C(I_n\lambda - A)^{-1}B + D]\boldsymbol{u}.$$

The matrix in square parentheses is the *transfer function* for system $\mathscr{S}$ (and gives the "output" $\boldsymbol{y}$ in terms of the "input" or "control vector" $\boldsymbol{u}$).

(b) Suppose, in addition, that $m = l$ and $D = I_m$ (the $m \times m$ identity). Show that

$$-C(I_n\lambda - A + BC)^{-1}B + I_m = [C(I_n\lambda - A)^{-1}B + I_m]^{-1},$$

and find a system whose transfer function is the inverse of that for $\mathscr{S}$.

CHAPTER 3

Linear, Euclidean, and Unitary Spaces

The notion of an abstract linear space has its origin in the familiar three-dimensional space with its well-known geometrical properties. Consequently, we start the discussion of linear spaces by keeping in mind properties of this particular and important space.

3.1 Definition of a Linear Space

Before making a formal definition of a linear space, we first collect some important properties of three-dimensional vectors and discuss them from a more general point of view.

Recall that a vector in three-dimensional space is referred to as a directed line segment. Thus, it is defined by its length, direction, and orientation.[†] So, the initial point of the vector is not essential: vectors having the same length, direction, and orientation are equivalent and thus describe the same vector.

This definition of a (three-dimensional) vector allows us to consider all vectors in the space as having the same initial point. If the space is equipped with a Cartesian coordinate system, then, putting this common initial point at the origin, we can deduce that a vector $\vec{\mathbf{a}}$ is completely described

[†] Direction can be defined as a set of parallel straight lines.

by three numbers, namely, the coordinates (a_1, a_2, a_3) of the terminal point of the corresponding vector with initial point at the origin. We write

$$\vec{\mathbf{a}} = (a_1, a_2, a_3).$$

and call this a *position vector*. Thus, there is a one-to-one correspondence between all position vectors in three-dimensional space and all three-tuples of real numbers.

In the set of all position vectors, the operation of addition is defined as in Section 1.2. Thus, to any two position vectors there corresponds another position vector according to a certain rule. This specific correspondence is a particular case of a binary operation for which we develop a formal definition.

Let $\mathscr{A}$, $\mathscr{B}$, and $\mathscr{S}$ be arbitrary sets, and let $\mathscr{A} \times \mathscr{B}$ denote the set of all ordered pairs of elements (a, b), where a and b are members of $\mathscr{A}$ and $\mathscr{B}$, respectively. A *binary operation* from $\mathscr{A} \times \mathscr{B}$ to $\mathscr{S}$ is a rule that associates a unique member of $\mathscr{S}$ with each ordered pair of $\mathscr{A} \times \mathscr{B}$.

Example 1. (a) If $\mathscr{A}$, $\mathscr{B}$, and $\mathscr{S}$ each refer to the set of all real numbers, then addition, subtraction, and multiplication of real numbers are binary operations from $\mathscr{A} \times \mathscr{A}$ to $\mathscr{A}$.

(b) If we take $\mathscr{A}_1$ to be the set of all real numbers excluding zero, then division is a binary operation from $\mathscr{A} \times \mathscr{A}_1$ to $\mathscr{A}$.

(c) Denoting the set of all position vectors (with real coordinates) as $\mathbb{R}_3$, we conclude that vector addition is a binary operation from $\mathbb{R}_3 \times \mathbb{R}_3$ to $\mathbb{R}_3$.

(d) In the above notation, scalar multiplication of position vectors by a real number is a binary operation from $\mathbb{R} \times \mathbb{R}_3$ to $\mathbb{R}_3$ □

A binary operation defined on $\mathscr{A} \times \mathscr{A}$ is said to be *closed* (on $\mathscr{A}$) if the operation is from $\mathscr{A} \times \mathscr{A}$ to $\mathscr{A}$. Thus, vector addition is closed on $\mathbb{R}_3$.

Let us collect the basic properties of position vectors regarding addition and scalar multiplication for our further consideration.

For any position vectors $\vec{\mathbf{a}}$, $\vec{\mathbf{b}}$, $\vec{\mathbf{c}}$ and any real numbers α, β, we have the following properties:

A1 $\vec{\mathbf{a}} + \vec{\mathbf{b}}$ is a unique vector of the set to which $\vec{\mathbf{a}}$, $\vec{\mathbf{b}}$ belong,
A2 $\vec{\mathbf{a}} + \vec{\mathbf{b}} = \vec{\mathbf{b}} + \vec{\mathbf{a}}$,
A3 $(\vec{\mathbf{a}} + \vec{\mathbf{b}}) + \vec{\mathbf{c}} = \vec{\mathbf{a}} + (\vec{\mathbf{b}} + \vec{\mathbf{c}})$,
A4 There exists a vector $\vec{\mathbf{0}}$ such that $\vec{\mathbf{a}} + \vec{\mathbf{0}} = \vec{\mathbf{a}}$,
A5 For every $\vec{\mathbf{a}}$ there exists a vector $-\vec{\mathbf{a}}$ such that $\vec{\mathbf{a}} + (-\vec{\mathbf{a}}) = \vec{\mathbf{0}}$.

S1 $\alpha\vec{\mathbf{a}}$ is a unique vector of the set to which $\vec{\mathbf{a}}$ belongs,
S2 $\alpha(\beta\vec{\mathbf{a}}) = (\alpha\beta)\vec{\mathbf{a}}$,
S3 $\alpha(\vec{\mathbf{a}} + \vec{\mathbf{b}}) = \alpha\vec{\mathbf{a}} + \alpha\vec{\mathbf{b}}$,

S4 $(\alpha + \beta)\vec{\mathbf{a}} = \alpha\vec{\mathbf{a}} + \beta\vec{\mathbf{a}}$,
S5 $1\vec{\mathbf{a}} = \vec{\mathbf{a}}$.

It is not difficult to see that these properties yield other fundamental properties of vector addition and scalar multiplication of position vectors. For example, $\alpha\vec{\mathbf{0}} = \vec{\mathbf{0}}$ for any real number α. To see this, first use A5 to obtain

$$\alpha\vec{\mathbf{0}} + (-\alpha\vec{\mathbf{0}}) = \vec{\mathbf{0}}, \tag{1}$$

Then by A4,

$$\alpha\vec{\mathbf{0}} + [\alpha\vec{\mathbf{0}} + (-\alpha\vec{\mathbf{0}})] = \alpha\vec{\mathbf{0}}.$$

Hence, applying A3,

$$[\alpha\vec{\mathbf{0}} + \alpha\vec{\mathbf{0}}] + (-\alpha\vec{\mathbf{0}}) = \alpha\vec{\mathbf{0}}$$

and, using S3 and A4,

$$\alpha(\vec{\mathbf{0}} + \vec{\mathbf{0}}) + (-\alpha\vec{\mathbf{0}}) = \alpha\vec{\mathbf{0}},$$
$$\alpha\vec{\mathbf{0}} + (-\alpha\vec{\mathbf{0}}) = \alpha\vec{\mathbf{0}}.$$

Comparing with (1) we get the desired result, $\vec{\mathbf{0}} = \alpha\vec{\mathbf{0}}$.

***Exercise* 2.** Deduce the following facts from A1–A5 and S1–S5 (α, β are real):

(a) $0\vec{\mathbf{a}} = \vec{\mathbf{0}}$.
(b) If $\alpha\vec{\mathbf{a}} = \vec{\mathbf{0}}$, then either $\alpha = 0$ or $\vec{\mathbf{a}} = \vec{\mathbf{0}}$, or both.
(c) $\alpha(-\vec{\mathbf{a}}) = -\alpha\vec{\mathbf{a}}$.
(d) $(-\alpha)\vec{\mathbf{a}} = -\alpha\vec{\mathbf{a}}$.
(e) $(\alpha - \beta)\vec{\mathbf{a}} = \alpha\vec{\mathbf{a}} - \beta\vec{\mathbf{a}}$.
(f) $\alpha(\vec{\mathbf{a}} - \vec{\mathbf{b}}) = \alpha\vec{\mathbf{a}} - \alpha\vec{\mathbf{b}}$. □

Thus, the properties A1–A5 and S1–S5 are basic in the sense that they are sufficient to obtain computational rules on the set of position vectors which extend those of arithmetic in a natural way.

It turns out that there are many sets with operations defined on them like vector addition and scalar multiplication and for which these conditions make sense. This justifies the important unifying concept called a "linear space." We now make a formal definition, which will be followed by a few examples.

Let $\mathscr{S}$ be a set on which a closed binary operation $(+)$ (like vector addition) is defined. Let $\mathscr{F}$ be a field and let a binary operation (scalar multiplication) be defined from $\mathscr{F} \times \mathscr{S}$ to $\mathscr{S}$. If, in this context, the axioms A1–A5 and S1–S5 are satisfied for any $a, b, c \in S$ and any $\alpha, \beta \in \mathscr{F}$, then $\mathscr{S}$ is said to be a *linear space over* $\mathscr{F}$.

In this definition we use the notion of a "field." The definition of a field is an issue we shall avoid. The reader who is not familiar with the concept need only understand that, in this book, wherever a general field $\mathscr{F}$ is introduced, $\mathbb{R}$ (the set of real numbers) or $\mathbb{C}$ (the set of complex numbers) may be substituted for $\mathscr{F}$.

Before proceeding to examples of linear spaces, let us introduce some notational devices. The set of all $m \times n$ matrices with elements from the field $\mathscr{F}$ is denoted by $\mathscr{F}^{m \times n}$. In particular, $\mathbb{R}^{m \times n}$ (respectively, $\mathbb{C}^{m \times n}$) stands for the set of real (respectively, complex) $m \times n$ matrices. Exploiting this notation, $\mathscr{F}^{1 \times n}$ (respectively, $\mathscr{F}^{m \times 1}$) denotes the set of all row- (respectively, column-) matrices of length n (respectively, m). For the sake of convenience, the set $\mathscr{F}^{m \times 1}$ will often be denoted by $\mathscr{F}^m$. In contrast, the symbol $\mathscr{F}_n$ stands for the set of all ordered n-tuples of numbers from the field $\mathscr{F}$. Borrowing some geometrical language, the elements $[a_1 \quad a_2 \quad \cdots \quad a_m]^{\mathrm{T}}$ of $\mathscr{F}^m$ and $(a_1, a_2, \ldots, a_m)$ from $\mathscr{F}_m$ are both referred to as *vectors of order* m, as is $[a_1 \quad a_2 \quad \cdots \quad a_m] \in \mathscr{F}^{1 \times m}$.

For solving the following exercises, it suffices to verify the conditions A1–A5 and S1–S5 (called the *linear space axioms*) for the elements of a set.

Exercise 3. Check that the following sets are linear spaces (with regard to the familiar operations defined on them) over the appropriate field:

(a) $\mathbb{R}^n$, $\mathbb{C}^n$, $\mathscr{F}^n$, $\mathscr{F}_n$.

(b) $\mathbb{R}^{m \times n}$, $\mathbb{C}^{m \times n}$, $\mathscr{F}^{m \times n}$.

(c) The sets of all upper triangular matrices and all symmetric matrices with real elements.

(d) The set of all polynomials of degree not exceeding n with real coefficients (including the identically zero function).

(e) The set of all real-valued continuous functions defined on $[0, 1]$.

(f) The set of all convergent sequences of real numbers. □

Thus, the concept of a linear space admits the study of structures such as those in Exercise 3 from a unified viewpoint and the development of a general approach for the investigation of such systems. This investigation will be performed in subsequent sections.

Before leaving this section, we make two remarks. The first concerns sets and operations that fail to generate linear spaces; the second concerns the geometrical approach to n-tuples.

Exercise 4. Confirm that the following sets fail to be linear spaces under the familiar operations defined on them over the appropriate field:

(a) the set of all position vectors of unit length in three-dimensional space;

(b) the set of all pairs of real numbers of the form $(1, a)$;
(c) the set of all real matrices of the form

$$\begin{bmatrix} a & b_0 \\ 0 & c \end{bmatrix}$$

with a fixed $b_0 \neq 0$;
(d) the set of all polynomials of a fixed degree n;
(e) the set of all Hermitian matrices over $\mathbb{C}$;
(f) the set of all rotations of a rigid body about a fixed point over $\mathbb{R}$.

In each case pick out at least one axiom that fails to hold. □

It should be noted that a set can be a linear space under certain operations of addition and scalar multiplication but may fail to generate a linear space (over the same field) under other operations. For instance, the set $\mathbb{R}_2$ is not a linear space over $\mathbb{R}$ under the operations defined by

$$(a_1, a_2) + (b_1, b_2) = (a_1 + b_1 - 1, a_2 + b_2),$$
$$\alpha(a_1, a_2) = (\alpha a_1, \alpha a_2)$$

for any $(a_1, a_2), (b_1, b_2) \in \mathbb{R}_2$ and any $\alpha \in \mathbb{R}$.

To avoid any such confusion, we shall always assume that the operations on the sets of n-tuples and matrices are defined as in Chapter 1.

In conclusion, note also that a triple (a_1, a_2, a_3), which was referred to as a vector with initial point at the origin and terminal point at the point P with Cartesian coordinates (a_1, a_2, a_3) (this approach was fruitful, for instance, for the definition of operations on triples), can be considered as a point in three-dimensional space. Similarly, an n-tuple $(a_1, a_2, \ldots, a_n)$ $(a_i \in \mathscr{F}, i = 1, 2, \ldots, n)$ may be referred to as a "point" in the "n-dimensional" linear space $\mathscr{F}_n$.

3.2 Subspaces

Let $\mathscr{S}$ denote a linear space over a field $\mathscr{F}$ and consider a subset $\mathscr{S}_0$ of elements from $\mathscr{S}$. The operations of addition and scalar multiplication are defined for all elements of $\mathscr{S}$ and, in particular, for those belonging to $\mathscr{S}_0$. The results of these operations on the elements of $\mathscr{S}_0$ are elements of $\mathscr{S}$. It may happen, however, that these operations are *closed* in $\mathscr{S}_0$, that is, when both operations are applied to arbitrary elements of $\mathscr{S}_0$, the resulting elements also belong to $\mathscr{S}_0$. In this case we say that $\mathscr{S}_0$ is a *subspace* of $\mathscr{S}$.

Thus, a nonempty subset $\mathscr{S}_0$ of a linear space $\mathscr{S}$ over $\mathscr{F}$ is a *subspace* of $\mathscr{S}$ if, for every $\boldsymbol{a}, \boldsymbol{b} \in \mathscr{S}_0$ and any $\alpha \in \mathscr{F}$,

(1) $\boldsymbol{a} + \boldsymbol{b} \in \mathscr{S}_0$.
(2) $\alpha\boldsymbol{a} \in \mathscr{S}_0$.

(Observe that here, and in general, bold face letters are used for members of a linear space if some other notation is not already established.)

It is not difficult to see that $\mathscr{S}_0$ is itself a linear space under the operations defined in $\mathscr{S}$ and hence the name "subspace."

If $\mathscr{S}$ is a linear space, it is easily seen that if $\boldsymbol{0}$ is the zero element of $\mathscr{S}$, then the singleton $\{\boldsymbol{0}\}$ and the whole space $\mathscr{S}$ are subspaces of $\mathscr{S}$. These are known as the *trivial subspaces* and any other subspace is said to be *nontrivial*. It is also important to note that the zero element of $\mathscr{S}$ is necessarily the zero element of any subspace $\mathscr{S}_0$ of $\mathscr{S}$. (This follows from Exercise 3.1.2(a).)

It is easy to verify that criteria (1) and (2) can be condensed into a single statement that characterizes a subspace. Thus, a *nonempty subset $\mathscr{S}_0$ of a linear space $\mathscr{S}$ over $\mathscr{F}$ is a subspace of $\mathscr{S}$ if and only if for every $\boldsymbol{a}, \boldsymbol{b} \in \mathscr{S}_0$ and ever $\alpha, \beta \in \mathscr{F}$, $\alpha\boldsymbol{a} + \beta\boldsymbol{b} \in \mathscr{S}_0$.*

Example 1. (a) Any straight line passing through the origin, that is, the set of triples

$$\{\boldsymbol{x} = (x_1, x_2, x_3): x_1 = at, x_2 = bt, x_3 = ct, (-\infty < t < \infty)\},$$

is a subspace of $\mathbb{R}_3$.

(b) Any plane in $\mathbb{R}_3$ passing through the origin, that is, the set of triples

$$\{\mathbf{x} = (x_1, x_2, x_3): ax_1 + bx_2 + cx_3 = 0, a^2 + b^2 + c^2 > 0\},$$

for fixed real numbers a, b, c, is a subspace of $\mathbb{R}_3$.

(c) The set of all $n \times n$ upper-(lower-) triangular matrices with elements from $\mathbb{R}$ is a subspace of $\mathbb{R}^{n \times n}$.

(d) The set of all real $n \times n$ matrices having zero main diagonal is a subspace of $\mathbb{R}^{n \times n}$.

(e) The set of all symmetric real $n \times n$ matrices is a subspace of $\mathbb{R}^{n \times n}$. □

Note again that any subspace is itself a linear space and therefore must contain the zero element. For this reason, any plane in $\mathbb{R}_3$ that does not pass through the origin fails to be a subspace of $\mathbb{R}_3$.

Exercise 2. Prove that if $\mathscr{S}_1$ and $\mathscr{S}_2$ are subspaces of a linear space $\mathscr{S}$, then the set $\mathscr{S}_1 \cap \mathscr{S}_2$ is also a subspace of $\mathscr{S}$. Give an example to show that $\mathscr{S}_1 \cup \mathscr{S}_2$ is not necessarily a subspace of $\mathscr{S}$.

Exercise 3. Let $A \in \mathscr{F}^{m \times n}$. Prove that the set of all solutions of the homogeneous equation $A\boldsymbol{x} = \boldsymbol{0}$ forms a subspace of $\mathscr{F}^n$. □

The set of all vectors $\boldsymbol{x}$ for which $A\boldsymbol{x} = \boldsymbol{0}$ is the *null space* or *kernel* of the matrix A and is written either $N(A)$ or Ker A. Observe that Exercise 3 simply states that Ker A is a subspace.

Exercise 4. Find Ker A (in $\mathbb{R}^3$) if

$$A = \begin{bmatrix} 1 & -1 & 0 \\ 1 & 1 & 1 \end{bmatrix}.$$

SOLUTION. A vector $[x_1 \quad x_2 \quad x_3]^{\mathrm{T}}$ belongs to Ker A if and only if

$$A \begin{bmatrix} x_1 \\ x_2 \\ x_3 \end{bmatrix} = \boldsymbol{0}$$

or, what is equivalent,

$$\begin{aligned} x_1 - x_2 \qquad &= 0, \\ x_1 + x_2 + x_3 &= 0. \end{aligned}$$

Hence, $x_1 = x_2$, $x_3 = -2x_2$, and every vector from Ker A is of the form

$$\begin{bmatrix} \alpha \\ \alpha \\ -2\alpha \end{bmatrix} = \alpha \begin{bmatrix} 1 \\ 1 \\ -2 \end{bmatrix}$$

for some (real) α. As in Example 1(a), this is a subspace of $\mathbb{R}^3$. □

Geometrically, a vector $\boldsymbol{x}$ from $\mathbb{R}^3$ belongs to the kernel of $A = [a_{ij}]_{i,j=1}^{m,3}$ if and only if it is orthogonal to the row-vectors of A. Thus for $i = 1, 2, \ldots, m$, the vector $\boldsymbol{x}$ must be perpendicular to $[a_{i1} \quad a_{i2} \quad a_{i3}]$. If $m = 2$, as in the previous example, then any $\boldsymbol{x} \in$ Ker A must be orthogonal to both of the vectors $[1 \quad -1 \quad 0]$ and $[1 \quad 1 \quad 1]$, that is, $\boldsymbol{x}$ must be orthogonal to the plane containing these vectors and passing through the origin. It is now obvious that Ker A represents a straight line through the origin orthogonal to this plane.

Now we introduce another subspace associated with a matrix, and this should be seen as a dual, or complementary, concept to that of the kernel.

If A is an $m \times n$ matrix, the set

$$R(A) = \operatorname{Im} A \triangleq \{\boldsymbol{y} \in \mathscr{F}^m : \boldsymbol{y} = A\boldsymbol{x} \text{ for some } \boldsymbol{x} \in \mathscr{F}^n\}$$

is said to be the *range* (or *image*) of A.

Exercise 5. The set

$$\left\{\begin{bmatrix}\alpha\\0\end{bmatrix} : \alpha \in \mathscr{F}\right\}$$

is the range of the matrix

$$A = \begin{bmatrix}1 & 0 & 0\\0 & 0 & 0\end{bmatrix}.$$

Exercise 6. Check that the range of any matrix $A \in \mathscr{F}^{m \times n}$ is a subspace of $\mathscr{F}^m$.

Exercise 7. Show that if the matrix product AB is defined, then

$$\operatorname{Ker}(AB) \supset \operatorname{Ker} B,$$

with equality when A is invertible, and

$$\operatorname{Im}(AB) \subset \operatorname{Im} A,$$

with equality when B is invertible. □

Note that the symbol $\subset$ (or $\supset$) is used consistently to denote either strict inclusion or equality.

3.3 Linear Combinations

We start with the following problem: Let $\boldsymbol{a}_1, \boldsymbol{a}_2, \ldots, \boldsymbol{a}_n$ be elements of a linear space $\mathscr{S}$ over the field $\mathscr{F}$. We want to find the minimal subspace $\mathscr{S}_0$ of $\mathscr{S}$ containing the given elements. The minimality of $\mathscr{S}_0$ is understood in the sense that if $\mathscr{S}_1$ is a subspace of $\mathscr{S}$ containing the elements $\boldsymbol{a}_1, \boldsymbol{a}_2, \ldots, \boldsymbol{a}_n$, then $\mathscr{S}_1 \supset \mathscr{S}_0$.

To solve the problem we first note that, according to the definition of a subspace, $\mathscr{S}_0$ must contain, along with $\boldsymbol{a}_i$ ($i = 1, 2, \ldots, n$), all the elements of the form $\alpha_i \boldsymbol{a}_i (\alpha_i \in \mathscr{F}, i = 1, 2, \ldots, n)$. Furthermore, the sums of elements belonging to $\mathscr{S}_0$ must also be elements of $\mathscr{S}_0$.

Thus, any subspace containing the elements $\boldsymbol{a}_1, \boldsymbol{a}_2, \ldots, \boldsymbol{a}_n$ must also contain all elements of the form

$$\alpha_1 \boldsymbol{a}_1 + \alpha_2 \boldsymbol{a}_2 + \cdots + \alpha_n \boldsymbol{a}_n \tag{1}$$

for any $\alpha_i \in \mathscr{F}$ $(i = 1, 2, \ldots, n)$. An expression of the form (1) is referred to as a *linear combination* of the elements $\boldsymbol{a}_1, \boldsymbol{a}_2, \ldots, \boldsymbol{a}_n$ over the field $\mathscr{F}$.

Example 1. (a) Since

$$a^{\mathrm{T}} = [2 \quad -3 \quad -4] = 2[1 \quad 0 \quad 1] - 3[0 \quad 1 \quad 2],$$

the vector $a^{\mathrm{T}} \in \mathbb{R}^{1 \times 3}$ is a linear combination of the vectors $a_1^{\mathrm{T}} = [1 \quad 0 \quad 1]$ and $a_2^{\mathrm{T}} = [0 \quad 1 \quad 2]$.

(b) Any vector in $\mathscr{F}^n$ is a linear combination of the *unit vectors*

$$e_1 = \begin{bmatrix} 1 \\ 0 \\ 0 \\ \vdots \\ 0 \end{bmatrix}, \quad e_2 = \begin{bmatrix} 0 \\ 1 \\ 0 \\ \vdots \\ 0 \end{bmatrix}, \ldots, \quad e_n = \begin{bmatrix} 0 \\ \vdots \\ 0 \\ 0 \\ 1 \end{bmatrix}.$$

Indeed, any $a = [\alpha_1 \quad \alpha_2 \quad \cdots \quad \alpha_n]^{\mathrm{T}} \in \mathscr{F}^n$ can be represented in the form $a = \sum_{i=1}^n \alpha_i e_i$.

(c) In the linear space $\mathscr{F}^{m \times n}$ of $m \times n$ matrices over $\mathscr{F}$, any matrix A is a linear combination of the mn matrices E_{ij} $(1 \le i \le m,\ 1 \le j \le n)$ with 1 in the i, jth position and zeros elsewhere.

(d) In the linear space of $n \times n$ Toeplitz matrices $A = [a_{i-j}]_{i,j=0}^{n-1}$ over $\mathscr{F}$, any matrix is a linear combination of $2n - 1$ matrices $A_k = [\delta_{i-j,k}]_{i,j=0}^{n-1}$ $(1 - n \le k \le n - 1)$, where the *Kronecker delta* δ_{ij} is defined by

$$\delta_{ij} = \begin{cases} 1 & \text{when} \quad i = j, \\ 0 & \text{when} \quad i \ne j. \end{cases}$$

(e) Any polynomial in x of degree not exceeding n can be represented as a linear combination of the monomials $1, x, x^2, \ldots, x^n$. □

Returning to the problem formulated above and making use of the new terminology, we conclude that any subspace containing the given elements $a_1, \ldots, a_n$, must contain all their linear combinations.

It is easily verified that the set of all linear combinations (over $\mathscr{F}$) of the elements $a_1, a_2, \ldots, a_n$ belonging to a linear space $\mathscr{S}$ *generates* a subspace $\mathscr{S}_0$ of $\mathscr{S}$. Obviously, this subspace solves the proposed problem: it contains the given elements themselves and is contained in any other subspace $\mathscr{S}_1$ of $\mathscr{S}$ such that $a_1, a_2, \ldots, a_n \in \mathscr{S}_1$.

The minimal subspace $\mathscr{S}_0$ generated by the elements $a_1, a_2, \ldots, a_n \in \mathscr{S}$ is referred to as the *linear hull* (or *span*) of the a_i $(i = 1, 2, \ldots, n)$ over $\mathscr{F}$ and is written $\operatorname{span}\{a_1, a_2, \ldots, a_n\}$. Thus by definition,

$$\operatorname{span}\{a_1, a_2, \ldots, a_n\} \triangleq \left\{ a \in \mathscr{S} : a = \sum_{i=1}^{n} \alpha_i a_i,\ \alpha_1, \alpha_2, \ldots, \alpha_n \in \mathscr{F} \right\}.$$

Note that a linear space $\operatorname{span}\{a_1, a_2, \ldots, a_n\}$ is determined uniquely by the *spanning elements* a_i $(i = 1, 2, \ldots, n)$. We have proved the following theorem.

Theorem 1. *If $\mathscr{S}$ is a linear space and $\boldsymbol{a}_1, \boldsymbol{a}_2, \ldots, \boldsymbol{a}_n \in \mathscr{S}$, then* $\operatorname{span}\{\boldsymbol{a}_1, \ldots. \boldsymbol{a}_n\}$ *is the minimal (in the sense of inclusion) subspace containing $\boldsymbol{a}_1, \boldsymbol{a}_2, \ldots, \boldsymbol{a}_n$.*

***Example* 2.** (a) If $\boldsymbol{a}_1$ and $\boldsymbol{a}_2$ stand for two nonparallel position vectors in $\mathbb{R}_3$, then their linear hull is the set of all vectors lying in the plane containing $\boldsymbol{a}_1$ and $\boldsymbol{a}_2$ and passing through the origin. This is obviously the minimal subspace containing the given vectors.

(b) Let $\boldsymbol{a}_1$ and $\boldsymbol{a}_2$ be parallel position vectors in $\mathbb{R}_3$. Then the linear hull of both of them, or of only one of them, is the same, namely, the line parallel to each of $\boldsymbol{a}_1$ and $\boldsymbol{a}_2$ and passing through the origin (in the language of $\mathbb{R}_3$). □

Example 2 shows that in some cases a subspace $\operatorname{span}\{\boldsymbol{a}_1, \boldsymbol{a}_2, \ldots, \boldsymbol{a}_n\}$ may be spanned by a proper subset of the given elements $\boldsymbol{a}_1, \boldsymbol{a}_2, \ldots, \boldsymbol{a}_n$. The problem of determining the minimal number of elements that span the subspace will be considered in the next section.

Now we note that the notion we introduced in the previous section of the image of a matrix $A \in \mathscr{F}^{m \times n}$ can also be defined using linear combinations. Let $\mathbf{x} = [x_1 \quad x_2 \quad \cdots \quad x_n]^{\mathrm{T}}$ be an arbitrary vector from $\mathscr{F}^n$ and let A_{*j} $(j = 1, 2, \ldots, n)$ denote the columns of A considered as vectors from $\mathscr{F}^m$. Observe that the vector $A\mathbf{x} \in \mathscr{F}^m$ can be written as the linear combination.

$$A\boldsymbol{x} = \boldsymbol{A}_{*1}x_1 + \boldsymbol{A}_{*2}x_2 + \cdots + \boldsymbol{A}_{*n}x_n \tag{2}$$

of columns of A. Representation (2) shows that every vector $\boldsymbol{y}$ in $\operatorname{Im} A$, the image of A, is a linear combination of the columns of A, and conversely. Thus, $\operatorname{Im} A$ can be defined as the linear hull of the columns of A.

Note that the span of the columns of A written as vectors from $\mathscr{F}^m$ is also referred as the *column space* of A, written $\mathscr{C}_A$. The *row space* $\mathscr{R}_A \subset \mathscr{F}^{1 \times n}$ of A is defined similarly. Thus

$$\mathscr{R}_A = \{\boldsymbol{y}^{\mathrm{T}} : \boldsymbol{y}^{\mathrm{T}} = \boldsymbol{a}^{\mathrm{T}}A \text{ for some } \boldsymbol{a} \in \mathscr{F}^m\} = \{\boldsymbol{y}^{\mathrm{T}} : \boldsymbol{y} \in \operatorname{Im} A^{\mathrm{T}}\}.$$

Although the column and row spaces of the matrix A are generally completely different (observe that $\mathscr{C}_A \subset \mathscr{F}^m$ and $\mathscr{R}_A \subset \mathscr{F}^{1 \times n}$), a characteristic common to them both will emerge in Section 3.7.

3.4 Linear Dependence and Independence

Let the elements $\boldsymbol{a}_1, \boldsymbol{a}_2, \ldots, \boldsymbol{a}_n$ from a linear space $\mathscr{S}$ over $\mathscr{F}$ span the subspace $\mathscr{S}_0 = \operatorname{span}\{\boldsymbol{a}_1, \boldsymbol{a}_2, \ldots, \boldsymbol{a}_n\}$, and suppose that $\mathscr{S}_0$ can be spanned by a proper subset of the given elements. For example, suppose that

$\mathscr{S}_0 = \text{span}\{\boldsymbol{a}_2, \boldsymbol{a}_3, \ldots, \boldsymbol{a}_n\}$. This means that the spaces consisting of linear combinations of the n elements $\boldsymbol{a}_1, \boldsymbol{a}_2, \ldots, \boldsymbol{a}_n$ and $n - 1$ elements $\boldsymbol{a}_2, \boldsymbol{a}_3, \ldots, \boldsymbol{a}_n$ are identical. In other words, for any fixed linear combination of $\boldsymbol{a}_1, \boldsymbol{a}_2, \ldots, \boldsymbol{a}_n$ there exists a linear combination of the last $n - 1$ elements such that they coincide.

For the investigation of this situation we take an element $\boldsymbol{a}$ of the subspace $\mathscr{S}_0$ expressed as:

$$\boldsymbol{a} = \sum_{i=1}^{n} \alpha_i \boldsymbol{a}_i,$$

where $\alpha_1, \alpha_2, \ldots, \alpha_n \in \mathscr{F}$, and $\alpha_1 \neq 0$. Since $\mathscr{S}_0$ is also the span of $\boldsymbol{a}_2, \boldsymbol{a}_3, \ldots, \boldsymbol{a}_n$, then necessarily $\boldsymbol{a} = \sum_{i=2}^{n} \beta_i \boldsymbol{a}_i$, where $\beta_2, \beta_3, \ldots, \beta_n \in \mathscr{F}$. Hence

$$\sum_{i=1}^{n} \alpha_i \boldsymbol{a}_i = \sum_{i=2}^{n} \beta_i \boldsymbol{a}_i$$

or, what is equivalent,

$$\alpha_1 \boldsymbol{a}_1 + (\alpha_2 - \beta_2)\boldsymbol{a}_2 + \cdots + (\alpha_n - \beta_n)\boldsymbol{a}_n = \boldsymbol{0}. \tag{1}$$

with $\alpha_1 \neq 0$.

Thus, in view of (1), a necessary condition that a set of n elements from $\mathscr{S}$ and a subset of $n - 1$ elements from them span the same subspace is that

$$\gamma_1 \boldsymbol{a}_1 + \gamma_2 \boldsymbol{a}_2 + \cdots + \gamma_n \boldsymbol{a}_n = \boldsymbol{0}. \tag{2}$$

where not all of $\gamma_1, \gamma_2, \ldots, \gamma_n$ are equal to zero. In general, elements $\boldsymbol{a}_1, \boldsymbol{a}_2, \ldots, \boldsymbol{a}_n$ satisfying (2) with at least one $\gamma_i \neq 0$ $(1 \leq i \leq n)$ are called *linearly dependent*. Thus, we conclude that n elements $\boldsymbol{a}_1, \boldsymbol{a}_2, \ldots, \boldsymbol{a}_n$ and a proper subset of those elements span the same subspace only if $\boldsymbol{a}_1, \boldsymbol{a}_2, \ldots, \boldsymbol{a}_n$ are linearly dependent.

It turns out that this condition is also sufficient: if $\boldsymbol{a}_1, \boldsymbol{a}_2, \ldots, \boldsymbol{a}_n$ are linearly dependent, then the subspace $\text{span}\{\boldsymbol{a}_1, \boldsymbol{a}_2, \ldots, \boldsymbol{a}_n\}$ can be spanned by some $n - 1$ elements of the set $\{\boldsymbol{a}_1, \boldsymbol{a}_2, \ldots, \boldsymbol{a}_n\}$. Indeed, let there exist scalars $\gamma_1, \gamma_2, \ldots, \gamma_n$ such that Eq. (2) holds and at least one of the γ_i, say γ_j, is not zero. Then, dividing Eq. (2) by γ_j, we obtain

$$\boldsymbol{a}_j = -\frac{\gamma_1}{\gamma_j}\boldsymbol{a}_1 - \cdots - \frac{\gamma_{j-1}}{\gamma_j}\boldsymbol{a}_{j-1} - \frac{\gamma_{j+1}}{\gamma_j}\boldsymbol{a}_{j+1} - \cdots - \frac{\gamma_n}{\gamma_j}\boldsymbol{a}_n,$$

and consequently, $\boldsymbol{a}_j$ is a linear combination of the remaining elements of $\{\boldsymbol{a}_1, \boldsymbol{a}_2, \ldots, \boldsymbol{a}_n\}$.

Furthermore, let $\boldsymbol{a} \in \text{span}\{\boldsymbol{a}_1, \boldsymbol{a}_2, \ldots, \boldsymbol{a}_n\}$, so that $\boldsymbol{a}$ is a linear combination of the spanning elements $\boldsymbol{a}_1, \boldsymbol{a}_2, \ldots, \boldsymbol{a}_n$. Since for some j, the element $\boldsymbol{a}_j$ is a linear combination of $a_1, \ldots, a_{j-1}, a_{j+1}, \ldots, a_n$, then so is a. Hence $a \in \operatorname{span}\{a_1, \ldots, a_{j-1}, a_{j+1}, \ldots, a_n\}$, and so

$$\operatorname{span}\{a_1, a_2, \ldots, a_n\} \subset \operatorname{span}\{a_1, \ldots, a_{j-1}, a_{j+1}, \ldots, a_n\}.$$

Since the reverse inclusion is obvious, the two subspaces must agree.

We have now established another characterization of a linearly dependent set of elements.

Theorem 1. *The subspace* $\operatorname{span}\{a_1, a_2, \ldots, a_n\}$ *is the span of a proper subset of* $\{a_i\}_{i=1}^n$ *if and only if the elements* $a_1, a_2, \ldots, a_n$ *are linearly dependent.*

A *linearly independent* set is now defined as one that is not linearly dependent. Thus, if $\mathscr{S}$ is a linear space, the set of elements $a_1, a_2, \ldots, a_n \in \mathscr{S}$ is *linearly independent* if the equation

$$\sum_{i=1}^{n} \gamma_i a_i = 0,$$

for $\gamma_1, \gamma_2, \ldots, \gamma_n \in \mathscr{F}$, implies that

$$\gamma_1 = \gamma_2 = \cdots = \gamma_n = 0.$$

We see that, in this case, it is no longer possible to delete an element of the set $\{a_1, a_2, \ldots, a_n\}$ in the construction of $\operatorname{span}\{a_1, a_2, \ldots, a_n\}$.

It turns out, therefore, that in this case n is the smallest number of spanning elements of the linear hull being considered. This and other related problems will be discussed in more detail in the next section, while the rest of this section will be devoted to some illustrations.

Exercise 1. Confirm the linear dependence of the following sets:

(a) any two vectors in $\mathbb{R}_2$ lying on the same line through the origin (*collinear vectors*);

(b) any three vectors in $\mathbb{R}_3$ lying in the same plane passing through the origin (*coplanar vectors*);

(c) the set $\{a_1, a_2, a_3\}$ of elements from $\mathbb{R}^3$ such that $a_1^T = [2 \ \ -3 \ \ 0]$, $a_2^T = [1 \ \ 0 \ \ -1]$, $a_3^T = [3 \ \ -6 \ \ 1]$;

(d) The set of polynomials $p_1(x) = 4 + 2x - 7x^2$, $p_2(x) = 2 - x^2$; $p_3(x) = 1 - x + 2x^2$;

(e) any set containing the zero element 0;

(f) any set containing an element that is a linear combination of the others;

(g) any set having a subset of linearly dependent elements.

Exercise 2. Confirm the linear independence of the following sets:

(a) the set of unit vectors $e_1, e_2, \ldots, e_n$ in $\mathscr{F}^n$;

(b) the set of matrices in $\mathbb{R}^{3\times 3}$

$$\begin{bmatrix} 1 & 2 & 0 \\ 0 & 1 & 0 \\ 0 & 0 & 0 \end{bmatrix}, \quad \begin{bmatrix} 0 & 0 & 0 \\ 0 & 0 & 1 \\ 0 & 0 & 1 \end{bmatrix}, \quad \begin{bmatrix} 0 & 0 & 0 \\ 0 & 0 & 0 \\ 0 & 1 & 0 \end{bmatrix};$$

(c) the set of $m \times n$ matrices E_{ij} $(1 \le i \le m,\ 1 \le j \le n)$ with 1 in the (i, j) position and zeros elsewhere;

(d) the set of polynomials

$$p_1(x) = 1 + 2x,\ p_2(x) = x^2,\ p_3(x) = 3 + 5x + 2x^2;$$

(e) the set of polynomials $1, x, x^2, \ldots, x^n$;

(f) any subset of a set of linearly independent elements. □

3.5 The Notion of a Basis

Let the elements $\boldsymbol{a}_1, \boldsymbol{a}_2, \ldots, \boldsymbol{a}_m$ (not all zero) belong to a linear space $\mathscr{S}$ over the field $\mathscr{F}$, and let

$$\mathscr{S}_0 = \operatorname{span}\{\boldsymbol{a}_1, \boldsymbol{a}_2, \ldots, \boldsymbol{a}_m\}.$$

If the elements $\boldsymbol{a}_1, \boldsymbol{a}_2, \ldots, \boldsymbol{a}_m$ are linearly independent, that is, if each of them fails to be a linear combination of others, then (Theorem 3.4.1) all $\mathbf{a}_i$ $(i = 1, 2, \ldots, m)$ are necessary for spanning $\mathscr{S}_0$. In the case of linearly dependent elements $\boldsymbol{a}_1, \boldsymbol{a}_2, \ldots, \boldsymbol{a}_m$ we may, according to the theorem, delete all vectors of the set that are linear combinations of the others so that the span of the remaining (linearly independent) n elements is $\mathscr{S}_0$. Thus we can always consider $\mathscr{S}_0$ as the linear hull of n linearly independent elements where $1 \le n \le m$.

Let $\mathscr{S}$ denote an arbitrary linear space over $\mathscr{F}$. A finite set of elements

$$\{\boldsymbol{a}_1, \boldsymbol{a}_2, \ldots, \boldsymbol{a}_n\} \tag{1}$$

is said to be a (finite) *basis* of $\mathscr{S}$ if they are linearly independent and every element $\boldsymbol{a} \in \mathscr{S}$ is a linear combination of the elements in (1):

$$\boldsymbol{a} = \sum_{i=1}^{n} \alpha_i \boldsymbol{a}_i \qquad \alpha_i \in \mathscr{F}, \quad i = 1, 2, \ldots, n. \tag{2}$$

In other words, the elements in (1) form a basis of $\mathscr{S}$ if

$$\mathscr{S} = \operatorname{span}\{\boldsymbol{a}_1, \boldsymbol{a}_2, \ldots, \boldsymbol{a}_n\} \tag{3}$$

and no $\mathbf{a}_i$ $(1 \le i \le n)$ in Eq. (3) can be discarded. In this case the elements in (1) are referred to as *basis elements* of $\mathscr{S}$.

Recalling our discussion of linear hulls in the first paragraph of this section, we conclude that any space generated by a set of m elements, not all zero, has a basis consisting of n elements, where $1 \le n \le m$. Some examples of bases are presented in the next exercise; the results of Example 3.3.1 and Exercise 3.4.2 are used.

Exercise 1. Verify the following statements:

(a) The set of n unit vectors $\boldsymbol{e}_1, \boldsymbol{e}_2, \ldots, \boldsymbol{e}_n \in \mathscr{F}^n$ defined in Example 3.3.1(b) is a basis for $\mathscr{F}^n$. This is known as the *standard basis* of the space.

(b) The set $\{1, x, x^2, \ldots, x^n\}$ is a basis for the space of polynomials (including zero) with degree less than or equal to n.

(c) The mn matrices E_{ij} $(1 \le i \le m,\ 1 \le j \le n)$ defined in Example 3.3.1(c) generate a basis for the space $\mathscr{F}^{m \times n}$. □

Note that a basis necessarily consists of only *finitely* many vectors. Accordingly, the linear spaces they generate are said to be *finite dimensional*; the properties of such spaces pervade the analysis of most of this book. If a linear space (containing nonzero vectors) has no basis, it is said to be of *infinite dimension*. These spaces are also of great importance in analysis and applications but are not generally within the scope of this book.

Proposition 1. *Any element of a (finite-dimensional) space $\mathscr{S}$ is expressed uniquely as a linear combination of the elements of a fixed basis.*

In other words, given $\boldsymbol{a} \in \mathscr{S}$ and a basis $\{\boldsymbol{a}_1, \boldsymbol{a}_2, \ldots, \boldsymbol{a}_n\}$ for $\mathscr{S}$, there is one and only one ordered set of scalars $\alpha_1, \alpha_2, \ldots, \alpha_n \in \mathscr{F}$ such that the representation (2) is valid.

PROOF. If

$$\boldsymbol{a} = \sum_{i=1}^{n} \alpha_i \boldsymbol{a}_i = \sum_{i=1}^{n} \beta_i \boldsymbol{a}_i,$$

then

$$\sum_{i=1}^{n} (\alpha_i - \beta_i)\boldsymbol{a}_i = \boldsymbol{0}.$$

The linear independence of the basis elements implies $\alpha_i = \beta_i$ for $i = 1, 2, \ldots, n$, and the assertion follows. ■

Let $\boldsymbol{a}$ belong to a linear space $\mathscr{S}$ over $\mathscr{F}$ having a basis $\{\boldsymbol{a}_1, \boldsymbol{a}_2, \ldots, \boldsymbol{a}_n\}$. The column matrix $\boldsymbol{\alpha} = [\alpha_1 \quad \alpha_2 \quad \cdots \quad \alpha_n]^{\mathrm{T}} \in \mathscr{F}^n$ such that the decomposition

$$\boldsymbol{a} = \sum_{i=1}^{n} \alpha_i \boldsymbol{a}_i$$

holds is said to be the *representation of* **a** *with respect to the basis* $\{a_i\}_{i=1}^n$. The scalars $\alpha_1, \alpha_2, \ldots, \alpha_n$ are referred to as the *coordinates* (or *components*) of **a** with respect to the basis. Proposition 1 thus means that the representation of an element with respect to a fixed basis is unique.

Example 2. If $\boldsymbol{\alpha} = [\alpha_1 \quad \alpha_2 \quad \cdots \quad \alpha_n]^{\mathrm{T}} \in \mathscr{F}^n$, then its representation with respect to the standard basis is again $[\alpha_1 \quad \alpha_2 \quad \cdots \quad \alpha_n]^{\mathrm{T}}$. □

The following example illustrates the fact that, in general, a linear space does not have a unique basis.

Exercise 3. Check that the vectors

$$\boldsymbol{a}_1 = \begin{bmatrix} 1 \\ 1 \\ 0 \end{bmatrix}, \qquad \boldsymbol{a}_2 = \begin{bmatrix} 0 \\ 1 \\ 1 \end{bmatrix}, \qquad \boldsymbol{a}_3 = \begin{bmatrix} 1 \\ 0 \\ 1 \end{bmatrix}$$

form a basis for $\mathbb{R}^3$, as do the unit vectors $\boldsymbol{e}_1, \boldsymbol{e}_2, \boldsymbol{e}_3$.

Hint. Observe that if $\boldsymbol{a} = [\alpha_1 \quad \alpha_2 \quad \alpha_3]^{\mathrm{T}} \in \mathbb{R}^3$, then its representation with respect to the basis $\{\boldsymbol{a}_1, \boldsymbol{a}_2, \boldsymbol{a}_3\}$ is

$$\tfrac{1}{2}[\alpha_1 + \alpha_2 - \alpha_3 \quad -\alpha_1 + \alpha_2 + \alpha_3 \quad \alpha_1 - \alpha_2 + \alpha_3]^{\mathrm{T}}$$

and differs from the representation $[\alpha_1 \quad \alpha_2 \quad \alpha_3]^{\mathrm{T}}$ of $\boldsymbol{a}$ with respect to the standard basis. □

In view of Exercise 3, the question of whether or not all bases for $\mathscr{S}$ have the same number of elements naturally arises. The answer is affirmative.

Theorem 1. *All bases of a finite-dimensional linear space $\mathscr{S}$ have the same number of elements.*

PROOF. Let $\{\boldsymbol{b}_1, \boldsymbol{b}_2, \ldots, \boldsymbol{b}_n\}$ and $\{\boldsymbol{c}_1, \boldsymbol{c}_2, \ldots, \boldsymbol{c}_m\}$ be two bases in $\mathscr{S}$. Our aim is to show that $n = m$. Let us assume, on the contrary, that $n < m$.

First, since $\{\boldsymbol{b}_1, \boldsymbol{b}_2, \ldots, \boldsymbol{b}_n\}$ constitutes a basis for $\mathscr{S}$ this system consists of linearly independent elements. Second, every element of $\mathscr{S}$ and, in particular, each $\mathbf{c}_j$ $(1 \le j \le m)$ is a linear combination of the $\mathbf{b}_i$ $(i = 1, 2, \ldots, n)$, say,

$$\boldsymbol{c}_j = \sum_{i=1}^{n} \beta_{ij} \boldsymbol{b}_i, \qquad 1 \le j \le m. \tag{4}$$

We shall show that Eq. (4) yields a contradiction of the assumed linear independence of the set $\{\boldsymbol{c}_1, \ldots, \boldsymbol{c}_m\}$. In fact, let

$$\sum_{j=1}^{m} \gamma_j \boldsymbol{c}_j = \boldsymbol{0}. \tag{5}$$

Substituting from Eq. (4), we obtain

$$\sum_{j=1}^{m} \gamma_j\left(\sum_{i=1}^{n} \beta_{ij}\boldsymbol{b}_i\right) = \sum_{i=1}^{n}\left(\sum_{j=1}^{m} \gamma_j\beta_{ij}\right)\boldsymbol{b}_i = \boldsymbol{0}.$$

The linear independence of the system $\{\boldsymbol{b}_1, \boldsymbol{b}_2, \ldots, \boldsymbol{b}_n\}$ now implies that

$$\sum_{j=1}^{m} \gamma_j\beta_{ij} = 0, \qquad i = 1, 2, \ldots, n, \tag{6}$$

and, in view of Exercise 2.9.5, our assumption $m > n$ yields the existence of a nontrivial solution $(\gamma_1, \gamma_2, \ldots, \gamma_m)$ of (6). This means [see Eq. (5)] that the elements $\boldsymbol{c}_1, \boldsymbol{c}_2, \ldots, \boldsymbol{c}_m$ are linearly dependent, and a contradiction with the assumption $m > n$ is derived.

If $m < n$, exchanging roles of the bases again leads to a contradiction. Thus $m = n$ and all bases of $\mathscr{S}$ possess the same number of elements. ■

The number of basis elements of a finite-dimensional space is, therefore, a characteristic of the space that is invariant under different choices of basis. We formally define the number of basis elements of a (finite-dimensional) space $\mathscr{S}$ to be in the *dimension* of the space, and write dim $\mathscr{S}$ for this number.

In other words, if dim $\mathscr{S} = n$ (in which case $\mathscr{S}$ is called an *n-dimensional space*), then $\mathscr{S}$ has n linearly independent elements and each set of $n + 1$ elements of $\mathscr{S}$ is necessarily linearly dependent. Note that if $\mathscr{S} = \{\boldsymbol{0}\}$ then, by definition, dim $\mathscr{S} = 0$.

The terminology "finite-dimensional space" can now be clarified as meaning a space having finite dimension. The "dimension" of an infinite-dimensional space is infinite in the sense that the span of any finite number of elements from the space is only a proper subspace of the space.

Exercise 4. Show that any n linearly independent elements $\boldsymbol{b}_1, \boldsymbol{b}_2, \ldots, \boldsymbol{b}_n$ in an n-dimensional linear space $\mathscr{S}$ constitutes a basis for $\mathscr{S}$.

SOLUTION. If, on the contrary, span$\{\boldsymbol{b}_1, \boldsymbol{b}_2, \ldots, \boldsymbol{b}_n\}$ is not the whole space, then there is an $\boldsymbol{a} \in \mathscr{S}$ such that $\boldsymbol{a}, \boldsymbol{b}_1, \boldsymbol{b}_2, \ldots, \boldsymbol{b}_n$ are linearly independent. Hence dim $\mathscr{S} \geq n + 1$ and a contradiction is achieved.

Exercise 5. Check that if $\mathscr{S}_0$ is a nontrivial subspace of the linear space $\mathscr{S}$, then

$$\dim \mathscr{S}_0 < \dim \mathscr{S}. \quad \square$$

Let $\mathscr{S}_0$ be a subspace of $\mathscr{F}^n$ and let $A \in \mathscr{F}^{m \times n}$. The subspace

$$\{A\boldsymbol{x} : \boldsymbol{x} \in \mathscr{S}_0\}$$

in $\mathscr{F}^m$ is known as *the image of $\mathscr{S}_0$ under A* and is written $A(\mathscr{S}_0)$.

Exercise 6. Prove that

$$\dim A(\mathcal{S}_0) \le \dim \mathcal{S}_0,$$

with equality when A is invertible.

Hint. Use Exercise 3.2.7 and Exercise 5 in this section. □

Let $\mathcal{S}$ be an n-dimensional linear space. It turns out that any system of r linearly independent elements from $\mathcal{S}$ $(1 \le r \le n-1)$ can be extended to form a basis for the space.

Proposition 2. *If the elements* $\{\boldsymbol{a}_i\}_{i=1}^r$ $(1 \le r \le n-1)$ *from a space* $\mathcal{S}$ *of dimension* n *are linearly independent, then there exist* $n-r$ *elements* $\boldsymbol{a}_{r+1}, \boldsymbol{a}_{r+2}, \ldots, \boldsymbol{a}_n$ *such that* $\{\boldsymbol{a}_i\}_{i=1}^n$ *constitutes a basis in* $\mathcal{S}$.

PROOF. Since $\dim \mathcal{S} = n > r$, the subspace $\mathcal{S}_1 = \operatorname{span}\{\boldsymbol{a}_1, \boldsymbol{a}_2, \ldots, \boldsymbol{a}_r\}$ is a proper subspace of $\mathcal{S}$. Hence there is an element $\boldsymbol{a}_{r+1} \in \mathcal{S}$ that does not belong to $\mathcal{S}_1$. The elements $\{\boldsymbol{a}_i\}_{i=1}^{r+1}$ are consequently linearly independent, since otherwise $\boldsymbol{a}_{r+1}$ would be a linear combination of $\{\boldsymbol{a}_i\}_{i=1}^r$ and would belong to $\mathcal{S}_1$. If $r+1=n$, then in view of Exercise 4 the proof is completed. If $r+1<n$ the argument can be applied repeatedly, beginning with $\operatorname{span}\{\boldsymbol{a}_1, \boldsymbol{a}_2, \ldots, \boldsymbol{a}_{r+1}\}$, until n linearly independent elements are constructed. ■

Let $\boldsymbol{a}_1, \boldsymbol{a}_2, \ldots, \boldsymbol{a}_r$ $(1 \le r \le n-1)$ belong to $\mathcal{S}$, with $\dim \mathcal{S} = n$. Denote $\mathcal{S}_1 = \operatorname{span}\{\boldsymbol{a}_1, \boldsymbol{a}_2, \ldots, \boldsymbol{a}_r\}$. Proposition 2 asserts the existence of basis elements $\boldsymbol{a}_{r+1}, \boldsymbol{a}_{r+2}, \ldots, \boldsymbol{a}_n$ spanning a subspace $\mathcal{S}_2$ in $\mathcal{S}$ such that the union of bases in $\mathcal{S}_1$ and $\mathcal{S}_2$ gives a basis in $\mathcal{S}$. In view of Eq. (2), this provides a representation of any element $\mathbf{a} \in \mathcal{S}$ as a sum of the elements

$$\boldsymbol{a}^{(1)} = \sum_{i=1}^{r} \alpha_i \boldsymbol{a}_i \quad (\in \mathcal{S}_1) \qquad \text{and} \qquad \boldsymbol{a}^{(2)} = \sum_{i=r+1}^{n} \alpha_i \boldsymbol{a}_i \quad (\in \mathcal{S}_2).$$

We write

$$\boldsymbol{a} = \boldsymbol{a}^{(1)} + \boldsymbol{a}^{(2)},$$

where $\boldsymbol{a}^{(1)} \in \mathcal{S}_1$ and $\boldsymbol{a}^{(2)} \in \mathcal{S}_2$. We will study this idea in more detail in the next section.

Exercise 7. Check that, with the notation of the last paragraph, $\mathcal{S}_1 \cap \mathcal{S}_2 = \{\boldsymbol{0}\}$. □

3.6 Sum and Direct Sum of Subspaces

Let $\mathcal{S}$ be a finite-dimensional linear space and let $\mathcal{S}_1$ and $\mathcal{S}_2$ be subspaces of $\mathcal{S}$. The *sum* $\mathcal{S}_1 + \mathcal{S}_2$ of the subspaces $\mathcal{S}_1$ and $\mathcal{S}_2$ is defined to be the set consisting of all sums of the form $\boldsymbol{a}_1 + \boldsymbol{a}_2$, where $\boldsymbol{a}_1 \in \mathcal{S}_1$ and $\boldsymbol{a}_2 \in \mathcal{S}_2$.

It is easily seen that $\mathscr{S}_1 + \mathscr{S}_2$ is also a subspace in $\mathscr{S}$ and that the operation of addition of subspaces satisfies the axioms A1–A4 of a linear space (see Section 3.1), but does not satisfy axiom A5.

Note also that the notion of the sum of several subspaces can be defined similarly.

Exercise 1. Show that if $\boldsymbol{a}_1 = [1 \quad 0 \quad 1]^T$, $\boldsymbol{a}_2 = [0 \quad 2 \quad -1]^T$, $\boldsymbol{b}_1 = [2 \quad 2 \quad 1]^T$, $\boldsymbol{b}_2 = [0 \quad 0 \quad 1]^T$, and

$$\mathscr{S}_1 = \operatorname{span}\{\boldsymbol{a}_1, \boldsymbol{a}_2\}, \qquad \mathscr{S}_2 = \operatorname{span}\{\boldsymbol{b}_1, \boldsymbol{b}_2\},$$

then $\mathscr{S}_1 + \mathscr{S}_2 = \mathbb{R}^3$. Observe that $\dim \mathscr{S}_1 = \dim \mathscr{S}_2 = 2$, while $\dim \mathbb{R}^3 = 3$. □

A general relationship between the dimension of a sum of subspaces and that of its summands is given below.

Proposition 1. *For arbitrary subspaces $\mathscr{S}_1$ and $\mathscr{S}_2$ of a finite-dimensional space,*

$$\dim(\mathscr{S}_1 + \mathscr{S}_2) + \dim(\mathscr{S}_1 \cap \mathscr{S}_2) = \dim \mathscr{S}_1 + \dim \mathscr{S}_2. \tag{1}$$

The proof of (1) is easily derived by extending a basis of $\mathscr{S}_1 \cap \mathscr{S}_2$ (see Proposition 3.5.2) to bases for $\mathscr{S}_1$ and $\mathscr{S}_2$.

Obviously, Eq. (1) is valid for the subspaces considered in Exercise 1. More generally, the following statement is a consequence of Eq. (1).

Exercise 2. Show that for any positive integer k,

$$\dim(\mathscr{S}_1 + \mathscr{S}_2 + \cdots + \mathscr{S}_k) \leq \sum_{i=1}^{k} \dim \mathscr{S}_i. \quad \square$$

Note that by definition of $\mathscr{S}_1 + \mathscr{S}_2$ every element **a** of $\mathscr{S}_1 + \mathscr{S}_2$ can be represented in the form

$$\boldsymbol{a} = \boldsymbol{a}_1 + \boldsymbol{a}_2, \tag{2}$$

where $\boldsymbol{a}_1 \in \mathscr{S}_1$, $\boldsymbol{a}_2 \in \mathscr{S}_2$, although in general, the decomposition in Eq. (2) is not unique.

Exercise 3. Let $\mathscr{S}_1$, $\mathscr{S}_2$, $\boldsymbol{a}_1, \boldsymbol{a}_2, \boldsymbol{b}_1$, and $\boldsymbol{b}_2$ be defined as in Exercise 1. Confirm the following representations of the element $\boldsymbol{a} = [1 \quad 0 \quad 0]^T$:

$$\boldsymbol{a} = \boldsymbol{a}_1 - \boldsymbol{b}_2 = -\tfrac{1}{2}\boldsymbol{a}_2 + \tfrac{1}{2}\boldsymbol{b}_1 - \boldsymbol{b}_2. \quad \square$$

It turns out that nonuniqueness in Eq. (2) is a result of the fact that $\mathscr{S}_1 \cap \mathscr{S}_2 \neq \{\boldsymbol{0}\}$.

Theorem 1. *Let $\mathscr{S}_1$ and $\mathscr{S}_2$ be subspaces of a finite-dimensional space and let $\boldsymbol{a}$ denote an element from $\mathscr{S}_1 + \mathscr{S}_2$. Then the decomposition (2) is unique if and only if $\mathscr{S}_1 \cap \mathscr{S}_2 = \{\boldsymbol{0}\}$.*

PROOF. Assume first that $\mathcal{S}_1 \cap \mathcal{S}_2 = \{0\}$. If simultaneously

$$a = a_1 + a_2 \quad \text{and} \quad a = a_1' + a_2',$$

where $a_1, a_1' \in \mathcal{S}_1$, $a_2, a_2' \in \mathcal{S}_2$, then

$$a_1 - a_1' = a_2 - a_2'.$$

Since $a_i - a_i' \in \mathcal{S}_i$ $(i = 1, 2)$, the elements $a_1 - a_1'$ and $a_2 - a_2'$ belong to $\mathcal{S}_1 \cap \mathcal{S}_2 = \{0\}$. Hence $a_1 = a_1'$, $a_2 = a_2'$ and the representation (2) is unique.

Conversely, suppose the representation (2) is unique but also $\mathcal{S}_1 \cap \mathcal{S}_2 \neq \{0\}$. Let $c \in \mathcal{S}_1 \cap \mathcal{S}_2$, $c \neq 0$. The second decomposition

$$a = (a_1 + c) + (a_2 - c)$$

leads to a contradiction. ■

This theorem reveals an important property of subspaces $\mathcal{S}_0 = \mathcal{S}_1 + \mathcal{S}_2$ for which $\mathcal{S}_1 \cap \mathcal{S}_2 = \{0\}$. In this case $\mathcal{S}_0$ is called the *direct sum* of the subspaces $\mathcal{S}_1$ and $\mathcal{S}_2$ and is written $\mathcal{S}_0 = \mathcal{S}_1 \dotplus \mathcal{S}_2$.

In view of Theorem 1, the direct sum $\mathcal{S}_0 = \mathcal{S}_1 \dotplus \mathcal{S}_2$ can also be defined (when it exists) as a subspace $\mathcal{S}_0$ such that for each $a \in \mathcal{S}_0$ there is a *unique* decomposition $a = a_1 + a_2$, with $a_1 \in \mathcal{S}_1$, $a_2 \in \mathcal{S}_2$. This remark admits an obvious generalization of the notion of direct sum to several subspaces. Thus, let $\mathcal{S}_0, \mathcal{S}_1, \mathcal{S}_2, \ldots, \mathcal{S}_k$ be subspaces of a (finite-dimensional) linear space $\mathcal{S}$. If any element a of $\mathcal{S}_0$ has a *unique* decomposition $a = \sum_{i=1}^k a_i$, with $a_i \in \mathcal{S}_i$ for $i = 1, 2, \ldots, k$, we say that $\mathcal{S}_0$ is the *direct sum* of the subspaces $\mathcal{S}_1, \mathcal{S}_2, \ldots, \mathcal{S}_k$ and write $\mathcal{S}_0 = \mathcal{S}_1 \dotplus \mathcal{S}_2 \dotplus \cdots \dotplus \mathcal{S}_k$ or, briefly, $\mathcal{S}_0 = \sum_{i=1}^k \cdot \mathcal{S}_i$. Note that the operation of taking the direct sum also satisfies the axioms A1–A4 of a linear space.

Exercise 4. Confirm that $\mathbb{R}^3 = \mathcal{S}_1 \dotplus \mathcal{S}_2$, where $\mathcal{S}_1 = \text{span}\{e_1, e_2\}$, $\mathcal{S}_2 = \text{span}\{e_3\}$, and e_1, e_2, e_3 are the unit vectors in $\mathbb{R}^3$. □

The rest of this section is devoted to the investigation of properties of direct sums.

Let the elements $b_1, b_2, \ldots, b_r$ and $c_1, c_2, \ldots, c_m$ be bases for the subspaces $\mathcal{S}_1$ and $\mathcal{S}_2$ in $\mathcal{S}$, respectively. The set

$$\{b_1, b_2, \ldots, b_r, c_1, c_2, \ldots, c_m\}$$

is referred to as the *union* of these bases. The union of several bases is defined in a similar way.

Theorem 2. *The subspace $\mathcal{S}_0$ is the direct sum of subspaces $\mathcal{S}_1, \mathcal{S}_2, \ldots, \mathcal{S}_k$ of a finite-dimensional space $\mathcal{S}$ if and only if the union of bases for $\mathcal{S}_1, \mathcal{S}_2, \ldots, \mathcal{S}_k$ constitutes a basis for $\mathcal{S}_0$.*

PROOF. We shall give details of the proof only for the case $k = 2$, from which the general statement is easily derived by induction.

Suppose, preserving the previous notation,

$$\mathscr{E} = \{\boldsymbol{b}_1, \boldsymbol{b}_2, \ldots, \boldsymbol{b}_r, \boldsymbol{c}_1, \boldsymbol{c}_2, \ldots, \boldsymbol{c}_m\}$$

is a basis of subspace $\mathscr{S}_0$. Then for every $\boldsymbol{a} \in \mathscr{S}_0$,

$$\boldsymbol{a} = \sum_{i=1}^{r} \beta_i \boldsymbol{b}_i + \sum_{j=1}^{m} \gamma_j \boldsymbol{c}_j = \boldsymbol{a}_1 + \boldsymbol{a}_2, \tag{3}$$

where $\boldsymbol{a}_1 \in \mathscr{S}_1$ and $\boldsymbol{a}_2 \in \mathscr{S}_2$. Thus, $\mathscr{S}_0 \subset \mathscr{S}_1 + \mathscr{S}_2$. It is easily seen that also $\mathscr{S}_1 + \mathscr{S}_2 \subset \mathscr{S}_0$ and hence $\mathscr{S}_0 = \mathscr{S}_1 + \mathscr{S}_2$. Since $\mathscr{S}_1 \cap \mathscr{S}_2 = \{\boldsymbol{0}\}$ (see Exercises 3.5.7), we have $\mathscr{S}_0 = \mathscr{S}_1 \dot{+} \mathscr{S}_2$.

Conversely, let $\mathscr{S}_0$ be a direct sum of $\mathscr{S}_1$ and $\mathscr{S}_2$. That is, for any $\mathbf{a} \in \mathscr{S}_0$ there is a unique representation

$$\boldsymbol{a} = \boldsymbol{a}_1 + \boldsymbol{a}_2,$$

where $\boldsymbol{a}_1 \in \mathscr{S}_1$, $\boldsymbol{a}_2 \in \mathscr{S}_2$. Let $\mathscr{E}$ be a union of bases $\{\boldsymbol{b}_i\}_{i=1}^{r}$ for $\mathscr{S}_1$ and $\{\boldsymbol{c}_j\}_{j=1}^{m}$ for $\mathscr{S}_2$. Our aim is to show that the system $\mathscr{E}$ is a basis for $\mathscr{S}_0$.

Indeed, any element $\boldsymbol{a}$ from $\mathscr{S}_0$ is a linear combination (3) of elements from $\mathscr{E}$. On the other hand, these $r + m$ elements of $\mathscr{E}$ are linearly independent. To see this, we first recall that the assumption $\mathscr{S}_0 = \mathscr{S}_1 \dot{+} \mathscr{S}_2$ implies (in fact, is equivalent to) the unique decomposition

$$\boldsymbol{0} = \boldsymbol{0} \dot{+} \boldsymbol{0} \tag{4}$$

of the zero element, since $\boldsymbol{0} \in \mathscr{S}_0 \cap \mathscr{S}_1 \cap \mathscr{S}_2$. Thus, if

$$\sum_{i=1}^{r} \alpha_i \boldsymbol{b}_i + \sum_{j=1}^{m} \delta_j \boldsymbol{c}_j = \boldsymbol{0},$$

where obviously $\sum_{i=1}^{r} \alpha_i \boldsymbol{b}_i \in \mathscr{S}_1$ and $\sum_{j=1}^{m} \delta_j \boldsymbol{c}_j \in \mathscr{S}_2$, then it follows from Eq. (4) that

$$\sum_{i=1}^{r} \alpha_i \boldsymbol{b}_i = \sum_{j=1}^{m} \delta_j \boldsymbol{c}_j = \boldsymbol{0}.$$

The linear independence of the basis elements for $\mathscr{S}_1$ and $\mathscr{S}_2$ now yields $\alpha_i = 0$ $(i = 1, 2, \ldots, r)$ and $\delta_j = 0$ $(j = 1, 2, \ldots, m)$ and, consequently, the linear independence of the elements of $\mathscr{E}$. Thus the system $\mathscr{E}$ is a basis for $\mathscr{S}_0$. ■

In the next three exercises it is assumed that $\mathscr{S}_0$ is the sum of subspaces $\mathscr{S}_1, \mathscr{S}_2, \ldots, \mathscr{S}_k$ and necessary and sufficient conditions are formulated for the sum to be direct.

Exercise 5. Show that $\mathscr{S}_0 = \sum_{i=1}^{k} {}^{\boldsymbol{\cdot}} \mathscr{S}_i$ if and only if

$$\dim \mathscr{S}_0 = \sum_{i=1}^{k} \dim \mathscr{S}_i. \quad \square$$

Now we can give a description of the direct sum of k subspaces comparable with the original one used in the case $k = 2$.

Exercise 6. Show that $\mathscr{S}_0 = \sum_{i=1}^{k} {}^{\cdot} \mathscr{S}_i$ if and only if for any i $(1 \le i \le k)$,

$$\mathscr{S}_i \cap (\mathscr{S}_1 + \cdots + \mathscr{S}_{i-1} + \mathscr{S}_{i+1} + \cdots + \mathscr{S}_k) = \{0\} \tag{5}$$

or, what is equivalent,

$$\mathscr{S}_i \cap (\mathscr{S}_1 + \cdots + \mathscr{S}_{i-1}) = \{0\}$$

for $i = 2, 3, \ldots, k$.

Exercise 7. Confirm that $\mathscr{S}_0 = \sum_{i=1}^{k} {}^{\cdot} \mathscr{S}_i$ if and only if for any nonzero $\boldsymbol{a}_i \in \mathscr{S}_i$ $(i = 1, 2, \ldots, k)$, the set $\{\boldsymbol{a}_1, \boldsymbol{a}_2, \ldots, \boldsymbol{a}_k\}$ is linearly independent.

Hint. Observe that linear dependence of $\{\boldsymbol{a}_1, \ldots, \boldsymbol{a}_k\}$ would conflict with (5). □

Let $\mathscr{S}_1$ be a subspace in the finite-dimensional space $\mathscr{S}$. The remark preceding Exercise 3.5.7 provides the existence of a subspace $\mathscr{S}_2$ of $\mathscr{S}$ such that $\mathscr{S}_1 \dot{+} \mathscr{S}_2 = \mathscr{S}$. The subspace $\mathscr{S}_2$ is said to be a *complementary subspace of* $\mathscr{S}_1$ *with respect to* $\mathscr{S}$.

It is worth mentioning that a complementary subspace of $\mathscr{S}_1$ with respect to $\mathscr{S}$ is not unique and depends on the (nonunique) extension of the basis of $\mathscr{S}_1$ to one for $\mathscr{S}$ (see the proof of Proposition 3.5.2).

Exercise 8. Let $\mathscr{S}_1$ be defined as in Exercise 1. Check that both of the subspaces

$$\mathscr{S}'_2 = \operatorname{span}\{\boldsymbol{e}_1\} \qquad \text{and} \qquad \mathscr{S}''_2 = \operatorname{span}\{\boldsymbol{e}_2\}$$

are complementary to $\mathscr{S}_1$ with respect to $\mathbb{R}^3$.

Exercise 9. Prove Proposition 1. □

3.7 Matrix Representation and Rank

Let $\mathscr{S}$ denote an m-dimensional linear space over $\mathscr{F}$ with a fixed basis $\{\boldsymbol{a}_1, \boldsymbol{a}_2, \ldots, \boldsymbol{a}_m\}$. An important relationship between a set of elements from $\mathscr{S}$ and a matrix associated with the set is revealed in this section. This will allow us to apply matrix methods to the investigation of the set, especially for determining the dimension of its linear hull.

Recall that if $\boldsymbol{b} \in \mathscr{S}$ and $\boldsymbol{b} = \sum_{i=1}^{m} \alpha_i \boldsymbol{a}_i$, then the vector $[\alpha_1 \quad \alpha_2 \quad \cdots \quad \alpha_m]^{\mathrm{T}}$ was referred to in Section 3.5 as the representation of $\boldsymbol{b}$ with respect to the basis $\{\boldsymbol{a}_i\}_{i=1}^{m}$. Now we generalize this notion for a system of several elements.

Let $\boldsymbol{b}_1, \boldsymbol{b}_2, \ldots, \boldsymbol{b}_n$ belong to $\mathscr{S}$. From the representations of these elements in the basis $\{\boldsymbol{a}_i\}_{i=1}^m$, we construct the $m \times n$ matrix

$$A_a = [\alpha_{ij}]_{i,j=1}^{m,n}, \tag{1}$$

where

$$\boldsymbol{b}_j = \sum_{i=1}^{m} \alpha_{ij}\boldsymbol{a}_i \qquad 1 \leq j \leq n. \tag{2}$$

The matrix (1) is referred to as the (*matrix*) *representation* of the ordered set $\{\boldsymbol{b}_1, \boldsymbol{b}_2, \ldots, \boldsymbol{b}_n\}$ with respect to the basis $\{\boldsymbol{a}_1, \boldsymbol{a}_2, \ldots, \boldsymbol{a}_m\}$. Obviously, the (matrix) representation of a set of vectors with respect to a basis is unique, while representations of the same set in different bases are generally distinct.

Exercise 1. Check (using Exercises 3.5.2 and 3.5.3) that the matrices

$$A_e = \begin{bmatrix} 1 & 0 \\ 2 & 1 \\ -1 & 2 \end{bmatrix} \qquad \text{and} \qquad A_a = \begin{bmatrix} 2 & -\frac{1}{2} \\ 0 & \frac{3}{2} \\ -1 & \frac{1}{2} \end{bmatrix}$$

are representations of the system $\boldsymbol{b}_1 = [1 \quad 2 \quad -1]^T$, $\boldsymbol{b}_2 = [0 \quad 1 \quad 2]^T$ with respect to the bases $\{\boldsymbol{e}_1, \boldsymbol{e}_2, \boldsymbol{e}_3\}$ and $\{\boldsymbol{a}_1, \boldsymbol{a}_2, \boldsymbol{a}_3\}$, respectively, where $\boldsymbol{a}_1 = [1 \quad 1 \quad 0]^T$, $\boldsymbol{a}_2 = [0 \quad 1 \quad 1]^T$, $\boldsymbol{a}_3 = [1 \quad 0 \quad 1]^T$. □

As in Exercise 1 we see that any matrix can be referred to as the representation of its column vectors with respect to the standard basis. It turns out that when A is a representation of $\boldsymbol{b}_1, \boldsymbol{b}_2, \ldots, \boldsymbol{b}_n$ with respect to some basis, there is a strong relationship between the dimension of the column (or row) space of A, that is, dim(Im A)(see Section 3.3), and that of span $\{\boldsymbol{b}_1, \boldsymbol{b}_2, \ldots, \boldsymbol{b}_n\}$.

We first observe that elements $\boldsymbol{b}_1, \ldots, \boldsymbol{b}_p$ $(1 \leq p \leq n)$ are linearly dependent if and only if their representations $\boldsymbol{\alpha}_j = [\alpha_{1j} \quad \alpha_{2j} \quad \cdots \quad \alpha_{mj}]^T$, $(j = 1, 2, \ldots, p)$ are linearly dependent. Indeed, the equality

$$\sum_{j=1}^{p} \beta_j \boldsymbol{b}_j = \boldsymbol{0} \tag{3}$$

implies, in view of (2),

$$\sum_{j=1}^{p} \beta_j \left(\sum_{i=1}^{m} \alpha_{ij}\boldsymbol{a}_i \right) = \sum_{i=1}^{m} \left(\sum_{j=1}^{p} \alpha_{ij}\beta_j \right) \boldsymbol{a}_i = \boldsymbol{0},$$

where $\{\boldsymbol{a}_i\}_{i=1}^m$ is the underlying basis. Consequently, the linear independence of $\{\boldsymbol{a}\}_{i=1}^m$ yields $\sum_{j=1}^{p} \alpha_{ij}\beta_j = 0$ for $i = 1, 2, \ldots, m$, or, what is equivalent,

$$\sum_{j=1}^{p} \beta_j \boldsymbol{\alpha}_j = \sum_{j=1}^{p} \beta_j [\alpha_{1j} \quad \alpha_{2j} \quad \cdots \quad \alpha_{mj}]^T = \boldsymbol{0}. \tag{4}$$

Thus, the linear dependence of $\boldsymbol{b}_1, \boldsymbol{b}_2, \ldots, \boldsymbol{b}_p$ implies that of $\boldsymbol{\alpha}_1, \boldsymbol{\alpha}_2, \ldots, \boldsymbol{\alpha}_p$. Conversely, starting with Eq. (4) and repeating the argument, we easily deduce the relation (3). Hence the systems $\{\boldsymbol{b}_j\}_{j=1}^p$ and $\{\boldsymbol{\alpha}_j\}_{j=1}^p$ are both linearly dependent or both linear independent. Rearranging the order of elements, this conclusion is easily obtained for any p elements from $\{\boldsymbol{b}_1, \boldsymbol{b}_2, \ldots, \boldsymbol{b}_n\}$ and their representations $(1 \le p \le n)$.

Hence, we have proved the following.

Proposition 1. *The number of linearly independent elements in the system* $\{\boldsymbol{b}_1, \boldsymbol{b}_2, \ldots, \boldsymbol{b}_n\}$ *is equal to the dimension of the column space of its (matrix) representation with respect to a fixed basis.*

In other words, if $\mathscr{C}_A$ is the column space of a matrix representation A of the system $\{\boldsymbol{b}_1, \boldsymbol{b}_2, \ldots, \boldsymbol{b}_n\}$ with respect to some basis, then

$$\dim(\operatorname{span}\{\boldsymbol{b}_1, \boldsymbol{b}_2, \ldots, \boldsymbol{b}_n\}) = \dim \mathscr{C}_A = \dim(\operatorname{Im} A). \tag{5}$$

The significance of this result lies in the fact that the problem of determining the dimension of the linear hull of elements from an arbitrary finite-dimensional space is reduced to that of finding the dimension of the column space of a matrix.

Note also that formula (5) is valid for a representation of the system with respect to *any* basis for the space $\mathscr{S}$. Consequently, all representations of the given system have column spaces of the same dimension.

The next proposition asserts a striking equality between the dimension of the column space $\mathscr{C}_A$ of a matrix A and its rank (as defined by the theory of determinants in Section 2.8).

Theorem 1. *For any matrix A,*

$$\dim \mathscr{C}_A = \operatorname{rank} A.$$

PROOF. We need to show that the dimension of the column space is equal to the order of the largest nonvanishing minor of $A \in \mathscr{F}^{m \times n}$ (see Theorem 2.8.1).

For convenience, let us assume that such a minor is the determinant of the submatrix of A lying in the upper left corner of A. (Otherwise, by rearranging columns and rows of A we proceed to an equivalent matrix having this property and the same rank.) Let the order of this submatrix be r. Consider the $(r+1) \times (r+1)$ submatrix

$$\tilde{A} = \left[\begin{array}{ccc|c} a_{11} & \cdots & a_{1r} & a_{1s} \\ a_{21} & & a_{2r} & a_{2s} \\ \vdots & & \vdots & \vdots \\ a_{r1} & \cdots & a_{rr} & a_{rs} \\ \hline a_{k1} & \cdots & a_{kr} & a_{ks} \end{array}\right]$$

and observe that $\det \tilde{A} = 0$ for all possible choices of k and s.

Indeed, if $k \leq r$ or $s \leq r$, then the minor is zero since there is a repeated row or column. If $k > r$ or $s > r$, then the minor is zero since the rank of A is r and any minor with order exceeding r is zero.

Now let A_{kj} ($j = 1, 2, \ldots, r$ or $j = s$) denote the cofactors of the elements of the last row of $\tilde{A}$. Expanding det $\tilde{A}$ by its last row, we obtain

$$a_{k1}A_{k1} + a_{k2}A_{k2} + \cdots + a_{kr}A_{kr} + a_{ks}A_{ks} = 0, \qquad 1 \leq k \geq m. \quad (6)$$

Now observe that the numbers A_{kj} ($1 \leq j \leq r$ or $j = s$) are independent of the choice of k, so we write $A_{kj} \equiv A_j$. Therefore, writing (6) for $k = 1, 2, \ldots, m$, we obtain

$$\mathbf{A}_{*1}A_1 + \mathbf{A}_{*2}A_2 + \cdots + \mathbf{A}_{*r}A_r + \mathbf{A}_{*s}A_s = \mathbf{0}. \quad (7)$$

where $\mathbf{A}_{*j}$ ($1 \leq j \leq r$ or $j = s$) denotes the jth column of A considered as a vector from $\mathscr{F}^m$. But A_s is a nonvanishing minor of A. Therefore Eq. (7) shows the linear dependence of the columns $\mathbf{A}_{*1}, \mathbf{A}_{*2}, \ldots, \mathbf{A}_{*r}, \mathbf{A}_{*s}$ and, consequently, of all the columns of A.

Now observe that the columns $\mathbf{A}_{*j}$ ($j = 1, 2, \ldots, r$) are linearly independent. Indeed, from

$$\alpha_1\mathbf{A}_{*1} + \alpha_2\mathbf{A}_{*2} + \cdots + \alpha_r\mathbf{A}_{*r} = \mathbf{0}, \qquad \alpha_j \in \mathscr{F}, \quad j = 1, 2, \ldots, r,$$

it follows that $\hat{A}\boldsymbol{\alpha} = \mathbf{0}$, where $\boldsymbol{\alpha} = [\alpha_1 \quad \alpha_2 \quad \cdots \quad \alpha_r]^{\mathrm{T}}$ and $\hat{A} = [a_{ij}]_{i,j=1}^{r}$. Since det $\hat{A} = A_s \neq 0$, it follows by Theorem 2.9.2 that $\boldsymbol{\alpha} = \mathbf{0}$ and the linear independence of $\{\mathbf{A}_{*1}, \ldots, \mathbf{A}_{*r}\}$ is established. Hence

$$\dim(\operatorname{span}\{\mathbf{A}_{*1}, \mathbf{A}_{*2}, \ldots, \mathbf{A}_{*n}\}) = r = \operatorname{rank} A$$

and the assertion follows. ■

Rephrasing the proof of Theorem 1 with respect to the rows of A, it is easily seen that

$$\dim \mathscr{R}_A = \operatorname{rank} A,$$

where $\mathscr{R}_A$ denotes the row space of A (see Section 3.3).

Observing that $A\mathbf{x} = \mathbf{0}$ for $\mathbf{x} \neq \mathbf{0}$ is equivalent to the linear dependence of the columns of A and using Theorem 2.9.2, we obtain a corollary to Theorem 1.

Corollary 1. *If $A \in \mathscr{F}^{n \times n}$, the equation $A\mathbf{x} = \mathbf{0}$ has a nontrivial solution if and only if A is singular.*

In conclusion, combining the assertions of Proposition 1 and Theorem 1, we immediately obtain the following result, which was the first objective of this section.

Theorem 2. *For any* $\boldsymbol{b}_1, \boldsymbol{b}_2, \ldots, \boldsymbol{b}_n \in \mathscr{S}$,

$$\dim(\operatorname{span}\{\boldsymbol{b}_1, \boldsymbol{b}_2, \ldots, \boldsymbol{b}_n\}) = \operatorname{rank} A, \tag{8}$$

where A *is a representation of* $\boldsymbol{b}_1, \boldsymbol{b}_2, \ldots, \boldsymbol{b}_n$ *with respect to a basis in* $\mathscr{S}$.

Exercise 2. Show that the representation of any system of m linearly independent elements from an m-dimensional space $\mathscr{S}$ with respect to another basis for the space is an $m \times m$ nonsingular matrix.

Hint. Use the result of Exercise 2.8.2. □

It follows from Eq, (8) that the ranks of all matrices that are representations of a fixed set $\{\mathbf{b}_1, \ldots, \mathbf{b}_n\}$ with respect to different bases are equal. Thus (see Section 2.8), all these matrices are equivalent. This will be discussed in Section 3.9 after some necessary preparations, and after a few important consequences of Theorem 1 are indicated in the next section.

3.8 Some Properties of Matrices Related to Rank

Let A be a matrix from $\mathscr{F}^{m \times n}$. It follows from Exercise 2.8.2 and Corollary 3.7.1 that, if $m = n$ and A is nonsingular, then rank $A = n$, Ker $A = \{0\}$, and hence, when det $A \neq 0$,

$$\operatorname{rank} A + \dim(\operatorname{Ker} A) = n.$$

This relation is, in fact, true for *any* $m \times n$ matrix.

Proposition 1. *For any* $A \in F^{m \times n}$,

$$\operatorname{rank} A + \dim(\operatorname{Ker} A) = n. \tag{1}$$

PROOF. Let rank $A = r$ and, without loss of generality, let it be assumed that the $r \times r$ submatrix A_{11} in the left upper corner of A is nonsingular:†

$$A = \begin{bmatrix} A_{11} & A_{12} \\ A_{21} & A_{22} \end{bmatrix},$$

where $A_{11} \in \mathscr{F}^{r \times r}$ and det $A_{11} \neq 0$. By virtue of Theorem 2.7.4, the given matrix A can be reduced to the form

$$\tilde{A} = \begin{bmatrix} I_r & 0 \\ 0 & 0 \end{bmatrix} = P^{-1}AQ^{-1}, \tag{2}$$

where P and Q are invertible $m \times m$ and $n \times n$ matrices, respectively.

† Recall (see Section 2.8) that the rank is not changed under, in particular, the elementary operations of interchanging columns or rows.

Note that $r = \text{rank } A = \text{rank } \tilde{A}$ and observe that Ker $\tilde{A}$ (being the linear hull of the unit vectors $e_{r+1}, e_{r+2}, \ldots, e_n$) has dimension $n - r$.

It remains now to show that the dimension of Ker $\tilde{A}$, or $\dim(\text{Ker } P^{-1}AQ^{-1})$ is equal to the dimension of Ker A. In fact, if $x \in \text{Ker } \tilde{A}$, then $Q^{-1}x \in \text{Ker } A$, since $AQ^{-1} = P\tilde{A}$. Similarly, $Q^{-1}x \in \text{Ker } A$ implies $x \in \text{Ker } A$. Using the terminology introduced just before Exercise 3.5.6, we have Ker $A = Q^{-1}$ (Ker $\tilde{A}$) and the desired result follows from Exercise 3.5.6. ■

We say that $A \in \mathscr{F}^{m \times n}$ is a matrix of *full rank* if rank $A = \min(m, n)$. In view of Theorem 3.7.1 and the remark thereafter, it is obvious that A is of full rank if and only if its column vectors are linearly independent in $\mathscr{F}^m$ in the case $m \geqslant n$, while for $m \leqslant n$, its row vectors must be linearly independent in $\mathscr{F}^n$. In particular, a square $n \times n$ matrix is of full rank if and only if both systems of column and row vectors are linearly independent in $\mathscr{F}^n$.

Appealing to Theorem 3.7.1 and Exercise 2.8.2, it is easily found that the following statement is valid:

Exercise 1. Let A be an $n \times n$ nonsingular matrix. Show that any submatrix of A consisting of r complete columns (or rows) of A $(1 \leq r \leq n)$ is of full rank.

Exercise 2. Let rank $A = r$. Show that at least one $r \times r$ submatrix lying in the "strip" of r linearly independent columns (rows) of A is nonsingular.

Exercise 3. Use the Binet–Cauchy formula to show that, if $A \in \mathbb{C}^{m \times n}$ and $m \leq n$, then $\det(AA^*) > 0$ if and only if A has full rank. (Compare with Exercise 2.5.2.) □

Proposition 2. *For any* $A, B \in \mathscr{F}^{m \times n}$,

$$\text{rank}(A + B) \leq \text{rank } A + \text{rank } B.$$

PROOF. Consider the subspace $\mathscr{S}_0$ spanned by all the column vectors of both A and B. Clearly, $\dim \mathscr{S}_0 \leq \text{rank } A + \text{rank } B$. On the other hand, the column space of $A + B$ is contained in $\mathscr{S}_0$ and hence, in view of Theorem 3.7.1,

$$\text{rank}(A + B) \leq \dim \mathscr{S}_0.$$

The assertion follows. ■

The next proposition shows that any nonzero matrix can be represented as a product of two matrices of full rank.

Proposition 3. *Any $m \times n$ matrix A from $\mathscr{F}^{m \times n}$ of* rank r $(r \geq 1)$ *can be written as a product*

$$A = P_1 Q_1, \tag{3}$$

where $P_1 \in \mathscr{F}^{m \times r}$, $Q_1 \in \mathscr{F}^{r \times n}$, and both matrices have rank r.

PROOF. Let two nonsingular matrices P and Q be defined as in Eq. (2). Writing

$$P = [P_1 \quad P_2] \qquad \text{and} \qquad Q = \begin{bmatrix} Q_1 \\ Q_2 \end{bmatrix},$$

where $P_1 \in \mathscr{F}^{m \times r}$, $P_2 \in \mathscr{F}^{m \times (m-r)}$, $Q_1 \in \mathscr{F}^{r \times n}$, $Q_2 \in \mathscr{F}^{(n-r) \times n}$, it is easy to see that the blocks P_2 and Q_2 do not appear in the final result:

$$A = P \begin{bmatrix} I_r & 0 \\ 0 & 0 \end{bmatrix} Q = P_1 I_r Q_1 = P_1 Q_1.$$

The desired properties of P_1 and Q_1 are provided by Exercise 1. ■

Note that Eq. (3) is known as a *rank decomposition* of A.

Exercise 4. Show that any matrix of rank r can be written as a sum of r matrices of rank 1.

Exercise 5. If A and B are two matrices such that the matrix product AB is defined, prove that

$$\operatorname{rank}(AB) \leq \min(\operatorname{rank} A, \operatorname{rank} B).$$

SOLUTION. Each column of AB is a linear combination of columns of A. Hence $\mathscr{C}_{AB} \subset \mathscr{C}_A$ and the inequality rank $(AB) \leq$ rank A follows from that of Exercise 3.5.5 and Theorem 3.7.1. Passing to the transpose and using Exercise 2.8.5, the second inequality is easily established.

Exercise 6. Let A be a nonsingular matrix from $\mathscr{F}^{n \times n}$. Prove that for any $B \in \mathscr{F}^{n \times k}$ and $C \in \mathscr{F}^{l \times n}$,

$$\operatorname{rank}(AB) = \operatorname{rank} B, \qquad \operatorname{rank}(CA) = \operatorname{rank} C.$$

Hint. Use Exercise 3.5.6 and Theorem 3.7.1. □

The next statement provides an extension of the Binet–Cauchy theorem (Theorem 2.5.1) for $m > n$.

Proposition 4. *Let A and B be $m \times n$ and $n \times m$ matrices, respectively. If $m > n$, then* det $AB = 0$.

PROOF. It follows from Exercise 5 that rank $(AB) \leq$ rank A. Since A has only n columns, it is clear that rank $A \leq n$ and therefore rank$(AB) \leq n < m$. But AB is an $m \times m$ matrix, hence Exercise 2.8.2 gives the required result. ■

3.9 Change of Basis and Transition Matrices

Let $\mathscr{S}$ denote an m-dimensional space and let $\boldsymbol{b} \in \mathscr{S}$. In this section we will study the relationship between the representations $\boldsymbol{\beta}$ and $\boldsymbol{\beta}'$ of $\boldsymbol{b}$ with respect to the bases $\mathscr{E} = \{\boldsymbol{a}_1, \boldsymbol{a}_2, \ldots, \boldsymbol{a}_m\}$ and $\mathscr{E}' = \{\boldsymbol{a}'_1, \boldsymbol{a}'_2, \ldots, \boldsymbol{a}'_m\}$ for $\mathscr{S}$. For this purpose, we first consider the representation $P \in \mathscr{F}^{m \times m}$ of the system $\mathscr{E}$ with respect to the basis $\mathscr{E}'$. Recall that by definition,

$$P \triangleq [\alpha_{ij}]_{i,j=1}^{m}$$

provided

$$\boldsymbol{a}_j = \sum_{i=1}^{m} \alpha_{ij} \boldsymbol{a}'_i, \qquad j = 1, 2, \ldots, m, \tag{1}$$

and that (by Exercise 3.7.2) the matrix P is nonsingular.

The matrix P is referred to as the *transition matrix* from the basis $\mathscr{E}$ to the basis $\mathscr{E}'$. Transition matrices are always square and nonsingular. The uniqueness of P (with respect to the ordered pair of bases) follows from Proposition 3.5.1.

Exercise 1. Check that the matrix

$$P = \begin{bmatrix} 1 & 0 & 1 \\ 1 & 1 & 0 \\ 0 & 1 & 1 \end{bmatrix}$$

is the transition matrix from the basis $\{\boldsymbol{a}_1, \boldsymbol{a}_2, \boldsymbol{a}_3\}$ defined in Exercise 3.7.1 to the standard one, while its inverse,

$$P^{-1} = \frac{1}{2} \begin{bmatrix} 1 & 1 & -1 \\ -1 & 1 & 1 \\ 1 & -1 & 1 \end{bmatrix},$$

is the transition matrix from the standard basis to $\{\boldsymbol{a}_1, \boldsymbol{a}_2, \boldsymbol{a}_3\}$. □

Now we are in a position to obtain the desired relation between the representations of a vector in two different bases. Preserving the previous notation, the representations $\boldsymbol{\beta}$ and $\boldsymbol{\beta}'$ of $\boldsymbol{b} \in \mathscr{S}$ are written

$$\boldsymbol{\beta} = [\beta_1 \quad \beta_2 \quad \cdots \quad \beta_m]^{\mathrm{T}}, \qquad \boldsymbol{\beta}' = [\beta'_1 \quad \beta'_2 \quad \cdots \quad \beta'_m]^{\mathrm{T}},$$

where

$$\boldsymbol{b} = \sum_{j=1}^{m} \beta_j \boldsymbol{a}_j = \sum_{j=1}^{m} \beta'_j \boldsymbol{a}'_j.$$

Substituting Eq. (1) for the $\boldsymbol{a}_j$, we obtain

$$\sum_{j=1}^{m} \beta_j \boldsymbol{a}_j = \sum_{j=1}^{m} \beta_j \left(\sum_{i=1}^{m} \alpha_{ij} \boldsymbol{a}'_i \right) = \sum_{i=1}^{m} \left(\sum_{j=1}^{m} \alpha_{ij} \beta_j \right) \boldsymbol{a}'_i.$$

Therefore

$$\boldsymbol{b} = \sum_{i=1}^{m} \beta'_i \boldsymbol{a}'_i = \sum_{i=1}^{m} \left(\sum_{j=1}^{m} \alpha_{ij} \beta_j \right) \boldsymbol{a}'_i$$

Since the representation of an element with respect to a basis is unique (Proposition 3.5.1), it follows that

$$\beta'_i = \sum_{j=1}^{m} \alpha_{ij} \beta_j, \qquad i = 1, 2, \ldots, m.$$

Rewriting the last equality in matrix form, we finally obtain the following:

Theorem 1. *Let* $\boldsymbol{b}$ *denote an element from the m-dimensional linear space* $\mathscr{S}$ *over* $\mathscr{F}$. *If* $\boldsymbol{\beta}$ *and* $\boldsymbol{\beta}'$ ($\in \mathscr{F}^m$) *are the representations of* $\boldsymbol{b}$ *with respect to bases* $\mathscr{E}$ *and* $\mathscr{E}'$ *for* $\mathscr{S}$, *respectively, then*

$$\boldsymbol{\beta}' = P\boldsymbol{\beta}, \tag{2}$$

where P is the transition matrix from $\mathscr{E}$ *to* $\mathscr{E}'$.

Exercise 2. Check the relation (2) for the representations of the vector $\boldsymbol{b} = [1 \quad 2 \quad -1]$ with respect to the bases $\{\boldsymbol{e}_1, \boldsymbol{e}_2, \boldsymbol{e}_3\}$ and $\{\boldsymbol{a}_1, \boldsymbol{a}_2, \boldsymbol{a}_3\}$ defined in Exercise 3.7.1.

Exercise 3. Let the $m \times n$ matrices A and A' denote the representations of the system $\{\boldsymbol{b}_1, \boldsymbol{b}_2, \ldots, \boldsymbol{b}_n\}$ with respect to the bases $\mathscr{E}$ and $\mathscr{E}'$ for $\mathscr{S}$ (dim $\mathscr{S} = m$), respectively. Show that

$$A' = PA, \tag{3}$$

where P is the transition matrix from $\mathscr{E}$ to $\mathscr{E}'$.

Hint. Apply Eq. (2) for each element of the system $\{\boldsymbol{b}_1, \boldsymbol{b}_2, \ldots, \boldsymbol{b}_n\}$. □

In Exercise 3 we again see the equivalence of the representations of the same system with respect to different bases. (See the remark after Theorem 3.7.2.)

Exercise 4. Check Eq. (3) for the matrices found in Exercise 1 in this section and Exercise 3.7.1 □

The relation (2) is characteristic for the transition matrix in the following sense:

Proposition 1. *Let Eq.* (2) *hold for the representations* $\boldsymbol{\beta}$ *and* $\boldsymbol{\beta}'$ *of every element* $\boldsymbol{b}$ *from* $\mathscr{S}$ (dim $\mathscr{S} = m$) *with respect to the fixed bases* $\mathscr{E}$ *and* $\mathscr{E}'$, *respectively. Then P is the transition matrix from* $\mathscr{E}$ *to* $\mathscr{E}'$.

PROOF. Let A' denote the representation of the elements of $\mathscr{E}$ with respect to $\mathscr{E}'$. Since the representation of these elements with respect to $\mathscr{E}$ is the identity matrix I_m and (2) for any $\boldsymbol{b} \in \mathscr{S}$ yields (3), it follows that $A' = P$. It remains to note that the representation A' of the elements of $\mathscr{E}$ with respect to $\mathscr{E}'$ coincides with the transition matrix from $\mathscr{E}$ to $\mathscr{E}'$. ■

Note that the phrase "every element $\boldsymbol{b}$ from $\mathscr{S}$" in the proposition can be replaced by "each element of a basis $\mathscr{E}$."

We have seen in Exercise 1 that the exchange of roles of bases yields the transition from the matrix P to its inverse. More generally, consider three bases $\{\boldsymbol{a}_i\}_{i=1}^m$, $\{\boldsymbol{a}_i'\}_{i=1}^m$, and $\{\boldsymbol{a}_j''\}_{i=1}^m$ for $\mathscr{S}$. Denote by P (respectively, P') the transition matrix from the first basis to the second (respectively, from the second to the third).

Proposition 2. *If P'' denotes the transition matrix from the first basis to the third then, in the previous notation,*

$$P'' = P'P.$$

PROOF. In fact, for the representations $\boldsymbol{\beta}$, $\boldsymbol{\beta}'$, and $\boldsymbol{\beta}''$ of an arbitrarily chosen element $\mathbf{b}$ from $\mathscr{S}$ with respect to the above bases, we have (by using Eq. (2) twice)

$$\boldsymbol{\beta}'' = P'\boldsymbol{\beta}' = P'(P\boldsymbol{\beta}) = P'P\boldsymbol{\beta}.$$

Hence, applying Proposition 1, the assertion follows. ■

Corollary 1. *If P is the transition matrix from the basis $\mathscr{E}$ to the basis $\mathscr{E}'$, then P^{-1} is the transition matrix from $\mathscr{E}'$ to $\mathscr{E}$.*

3.10 Solution of Equations

Consider the following system of m linear equations with n unknowns written in matrix form:

$$A\boldsymbol{x} = \boldsymbol{b}, \qquad A \in \mathscr{F}^{m \times n}, \tag{1}$$

where, as before, $\mathscr{F}$ may be read as $\mathbb{C}$ or $\mathbb{R}$. Investigation of such a system began in Section 2.9 and will continue here. Note that Eq. (1) is also referred to as a *matrix equation.*

We start with the question of *consistency* (or *solvability*) for the given equations. In other words, we are interested in conditions under which Eq. (1) is sensible, meaning that at least one solution exists.

Theorem 1. *The system* (1) *is solvable if and only if*

$$\operatorname{rank}[A \quad b] = \operatorname{rank} A. \tag{2}$$

PROOF. If there is a solution $\mathbf{x} = [x_1 \quad x_2 \quad \cdots \quad x_n]^T \in \mathbb{C}^n$ of Eq. (1), then $\sum_{i=1}^{n} A_{*i}x_i = \boldsymbol{b}$, where $A_{*1}, \ldots, A_{*n}$ stand for the columns of A. Hence $\boldsymbol{b}$ is a linear combination of the columns of A and, therefore, by virtue of Theorem 3.7.1, $\operatorname{rank}[A \quad \boldsymbol{b}] = \operatorname{rank} A$.

Conversely, if Eq. (2) holds, then $\boldsymbol{b}$ is a linear combination of the columns of A and the coefficients of this combination provide a solution of (1). This completes the proof. ■

Exercise 1. Observe that the matrix equation

$$A\boldsymbol{x} = \begin{bmatrix} 1 & 2 & 0 \\ 2 & -3 & 7 \\ 1 & 4 & -2 \end{bmatrix} \begin{bmatrix} x_1 \\ x_2 \\ x_3 \end{bmatrix} = \begin{bmatrix} 11 \\ -2 \\ -7 \end{bmatrix} = \boldsymbol{b}$$

has no solution since $\operatorname{rank} A = 2$ but $\operatorname{rank}[A \quad \boldsymbol{b}] = 3$.

Exercise 2. Confirm the existence of solutions of the linear system

$$\begin{aligned} x_1 + 2x_2 &= 1, \\ 2x_1 - 3x_2 + 7x_3 &= 2, \\ x_1 + 4x_2 - 2x_3 &= 1. \quad \square \end{aligned}$$

Note that Theorem 1 can be interpreted as saying that the system (1) is consistent if and only if $\boldsymbol{b}$ belongs to the column space (range) of A.

The next result allows us to describe the set of all solutions of (1) by combining a particular solution with the solutions of the corresponding homogeneous matrix equation $A\boldsymbol{x} = \boldsymbol{0}$.

Theorem 2. *Let $\boldsymbol{x}_0$ be a particular solution of the matrix equation $A\boldsymbol{x} = \boldsymbol{b}$. Then*

(a) *If $\boldsymbol{x}' \in \operatorname{Ker} A$, the vector $\boldsymbol{x}_0 + \boldsymbol{x}'$ is a solution of* (1).

(b) *For every solution $\boldsymbol{x}$ of* (1) *there exists a vector $\boldsymbol{x}'$ such that $A\boldsymbol{x}' = \boldsymbol{0}$ and $\boldsymbol{x} = \boldsymbol{x}_0 + \boldsymbol{x}'$.*

In other words, the solution set of (1) consists of vectors of the form $\boldsymbol{x}_0 + \boldsymbol{x}'$, where $\boldsymbol{x}_0$ is a particular solution of (1) and $\boldsymbol{x}'$ is some solution of the corresponding homogeneous equation.

PROOF. For (a) we have

$$A(\boldsymbol{x}_0 + \boldsymbol{x}') = A\boldsymbol{x}_0 + A\boldsymbol{x}' = \boldsymbol{b} + \boldsymbol{0} = \boldsymbol{b},$$

and hence $\boldsymbol{x}_0 + \boldsymbol{x}'$ is a solution of (1). For part (b), if $A\boldsymbol{x} = \boldsymbol{b}$ and $A\boldsymbol{x}_0 = \boldsymbol{b}$, then $A(\boldsymbol{x} - \boldsymbol{x}_0) = \boldsymbol{0}$ and $\boldsymbol{x}' = \boldsymbol{x} - \boldsymbol{x}_0 \in \operatorname{Ker} A$. Thus, $\boldsymbol{x} = \boldsymbol{x}_0 + \boldsymbol{x}'$. ■

Exercise 3. The vector $\mathbf{x}_0 = [-1 \quad 1 \quad 1]^T$ is a solution vector of the system considered in Exercise 2. A solution vector of the homogeneous equation is $\mathbf{x}' = [2 \quad -1 \quad -1]^T$. Now it is easy to observe that any solution of the nonhomogeneous system is of the form $\mathbf{x}_0 + \alpha\mathbf{x}'$ ($\alpha \in \mathbb{R}$).

Corollary 1. *If A is a square matrix, then the matrix equation $A\mathbf{x} = \mathbf{b}$ has a unique solution if and only if A is nonsingular.*

PROOF. If Eq. (1) has a unique solution, then by Theorem 2 we have $\mathbf{x}' = \mathbf{0}$ and $A\mathbf{x} = \mathbf{0}$ has only the trivial solution. Hence, by Corollary 3.7.1, the matrix A is nonsingular. The converse statement follows from Theorem 2.9.2. ■

There are some other important consequences of Theorem 2. When solutions exist, the system $A\mathbf{x} = \mathbf{b}$ has as many solutions as the corresponding homogeneous system; and the number of solutions is either 1 or ∞ in both cases. Since any solution vector of $A\mathbf{x} = \mathbf{0}$ belongs to Ker A, the dimension of Ker A provides the maximal number of linearly independent solutions (as vectors of $\mathscr{F}^n$) of this homogeneous equation. The set of solutions of a nonhomogeneous equation is not a linear space. The next theorem suggests that it should be regarded as a subspace (Ker A) shifted bodily from the origin through $\mathbf{x}_0$ (a particular solution).

Theorem 3. *Let $A \in \mathscr{F}^{m \times n}$ have rank $r < n$. The number of independent parameters required to describe the solution set of the solvable matrix equation $A\mathbf{x} = \mathbf{b}$ is equal to $n - r$.*

The proof follows immediately from Theorem 2 and Proposition 3.8.1.

Theorem 3 merely restates the fact that any solution of Eq. (1) is a linear combination of $n - r$ linearly independent solutions of $A\mathbf{x} = \mathbf{0}$ (hence the $n - r$ parameters) plus a particular solution.

Exercise 4. Verify the assertion of Theorem 3 for the system considered in Exercises 2 and 3.

Example 5. Describe the solution set of the system given in the matrix form

$$A\mathbf{x} = \begin{bmatrix} 1 & -2 & 1 & 1 \\ 2 & 0 & 3 & -1 \\ 3 & -2 & 4 & 0 \\ -1 & -2 & -2 & 2 \end{bmatrix} \begin{bmatrix} x_1 \\ x_2 \\ x_3 \\ x_4 \end{bmatrix} = \begin{bmatrix} 0 \\ -1 \\ -1 \\ 1 \end{bmatrix} = \mathbf{b}.$$

SOLUTION. It is easily seen that the last two rows of A are linear combinations of the first two (linearly independent) rows. Hence rank $A = 2$. Since $\boldsymbol{b}$ is a linear combination of the first and the third columns of A, then rank $[A \quad \boldsymbol{b}] = 2$ also, and the given equation is solvable by Theorem 1. Observe, furthermore, that $\boldsymbol{x}_0 = [1 \quad 0 \quad -1 \quad 0]$ is a particular solution of the system and that $A\boldsymbol{x}' = \boldsymbol{0}$ has $n - r = 2$ linearly independent solutions. To find them, first observe that the 2×2 minor in the left upper corner of A is not zero, and so we consider the first two equations of $A\boldsymbol{x}' = \boldsymbol{0}$:

$$\begin{bmatrix} 1 & -2 \\ 2 & 0 \end{bmatrix} \begin{bmatrix} x'_1 \\ x'_2 \end{bmatrix} + \begin{bmatrix} 1 & 1 \\ 3 & -1 \end{bmatrix} \begin{bmatrix} x'_3 \\ x'_4 \end{bmatrix} = \begin{bmatrix} 0 \\ 0 \end{bmatrix},$$

or

$$\begin{bmatrix} 1 & -2 \\ 2 & 0 \end{bmatrix} \begin{bmatrix} x'_1 \\ x'_2 \end{bmatrix} = \begin{bmatrix} -x'_3 - x'_4 \\ -3x' + x'_4 \end{bmatrix}.$$

Hence, inverting the 2×2 matrix on the left, we have

$$\begin{bmatrix} x'_1 \\ x'_2 \end{bmatrix} = \frac{1}{4} \begin{bmatrix} -6x'_3 + 2x'_4 \\ -x'_3 + 3x'_4 \end{bmatrix}.$$

Write $x'_3 = u$, $x'_4 = v$ (to suggest that they will be free parameters), then we have $x'_1 = \frac{1}{4}(-6u + 2v)$, $x'_2 = \frac{1}{4}(-u + 3v)$, $x'_3 = u$, $x'_4 = v$.

Setting first $v = 0$ and $u = 1$, and then $u = 0$ and $v = 1$, it is easily concluded that $[-\frac{3}{2} \quad -\frac{1}{4} \quad 1 \quad 0]^{\mathrm{T}}$ and $[\frac{1}{2} \quad \frac{3}{4} \quad 0 \quad 1]^{\mathrm{T}}$ are linearly independent solution vectors of $A\boldsymbol{x} = \boldsymbol{0}$. Thus, the general solution of the inhomogeneous system can be described by

$$\boldsymbol{x} = \boldsymbol{x}_0 + \boldsymbol{x}' = [1 \quad 0 \quad -1 \quad 0]^{\mathrm{T}} + u[-\tfrac{3}{2} \quad -\tfrac{1}{4} \quad 1 \quad 0]^{\mathrm{T}} + v[\tfrac{1}{2} \quad \tfrac{3}{4} \quad 0 \quad 1]^{\mathrm{T}}$$

for any scalars u and v.

Note that the number of free (mutually independent) parameters in the general solution is (as indicated by Theorem 3) just $n - r = 2$.

Exercise 6. Use Proposition 3.8.1 to show that if $A \in \mathscr{F}^{m \times n}$, then $A\boldsymbol{x} = \boldsymbol{0}$ implies $\boldsymbol{x} = \boldsymbol{0}$ if and only if $m \geq n$ and A has full rank.

Exercise 7. If $A \in \mathscr{F}^{m \times n}$, show that the system $A\boldsymbol{x} = \boldsymbol{b}$ has a unique solution if and only if $m \geq n$, A has full rank, and the equation is solvable.

Exercise 8. Consider the equation $A\boldsymbol{x} = \boldsymbol{b}$, where $A \in \mathscr{F}^{m \times n}$ and $\boldsymbol{b} \in \mathscr{F}^{m}$, and let $M \in \mathscr{F}^{p \times m}$ be a matrix for which $\operatorname{rank}(MA) = \operatorname{rank} A$. Show that the Solution sets of $A\boldsymbol{x} = \boldsymbol{b}$ and $MA\boldsymbol{x} = M\boldsymbol{b}$ coincide.

Hint. Show first that Ker A = Ker MA. □

3.11 Unitary and Euclidean Spaces

We started Chapter 3 by formalizing some familiar concepts in $\mathbb{R}_3$ and then extending them to define a linear space. Now we go a step further in this direction and generalize the important concept of the scalar product in $\mathbb{R}_3$ (see Section 1.3) to include a wide class of linear spaces. This allows us to generalize the familiar notions of length, angle, and so on.

Let $\mathscr{S}$ be a (finite-dimensional) linear space over the field $\mathscr{F}$ of real or complex numbers and let $\boldsymbol{x}, \boldsymbol{y} \in \mathscr{S}$. A binary operation $(\boldsymbol{x}, \boldsymbol{y})$ from $\mathscr{S} \times \mathscr{S}$ to $\mathscr{F}$ is said to be an *inner* (or scalar) *product on* $\mathscr{S}$ if the following properties are satisfied for all $\boldsymbol{x}, \boldsymbol{y}, \boldsymbol{z} \in \mathscr{S}$ and $\alpha, \beta \in \mathscr{F}$:

I1 $(\boldsymbol{x}, \boldsymbol{x}) \geq 0$ and $(\boldsymbol{x}, \boldsymbol{x}) = 0$ if and only if $\boldsymbol{x} = \boldsymbol{0}$,

I2 $(\alpha\boldsymbol{x} + \beta\boldsymbol{y}, \boldsymbol{z}) = \alpha(\boldsymbol{x}, \boldsymbol{z}) + \beta(\boldsymbol{y}, \boldsymbol{z})$,

I3 $(\boldsymbol{x}, \boldsymbol{y}) = \overline{(\boldsymbol{y}, \boldsymbol{x})}$ (where the bar denotes complex conjugate).

These three properties may be described as *positivity*, *linearity in the first argument*, and *antisymmetry*, respectively. It is easily verified that they are all satisfied by the familiar scalar product in $\mathbb{R}_3$. The next example is the natural extension of that case to $\mathbb{C}^n$.

Exercise 1. Confirm that the binary operation

$$(\boldsymbol{x}, \boldsymbol{y}) = x_1\bar{y}_1 + x_2\bar{y}_2 + \cdots + x_n\bar{y}_n = \boldsymbol{y}^*\boldsymbol{x}, \tag{1}$$

defined for any pair of elements

$$\boldsymbol{x} = [x_1 \quad x_2 \quad \cdots \quad x_n]^{\mathrm{T}}, \quad \boldsymbol{y} = [y_1 \quad y_2 \quad \cdots \quad y_n]^{\mathrm{T}}$$

from $\mathbb{C}^n$, is a scalar product on $\mathbb{C}^n$. □

The inner product given by Eq. (1) will be called the *standard inner product* for $\mathbb{C}^n$, it will be the one applied if no other is specified. The standard inner product for $\mathbb{R}_3$ also is defined by Eq. (1), where obviously the bars may be omitted. Thus, in this case the inner product is *bilinear*, that is, linear in both arguments.

The next exercise shows how an inner product can be defined on *any* finite-dimensional space $\mathscr{S}$ over a field $\mathscr{F}$ in which complex conjugates are defined.

Exercise 2. (a) Let $\{\boldsymbol{a}_1, \boldsymbol{a}_2, \ldots, \boldsymbol{a}_n\}$ be a basis for $\mathscr{S}$ over $\mathscr{F}$. Check that the binary operation

$$(\boldsymbol{x}, \boldsymbol{y}) = \sum_{i=1}^{n} x_i\bar{y}_i \tag{2}$$

is an inner product on $\mathscr{S}$, where $\boldsymbol{x} = \sum_{i=1}^{n} x_i\boldsymbol{a}_i$ and $\boldsymbol{y} = \sum_{i=1}^{n} y_i\boldsymbol{a}_i$.

(b) Let $\delta_1, \delta_2, \ldots, \delta_n$ be positive numbers and define another binary operation on $\mathscr{S}$ by

$$(\mathbf{x}, \mathbf{y}) = \sum_{i=1}^{n} \delta_i x_i \bar{y}_i. \tag{3}$$

Show that this is also an inner product on $\mathscr{S}$. □

The inner product possesses some additional properties that are simple consequences of the axioms I1–I3.

Exercise 3. Check that

(a) $(\mathbf{x}, \alpha \mathbf{y} + \beta \mathbf{z}) = \bar{\alpha}(\mathbf{x}, \mathbf{y}) + \bar{\beta}(\mathbf{x}, \mathbf{z})$,
(b) $(\mathbf{x}, \mathbf{0}) = (\mathbf{0}, \mathbf{x}) = 0$,

for any $\mathbf{x}, \mathbf{y}, \mathbf{z} \in \mathscr{S}$ and $\alpha, \beta \in \mathscr{F}$. □

Linear spaces with inner products are important enough to justify their own name: A complex (respectively, real) linear space $\mathscr{S}$ together with an inner product from $\mathscr{S} \times \mathscr{S}$ to $\mathbb{C}$ (respectively, $\mathbb{R}$) is referred to as a *unitary* (respectively, *Euclidean*) *space.*

To distinguish between unitary spaces generated by the same $\mathscr{S}$ but different inner products $(\mathbf{x}, \mathbf{y})$ and $\langle \mathbf{x}, \mathbf{y} \rangle$, we shall write $\mathscr{U}_1 = \mathscr{S}(\ ,\ )$ and $\mathscr{U}_2 = \mathscr{S}\langle\ ,\ \rangle$, respectively. Recall that any n-dimensional space over the field of complex (or real) numbers can be considered a unitary (or Euclidean) space by applying an inner product of the type in Eq. (2), or of the type in Eq. (3), or of some other type.

Exercise 4. Let $\mathscr{U} = \mathscr{S}(\ ,\ )$ be a unitary space and let $\mathscr{S}_0$ be a subspace of $\mathscr{S}$. Show that $\mathscr{U}_0 = \mathscr{S}_0(\ ,\ )$ is also a unitary space. (When discussing subspaces of unitary spaces, this property is generally tacitly understood.) □

Let $\mathscr{U}$ denote the unitary space $\mathscr{S}(\ ,\ )$ and let $\mathbf{x} \in \mathscr{U}$. The inner product $(\mathbf{x}, \mathbf{y})$ on $\mathscr{S}$ allows us to define the *length* (or *norm*) of the element $\mathbf{x}$ as the number $\sqrt{(\mathbf{x}, \mathbf{x})}$, written $\|\mathbf{x}\|$. Axiom I1 of the inner product says that $\|\mathbf{x}\| \geq 0$ and only the zero element has zero length. It should be noted that the length of the element $\mathbf{x}$ depends on the chosen inner product on $\mathscr{S}$ and that length in another unitary space $\mathscr{S}\langle\ ,\ \rangle$ would be given by $\|\mathbf{x}\|_1 = \sqrt{\langle \mathbf{x}, \mathbf{x} \rangle}$.

It is clear that if $(\ ,\ )$ denotes the familiar scalar product in $\mathbb{R}^3$, then the length of $\mathbf{x} \in \mathbb{R}^3(\ ,\ )$ defined above coincides with the classical Euclidean length of the vector. The next exercise can be seen as a generalization of this statement.

Exercise 5. Check that the length of the element $x \in \mathscr{S}$ having the representation $[\alpha_1 \quad \alpha_2 \quad \cdots \quad \alpha_n]^T$ with respect to a basis $\{a_1, \ldots, a_n\}$ for $\mathscr{S}$ is given by

$$\|x\|^2 = \sum_{j,k=1}^{n} \alpha_j \bar{\alpha}_k (a_j, a_k) \tag{4}$$

in the standard inner product on $\mathscr{S}$. □

An element $x \in \mathscr{U}$ with length (norm) 1 is referred to as a *normalized* element. It is evident that any nonzero element x can be normalized by transition to the element λx, where $\lambda = 1/\|x\|$.

The following inequality, concerning the norms and inner product of any two elements in a unitary space, is known as the (generalized) *Cauchy–Schwarz inequality* and contains Exercise 2.5.4 as a special case (in $\mathbb{R}^n$).

Theorem 1. *If* x, y *are members of a unitary space, then*

$$|(x, y)| \le \|x\| \, \|y\|. \tag{5}$$

PROOF. Let $\alpha = -(y, x)$, $\beta = (x, x)$, and $z = \alpha x + \beta y$. To prove the equivalent inequality $|\alpha|^2 \le \beta(y, y)$, we compute, using I2,

$$(z, z) = \alpha(x, \alpha x + \beta y) + \beta(y, \alpha x + \beta y).$$

Appealing to I1 and Exercise 3(a), we obtain

$$0 \le |\alpha|^2(x, x) + \alpha\bar{\beta}(x, y) + \bar{\alpha}\beta(y, x) + |\beta|^2(y, y).$$

Recalling the definitions of α and β, this reduces to

$$0 \le \beta(-|\alpha|^2 + \beta(y, y)).$$

If $\beta = 0$, then $x = 0$ and (5) is trivially true. If $\beta > 0$, then $|\alpha|^2 \le \beta(y, y)$. ■

Exercise 6. Check that equality holds in (5) for nonzero x and y if and only if $x = \lambda y$ for some $\lambda \in \mathbb{C}$. □

Proceeding to the definition of the angle θ between two nonzero elements x and y of a Euclidean space $\mathscr{E}$, we first recall that for $\mathscr{E} = \mathbb{R}^3$ with the familiar scalar product,

$$\cos\theta = \frac{(x, y)}{\|x\| \, \|y\|}, \qquad 0 \le \theta \le \pi. \tag{6}$$

Generalizing this, we say that the *angle* θ between nonzero x and y from the Euclidean space $\mathscr{E} = \mathscr{S}(\ ,\ )$ is given by Eq. (6). Note that the expression on the right in (6) is, because of the Cauchy–Schwarz inequality, less than or equal to 1 in absolute value and therefore the cosine exists. Furthermore, it is defined uniquely since, in particular, $\cos\theta$ is not changed if $\boldsymbol{x}$, $\boldsymbol{y}$ are replaced by $\alpha\boldsymbol{x}$, $\beta\boldsymbol{y}$ for any positive α, β.

Exercise 7. Check that the angle θ between the vectors $[0 \quad 3 \quad -4 \quad 0]^{\mathrm{T}}$ and $[1 \quad 1 \quad 1 \quad 1]^{\mathrm{T}}$ from $\mathbb{R}^4$ is equal to $\pi - \arccos(0.1)$ (with respect to the standard inner product). □

The notion of angle is not usually considered in unitary spaces for which the expression on the right side of (6) is complex-valued. Nevertheless, extending some geometrical language from the Euclidean situation, we say that nonzero elements $\boldsymbol{x}$ and $\boldsymbol{y}$ from a unitary (or Euclidean) space are *orthogonal* if and only if $(\boldsymbol{x}, \boldsymbol{y}) = 0$. In particular, using the standard scalar product in $\mathbb{C}^n$ of Exercise 1, nonzero vectors $\boldsymbol{x}$, $\boldsymbol{y}$ in $\mathbb{C}^n$ are orthogonal if and only if $\boldsymbol{y}^*\boldsymbol{x} = \boldsymbol{x}^*\boldsymbol{y} = 0$.

Exercise 8. Check that any two unit vectors $\boldsymbol{e}_i$ and $\boldsymbol{e}_j$ $(1 \le i < j \le n)$ in $\mathbb{C}^n$ (or $\mathbb{R}^n$) are orthogonal in the standard inner product.

Exercise 9. (Pythagoras's Theorem) Prove that if $\boldsymbol{x}$ and $\boldsymbol{y}$ are orthogonal members of a unitary space, then

$$\|\boldsymbol{x} \pm \boldsymbol{y}\|^2 = \|\boldsymbol{x}\|^2 + \|\boldsymbol{y}\|^2. \quad \square \tag{7}$$

Note that if $\boldsymbol{x}$, $\boldsymbol{y}$ denote position vectors in $\mathbb{R}^3$, then $\|\boldsymbol{x} - \boldsymbol{y}\|$ is the length of the third side of the triangle determined by $\boldsymbol{x}$ and $\boldsymbol{y}$. Thus, the customary Pythagoras's theorem follows from Eq. (7). The next result is sometimes described as Appollonius's theorem or as the parallelogram theorem.

Exercise 10 Prove that for any $\boldsymbol{x}, \boldsymbol{y} \in \mathscr{U}$,

$$\|\boldsymbol{x} + \boldsymbol{y}\|^2 + \|\boldsymbol{x} - \boldsymbol{y}\|^2 = 2(\|\boldsymbol{x}\|^2 + \|\boldsymbol{y}\|^2)$$

and interpret this result for $\mathbb{R}^3$.

Exercise 11 Let $\mathscr{S}$ be the complex linear space $\mathbb{C}^{m \times n}$ (see Exercise 3.1.3(b)). Show that the binary operation defined on $\mathscr{S}$ by

$$(X, Y) = \operatorname{tr}(XY^*),$$

for all $X, Y \in \mathscr{S}$, defines an inner product on $\mathscr{S}$ (see Exercise 1.8.3). □

3.12 Orthogonal Systems

Any set of nonzero elements from a unitary space is said to be an *orthogonal set*, or *system*, if any two members of the set are orthogonal, or if the set consists of only one element.

Exercise 1. Show that an orthogonal set in a unitary space is necessarily linearly independent. □

An orthogonal system in which each member is normalized is described as an *orthonormal system.* Clearly, any subset of the set of unit vectors in $\mathbb{C}^n$ (or $\mathbb{R}^n$) with the standard inner product constitutes the prime example of an orthonormal system.

Consider the possibility of generating a converse statement to Exercise 1. Given a linearly independent set $\{\mathbf{x}_1, \mathbf{x}_2, \ldots, \mathbf{x}_r\}$ in a unitary space, can we construct an orthogonal set from them in the subspace $\text{span}\{\mathbf{x}_1, \ldots, \mathbf{x}_r\}$? The answer is affirmative and, once such a system is obtained, an orthonormal system can be found simply by normalizing each element of the orthogonal set. One procedure for completing this process is known as the *Gram–Schmidt orthogonalization process*, which we now describe.

Let $\{\mathbf{x}_1, \ldots, \mathbf{x}_r\}$ be a linearly independent set in a unitary space $\mathscr{U}$. We shall construct an orthogonal set $\{\mathbf{y}_1, \mathbf{y}_2, \ldots, \mathbf{y}_r\}$ with the property that for $k = 1, 2, \ldots, r$,

$$\text{span}\{\mathbf{y}_1, \ldots, \mathbf{y}_k\} = \text{span}\{\mathbf{x}_1, \ldots, \mathbf{x}_k\}. \tag{1}$$

First assign $\mathbf{y}_1 = \mathbf{x}_1$ and seek an element $\mathbf{y}_2 \in \text{span}\{\mathbf{x}_1, \mathbf{x}_2\}$ that is orthogonal to $\mathbf{y}_1$. Write $\mathbf{y}_2 = \alpha_1 \mathbf{y}_1 + \mathbf{x}_2$ and determine α_1 in such a way that $(\mathbf{y}_1, \mathbf{y}_2) = 0$. It is easily seen that $\alpha_1 = -(\mathbf{y}_1, \mathbf{x}_2)/(\mathbf{y}_1, \mathbf{y}_1)$. Note that the linear independence of $\mathbf{x}_1$ and $\mathbf{x}_2$ ensures that $\mathbf{y}_2 \neq \mathbf{0}$. Now the process is complete if $r = 2$. More generally, an induction argument establishes the result, as indicated in the next exercise.

Exercise 2. Let $\mathbf{x}_1, \mathbf{x}_2, \ldots, \mathbf{x}_r$ be linearly independent elements of a unitary space $\mathscr{U} = \mathscr{S}(\ ,\ )$. Prove that a set $\{\mathbf{y}_1, \ldots, \mathbf{y}_r\}$ is orthogonal and satisfies Eq. (1) if $\mathbf{y}_1 = \mathbf{x}_1$ and, for $p = 2, 3, \ldots, r$.

$$\mathbf{y}_p = \mathbf{x}_p - \sum_{j=1}^{p-1} \frac{(\mathbf{y}_j, \mathbf{x}_p)}{(\mathbf{y}_j, \mathbf{y}_j)} \mathbf{y}_j. \quad \square$$

It is quite clear that the procedure we have just described can be used to generate a basis of orthonormal vectors for a unitary space, that is, an *orthonormal basis.* In fact, we have the following important result.

Theorem 1. *Every finite-dimensional unitary space has an orthonormal basis.*

PROOF. Starting with any basis for the space $\mathscr{U}$, an orthogonal set of n elements where $n = \dim \mathscr{U}$ can be obtained by the Gram–Schmidt process. Normalizing the set and using Exercises 1 and 3.5.4, an orthonormal basis for $\mathscr{U}$ is constructed. ■

Compared to other bases, orthonormal bases have the great advantage that the representation of an arbitrary element from the space is easily found, as is the inner product of any two elements. The next exercises give the details.

Exercise 3. Let $\{\boldsymbol{a}_1, \boldsymbol{a}_2, \ldots, \boldsymbol{a}_n\}$ be an orthonormal basis for the unitary space $\mathscr{U}$ and let $\boldsymbol{x} = \sum_{i=1}^{n} \alpha_i \boldsymbol{a}_i$ be the decomposition of an $\boldsymbol{x} \in \mathscr{U}$ in this basis. Show that the expansion coefficients α_i are given by $\alpha_i = (\boldsymbol{x}, \boldsymbol{a}_i)$, $i = 1, 2, \ldots, n$.

Exercise 4. Let $\{\boldsymbol{a}_1, \boldsymbol{a}_2, \ldots, \boldsymbol{a}_n\}$ be a basis for the unitary space $\mathscr{U}$ and let

$$\boldsymbol{x} = \sum_{i=1}^{n} \alpha_i \boldsymbol{a}_i, \qquad \boldsymbol{y} = \sum_{i=1}^{n} \beta_i \boldsymbol{a}_i.$$

Prove that $(\boldsymbol{x}, \boldsymbol{y}) = \sum_{i=1}^{n} \alpha_i \bar{\beta}_i$ for any $\boldsymbol{x}, \boldsymbol{y} \in \mathscr{U}$ if and only if the basis $\{\boldsymbol{a}_i\}_{i=1}^{n}$ is orthonormal. □

Recall that any system of linearly independent elements in a space can be extended to a basis for that space (Proposition 3.5.2) that can then be orthogonalized and normalized. Hence the result of the next exercise is to be expected.

Exercise 5. Prove that any orthonormal system of elements from a unitary space $\mathscr{U}$ can be extended to an orthonormal basis for $\mathscr{U}$.

Exercise 6. (Bessel's inequality) Let $\{\boldsymbol{a}_1, \boldsymbol{a}_2, \ldots, \boldsymbol{a}_r\}$ be an orthonormal basis for $\mathscr{U}_0 = \mathscr{S}_0(\ ,\ )$ and let $\boldsymbol{x}$ belong to $\mathscr{U} = \mathscr{S}(\ ,\ )$. Prove that if $\mathscr{S}_0 \subset \mathscr{S}$, then

$$\sum_{i=1}^{r} |(\boldsymbol{x}, \boldsymbol{a}_i)|^2 \le (\boldsymbol{x}, \boldsymbol{x}), \tag{2}$$

with equality (Parseval's equality) if and only if $\boldsymbol{x} \in \mathscr{S}_0$.

Hint. Use Exercises 3–5, and recall the comment on Exercise 3.11.4. □

It is instructive to consider another approach to the orthogonalization problem. Once again a linearly independent set $\{\boldsymbol{x}_1, \boldsymbol{x}_2, \ldots, \boldsymbol{x}_r\}$ is given and we try to construct an orthogonal set $\{\boldsymbol{y}_1, \ldots, \boldsymbol{y}_r\}$ with property (1) We again take $\boldsymbol{y}_1 = \boldsymbol{x}_1$, and for $p = 2, 3, \ldots, r$ we write

$$\boldsymbol{y}_p = \boldsymbol{x}_p + \sum_{j=1}^{p-1} \alpha_{jp} \boldsymbol{x}_j,$$

where $\alpha_{1p}, \ldots, \alpha_{p-1,p}$ are coefficients to be determined. We then impose the orthogonality conditions in the form $(\boldsymbol{x}_i, \boldsymbol{y}_p) = 0$ for $i = 1, 2, \ldots, p-1$, and note that these conditions also ensure the orthogonality of $\boldsymbol{y}_p$ with $\boldsymbol{y}_1, \ldots, \boldsymbol{y}_{p-1}$. The result is a set of equations for the α's that can be written as one matrix equation:

$$G\boldsymbol{\alpha}_p = -\boldsymbol{\beta}_p, \tag{3}$$

where $\boldsymbol{\alpha}_p = [\alpha_{1p} \quad \alpha_{2p} \quad \cdots \quad \alpha_{p-1,p}]^{\mathrm{T}}$, $\boldsymbol{\beta}_p = [(\boldsymbol{x}_1, \boldsymbol{x}_p) \quad \cdots \quad (\boldsymbol{x}_{p-1}, \boldsymbol{x}_p)]^{\mathrm{T}}$, and the coefficient matrix is

$$G = [(\boldsymbol{x}_i, \boldsymbol{x}_j)]_{i,j=1}^{p-1}. \tag{4}$$

Thus, the orthogonalization process can be continued with the calculation of unique elements $\boldsymbol{y}_2, \ldots, \boldsymbol{y}_r$ if and only if eq. (3) has a unique solution vector $\boldsymbol{\alpha}_p$ for each p. We know from Corollary 3.10.1 that this is the case if and only if matrix G is nonsingular. We are about to show that, because of the special form of G illustrated in Eq. (4), this follows immediately from the linear independence of the $\boldsymbol{x}$'s.

We turn now to the special structure of G. More generally, let $\boldsymbol{x}_1, \boldsymbol{x}_2, \ldots, \boldsymbol{x}_r$ be *any* set of elements from a unitary space $\mathscr{U} = \mathscr{S}(\ ,\ )$. The $r \times r$ matrix

$$G = [(\boldsymbol{x}_j, \boldsymbol{x}_k)]_{j,k=1}^{r}$$

is said to be the *Gram matrix* of $\{\boldsymbol{x}_1, \ldots, \boldsymbol{x}_r\}$ and its determinant, written $\det G = g(\boldsymbol{x}_1, \ldots, \boldsymbol{x}_r)$ is the *Gramian* or *Gram determinant* of $\{\boldsymbol{x}_1, \ldots, \boldsymbol{x}_r\}$. Our experience with the Gram–Schmidt process suggests that the Gramian is nonzero provided the **x**'s are linearly independent. This is the case, but more is true.

Theorem 2. (Gram's criterion) *Elements $\boldsymbol{x}_1, \ldots, \boldsymbol{x}_r$ from a unitary space are linearly independent, or linearly dependent, according as their Gramian $g(\boldsymbol{x}_1, \ldots, \boldsymbol{x}_r)$ is positive or zero.*

PROOF. Observe first that the Gram matrix is Hermitian and, consequently, the Gramian is real (Exercise 2.2.6).

Taking advantage of Theorem 1, let $\boldsymbol{y}_1, \boldsymbol{y}_2, \ldots, \boldsymbol{y}_n$ be an orthonormal basis for the space and for $j = 1, 2, \ldots, r$, let $\boldsymbol{x}_j = \sum_{k=1}^{n} \alpha_{jk} \boldsymbol{y}_k$. Define the $r \times n$ matrix $A = [\alpha_{jk}]$ and note that A has full rank if and only if $\boldsymbol{x}_1, \ldots, \boldsymbol{x}_r$ are linearly independent (see Theorem 3.7.2).

Now observe that

$$(\boldsymbol{x}_j, \boldsymbol{x}_k) = \sum_{u,v=1}^{n} \alpha_{ju} \bar{\alpha}_{kv} (\boldsymbol{y}_u, \boldsymbol{y}_v) = \sum_{u=1}^{n} \alpha_{ju} \bar{\alpha}_{ku},$$

where we have used the fact that $(\boldsymbol{y}_u, \boldsymbol{y}_v) = \delta_{uv}$, the Kronecker delta. Now it is easily verified that the Gram matrix $G = [(\boldsymbol{x}_j, \boldsymbol{x}_k)] = AA^*$ and, using Exercise 3.8.3, the result follows. ■

The Gram–Schmidt process leads to the following decomposition of matrices of full rank, which should be compared with the rank decomposition of proposition 3.8.3. This result is of considerable importance in numerical work. Surprisingly, it forms the first step in a useful technique for computing eigenvalues (an important concept to be introduced in the next chapter.)

Exercise 7. Let $A \in \mathbb{C}^{m \times n}$ with $m \geq n$ and let rank $A = n$. Confirm the representation of $A = QR$, where R is an $n \times n$ nonsingular upper triangular matrix, $Q \in \mathbb{C}^{m \times n}$, and $Q^*Q = I_n$ (i.e., the columns of Q form an orthonormal system).

Hint. If $z_p = \sum_{j=1}^{p} t_j^{(p)} A_{*j}$ $(p = 1, 2, \ldots, n)$ are the orthonormal vectors constructed from the columns A_{*j} of A, show that $Q = [z_1 \quad z_2 \quad \cdots \quad z_n]$ and $R = ([t_{jp}]_{j,p=1}^{n})^{-1}$, where $t_{jp} = t_j^{(p)}$, satisfy the required conditions. □

Observe that when the QR decomposition of a square matrix A is known, the equation $Ax = b$ is then equivalent to the equation $Rx = Q^*b$ in which the coefficient matrix is triangular. Thus, in some circumstances the QR decomposition is the basis of a useful equation-solving algorithm.

Exercise 8. Let $\{u_1, \ldots, u_n\}$ be a basis for a unitary space $\mathscr{U}$ and let $v \in \mathscr{U}$. Show that v is uniquely determined by the n numbers (v, u_j), $j = 1, 2, \ldots, n$. □

3.13 Orthogonal Subspaces

Let $\mathscr{S}_1$ and $\mathscr{S}_2$ be two subsets of elements from the unitary space $\mathscr{U} = \mathscr{S}$ $(\ ,\ )$. We say that $\mathscr{S}_1$ and $\mathscr{S}_2$ are *(mutually) orthogonal*, written $\mathscr{S}_1 \perp \mathscr{S}_2$, if and only if $(x, y) = 0$ for every $x \in \mathscr{S}_1$, $y \in \mathscr{S}_2$.

Exercise 1. Prove that two nontrivial subspaces $\mathscr{S}_1$ and $\mathscr{S}_2$ of $\mathscr{U}$ are orthogonal if and only if each basis element of one of the subspaces is orthogonal to all basis elements of the second.

Exercise 2. Show that $(x, y) = 0$ for all $y \in \mathscr{U}$ implies $x = 0$.

Exercise 3. Let $\mathscr{S}_0$ be a subspace of the unitary space $\mathscr{U}$. Show that the set of all elements of $\mathscr{U}$ orthogonal to $\mathscr{S}_0$ is a subspace of $\mathscr{U}$. □

The subspace defined in Exercise 3 is said to be the *complementary orthogonal subspace* of $\mathscr{S}_0$ in $\mathscr{U}$ and is written $\mathscr{S}_0^{\perp}$. Note that Exercise 2 shows that $\mathscr{U}^{\perp} = \{0\}$ and $\{0\}^{\perp} = \mathscr{U}$ (in $\mathscr{U}$).

Exercise 4. Confirm the following properties of complementary orthogonal subspaces in $\mathscr{U}$:

(a) $(\mathscr{S}_0^{\perp})^{\perp} = \mathscr{S}_0$,
(b) If $\mathscr{S}_1 \subset \mathscr{S}_2$, then $\mathscr{S}_1^{\perp} \supset \mathscr{S}_2^{\perp}$,
(c) $(\mathscr{S}_1 + \mathscr{S}_2)^{\perp} = \mathscr{S}_1^{\perp} \cap \mathscr{S}_2^{\perp}$,
(d) $(\mathscr{S}_1 \cap \mathscr{S}_2)^{\perp} = \mathscr{S}_1^{\perp} + \mathscr{S}_2^{\perp}$. □

The sum of a subspace $\mathscr{S}_0$ in $\mathscr{U}$ and its complementary orthogonal subspace $\mathscr{S}_0^\perp$ is of special interest. Note first that this sum is direct: $\mathscr{S}_0 \cap \mathscr{S}_0^\perp = \{0\}$, so that $\mathscr{S}_0 \dotplus \mathscr{S}_0^\perp$ is a direct sum of mutually orthogonal subspaces, that is, an *orthogonal sum*. More generally, the direct sum of k linear subspaces $\mathscr{S}_i$ $(i = 1, 2, \ldots, k)$ of $\mathscr{U}$ is said to be the *orthogonal sum* of the subspaces if $\mathscr{S}_i \perp \mathscr{S}_j$ for $i \neq j$ $(1 \leq i, j \leq k)$ and is indicated by $\mathscr{S}_1 \oplus \mathscr{S}_2 \oplus \cdots \oplus \mathscr{S}_k$ or, briefly, by $\sum_{i=1}^k \oplus \mathscr{S}_i$. Note the contrast with $\mathscr{S}_1 \dotplus \mathscr{S}_2 \dotplus \cdots \dotplus \mathscr{S}_k$ used for a customary (nonorthogonal) direct sum.

Exercise 5. Let $\mathscr{U}$ be a unitary space of dimension n. Show that for a subspace $\mathscr{S}_0$ of $\mathscr{U}$,

$$\mathscr{U} = \mathscr{S}_0 \oplus \mathscr{S}_0^\perp \tag{1}$$

and in particular, $\dim \mathscr{S}_0 + \dim \mathscr{S}_0^\perp = n$. □

In view of Eq. (1), each element $\boldsymbol{x}$ of $\mathscr{U}$ can be written uniquely in the form

$$\boldsymbol{x} = \boldsymbol{x}_1 + \boldsymbol{x}_2, \tag{2}$$

where $\boldsymbol{x}_1 \in \mathscr{S}_0$ and $\boldsymbol{x}_1 \perp \boldsymbol{x}_2$. The representation (2) reminds one of the decomposition of a vector in $\mathbb{R}^3$ into the sum of two mutually orthogonal vectors. Hence, the element $\boldsymbol{x}_1$ in Eq. (2) is called the *orthogonal projection* of $\boldsymbol{x}$ onto $\mathscr{S}_0$.

Exercise 6. Show that if $\mathscr{U}$ is a unitary space,

$$\mathscr{U} = \sum_{i=1}^{n} \oplus \mathscr{S}_i,$$

where $\mathscr{S}_i = \text{span}\{\boldsymbol{a}_i\}$, $1 \leq i \leq n$, and $\{\boldsymbol{a}_1, \boldsymbol{a}_2, \ldots, \boldsymbol{a}_n\}$ is an orthonormal basis for $\mathscr{U}$.

Exercise 7. Let

$$\mathscr{U} = \sum_{i=1}^{k} \oplus \mathscr{S}_i$$

and let $\boldsymbol{x} = \sum_{i=1}^k \boldsymbol{x}_i$, where $\boldsymbol{x}_i \in \mathscr{S}_i$ $(1 \leq i \leq k)$. Prove that

$$(\boldsymbol{x}, \boldsymbol{x}) = \sum_{i=1}^{k} (\boldsymbol{x}_i, \boldsymbol{x}_i);$$

compare this with Parseval's equality, Eq. (3.12.2). □

A converse statement to Exercise 7 is also true in the following sense.

Exercise 8. Check that if $(\sum_{i=1}^k \boldsymbol{x}_i, \sum_{i=1}^k \boldsymbol{x}_i) = \sum_{i=1}^k (\boldsymbol{x}_i, \boldsymbol{x}_i)$ for any elements $\boldsymbol{x}_i \in \mathscr{S}_i$, $1 \leq i \leq n$, then the sum $\sum_{i=1}^k \mathscr{S}_i$ is orthogonal. □

Two systems of elements $\{x_1, x_2, \ldots, x_k\}$ and $\{y_1, y_2, \ldots, x_k\}$ in a unitary space are said to be *biorthogonal* if $(x_i, y_j) = \delta_{ij}$ $(1 \le i, j \le k)$.

Exercise 9. Prove that if $\{x_j\}_{i=1}^k$ and $\{y\}_{j=1}^k$ generate a biorthogonal system, then each system consists of linearly independent elements.

Exercise 10. Show that if $\{x_i\}_{i=1}^n$ and $\{y_j\}_{j=1}^n$ are biorthogonal bases for $\mathscr{U}$, then $(\text{span}\{x_1, \ldots, x_k\})^{\perp} = \text{span}\{y_{k+1}, y_{k+2}, \ldots, y_n\}$ for any positive integer $k < n$.

Exercise 11. For biorthogonal systems $\{x_i\}_{i=1}^n$ and $\{y_j\}_{j=1}^n$ in $\mathscr{U}$, show that

$$(x, x) = \sum_{i=1}^{n} (x, x_i)(y_i, x)$$

for any $x \in \mathscr{U}$ if and only if the system $\{x_i\}_{i=1}^n$ consistutes a basis in $\mathscr{U}$. Deduce the Parseval equality (Eq. (3.12.2)) from this.

Exercise 12. Let $\mathscr{U}$ be a unitary space.

(a) (cf. Ex. 3.11.9). Let $x, y \in \mathscr{U}$. Show that

$$\|x + y\|^2 = \|x\|^2 + \|y\|^2 \tag{3}$$

if and only if $\mathscr{Re}(x, y) = 0$.

(b) In contrast, let $\mathscr{X}$, $\mathscr{Y}$ be subspaces of $\mathscr{U}$, and show that these subspaces are orthogonal if and only if Eq. (3) holds *for all* $x \in \mathscr{X}$ and $y \in \mathscr{Y}$. □

3.14 Miscellaneous Exercises

1. Find a basis and the dimension of the subspace $\mathscr{S}_1$ spanned by $a_1 = [1 \;\; 0 \;\; 2 \;\; -1]^{\mathrm{T}}$, $a_2 = [0 \;\; -1 \;\; 2 \;\; 0]^{\mathrm{T}}$, and $a_3 = [2 \;\; -1 \;\; 6 \;\; -2]^{\mathrm{T}}$.

   Answer. The set $\{a_1, a_2\}$ is a basis for $\mathscr{S}_1$, and $\dim \mathscr{S}_1 = 2$.

2. Let

   $$b_1 = [1 \;\; -1 \;\; 4 \;\; -1]^{\mathrm{T}}, \; b_2 = [1 \;\; 0 \;\; 0 \;\; 1]^{\mathrm{T}}, \; b_3 = [-1 \;\; -2 \;\; 2 \;\; 1]^{\mathrm{T}},$$

   and let $\mathscr{S}_2 = \text{span}\{b_1, b_2, b_3\}$. If $\mathscr{S}_1$ is defined as in Exercise 1, show that $\dim(\mathscr{S}_1 \cap \mathscr{S}_2) = 2$, $\dim(\mathscr{S}_1 + \mathscr{S}_2) = 3$.
3. Let $\mathscr{S}_1$, $\mathscr{S}_2$ be subspaces of the linear space $\mathscr{S}$ of dimension n. Show that if $\dim \mathscr{S}_1 + \dim \mathscr{S}_2 > n$, then $\mathscr{S}_1 \cap \mathscr{S}_2 \neq \{0\}$.
4. Two linear spaces $\mathscr{S}$ and $\mathscr{S}_1$ over $\mathscr{F}$ are *isomorphic* if and only if there is a one-to-one correspondence $x \leftrightarrow x_1$ between the elements $x \in \mathscr{S}$ and

$\mathbf{x}_1 \in \mathscr{S}_1$ such that if $\mathbf{x} \leftrightarrow \mathbf{x}_1$ and $\mathbf{y} \leftrightarrow \mathbf{y}_1$, then $\mathbf{x} + \mathbf{y} \leftrightarrow \mathbf{x}_1 + \mathbf{y}_1$ and $\alpha\mathbf{x} \leftrightarrow \alpha\mathbf{x}_1$ ($\mathbf{y} \in \mathscr{S}$, $\mathbf{y}_1 \in \mathscr{S}_1$, $\alpha \in \mathscr{F}$). Prove that two finite-dimensional spaces are isomorphic if and only if they are of the same dimension. (The correspondence, or mapping, defining isomorphic linear spaces is called an *isomorphism*.)

Hint. Consider first a mapping of linearly independent elements from $\mathscr{S}$ onto a system of linearly independent elements in $\mathscr{S}_1$.

5. Show that if all the row sums of a matrix $A \in \mathbb{C}^{n \times m}$ are zeros, then A is singular.

Hint. Observe that $A\mathbf{x} = \mathbf{0}$ for $\mathbf{x} = [1 \quad 1 \quad \cdots \quad 1]^{\mathrm{T}}$.

6. Check that, for any $n \times n$ matrices A, B,

$$\operatorname{rank}(AB) \geq \operatorname{rank} A + \operatorname{rank} B - n. \quad \square$$

In Exercises 7 and 8, let (,) denote the standard inner product on $\mathbb{C}^n$.

7. If $A \in \mathbb{C}^{n \times n}$, prove that if $(\mathbf{x}, A\mathbf{y}) = 0$ for all $\mathbf{x}, \mathbf{y} \in \mathbb{C}^n$, then $A = 0$.

8. (a) If $A \in \mathbb{C}^{n \times n}$, show that $(\mathbf{x}, A\mathbf{y}) = (A^*\mathbf{x}, \mathbf{y})$ for all $\mathbf{x}, \mathbf{y} \in \mathbb{C}^n$.
(b) If A is an $n \times n$ matrix and $(\mathbf{x}, A\mathbf{y}) = (B\mathbf{x}, \mathbf{y})$ for all $\mathbf{x}, \mathbf{y} \in \mathbb{C}^n$, then $B = A^*$.

9. Let $\mathscr{S}$ denote the space of all real polynomials of degree not exceeding n, with an inner product on $\mathscr{S}$ defined by

$$(P, Q) \triangleq \int_{-1}^{1} P(x)Q(x)\,dx \qquad P, Q \in \mathscr{S}.$$

(a) Verify that this is indeed an inner product on $\mathscr{S}$.
(b) Check that the Cauchy–Schwartz inequality, Eq. (3.11.5), becomes

$$\left| \int_{-1}^{1} P(x)Q(x)\,dx \right| \leq \left[\int_{-1}^{1} P^2(x)\,dx \right]^{1/2} \left[\int_{-1}^{1} Q^2(x)\,dx \right]^{1/2}.$$

(c) Check that the Gram–Schmidt orthogonalization process applied to the basis $1, x, x^2, \ldots x^n$ of $\mathscr{S}$ leads to the polynomials $\alpha_k L_k(x)$ ($k = 0, 1, \ldots, n$), where

$$L_k(x) = \frac{1}{2^k k!} \frac{d^k (x^2 - 1)^k}{dx^k} \qquad \text{(the Legendre polynomials)}$$

and α_k denote some scalars.

10. Prove the Fredholm theorem: The equation $A\mathbf{x} = \mathbf{b}$ ($A \in \mathbb{C}^{m \times n}$) is solvable if and only if the vector $\mathbf{b} \in \mathbb{C}^m$ is orthogonal to all solutions of the homogeneous equation $A^*\mathbf{y} = \mathbf{0}$.

Hint. Use the result of Exercise 8.

11. Check that the rank of a symmetric (or skew-symmetric) matrix is equal to the order of its largest nonzero *principal* minor. In particular, show that the rank of a skew-symmetric matrix cannot be odd.

12. Show that the transition matrix P_n from one orthonormal basis to another is unitary, that is, $P_n^* P_n = I_n$.

Hint. Observe that the columns of P viewed as elements in $\mathscr{F}^n$ generate an orthonormal system (with respect to the standard inner product).

CHAPTER 4

Linear Transformations and Matrices

The notion of a function of a real variable as a map of one set of real numbers into another can be extended in a natural way to functions defined on arbitrary sets. Thus, if F_1 and F_2 denote any two nonempty sets of elements, a rule T that assigns to each element $\mathbf{x} \in F_1$ some unique element $\mathbf{y} \in F_2$ is called a *transformation* (or *operator*, or *mapping*) of F_1 *into* F_2, and we write $T: F_1 \overset{\text{into}}{\longrightarrow} F_2$ or just $T: F_1 \to F_2$. The set F_1 is the *domain* of T. If T assigns $\mathbf{y} \in F_2$ to $\mathbf{x} \in F_1$, then $\mathbf{y}$ is said to be the *image of* $\mathbf{x}$ *under* T and is denoted $\mathbf{y} = T(\mathbf{x})$. The totality of all images of the elements of F_1 under T is written

$$T(F_1) = \{\mathbf{y} \in F_2 : \mathbf{y} = T(\mathbf{x}) \text{ for some } \mathbf{x} \in F_1\}$$

and is referred to as the *image* of the set F_1 under the transformation T or as the *range* of T.

Generally, $T(F_1)$ is a subset of F_2. In the case $T(F_1) = F_2$, we say that T maps (or transforms) F_1 *onto* F_2; we shall indicate that by writing $T: F_1 \overset{\text{onto}}{\longrightarrow} F_2$. Obviously, $T: F_1 \overset{\text{onto}}{\longrightarrow} T(F_1)$. In the important case $F_1 = F_2 = F$ we also say that the transformation T *acts* on F.

In this book our main concern is transformations with the property of "linearity," which will be discussed next. Many properties of matrices become vivid and clear if they are considered from the viewpoint of linear transformations.

4.1 Linear Transformations

Let $\mathscr{S}_1$ and $\mathscr{S}_2$ denote linear spaces over a field $\mathscr{F}$. A transformation $T: \mathscr{S}_1 \to \mathscr{S}_2$ is said to be a *linear transformation* if, for any two elements $\boldsymbol{x}_1$ and $\boldsymbol{x}_2$ from $\mathscr{S}_1$ and any scalar $\alpha \in \mathscr{F}$,

$$\begin{aligned} T(\boldsymbol{x}_1 + \boldsymbol{x}_2) &= T(\boldsymbol{x}_1) + T(\boldsymbol{x}_2), \\ T(\alpha \boldsymbol{x}_1) &= \alpha T(\boldsymbol{x}_1), \end{aligned} \tag{1}$$

that is, if T is *additive* and *homogeneous.*

Note that by assuming $\mathscr{S}_1$ to be a linear space over $\mathscr{F}$ we guarantee that the elements $\boldsymbol{x}_1 + \boldsymbol{x}_2$ and $\alpha\boldsymbol{x}_1$ also belong to the space $\mathscr{S}_1$ on which T is defined. Furthermore, since $\mathscr{S}_2$ is a linear space over $\mathscr{F}$, the images $T(\boldsymbol{x}_1 + \boldsymbol{x}_2)$ and $T(\alpha\boldsymbol{x}_1)$ are elements of $\mathscr{S}_2$ provided $T(\boldsymbol{x}_1)$ and $T(\boldsymbol{x}_2)$ are.

An isomorphism between two linear spaces of the same dimension (see Exercise 3.14.4) is an example of a linear transformation of a special kind. Other examples are presented below.

Example 1. Consider the two-dimensional space $\mathbb{R}_2$ of position vectors and let $\mathscr{S}_1 = \mathscr{S}_2 = \mathbb{R}_2$. If T denotes a transformation of this space into itself such that each vector $\boldsymbol{x}$ is mapped (see Fig. 4.1) into a collinear vector $\alpha_0 \boldsymbol{x}$, where $\alpha_0 \in \mathscr{F}$ is fixed ($\alpha_0 \neq 0$),

$$T(\boldsymbol{x}) = \alpha_0 \boldsymbol{x},$$

then it is easily seen that T is a linear transformation of $\mathbb{R}_2$ onto itself. Indeed,

$$T(\boldsymbol{x}_1 + \boldsymbol{x}_2) = \alpha_0(\boldsymbol{x}_1 + \boldsymbol{x}_2) = \alpha_0 \boldsymbol{x}_1 + \alpha_0 \boldsymbol{x}_2 = T(\boldsymbol{x}_1) + T(\boldsymbol{x}_2),$$

and for all $\alpha \in \mathbb{R}$,

$$T(\alpha\boldsymbol{x}) = \alpha_0 \alpha\boldsymbol{x} = \alpha(\alpha_0 \boldsymbol{x}) = \alpha T(\boldsymbol{x}).$$

Obviously, for any $\boldsymbol{y} \in \mathbb{R}_2$ there is an $\boldsymbol{x}$, namely, $\boldsymbol{x} = (1/\alpha_0)\boldsymbol{y}$, such that $T(\boldsymbol{x}) = \boldsymbol{y}$ and hence $T: \mathbb{R}_2 \xrightarrow{\text{onto}} \mathbb{R}_2$.

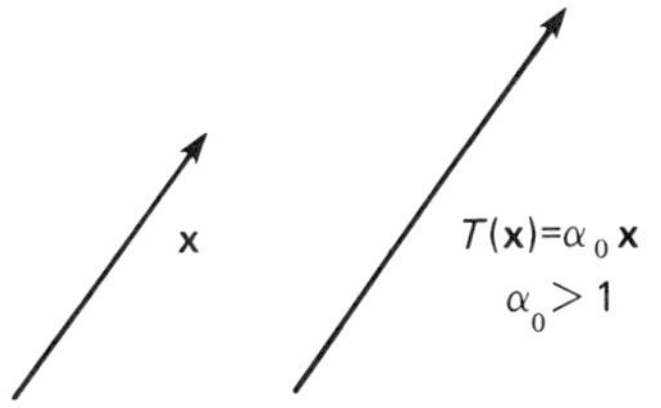

Fig. 4.1 Collinear mapping.

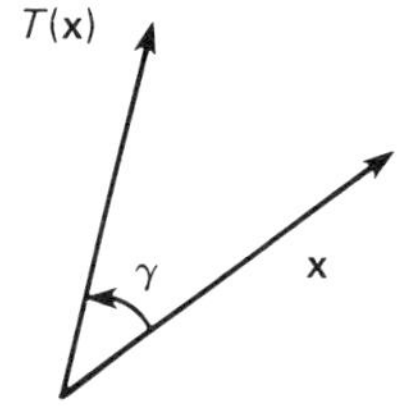

Fig. 4.2 Rotation.

***Example* 2.** Let T be the transformation of anticlockwise rotation of vectors (see Fig. 4.2) from $\mathbb{R}_2$ through a fixed angle γ $(0 < \gamma < 2\pi)$. If $\boldsymbol{x} = (x_1, x_2)$, then it is easily seen that the resulting vector is $\boldsymbol{x}' = (x'_1, x'_2)$, where

$$x'_1 = (\cos\gamma)x_1 - (\sin\gamma)x_2,$$
$$x'_2 = (\sin\gamma)x_1 + (\cos\gamma)x_2.$$

Using matrix notation, we can write

$$\begin{bmatrix} x'_1 \\ x'_2 \end{bmatrix} = \begin{bmatrix} \cos\gamma & -\sin\gamma \\ \sin\gamma & \cos\gamma \end{bmatrix} \begin{bmatrix} x_1 \\ x_2 \end{bmatrix},$$

or, representing $\boldsymbol{x}$ and $\boldsymbol{x}'$ as column vectors,

$$T(\boldsymbol{x}) = \boldsymbol{x}' = A\boldsymbol{x},$$

where

$$A = \begin{bmatrix} \cos\gamma & -\sin\gamma \\ \sin\gamma & \cos\gamma \end{bmatrix}.$$

Now the linearity of the transformation T easily follows from the appropriate properties of matrices (see Section 1.2):

$$T(\boldsymbol{x}_1 + \boldsymbol{x}_2) = A(\boldsymbol{x}_1 + \boldsymbol{x}_2) = A\boldsymbol{x}_1 + A\boldsymbol{x}_2 = T(\boldsymbol{x}_1) + T(\boldsymbol{x}_2),$$
$$T(\alpha\boldsymbol{x}) = A(\alpha\boldsymbol{x}) = \alpha(A\boldsymbol{x}) = \alpha T(\boldsymbol{x}).$$

***Example* 3.** Let T_A denote a transformation of the n-dimensional space $\mathscr{F}^n$ into $\mathscr{F}^m$ given by the rule

$$T_A(\boldsymbol{x}) = A\boldsymbol{x}, \qquad \boldsymbol{x} \in \mathscr{F}^n,$$

where A is a fixed $m \times n$ matrix. Obviously T_A is linear (see Section 1.2 and Example 2). □

Thus, any transformation from $\mathscr{F}^n$ into $\mathscr{F}^m$ determined by an $m \times n$ matrix, together with matrix–vector multiplication, is linear. It is remarkable that, in an appropriate sense, *any* linear transformation can be described in this way. It will be shown later that *any* linear transformation between *any* finite-dimensional spaces can be represented by a matrix together with matrix-vector multiplication. Before presenting one more example, note that the transformation of Example 1 is determined by the scalar matrix $A = \alpha_0 I$ in the sense of Example 3.

Example 4. Let T be the transformation mapping the linear space P^n of polynomials over $\mathscr{F}$ with degree not exceeding n into P^{n-1} according to the rule

$$T(p_n(x)) = p_n'(x).$$

That is, T maps a polynomial in P^n onto its derivative (a polynomial in P^{n-1}).

Recalling the appropriate properties of derivatives, it is easily seen that the conditions in (1) hold, so the transformation is linear. Identifying the polynomial $p_n(x) = \sum_{i=0}^{n} a_i x^i$ with the vector $\boldsymbol{p} = [a_0 \quad a_1 \quad \cdots \quad a_n]^{\mathrm{T}}$, we can think of the transformation as a mapping $\tilde{T}$ from $\mathscr{F}^{n+1}$ onto $\mathscr{F}^n$, given by the rule

$$\tilde{T}(\boldsymbol{p}) = \begin{bmatrix} 0 & 1 & 0 & \cdots & 0 \\ \vdots & 0 & 2 & \ddots & \vdots \\ 0 & \cdots & & 0 & n \end{bmatrix} \boldsymbol{p} = \boldsymbol{p}_1,$$

where the matrix is $n \times (n+1)$ and $\boldsymbol{p}_1 = [a_1 \quad 2a_2 \quad \cdots \quad na_n]^{\mathrm{T}}$. □

Let us denote by $\mathscr{L}(\mathscr{S}_1, \mathscr{S}_2)$ the set of all linear transformations mapping the linear space $\mathscr{S}_1$ into the linear space $\mathscr{S}_2$. Both spaces are assumed throughout the book to be finite-dimensional and over the same field $\mathscr{F}$. When $\mathscr{S}_1 = \mathscr{S}_2 = \mathscr{S}$, we write $\mathscr{L}(\mathscr{S})$ instead of $\mathscr{L}(\mathscr{S}, \mathscr{S})$.

Note that, combining the conditions in (1), a transformation $T \in \mathscr{L}(\mathscr{S}_1, \mathscr{S}_2)$ can be characterized by a single condition:

$$T(\alpha \boldsymbol{x}_1 + \beta \boldsymbol{x}_2) = \alpha T(\boldsymbol{x}_1) + \beta T(\boldsymbol{x}_2) \tag{2}$$

for any elements $\boldsymbol{x}_1, \boldsymbol{x}_2$ from $\mathscr{S}_1$ and $\alpha, \beta \in \mathscr{F}$. This implies a more general relation:

$$T\left(\sum_{i=1}^{k} \alpha_i \boldsymbol{x}_i\right) = \sum_{i=1}^{k} \alpha_i T(\boldsymbol{x}_i),$$

where $\boldsymbol{x}_i \in \mathscr{S}$, $\alpha_i \in \mathscr{F}$ $(i = 1, 2, \ldots, k)$. In particular,

$$T(\boldsymbol{0}) = \boldsymbol{0},$$
$$T(-\boldsymbol{x}) = -T(\boldsymbol{x}),$$
$$T(\boldsymbol{x}_1 - \boldsymbol{x}_2) = T(\boldsymbol{x}_1) - T(\boldsymbol{x}_2).$$

Exercise 5. Let $T \in \mathscr{L}(\mathscr{S}_1, \mathscr{S}_2)$. Show that if the elements $\boldsymbol{x}_i \in \mathscr{S}_1$ $(i = 1, 2, \ldots, k)$ are linearly dependent, so are their images $T(\boldsymbol{x}_i)$ $(i = 1, 2, \ldots, k)$. □

Exercise 5 shows that a linear transformation carries a linearly dependent system into a linear dependent one. However, a linear transformation may also transform linearly independent elements into a linearly dependent set. For example, the transformation mapping every element of $\mathscr{S}_1$ onto the zero element of $\mathscr{S}_2$.

Example 6. The transformation $T \in \mathscr{L}(\mathbb{R}^4, \mathbb{R}^3)$ defined by

$$T\left(\begin{bmatrix} x_1 \\ x_2 \\ x_3 \\ x_4 \end{bmatrix}\right) = \begin{bmatrix} x_1 - x_2 + x_3 \\ x_2 - x_3 + x_4 \\ 0 \end{bmatrix}$$

maps the linearly independent vectors $[-1 \quad 0 \quad 1 \quad 1]^{\mathrm{T}}$ and $[0 \quad 0 \quad 0 \quad 1]^{\mathrm{T}}$ into the (linearly dependent) vectors $\boldsymbol{0}$ and $[0 \quad 1 \quad 0]^{\mathrm{T}}$, respectively. □

The rest of this section will focus on the algebraic properties of the set $\mathscr{L}(\mathscr{S}_1, \mathscr{S}_2)$. It turns out that $\mathscr{L}(\mathscr{S}_1, \mathscr{S}_2)$ is itself a linear space over $\mathscr{F}$ if *addition* and *scalar multiplication* of linear transformations are defined as follows:

For any $T_1, T_2 \in \mathscr{L}(\mathscr{S}_1, \mathscr{S}_2)$ and any $\boldsymbol{x} \in \mathscr{S}_1$,

$$(T_1 + T_2)(\boldsymbol{x}) \triangleq T_1(\boldsymbol{x}) + T_2(\boldsymbol{x}).$$

For any $T \in \mathscr{L}(\mathscr{S}_1, \mathscr{S}_2)$, $\alpha \in \mathscr{F}$, and $\boldsymbol{x} \in \mathscr{S}_1$,

$$(\alpha T)(\boldsymbol{x}) \triangleq \alpha T(\boldsymbol{x}).$$

Note that two transformations T_1 and T_2 from $\mathscr{L}(\mathscr{S}_1, \mathscr{S}_2)$ are *equal* or *identical* if $T_1(\boldsymbol{x}) = T_2(\boldsymbol{x})$ for every $\boldsymbol{x} \in \mathscr{S}_1$. Obviously, the zero *transformation* 0 such that $0(\boldsymbol{x}) = \boldsymbol{0}$ for all $\boldsymbol{x} \in \mathscr{S}_1$ plays the role of the zero element in $\mathscr{L}(\mathscr{S}_1, \mathscr{S}_2)$.

Exercise 7. Check that, with respect to the operations introduced, the set $\mathscr{L}(\mathscr{S}_1, \mathscr{S}_2)$ is a linear space over $\mathscr{F}$.

Exercise 8. Let $\{\boldsymbol{x}_1, \boldsymbol{x}_2, \ldots, \boldsymbol{x}_n\}$ and $\{\boldsymbol{y}_1, \boldsymbol{y}_2, \ldots, \boldsymbol{y}_m\}$ denote bases in $\mathscr{S}_1$ and $\mathscr{S}_2$, respectively. The transformations $T_{ij} \in \mathscr{L}(\mathscr{S}_1, \mathscr{S}_2)$, $i = 1, 2, \ldots, n$ and $j = 1, 2, \ldots, m$, are defined by the rule

$$T_{ij}(\boldsymbol{x}_i) = \boldsymbol{y}_j \qquad \text{and} \qquad T_{ij}(\boldsymbol{x}_k) = \boldsymbol{0} \quad \text{if } k \neq i,$$

then $T_{ij}(\boldsymbol{x})$ is defined for any $\boldsymbol{x} \in \mathscr{S}_1$ by linearity. Show that the T_{ij}'s constitute a basis in $\mathscr{L}(\mathscr{S}_1, \mathscr{S}_2)$. In particular, $\dim \mathscr{L}(\mathscr{S}_1, \mathscr{S}_2) = mn$. □

Consider three linear spaces $\mathcal{S}_1$, $\mathcal{S}_2$, and $\mathcal{S}_3$ over the field $\mathcal{F}$, and let $T_1 \in \mathcal{L}(\mathcal{S}_1, \mathcal{S}_2)$, $T_2 \in \mathcal{L}(\mathcal{S}_2, \mathcal{S}_3)$. If $\boldsymbol{x} \in \mathcal{S}_1$, then T_1 maps $\boldsymbol{x}$ onto an element $\boldsymbol{y} \in \mathcal{S}_2$ and, subsequently, T_2 maps $\boldsymbol{y}$ onto an element $\boldsymbol{z} \in \mathcal{S}_3$. Hence a successive application of T_1 and T_2 performs a transformation of $\mathcal{S}_1$ into $\mathcal{S}_3$. This transformation T is referred to as the *composition* of T_2 and T_1, and we write $T = T_2 T_1$. The formal definition is

$$T(\boldsymbol{x}) \triangleq T_2(T_1(\boldsymbol{x})) \tag{3}$$

for all $\boldsymbol{x} \in \mathcal{S}_1$.

Exercise 9. Show that T given by Eq. (3) is a linear transformation from $\mathcal{S}_1$ to $\mathcal{S}_3$, that is, $T \in \mathcal{L}(\mathcal{S}_1, \mathcal{S}_3)$. □

Similarly, a composition of several linear transformations between appropriate spaces over the same field, $T = T_k T_{k-1} \cdots T_1$, is defined by the rule

$$T(\boldsymbol{x}) \triangleq T_k(T_{k-1}(\cdots(T_2(T_1(\boldsymbol{x}))\cdots)) \qquad \text{for all } \boldsymbol{x} \in \mathcal{S}.$$

In particular, if $T \in \mathcal{L}(\mathcal{S})$, T^k denotes the composition of the transformation T with itself k times.

Exercise 10. Check the following properties of composition of linear transformations:

(a) $T_1(T_2 T_3) = (T_1 T_2)T_3$;
(b) $\alpha(T_1 T_2) = (\alpha T_1)T_2 = T_1(\alpha T_2)$;
(c) $T_1(T_2 + T_3) = T_1 T_2 + T_1 T_3$;
(d) $(T_1 + T_2)T_3 = T_1 T_3 + T_2 T_3$.

Hint. Consider the image of an element under the transformations on both sides of the equations.

Exercise 11. Show that if transformations are defined by matrices as in Example 3, then the properties of compositions of transformations in Exercise 10 are equivalent to the corresponding properties of matrices under matrix multiplication. □

Consider the linear space $\mathcal{L}(\mathcal{S})$. The operation of composition defined above, with properties indicated in Exercise 10, is an additional operation defined on the elements of $\mathcal{L}(\mathcal{S})$. A linear space with such an operation is referred to as an *algebra*. Thus, we may describe $\mathcal{L}(\mathcal{S})$ as an algebra of linear transformations acting on $\mathcal{S}$. The *identity transformation* I defined by $I(\mathbf{x}) \triangleq \mathbf{x}$ for all $\mathbf{x} \in \mathcal{S}$ has the property that $IT = TI = T$ for all $T \in \mathcal{L}(\mathcal{S})$.

It was shown in Exercise 3.1.3(b) that $\mathcal{F}^{n \times n}$ is a linear space. The space $\mathcal{F}^{n \times n}$ together with the operation of matrix multiplication is another example of an algebra.

Polynomials in a transformation are defined as for those in square matrices (see Section 1.7).

Exercise 12. Show that for any $T \in \mathcal{L}(\mathcal{S})$ there is a nonzero scalar polynomial $p(\lambda) = \sum_{i=0}^{l} p_i \lambda^i$ with coefficients from $\mathcal{F}$ such that $p(T) \triangleq \sum_{i=0}^{l} p_i T^i$ is the zero transformation.

SOLUTION. Since (by Exercise 8) the space $\mathcal{L}(\mathcal{S})$ is finite-dimensional, there is a sufficiently large positive integer l such that the transformations I, $T, T^2, \ldots, T^l$ viewed as elements of $\mathcal{L}(\mathcal{S})$ are linearly dependent. Hence there are scalars $p_1, p_2, \ldots, p_l$ from $\mathcal{F}$, not all zero, such that $\sum_{i=0}^{l} p_i T^i = 0$. □

4.2 Matrix Representation of Linear Transformations

Let $T \in \mathcal{L}(\mathcal{S}_1, \mathcal{S}_2)$, where $\dim \mathcal{S}_1 = n$ and $\dim \mathcal{S}_2 = m$. We have seen in Example 4.1.3 that, in the particular case $\mathcal{S}_1 = \mathcal{F}^n$ and $\mathcal{S}_2 = \mathcal{F}^m$, any $m \times n$ matrix defines a transformation $T \in \mathcal{L}(\mathcal{F}^n, \mathcal{F}^m)$. Now we show that any transformation $T \in \mathcal{L}(\mathcal{S}_1, \mathcal{S}_2)$ is associated with a set of $m \times n$ matrices.

Let $\mathcal{E} = \{\mathbf{x}_1, \mathbf{x}_2, \ldots, \mathbf{x}_n\}$ and $\mathcal{G} = \{\mathbf{y}_1, \mathbf{y}_2, \ldots, \mathbf{y}_m\}$ be any bases in the spaces $\mathcal{S}_1$ and $\mathcal{S}_2$ over $\mathcal{F}$, respectively. Evaluating the images $T(\mathbf{x}_j)$ $(j = 1, 2, \ldots, n)$ and finding the representation of the system $\{T(\mathbf{x}_1), T(\mathbf{x}_2), \ldots, T(\mathbf{x}_n)\}$ with respect to the basis $\mathcal{G}$ (see Section 3.7), we obtain

$$T(\mathbf{x}_j) = \sum_{i=1}^{m} a_{ij} \mathbf{y}_i, \qquad j = 1, 2, \ldots, n. \tag{1}$$

The matrix $A_{\mathcal{E},\mathcal{G}} = [a_{ij}]_{i,j=1}^{m,n} \in \mathcal{F}^{m \times n}$ is defined by this relation and is said to be the (*matrix*) *representation of T with respect to the bases* $(\mathcal{E}, \mathcal{G})$. Note that the *columns* of the representation $A_{\mathcal{E},\mathcal{G}}$ of T are just the coordinates of $T(\mathbf{x}_1), T(\mathbf{x}_2), \ldots, T(\mathbf{x}_n)$ with respect to the basis $\mathcal{G}$ and that, as indicated in Section 3.7, the representation of the system $\{T(\mathbf{x}_j)\}_{j=1}^{n}$ with respect to the basis $\mathcal{G}$ is unique. Hence, also fixing the basis $\mathcal{E}$, the (matrix) representation of T with respect to the pair $(\mathcal{E}, \mathcal{G})$ is uniquely determined.

Exercise 1. Find the matrix representation of the transformation $T \in \mathscr{L}(\mathscr{F}^3, \mathscr{F}^4)$ such that

$$T\left(\begin{bmatrix} x_1 \\ x_2 \\ x_3 \end{bmatrix}\right) = \begin{bmatrix} x_2 + x_3 \\ x_1 + x_2 \\ x_1 \\ x_1 - x_2 \end{bmatrix}, \qquad \mathbf{x} = [x_1 \quad x_2 \quad x_3]^{\mathrm{T}} \in \mathscr{F}^3,$$

with respect to the pair of standard bases.

SOLUTION. Computing $T(\mathbf{e}_j), j = 1, 2, 3$, where $\mathscr{E} = \{\mathbf{e}_1, \mathbf{e}_2, \mathbf{e}_3\}$ denotes the standard basis in $\mathscr{F}^3$, we obtain the system of elements

$$T(\mathbf{e}_1) = \begin{bmatrix} 0 \\ 1 \\ 1 \\ 1 \end{bmatrix}, \qquad T(\mathbf{e}_2) = \begin{bmatrix} 1 \\ 1 \\ 0 \\ -1 \end{bmatrix}, \qquad T(\mathbf{e}_3) = \begin{bmatrix} 1 \\ 0 \\ 0 \\ 0 \end{bmatrix}.$$

For $\mathbf{y} = [y_1 \quad y_2 \quad y_3 \quad y_4]^{\mathrm{T}}$ and the standard basis $\mathscr{G} = \{\mathbf{e}'_1, \mathbf{e}'_2, \mathbf{e}'_3, \mathbf{e}'_4\}$ in $\mathscr{F}^4$, we have $\mathbf{y} = \sum_{i=1}^{4} y_i \mathbf{e}'_i$. Hence the representation of $\mathbf{y}$ is the vector $[y_1 \quad y_2 \quad y_3 \quad y_4]^{\mathrm{T}}$ itself, and the required representation of T is the matrix

$$A = A_{\mathscr{E},\mathscr{G}} = \begin{bmatrix} 0 & 1 & 1 \\ 1 & 1 & 0 \\ 1 & 0 & 0 \\ 1 & -1 & 0 \end{bmatrix}. \quad \square$$

The representation of a linear transformation $T \in \mathscr{L}(\mathscr{F}^n, \mathscr{F}^m)$ with respect to the standard bases of the two spaces is referred to as the *standard (matrix) representation of T*.

In view of Exercise 3.7.1, it is clear that the transition from one basis in $\mathscr{S}_2$ to another changes the representation of the system $\{T(\mathbf{x}_j)\}_{j=1}^{n}$. There is also a change of the representation when a basis in $\mathscr{S}_1$ is replaced by another.

Exercise 2. For the same $T \in \mathscr{L}(\mathscr{F}^3, \mathscr{F}^4)$ defined in Exercise 1, verify that the representation with respect to the bases

$$\mathbf{x}_1 = \begin{bmatrix} 1 \\ 1 \\ 0 \end{bmatrix}, \qquad \mathbf{x}_2 = \begin{bmatrix} 1 \\ 0 \\ -1 \end{bmatrix}, \qquad \mathbf{x}_3 = \begin{bmatrix} 0 \\ 1 \\ 0 \end{bmatrix}$$

for $\mathscr{F}^3$, and

$$\mathbf{y}_1 = \begin{bmatrix} 1 \\ -1 \\ 0 \\ 0 \end{bmatrix}, \qquad \mathbf{y}_2 = \begin{bmatrix} 1 \\ 0 \\ 1 \\ 0 \end{bmatrix}, \qquad \mathbf{y}_3 = \begin{bmatrix} 0 \\ 1 \\ 0 \\ 0 \end{bmatrix}, \qquad \mathbf{y}_4 = \begin{bmatrix} 0 \\ 0 \\ 1 \\ 1 \end{bmatrix}.$$

for $\mathscr{F}^4$, is

$$A' = \begin{bmatrix} 0 & -1 & 0 \\ 1 & 0 & 1 \\ 2 & 0 & 1 \\ 0 & 1 & -1 \end{bmatrix}.$$

Hint. Calculate $T(\mathbf{x}_j)$ $(j = 1, 2, 3)$. The solution of the jth system of type (1) (with $m = 4$) gives the jth column of A'. □

Thus, the same linear transformation generally has different matrix representations in distinct pairs of bases. It is natural to ask which properties all of these matrices have in common; this will be our main concern in the next section. Now we are focusing on a relationship between the set of linear transformations $\mathscr{L}(\mathscr{S}_1, \mathscr{S}_2)$ and the set of their representations with respect to a fixed pair of bases $(\mathscr{E}, \mathscr{G})$. We already know that any linear transformation produces a unique $m \times n$ matrix, which is its representation with respect to $(\mathscr{E}, \mathscr{G})$. The converse is also true.

Exercise 3. Let $\mathscr{S}_1$, $\mathscr{S}_2$, $\mathscr{E}$, and $\mathscr{G}$ be defined as before. Given $A \in \mathscr{F}^{m \times n}$, confirm the existence of a unique $T \in \mathscr{L}(\mathscr{S}_1, \mathscr{S}_2)$ such that the representation $A_{\mathscr{E},\mathscr{G}}$ of T with respect to $(\mathscr{E}, \mathscr{G})$ is A.

Hint. Define a transformation T on the elements of $\mathscr{E}$ by Eq. (1) and then extend T to $\mathscr{S}_1$ by linearity:

$$T(\boldsymbol{x}) = \sum_{j=1}^{n} \alpha_j T(\boldsymbol{x}_j),$$

provided $\boldsymbol{x} = \sum_{j=1}^{n} \alpha_j \boldsymbol{x}_j$. □

Thus, given the bases $\mathscr{E}$ and $\mathscr{G}$ there is a one-to-one correspondence between the sets $\mathscr{L}(\mathscr{S}_1, \mathscr{S}_2)$ and $\mathscr{F}^{m \times n}$ in the sense that each matrix A from $\mathscr{F}^{m \times n}$ is a representation of a transformation $T \in \mathscr{L}(\mathscr{S}_1, \mathscr{S}_2)$ with respect to the bases $\mathscr{E}$ and $\mathscr{G}$, and vice versa. *This correspondence $T \leftrightarrow A$ is an isomorphism between the linear spaces $\mathscr{L}(\mathscr{S}_1, \mathscr{S}_2)$ and $\mathscr{F}^{m \times n}$.* (See Exercise 3.14.4 for the definition of an isomorphism). Indeed, if A_1 (respectively, A_2) is the representation of T_1 (respectively, T_2) from $\mathscr{L}(\mathscr{S}_1, \mathscr{S}_2)$ with respect to the bases $\mathscr{E} = \{\boldsymbol{x}_j\}_{j=1}^{n}$ and $\mathscr{G} = \{\boldsymbol{y}_i\}_{i=1}^{m}$, and if we write $A_k = [a_{ij}^{(k)}]$, $k = 1, 2$, then (see Eq. (1))

$$T_k(\boldsymbol{x}_j) = \sum_{i=1}^{m} a_{ij}^{(k)} \boldsymbol{y}_i$$

for $j = 1, 2, \ldots, n$ and $k = 1, 2$. Hence for any $\alpha \in \mathscr{F}$,

$$\alpha T_k(\boldsymbol{x}_j) = \sum_{i=1}^{m} \alpha a_{ij}^{(k)} \boldsymbol{y}_i,$$

$$(T_1 + T_2)(\boldsymbol{x}_j) = \sum_{i=1}^{m} (a_{ij}^{(1)} + a_{ij}^{(2)}) \boldsymbol{y}_i.$$

Thus, if $T_1 \leftrightarrow A_1$ and $T_2 \leftrightarrow A_2$, then, for $\alpha \in \mathscr{F}$,

$$\alpha T_1 \leftrightarrow \alpha A_1, \qquad \alpha T_2 \leftrightarrow \alpha A_2, \tag{2}$$

$$T_1 + T_2 \quad \leftrightarrow \quad A_1 + A_2, \tag{3}$$

and the isomorphism is established.

We consider now the operation of matrix multiplication and its connection with the composition of transformations. Assume that $T_1 \in \mathscr{L}(\mathscr{S}_1, \mathscr{S}_2)$ and $T_2 \in \mathscr{L}(\mathscr{S}_2, \mathscr{S}_3)$. Let $\mathscr{R} = \{\boldsymbol{z}_s\}_{s=1}^{l}$ denote a basis in $\mathscr{S}_3$ and let $\mathscr{E}$ and $\mathscr{G}$ be defined as before. If $A = [a_{ij}]_{i,j=1}^{m,n}$ and $B = [b_{si}]_{s,i=1}^{l,m}$ are representations of T_1 and T_2 with respect to the pairs $(\mathscr{E}, \mathscr{G})$ and $(\mathscr{G}, \mathscr{R})$, respectively, then

$$\begin{aligned} T_2 T_1(\boldsymbol{x}_j) = T_2(T_1(\boldsymbol{x}_j)) &= \sum_{i=1}^{m} a_{ij} T_2(\boldsymbol{y}_i) \\ &= \sum_{i=1}^{M} a_{ij} \sum_{s=1}^{l} b_{si} \boldsymbol{z}_s = \sum_{s=1}^{l} \left(\sum_{i=1}^{m} b_{si} a_{ij} \right) \boldsymbol{z}_s \end{aligned}$$

for all $j = 1, 2, \ldots, n$. Hence, the representation of $T_2 T_1$ with respect to the bases $(\mathscr{E}, \mathscr{R})$ is the matrix BA. Briefly, if $T_1 \leftrightarrow A$, $T_2 \leftrightarrow B$, then

$$T_2 T_1 \leftrightarrow BA. \tag{4}$$

Note that the relation in (4) obviously holds for transformations T_1 and T_2 both acting on $\mathscr{S}_1$ and their representations with respect to the pairs of bases $(\mathscr{E}, \mathscr{G})$ and $(\mathscr{G}, \mathscr{R})$, where $\mathscr{E}$, $\mathscr{G}$, and $\mathscr{R}$ are all bases for $\mathscr{S}_1$. This general approach simplifies considerably when we consider transformations acting on a single space and their representations with respect to *one* basis (that is, from the general point of view, with respect to a pair of identical bases). Clearly, the correspondences in Eqs. (2), (3), and (4) are valid for transformations T_1 and T_2 acting on the linear space $\mathscr{S}$ and their representations with respect to one basis $\mathscr{E}$ $(= \mathscr{G} = \mathscr{R})$ for $\mathscr{S}$. Using Exercise 3 and supposing the basis $\mathscr{E}$ to be fixed, we say that there is a one-to-one correspondence between algebraic operations involving linear transformations in $\mathscr{L}(\mathscr{S})$ and those involving square matrices. Such a correspondence, given by

Eqs. (2)–(4), is known as *an isomorphism between algebras* ($\mathscr{L}(\mathscr{S})$ and $\mathscr{F}^{n\times n}$, in this case).

Theorem 1. *(a) There is an isomorphism between the linear spaces* $\mathscr{L}(\mathscr{S}_1, \mathscr{S}_2)$ *and* $\mathscr{F}^{m\times n}$ *given by the correspondence* $T \leftrightarrow A_{\mathscr{E},\mathscr{G}}$, *where* $T \in \mathscr{L}(\mathscr{S}_1, \mathscr{S}_2)$ *and* $A_{\mathscr{E},\mathscr{G}} \in \mathscr{F}^{m\times n}$ *is the representation of* T *with respect to a fixed pair of bases* $(\mathscr{E}, \mathscr{G})$.

(b) There is an isomorphism between the algebras $\mathscr{L}(\mathscr{S})$ *and* $\mathscr{F}^{n\times n}$ *given by the correspondence* $T \leftrightarrow A_{\mathscr{E}}$, *where* $T \in \mathscr{L}(\mathscr{S})$ *and* $A_{\mathscr{E}} \in \mathscr{F}^{n\times n}$ *is the representation of* T *with respect to a fixed basis* $\mathscr{E}$ *for* $\mathscr{S}$.

Along with the established isomorphism $\mathscr{L}(\mathscr{S}_1, \mathscr{S}_2) \leftrightarrow \mathscr{F}^{m\times n}$ between the spaces $\mathscr{L}(\mathscr{S}_1, \mathscr{S}_2)$ and $\mathscr{F}^{m\times n}$, there are isomorphisms $\mathscr{S}_1 \leftrightarrow \mathscr{F}^n$ and $\mathscr{S}_2 \leftrightarrow \mathscr{F}^m$ between pairs of linear spaces of the same dimension (see Exercise 3.14.4). For instance, these isomorphisms are given by the correspondence $\boldsymbol{x} \leftrightarrow \boldsymbol{\alpha}$ (respectively, $\boldsymbol{y} \leftrightarrow \boldsymbol{\beta}$), where $\boldsymbol{\alpha}$ (respectively, $\boldsymbol{\beta}$) denotes the representation of $\boldsymbol{x} \in \mathscr{S}_1$ (respectively, $\boldsymbol{y} \in \mathscr{S}_2$) with respect to a fixed basis $\mathscr{E}$ for $\mathscr{S}_1$ (respectively, $\mathscr{G}$ for $\mathscr{S}_2$). It turns out that these isomorphic correspondences carry over to the relations $\boldsymbol{y} = T(\boldsymbol{x})$ and $\boldsymbol{\beta} = A_{\mathscr{E},\mathscr{G}}\boldsymbol{\alpha}$.

Theorem 2. *In the previous notation, let the bases* $\mathscr{E}, \mathscr{G}$ *be fixed and* $\boldsymbol{x} \leftrightarrow \boldsymbol{\alpha}$, $\boldsymbol{y} \leftrightarrow \boldsymbol{\beta}$, $T \leftrightarrow A_{\mathscr{E},\mathscr{G}}$. *Then*

$$\boldsymbol{y} = T(\boldsymbol{x}) \tag{5}$$

if and only if

$$\boldsymbol{\beta} = A_{\mathscr{E},\mathscr{G}}\boldsymbol{\alpha}. \tag{6}$$

PROOF. Let $T \in \mathscr{L}(\mathscr{S}_1, \mathscr{S}_2)$. Assume Eq. (5) holds and that $\boldsymbol{x} \leftrightarrow \boldsymbol{\alpha}$ and $\boldsymbol{y} \leftrightarrow \boldsymbol{\beta}$, where $\boldsymbol{\alpha} = [\alpha_1 \ \ \alpha_2 \ \ \cdots \ \ \alpha_n]^T$ and $\boldsymbol{\beta} = [\beta_1 \ \ \beta_2 \ \ \cdots \ \ \beta_m]^T$. Then, using Eq. (1),

$$\sum_{i=1}^{m} \beta_i \boldsymbol{y}_i = \boldsymbol{y} = T(\boldsymbol{x}) = \sum_{j=1}^{n} \alpha_j T(\boldsymbol{x}_j) = \sum_{i=1}^{m} \left(\sum_{j=1}^{n} a_{ij}\alpha_j \right) \boldsymbol{y}_i. \tag{7}$$

Therefore

$$\beta_i = \sum_{j=1}^{n} a_{ij}\alpha_j, \qquad i = 1, 2, \ldots, m, \tag{8}$$

and Eq. (6) holds.

Conversely, if (6) (or the equivalent (8)) is given, then by defining T as suggested in Exercise 3 and using (7), it is easily seen that (8) implies Eq. (5).

Exercise 4. Check the statement of Theorem 2 for the transformation T considered in Exercise 1 and its standard representation.

SOLUTION. Preserving the previous notation, if $\boldsymbol{x} = [x_1 \quad x_2 \quad x_3]^T \in \mathscr{F}^3$, then the representation $\boldsymbol{\alpha}$ of $\boldsymbol{x}$ with respect to the standard basis is $\boldsymbol{\alpha} = [x_1 \quad x_2 \quad x_3]^T$. Noting a similar result for $\boldsymbol{y} \in \mathscr{F}^4$ and its representation $\boldsymbol{\beta}$, we have $\boldsymbol{y} = T(\boldsymbol{x})$ if and only if (see the solution of Exercise 1)

$$\begin{bmatrix} y_1 \\ y_2 \\ y_3 \\ y_4 \end{bmatrix} = \begin{bmatrix} 0 & 1 & 1 \\ 1 & 1 & 0 \\ 1 & 0 & 0 \\ 1 & -1 & 0 \end{bmatrix} \begin{bmatrix} x_1 \\ x_2 \\ x_3 \end{bmatrix}.$$

Exercise 5. Verify that the matrices

$$\begin{bmatrix} 2 & 0 & 0 \\ 1 & 0 & 0 \\ 0 & 1 & 1 \end{bmatrix} \quad \text{and} \quad \begin{bmatrix} 1 & -1 & 1 \\ 0 & 1 & 0 \\ 1 & 1 & 1 \end{bmatrix}$$

are representations of the same transformation $T \in \mathscr{L}(\mathbb{R}^3)$ with respect to the standard basis and the basis $[1 \quad 1 \quad 0]^T$, $[0 \quad 1 \quad 1]^T$, $[1 \quad 0 \quad 1]^T$, respectively. Check the statement of Theorem 2 for these cases.

Hint. Use Exercise 3.5.3. and calculate $T(\boldsymbol{e}_j)$, $j = 1, 2, 3$. □

4.3 Matrix Representations, Equivalence, and Similarity

Let $T \in \mathscr{L}(\mathscr{S}_1, \mathscr{S}_2)$. To describe the set $\mathscr{A}_T$ of all matrices that are representations of T in different pairs of bases, we shall use the results of Section 3.9 about how a change of bases influences the representation of a system. As before, assume that $\dim \mathscr{S}_1 = n$ and $\dim \mathscr{S}_2 = m$.

Let $\mathscr{G}$ and $\mathscr{G}'$ be two bases in $\mathscr{S}_2$. If $\mathscr{E}$ is a basis in $\mathscr{S}_1$, then the representations $A_{\mathscr{E},\mathscr{G}}$ and $A_{\mathscr{E},\mathscr{G}'}$ of T are related by Exercise 3.9.3 as

$$A_{\mathscr{E},\mathscr{G}'} = PA_{\mathscr{E},\mathscr{G}}, \tag{1}$$

where $P \in \mathscr{F}^{m \times m}$ is the transition matrix from the basis $\mathscr{G}$ to the basis $\mathscr{G}'$.

Now consider along with $\mathscr{E}$ a basis $\mathscr{E}'$ in $\mathscr{S}_1$ and let Q denote the transition matrix from $\mathscr{E}$ to $\mathscr{E}'$. Then we will show that

$$A_{\mathscr{E},\mathscr{G}'} = A_{\mathscr{E}',\mathscr{G}'}Q. \tag{2}$$

Indeed, if $\boldsymbol{\alpha}$ and $\boldsymbol{\alpha}'$ denote the representations of $\mathbf{x} \in \mathscr{S}_1$ with respect to the bases $\mathscr{E}$ and $\mathscr{E}'$, respectively, then, by Theorem 3.9.1,

$$\boldsymbol{\alpha}' = Q\boldsymbol{\alpha}. \tag{3}$$

Let $y = T(x)$ and let $\boldsymbol{\beta}$ denote the representation of y with respect to the basis $\mathscr{G}'$. By Theorem 4.2.2 we have

$$\boldsymbol{\beta} = A_{\mathscr{E},\mathscr{G}'}\boldsymbol{\alpha} \qquad \text{and} \qquad \boldsymbol{\beta} = A_{\mathscr{E}',\mathscr{G}'}\boldsymbol{\alpha}'. \tag{4}$$

Hence, using Eq. (3), the second relation in (4) yields $\boldsymbol{\beta} = A_{\mathscr{E}',\mathscr{G}'}Q\boldsymbol{\alpha}$. Comparing this with the first relation in (4) and noting that $x \in \mathscr{S}_1$ is arbitrary, the required equality (2) follows.

Combining Eqs. (1) and (2) and recalling (Exercise 3.7.2) that transition matrices are nonsingular, we arrive at the following theorem.

Theorem 1. *Let $T \in \mathscr{L}(\mathscr{S}_1\ \mathscr{S}_2)$. If $A_{\mathscr{E},\mathscr{G}}$ and $A_{\mathscr{E}',\mathscr{G}'}$ denote the representations of T with respect to the pairs of bases $(\mathscr{E}, \mathscr{G})$ and $(\mathscr{E}', \mathscr{G}')$, respectively, then*

$$A_{\mathscr{E}',\mathscr{G}'} = PA_{\mathscr{E},\mathscr{G}}Q^{-1}, \tag{5}$$

where P and Q denote the transition matrices from $\mathscr{G}$ to $\mathscr{G}'$ and from $\mathscr{E}$ to $\mathscr{E}'$, respectively.

Example 1. Consider the matrices A and A' obtained in Exercises 4.2.1 and 4.2.2, respectively. We already know that they are equivalent: $A' = PAQ^{-1}$ where P and Q are transition matrices from the standard bases to the bases in Exercise 4.2.2. In fact, calculating these transition matrices as in Exercise 3.9.1 (see also Corollary 3.9.1),

$$P = \begin{bmatrix} 1 & 1 & 0 & 0 \\ -1 & 0 & 1 & 0 \\ 0 & 1 & 0 & 1 \\ 0 & 0 & 0 & 1 \end{bmatrix}^{-1}, \qquad Q^{-1} = \begin{bmatrix} 1 & 1 & 0 \\ 1 & 0 & 1 \\ 0 & -1 & 0 \end{bmatrix}. \quad \square$$

Recalling the definition of equivalent matrices (Section 2.7), the last theorem can be restated in another way:

Theorem 1′. *Any two matrices that are representations of the same linear transformation with respect to different pairs of bases are equivalent.*

It is important to note that the last proposition admits a converse statement.

Exercise 2. Show that if A is a representation of $T \in \mathscr{L}(\mathscr{S}_1, \mathscr{S}_2)$, then any matrix A' that is equivalent to A is also a representation of T (with respect to some other pair of bases).

Hint. Given $A' = PAQ^{-1}$, use the matrices P and Q as transition matrices and construct a pair of bases with respect to which A' is a representation of T. ☐

Thus, any transformation $T \in \mathscr{L}(\mathscr{S}_1, \mathscr{S}_2)$ ($\dim \mathscr{S}_1 = n$, $\dim \mathscr{S}_2 = m$) generates a whole class $\mathscr{A}_T$ of mutually equivalent $m \times n$ matrices. Furthermore, this is an equivalence class of matrices within $\mathscr{F}^{m \times n}$ (see Section 2.7). Given $T \in \mathscr{L}(\mathscr{S}_1, \mathscr{S}_2)$, it is now our objective to find the "simplest" matrix in this class since, if it turns out to be simple enough, it should reveal at a glance some important features of the "parent" transformation.

Let $T \in \mathscr{L}(\mathscr{S}_1, \mathscr{S}_2)$, $T \neq 0$. The class of equivalent matrices generated by T is uniquely defined by the rank r of an arbitrary matrix belonging to the class (see Section 2.8). Referring to Section 2.7, we see that the simplest matrix of this equivalence class is of one of the forms

$$\begin{bmatrix} I_r & 0 \\ 0 & 0 \end{bmatrix} \quad \text{or} \quad [I_m \quad 0], \quad \begin{bmatrix} I_n \\ 0 \end{bmatrix}, \quad I_n, \tag{6}$$

depending on the relative sizes of r, m, and n.

A matrix in (6) is referred to as the *canonical form of the representations of T with respect to equivalence.* This canonical form (also referred to as the canonical form for any matrix in $\mathscr{A}_T$ with respect to equivalence) instantly gives the rank of all the matrices in the corresponding equivalence class.

Exercise 3. Check that the canonical form with respect to equivalence of the representations of $T \in \mathscr{L}(\mathscr{F}^3, \mathscr{F}^4)$ defined in Exercise 4.2.1 is

$$A_0 = \begin{bmatrix} 1 & 0 & 0 \\ 0 & 1 & 0 \\ 0 & 0 & 1 \\ 0 & 0 & 0 \end{bmatrix}. \quad \square$$

Finding the bases with respect to which the representation of a transformation has the canonical form can be performed with the help of transition matrices. In more detail, let $T \in \mathscr{L}(\mathscr{S}_1, \mathscr{S}_2)$ and let $A_{\mathscr{E},\mathscr{G}}$ denote the representation of T with respect to the bases $\mathscr{E}$ and $\mathscr{G}$ in $\mathscr{S}_1$ and $\mathscr{S}_2$, respectively. If A_0 is the canonical form of the equivalence class generated by T, then there are nonsingular matrices $P \in \mathscr{F}^{m \times m}$ and $Q \in \mathscr{F}^{n \times n}$ such that (Theorem 1)

$$A_0 = P^{-1} A_{\mathscr{E},\mathscr{G}} Q,$$

where P, Q are transition matrices from the bases $\mathscr{E}$, $\mathscr{G}$ to the bases with respect to which the representation of T is in the canonical form A_0. By use of Eq. (3.9.1), the required bases are easily found.

There are no essentially new features in Theorems 1 and 1′, or in the related discussion, when the special case of transformations from $\mathscr{L}(\mathscr{S})$ is considered. That is, transformations from $\mathscr{S}$ into itself. The associated matrix representations $A_{\mathscr{E},\mathscr{G}}$ are square, of course. This means that the possible canonical forms under equivalence are confined to the first and last of those displayed in (6).

A new phenomenon arises, however, if we insist that $\mathscr{G} = \mathscr{E}$. As might be expected, the canonical forms of (6) cannot generally be achieved under this additional constraint. In the context of Theorem 4.2.1(b), we now seek a single basis $\mathscr{E}$ for $\mathscr{S}$ in which the representation $A_{\mathscr{E}}$ of T is as simple as possible. The next theorem is a first step in the analysis of this problem. Keeping in mind the meaning of the matrices P and Q in Eq. (5), it is an immediate consequence of Theorem 1.

Theorem 2. *Let $T \in \mathscr{L}(\mathscr{S})$. If $A_{\mathscr{E}}$ and $A_{\mathscr{E}'}$ denote the representations of T with respect to the bases $\mathscr{E}$ and $\mathscr{E}'$ in $\mathscr{S}$, respectively, then*

$$A_{\mathscr{E}'} = PA_{\mathscr{E}}P^{-1}, \tag{7}$$

where P is the transition matrix from the basis $\mathscr{E}$ to the basis $\mathscr{E}'$.

The relation (7) between matrices $A_{\mathscr{E}'}$ and $A_{\mathscr{E}}$ turns out to be so important in the theory of matrices and linear transformations that a special name is justified: Square matrices A and B are said to be *similar*, written $A \approx B$, if there is a nonsingular matrix P such that

$$A = PBP^{-1}. \tag{8}$$

It is easy to verify that similarity determines an equivalence relation on $\mathscr{F}^{n \times n}$. The next result is then a consequence of Theorem 2.

Theorem 3. *Let $T \in \mathscr{L}(\mathscr{S})$ and $\dim \mathscr{S} = n$. Then the set of all matrix representations of T (with respect to different bases in $\mathscr{S}$) is an equivalence class of similar matrices in $\mathscr{F}^{n \times n}$.*

Exercise 4. Check that the representations of the transformation T defined in Exercise 4.2.5, with respect to the bases indicated there, are similar matrices. □

Denote the set of matrix representations of T (of the last theorem) by $\mathscr{A}_T^0$. In contrast to the problem of finding the "simplest" matrix in the equivalence class $\mathscr{A}_T$ of *equivalent* matrices, the problem of determining the "simplest" matrix in the equivalence class $\mathscr{A}_T^0$ of *similar* matrices (that is, the *canonical form of the representations of T with respect to similarity*) is much more complicated. For the general case, this problem will be solved in Chapter 6, while its solution for an especially simple but nonetheless important kind of transformation is found in Section 4.8.

4.4 Some Properties of Similar Matrices

Throughout this section all matrices are assumed to be square.

Exercise 1. Let $A \approx B$; that is, let A and B be similar matrices. Show that

(a) $\operatorname{rank} A = \operatorname{rank} B$;
(b) $\det A = \det B$;
(c) $A - \lambda I \approx B - \lambda I$;
(d) $A^k \approx B^k$ for $k = 0, 1, 2, \ldots$;
(e) $p(A) \approx p(B)$ for any scalar polynomial $p(\lambda)$;
(f) $A^{\mathrm{T}} \approx B^{\mathrm{T}}$.

SOLUTIONS (b) If $A = PBP^{-1}$, then $\det A = (\det P)(\det B)(\det P^{-1})$ and the result follows from Exercise 2.6.11.

(e) For $p(\lambda) = \sum_{i=0}^{l} p_i \lambda^i$ we have

$$p(A) = \sum_{i=0}^{l} p_i(PBP^{-1})^i$$
$$= P\left(\sum_{i=0}^{l} p_i B^i\right)P^{-1} = Pp(B)P^{-1}. \quad \square$$

When A and B are similar the matrix P of the relation $A = PBP^{-1}$ is said to be a *transforming matrix* (of B to A). It should be pointed out that this transforming matrix P is not unique. For example, P may be replaced by PP_1 where P_1 is any nonsingular matrix that commutes with B.

Exercise 2. Show that

$$\begin{bmatrix} A_1 & 0 \\ 0 & A_2 \end{bmatrix} \approx \begin{bmatrix} A_2 & 0 \\ 0 & A_1 \end{bmatrix}, \qquad A_1 \in \mathscr{F}^{k \times k}, \quad A_2 \in \mathscr{F}^{m \times m},$$

and that nonsingular matrices of the form

$$\begin{bmatrix} 0 & P_1 \\ P_2 & 0 \end{bmatrix}, \qquad P_1 \in \mathscr{F}^{k \times k}, \quad P_2 \in \mathscr{F}^{m \times m},$$

where P_1 commutes with A_1 and P_2 commutes with A_2, are some of the transforming matrices.

Exercise 3. Confirm the following generalization of the previous result: A square matrix A is similar to a matrix obtained from A by reflection in the secondary diagonal.

Exercise 4. Prove that if $(i_1, i_2, \ldots, i_n)$ is a permutation of $(1, 2, \ldots, n)$, then

$$[a_{i_k i_m}]_{k,m=1}^{n} \approx [a_{sj}]_{s,j=1}^{n}$$

for any $A = [a_{sj}]_{s,j=1}^{n}$.

Hint. Observe that the matrix $\tilde{A} = [a_{i_k i_m}]_{k,m=1}^{n}$ is obtained from A by interchanging columns accompanied by a corresponding interchanging of rows. Deduce from this the existence of a permutation matrix P such that $\tilde{A} = PAP^{-1}$. □

Exercise 4 gives rise to the problem of determining the operations on elements of a matrix that transform it into a similar one. Recalling that any nonsingular matrix is a product of elementary matrices (Exercise 2.7.7), we write the matrix P in Eq. (4.3.8) in the form $P = E_k E_{k-1} \cdots E_1$. Hence the transformation of B into a similar matrix A can be interpreted as a result of successive applications of elementary matrices:

$$A = E_k(E_{k-1} \cdots (E_2(E_1 B E_1^{-1})E_2^{-1}) \cdots E_{k-1}^{-1})E_k^{-1}.$$

For any elementary matrix E, the result of the operation EBE^{-1} can be easily described in terms of row and column operations depending on the type (see Section 2.7) of elementary matrix. The following exercise is established in this way.

Exercise 5. Show that a matrix is transformed into a similar one if and only if elementary operations of the following kind are performed:

(a) The ith column (row) is interchanged with the jth column (row) and subsequently, the ith row (column) and jth row (column) are interchanged;

(b) The jth column (row) is multiplied by a nonzero scalar α and, subsequently, the jth row (column) is multiplied by $1/\alpha$;

(c) The jth column (row), multiplied by α, is added to the ith column (row), while the ith column (row), multiplied by $-\alpha$, is added to the jth column (row). □

Observing that the trace of a matrix is not influenced by the operations of Exercise 5, the next exercise can be completed.

Exercise 6. Check that if $A \approx B$, then $\operatorname{tr} A = \operatorname{tr} B$. □

Note that the common characteristics we have found (rank, determinant, trace) for an equivalence class of similar matrices fail to provide complete information about the class.

Exercise 7. Check that the matrices

$$A = \begin{bmatrix} 1 & 1 \\ 0 & 1 \end{bmatrix} \quad \text{and} \quad I = \begin{bmatrix} 1 & 0 \\ 0 & 1 \end{bmatrix}$$

are not similar, although they have the same rank, determinant, and trace. □

The similarity relation preserves some other properties of matrices.

Exercise 8. Show that if $A \approx B$, then

(a) if B is idempotent, so is A;
(b) if B is nilpotent of degree k, so is A.

Exercise 9. Show that the scalar matrices are similar to themselves using any transforming matrix P. Conversely, prove that a matrix having this property must be scalar.

Hint. Consider first diagonal transforming matrices. □

4.5 Image and Kernel of a Linear Transformation

Theorem 1. *Let $T \in \mathscr{L}(\mathscr{S}_1, \mathscr{S}_2)$. If $\mathscr{S}$ is a subspace in $\mathscr{S}_1$, then $T(\mathscr{S})$ is a subspace in $\mathscr{S}_2$.*

PROOF. Let $\mathbf{y}_1, \mathbf{y}_2 \in T(\mathscr{S})$. Then there are elements $\mathbf{x}_1, \mathbf{x}_2 \in \mathscr{S}$ such that $\mathbf{y}_i = T(\mathbf{x}_i)$ $(i = 1, 2)$. Since T is linear, it follows that

$$\alpha \mathbf{y}_1 + \beta \mathbf{y}_2 = \alpha T(\mathbf{x}_1) + \beta T(\mathbf{x}_2) = \mathrm{T}(\alpha \mathbf{x}_1 + \beta \mathbf{x}_2).$$

Hence for any $\alpha, \beta \in \mathscr{F}$, the element $\alpha \mathbf{y}_1 + \beta \mathbf{y}_2$ is the image of $\alpha \mathbf{x}_1 + \beta \mathbf{x}_2 \in \mathscr{S}$, that is, it belongs to $T(\mathscr{S})$. This proves that the image $T(\mathscr{S})$ is a linear subspace in $\mathscr{S}_2$. ■

In particular, the image $T(\mathscr{S}_1)$ of the whole space $\mathscr{S}_1$, called the *image of the transformation* T and usually denoted by Im T, is a subspace in $\mathscr{S}_2$. Its dimension is called the *rank* of the transformation T and is written rank T. Thus,

$$\operatorname{rank} T \triangleq \dim(\operatorname{Im} T).$$

We have seen in Exercise 4.1.3 that an $m \times n$ matrix A over $\mathscr{F}$ determines a linear transformation T_A from $\mathscr{F}^n$ to $\mathscr{F}^m$ in a natural way. Observe that Im T_A as defined here is just the subspace Im A of $\mathscr{F}^m$, as introduced in Section 3.2.

Exercise 1. Describe the image of the transformation $T \in \mathscr{L}(\mathbb{R}^3, \mathbb{R}^4)$ defined in Exercise 4.2.1 and determine its rank.

SOLUTION. Representing T as

$$\begin{bmatrix} x_2 + x_3 \\ x_1 + x_2 \\ x_1 \\ x_1 - x_2 \end{bmatrix} = x_1 \begin{bmatrix} 0 \\ 1 \\ 1 \\ 1 \end{bmatrix} + x_2 \begin{bmatrix} 1 \\ 1 \\ 0 \\ -1 \end{bmatrix} + x_3 \begin{bmatrix} 1 \\ 0 \\ 0 \\ 0 \end{bmatrix},$$

it is not difficult to see that Im T coincides with span$\{\mathbf{v}_1, \mathbf{v}_2, \mathbf{v}_3\}$, where $\mathbf{v}_3 = [0 \quad 1 \quad 1 \quad 1]^T$, $\mathbf{v}_2 = [1 \quad 1 \quad 0 \quad -1]^T$, $\mathbf{v}_3 = [1 \quad 0 \quad 0 \quad 0]^T$. The vectors $\mathbf{v}_i$ $(i = 1, 2, 3)$ are linearly independent and, therefore, rank $T = 3$.

Exercise 2. Show that, for any linear subspace $\mathscr{S}$ of $\mathscr{S}_1$,

$$\dim T(\mathscr{S}) \leq \dim \mathscr{S} \tag{1}$$

and hence, that rank $T \leq \dim \mathscr{S}_1$.

Hint. Use Exercise 4.1.5. Compare this result with that of Exercise 3.5.6.

Exercise 3. Check that for any $T_1, T_2 \in \mathscr{L}(\mathscr{S}_1, \mathscr{S}_2)$,

$$\text{Im}(T_1 + T_2) \subset \text{Im } T_1 + \text{Im } T_2.$$

Then prove the inequalities

$$|\text{rank } T_1 - \text{rank } T_2| \leq \text{rank}(T_1 + T_2) \leq \text{rank } T_1 + \text{rank } T_2.$$

Hint. Use Exercise 3.6.2; compare this with the assertion of Proposition 3.8.2.

Exercise 4. Let $T_1 \in \mathscr{L}(\mathscr{S}_1, \mathscr{S}_2)$ and $T_2 \in \mathscr{L}(\mathscr{S}_2, \mathscr{S}_3)$. Show that

$$\text{Im}(T_2 T_1) = T_2(\text{Im } T_1)$$

and that

$$\text{rank}(T_2 T_1) \leq \min(\text{rank } T_1, \text{rank } T_2).$$

Hint. Use Eq. (1). Compare this with the result of Exercise 3.8.5. □

The notion of the rank of a transformation can be defined in a different way. Recall (Section 4.3) that all matrices from the equivalence class $\mathscr{A}_T$ have the same rank and so this can be viewed as a characteristic of the transformation T itself. It is thus natural to consider this characteristic as the *rank of T*. Note that both definitions are equivalent.

Theorem 2. *With the previous notation,*

$$\text{rank } T = \text{rank } A$$

for any $A \in \mathscr{A}_T$.

PROOF. Let A be the representation of $T \in \mathscr{L}(\mathscr{S}_1, \mathscr{S}_2)$ with respect to the bases $\mathscr{E} = \{\mathbf{y}_j\}_{j=1}^n$ and $\mathscr{G} = \{\mathbf{y}_i\}_{i=1}^m$ for $\mathscr{S}_1$ and $\mathscr{S}_2$, respectively. Since rank $T = \dim(\text{span}\{T(\mathbf{x}_j)_{j=1}^n)$, it follows from Proposition 3.7.1 that rank $T = \dim(\text{span}\{\boldsymbol{\beta}_j\}_{j=1}^n)$, where $\boldsymbol{\beta}_j$ $(1 \leq j \leq n)$ is the representation of the element $T(\mathbf{x}_j)$ with respect to the basis $\mathscr{G}$. Recalling that $A = [\boldsymbol{\beta}_1 \ \ \boldsymbol{\beta}_2 \ \ \cdots \ \ \boldsymbol{\beta}_n]$ and using the result of Theorem 3.7.2, the proof is completed. ■

Generalizing the definition of the kernel of a matrix, as introduced in Section 3.2, we now define the *kernel* of any transformation $T \in \mathscr{L}(\mathscr{S}_1, \mathscr{S}_2)$ by

$$\text{Ker } T \triangleq \{\mathbf{x} \in \mathscr{S}_1 : T(\mathbf{x}) = \mathbf{0}\}.$$

In the notation of Exercise 4.1.3 we see that for any matrix A, Ker A = Ker T_A.

Theorem 3. *Let* $T \in \mathscr{L}(\mathscr{S}_1, \mathscr{S}_2)$. *The set* Ker T *is a linear subspace in* $\mathscr{S}_1$.

PROOF. For any $\mathbf{x}_1, \mathbf{x}_2 \in$ Ker T and any $\alpha, \beta \in \mathscr{F}$, the element $\alpha\mathbf{x}_1 + \beta\mathbf{x}_2$ also belongs to the kernel of T:

$$T(\alpha\mathbf{x}_1 + \beta\mathbf{x}_2) = \alpha T(\mathbf{x}_1) + \beta T(\mathbf{x}_2) = \alpha\mathbf{0} + \beta\mathbf{0} = \mathbf{0}.$$

This completes the proof. ■

The dimension of the subspace Ker T is referred to as the *defect* of the transformation and is denoted by def T.

Exercise 5. Describe the kernel of the transformation defined in Exercise 4.2.1 and determine its defect.

SOLUTION. To find the elements $\mathbf{x} \in \mathscr{S}_1$ such that $T(\mathbf{x}) = \mathbf{0}$, we solve the system

$$\begin{bmatrix} x_2 + x_3 \\ x_1 + x_2 \\ x_1 \\ x_1 - x_2 \end{bmatrix} = \mathbf{0},$$

which gives $x_1 = x_2 = x_3 = 0$ and Ker $T = \{\mathbf{0}\}$. Thus, def $T = 0$.

Exercise 6. Describe the kernel and determine the defect of the transformation considered in Exercise 4.1.6.

Answer. $\operatorname{Ker} T = \operatorname{span}\{[0 \ \ 1 \ \ 1 \ \ 0]^{\mathrm{T}}, [1 \ \ 1 \ \ 0 \ \ -1]^{\mathrm{T}}\}$; $\operatorname{def} T = 2$. □

Theorem 4. *For any matrix representation A of a linear transformation T,*

$$\dim(\operatorname{Ker} T) = \dim(\operatorname{Ker} A).$$

PROOF. Let us use the notations introduced in the proof of Theorem 2. An element $\boldsymbol{x}$ belongs to $\operatorname{Ker} T$ if and only if its representation $\boldsymbol{\alpha} = [\alpha_1 \ \ \alpha_2 \ \ \cdots \ \ \alpha_n]^{\mathrm{T}}$ with respect to a basis $\mathscr{E}$ belongs to the kernel of $A_{\mathscr{E},\mathscr{G}}$. Indeed, if $\boldsymbol{x} = \sum_{j=1}^{n} \alpha_j \boldsymbol{x}_j$, then $\boldsymbol{x} \in \operatorname{Ker} T$ if and only if $\sum_{j=1}^{n} \alpha_j T(\boldsymbol{x}_j) = \boldsymbol{0}$ or, what is equivalent,

$$\sum_{j=1}^{n} \alpha_j \sum_{i=1}^{m} a_{ij}\boldsymbol{y}_i = \sum_{i=1}^{m} \left(\sum_{j=1}^{n} \alpha_j a_{ij}\right)\boldsymbol{y}_i = \boldsymbol{0}.$$

The linear independence of the elements $\boldsymbol{y}_1, \boldsymbol{y}_2, \ldots, \boldsymbol{y}_m$ from the basis $\mathscr{G}$ implies that $A_{\mathscr{E},\mathscr{G}}\boldsymbol{\alpha} = \boldsymbol{0}$, where $A_{\mathscr{E},\mathscr{G}} = [a_{ij}]_{i,j=1}^{m,n}$. It remains only to apply Proposition 3.7.1. ■

Using the notion of the defect for matrices, we say that $\operatorname{def} T = \operatorname{def} A$ for any A that is a representation of T. It should be emphasized that, in general, the subspaces $\operatorname{Im} T$ and $\operatorname{Im} A$, as well as $\operatorname{Ker} T$ and $\operatorname{Ker} A$, lie in different spaces and are essentially different.

Theorems 2 and 4, combined with the relation (3.8.1), yield the following useful result:

Theorem 5. *For any $T \in \mathscr{L}(\mathscr{S}_1, \mathscr{S}_2)$,*

$$\operatorname{rank} T + \operatorname{def} T = \dim \mathscr{S}_1. \tag{2}$$

This result is in fact a particular case of the next theorem, which is inspired by the observation that $\operatorname{Ker} T$ consists of all elements of $\mathscr{S}_1$ that are carried by T onto the zero element of $\mathscr{S}_2$. Hence, it is reasonable to expect that $\operatorname{Im} T$ is spanned by images of the basis elements in a complement to $\operatorname{Ker} T$.

Theorem 6. *Let $T \in \mathscr{L}(\mathscr{S}_1, \mathscr{S}_2)$ and let $\mathscr{S}_0$ be a direct complement to $\operatorname{Ker} T$ in $\mathscr{S}_1$:*

$$\mathscr{S}_1 = \operatorname{Ker} T \dotplus \mathscr{S}_0. \tag{3}$$

If $\boldsymbol{x}_1, \boldsymbol{x}_2, \ldots, \boldsymbol{x}_r$ form a basis for $\mathscr{S}_0$, then $T(\boldsymbol{x}_1), T(\boldsymbol{x}_2), \ldots, T(\boldsymbol{x}_r)$ form a basis for $\operatorname{Im} T$.

PROOF. We first observe that the elements $T(\boldsymbol{x}_1), \ldots, T(\boldsymbol{x}_r)$ are linearly independent. Indeed, if

$$\sum_{i=1}^{r} \alpha_i T(\boldsymbol{x}_i) = T\left(\sum_{i=1}^{r} \alpha_i \boldsymbol{x}_i\right) = \boldsymbol{0},$$

then the element $\sum_{i=1}^{r} \alpha_i \boldsymbol{x}_i$ is in $(\operatorname{Ker} T) \cap \mathscr{S}_0$ and hence is the zero element of $\mathscr{S}_1$. Now the linear independence of $\{\boldsymbol{x}_i\}_{i=1}^{r}$ yields $\alpha_i = 0$ $(i = 1, 2, \ldots, r)$ and the linear independence of $T(\boldsymbol{x}_1), \ldots, T(\boldsymbol{x}_r)$.

Furthermore, every element $\boldsymbol{y} \in \operatorname{Im} T$ can be represented as a linear combination of $T(\boldsymbol{x}), \ldots, T(\boldsymbol{x}_r)$. To see this, we first write $\boldsymbol{y} = T(\boldsymbol{x})$ for some $\boldsymbol{x} \in \mathscr{S}_1$. Let $\{\boldsymbol{x}_i'\}_{i=1}^{k}$ be a basis for Ker T, where $k = \operatorname{def} T$. If the basis representation of $\boldsymbol{x}$ is

$$\boldsymbol{x} = \sum_{i=1}^{k} \beta_i \boldsymbol{x}_i' + \sum_{i=1}^{r} \gamma_i \boldsymbol{x}_i,$$

we obtain $T(\boldsymbol{x}) = \sum_{i=1}^{r} \gamma_i T(\boldsymbol{x}_i)$, which completes the proof. ■

Exercise 7. Deduce Eq. (2) from Theorem 6. □

The assertion of Theorem 6 is seen from a different point of view in the next exercise.

Exercise 8. $T \in \mathscr{L}(\mathscr{S}_1, \mathscr{S}_2)$ and let $\operatorname{Im} T = \operatorname{span}\{\boldsymbol{y}_1, \boldsymbol{y}_2, \ldots, \boldsymbol{y}_r\}$, where $\boldsymbol{y}_i = T(\boldsymbol{x}_i)$ for $i = 1, 2, \ldots, r$. (We say $\boldsymbol{x}_i$ is the *preimage* of $\boldsymbol{y}_i$.) Confirm that $\mathscr{S}_0 = \operatorname{span}\{\boldsymbol{x}_1, \boldsymbol{x}_2, \ldots, \boldsymbol{x}_r)$ satisfies Eq. (3). □

Using the same technique as in the proof of Theorem 6, it is not difficult to prove the next exercise.

Exercise 9. If $T \in \mathscr{L}(\mathscr{S}_1, \mathscr{S}_2)$ and $\mathscr{S}$ is a linear subspace in $\mathscr{S}_1$, show that

$$\dim T(\mathscr{S}) = \dim \mathscr{S} - \dim(\mathscr{S} \cap \operatorname{Ker} T). \tag{4}$$

Hint. Extend a basis in $\mathscr{S} \cap \operatorname{Ker} T$ to a basis in $\mathscr{S}$. □

Note that Eq. (4) implies Eqs. (1) and (2) (for $\mathscr{S} = \mathscr{S}_1$) and also the inequality

$$\dim T(\mathscr{S}) \geq \dim \mathscr{S} - \operatorname{def} T. \tag{5}$$

The next exercises concern properties of sums and compositions of linear transformations.

Exercise 10. Let $T_1, T_2 \in \mathscr{L}(\mathscr{S}_1, \mathscr{S}_2)$. Check that

$$\operatorname{Ker} T_1 \cap \operatorname{Ker} T_2 \subset \operatorname{Ker}(T_1 + T_2).$$

Exercise 11. Let $T_1 \in \mathscr{L}(\mathscr{S}_1, \mathscr{S}_2)$ and $T_2 \in \mathscr{L}(\mathscr{S}_2, \mathscr{S}_3)$

(a) Show that

$$\operatorname{Ker} T_1 \subset \operatorname{Ker}(T_2 T_1) = \{\mathbf{x} \in \mathscr{S}_1 : T_1(\mathbf{x}) \in \operatorname{Ker} T_2\}.$$

(b) Show that $T_1(\operatorname{Ker} T_2 T_1) \subset \operatorname{Ker} T_2$ with equality if $\operatorname{Ker} T_2 \subset \operatorname{Im} T_1$.

(c) Apply the inequality (5) with $\mathcal{S} = \operatorname{Ker}(T_2 T_1)$ and $T = T_1$ to show that

$$\operatorname{def}(T_2 T_1) \leq \operatorname{def} T_1 + \operatorname{def} T_2. \tag{6}$$

Exercise 12. Check that for any $T \in \mathcal{L}(\mathcal{S})$,

$$\operatorname{Ker} T \subset \operatorname{Ker} T^2 \subset \operatorname{Ker} T^3 \subset \cdots,$$
$$\operatorname{Im} T \supset \operatorname{Im} T^2 \supset \operatorname{Im} T^3 \supset \cdots.$$

Exercise 13. Show that if

$$\operatorname{Ker} T^k = \operatorname{Ker} T^{k+1}, \tag{7}$$

then $\operatorname{Ker} T^j = \operatorname{Ker} T^{j+1}$ for all $j \geq k$.

SOLUTION. Suppose on the contrary that for some $j > k$, we have $\operatorname{Ker} T^j \neq \operatorname{Ker} T^{j+1}$. In other words, $T^{j+1}(\boldsymbol{x}) = \boldsymbol{0}$ but $T^j(\boldsymbol{x}) \neq \boldsymbol{0}$ for some $\boldsymbol{x} \in \mathcal{S}_1$. Writing $j = k + i$, we obtain $T^{k+1}(T^i(\boldsymbol{x})) = \boldsymbol{0}$ while $T^k(T^i(\boldsymbol{x})) \neq \boldsymbol{0}$. This contradicts the assumption that $\operatorname{Ker} T^k = \operatorname{Ker} T^{k+1}$. □

Since *strict* inclusion between subspaces, $\mathcal{S}_0 \subset \mathcal{S}_1$, implies $\dim \mathcal{S}_0 < \dim \mathcal{S}_1$, then, in a chain of subspaces of a finite-dimensional space $\mathcal{S}$ such that

$$\mathcal{S}_0 \subset \mathcal{S}_1 \subset \cdots \subset \mathcal{S}_k \subset \cdots \subset \mathcal{S},$$

there exists a positive integer $k \leq \dim \mathcal{S}$ such that $\mathcal{S}_k = \mathcal{S}_{k+1} = \cdots$. Using this observation and combining Exercises 12 and 13 will yield the conclusion of the next exercise.

Exercise 14. Check that for any $T \in \mathcal{L}(\mathcal{S})$,

$$\operatorname{Ker} T \subset \operatorname{Ker} T^2 \subset \cdots \subset \operatorname{Ker} T^k = \operatorname{Ker} T^{k+1} = \cdots \subset \mathcal{S},$$

where k denotes a positive integer between 1 and n, $n = \dim \mathcal{S}$, and the inclusions are strict. □

4.6 Invertible Transformations

A transformation $T \in \mathcal{L}(\mathcal{S}_1, \mathcal{S}_2)$ is said to be *left invertible* if there exists a transformation $T_1 \in \mathcal{L}(\mathcal{S}_2, \mathcal{S}_1)$ such that

$$T_1 T = I_1, \tag{1}$$

where I_1 denotes the identity transformation in $\mathcal{S}_1$. If Eq. (1) holds, then the transformation T_1 is called a *left inverse* of T.

Theorem 1. *Let* $T \in \mathscr{L}(\mathscr{S}_1, \mathscr{S}_2)$ *and let* $\dim \mathscr{S}_1 = n$, $\dim \mathscr{S}_2 = m$. *The following statements are equivalent*:

(a) *T is left invertible*;
(b) $\operatorname{Ker} T = \{\boldsymbol{0}\}$;
(c) *T defines an isomorphism between* $\mathscr{S}_1$ *and* $\operatorname{Im} T$;
(d) $n \le m$ *and* $\operatorname{rank} T = n$;
(e) *Every representation of T is an* $m \times n$ *matrix with* $n \le m$ *and full rank.*

PROOF. If Eq. (1) holds and $T(\boldsymbol{x}) = \boldsymbol{0}$, then $\boldsymbol{x} = I_1(\boldsymbol{x}) = T_1(T(\boldsymbol{x})) = \boldsymbol{0}$; hence $\operatorname{Ker} T = \{\boldsymbol{0}\}$ and $(a) \Rightarrow (b)$. To deduce (c) from (b) it suffices, in view of the linearity of T, to establish a one-to-one correspondence between $\mathscr{S}_1$ and $\operatorname{Im} T$. Let $\boldsymbol{y} = T(\boldsymbol{x}_1) = T(\boldsymbol{x}_2)$, where $\boldsymbol{x}_i \in \mathscr{S}_1$ $(i = 1, 2)$. Then $T(\boldsymbol{x}_1 - \boldsymbol{x}_2) = \boldsymbol{0}$ and the condition $\operatorname{Ker} T = \{\boldsymbol{0}\}$ yields $\boldsymbol{x}_1 = \boldsymbol{x}_2$ and, consequently, the required uniqueness of $\boldsymbol{x}$. To prove the implication $(c) \Rightarrow (a)$, we observe that the existence of a one-to-one map (namely, T) from $\mathscr{S}_1$ onto $\operatorname{Im} T$ permits the definition of the transformation $T_1 \in \mathscr{L}(\operatorname{Im} T, \mathscr{S}_1)$ by the rule: If $\boldsymbol{y} = T(\boldsymbol{x})$ then $T_1(\boldsymbol{y}) = \boldsymbol{x}$. Obviously $T_1(T(\boldsymbol{x})) = T_1(\boldsymbol{y}) = \boldsymbol{x}$ for all $\boldsymbol{x} \in \mathscr{S}_1$, or, what is equivalent, $T_1 T = I_1$. Thus, $(c) \Rightarrow (a)$. We have shown that (a), (b) and (c) are equivalent.

Now we note that the equivalence $(b) \Leftrightarrow (d)$ follows from Eq. (4.5.2), and Theorem 4.5.2 gives the equivalence $(d) \Leftrightarrow (e)$. This completes the proof. ■

Exercise 1. Check the equivalence of the following statements:

(a) $T \in \mathscr{L}(\mathscr{S}_1, \mathscr{S}_2)$ is left invertible.
(b) For any linear subspace $\mathscr{S} \subset \mathscr{S}_1$,

$$\dim T(\mathscr{S}) = \dim \mathscr{S},$$

(c) Any linearly independent system in $\mathscr{S}_1$ is carried by T into another linearly independent system in $\mathscr{S}_2$. □

The notion of right invertibility of a transformation is defined similarly by saying that $T \in \mathscr{L}(\mathscr{S}_1, \mathscr{S}_2)$ is *right invertible* if one can find a transformation $T_2 \in \mathscr{L}(\mathscr{S}_2, \mathscr{S}_1)$ such that

$$TT_2 = I_2, \tag{2}$$

where I_2 denotes the identity transformation in $\mathscr{S}_2$. The transformation T_2 in Eq. (2) is referred to as a *right inverse* of T.

Theorem 2. *Let* $T \in \mathscr{L}(\mathscr{S}_1, \mathscr{S}_2)$ *and let* $\dim \mathscr{S}_1 = n$, $\dim \mathscr{S}_2 = m$. *The following statements are equivalent*:

(a) *T is right invertible*;
(b) $\operatorname{rank} T = m$;

(*c*) *T defines an isomorphism between any direct complement to* Ker T and $\mathscr{S}_2$;

(*d*) $n \geq m$ *and* def $T = n - m$;

(*e*) *Every representation of T is an $m \times n$ matrix with $n \geq m$ and full rank.*

PROOF. (*a*) ⇒ (*b*). Let Eq. (2) hold and let $\mathbf{y} \in \mathscr{S}_2$. Then $\mathbf{y} = TT_2(\mathbf{y}) = T(T_2(\mathbf{y}))$ and the element $\mathbf{x} = T_2(\mathbf{y}) \in \mathscr{S}_1$ is a pre-image of $\mathbf{y}$. Thus $\mathscr{S}_2 = \text{Im } T$ and part (*b*) follows from part (*a*). To prove the implication (*b*) ⇒ (*c*), we consider a subspace $\mathscr{S}_0$ such that $\mathscr{S}_1 = \text{Ker } T \dotplus \mathscr{S}_0$ and show that T provides a one-to-one correspondence between $\mathscr{S}_0$ and $\mathscr{S}_2$. Indeed, if $\mathbf{y} = T(\mathbf{x}_1) = T(\mathbf{x}_2)$, where $\mathbf{x}_i \in \mathscr{S}_0$ $(i = 1, 2)$, then $\mathbf{x}_1 - \mathbf{x}_2 = \mathscr{S}_0 \cap \text{Ker } T = \{\mathbf{0}\}$. Hence $\mathbf{x}_1 = \mathbf{x}_2$ and the required isomorphism between $\mathscr{S}_0$ and $\mathscr{S}_2$ follows from the linearity of T. Now observe that this isomorphism guarantees the existence of a transformation $T_2 \in \mathscr{L}(\mathscr{S}_2, \mathscr{S}_0)$ defined by the rule:

$$\text{If } \mathbf{y} = T(\mathbf{x}) \text{ where } \mathbf{x} \in \mathscr{S}_0, \text{ then } T_2(\mathbf{y}) = \mathbf{x}.$$

Clearly, $T(T_2(\mathbf{y})) = T(\mathbf{x}) = \mathbf{y}$ for all $\mathbf{y} \in \mathscr{S}_2$ and the relation (2) holds. Thus, the equivalence of (*a*), (*b*), and (*c*) is established.

Note that the equivalences (*b*) ⇔ (*d*) and (*d*) ⇔ (*e*) are immediate consequences of Theorems 4.5.5 and 4.5.2, respectively. ■

The notions of left (or right) invertible matrices and their one-sided inverses are defined similarly to those for transformations. As pointed out in Section 2.6, a one-sided inverse of a *square* matrix is necessarily an inverse matrix, that is, both a left and right inverse.

Exercise 2. Show that a transformation $T \in \mathscr{L}(\mathscr{S}_1, \mathscr{S}_2)$ is left (respectively, right) invertible if and only if any representation A is a left (respectively, right) invertible matrix.

Hint. Use the correspondence (4.2.4). □

Theorems 1 and 2 applied to matrices state, in particular, that a rectangular $m \times n$ matrix of full rank is left invertible if $m \geq n$, and it is right invertible if $m \leq n$. The treatment of one-sided invertible matrices will be continued in the chapter on generalized inverses, while the invertibility properties of square matrices are discussed later in this section after some further preparations.

Let $T \in \mathscr{L}(\mathscr{S}_1, \mathscr{S}_2)$. The transformation T is said to be *invertible* if it is both left and right invertible. We combine Theorems 1 and 2 and the corollary to Theorem 3.7.1 to derive the next theorem.

Theorem 3. *Let* $T \in \mathscr{L}(\mathscr{S}_1\ \mathscr{S}_2)$ *and let* $\dim \mathscr{S}_1 = n$, $\dim \mathscr{S}_2 = m$. *The following statements are equivalent*:

(*a*) *T is invertible*;
(*b*) Ker $T = \{0\}$, rank $T = m = n$;
(*c*) *T defines an isomorphism between* $\mathscr{S}_1$ *and* $\mathscr{S}_2$;
(*d*) *Every representation of T is an* $n \times n$ *nonsingular matrix.*

Using Theorem 4.5.5, we deduce three corollaries.

Corollary 1. *Let* $T \in \mathscr{L}(\mathscr{S}_1, \mathscr{S}_2)$ *be left invertible. Then T is invertible if and only if* $\dim \mathscr{S}_1 = \dim \mathscr{S}_2$. *In particular, a left invertible transformation acting on* $\mathscr{S}$ *is invertible and defines a one-to-one map of* $\mathscr{S}$ *onto itself.*

Corollary 2. *Let* $T \in \mathscr{L}(\mathscr{S}_1, \mathscr{S}_2)$ *be right invertible. Then T is invertible if and only if* def $T = 0$. *In particular, a right invertible transformation acting on* $\mathscr{S}$ *is invertible and produces a one-to-one map of* $\mathscr{S}$ *onto itself.*

Corollary 3. *Any representation of an invertible transformation is an invertible matrix.*

Consider an invertible transformation $T \in \mathscr{L}(\mathscr{S}_1, \mathscr{S}_2)$. By definition there are transformations $T_1, T_2 \in \mathscr{L}(\mathscr{S}_2, \mathscr{S}_1)$ such that simultaneously Eqs. (1) and (2) hold. In this case, the equalities

$$T_1 = T_1 I_2 = T_1(TT_2) = (T_1 T)T_2 = I_1 T_2 = T_2$$

show that the one-sided inverses of T coincide. Hence there is a transformation $T^{-1} \in \mathscr{L}(\mathscr{S}_2, \mathscr{S}_1)$ satisfying the conditions

$$T^{-1}T = I_1 \qquad \text{and} \qquad TT^{-1} = I_2. \tag{3}$$

The transformation T^{-1} is called the *inverse* of T and is easily seen to be unique. It is clear, comparing Eq. (4.2.4) with Eq. (3), that if A is a representation of an invertible transformation $T \in \mathscr{L}(\mathscr{S}_1, \mathscr{S}_2)$ with respect to the bases $(\mathscr{E}, \mathscr{G})$, then A^{-1} is the representation of T^{-1} with respect to $(\mathscr{G}, \mathscr{E})$. In particular, if $T, T^{-1} \in \mathscr{L}(\mathscr{S})$, then their representations with respect to a fixed basis in $\mathscr{S}$ are correspondingly A and A^{-1}.

Note that part (*d*) of Theorem 3 asserts that a transformation T is invertible if and only if $\det A \neq 0$ for every $A \in \mathscr{A}_T$. When T acts on a space $\mathscr{S}$, then by Exercise 4.4.1(b) all its representations have the same determinant. In this case it is then reasonable to define the *determinant of the transformation* $T \in \mathscr{L}(\mathscr{S})$, written det T, as the determinant of any of its representations in $\mathscr{A}_T^0$. Hence we now conclude that a *transformation* $T \in \mathscr{L}(\mathscr{S})$ *is invertible if and only if* $\det T \neq 0$. This result for matrices was established in Section 2.6.

Exercise 3. Let $T_i: \mathscr{S}_i \xrightarrow{\text{onto}} \mathscr{S}_{i+1}$ $(i = 1, 2, \ldots, k)$ be invertible linear transformations. Show that $T = T_k T_{k-1} \cdots T_1 \in \mathscr{L}(\mathscr{S}_1, \mathscr{S}_{k+1})$ is also invertible and that $T^{-1} = T_1^{-1} T_2^{-1} \cdots T_k^{-1}$. □

The result in Exercise 3 can be reversed in the following sense.

Exercise 4. Let $T_i \in \mathscr{L}(\mathscr{S})$, $i = 1, 2, \ldots, k$. Check that all T_i are invertible if and only if, for any permutation $(j_1, j_2, \ldots, j_k)$ of $(1, 2, \ldots, k)$, the transformation $T_{j_1} T_{j_2} \cdots T_{j_k}$ is invertible.

Exercise 5. Check that if $\sum_{i=0}^{k} p_i T^i = 0$, $p_0 \neq 0$, and $T \in \mathscr{L}(\mathscr{S})$, then T is invertible. □

4.7 Restrictions, Invariant Subspaces, and Direct Sums of Transformations

Let $T \in \mathscr{L}(\mathscr{S}_1, \mathscr{S}_2)$ and let $\mathscr{S}_0$ be a linear subspace in $\mathscr{S}_1$. The transformation $\tilde{T} : \mathscr{S}_0 \to \mathscr{S}_2$ defined by $\tilde{T}(\mathbf{x}) = T(\mathbf{x})$ for every $\mathbf{x} \in \mathscr{S}_0$ is referred to as the *restriction* of T to $\mathscr{S}_0$ and is denoted by $T|_{\mathscr{S}_0}$. Obviously, $T|_{\mathscr{S}_0} \in \mathscr{L}(\mathscr{S}_0, \mathscr{S}_2)$. In contrast, a transformation $T \in \mathscr{L}(\mathscr{S}_1, \mathscr{S}_2)$ coinciding with $\tilde{T}$ on $\mathscr{S}_0 \subset \mathscr{S}_1$ is called an *extension of* $\tilde{T}$.

Exercise 1. Let $T \in \mathscr{L}(\mathscr{S}_1, \mathscr{S}_2)$. Check that for any linear subspace $\mathscr{S}_0 \subset \mathscr{S}_1$,

$$\operatorname{Im}(T|_{\mathscr{S}_0}) = T(\mathscr{S}_0), \qquad \operatorname{Ker}(T|_{\mathscr{S}_0}) = \mathscr{S}_0 \cap \operatorname{Ker} T: \quad \square$$

The second relation of Exercise 1 shows that if $\mathscr{S}_0$ is any direct complement to Ker T in $\mathscr{S}_1$ (thus $\mathscr{S}_0 \dotplus \operatorname{Ker} T = \mathscr{S}_1$), then $T|_{\mathscr{S}_0}$ has the trivial subspace $\{\mathbf{0}\}$ as its kernel. In this case, we deduce from Theorem 4.5.6 that $\operatorname{Im} T = T(\mathscr{S}_0)$. So the action of a linear transformation T is completely determined by its action on a direct complement to Ker T, that is, by the restriction of T to such a subspace.

We now consider an arbitrary decomposition of the domain into a direct sum of subspaces (as defined in Section 3.6).

Exercise 2. Let $T \in \mathscr{L}(\mathscr{S}_1, \mathscr{S}_2)$ and let $\mathscr{S}_1 = \mathscr{S}_0 \dotplus \tilde{\mathscr{S}}$. Write $\mathbf{x} = \mathbf{x}_1 + \mathbf{x}_2$ where $\mathbf{x}_1 \in \mathscr{S}_0$, $\mathbf{x}_2 \in \tilde{\mathscr{S}}$, and show that, if $T_1 \in \mathscr{L}(\mathscr{S}_0, \mathscr{S}_2)$ and $T_2 \in \mathscr{L}(\tilde{\mathscr{S}}, \mathscr{S}_2)$, the equality

$$T(\mathbf{x}) = T_1(\mathbf{x}_1) + T_2(\mathbf{x}_2) \tag{1}$$

holds for any $\mathbf{x} \in \mathscr{S}_1$ if and only if $T_1 = T|_{\mathscr{S}_0}$ and $T_2 = T|_{\tilde{\mathscr{S}}}$. □

A similar result can be easily formulated for a direct sum of several subspaces as defined in Section 3.6.

Suppose now that $\mathscr{S}_1 = \mathscr{S}_2$. Thus we investigate transformations $T \in \mathscr{L}(\mathscr{S})$. The following concept, which is important in its own right, is helpful in this context.

Let $T \in \mathscr{L}(\mathscr{S})$. A subspace $\mathscr{S}_0 \subset \mathscr{S}$ is called *invariant under* T (or *T-invariant*), if for any $\boldsymbol{x} \in \mathscr{S}_0$, the image $T(\boldsymbol{x})$ also belongs to $\mathscr{S}_0$:

$$T(\boldsymbol{x}) \in \mathscr{S}_0 \qquad \text{for all } \boldsymbol{x} \in \mathscr{S}_0. \tag{2}$$

In other words, if $\mathscr{S}_0$ is T-invariant, then every element $\boldsymbol{x} \in \mathscr{S}_0$ is transformed by T into an element of the same subspace $\mathscr{S}_0$. Hence the restriction $T|_{\mathscr{S}_0}$ in this case is a transformation acting on $\mathscr{S}_0$ and therefore can be viewed as a "part" of T in $\mathscr{S}_0$.

Obviously, $\{\boldsymbol{0}\}$ and the whole space are invariant under any transformation. These spaces are referred to as the *trivial* invariant subspaces and usually are excluded from further considerations.

The next exercise shows that linear transformations that are not invertible always have nontrivial invariant subspaces.

Exercise 3. Let $T \in \mathscr{L}(\mathscr{S})$. Show that Ker T and Im T are T-invariant. Show also that any subspace containing Im T is T-invariant.

SOLUTION. For the first subspace, let $\boldsymbol{x} \in$ Ker T. Then $T(\boldsymbol{x}) = \boldsymbol{0}$ and, since $\boldsymbol{0} \in$ Ker T, the condition (2) is valid. The proof of T-invariance of Im T is similar.

Exercise 4. Show that $\mathscr{S}_0$ is T-invariant if and only if $T(\mathbf{x}_i) \in \mathscr{S}_0$ ($i = 1, 2, \ldots, k$), where $\{\boldsymbol{x}_1, \boldsymbol{x}_2, \ldots, \boldsymbol{x}_k\}$ is a basis for $\mathscr{S}_0$.

Exercise 5. Check that if $T_1 T_2 = T_2 T_1$, then Ker T_2 and Im T_2 are T_1-invariant.

Exercise 6. Show that sums and intersections of T-invariant subspaces are T-invariant. □

The notion of an invariant subspace for a matrix A is obtained from that for transformations by treating A as the transformation T_A introduced in Exercise 4.1.3.

Proposition 1. *Let $T \in \mathscr{L}(\mathscr{S})$ and suppose that T has an invariant subspace in $\mathscr{S}$ of dimension k. Then there exists a matrix $A \in \mathscr{A}_T^0$ such that one of its invariant subspaces is also of dimension k.*

PROOF. Let the subspace $\mathscr{S}_0 \subset \mathscr{S}$ be T-invariant of dimension k. Let $\mathscr{E}_0$ be a basis in $\mathscr{S}_0$ and let the basis $\mathscr{E}$ in $\mathscr{S}$ be the union of $\mathscr{E}_0$ and a basis in a direct complement to $\mathscr{S}_0$ in $\mathscr{S}$. If $\boldsymbol{y} = T(\boldsymbol{x})$, $\boldsymbol{x} \in \mathscr{S}_0$, then, because of the invariance of $\mathscr{S}_0$ under T, the representation $\boldsymbol{\beta}$ of $\boldsymbol{y}$ with respect to $\mathscr{E}$ is of the form $\boldsymbol{\beta} = [\beta_1 \ \ \beta_2 \ \ \cdots \ \ \beta_k \ \ 0 \ \ \cdots \ \ 0]^{\mathrm{T}}$. The representation $\boldsymbol{\alpha}$ of $\boldsymbol{x}$ with respect to the *same* basis $\mathscr{E}$ is also of this form, and then Theorem 4.2.2

implies that the subspace of the representations of the elements of $\mathscr{S}_0$ with respect to $\mathscr{E}$ is $A_{\mathscr{E}}$-invariant. Clearly, the dimension of this subspace is equal to k. ■

Let $T \in \mathscr{L}(\mathscr{S})$ and let $\mathscr{S}_0 \subset \mathscr{S}$ be a T-invariant subspace. Denote by $\tilde{\mathscr{S}}$ a direct complement to $\mathscr{S}_0$ in $\mathscr{S}$: $\mathscr{S} = \mathscr{S}_0 \dotplus \tilde{\mathscr{S}}$. (For the existence of such a space $\tilde{\mathscr{S}}$, see Section 3.5.) Note that $\tilde{\mathscr{S}}$ is not necessarily T-invariant.

***Example* 7.** Let $T \in \mathscr{L}(\mathscr{F}^2)$ be defined as follows:

$$T\left(\begin{bmatrix} x_1 \\ x_2 \end{bmatrix}\right) = \begin{bmatrix} x_1 + x_2 \\ x_2 \end{bmatrix}.$$

Obviously, $\mathscr{S}_0 \triangleq \text{span}\{[1 \quad 0]^{\text{T}}\} = \{[\alpha \quad 0]^{\text{T}} : \alpha \in \mathscr{F}\}$ is T-invariant: $T([\alpha \quad 0]^{\text{T}}) = [\alpha \quad 0]^{\text{T}} \in \mathscr{S}_0$. However, the subspace $\tilde{\mathscr{S}} \triangleq \text{span}\{[0 \quad 1]^{\text{T}}\}$ is a direct complement to $\mathscr{S}_0$, but it fails to be T-invariant: $T([0 \quad 1]^{\text{T}}) = [1 \quad 1]^{\text{T}} \notin \tilde{\mathscr{S}}$. Observing that $\mathscr{S}_0$ is the only nontrivial T-invariant subspace of $\mathscr{F}^2$. we deduce that it is impossible to decompose $\mathscr{F}^2$ into a direct sum of T-invariant subspaces. □

If the space $\mathscr{S}$ can be decomposed into a direct sum of k subspaces $\mathscr{S}_i$, all of which are T-invariant, we may write, for every $\mathbf{x} \in \mathscr{S}$,

$$T(\mathbf{x}) = \sum_{i=1}^{k} {}^{\boldsymbol{\cdot}}\, T_i(\mathbf{x}_i), \tag{3}$$

where $\mathbf{x} = \sum_{i=1}^{k} \mathbf{x}_i$ ($\mathbf{x}_i \in \mathscr{S}_i$, $i = 1, 2, \ldots, k$) and $T_i = T|_{\mathscr{S}_i}$ for each i. (See the remark immediately following Exercise 2.) Note that when $k = 2$ Eq. (3) differs from Eq. (1) to the extent that we may now interpret T_i as being in $\mathscr{L}(\mathscr{S}_i)$ for $i = 1$ and 2. Bearing Eq. (3) in mind, we give the following definition.

Let $T \in \mathscr{L}(\mathscr{S})$ and let $\mathscr{S} = \sum_{i=1}^{k} {}^{\boldsymbol{\cdot}}\, \mathscr{S}_i$, where each $\mathscr{S}_i$ is T-invariant. Then the transformation T is referred to as *the direct sum* of the transformations $T_i = T|_{\mathscr{S}_i}$ ($i = 1, 2, \ldots, k$) and is written

$$T = \sum_{i=1}^{k} {}^{\boldsymbol{\cdot}}\, T_i \triangleq T_1 \dotplus T_2 \dotplus \cdots \dotplus T_k. \tag{4}$$

Note that Eq. (4) now implies Eq. (3) for every $\mathbf{x} \in \mathscr{S}$, and vice versa. The significance of the representation (4) lies in the possibility of reducing the study of the transformation T to that of its "parts" T_i acting on smaller subspaces. This is the basic strategy adopted in the next section for the analysis of the class of matrix representations of T.

***Exercise* 8.** Let $T \in \mathscr{L}(\mathscr{S})$ and $p(\lambda)$ be a scalar polynomial. Show that if a subspace is T-invariant, then it is $p(T)$-invariant. □

4.8 Direct Sums and Matrices

Let $T \in \mathscr{L}(\mathscr{S})$ and let $\mathscr{A}_T^0$ denote the equivalence class of similar matrices associated with T. In searching for the simplest matrices in this class, we first study the representations of T in the case when the space $\mathscr{S}$ is decomposed into a direct sum of subspaces, at least one of which is T-invariant.

Let $\mathscr{S} = \mathscr{S}_1 \dotplus \mathscr{S}_2$, where $\mathscr{S}_1$ is T-invariant, assuming for the moment that such a decomposition is always possible. Now we construct a representation of T that is most "appropriate" for the given decomposition of the space, namely, we build up the representation of T with respect to the basis

$$\{\mathbf{x}_1, \mathbf{x}_2, \ldots, \mathbf{x}_k, \mathbf{x}_{k+1}, \ldots, \mathbf{x}_n\} \tag{1}$$

in $\mathscr{S}$, where $k = \dim \mathscr{S}_1$ and the first k elements constitute a basis in $\mathscr{S}_1$. Since $\mathscr{S}_1$ is T-invariant, $T(\mathbf{x}_j) \in \mathscr{S}_1$ $(j = 1, 2, \ldots, k)$ and Eq. (4.2.1) gives

$$T(\mathbf{x}_j) = \sum_{i=1}^{k} \alpha_{ij}\mathbf{x}_i, \qquad j = 1, 2, \ldots, k. \tag{2}$$

Hence the matrix $A = [\alpha_{ij}]_{i,j=1}^n$, where the α_{ij} $(1 \le i, j \le n)$ are found from the equation $T(\mathbf{x}_j) = \sum_{i=1}^n \alpha_{ij}\mathbf{x}_i$ $(j = 1, 2, \ldots, n)$, is of the form:

$$A = \begin{bmatrix} A_1 & A_3 \\ 0 & A_2 \end{bmatrix}, \tag{3}$$

where $A_1 \in \mathscr{F}^{k \times k}$ and $A_2 \in \mathscr{F}^{(n-k)\times(n-k)}$. By definition, the matrix (3) is the representation of T with respect to the basis (1). Thus, a decomposition of the space into a direct sum of subspaces, one of which is T-invariant, admits a representation of T in block-triangular form (3).

Exercise 1. Let $T \in \mathscr{L}(\mathscr{S})$, where $\mathscr{S} = \mathscr{S}_1 \dotplus \mathscr{S}_2$ and $\mathscr{S}_1$ is T-invariant. Check that the representation of T with respect to a basis in which the basis elements of $\mathscr{S}_1$ are in the last places is a lower-block-triangular matrix. □

If $\mathscr{S} = \mathscr{S}_1 \dotplus \mathscr{S}_2$ and both subspaces $\mathscr{S}_1$ and $\mathscr{S}_2$ are T-invariant, then a simpler representation of T can be found. Namely, reasoning as before and constructing the representation of T with respect to the basis $\{\mathbf{x}_1, \mathbf{x}_2, \ldots, \mathbf{x}_k, \mathbf{x}_{k+1}, \ldots, \mathbf{x}_n\}$, in which the first k elements constitute a basis in $\mathscr{S}_1$ and the others generate a basis in $\mathscr{S}_2$, we obtain the matrix

$$A = \begin{bmatrix} A_1 & 0 \\ 0 & A_2 \end{bmatrix}, \tag{4}$$

where $A_1 \in \mathscr{F}^{k \times k}$ and $A_2 \in \mathscr{F}^{(n-k)\times(n-k)}$. Hence, if $T \in \mathscr{L}(\mathscr{S})$ can be decomposed into a direct sum of transformations $T_1 \in \mathscr{L}(\mathscr{S}_1)$ and $T_2 \in \mathscr{L}(\mathscr{S}_2)$, that is, $T = T_1 \dotplus T_2$, then there is a representation A of T in a block-diagonal form (4), where obviously A_1 and A_2 are representations of T_1 and T_2, respectively. Note that, conversely, if there exists a matrix A in $\mathscr{A}_T^0$ of the form (4), then there is a basis in $\mathscr{S}$ such that the Eq. (2) holds and

$$T(\mathbf{x}_j) = \sum_{i=k+1}^{n} \alpha_{ij}\mathbf{x}_i \qquad j = k+1, k+2 \ldots, n.$$

In view of Exercise 4.7.4, the subspaces $\mathscr{S}_1 = \text{span}\{\mathbf{x}_j\}_{j=1}^k$ and $\mathscr{S}_2 = \text{span}\,\{\mathbf{x}_j\}_{j=k+1}^n$ are T-invariant and, obviously, $\mathscr{S}_1 \dotplus \mathscr{S}_2 = \mathscr{S}$. By analogy, the matrix A in (4) is said to be a *direct sum* of the matrices A_1 and A_2 and is written $A = A_1 \dotplus A_2$. A similar definition and notation, $A = A_1 \dotplus A_2 \dotplus \cdots \dotplus A_p$ or, briefly, $A = \sum_{i=1}^{p}{}^{\boldsymbol{\cdot}} A_i$, is given for the direct sum of p matrices. We summarize this discussion.

Theorem 1. *A transformation $T \in \mathscr{L}(\mathscr{S})$ has a representation $A \in \mathscr{A}_T^0$ in block-diagonal form, consisting of p blocks, if and only if T can be decomposed into a direct sum of p linear transformations.*

In symbols, if $A \in \mathscr{A}_T^0$, then $A = \sum_{i=1}^{p}{}^{\boldsymbol{\cdot}}\ A_i$ if and only if $T = \sum_{i=1}^{p}{}^{\boldsymbol{\cdot}}\ T_i$. In this case the matrix A_i is the representation of T_i $(1 \le i \le p)$.

Recall that Example 4.7.7 provides an example of a transformation T from $\mathscr{L}(\mathscr{F}^2)$ that cannot be written $T = T_1 \dotplus T_2$. Hence, there are linear transformations having no block-diagonal representations (with more than one block). Such transformations can be referred to as *one-block transformations.* In contrast, a particularly important case arises when a transformation from $\mathscr{L}(\mathscr{S})$ can be decomposed into a direct sum of transformations acting on one-dimensional spaces. Such transformations are called *simple.* Note the following consequence from Theorem 1 regarding simple transformations.

Theorem 2. *Let $\mathscr{A}_T^0$ denote the equivalence class of matrix representations of the transformation $T \in \mathscr{L}(\mathscr{S})$. Any matrix $A \in \mathscr{A}_T^0$ is similar to a diagonal matrix D if and only if T is simple.*

By analogy, a matrix $A \in \mathscr{F}^{n \times n}$ is *simple* if it is a representation of a simple transformation T acting on an n-dimensional space. Using this notion, the previous result can be partially restated.

Theorem 2′. *Any matrix $A \in \mathscr{F}^{n \times n}$ is simple if and only if it is similar to a diagonal matrix D.*

The meaning of the diagonal elements in D, as well as the structure of the matrix that transforms A into D, will be revealed later (Theorem 4.10.2), after the effect of a linear transformation on a one-dimensional space is studied.

Exercise 2. Check that any circulant is a simple matrix.

Exercise 3. Show that if the nonsingular matrix A is simple, so are A^{-1} and adj A.

Exercise 4. Let $A \in \mathcal{F}^{n \times n}$ and let there be a k-dimensional subspace of $\mathcal{F}^n$ that is A-invariant $(1 \leq k < n)$. Show that the relation

$$A = P\begin{bmatrix} A_1 & A_3 \\ 0 & A_2 \end{bmatrix} P^{-1} \tag{5}$$

is valid for some $A_1 \in \mathcal{F}^{k \times k}$ and a nonsingular $P \in \mathcal{F}^{n \times n}$.

Hint. Choose a basis $\boldsymbol{b}_1, \boldsymbol{b}_2, \ldots, \boldsymbol{b}_n$ for $\mathcal{F}^n$ by extending a basis $\boldsymbol{b}_1, \ldots, \boldsymbol{b}_k$ for the A-invariant subspace. Then define P to be the matrix whose columns are $\boldsymbol{b}_1, \boldsymbol{b}_2, \ldots, \boldsymbol{b}_n$.

4.9 Eigenvalues and Eigenvectors of a Transformation

Let $T \in \mathcal{L}(\mathcal{S})$. We start the investigation of one-dimensional T-invariant subspaces with a description of their elements in terms of T.

Proposition 1. *Let $T \in \mathcal{L}(\mathcal{S})$ and let $\mathcal{S}_0 \subset \mathcal{S}$ be a T-invariant subspace of dimension* 1. *Then for all nonzero $\boldsymbol{x} \in \mathcal{S}_0$ there is a scalar $\lambda \in \mathcal{F}$ (independent of $\boldsymbol{x}$), such that*

$$T(\boldsymbol{x}) = \lambda \boldsymbol{x}. \tag{1}$$

PROOF. Since $\mathcal{S}_0$ is one-dimensional, it is of the form

$$\mathcal{S}_0 = \{\alpha \boldsymbol{x}_0 : \boldsymbol{0} \neq \boldsymbol{x}_0 \in \mathcal{S}_0 \text{ and } \alpha \in \mathcal{F}\}.$$

The T-invariance of $\mathcal{S}_0$ means $T(\boldsymbol{x}) \in \mathcal{S}_0$ for every nonzero $\boldsymbol{x} \in \mathcal{S}_0$ and, therefore, representing $\boldsymbol{x}$ as $\alpha_0 \boldsymbol{x}_0$, we have $T(\alpha_0 \boldsymbol{x}_0) = \alpha \boldsymbol{x}_0$. Thus, the condition (1) holds for $\lambda = \alpha/\alpha_0$ $(\alpha_0 \neq 0)$. If $\boldsymbol{x}_1$ denotes another element from $\mathcal{S}_0$, then $\boldsymbol{x}_1 = \gamma \boldsymbol{x}_0$ for some $\gamma \in \mathcal{F}$ and $T(\boldsymbol{x}_1) = \gamma T(\boldsymbol{x}_0) = \gamma \lambda \boldsymbol{x}_0 = \lambda \boldsymbol{x}_1$, which shows the uniqueness of λ in Eq. (1). ■

Note that a one-dimensional T-invariant subspace exists provided Eq. (1) holds true for at least one nonzero element $\boldsymbol{x}$.

Exercise 1. Check that if for some nonzero $\boldsymbol{x}_0 \in \mathcal{S}$,

$$T(\boldsymbol{x}_0) = \lambda \boldsymbol{x}_0, \tag{2}$$

then the subspace $\mathcal{S}_0 = \text{span}\{\boldsymbol{x}_0\}$ is T-invariant. □

Thus, the nonzero elements x satisfying Eq. (2) play a crucial role in the problem of existence and construction of one dimensional T-invariant subspaces. We need a formal definition. If $T \in \mathscr{L}(\mathscr{S})$, a nonzero element $x_0 \in \mathscr{S}$ satisfying Eq. (2) is said to be an *eigenvector* of T corresponding to the *eigenvalue* λ of T. Thus, the existence of T-invariant subspaces of dimension 1 is equivalent to the existence of eigenvectors of T.

If we now observe that Eq. (2) is equivalent to

$$(T - \lambda I)x_0 = 0, \qquad x_0 \neq 0,$$

we can characterize eigenvalues in another way.

***Exercise* 2.** Show that $\lambda \in \mathscr{F}$ is an eigenvalue of T if and only if the transformation $T - \lambda I$ fails to be invertible. □

The next proposition uses the concept of an algebraically closed field, which is introduced in Appendix 1. For our purposes, it is important to note that $\mathbb{C}$ is algebraically closed and $\mathbb{R}$ is not.

Proposition 2. *Let $\mathscr{S}$ be a finite-dimensional linear space over an algebraically closed field. Then any linear transformation $T \in \mathscr{L}(\mathscr{S})$ has at least one eigenvalue.*

PROOF. By virtue of Exercise 4.1.12, there is a monic polynomial $p(\lambda)$ with coefficients from $\mathscr{F}$ such that $p(T) = 0$. Since $\mathscr{F}$ is an algebraically closed field, there exists a decomposition of $p(\lambda)$ into a product of linear factors:

$$p(\lambda) = \sum_{i=1}^{l} (\lambda - \lambda_i),$$

where $\lambda_i \in \mathscr{F}$; $(i = 1, 2, \ldots, l)$. Then $p(T) = \prod_{i=1}^{l} (T - \lambda_i I)$ is the zero transformation, hence at least one of the factors $T - \lambda_i I$ $(1 \leq i \leq l)$ must be noninvertible (see Exercise 4.6.4). Now the required assertion follows from Exercise 2. ■

Obviously, if x is an eigenvector of T corresponding to λ, then it follows from $T(\alpha x) = \lambda(\alpha x)$ for any $\alpha \in \mathscr{F}$ that any nonzero element from the one-dimensional space spanned by x is an eigenvector of T corresponding to the same eigenvalue λ. This also follows from the results of Proposition 1 and Exercise 2. Moreover, every nonzero linear combination of (linearly dependent or independent) eigenvectors corresponding to the same eigenvalue λ is an eigenvector associated with λ. Indeed, if $T(x_i) = \lambda x_i$ $(x_i \neq x,\ i = 1, 2, \ldots, k)$, then the linearity of T implies

$$T\left(\sum_{i=1}^{k} \alpha_i x_i\right) = \sum_{i=1}^{k} \alpha_i T(x_i) = \lambda\left(\sum_{i=1}^{k} \alpha_i x_i\right),$$

and if $\sum_{i=1}^{k} \alpha_i x_i \neq 0$, it is an eigenvector of T corresponding to λ.

Consider the linear hull $\mathscr{S}_0$ of all eigenvectors of T corresponding to λ. In view of the last observation, every nonzero element of this subspace is an eigenvector of T associated with λ and hence it makes sense to call $\mathscr{S}_0$ the *eigenspace* of T corresponding to the eigenvalue λ of the transformation.

Exercise 3. Let λ denote an eigenvalue of $T \in \mathscr{L}(\mathscr{S})$. Show that the eigenspace associated with λ coincides with the subspace $\operatorname{Ker}(T - \lambda I)$.

Exercise 4. Find eigenvectors, eigenspaces, and eigenvalues of the transformation $T \in \mathscr{L}(\mathbb{R}^3)$ defined as follows:

$$T\left(\begin{bmatrix} x_1 \\ x_2 \\ x_3 \end{bmatrix}\right) = \begin{bmatrix} 2x_1 \\ x_1 \\ x_2 + x_3 \end{bmatrix}.$$

SOLUTION Solving the system $T(\boldsymbol{x}) = \lambda \boldsymbol{x}$, which becomes in this case

$$2x_1 = \lambda x_1, \qquad x_1 = \lambda x_2, \qquad x_2 + x_3 = \lambda x_3,$$

we obtain

$$\begin{aligned} \lambda &= 2 \quad \text{and} \quad x_1 = 2x_2 = 2x_3, \\ \lambda &= 0 \quad \text{and} \quad x_1 = 0, \ \ x_2 = -x_3, \\ \lambda &= 1 \quad \text{and} \quad x_1 = 0, \ \ x_2 = 0. \end{aligned}$$

Thus, the transformation has three one-dimensional eigenspaces spanned by the elements $[2 \ \ 1 \ \ 1]^{\mathrm{T}}$, $[0 \ \ 1 \ \ -1]^{\mathrm{T}}$, $[0 \ \ 0 \ \ 1]^{\mathrm{T}}$ and corresponding to the eigenvalues 2, 0, and 1, respectively. □

It is important to note that, in contrast to Exercise 4, a transformation acting on an n-dimensional space may have less than n distinct eigenvalues.

Exercise 5. Check that the transformation $T \in \mathscr{L}(\mathscr{F}^2)$ defined in Example 4.7.7 has only one eigenvalue, $\lambda = 1$, associated with the eigenspace spanned by $[1 \ \ 0]^{\mathrm{T}}$. □

Furthermore, it may happen that the transformation has two or more linearly independent eigenvectors corresponding to the same eigenvalue, as the next exercise shows.

Exercise 6. Check that the transformation $T \in \mathscr{L}(\mathscr{F}^3)$ defined by

$$T\left(\begin{bmatrix} x_1 \\ x_2 \\ x_3 \end{bmatrix}\right) = \begin{bmatrix} 2x_1 + x_3 \\ 2x_2 \\ 3x_3 \end{bmatrix}$$

has eigenvalues $\lambda_1 = \lambda_2 = 2$ and $\lambda_3 = 3$ associated with eigenspaces $\operatorname{span}\{[1 \ \ 0 \ \ 0]^{\mathrm{T}}, [0 \ \ 1 \ \ 0]^{\mathrm{T}}\}$ and $\operatorname{span}\{[1 \ \ 0 \ \ 1]^{\mathrm{T}}\}$, respectively. □

Thus, two eigenvectors of T corresponding to the same eigenvalue may be linearly dependent or independent. However, the situation is clear when the eigenvalues are distinct.

Theorem 1. *Eigenvectors corresponding to distinct eigenvalues are linearly independent.*

PROOF. Let $\lambda_1, \lambda_2, \ldots, \lambda_s$ be the distinct eigenvalues of the transformation $T \in \mathscr{L}(\mathscr{S})$ and let $\boldsymbol{x}_1, \boldsymbol{x}_2, \ldots, \boldsymbol{x}_s$ denote corresponding eigenvectors.

Suppose that there exist numbers $\alpha_1, \alpha_2, \ldots, \alpha_s$ such that

$$\sum_{i=1}^{s} \alpha_i \boldsymbol{x}_i = \boldsymbol{0}. \tag{3}$$

To prove that $\alpha_i = 0$ $(i = 1, 2, \ldots, s)$, we first multiply both sides of Eq. (3) on the left by $T - \lambda_1 I$, noting that $(T - \lambda_1 I)\boldsymbol{x}_1 = \boldsymbol{0}$ and $T(\boldsymbol{x}_i) = \lambda_i \boldsymbol{x}_i$, $i = 1, 2, \ldots, s$. Then

$$\sum_{i=2}^{s} (\lambda_1 - \lambda_i)\alpha_i \boldsymbol{x}_i = \boldsymbol{0}. \tag{4}$$

After premultiplication by $T - \lambda_2 I$, Eq. (4) becomes

$$\sum_{i=3}^{s} (\lambda_1 - \lambda_i)(\lambda_2 - \lambda_i)\alpha_i \boldsymbol{x}_i = \boldsymbol{0}.$$

After $s - 1$ such operations we finally have

$$(\lambda_1 - \lambda_s)(\lambda_2 - \lambda_s) \cdots (\lambda_{s-1} - \lambda_s)\alpha_s \boldsymbol{x}_s = \boldsymbol{0}.$$

which implies $\alpha_s = 0$. But the ordering of the eigenvalues and eigenvectors is arbitrary, so we can also prove $\alpha_1 = \alpha_2 = \cdots = \alpha_{s-1} = 0$. Hence the elements $\boldsymbol{x}_1, \boldsymbol{x}_2, \ldots, \boldsymbol{x}_s$ are linearly independent. ■

The *spectrum* of a transformation $T \in \mathscr{L}(\mathscr{S})$, written $\sigma(T)$, is defined as the set of all distinct eigenvalues of T. Note that Proposition 2 can be restated in this terminology as follows: Any linear transformation acting on a finite-dimensional space over an algebraically closed field has a nonempty spectrum. Now Theorem 1 implies a limit to the size of the spectrum.

Corollary 1. *The spectrum of a linear transformation acting on an n-dimensional space consists of at most n distinct eigenvalues.*

PROOF. If $T \in \mathscr{L}(\mathscr{S})$ and T has k distinct eigenvalues, $k > n = \dim \mathscr{S}$, then the n-dimensional space $\mathscr{S}$ has more than n linearly independent elements (eigenvectors of T, in this case). This is a contradiction. ■

Corollary 2. *Let $T \in \mathcal{L}(\mathcal{S})$ and $\dim \mathcal{S} = n$. If T has n distinct eigenvalues, then there is a basis for $\mathcal{S}$ consisting of eigenvectors of T.*

In this event we also say that the eigenvectors of T *span* the space $\mathcal{S}$ or that T has an *eigenbasis* in $\mathcal{S}$. Note that Exercise 6 shows that a transformation with $n_0 < n$ distinct eigenvalues can still have an eigenbasis. On the other hand, not every transformation (see Exercise 5) has enough eigenvectors to span the space. It turns out that the existence of an eigenbasis of T in $\mathcal{S}$ is equivalent to saying that T is a simple transformation.

Theorem 2. *A transformation $T \in \mathcal{L}(\mathcal{S})$ is simple if and only if it has an eigenbasis in $\mathcal{S}$.*

PROOF. If $\{\boldsymbol{x}_j\}_{j=1}^n$ is an eigenbasis in $\mathcal{S}$, then, putting $\mathcal{S}_i = \operatorname{span}\{\boldsymbol{x}_i\}$ for $1 \le i \le n$, we obtain the decomposition $T = \sum_{i=1}^{n} \dot{} \, T_i$, where T_i acts on $\mathcal{S}_i$ and $\dim \mathcal{S}_i = 1$ $(1 \le i \le n)$. Hence T is simple. Conversely, let $\mathcal{S} = \sum_{i=1}^{n} \dot{} \, \mathcal{S}_i$, where subspaces $\mathcal{S}_i$ of dimension one are T-invariant, $i = 1, 2, \ldots, n$. By Proposition 1, any nonzero element of $\mathcal{S}_i$, say $\boldsymbol{x}_i$, is an eigenvector of T, $1 \le i \le n$. Clearly, the set of all such $\boldsymbol{x}_i$ $(i = 1, 2, \ldots, n)$ gives an eigenbasis of T in $\mathcal{S}$. ■

In view of Theorem 2, Corollary 2 provides a sufficient (but not necessary) condition for T to be simple (see Exercise 5).

Exercise 7. Check that the transformation $T \in \mathcal{L}(\mathcal{S})$ is invertible if and only if all eigenvalues of T are nonzero.

Exercise 8. Prove that each T-invariant subspace $\mathcal{S}_0$ contains at least one eigenvector of T (the field is algebraically closed).

Hint. Consider $T|_{\mathcal{S}_0}$ and Proposition 2.

Exercise 9. Let T be a simple transformation from $\mathcal{L}(\mathcal{S})$. Show that

(a) Im T is spanned by the eigenvectors of T corresponding to the nonzero eigenvalues,

(b) $\operatorname{Im} T \cap \operatorname{Ker} T = \{\boldsymbol{0}\}$.

Exercise 10. Let $T \in \mathcal{L}(\mathcal{S})$ and $\sigma(T) = \{\lambda_1, \lambda_2, \ldots, \lambda_s\}$. Show that if $\mathcal{S}_j$ denotes the eigenspace of T associated with λ_j, then

(a) The sum $\sum_{j=1}^{s} \mathcal{S}_j$ is direct;

(b) The transformation T is simple if and only if

$$\sum_{j=1}^{s} \cdot \mathcal{S}_j = \mathcal{S}. \quad \square$$

4.10 Eigenvalues and Eigenvectors of a Matrix

Let $A \in \mathscr{F}^{n \times n}$. The matrix A can be viewed as a linear transformation acting in $\mathscr{F}^n$ (see Exercise 4.1.3). Hence we define a nonzero vector $\boldsymbol{\alpha} \in \mathscr{F}^n$ to be an *eigenvector* of A corresponding to the *eigenvalue* λ if

$$A\boldsymbol{\alpha} = \lambda\boldsymbol{\alpha}, \qquad \boldsymbol{\alpha} \neq \boldsymbol{0}. \tag{1}$$

The set of all distinct eigenvalues of A is called the *spectrum* of A and is denoted by $\sigma(A)$. Observe that the spectrum is invariant under similarity transformations.

Proposition 1. *Similar matrices have the same spectrum.*

PROOF. Let Eq. (1) hold and suppose $A = PBP^{-1}$. Then

$$B(P^{-1}\boldsymbol{\alpha}) = P^{-1}A\alpha = \lambda(P^{-1}\boldsymbol{\alpha})$$

and since $\boldsymbol{\alpha} \neq \boldsymbol{0}$, the vector $P^{-1}\boldsymbol{\alpha}$ is an eigenvector of B corresponding to the eigenvalue λ. Hence $\lambda \in \sigma(B)$ and $\sigma(A) \subset \sigma(B)$. The reverse inclusion is shown similarly. ■

Note that, in fact, we proved more.

Proposition 2. *If $\boldsymbol{\alpha}$ is an eigenvector of A associated with the eigenvalue λ, then $\boldsymbol{\alpha}' = P^{-1}\boldsymbol{\alpha}$ is an eigenvector of the matrix $B = P^{-1}AP$ corresponding to the same eigenvalue λ.*

Thus, all matrices from the equivalence class $\mathscr{A}_T^0$ of matrix representations for a transformation $T \in \mathscr{L}(\mathscr{S})$ have the same eigenvalues associated with eigenvectors of the form $P^{-1}\mathbf{x}$, where P is a transforming matrix. Hence, in particular, the spectrum of the matrices from $\mathscr{A}_T^0$ is uniquely determined by the transformation T. Consequently, a strong connection between the spectrum of T and the representations of T is to be expected.

Theorem 1. *Let $T \in \mathscr{L}(\mathscr{S})$ and let $\mathscr{A}_T^0$ denote the corresponding equivalence class of matrix representations for T. Then*

$$\sigma(T) = \sigma(A) \tag{2}$$

for any $A \in \mathscr{A}_T^0$.

PROOF. In view of Proposition 1; it suffices to show that Eq. (2) holds for at least one matrix A from $\mathscr{A}_T^0$. Let $\lambda \in \sigma(T)$ and

$$T(\boldsymbol{x}) = \lambda\boldsymbol{x}, \qquad \boldsymbol{x} \neq \boldsymbol{0}. \tag{3}$$

If $\boldsymbol{\alpha} \in \mathscr{F}^n$ is a representation of $\boldsymbol{x}$ with respect to the same basis in $\mathscr{S}$ in which A is a representation of T, then Theorem 4.2.2 and the isomorphism between elements and their representations show that the relation (3) is equivalent to $A\boldsymbol{\alpha} = \lambda\boldsymbol{\alpha}$. Note that $\boldsymbol{\alpha} \neq \boldsymbol{0}$ since $\boldsymbol{x} \neq \boldsymbol{0}$ and, therefore, $\lambda \in \sigma(A)$. ■

Observe that we have also found a relation between the eigenvectors of a matrix and the eigenvectors of the transformation which the matrix represents.

Proposition 3. *An element $\boldsymbol{x} \in \mathscr{S}$ is an eigenvector of $T \in \mathscr{L}(\mathscr{S})$ associated with $\lambda \in \sigma(T)$ if and only if the representation $\boldsymbol{\alpha} \in \mathscr{F}^n$ of $\boldsymbol{x}$ with respect to a basis in $\mathscr{S}$ is an eigenvector of the matrix representation of T with respect to the same basis.*

By varying the bases in $\mathscr{S}$, we obtain a set of vectors $\{P^{-1}\boldsymbol{\alpha}\colon \det P \neq 0\}$ that are eigenvectors of the corresponding representations of T (see Proposition 2). Note also that Theorem 1 and Proposition 3 provide an easy transition from the eigenvalues and eigenvectors of a transformation to those of matrices. In particular, they admit an analogous reformulation for matrices of the results concerning eigenvalues and eigenvectors of transformations. For instance, recalling that the representations of linearly independent elements also are linearly independent (see Exercise 3.7.2), we obtain an analog of Theorem 4.9.1.

Proposition 4. *Eigenvectors of a matrix corresponding to distinct eigenvalues are linearly independent.*

We now use this together with Proposition 4.9.2 and Theorem 4.9.2.

Proposition 5. *The spectrum $\sigma(A)$ of a matrix $A \in \mathbb{C}^{n \times n}$ is not empty and consists of at most n distinct eigenvalues. If A has n distinct eigenvalues, then it is simple.*

We can now complement the statement of Theorem 4.8.2′ by indicating the structure of the diagonal matrix D and the corresponding transforming matrix P.

Theorem 2. *Let $A \in \mathscr{F}^{n \times n}$ be a simple matrix and let D be the diagonal matrix occurring in Theorem 4.8.2′:*

$$A = PDP^{-1}, \qquad \det P \neq 0. \tag{4}$$

Then $D = \operatorname{diag}[\lambda_j]_{j=1}^n$ with $\lambda_j \in \sigma(A)$, $j = 1, 2, \ldots, n$, and

$$P = [\boldsymbol{x}_1 \quad \boldsymbol{x}_2 \quad \cdots \quad \boldsymbol{x}_n],$$

in which $\boldsymbol{x}_j$ $(1 \leq j \leq n)$ is an eigenvector of A corresponding to the eigenvalue λ_j $(1 \leq j \leq n)$.

PROOF. Rewriting Eq. (4) in the form $AP = PD$ and comparing the columns of the matrices AP and PD, it is easily found that

$$A\mathbf{x}_j = \lambda_j \mathbf{x}_j, \qquad j = 1, 2, \ldots, n, \tag{5}$$

where $\mathbf{x}_j$ $(1 \le j \le n)$ is the jth column of P and λ_j $(1 \le j \le n)$ is the jth diagonal element of D. Since P is nonsingular, none of its columns can be the zero element in $\mathscr{F}^n$ and hence $\mathbf{x}_j$ in (5) is an eigenvector of A corresponding to λ_j. ■

Theorem 2 can be reformulated in an illuminating way. Define the matrix $Q = (P^{-1})^{\mathrm{T}}$ and write $Q = [\mathbf{y} \quad \mathbf{y}_2 \quad \cdots \quad \mathbf{y}_n]$. Then $Q^{\mathrm{T}}P = P^{-1}P = I$ and, comparing elements, we obtain

$$\mathbf{y}_j^{\mathrm{T}} \mathbf{x}_k = \delta_{jk} \tag{6}$$

for $1 \le j, k \le n$. Note that in the sense of Section 3.13, the systems $\{\mathbf{x}_j\}_{j=1}^n$ and $\{\bar{\mathbf{y}}_j\}_{j=1}^n$ are biorthogonal with respect to the standard inner product in $\mathbb{C}^n$.

Take transposes in Eq. (4) and we have $A^{\mathrm{T}} = QDQ^{-1}$, so that

$$A^{\mathrm{T}}Q = QD.$$

This implies $A^{\mathrm{T}}\mathbf{y}_j = \lambda_j \mathbf{y}_j$ for $j = 1, 2, \ldots, n$. Thus, the columns of $(P^{-1})^{\mathrm{T}}$ are eigenvectors for A^{T}. They are also known as *left eigenvectors* of A since they satisfy $\mathbf{y}_j^{\mathrm{T}} A = \lambda_j \mathbf{y}_j^{\mathrm{T}}$.

Now, to recast Theorem 2, observe that Eq. (4) implies, using Eq. (1.3.6),

$$A = PDQ^{\mathrm{T}} = [\mathbf{x}_1 \quad \mathbf{x}_2 \quad \cdots \quad \mathbf{x}_n] \begin{bmatrix} \lambda_1 \mathbf{y}_1^{\mathrm{T}} \\ \vdots \\ \lambda_n \mathbf{y}_n^{\mathrm{T}} \end{bmatrix}$$

$$= \sum_{j=1}^{n} \lambda_j \mathbf{x}_j \mathbf{y}_j^{\mathrm{T}}.$$

Thus, A is expressed as a sum of matrices of rank one (or zero if $\lambda_j = 0$), and each of these is defined in terms of the *spectral properties* of A, that is, in terms of its eigenvalues, eigenvectors, and left eigenvectors. Consequently, the next result is called a *spectral theorem.*

Theorem 3. *Let $A \in \mathscr{F}^{n \times n}$ be a simple matrix with eigenvalues $\lambda_1, \ldots, \lambda_n$ and associated eigenvectors $\mathbf{x}_1, \mathbf{x}_2, \ldots, \mathbf{x}_n$. Then there are left eigenvectors $\mathbf{y}_1, \mathbf{y}_2, \ldots, \mathbf{y}_n$ for which $\mathbf{y}_j^{\mathrm{T}} \mathbf{x}_k = \delta_{jk}$, $1 \le j\ k \le n$, and*

$$A = \sum_{j=1}^{n} \lambda_j \mathbf{x}_j \mathbf{y}_j^{\mathrm{T}}. \tag{7}$$

Exercise 1. Consider a matrix $A \in \mathscr{F}^{n \times n}$ with eigenvectors $x_1, x_2, \ldots, x_n$ corresponding to the eigenvalues $\lambda_1, \lambda_2, \ldots, \lambda_n$, respectively. Show that any subspace of $\mathscr{F}^n$ generated by a set of eigenvectors

$$x_{j_1}, x_{j_2}, \ldots, x_{j_k} (1 \leq j_1 < j_2 < \cdots < j_k \leq n)$$

of A is A-invariant.

Exercise 2. Show that the spectrum of the square matrix

$$A = \begin{bmatrix} A_1 & A_3 \\ 0 & A_2 \end{bmatrix},$$

where A_1 and A_2 are square, is the union of the spectra of A_1 and A_2.

Exercise 3. Check that the matrix $A \in \mathscr{F}^{n \times n}$ is singular if and only if it has a zero eigenvalue.

Exercise 4. Let $\lambda \in \sigma(A)$, A be nonsingular, and $A\mathbf{x} = \lambda \mathbf{x}$, $\mathbf{x} \neq \mathbf{0}$. Prove that λ^{-1} is an eigenvalue of A^{-1} associated with eigenvector $\mathbf{x}$. □

4.11 The Characteristic Polynomial

We now bring the ideas of the theory of determinants to bear on the problem of finding the eigenvalues of an $n \times n$ matrix $A \in \mathscr{F}^{n \times n}$. As in the situation for transformations (see Exercise 4.9.3), it is easily shown that $\lambda \in \sigma(A)$ if and only if the matrix $\lambda I - A$ is not invertible, or equivalently (Theorem 2.6.2),

$$\det(\lambda I - A) = 0. \tag{1}$$

Observe now that the expression on the left side of Eq. (1) is a monic polynomial of degree n in λ:

$$\det(\lambda I - A) = c_0 + c_1\lambda + \cdots + c_{n-1}\lambda^{n-1} + c_n\lambda^n, \qquad c_n = 1. \tag{2}$$

The polynomial $c(\lambda) \triangleq \sum_{i=0}^{n} c_i \lambda^i$ in Eq. (2) is referred to as the *characteristic polynomial* of the matrix A. These ideas are combined to give Theorem 1.

Theorem 1. *A scalar $\lambda \in \mathscr{F}$ is an eigenvalue of $A \in \mathscr{F}^{n \times n}$ if and only if λ is a zero of the characteristic polynomial of A.*

The significance of this result is that the evaluation of eigenvalues of a matrix is reduced to finding the zeros of a scalar polynomial and is separated from the problem of finding (simultaneously) its eigenvectors.

Exercise 1. Check that the eigenvalues of the matrix

$$A = \begin{bmatrix} 2 & 0 & 0 \\ 11 & 4 & 5 \\ 7 & -1 & -2 \end{bmatrix}$$

are $\lambda_1 = 2, \lambda_2 = 3, \lambda_3 = -1$.

Exercise 2. Show that the diagonal elements of a triangular matrix are its eigenvalues. □

Note that if the field $\mathscr{F}$ is algebraically closed (for definition and discussion see Appendix 1), then all the zeros of the scalar polynomial $c(\lambda)$ belong to $\mathscr{F}$. Hence the spectrum of any $A \in \mathscr{F}^{n \times n}$ is nonempty and consists of at most n points; properties that are already familiar. They are established in Theorem 4.10.5 and, as indicated in Theorem 4.10.1, are also properties of the underlying linear transformation.

Observe also that, if $A = PBP^{-1}$, then

$$c_A(\lambda) = \det(\lambda I - A) = \det(P(\lambda I - B)P^{-1}) = \det(\lambda I - B) = c_B(\lambda).$$

Thus the characteristic polynomial is invariant under similarity and hence all matrices in the equivalence class $\mathscr{A}_T^0$ have the same characteristic polynomial. In other words, this polynomial is a characteristic of the transformation T itself and therefore can be referred to as the *characteristic polynomial of the transformation*.

Proceeding to an investigation of the coefficients of the characteristic polynomial of a matrix A or, equivalently, of a transformation T having A as a representation, we first formulate a lemma.

Lemma 1. *If $A \in \mathscr{F}^{n \times n}$ and distinct columns $i_1, i_2, \ldots, i_r$ $(1 \le r < n)$ of A are the unit vectors $\mathbf{e}_{i_1}, \mathbf{e}_{i_2}, \ldots, \mathbf{e}_{i_r}$, then* det A *is equal to the principal minor of A obtained by striking out rows and columns $i_1, i_2, \ldots, i_r$.*

PROOF. Expanding det A by column i_1, we obtain

$$\det A = (-1)^{i_1+i_1} A\begin{pmatrix} 1 & 2 & \cdots & i_1 - 1 & i_1 + 1 & \cdots & n \\ 1 & 2 & \cdots & i_1 - 1 & i_1 + 1 & \cdots & n \end{pmatrix}$$

$$= A\begin{pmatrix} 1 & 2 & \cdots & i_1 - 1 & i_1 + 1 & \cdots & n \\ 1 & 2 & \cdots & i_1 - 1 & i_1 + 1 & \cdots & n \end{pmatrix}.$$

Now evaluate this minor by expansion on the column numbered i_2. It is found that det A is the principal minor of A obtained by striking out rows and columns i_1 and i_2. Continuing in this way with columns $i_3, i_4, \ldots, i_r$, the result is obtained. ■

Theorem 2. *Let $A \in \mathcal{F}^{n \times n}$ and let $c(\lambda) = \sum_{i=0}^{n} c_i \lambda^i$ denote the characteristic polynomial of A. Then the scalar $c_r (0 \leq r < n)$ is equal to the sum of all principal minors of order $n - r$ of A multiplied by $(-1)^{n-r}$.*

PROOF. Let $\boldsymbol{a}_1, \boldsymbol{a}_2, \ldots, \boldsymbol{a}_n$ be the columns of A. Then, using the unit vectors in $\mathcal{F}^n$, we write

$$(-1)^n c(\lambda) = \det(A - \lambda I) = \det[\boldsymbol{a}_1 - \lambda \boldsymbol{e}_1 \quad \boldsymbol{a}_2 - \lambda \boldsymbol{e}_2 \quad \cdots \quad \boldsymbol{a}_n - \lambda \boldsymbol{e}_n].$$

Since the determinant is a homogeneous linear function of its columns, we can express the latter determinant as the sum of 2^n determinants of matrices having columns of the form $\boldsymbol{a}_i$ or $-\lambda \boldsymbol{e}_i$. Let us select those determinants involving just r columns of the form $-\lambda \boldsymbol{e}_i$ for some i. There are precisely $C_{n,r}$ of these obtained by replacing r of the columns of A by $(-\lambda) \times$ (a unit vector) in every possible way. Thus, from the lemma, each of these determinants is of the form $(-\lambda)^r$ times a principal minor of A of order $n - r$. Furthermore, every principal minor of order $n - r$ appears in this summation. The result follows on putting $r = 0, 1, 2, \ldots, n - 1$. ■

In particular, by setting $r = n - 1$ and $r = 0$ in Theorem 2, we obtain

$$-c_{n-1} = \sum_{i=1}^{n} a_{ii} = \operatorname{tr} A \qquad \text{and} \qquad c_0 = (-1)^n \det A,$$

respectively. The last result can be obtained directly by putting $\lambda = 0$ in the definition of the characteristic polynomial. A fundamental result of the theory of scalar polynomials tells us that $(-1)^{n-r} c_r$ is also the sum of all products of the n zeros of $c(\lambda)$ taken $n - r$ at a time $(0 \leq r < n)$. Thus, if $\lambda_1, \lambda_2, \ldots, \lambda_n$ are eigenvalues of A (not necessarily distinct) then Theorem 2 gives the important special cases:

$$\operatorname{tr} A = \sum_{r=1}^{n} \lambda_r \qquad \text{and} \qquad \det A = \prod_{r=1}^{n} \lambda_r. \tag{3}$$

We have seen that any $n \times n$ matrix has a monic polynomial for its characteristic polynomial. It turns out that, conversely, if $c(\lambda)$ is a monic polynomial over $\mathcal{F}$ of degree n, then there exists a matrix $A \in \mathcal{F}^{n \times n}$ such that its characteristic polynomial is $c(\lambda)$.

Exercise 3. Check that the $n \times n$ companion matrix

$$\begin{bmatrix} 0 & 1 & 0 & \cdots & 0 \\ 0 & 0 & 1 & \ddots & \vdots \\ \vdots & \vdots & \ddots & \ddots & 0 \\ 0 & 0 & & 0 & 1 \\ -c_0 & -c_1 & -c_2 & \cdots & -c_{n-1} \end{bmatrix}$$

has $c(\lambda) = \lambda^n + \sum_{i=0}^{n-1} c_i \lambda^i$ for its characteristic polynomial. (See Exercise 2.2.9.) □

The last result implies, in particular, that for any set σ_0 of k points in $\mathbb{C}$ $(1 \le k \le n)$, there is an $n \times n$ matrix A such that $\sigma(A) = \sigma_0$.

It was mentioned earlier that similar matrices have the same characteristic polynomial. The converse is not true.

Exercise 4. Check that the matrices

$$\begin{bmatrix} 1 & 2 \\ 0 & 1 \end{bmatrix} \quad \text{and} \quad \begin{bmatrix} 1 & 0 \\ 0 & 1 \end{bmatrix}$$

are not similar and that they have the same characteristic polynomial, $c(\lambda) = (\lambda - 1)^2$.

Exercise 5. Prove that two simple matrices are similar if and only if their characteristic polynomials coincide.

Exercise 6. Show that

(a) The characteristic polynomial of the matrix

$$\begin{bmatrix} A_1 & A_3 \\ 0 & A_2 \end{bmatrix}$$

is equal to the product of the characteristic polynomials of A_1 and A_2.

(b) If $\mathscr{S}_0$ is T-invariant, then the characteristic polynomial of $T|_{\mathscr{S}_0}$ divides the characteristic polynomial of T. □

An important relation between the eigenvalues of the matrices A and $p(A)$, where $p(\lambda)$ is a scalar polynomial, is considered next.

Theorem 3. *Let λ_0 be an eigenvalue of $A \in \mathscr{F}^{n \times n}$ and let $p(\lambda)$ be any scalar polynomial. Then $p(\lambda_0)$ is an eigenvalue of $p(A)$.*

PROOF. The reader can easily verify, using the commutativity of powers of A, that

$$p(A) - p(\lambda_0)I = (A - \lambda_0 I)q(A)$$

for some polynomial $q(\lambda)$. Thus $\det(A - \lambda_0 I) = 0$ yields $\det(p(A) - p(\lambda_0)I) = 0$ and the required result follows from Theorem 1. ■

Exercise 7. Prove that the eigenvalues of an idempotent matrix are all equal to zero or one. Using Theorems 4.5.6 and 4.9.2, show that an idempotent is a simple matrix.

Exercise 8. Prove that the eigenvalues of nilpotent matrices are all equal to zero and that nonzero nilpotent matrices cannot be simple.

Exercise 9. Verify that the eigenvalues of A and A^{T} coincide, while the eigenvalues of A and A^* are complex conjugates.

Exercise 10. Show that the eigenvalues of the real symmetric matrix

$$\begin{bmatrix} 13 & 4 & -2 \\ 4 & 13 & -2 \\ -2 & -2 & 10 \end{bmatrix}$$

are 9, 9, and 18.

Exercise 11. Show that the following matrices are simple and find corresponding matrices D and P as described in Theorem 4.10.2:

$$\text{(i)} \quad \begin{bmatrix} 1 & -1 & 0 \\ 2 & 3 & 2 \\ 1 & 1 & 2 \end{bmatrix}, \qquad \text{(ii)} \quad \begin{bmatrix} 0 & 0 & 1 \\ 1 & 0 & -1 \\ 0 & 1 & 1 \end{bmatrix}, \qquad \text{(iii)} \quad \begin{bmatrix} 5 & 8 & 16 \\ 4 & 1 & 8 \\ -4 & -4 & -11 \end{bmatrix}.$$

Hint. Show first that the characteristic polynomials are

$$(\lambda - 1)(\lambda - 2)(\lambda - 3), \quad (\lambda - 1)(\lambda^2 + 1), \quad \text{and} \quad (\lambda - 1)(\lambda + 3)^2,$$

respectively. □

4.12 The Multiplicities of an Eigenvalue

Let $A \in \mathcal{F}^{n \times n}$ be a representation of the transformation $T \in \mathcal{L}(\mathcal{S})$, and let $\mathcal{F}$ be algebraically closed. Then the characteristic polynomial $c(\lambda)$ of A (or of T) can be factorized into a product of n linear factors:

$$c(\lambda) = \prod_{i=1}^{n} (\lambda - \lambda_i), \tag{1}$$

where $\lambda_i \in \sigma(A)$, $i = 1, 2, \ldots, n$. It may happen that some of these eigenvalues are equal, and we define the *algebraic multiplicity* of the eigenvalue λ_i of A (or of T) to be the number of times the factor $(\lambda - \lambda_i)$ appears in Eq. (1). In other words, the algebraic multiplicity of $\lambda_i \in \sigma(A)$ is its multiplicity as a zero of the characteristic polynomial.

We have seen that $\lambda_0 \in \sigma(A)$ if and only if $\text{Ker}(A - \lambda_0 I) \neq \{0\}$, and the *geometric multiplicity* of the eigenvalue λ_0 of A (or of T) is defined as the dimension of the subspace $\text{Ker}(A - \lambda_0 I)$. Note that by referring to the subspace $\text{Ker}(A - \lambda I)$ as the *eigenspace of the matrix A associated with the eigenvalue λ*, we can restate for matrices all the results concerning eigenspaces for transformations. We also point out that scalars that fail to be eigenvalues of A can be viewed as having zero (algebraic and geometric) multiplicities. However, the algebraic and geometric multiplicities are not generally equal.

Example 1. Consider the matrix

$$A = \begin{bmatrix} 0 & 1 \\ 0 & 0 \end{bmatrix}.$$

The characteristic polynomial $c(\lambda)$ of A is $c(\lambda) = \lambda^2$ and therefore the algebraic multiplicity of the zero eigenvalue $\lambda_0 = 0$ is 2. On the other hand, $\dim(\text{Ker}(A - \lambda_0 I)) = \dim(\text{Ker } A) = 1$ and the geometric multiplicity of $\lambda_0 = 0$ is 1. □

Theorem 1. *The geometric multiplicity of an eigenvalue does not exceed its algebraic multiplicity.*

PROOF. Let λ be an eigenvalue of the $n \times n$ matrix A and let

$$m_\lambda = \dim(\text{Ker}(A - \lambda I)).$$

If r denotes the rank of $A - \lambda I$ then, by Eq. (4.5.2), we have $r = n - m_\lambda$ and all the minors of $A - \lambda I$ of order greater than $n - m_\lambda$ are zero. Hence if $\tilde{c}(\mu) = \sum_{i=0}^{n} \tilde{c}_i \mu^i$ is the characteristic polynomial of $A - \lambda I$, Theorem 4.11.2 implies that $\tilde{c}_0 = \tilde{c}_1 = \cdots = \tilde{c}_{m_\lambda - 1} = 0$. Therefore

$$\tilde{c}(\mu) = \sum_{i=m_\lambda}^{n} \tilde{c}_i \mu^i$$

and must have a zero of multiplicity greater than or equal to m_λ. It remains to observe that, if $c(\mu)$ is the characteristic polynomial of A, then

$$c(\mu) = \tilde{c}(\mu - \lambda). \quad ■$$

Denote by $\sum_a$ (respectively, $\sum_g$) the sum of the algebraic (respectively, geometric) multiplicities of the eigenvalues of an $n \times n$ complex matrix A. Then Theorem 1 implies

$$\textstyle\sum_g \le \sum_a = n. \tag{2}$$

In the case $\sum_g = n$, the relations (2) imply that the algebraic and geometric multiplicities of each eigenvalue of A must coincide. On the other hand, if we use Proposition 4.10.4, the equality $\sum_g = n$ also means that the space $\mathscr{F}^n$ can be spanned by nonzero vectors from the eigenspaces of the matrix, that is, by their eigenvectors. Hence if $\sum_g = n$, then the matrix A is simple. It is easily seen that the converse also holds and a criterion for a matrix to be simple is obtained.

Theorem 2. *Any matrix A with entries from an algebraically closed field is simple if and only if the algebraic and geometric multiplicities of each eigenvalue of A coincide.*

4.13 First Applications to Differential Equations

Consider a system of ordinary differential equations with constant coefficients of the following form:

$$\begin{aligned}\dot{x}_1 &= a_{11}x_1 + a_{12}x_2 + \cdots + a_{1n}x_n,\\ \dot{x}_2 &= a_{21}x_1 + a_{22}x_2 + \cdots + a_{2n}x_n,\\ &\vdots\\ \dot{x}_n &= a_{n1}x_1 + a_{n2}x_2 + \cdots + a_{nn}x_n,\end{aligned}$$

where $x_1, \ldots, x_n$ are functions of the independent variable t, the symbol $\dot{x}_j$ denotes the derivative of x_j with respect to t; and the coefficients a_{ij} are independent of t. We can obviously abbreviate this set of equations to one matrix equation:

$$\dot{\mathbf{x}}(t) = A\mathbf{x}(t), \tag{1}$$

where $\mathbf{x}(t) = [x_1 \quad x_2 \quad \cdots \quad x_n]^{\mathrm{T}}$ and $A = [a_{ij}]_{i,j=1}^{n}$. Note that we may write $\mathbf{x}$ or $\dot{\mathbf{x}}$ instead of $\mathbf{x}(t)$ or $\dot{\mathbf{x}}(t)$ if their dependence on t is clear from the context.

If A has at least one eigenvalue, we can easily demonstrate the existence of a solution of $\dot{\mathbf{x}} = A\mathbf{x}$. We look for a solution of the form $\mathbf{x}(t) = \mathbf{x}_0 e^{\lambda_0 t}$, where $\mathbf{x}_0$ and λ_0 are independent of t. Then $\dot{\mathbf{x}} = \lambda_0 \mathbf{x}$ and Eq. (1) can be written $(A - \lambda_0 I)\mathbf{x}_0 = \mathbf{0}$. Thus, if we choose $\mathbf{x}_0$ to be a (right) eigenvector of A with eigenvalue λ_0, then $\mathbf{x}(t)$ is indeed a solution. Furthermore, if $A \in \mathbb{C}^{n \times n}$, such a solution always exists because every A then has at least one eigenvalue with an associated (right) eigenvector.

Proposition 1. *The vector-valued function* $\mathbf{x}(t) = \mathbf{x}_0 e^{\lambda_0 t}$ *is a solution of Eq. (1) if and only if* λ_0 *is an eigenvalue of* A *associated with an eigenvector* $\mathbf{x}_0$.

Hence the problem of determining the eigenvalues and eigenvectors of a matrix is closely related to finding the solutions of the corresponding system of differential equations. Can we characterize the *solution set* $\mathscr{S}$ of all solutions of Eq. (1)? It is easily verified that $\mathscr{S}$ is a linear space and it will be shown later that the dimension of this space is n (see Section 7.10). Here, the result will be proved for simple matrices only. The first exercise shows how a subspace of $\mathscr{S}$ can always be generated.

Exercise 1. Let $A \in \mathbb{C}^{n \times n}$ and let $\mathscr{S}$ denote the solution space of Eq. (1). If $\lambda_1, \lambda_2, \ldots, \lambda_s$ are the distinct eigenvalues of A associated with the eigenvectors $\mathbf{x}_1, \mathbf{x}_2, \ldots, \mathbf{x}_s$, respectively, show that the set

$$\mathscr{S}_0 = \operatorname{span}\{\mathbf{x}_1 e^{\lambda_1 t}, \mathbf{x}_2 e^{\lambda_2 t}, \ldots, \mathbf{x}_s e^{\lambda_s t}\}$$

is a linear subspace in $\mathscr{S}$ of dimension s.

Hint. Observe that the linear independence of $\mathbf{x}_1, \mathbf{x}_2, \ldots, \mathbf{x}_s$ implies that of $\mathbf{x}_1 e^{\lambda_1 t}, \mathbf{x}_2 e^{\lambda_2 t}, \ldots, \mathbf{x}_s e^{\lambda_s t}$ and use Proposition 1. □

For a simple matrix A we can say rather more than the conclusion of Exercise 1, for in this case we have a basis for $\mathbb{C}^n$ consisting of eigenvectors of A. In fact, if $\mathbf{x}_1, \ldots, \mathbf{x}_n$ is such a basis with associated eigenvalues $\lambda_1, \ldots, \lambda_n$, respectively, it is easily verified that the n vector-valued functions $\mathbf{x}_j e^{\lambda_j t}$, are solutions, $j = 1, 2, \ldots, n$, and they generate an n-dimensional subspace of solutions, $\mathcal{S}_0$. We are to show that $\mathcal{S}_0$ is in fact, all of $\mathcal{S}$.

More generally, let us first consider the *inhomogeneous system*

$$\dot{\mathbf{x}}(t) = A\mathbf{x}(t) + \mathbf{f}(t) \tag{2}$$

(where $\mathbf{f}(t)$ is a given integrable function). It will be shown that Eq. (2) can be reduced to a set of n independent scalar equations. To see this, let eigenvectors, $\mathbf{x}_1, \ldots, \mathbf{x}_n$ and left eigenvectors $\mathbf{y}_1, \ldots, \mathbf{y}_n$ be defined for A as required by Theorem 4.10.3. If $P = [\mathbf{x}_1 \ \cdots \ \mathbf{x}_n]$ and $Q = [\mathbf{y}_1 \ \cdots \ \mathbf{y}_n]$, then the biorthogonality properties of the eigenvectors (see Eq. (4.10.6)) are summarized in the relation

$$Q^{\mathrm{T}} P = I, \tag{3}$$

and the representation for A itself (see Eq. (4.10.7)) is

$$A = PDQ^{\mathrm{T}}, \tag{4}$$

where $D = \operatorname{diag}[\lambda_1, \ldots, \lambda_n]$.

The reduction of the system (2) to a set of scalar equations is now accomplished if we substitute Eq. (4) in Eq. (2), multiply on the left by Q^{T}, and take advantage of Eq. (3) to obtain

$$Q^{\mathrm{T}}\dot{\mathbf{x}}(t) = DQ^{\mathrm{T}}\mathbf{x}(t) + Q^{\mathrm{T}}\mathbf{f}(t).$$

Introducing new variables $z_j(t) = \mathbf{y}_j^{\mathrm{T}}\mathbf{x}(t)$ for $j = 1, 2, \ldots, n$ and transformed prescribed functions $\phi_j(t) = \mathbf{y}_j^{\mathrm{T}}\mathbf{f}(t)$, the equations become $\dot{\mathbf{z}}(t) = D\mathbf{z}(t) + \boldsymbol{\varphi}(t)$, where

$$\mathbf{z}(t) = [z_1(t) \ \ z_2(t) \ \ \cdots \ \ z_n(t)]^{\mathrm{T}}, \quad \boldsymbol{\varphi}(t) = [\varphi_1(t) \ \ \varphi_2(t) \ \ \cdots \ \ \varphi_n(t)]^{\mathrm{T}}.$$

Or, equivalently, we have n "uncoupled" scalar equations

$$\dot{z}_j(t) = \lambda_j z_j(t) + \varphi_j(t), \tag{5}$$

for $j = 1, 2, \ldots, n$. These can be solved by elementary methods and the solution of the original system (2) is recovered by observing that $\mathbf{z}(t) = Q^{\mathrm{T}}\mathbf{x}(t)$ and, using Eq. (3), $\mathbf{x}(t) = P\mathbf{z}(t)$.

For the *homogeneous* Eq. (1) (when $\mathbf{f}(t) = \mathbf{0}$ for all t in Eq. (2)), we simply have $\varphi_j(t) = 0$ in Eq. (5) and the solution of (5) is $z_j(t) = c_j e^{\lambda_j t}$ for some constant c_j. Thus, in this case, $z(t) = \operatorname{diag}[e^{\lambda_1 t}, \ldots, e^{\lambda_n t}]\boldsymbol{c}$ and so *every* solution of Eq. (1) has the form

$$\boldsymbol{x}(t) = P \operatorname{diag}[e^{\lambda_1 t}, \ldots, e^{\lambda_n t}]\boldsymbol{c} \tag{6}$$

for some $\boldsymbol{c} \in \mathbb{C}^n$, $\boldsymbol{c} = [c_1 \quad c_2 \quad \cdots \quad c_n]^{\mathrm{T}}$.

Observe that Eq. (6) just means that any solution of Eq. (1) is a linear combination of the *primitive solutions* $\boldsymbol{x}_j e^{\lambda_j t}$, $j = 1, 2, \ldots, n$. Since these primitive solutions are linearly independent (see Exercise 1), it follows that if $\mathscr{S}$ is the solution set of Eq. (1) and A is a simple $n \times n$ matrix, then $\dim \mathscr{S} = n$.

In Chapter 9 we will return to the discussion of Eq. (2) and remove our hypothesis that A be simple; also we will consider the case when $\boldsymbol{f}(t) \neq \boldsymbol{0}$ more closely.

Exercise 2. Describe the solutions of the following system of differential equations with constant coefficients ($x_j = x_j(t), j = 1, 2, 3$)

$$\begin{aligned} \dot{x}_1 &= 2x_1 \\ \dot{x}_2 &= 11x_1 + 4x_2 + 5x_3, \\ \dot{x}_3 &= 7x_1 - x_2 - 2x_3. \end{aligned} \tag{7}$$

SOLUTION. Rewrite the system in matrix form:

$$\dot{\boldsymbol{x}} = A\boldsymbol{x}, \tag{8}$$

where

$$A = \begin{bmatrix} 2 & 0 & 0 \\ 11 & 4 & 5 \\ 7 & -1 & -2 \end{bmatrix}, \qquad \boldsymbol{x} = \begin{bmatrix} x_1 \\ x_2 \\ x_3 \end{bmatrix}, \qquad \dot{\boldsymbol{x}} = \begin{bmatrix} \dot{x}_1 \\ \dot{x}_2 \\ \dot{x}_3 \end{bmatrix},$$

and recall (see Exercise 4.11.1) that A has eigenvalues 2, 3, and -1 corresponding to the eigenvectors $\boldsymbol{x}_1 = [3 \quad -79 \quad 25]^{\mathrm{T}}$, $\boldsymbol{x}_2 = [0 \quad 5 \quad -1]^{\mathrm{T}}$, and $\boldsymbol{x}_3 = [0 \quad 1 \quad -1]^{\mathrm{T}}$. Thus, by Eq. (6), any solution $\boldsymbol{x} = [x_1 \quad x_2 \quad x_3]^{\mathrm{T}}$ of Eq. (8) is of the form

$$\boldsymbol{x} = c_1\boldsymbol{x}_1 e^{2t} + c_2\boldsymbol{x}_2 e^{3t} + c_3\boldsymbol{x}_3 e^{-t} = [\boldsymbol{x}_1 \quad \boldsymbol{x}_2 \quad \boldsymbol{x}_3] \begin{bmatrix} e^{2t} & 0 & 0 \\ 0 & e^{3t} & 0 \\ 0 & 0 & e^{-t} \end{bmatrix} \begin{bmatrix} c_1 \\ c_2 \\ c_3 \end{bmatrix}$$

or, what is equivalent, comparing the components of the vectors,

$$\begin{aligned} x_1 &= 3c_1 e^{2t}, \\ x_2 &= -79c_1 e^{2t} + 5c_2 e^{3t} + c_3 e^{-t}, \\ x_3 &= 25c_1 e^{2t} - c_2 e^{3t} - c_3 e^{-t}, \end{aligned}$$

for some $c_1, c_2, c_3 \in \mathbb{C}$. □

4.14 Miscellaneous Exercises

1. Let $\mathscr{S} = \mathscr{S}_1 \dotplus \mathscr{S}_2$. Define a transformation T on $\mathscr{S}$ by

$$T(\boldsymbol{x}) = \boldsymbol{x}_1,$$

where $\boldsymbol{x} = \boldsymbol{x}_1 + \boldsymbol{x}_2$ and $\boldsymbol{x}_1 \in \mathscr{S}_1$, $\boldsymbol{x}_2 \in \mathscr{S}_2$. Prove that

(a) $T \in \mathscr{L}(\mathscr{S})$;
(b) $\mathscr{S}_1$ and $\mathscr{S}_2$ are T-invariant;
(c) T satisfies the condition $T^2 = T$ (i.e., T is *idempotent*).

The transformation T is called a *projector* of $\mathscr{S}$ on $\mathscr{S}_1$ *parallel* to $\mathscr{S}_2$. Observe that $T|_{\mathscr{S}_1}$ is the identity transformation on $\mathscr{S}_1$.

2. Consider a simple linear transformation $T \in \mathscr{L}(\mathscr{S})$ where $\dim \mathscr{S} = n$, $\lambda_1, \lambda_2, \ldots, \lambda_n$ are its eigenvalues, and $\boldsymbol{x}_1, \boldsymbol{x}_2, \ldots, \boldsymbol{x}_n$ are corresponding eigenvectors which span $\mathscr{S}$. Define $\mathscr{S}_1 = \operatorname{span}(\boldsymbol{x}_i)$ for $i = 1, 2, \ldots, n$. Show that $T = \sum_{i=1}^n \cdot \lambda_i I_i$, where I_i is the identity transformation on $\mathscr{S}_i$ $(1 \le i \le n)$ and that $\mathscr{S} = \sum_{i=1}^n \cdot \mathscr{S}_i$; this is the spectral theorem for simple transformations.

3. Consider the spectral decomposition presented as Theorem 4.10.3. Let

$$G_j = \boldsymbol{x}_j \boldsymbol{y}_j^{\mathrm{T}} \text{ for } j = 1, 2, \ldots, n, \text{ so that } A = \sum_{j=1}^{n} \lambda_j G_j.$$

(a) Show that each G_j has rank one and that $G_1, \ldots, G_n$ are mutually orthogonal in the sense that $G_k G_j = \delta_{kj} G_k$ for $1 \le j, k \le n$.

(b) If $p(\lambda)$ is any scalar polynomial, show that $p(A)$ is simple and $p(A) = \sum_{j=1}^n p(\lambda_j) G_j$.

The matrix G_j, $1 \le j \le n$, is referred to as the *constituent matrix* of A associated with λ_j.

4. The matrix $R_\lambda = (\lambda I - A)^{-1}$, called the *resolvent* of A, is defined at the points $\lambda \notin \sigma(A)$. Verify the following statements:

(a) $R_\lambda - R_\mu = (\mu - \lambda) R_\lambda R_\mu$;
(b) If A is simple and $G_1, G_2, \ldots, G_n$ are defined as in Exercise 3, then

$$R_\lambda = \sum_{i=1}^{n} \frac{1}{\lambda - \lambda_i} G_i, \qquad \lambda \notin \sigma(A).$$

Hint. First use Eqs. (4.13.3) and (4.13.4) to write $R_\lambda = P(\lambda I - D)^{-1} Q^{\mathrm{T}}$.

5. Prove that if $\dim \mathscr{S} = n$, any transformation $T \in \mathscr{L}(\mathscr{S})$ has an invariant subspace of dimension $n - 1$.

Hint. Let $\lambda \in \sigma(T)$ and use Eq. (4.5.2) for $T - \lambda I$ to confirm the existence of the k-dimensional subspace $\operatorname{Im}(T - \lambda I)$ invariant under T, where $k \leq n - 1$. Then observe that any subspace in $\mathscr{S}$ containing $\operatorname{Im}(T - \lambda I)$ is T-invariant.

6. Prove that for any $T \in \mathscr{L}(\mathscr{S})$ there is a set of T-invariant subspaces $\mathscr{S}_i$ $(i = 1, 2, \ldots, n = \dim \mathscr{S})$ such that

$$\mathscr{S}_1 \subset \mathscr{S}_2 \subset \cdots \subset \mathscr{S}_{n-1} \subset \mathscr{S}_n = \mathscr{S}$$

and $\dim \mathscr{S}_i = i$ for $i = 1, 2, \ldots, n$.

Hint. Use an inductive argument based on the previous result.

7. Interpret the result of Exercise 6 for matrices. Show that the representation of T with respect to the basis $\mathscr{E} = \{\mathbf{x}_1, \mathbf{x}_2, \ldots, \mathbf{x}_n\}$, where $\mathbf{x}_i \in \mathscr{S}_i$ $(1 \leq i \leq n)$ (see Exercise 6) is an upper-triangular matrix, while with respect to $\mathscr{E}' = \{\mathbf{x}_n, \mathbf{x}_{n-1}, \ldots, \mathbf{x}_1\}$, it is lower-triangular (*Schur's theorem*).

8. Prove that if A is a simple matrix with characteristic polynomial $c(\lambda)$, then $c(A) = O$. (This is a special case of the Cayley–Hamilton theorem, to be proved in full generality in Sections 6.2 and 7.2).

Hint. Use theorem 4.10.2.

9. Show that the inverse of a nonsingular simple matrix $A \in \mathbb{C}^{n \times n}$ is given by the formula

$$A^{-1} = -\frac{1}{c_0}(A^{n-1} + c_{n-1}A^{n-2} + \cdots + c_1 I),$$

where $c(\lambda) = \lambda^n + \sum_{i=0}^{n-1} c_i \lambda^i$ is the characteristic polynomial of A.

Hint. Observe that $A^{-1}c(A) = O$.

10. Check the following statements.

(a) If A or B is nonsingular, then $AB \approx BA$. Consider the example

$$A = \begin{bmatrix} 1 & 0 \\ 0 & 0 \end{bmatrix}, \qquad B = \begin{bmatrix} 0 & 1 \\ 0 & 0 \end{bmatrix}$$

to confirm that in the case of singular matrices, AB and BA may not be similar.

(b) For any $A, B \in \mathscr{F}^{n \times n}$, the matrices AB and BA have the same characteristic polynomial and hence the same eigenvalues.

Hint. For part (b) use the following polynomial identity in λ:

$$\det(\lambda I - (A - \mu I)B) = \det(\lambda I - B(A - \mu I))$$

for $\mu \in \mathscr{F}$ such that $A - \mu I$ is nonsingular.

11. Prove that if $AB = BA$, then A and B have at least one common eigenvector. Furthermore, if A and B are simple, then there exists a basis in the space consisting of their common eigenvectors.

Hint. Consider span$\{\mathbf{x}, B\mathbf{x}, \ldots, B^s\mathbf{x}, \ldots\}$, where $A\mathbf{x} = \lambda\mathbf{x}$, $\mathbf{x} \neq \mathbf{0}$, and use Exercise 4.9.8 to establish the first statement.

12. Let $A \in \mathscr{F}^{n \times n}$. Show that, in comparison with Exercise 4.5.12,

(a) If the zero eigenvalue of A has the same algebraic and geometric multiplicities, then Ker A = Ker A^2;

(b) If A is simple, then Im A^2 = Im A.

13. If A is an idempotent matrix of rank r, show that tr $A = r$.

Hint. Use Exercise 4.11.7, Theorem 4.10.3, and Eq. (4.11.3).

14. Let $A \in \mathscr{F}^{n \times n}$ and let $A = [\mathbf{a} \quad \mathbf{a} \quad \cdots \quad \mathbf{a}]$, where $\mathbf{a} \in \mathscr{F}^n$. Show that at least $n - 1$ eigenvalues of A are zero.

Hint. Use Theorem 4.11.2.

15. (a) Let A_n be the $n \times n$ matrix

$$\begin{bmatrix} 2c & 1 & 0 & \cdots & 0 \\ 1 & 2c & 1 & & \vdots \\ 0 & 1 & 2c & & 0 \\ \vdots & & \ddots & \ddots & 1 \\ 0 & \cdots & 0 & 1 & 2c \end{bmatrix},$$

where $c = \cos\theta$. If $D_n = \det A_n$, show that

$$D_n = \frac{\sin(n+1)\theta}{\sin n\theta}$$

(b) Find the eigenvalues of A_n.

Note that $D_n(c)$ is the *Chebyshev polynomial* of the second kind, $U_n(c)$. The matrix A_n with $c = -1$ arises in Rayleigh's finite-dimensional approximation to a vibrating string.

Hint. For part (b), observe that $A_n - \lambda I$ preserves the form of A_n for small λ to obtain the eigenvalues of A_n:

$$\lambda_k = 2\left(c - \cos\frac{k\pi}{n+1}\right), \qquad k = 1, 2, \ldots, n.$$

16. Find the eigenvalues and eigenvectors of the $n \times n$ circulant matrix with the first row $[a_0 \quad a_1 \quad \cdots \quad a_{n-1}]$.

Hint. Use Exercise 2.11.14 to determine the eigenvalues $\lambda_k = a(\varepsilon_k)$, and the eigenvectors $[1 \quad \varepsilon_k \quad \varepsilon_k^2 \quad \cdots \quad \varepsilon_k^{n-1}]^T$, where

$$a(\lambda) = \sum_{i=0}^{n-1} a_i \lambda^i$$

and

$$\varepsilon_k = \cos\left(\frac{2\pi k}{n}\right) + i \sin\left(\frac{2\pi k}{n}\right), \quad k = 0, 1, \ldots, n-1.$$

17. Let $V \in \mathcal{L}(\mathcal{S}_1, \mathcal{S}_2)$, $X \in \mathcal{L}(\mathcal{S}_2, \mathcal{S}_1)$ and satisfy $XV = I$, the identity on $\mathcal{S}_1$. Show that $\mathcal{S}_2 = \operatorname{Im} V \dotplus \operatorname{Ker} X$.

Hint. To show that $\mathcal{S}_2 = \operatorname{Im} V + \operatorname{Ker} X$, let $\boldsymbol{x} \in \mathcal{S}_2$ and write $\boldsymbol{x} = \boldsymbol{v} + \boldsymbol{w}$, where $\boldsymbol{v} = VX\boldsymbol{x}$. Then show that the sum is direct.

CHAPTER 5

Linear Transformations in Unitary Spaces and Simple Matrices

The results discussed in Chapter 4 are obviously applicable to linear transformations mapping one unitary space into another. Some additional properties of transformations resulting from the existence of the inner product are considered in this chapter. In particular, it will be shown that any linear transformation acting on a unitary space has a dual transformation, known as its "adjoint." Normal matrices and transformations are defined and studied in terms of adjoints prior to the analysis of the important subclasses of Hermitian matrices (and self-adjoint transformations) and unitary matrices and transformations. All of these special classes are subsets of the simple matrices and transformations introduced in Chapter 4.

Other important ideas associated with simple matrices are developed, for example, projectors and the reduction of quadratic forms. The chapter concludes with an important application (Section 5.12) and some preparatory ideas for subsequent applications in systems theory (Section 5.13). Note that throughout this chapter the underlying field is assumed to be $\mathscr{F} = \mathbb{C}$ or $\mathscr{F} = \mathbb{R}$.

5.1 Adjoint Transformations

Let $\mathscr{U}$, $\mathscr{V}$ be unitary spaces with inner products $(\ ,\ )_1$ and $(\ ,\ )_2$, respectively, and let $T \in \mathscr{L}(\mathscr{U}, \mathscr{V})$. Observe first that any vector $\mathbf{v} \in \mathscr{V}$ is uniquely determined if $(\mathbf{v}, \mathbf{y})_2$ is known for all $\mathbf{y} \in \mathscr{V}$. Indeed, as indicated in Exercise 3.12.8, it is sufficient to know these inner products for all vectors $\boldsymbol{y}$ in a basis for $\mathscr{V}$. It follows that the action of T is determined by the values of all the numbers $(T\boldsymbol{x}, \boldsymbol{y})_2$, where $\boldsymbol{x}, \boldsymbol{y}$ are any vectors from $\mathscr{U}$ and $\mathscr{V}$, respectively. We ask whether this same set of numbers is also attained by first forming the image of $\boldsymbol{y}$ under some $T_1 \in \mathscr{L}(\mathscr{V}, \mathscr{U})$, and then taking the $(\ ,\ )_1$ inner product with $\boldsymbol{x}$. In other words, is there a $T_1 \in \mathscr{L}(\mathscr{V}, \mathscr{U})$ such that

$$(T(\boldsymbol{x}), \boldsymbol{y})_2 = (\boldsymbol{x}, T_1(\boldsymbol{y}))_1 \tag{1}$$

for all $\boldsymbol{x} \in \mathscr{U}$ and $\boldsymbol{y} \in \mathscr{V}$? The answer is yes, and the "partner" for T, written T^*, is known as the *adjoint* of T.

To define the adjoint we consider an orthonormal basis $\boldsymbol{u}_1, \ldots, \boldsymbol{u}_n$ for $\mathscr{U}$. Then define a map $T^*: \mathscr{V} \to \mathscr{U}$ in terms of T by

$$T^*(\boldsymbol{y}) = \sum_{j=1}^{n} (\boldsymbol{y}, T(\boldsymbol{u}_j))_2 \boldsymbol{u}_j \tag{2}$$

for any $\boldsymbol{y} \in \mathscr{V}$. We check first that $T^* \in \mathscr{L}(\mathscr{V}, \mathscr{U})$. If $\boldsymbol{y}, \boldsymbol{z} \in \mathscr{V}$ and $\alpha, \beta \in \mathscr{F}$ then, using the linearity of the inner product (Section 3.11),

$$\begin{aligned} T^*(\alpha\boldsymbol{y} + \beta\boldsymbol{z}) &= \alpha \sum_{j=1}^{n} (\boldsymbol{y}, T(\boldsymbol{u}_j))_2 \boldsymbol{u}_j + \beta \sum_{j=1}^{n} (\boldsymbol{z}, T(\boldsymbol{u}_j))_2 \boldsymbol{u}_j \\ &= \alpha T^*(\boldsymbol{y}) + \beta T^*(\boldsymbol{z}), \end{aligned}$$

by definition (2) of T^*. Thus $T^* \in \mathscr{L}(\mathscr{V}, \mathscr{U})$. Our first theorem will show that T^*, so defined, does indeed have the property (1) that we seek. It will also show that our definition of T^* is, in fact, independent of the choice of orthonormal basis for $\mathscr{U}$.

First we should note the important special case of this discussion in which $\mathscr{U} = \mathscr{V}$ and the inner products also coincide. In this case, the definition (2) reads

$$T^*(\boldsymbol{x}) = \sum_{j=1}^{n} (\boldsymbol{x}, T(\boldsymbol{u}_j))\boldsymbol{u}_j, \tag{3}$$

and T^*, as well as T, belongs to $\mathscr{L}(\mathscr{U})$.

Theorem 1. *Let $T \in \mathscr{L}(\mathscr{U}, \mathscr{V})$, where $\mathscr{U}$ is a unitary space with inner product $(\ ,\ )_1$ and $\mathscr{V}$ is a unitary space with inner product $(\ ,\ )_2$. The adjoint transformation $T^* \in \mathscr{L}(\mathscr{V}, \mathscr{U})$ satisfies the condition*

$$(T(\boldsymbol{x}), \boldsymbol{y})_2 = (\boldsymbol{x}, T^*(\boldsymbol{y}))_1 \tag{4}$$

for every $\boldsymbol{x} \in \mathscr{U}, \boldsymbol{y} \in \mathscr{V}$ and is the unique member of $\mathscr{L}(\mathscr{V}, \mathscr{U})$ for which Eq. (4) holds. Conversely, if Eq. (4) holds, then T^* is the adjoint of T.

PROOF. It has been shown that $T^* \in \mathscr{L}(\mathscr{V}, \mathscr{U})$. To prove the equality in Eq. (4) we use Exercise 3.12.3 to write $x = \sum_{i=1}^{n} (x, u_i)_1 u_i$. Then, using the linearity of T,

$$(T(x), y)_2 = \left(\sum_{i=1}^{n} (x, u_i)_1 T(u_i), y \right)_2 = \sum_{i=1}^{n} (x, u_i)_1 (T(u_i), y)_2. \tag{5}$$

On the other hand, by Eq. (2) and the orthonormality of the basis,

$$\begin{aligned}(x, T^*(y))_1 &= \left(\sum_{i=1}^{n} (x, u_i)_1 u_i, \sum_{j=1}^{n} (y, T(u_j))_2 u_j \right)_1 \\ &= \sum_{i=1}^{n} (x, u_i)_1 \overline{(y, T(u_i))_2}. \end{aligned} \tag{6}$$

Since $\overline{(y, T(u_i))_2} = (T(u_i), y)_2$, Eq. (4) follows on comparing Eq. (6) with Eq. (5).

For the uniqueness, let $(T(x), y)_2 = (x, T_1(y))_1$ for some $T_1 \in \mathscr{L}(\mathscr{V}, \mathscr{U})$ and all $x \in \mathscr{U}$, $y \in \mathscr{V}$. Then, on comparison with Eq. (4),

$$(x, T_1(y))_1 = (x, T^*(y))_1$$

for all $x \in \mathscr{U}$ and $y \in \mathscr{V}$. But this clearly means that $(x, (T_1 - T^*)(y))_1 = 0$ for all $x \in \mathscr{U}$ and $y \in \mathscr{V}$, and this implies (by Exercise 3.13.2) that $T_1 - T^* = O$, that is, $T_1 = T^*$.

The proof of the converse statement uses similar ideas and is left as an exercise. ∎

It is interesting to compare Eq. (4) with the result of Exercise 3.14.8. The similarity between these statements will be explained in our next theorem.

Exercise 1. Let $T \in \mathscr{L}(\mathscr{F}^3, \mathscr{F}^4)$ be defined as in Exercise 4.2.1. Check that the transformation

$$T_1(y) = \begin{bmatrix} y_2 + y_3 + y_4 \\ y_1 + y_2 - y_4 \\ y_1 \end{bmatrix},$$

for all $y = [y_1 \quad y_2 \quad y_3 \quad y_4]^{\mathrm{T}} \in \mathscr{F}^4$, is the adjoint of T with respect to the standard inner products in $\mathscr{F}^3$ and $\mathscr{F}^4$. Confirm Eq. (4) for the given transformations. □

Some important properties of the adjoint are indicated next.

Exercise 2. Prove the following statements:

(a) $I^* = I, \quad O^* = O$;
(b) $(T^*)^* = T$;
(c) $(\alpha T)^* = \bar{\alpha} T^*$ for any $\alpha \in \mathscr{F}$;

(d) $(T_1 + T_2)^* = T_1^* + T_2^*$;
(e) $(T_2 T_1)^* = T_1^* T_2^*$ for every $T_1 \in \mathscr{L}(\mathscr{U}, \mathscr{W})$, $T_2 \in \mathscr{L}(\mathscr{W}, \mathscr{V})$.

SOLUTION. We shall prove only the last equality, since all of them are established analogously. Successive use of Eq. (4) gives

$$(\mathbf{x}, (T_2 T_1)^*(\mathbf{y})) = (T_2 T_1(\mathbf{x}), \mathbf{y}) = (T_1(\mathbf{x}), T_2^*(\mathbf{y})) = (\mathbf{x}, T_1^* T_2^*(\mathbf{y}))$$

for all $\mathbf{x} \in \mathscr{U}$, $\mathbf{y} \in \mathscr{V}$. Hence (see Exercise 3.13.2), $(T_2 T_1)^*(\mathbf{y}) = T_1^* T_2^*(\mathbf{y})$ for all $\mathbf{y} \in \mathscr{V}$ and therefore part (e) holds.

Exercise 3. Check that $T \in \mathscr{L}(\mathscr{U})$ is invertible if and only if $T^* \in \mathscr{L}(\mathscr{U})$ is invertible and that in this case,

$$(T^*)^{-1} = (T^{-1})^*. \tag{7}$$

SOLUTION. Let T be invertible. If $T^*(\mathbf{x}) = \mathbf{0}$, then by Eq. (3),

$$\sum_{i=1}^{n} (\mathbf{x}, T(\mathbf{u}_i))\mathbf{u}_i = \mathbf{0}$$

and consequently $(\mathbf{x}, T(\mathbf{u}_i)) = 0$, $i = 1, 2, \ldots, n$. The invertibility of T means that since all the elements $T(\mathbf{u}_i)$ are images of the basis elements in $\mathscr{U}$, they generate another basis in $\mathscr{U}$. Hence by Exercise 3.13.2, we must have $\mathbf{x} = \mathbf{0}$, and by Theorem 4.6.3, the transformation T^* is invertible. The converse follows by applying part (b) of Exercise 2. To establish Eq. (7), write

$$(\mathbf{x}, (T^{-1})^*(\mathbf{y})) = (T^{-1}(\mathbf{x}), \mathbf{y})$$

and use the invertibility of T^* to find an element $\mathbf{z} \in \mathscr{U}$ such that $T^*(\mathbf{z}) = \mathbf{y}$. Hence, continuing, we have by Exercise 2(b)

$$(T^{-1}(\mathbf{x}), \mathbf{y}) = (T^{-1}(\mathbf{x}), T^*(\mathbf{z})) = (TT^{-1}(\mathbf{x}), \mathbf{z}) = (\mathbf{x}, \mathbf{z}) = (\mathbf{x}, (T^*)^{-1}(\mathbf{y}))$$

for any $\mathbf{x}, \mathbf{y} \in \mathscr{U}$. Thus, reasoning as in the solution of Exercise 2, we deduce the required relation (7). □

Proceeding to the study of matrix representations for a transformation $T \in \mathscr{L}(\mathscr{U}, \mathscr{V})$ and its adjoint, we discover the reason for the use of the "star" notation for both the adjoint of a linear transformation and the conjugate transpose of a matrix.

Theorem 2. *Let $T \in \mathscr{L}(\mathscr{U}, \mathscr{V})$ and let T^* denote the adjoint of T. If $A \in \mathscr{F}^{m \times n}$ is the representation of T with respect to the orthonormal bases $(\mathscr{E}, \mathscr{G})$, then the conjugate transpose matrix $A^* \in \mathscr{F}^{n \times m}$ is the representation of T^* with respect to the pair of bases $(\mathscr{G}, \mathscr{E})$.*

PROOF. Let $\mathscr{E} = \{u_1, u_2, \ldots, u_n\}$ and $\mathscr{G} = \{v_1, v_2, \ldots, v_m\}$ denote orthonormal bases in $\mathscr{U}$ and $\mathscr{V}$, respectively. Then the representation $A = [a_{ij}]_{i,j=1}^{m,n}$ is found from the equations

$$T(u_j) = \sum_{i=1}^{m} a_{ij} v_i, \qquad j = 1, 2, \ldots, n,$$

where $a_{ij} = (T(u_j), v_i)_2$, and $i = 1, 2, \ldots, m, j = 1, 2, \ldots, n$. Similarly, for the element b_{ji} $(1 \le i \le m,\ 1 \le j \le n)$ of the representation of T^* with respect to the bases $\mathscr{G}$ and $\mathscr{E}$, we have

$$T^*(v_i) = \sum_{j=1}^{n} b_{ji} u_j, \qquad i = 1, 2, \ldots, m,$$

where $b_{ji} = (T^*(v_i), u_j)_1$, for $1 \le i \le m$ and $1 \le j \le n$. Hence Theorem 1 implies

$$b_{ji} = (v_i, T(u_j))_2 = \overline{(T(u_j), v_i)_2} = \bar{a}_{ij} \qquad 1 \le i \le m, \quad 1 \le j \le n,$$

and the assertion follows. ■

Exercise 4. Check the assertion of Theorem 2 for the transformations T and $T_1 = T^*$ considered in Exercise 1.

Exercise 5. Deduce from Theorem 2 properties of the conjugate transpose similar to those of the adjoint presented in Exercises 2 and 3.

Exercise 6. Check that for any $T \in \mathscr{L}(\mathscr{U}, \mathscr{V})$,

$$\text{rank } T = \text{rank } T^*,$$

and that, if $\mathscr{U} = \mathscr{V}$, then in addition

$$\text{def } T = \text{def } T^*.$$

Exercise 7. Let $T \in \mathscr{L}(\mathscr{U}, \mathscr{V})$ and let T^* denote the adjoint of T. Show that

$$\mathscr{U} = \text{Ker } T \oplus \text{Im } T^*, \qquad \mathscr{V} = \text{Ker } T^* \oplus \text{Im } T.$$

(See the remark at the end of Section 5.8.)

Hint. Note that $(x, T(y))_2 = 0$ for all $y \in \mathscr{U}$ yields $x \in \text{Ker } T^*$.

Exercise 8. Let $T \in \mathscr{L}(\mathscr{U})$. Check that if a subspace $\mathscr{U}_0 \subset \mathscr{U}$ is T-invariant, then the orthogonal complement to $\mathscr{U}_0$ in $\mathscr{U}$ is T^*-invariant. □

The next example indicates that the relation between the representations of T and T^* with respect to a pair of nonorthonormal bases is more complicated than the one pointed out in Theorem 2.

Exercise 9. Prove that if $T \in \mathscr{L}(\mathscr{U})$, the set $\mathscr{E} = \{u_1, u_2, \ldots, u_n\}$ is an orthogonal basis in $\mathscr{U}$, and A is the representation of T with respect to $\mathscr{E}$, then the representation A' of T^* with respect to the same basis is

$$A' = D^{-1}A^*D, \tag{8}$$

where $D = \operatorname{diag}\{(u_i, u_i)\}_{i=1}^{n}$. □

In particular, the relation (8) shows that $A' = A^*$ provided the basis $\mathscr{E}$ is orthonormal. However, it turns out that the assertion of Theorem 2 remains true if $\mathscr{U} = \mathscr{V}$ and the two sets of basis vectors are biorthogonal in $\mathscr{U}$ (see Section 3.13).

Theorem 3. *Let $T \in \mathscr{L}(\mathscr{U})$ and let $(\mathscr{E}, \mathscr{E}')$ denote a pair of biorthogonal bases in $\mathscr{U}$. If A is the representation of T with respect to $\mathscr{E}$, then A^* is the representation of T^* with respect to $\mathscr{E}'$.*

PROOF. Let $\mathscr{E} = \{u_1, u_2, \ldots, u_n\}$, $\mathscr{E}' = \{u'_1, u'_2, \ldots, u'_n\}$, and $A = [a_{ij}]_{i,j=1}^{n}$. Since $T(u_j) = \sum_{i=1}^{n} a_{ij}u_i$ $(1 \le j \le n)$ and $(u_k, u'_i) = \delta_{ki}$ $(1 \le i, k \le n)$, it follows that $(T(u_j), u'_i) = a_{ij}$. Computing the entries b_{ij} of the representation B of T^* with respect to $\mathscr{E}'$ in a similar way, we obtain $b_{ij} = (T^*(u'_j), u_i)$ $(1 \le i, j \le n)$. Hence for all possible i and j,

$$b_{ij} = (u'_j, T(u_i)) = \overline{(T(u_i), u'_j)} = \overline{a_{ji}},$$

and therefore $B = A^*$. ■

Exercise 10. Check that if $T \in \mathscr{L}(\mathscr{U})$ then $\lambda \in \sigma(T)$ if and only if $\bar{\lambda} \in \sigma(T^*)$.

Hint. See Exercise 4.11.9.

Exercise 11. Prove that the transformation $T \in \mathscr{L}(\mathscr{U})$ is simple if and only if T^* is simple.

Hint. Use Theorem 3.

Exercise 12. If $T \in \mathscr{L}(\mathscr{U})$ is simple, show that there exist eigenbases of T and T^* in $\mathscr{U}$ that are biorthogonal.

Hint. Consider $\operatorname{span}\{x_1, x_2, \ldots, x_{j-1}, x_{j+1}, \ldots, x_n\}$, $1 \le j \le n$, where $\{x_i\}_{i=1}^{n}$ is an eigenbasis of T in $\mathscr{U}$, and use Exercise 8.

Exercise 13. If $T \in \mathscr{L}(\mathscr{U})$ is simple and $\{x_1, x_2, \ldots, x_n\}$ is its eigenbasis in $\mathscr{U}$, prove that a system $\{y_1, y_2, \ldots, y_n\}$ in $\mathscr{U}$ that is biorthogonal to $\{x_i\}_{i=1}^{n}$ is an eigenbasis of T^* in $\mathscr{U}$. Moreover, show that $T(x_i) = \lambda_i x_i$ yields $T^*(y_i) = \bar{\lambda}_i y_i$, $1 \le i \le n$.

Hint. Use Eq. (4).

Exercise 14. Check that the eigenspaces of T and T^* associated with non-complex-conjugate eigenvalues are orthogonal. □

We conclude this section with an important result concerning the general linear equation $T(\mathbf{x}) = \mathbf{b}$, where $T \in \mathscr{L}(\mathscr{U}, \mathscr{V})$ and $\mathbf{b} \in \mathscr{V}$ are given and solutions $\mathbf{x}$ in $\mathscr{U}$ are to be found. This is a general result giving a condition for existence of solutions for *any* choice of $\mathbf{b} \in \mathscr{V}$ in terms of the adjoint operator T^*. It is known as the *Fredholm alternative*. (Compare with Exercise 3.14.10.)

Theorem 4. *Let $T \in \mathscr{L}(\mathscr{U}, \mathscr{V})$. Either the equation $T(\mathbf{x}) = \mathbf{b}$ is solvable for any $\mathbf{b} \in \mathscr{V}$ or the equation $T^*(\mathbf{y}) = \mathbf{0}$ has nonzero solutions.*

PROOF. Since T^* is a linear transformation it is enough to show that $T(\mathbf{x}) = \mathbf{b}$ solvable for any $\mathbf{b} \in \mathscr{V}$ is equivalent to $T^*(\mathbf{y}) = \mathbf{0}$ implies $\mathbf{y} = \mathbf{0}$.

Now the solvability condition clearly implies that $\operatorname{Im} T = \mathscr{V}$. But, by Exercise 7,

$$\mathscr{V} = \operatorname{Im} T \oplus \operatorname{Ker} T^*,$$

consequently $\operatorname{Ker} T^* = \{\mathbf{0}\}$, i.e., $T^*(\mathbf{y}) = \mathbf{0}$ implies $\mathbf{y} = \mathbf{0}$.

Clearly, the argument can be reversed to establish the converse statement. ■

5.2 Normal Transformations and Matrices

It was shown in Section 4.9 that a simple transformation T acting on a linear space $\mathscr{S}$ has an eigenbasis in $\mathscr{S}$. If the space is unitary, then the possibility arises of an *orthonormal eigenbasis* of T, that is, a basis consisting of mutually orthogonal normalized eigenvectors of the transformation. Obviously (see Theorem 4.9.2) transformations T with this property are necessarily simple. Thus, a transformation $T \in \mathscr{L}(\mathscr{U})$, where $\mathscr{U}$ is unitary, is said to be *normal* if there exists an orthonormal basis in $\mathscr{U}$ consisting entirely of eigenvectors of T. The *normal matrices* are defined similarly.

Exercise 1. Show that a transformation $T \in \mathscr{L}(\mathscr{U})$ is normal if and only if the representation of T with respect to any orthonormal basis in $\mathscr{U}$ is a normal matrix.

Hint. See Theorem 4.2.2. □

For a further investigation of normal matrices and transformations, we need some preparations. First, let us agree to use the notation $\mathscr{A}_T^0$ for the set of all matrices $A \in \mathscr{F}^{n \times n}$ such that A is a representation of $T \in \mathscr{L}(\mathscr{U})$ with respect to an *orthonormal* basis in $\mathscr{U}$. Furthermore, recall (Exercise 3.14.12) that the transition matrix $P \in \mathscr{F}^{n \times n}$ from one orthonormal basis in $\mathscr{U}$ into another is a unitary matrix, that is, P satisfies the condition $P^*P = I$. Hence Theorem 4.3.2 implies that any two representations A and A' from $\mathscr{A}_T^0$ are related by the equation $A' = PAP^*$, where P is some unitary matrix.

In general, if the matrix B is similar to A and there exists a unitary transforming matrix U so that

$$B = UAU^* = UAU^{-1},$$

then we say that B is *unitarily similar to* A, and observe that B and A are simultaneously similar and congruent.† It is easily seen that unitary similarity is an equivalence relation. In other words, $\mathscr{A}_T^0$ coincides with the equivalence class of matrix representations of $T \in \mathscr{L}(\mathscr{U})$ with respect to orthonormal bases in $\mathscr{U}$, and all matrices in this class are unitarily similar. Note that the simplest (canonical) matrix in this class is a diagonal matrix of the eigenvalues of T. We now summarize the discussion of this paragraph in a formal statement.

Theorem 1. *An $n \times n$ matrix A is normal if and only if it is unitarily similar to a diagonal matrix of its eigenvalues;*

$$A = UDU^*, \tag{1}$$

where $D = \operatorname{diag}[\lambda_i]_{i=1}^n$, $\lambda_i \in \sigma(A)$, $i = 1, 2, \ldots, n$, and U is unitary.

Note (see Theorem 4.10.2) that the columns of U viewed as vectors from $\mathscr{F}^n$ give the (orthonormal) eigenvectors of A and that D in Eq. (1) is a *canonical form* for matrices in $\mathscr{A}_T^0$ *with respect to unitary similarity.*

Recall that in $\mathbb{R}^{n \times n}$ the unitary matrices are known as *orthogonal* matrices (ref. Exercise 2.6.16) and satisfy the condition $U^{\mathrm{T}}U = I$. Naturally, real matrices A and B from $\mathbb{R}^{n \times n}$ are said to be *orthogonally similar* when $B = UAU^{\mathrm{T}}$ for some orthogonal matrix $U \in \mathbb{R}^{n \times n}$.

Exercise 2. Let A denote a normal matrix. Prove that the matrix A^p is normal for any positive integer p (and, if $\det A \neq 0$, for any integer p).

Exercise 3. Show that the matrix A is normal if and only if A^* is normal and that the eigenspaces associated with mutually conjugate eigenvalues, of A and A^* coincide.

Exercise 4. Check that if A is normal then, for any polynomial $p(\lambda)$ over $\mathbb{C}$, the matrix $p(A)$ is normal.

Exercise 5. Prove a spectral theorem for normal matrices (see Theorem 4.10.3): If A is an $n \times n$ normal matrix with eigenvalues $\lambda_1, \lambda_2, \ldots, \lambda_n$, then there is an associated set of orthonormal eigenvectors $\{\boldsymbol{x}_1, \boldsymbol{x}_2, \ldots, \boldsymbol{x}_n\}$ such that $A = \sum_{j=1}^n \lambda_j \boldsymbol{x}_j \boldsymbol{x}_j^*$. □

If a matrix fails to be normal then by Theorem 1 it cannot be unitarily similar to a diagonal matrix. However, it is interesting to ask how far an

† Congruence of matrices will be studied in Section 5.5.

arbitrary matrix *can be* reduced by means of a unitary similarity. The next theorem, which is due to Schur[†] and Toeplitz[‡], casts some light on this more general question.

Theorem 2. *Any square matrix is unitarily similar to an upper triangular matrix.*

PROOF. Let $A \in \mathscr{F}^{n \times n}$ and let $\mathscr{U}$ be the unitary space of $\mathscr{F}^n$ equipped with the standard inner product. Then, in the natural way, A determines a transformation $T \in \mathscr{L}(\mathscr{U})$. By Exercise 4.14.6 there exists a chain of T-invariant subspaces

$$\{\mathbf{0}\} = \mathscr{U}_0 \subset \mathscr{U}_1 \subset \cdots \subset \mathscr{U}_n = \mathscr{U}$$

such that $\mathscr{U}_i$ has dimension i for each i. It follows that, using the Gram–Schmidt procedure, an orthonormal basis $\{\mathbf{u}_1, \ldots, \mathbf{u}_n\}$ can be constructed for $\mathscr{U}$ with the properties that $\mathbf{u}_i \in \mathscr{U}_i$ and $\mathbf{u}_i \notin \mathscr{U}_{i-1}$ for $i = 1, 2, \ldots, n$. Then the T-invariance of the chain of subspaces implies that the representation of T with respect to this orthonormal basis must be an upper-triangular matrix, B say (see Section 4.8).

But the matrix A is simply the representation of T with respect to the standard basis for $\mathscr{F}^n$ (which is also orthonormal), and so A and B must be unitarily similar. ■

The Schur–Toeplitz theorem plays a crucial role in proving the next beautiful characterization of normal matrices; it is, in fact, often used as the definition of a normal matrix.

Theorem 3. *An $n \times n$ matrix A is normal if and only if*

$$AA^* = A^*A. \tag{2}$$

PROOF. If A is normal, then by Eq. (1) there is a unitary matrix U and a diagonal matrix D for which $A = UDU^*$. Then

$$AA^* = (UDU^*)(U\bar{D}U^*) = UD\bar{D}U^* = U\bar{D}DU^* = (U\bar{D}U^*)(UDU^*) = A^*A,$$

so that A satisfies Eq. (2).

Conversely, we suppose $AA^* = A^*A$ and apply Theorem 2. Thus, there is a unitary matrix U and an upper triangular matrix $S = [s_{ij}]_{i,j=1}^n$ for which $A = USU^*$. It is easily seen that $AA^* = A^*A$ if and only if $SS^* = S^*S$. The 1,1 element of the last equation now reads

$$|s_{11}|^2 + |s_{12}|^2 + \cdots + |s_{1n}|^2 = |s_{11}|^2.$$

† *Math. Annalen* **66** (1909), 488–510.

‡ *Math. Zeitschrift* **2** (1918), 187–197.

Hence $s_{1j} = 0$ for $j = 2, 3, \ldots, n$. The 2,2 element of the equation $SS^* = S^*S$ yields

$$|s_{22}|^2 + |s_{23}|^2 + \cdots + |s_{2n}|^2 = |s_{12}|^2 + |s_{22}|^2,$$

and since we have proved $s_{12} = 0$ it follows that $s_{2j} = 0$ for $j = 3, 4, \ldots, n$. Continuing in this way we find that $s_{jk} = 0$ whenever $j \neq k$ and $1 \leq j, k \leq n$, Thus, S must be a diagonal matrix and $A = USU^*$. Now use Theorem 1. ■

Note that by virtue of Exercise 1 and the relation (4.2.4), a similar result holds for transformations.

Theorem 3′. *A transformation $T \in \mathscr{L}(\mathscr{U})$ is normal if and only if*

$$TT^* = T^*T. \tag{3}$$

Proceeding to special kinds of normal transformations, we first define a transformation $T \in \mathscr{L}(\mathscr{U})$ to be *self-adjoint* if $T = T^*$, and to be *unitary* if $TT^* = T^*T = I$.

Exercise 6. Check that a transformation $T \in \mathscr{L}(\mathscr{U})$ is self-adjoint (respectively, unitary) if and only if its representation with respect to any orthonormal basis is a Hermitian (respectively, unitary) matrix. □

The important results of Exercise 6 allow us to deduce properties of self-adjoint and unitary transformations from those of the corresponding matrices studied in the subsequent sections. If not indicated otherwise, the inner product in $\mathscr{F}^n$ is assumed to be standard, that is, $(\mathbf{x}, \mathbf{y}) = \mathbf{y}^*\mathbf{x}$ for any $\mathbf{x}, \mathbf{y} \in \mathscr{F}^n$.

Exercise 7. If $T \in \mathscr{L}(\mathscr{U})$ is normal and $\{\mathbf{u}_1, \mathbf{u}_2, \ldots, \mathbf{u}_n\}$ is an orthonormal eigenbasis in $\mathscr{U}$, prove that $\{\mathbf{u}_i\}_{i=1}^n$ is also an (orthonormal) eigenbasis of T^* in $\mathscr{U}$. Furthermore, if $T(\mathbf{u}_i) = \lambda_i \mathbf{u}_i$, then $T^*(\mathbf{u}_i) = \bar{\lambda}_i \mathbf{u}_i$ for all $i = 1, 2, \ldots, n$. Obviously, a similar result holds for normal matrices.

Hint. See Exercise 5.1.13.

Exercise 8. Find a real unitary matrix U and a diagonal matrix D for which $A = UDU^{\mathrm{T}}$ if

$$A = \begin{bmatrix} 1 & 0 & -4 \\ 0 & 5 & 4 \\ -4 & 4 & 3 \end{bmatrix}.$$

Exercise 9. If $T \in \mathscr{L}(\mathscr{U})$ is normal and $\mathscr{U}_0$ is a T-invariant subspace in $\mathscr{U}$, show that $T|_{\mathscr{U}_0}$ is a normal transformation.

Exercise 10. Consider a *simple* transformation $T \in \mathscr{L}(\mathscr{S})$. Show that an inner product (,) can be defined on $\mathscr{S}$ such that the transformation $T \in \mathscr{L}(\mathscr{U})$ is normal, where $\mathscr{U} = \mathscr{S}, (\, , \,)$. □

5.3 Hermitian, Skew-Hermitian, and Definite Matrices

The structure and properties of Hermitian matrices are particularly simple and elegant. Because such matrices arise frequently in applications, it is important that we fully understand them. In particular, A is Hermitian if $A^* = A$. Thus, using Theorem 5.2.3, it is apparent that a Hermitian matrix is normal and so they have all the properties of normal matrices established in the previous section.

Theorem 1. *The spectrum of a Hermitian matrix is real.*

PROOF. A Hermitian matrix H is normal and so, by Eq. (5.2.1), $H = UDU^*$, where U is unitary and D is a diagonal matrix of the eigenvalues of H. But $H^* = H$ implies $U\bar{D}U^* = UDU^*$ and hence $\bar{D} = D$, from which the result follows. ■

More can be said: the existence of a real spectrum for a normal matrix A is a necessary and sufficient condition for A to be Hermitian.

Theorem 2. *A normal matrix A is Hermitian if and only if its spectrum lies on the real line.*

PROOF. Let $A^*A = AA^*$ and suppose $\sigma(A) \subset (-\infty, \infty)$. Since $A = UDU^*$ and $A^* = UD^*U^* = UDU^*$, it follows that $A = A^*$ and hence A is Hermitian.

The second part of this theorem is Theorem 1. ■

Thus Hermitian matrices, and only they, have an orthonormal eigenbasis along with a real spectrum. Another important description of Hermitian matrices follows from their definition and Theorem 5.1.1.

Exercise 1. Prove that an $n \times n$ matrix A is Hermitian if and only if $(A\mathbf{x}, \mathbf{y}) = (\mathbf{x}, A\mathbf{y})$ for all $\mathbf{x}, \mathbf{y} \in \mathcal{F}^n$. □

We know that for an arbitrary square matrix, the eigenvectors associated with distinct eigenvalues are necessarily linearly independent (Proposition 4.10.4). If the matrix is Hermitian, it is also normal and, by definition of a normal transformation, it has an *orthonormal* eigenbasis.

Exercise 2. Show (without using the fact that a Hermitian matrix is normal) that the eigenvectors associated with distinct eigenvalues of a Hermitian matrix are orthogonal. □

It turns out that any square matrix A can be represented in terms of a pair of Hermitian matrices. Indeed, it is easy to verify that for $A \in \mathcal{F}^{n \times n}$,

$$A = H_1 + iH_2, \tag{1}$$

where $i^2 = -1$ and H_1 and H_2 are Hermitian matrices defined by

$$H_1 = \mathcal{R}e\, A \triangleq \frac{1}{2}(A + A^*), \qquad H_2 = \mathcal{I}m\, A \triangleq \frac{1}{2i}(A - A^*). \tag{2}$$

The representation (1) is called the *Cartesian decomposition* of A. Its analog in the case $n = 1$ is the Cartesian decomposition of a complex number in the form $z = a + ib$, where a and b are real.

Exercise 3. Check that a matrix A is normal if and only if its *real part* ($\mathcal{R}e\, A$) and *imaginary part* ($\mathcal{I}m\, A$) commute.

Exercise 4. Prove that if A is normal and $\lambda \in \sigma(A)$, then $\lambda = \lambda_1 + i\lambda_2$, where $\lambda_1 \in \sigma(\mathcal{R}e\, A)$ and $\lambda_2 \in \sigma(\mathcal{I}m\, A)$. □

Consider a Hermitian matrix H of order n. Since all eigenvalues of H are real, they can be ordered: $\lambda_1 \le \lambda_2 \le \cdots \le \lambda_n$. Those cases in which all eigenvalues are of one sign are particularly important. A Hermitian matrix A is said to be *positive* (respectively, *negative*) *definite* if all its eigenvalues are positive (respectively, negative). Obviously, a *definite* (i.e., either positive or negative definite) matrix is nonsingular (see Exercise 4.10.3). We shall also need some refinements: A Hermitian matrix is called *positive* (respectively, *negative*) *semi-definite* or *nonnegative* (respectively, *nonpositive*) *definite*, if all its eigenvalues are nonnegative (respectively, nonpositive). A *semi-definite* matrix is understood to be one of the two varieties defined above. We shall write $H > 0$ and $H \ge 0$ for positive definite and positive semi-definite matrices, respectively. Similarly, the notations $H < 0$ and $H \le 0$ will be used for negative definite and negative semi-definite matrices.

The following result is very important and is frequently used in the definition of definite matrices. It has the great advantage of being free of ideas associated with the spectrum of A.

Theorem 3. *Given a Hermitian matrix $H \in \mathcal{F}^{n \times n}$, $H \ge 0$ (or $H > 0$) if and only if $(H\mathbf{x}, \mathbf{x}) \ge 0$ (or $(H\mathbf{x}, \mathbf{x}) > 0$) for all nonzero $\mathbf{x} \in \mathcal{F}^n$.*

PROOF. The necessity of the condition will be shown in Corollary 5.4.1.

Let $(H\mathbf{x}, \mathbf{x}) \ge 0$ for all $\mathbf{x} \in \mathcal{F}^n$ and let the eigenvectors $\mathbf{u}_1, \mathbf{u}_2, \ldots, \mathbf{u}_n$ of H associated with the eigenvalues $\lambda_1, \lambda_2, \ldots, \lambda_n$, respectively, form an orthonormal basis in $\mathcal{F}^n$. Since

$$0 \le (H\mathbf{u}_i, \mathbf{u}_i) = \lambda_i(\mathbf{u}_i, \mathbf{u}_i) = \lambda_i$$

for $i = 1, 2, \ldots, n$, the matrix H is positive semi-definite.

The positive definite case is proved similarly. ■

Note that results similar to Theorem 3 hold for negative definite and semi-definite matrices.

Exercise 5. Show that for any (even rectangular) matrix A, the matrices A^*A and AA^* are positive semi-definite.

Hint. Let A be $m \times n$ and consider $(A^*A\mathbf{x}, \mathbf{x})$ for $\mathbf{x} \in \mathcal{F}^n$.

Exercise 6. Confirm that for any $A \in \mathcal{F}^{m \times n}$,

$$\operatorname{Ker}(A^*A) = \operatorname{Ker} A, \qquad \operatorname{Im}(A^*A) = \operatorname{Im} A^*.$$

Exercise 7. Deduce from Exercise 6 that the matrix A^*A (where $A \in \mathcal{F}^{m \times n}$ and $m \geq n$) is positive definite if and only if A is of full rank.

Exercise 8. Check that if H_1 and H_2 are positive definite and $\alpha, \beta \in \mathbb{R}$ are nonnegative real numbers, not both zero, then the matrices H_1^{-1} and $\alpha H_1 + \beta H_2$ are positive definite.

Exercise 9. Show that if H is positive semi-definite and rank $H = r$, then H has exactly r positive eigenvalues.

Exercise 10. Show that if H is any positive definite matrix, then the function $(\ ,\ )$ defined on $\mathbb{C}^n \times \mathbb{C}^n$ by $(\mathbf{x}, \mathbf{y}) = \mathbf{y}^*H\mathbf{x}$ is an inner product on $\mathbb{C}^n$. □

Skew-Hermitian matrices have a number of properties that can be seen as analogs to those of Hermitian matrices. The next two exercises give results of this kind.

Exercise 11. Prove that the spectrum of a skew-Hermitian matrix is pure imaginary.

Exercise 12. Show that a normal matrix is skew-Hermitian if and only if its spectrum lies on the imaginary axis.

Exercise 13. Let A be Hermitian and let H be a positive definite matrix of the same size. Show that AH has only real eigenvalues.

Exercise 14. Use Exercise 2.10.2 to show that, if H is positive definite, there is a unique lower triangular matrix G with positive diagonal elements such that $H = GG^*$. (This is known as the *Cholesky factorization* of H.) □

5.4 Square Root of a Definite Matrix and Singular Values

An analogy between Hermitian matrices and real numbers can be seen in the following result, which states the existence of a *square root* H_0 of a positive semi-definite matrix H, that is, a matrix H_0 such that $H_0^2 = H$.

Theorem 1. *A matrix H is positive definite (or semi-definite) if and only if it has a positive definite (respectively, semi-definite) square root H_0. Moreover,* rank H_0 = rank H.

PROOF. If $H \geq 0$ then, by definition, its eigenvalues $\{\lambda_i\}_{i=1}^n$ are nonnegative and we can define a real matrix $D_0 = \text{diag}[\sqrt{\lambda_1}, \sqrt{\lambda_2}, \ldots, \sqrt{\lambda_n}]$. The matrix H is normal and hence there is a unitary matrix U such that $H = UDU^*$, where $D = D_0^2$. The required square root of H is the matrix

$$H_0 = UD_0U^*, \tag{1}$$

since $H_0^2 = UD_0U^*UD_0U^* = UD_0^2U^* = H$. Note that the representation (1) shows that the eigenvalues of H_0 are the (arithmetic) square roots of the eigenvalues of H. This proves that rank H_0 = rank H (Exercise 5.3.9), and $H \geq 0$ yields $H_0 \geq 0$.

Conversely, if $H = H_0^2$ and $H_0 \geq 0$, then the eigenvalues of H, being the squares of those of H_0, are nonnegative. Hence $H \geq 0$. This fact also implies the equality of ranks.

The same argument proves the theorem for positive definite matrices. ∎

The following simple corollary gives a proof of the necessary part of Theorem 5.3.3.

Corollary 1. *If $H \geq 0$ (or $H > 0$), then $(Hx, x) \geq 0$ (or $(Hx, x) > 0$) for all $x \in \mathscr{F}^n$.*

PROOF. Representing $H = H_0^2$ and using the matrix version of Theorem 5.1.1, we have

$$(Hx, x) = (H_0^2x, x) = (H_0x, H_0x) \geq 0 \qquad \text{for all } x \in \mathscr{F}^n. \quad ∎$$

Exercise 1. Prove that if H is positive semi-definite and $(Hx, x) = 0$ for some $x \in \mathscr{F}^n$, then $Hx = \mathbf{0}$. □

Note that the square root of a positive semi-definite matrix cannot be unique since any matrix of the form (1) with a diagonal matrix

$$D = \text{diag}[\pm\sqrt{\lambda_1}, \pm\sqrt{\lambda_2}, \ldots, \pm\sqrt{\lambda_n}]$$

is a square root of H. However, the positive semi-definite square root H_0 of H is unique.

Proposition 1. *Let H be positive semi-definite. The positive semi-definite square root H_0 of H is unique and is given by Eq.* (1).

PROOF. Let H_1 satisfy $H_1^2 = H$. By Theorem 4.11.3, the eigenvalues of H_1 are square roots of those of H and, since $H_1 \geq 0$, they must be nonnegative. Thus, the eigenvalues of H_1 and H_0 of Eq. (1) coincide. Furthermore, H_1

is Hermitian and, therefore (Theorem 5.2.1), $H_1 = VD_0V^*$ for some unitary V. Now $H_1^2 = H_0^2 = H$, so $VD_0^2V^* = UD_0^2U^*$ and hence $(U^*V)D_0^2 = D_0^2(U^*V)$. It is easily seen that, because $D_0 \geq 0$, this implies $(U^*V)D_0 = D_0(U^*V)$ and, consequently, $H_1 = H_0$, as required. ■

In the sequel the unique positive semi-definite (or definite) square root of a positive semi-definite (or definite) matrix H is denoted by $H^{1/2}$. Summarizing the above discussion, note that $\lambda \in \sigma(H^{1/2})$ if and only if $\lambda^2 \in \sigma(H)$, and the corresponding eigenspaces of $H^{1/2}$ and H coincide. The concept of a square root of a positive semi-definite matrix allows us to introduce a spectral characteristic for *rectangular* matrices.

Consider an arbitrary $m \times n$ matrix A. The $n \times n$ matrix A^*A is (generally) positive semi-definite (see Exercise 5.3.5). Therefore by Theorem 1 the matrix A^*A has a positive semi-definite square root H_1 such that $A^*A = H_1^2$. The eigenvalues $\lambda_1, \lambda_2, \ldots, \lambda_n$ of the matrix $H_1 = (A^*A)^{1/2}$ are referred to as the *singular values* $s_1, s_2, \ldots, s_n$ of the (generally rectangular) matrix A. Thus, for $i = 1, 2, \ldots, n$,

$$s_i(A) \triangleq \lambda_i((A^*A)^{1/2}).$$

Obviously, the singular values of a matrix are nonnegative numbers.

Exercise 2. Check that $s_1 = \sqrt{2}$, $s_2 = 1$ are singular values of the matrix

$$A = \begin{bmatrix} 1 & 0 \\ 0 & 1 \\ 1 & 0 \end{bmatrix}. \quad \square$$

Note that the singular values of A are sometimes defined as eigenvalues of the matrix $(AA^*)^{1/2}$ of order m. It follows from the next fact that the difference in definition is not highly significant.

Theorem 2. *The nonzero eigenvalues of the matrices $(A^*A)^{1/2}$ and $(AA^*)^{1/2}$ coincide.*

PROOF. First we observe that it suffices to prove the assertion of the theorem for the matices A^*A and AA^*. Furthermore, we select eigenvectors $\mathbf{x}_1, \mathbf{x}_2, \ldots, \mathbf{x}_n$ of A^*A corresponding to the eigenvalues $\lambda_1, \lambda_2, \ldots, \lambda_n$ such that $\{\mathbf{x}_1, \mathbf{x}_2, \ldots, \mathbf{x}_n\}$ forms an orthonormal basis in $\mathscr{F}^n$. We have

$$(A^*A\mathbf{x}_i, \mathbf{x}_j) = \lambda_i(\mathbf{x}_i, \mathbf{x}_j) = \lambda_i\delta_{ij}, \qquad 1 \leq i, j \leq n.$$

On the other hand, $(A^*A\mathbf{x}_i, \mathbf{x}_j) = (A\mathbf{x}_i, A\mathbf{x}_j)$, and comparison shows that $(A\mathbf{x}_i, A\mathbf{x}_i) = \lambda_i$, $i = 1, 2, \ldots, n$. Thus, $A\mathbf{x}_i = \mathbf{0}$ $(1 \leq i \leq n)$ if and only if $\lambda_i = 0$. Since

$$AA^*(A\mathbf{x}_i) = A(A^*A\mathbf{x}_i) = \lambda_i(A\mathbf{x}_i), \qquad 1 \leq i \leq n,$$

the preceding remark shows that for $\lambda_i \neq 0$, the vector $A\mathbf{x}_i$ is an eigenvector of AA^*. Hence if a nonzero $\lambda_i \in \sigma(A^*A)$ is associated with the eigenvector

$\boldsymbol{x}_i$, then $\lambda_i \in \sigma(AA^*)$ and is associated with the eigenvector $A\boldsymbol{x}_i$. In particular, $\sigma(A^*A) \subset \sigma(AA^*)$. The opposite inclusion is obtained by exchanging the roles of A and A^*. ■

Thus the eigenvalues of A^*A and AA^*, as well as of $(A^*A)^{1/2}$ and $(AA^*)^{1/2}$, differ only by the geometric multiplicity of the zero eigenvalue, which is $n - r$ for A^*A and $m - r$ for AA^*, where $r = \text{rank}(A^*A) = \text{rank}(AA^*)$. Also, for a square matrix A it follows immediately that the eigenvalues of A^*A and AA^* coincide and have the same multiplicities. [Compare this with the result of Exercise 4.14.10(b).]

Note that we proved more than was stated in Theorem 2.

Exercise 3. Show that if $\boldsymbol{x}_1, \boldsymbol{x}_2, \ldots, \boldsymbol{x}_k$ are orthonormal eigenvectors of A^*A corresponding to nonzero eigenvalues, then $A\boldsymbol{x}_1, A\boldsymbol{x}_2, \ldots, A\boldsymbol{x}_k$ are orthogonal eigenvectors of AA^* corresponding to the same eigenvalues. A converse statement is obtained by replacing A by A^*.

Exercise 4. Verify that the nonzero singular values of the matrices A and A^* are the same. If, in addition, A is a square $n \times n$ matrix, show that $s_i(A) = s_i(A^*)$ for $i = 1, 2, \ldots, n$. □

In the rest of this section only square matrices are considered.

Proposition 2. *The singular values of a square matrix are invariant under unitary transformation.*

In other words, for any unitary matrix $U \in \mathbb{C}^{n \times n}$ and any $A \in \mathbb{C}^{n \times n}$,

$$s_i(A) = s_i(UA) = s_i(AU), \qquad i = 1, 2, \ldots, n. \tag{2}$$

PROOF. By definition,

$$s_i(UA) = \lambda_i((A^*U^*UA)^{1/2}) = \lambda_i((A^*A)^{1/2}) = s_i(A).$$

To prove the second equality in Eq. (2), use Exercise 4 and the part of the proposition already proved. ■

Theorem 3. *For an $n \times n$ normal matrix $A \in \mathscr{F}^{n \times n}$,*

$$s_i(A) = |\lambda_i(A)|, \qquad i = 1, 2, \ldots, n.$$

PROOF. Let λ_i denote an eigenvalue of A associated with the eigenvector $\boldsymbol{x}_i$. Since $A^*A = AA^*$, it follows (see Exercise 5.2.7) that

$$A^*A\boldsymbol{x}_i = \lambda_i A^*\boldsymbol{x}_i = \lambda_i \bar{\lambda}_i \boldsymbol{x}_i = |\lambda_i|^2 \boldsymbol{x}_i. \tag{3}$$

Hence $\boldsymbol{x}_i$ is an eigenvector of A^*A corresponding to the eigenvalue $|\lambda_i|^2$. By definition, $|\lambda_i|$ is a singular value of A. Note that A has no other singular values, since in varying i from 1 to n in Eq. (3), all the eigenvalues of A^*A are obtained. ■

Exercise 5. Check that for any $n \times n$ matrix A,

$$|\det A| = s_1(A)s_2(A)\cdots s_n(A).$$

Hint. If $H^2 = A^*A$, where $H \geq 0$, then $\det H = |\det A|$.

Exercise 6. Confirm that a square matrix is unitary if and only if all its singular values are equal to one.

Exercise 7. Let $A \in \mathbb{C}^{m \times n}$ and $A = QR$ be the "Q–R decomposition" of A described in Exercise 3.12.7. Show that A and R have the same singular values. □

5.5 Congruence and the Inertia of a Matrix

The notion of unitary similarity discussed in Section 5.2 can be generalized in the following way. Two square matrices A and B are said to be *congruent* if there exists a nonsingular matrix P such that

$$A = PBP^*. \tag{1}$$

It is clear that for a unitary P the matrix A is unitarily similar to B. Also, congruence is an equivalence relation on $\mathcal{F}^{n \times n}$. It is also obvious from Eq. (1) that if A is Hermitian then all matrices in the equivalence class of A are Hermitian. In Theorem 1 it will be shown that there is a particularly simple canonical matrix in each such class of Hermitian matrices.

Theorem 1. *A Hermitian matrix $H \in \mathcal{F}^{n \times n}$ is congruent to the matrix*

$$D_0 = \begin{bmatrix} I_s & 0 & 0 \\ 0 & -I_{r-s} & 0 \\ 0 & 0 & 0_{n-r} \end{bmatrix}, \tag{2}$$

where $r =$ rank H and s is the number of positive eigenvalues of H counted according to multiplicities.

In other words, the matrix D_0 in Eq. (2) is the *canonical form of H with respect to congruence*; it is the simplest representative of the equivalence class of Hermitian matrices congruent to H.

PROOF. By Theorem 5.2.1,

$$H = UDU^*, \tag{3}$$

where D is a diagonal matrix of the eigenvalues of H, and U is unitary.

Ordering the eigenvalues so that the first s scalars $\lambda_1, \lambda_2, \ldots, \lambda_s$ on the main diagonal of D are positive and the next $r - s$ numbers $\lambda_{s+1}, \lambda_{s+2}, \ldots, \lambda_r$ are negative, we may write $D = U_1 D_1 D_0 D_1 U_1^*$, where

$$D_1 = \operatorname{diag}[\sqrt{\lambda_1}, \sqrt{\lambda_2}, \ldots, \sqrt{\lambda_s}, \sqrt{|\lambda_{s+1}|}, \ldots, \sqrt{|\lambda_r|}, 0, \ldots, 0],$$

D_0 is given by Eq. (2), and U_1 is a permutation (and therefore a unitary) matrix. Hence, substituting into (3), we obtain $H = PD_0P^*$, where $P = UU_1D_1$. The theorem is established. ■

Corollary 1. *A Hermitian matrix $H \in \mathscr{F}^{n \times n}$ is positive definite if and only if it is congruent to the identity matrix.*

In other words, $H > 0$ if and only if $H = PP^*$ for some nonsingular matrix P.

Corollary 2. *A Hermitian matrix $H \in \mathscr{F}^{n \times n}$ of rank r is positive semi-definite if and only if it is congruent to a matrix of the form*

$$\begin{bmatrix} I_r & 0 \\ 0 & 0_{n-r} \end{bmatrix}.$$

Thus $H \geq O$ if and only if $H = P^*P$ for some P.

Corollary 3. *Two Hermitian matrices $A, B \in \mathscr{F}^{n \times n}$ are congruent if and only if* rank A = rank B *and the number of positive eigenvalues (counting multiplicities) of both matrices is the same.*

We already know that a Hermitian matrix H can be reduced by a congruence transformation to a diagonal matrix. If several Hermitian matrices are given, then each of them can be reduced to diagonal form by means of (generally distinct) congruence transformations.

The important problem of the simultaneous reduction of *two* Hermitian matrices to diagonal form using the *same* congruence transformation will now be discussed. This turns out to be possible provided one of the Hermitian matrices is positive (or negative) definite. If this condition is relaxed, the question of canonical reduction becomes significantly more complicated.

Theorem 2. *If H_1 and H_2 are two $n \times n$ Hermitian matrices and at least one of them, say H_1, is positive definite, then there exists a nonsingular matrix Q such that*

$$Q^*H_1Q = I \qquad \text{and} \qquad Q^*H_2Q = D, \tag{4}$$

where $D = \operatorname{diag}[\lambda_1, \ldots, \lambda_n]$ and $\lambda_1, \lambda_2, \ldots, \lambda_n$ are the eigenvalues of $H_1^{-1}H_2$, all of which are real.

Furthermore, if $\boldsymbol{q}_j$ denotes the jth *column of Q, then for $j = 1, 2, \ldots, n$,*

$$(\lambda_j H_1 - H_2)\boldsymbol{q}_j = \mathbf{0}. \tag{5}$$

PROOF. Let $H_1^{1/2}$ be the positive definite square root of H_1 and define $H_1^{-1/2} = (H_1^{1/2})^{-1}$. Observe that $H = H_1^{-1/2}H_2H_1^{-1/2}$ is Hermitian, and so there is a unitary matrix U such that $U^*HU = \operatorname{diag}[\lambda_1, \lambda_2, \ldots, \lambda_n] \triangleq D$, a real diagonal matrix, where $\lambda_1, \lambda_2, \ldots, \lambda_n$ are the eigenvalues of H. Since $H = H_1^{1/2}(H_1^{-1}H_2)H_1^{-1/2}$, the matrices H and $H_1^{-1}H_2$ are similar. It follows that $\lambda_1, \lambda_2, \ldots, \lambda_n$ are also the eigenvalues of $H_1^{-1}H_2$ and are real.

Now define $Q = H_1^{-1/2}U$; we have

$$\begin{aligned} Q^*H_1Q &= U^*H_1^{-1/2}H_1H_1^{-1/2}U = U^*U = I, \\ Q^*H_2Q &= U^*H_1^{-1/2}H_2H_1^{-1/2}U = U^*HU = D, \end{aligned} \tag{6}$$

as required for the first part of the theorem.

It is clear from Eqs. (6) that

$$(Q^*H_1Q)D - Q^*H_2Q = 0,$$

and this implies $H_1QD - H_2Q = O$. The jth column of this relation is Eq. (5). ■

Exercise 1. Show that if the matrices H_1 and H_2 of Theorem 2 are real, then there is a real transforming matrix Q with the required properties.

Exercise 2. Prove that the eigenvalues of $H_1^{-1}H_2$ in Theorem 2 are real by first defining an inner product $\langle \mathbf{x}, \mathbf{y} \rangle = \mathbf{y}^*H_1\mathbf{x}$ and then showing that $H_1^{-1}H_2$ is Hermitian in this inner product. □

We conclude our discussion on the simultaneous reduction of a pair of Hermitian matrices to a diagonal form by noting that this is not always possible.

Exercise 3. Show that there is no nonsingular matrix Q such that both matrices

$$Q^*\begin{bmatrix} 0 & 1 \\ 1 & 0 \end{bmatrix}Q \quad \text{and} \quad Q^*\begin{bmatrix} 0 & 0 \\ 0 & 1 \end{bmatrix}Q$$

are diagonal. □

In the study of differential equations, the analysis of stability of solutions requires information on the location of the eigenvalues of matrices in relation to the imaginary axis. Many important conclusions can be deduced by knowing only how many eigenvalues are located in the right and left half-planes and on the imaginary axis itself. These numbers are summarized in the notion of "inertia." The *inertia* of a square matrix $A \in \mathcal{F}^{n \times n}$, written In A, is the triple of integers

$$\operatorname{In} A = \{\pi(A), \nu(A), \delta(A)\},$$

where $\pi(A)$ $\nu(A)$ and $\delta(A)$ denote the number of eigenvalues of A, counted with their algebraic multiplicities, lying in the open right half-plane, in the open left half-plane, and on the imaginary axis, respectively. Obviously,

$$\pi(A) + \nu(A) + \delta(A) = n$$

and the matrix A is nonsingular if $\delta(A) = 0$.

In the particular case of a Hermitian matrix H, the number $\pi(H)$ (respectively, $\nu(H)$) merely denotes the number of positive (respectively, negative) eigenvalues counted with their multiplicities. Furthermore, $\delta(H)$ is equal to the number of zero eigenvalues of H, and hence, H is nonsingular if and only if $\delta(H) = 0$. Note also that for a Hermitian matrix H,

$$\pi(H) + \nu(H) = \text{rank } H.$$

The difference $\pi(H) - \nu(H)$, written sig H, is referred to as the *signature* of H.

The next result gives a useful sufficient condition for two matrices to have the same inertia. The corollary is a classical result that can be proved more readily by a direct calculation. The interest of the theorem lies in the fact that matrix M of the hypothesis may be singular.

Theorem 3. *Let A and B be $n \times n$ Hermitian matrices of the same rank r. If $A = MBM^*$ for some matrix M, then* In A = In B.

PROOF. By Theorem 1 there exist nonsingular matrices P and Q such that

$$PAP^* = \text{diag}[I_t, \quad -I_{r-t}, \quad 0] \triangleq D_0^{(1)}$$

and

$$Q^{-1}B(Q^{-1})^* = \text{diag}[I_s, \quad -I_{r-s}, \quad 0] \triangleq D_0^{(2)},$$

where t and s denote the number of positive eigenvalues of A and B, respectively; $t = \pi(A)$ and $s = \pi(B)$.

To prove the theorem it suffices to show that $s = t$. Observe that, since $A = MBM^*$,

$$D_0^{(1)} = PAP^* = PM(QD_0^{(2)}Q^*)M^*P^*$$

or, writing $R = PMQ$,

$$D_0^{(1)} = RD_0^{(2)}R^*. \tag{7}$$

Now suppose $s < t$ and let us seek a contradiction. Let $\mathbf{x} \in \mathscr{F}^n$ be given by $\boldsymbol{x} = \begin{bmatrix} \boldsymbol{x}_1 \\ \mathbf{0} \end{bmatrix}$, where $\boldsymbol{x}_1 = [x_1 \quad x_2 \quad \cdots \quad x_t]^{\mathrm{T}} \in \mathscr{F}^t$ and $\boldsymbol{x}_1 \neq \mathbf{0}$. Then

$$\boldsymbol{x}^* D_0^{(1)} \boldsymbol{x} = \sum_{i=1}^{t} |x_i|^2 > 0. \tag{8}$$

Partition R^* in the form

$$R^* = \begin{bmatrix} R_1 & R_2 \\ R_3 & R_4 \end{bmatrix},$$

where R_1 is $s \times t$. Since $s < t$, an x_1 can be chosen so that $x_1 \neq 0$ and $R_1 x_1 = 0$. Then define $y = R_3 x_1 \in \mathscr{F}^{n-s}$ and we have $R^* x = \begin{bmatrix} 0_s \\ y \end{bmatrix}$. Consequently, using Eq. (7) and writing $y = [y_1, y_2, \ldots, y_{n-s}]^{\mathrm{T}}$,

$$x^* D_0^{(1)} x = (R^* x)^* D_0^{(2)} (R^* x) = -\sum_{j=1}^{r-s} |y_j|^2 \leq 0,$$

and a contradiction with Eq. (8) is achieved.

Similarly, interchanging the roles of $D_0^{(1)}$ and $D_0^{(2)}$, it is found that $t < s$ is impossible. Hence $s = t$ and the theorem is proved. ■

Corollary 1. (Sylvester's law of inertia). *Congruent Hermitian matrices have the same inertia characteristics.*

PROOF. If $A = PBP^*$ and P is nonsingular, then rank A = rank B and the result follows. ■

Exercise 4. Show that the inertia of the Hermitian matrix

$$\begin{bmatrix} 0 & iI_n \\ -iI_n & 0 \end{bmatrix}$$

is $(n, n, 0)$. □

5.6 Unitary Matrices

We now continue the study of unitary matrices. Recall that a square matrix U is unitary if it satisfies the condition $U^*U = UU^* = I$.

Theorem 1. *The spectrum of a unitary matrix lies on the unit circle.*

PROOF. If $Ux = \lambda x$ and $(x, x) = 1$, we obtain

$$(Ux, Ux) = (U^*Ux, x) = (x, x) = 1.$$

On the other hand,

$$(Ux, Ux) = (\lambda x, \lambda x) = |\lambda|^2 (x, x) = |\lambda|^2.$$

The comparison yields $|\lambda|^2 = 1$ and the desired result. ■

The converse of this statement is also true for normal matrices. (It is an analog of Theorem 5.3.2 for Hermitian matrices).

Theorem 2. *A normal matrix $A \in \mathscr{F}^{n \times n}$ is unitary if all its eigenvalues lie on the unit circle.*

PROOF. Let $Ax_i = \lambda_i x_i$, where $|\lambda_i| = 1$ and $(x_i, x_j) = \delta_{ij}$ $(1 \le i, j \le n)$. For any vector $x \in \mathscr{F}^n$, write $x = \sum_{i=1}^n \alpha_i x_i$; then it is found that

$$A^*Ax = A^*\left(\sum_{i=1}^n \alpha_i Ax_i\right) = A^*\left(\sum_{i=1}^n \alpha_i \lambda_i x_i\right). \tag{1}$$

From Exercise 5.2.7 we know that $A^*x_i = \bar{\lambda}_i x_i$ $(i = 1, 2, \ldots, n)$, so Eq. (1) becomes

$$A^*Ax = \sum_{i=1}^n \alpha_i \lambda_i A^*x_i = \sum_{i=1}^n \alpha_i |\lambda_i|^2 x_i = \sum_{i=1}^n \alpha_i x_i = x.$$

Thus $A^*Ax = x$ for every $x \in \mathscr{F}^n$ and therefore $A^*A = I$. The relation, $AA^* = I$ now follows since a left inverse is also a right inverse (see Section 2.6). Hence the matrix A is unitary. ■

Thus, a normal matrix is unitary if and only if its eigenvalues lie on the unit circle. Unitary matrices can also be described in terms of the standard inner product. In particular, the next result shows that if U is unitary, then for any $\mathbf{x}$ the vectors $\mathbf{x}$ and $U\mathbf{x}$ have the same Euclidean length or norm.

Theorem 3. *A matrix $U \in \mathscr{F}^{n \times n}$ is unitary if and only if*

$$(Ux, Uy) = (x, y) \tag{2}$$

for all $x, y \in \mathscr{F}^n$.

PROOF. If $U^*U = I$, then

$$(Ux, Uy) = (U^*Ux, y) = (x, y)$$

and Eq. (2) holds. Conversely, from Eq. (2) it follows that $(U^*Ux, y) = (x, y)$ or, equivalently, $((U^*U - I)x, y) = 0$ for all $x, y \in \mathscr{F}^n$. Referring to Exercise 3.13.2, we immediately obtain $U^*U = I$. Consequently, $UU^* = I$ also, and U is a unitary matrix. ■

Observe that Eq. (2) holds for every $x, y \in \mathscr{F}^n$ if and only if it is true for the vectors of a basis in $\mathscr{F}^n$. (Compare the statement of the next Corollary with Exercise 3.14.12.)

Corollary 1. *A matrix $U \in \mathscr{F}^{n \times n}$ is unitary if and only if it transforms an orthonormal basis for $\mathscr{F}^n$ into an orthonormal basis.*

Now take the elements e_i ($i = 1, 2, \ldots, n$) of the standard basis in $\mathscr{F}^n$ for x and y in Eq. (2).

Corollary 2. *A matrix $U \in \mathscr{F}^{n \times n}$ is unitary if and only if the columns of U, viewed as vectors of $\mathscr{F}^n$, constitute an orthonormal basis in $\mathscr{F}^n$.*

Note, in conclusion, that some properties of unitary matrices are reminiscent of those of the unimodular complex numbers $e^{i\gamma}$, where $0 \leq \gamma < 2\pi$. In particular, it follows from Theorem 1 that all eigenvalues of a unitary matrix are of this form. The polar form of more general complex numbers is the starting point for our next topic.

Exercise 1. (Householder transformations). Let x be a nonzero vector in $\mathbb{C}^n$ and define the $n \times n$ matrix

$$U = I - \alpha x x^*,$$

where $\alpha = 2(x^*x)^{-1}$. Show that U is both Hermitian and unitary. Find the n eigenvalues of U and an associated orthonormal eigenbasis. □

5.7 Polar and Singular-Value Decompositions

The following result has its origin in the familiar polar form of a complex number: $\lambda = \lambda_0 e^{i\gamma}$, where $\lambda_0 \geq 0$ and $0 \leq \gamma < 2\pi$.

Theorem 1. *Any matrix $A \in \mathscr{F}^{n \times n}$ can be represented in the form*

$$A = HU, \tag{1}$$

where $H \geq 0$ and U is unitary. Moreover, the matrix H in Eq. (1) *is unique and is given by $H = (AA^*)^{1/2}$.*

PROOF. Let

$$\lambda_1 \geq \lambda_2 \geq \cdots \geq \lambda_r > 0 = \lambda_{r+1} = \cdots = \lambda_n$$

denote the eigenvalues of A^*A (see Theorem 5.4.2) with corresponding eigenvectors $x_1, x_2, \ldots, x_n$ that comprise an orthonormal basis in $\mathscr{F}^n$.

Then, by Exercise 5.4.3, the normalized elements

$$y_i = \frac{1}{\|Ax_i\|} Ax_i = \frac{1}{\sqrt{\lambda_i}} Ax_i, \qquad i = 1, 2, \ldots, r, \tag{2}$$

are orthonormal eigenvectors of AA^* corresponding to the eigenvalues $\lambda_1, \lambda_2, \ldots, \lambda_r$, respectively. Choosing an orthonormal basis $\{\boldsymbol{y}_{r+1}, \ldots, \boldsymbol{y}_n\}$ in Ker AA^*, we extend it to an orthonormal eigenbasis $\{\boldsymbol{y}_i\}_{i=1}^n$ for AA^*.

Proceeding to a construction of the matrices H and U in Eq. (1), we write $H = (AA^*)^{1/2}$ and note (see Section 5.4) that $H\boldsymbol{y}_i = (\lambda_i)^{1/2}\boldsymbol{y}_i$ for $i = 1, 2, \ldots, n$. Also, we introduce an $n \times n$ (transition) matrix U by $U\boldsymbol{x}_i = \boldsymbol{y}_i, i = 1, 2, \ldots, n$. Note that Corollary 5.6.1 asserts that U is unitary. We have

$$HU\boldsymbol{x}_i = H\boldsymbol{y}_i = \sqrt{\lambda_i}\,\boldsymbol{y}_i = A\boldsymbol{x}_i, \qquad i = 1, 2, \ldots, r. \tag{3}$$

Since $(A\boldsymbol{x}_i, A\boldsymbol{x}_i) = (A^*A\boldsymbol{x}_i, \boldsymbol{x}_i) = \lambda_i = 0$ for $i = r + 1, r + 2, \ldots, n$, it follows that $A\boldsymbol{x}_i = \boldsymbol{0}$ $(r + 1 \le i \le n)$. Furthermore, as observed in Section 5.4, $AA^*\boldsymbol{y}_i = \boldsymbol{0}$ implies $H\boldsymbol{y}_i = \boldsymbol{0}$. Thus

$$HU\boldsymbol{x}_i = H\boldsymbol{y}_i = \boldsymbol{0} = A\boldsymbol{x}_i, \qquad i = r + 1, \ldots, n. \tag{4}$$

The equalities (3) and (4) for basis elements clearly give $HU\boldsymbol{x} = A\boldsymbol{x}$ for all $\boldsymbol{x} \in \mathscr{F}^n$, proving Eq. (1). ■

Note that if A is nonsingular, so is $(AA^*)^{1/2}$ and in the *polar decomposition* (1), the matrix $H = (AA^*)^{1/2}$ is positive definite. Observe that in this case the unitary matrix U can be chosen to be $H^{-1}A$ and the representation (1) is unique.

A dual polar decomposition can be established similarly.

Exercise 1. Show that any $A \in \mathscr{F}^{n \times n}$ can be represented in the form

$$A = U_1 H_1, \tag{5}$$

where $H_1 = (A^*A)^{1/2}$ and U_1 is unitary. □

For normal matrices both decompositions coincide in the following sense.

Proposition 1. *A matrix $A \in \mathscr{F}^{n \times n}$ is normal if and only if the matrices H and U in Eq.* (1) *commute.*

PROOF. If $A = HU = UH$, then

$$A^*A = (HU^*)(UH) = H^2 = (HU)(U^*H) = AA^*$$

and A is normal.

Conversely, let A be normal and adopt the notations used in the proof of Theorem 1. Since $AA^* = A^*A$, the system $\{\boldsymbol{x}_i\}_{i=1}^n$ is an orthonormal eigenbasis for AA^* and, consequently, for $H = (AA^*)^{1/2}$. Hence, for $i = 1, 2, \ldots, n$,

$$UH\boldsymbol{x}_i = \sqrt{\lambda_i}\, U\boldsymbol{x}_i. \tag{6}$$

Furthermore, from $A = HU$ it follows that $\lambda_i\boldsymbol{x}_i = A^*A\boldsymbol{x}_i = U^*H^2U\boldsymbol{x}_i$, and consequently $H^2U\boldsymbol{x}_i = \lambda_i U\boldsymbol{x}_i$. Thus the vectors $U\boldsymbol{x}_i$ are eigenvectors of

H^2 corresponding to the eigenvalues λ_i $(i = 1, 2, \ldots, n)$. Then the matrix H has the same eigenvectors corresponding to the eigenvalues $\sqrt{\lambda_i}$, respectively:

$$HU\mathbf{x}_i = \sqrt{\lambda_i}\, U\mathbf{x}_i, \qquad i = 1, 2, \ldots, n.$$

Comparing this with Eq. (6), it is deduced that $UH = HU$, as required. ■

Exercise 2. Check that if $\lambda = \lambda_0 e^{i\gamma}$ $(\lambda_0 > 0)$ is a nonzero eigenvalue of a normal matrix A, then λ_0 is an eigenvalue of H while $e^{i\gamma}$ is an eigenvalue of U in the polar decomposition (1).

Exercise 3. Check that the representation

$$A \triangleq \begin{bmatrix} 0 & 2 \\ 0 & 2 \end{bmatrix} = \begin{bmatrix} \alpha^{-1} & \alpha^{-1} \\ \alpha^{-1} & \alpha^{-1} \end{bmatrix} \begin{bmatrix} \alpha & \alpha \\ -\alpha & \alpha \end{bmatrix}, \qquad \alpha = \frac{1}{\sqrt{2}},$$

is a polar decomposition of A. Find its dual polar decomposition. □

The procedure used in proving Theorem 1 can be developed to prove another important result.

Theorem 2. *Let A denote an arbitrary matrix from $\mathscr{F}^{m \times n}$ and let $\{s_i\}_{i=1}^r$ be the nonzero singular values of A. Then A can be represented in the form*

$$A = UDV^*, \tag{7}$$

where $U \in \mathscr{F}^{m \times m}$ and $V \in \mathscr{F}^{n \times n}$ are unitary and the $m \times n$ matrix D has s_i in the i, i position $(1 \le i \le r)$ and zeros elsewhere.

The representation (7) is referred to as a *singular-value decomposition* of the matrix A.

PROOF. Let us agree to preserve the notation introduced in the proof of Theorem 1. Note only that now the matrix A is generally a rectangular matrix and the systems of eigenvectors $\{\mathbf{x}_i\}_{i=1}^n$ for A^*A and $\{\mathbf{y}_j\}_{j=1}^m$ for AA^* are now orthonormal bases in $\mathscr{F}^n$ and $\mathscr{F}^m$, respectively. We have [see Eq. (2)]

$$A\mathbf{x}_i = \sqrt{\lambda_i}\,\mathbf{y}_i, \qquad i = 1, 2, \ldots, r, \tag{8}$$

and, by definition of $\mathbf{x}_i$ (see Exercise 5.3.6), $A\mathbf{x}_i = \mathbf{0}$ $(r + 1 \le i \le n)$.

Now we construct matrices

$$V = [\mathbf{x}_1 \quad \mathbf{x}_2 \quad \cdots \quad \mathbf{x}_n] \qquad \text{and} \qquad U = [\mathbf{y}_1 \quad \mathbf{y}_2 \quad \cdots \quad \mathbf{y}_m] \tag{9}$$

and note that, according to Corollary 5.6.2, they are unitary. Since by definition $s_i = \sqrt{\lambda_i}$ $(i = 1, 2, \ldots, n)$, the relation (8) implies

$$AV = [s_1\mathbf{y}_1 \quad s_2\mathbf{y}_2 \quad \cdots \quad s_r\mathbf{y}_r \quad \mathbf{0} \quad \cdots \quad \mathbf{0}] = UD,$$

where D is the matrix in Eq. (7). Thus, the representation (7) is established. ■

Note the structure of the matrices U and V in Eqs. (7) and (9): the columns of U (respectively, V) viewed as vectors from $\mathscr{F}^m$ (respectively, $\mathscr{F}^n$) constitute an orthonormal eigenbasis of AA^* in $\mathscr{F}^m$ (respectively, of A^*A in $\mathscr{F}^n$). Thus, a singular-value decomposition of A can be obtained by solving the eigenvalue–eigenvector problem for the matrices AA^* and A^*A.

Example 4. Consider the matrix A defined in Exercise 5.4.2; it has singular values $s_1 = \sqrt{2}$ and $s_2 = 1$. To find a singular-value decomposition of A, we compute A^*A and AA^* and construct orthonormal eigenbases for these matrices. The standard basis in $\mathscr{F}^2$ can be used for A^*A and the system $\{[\alpha \;\; 0 \;\; \alpha]^{\mathrm{T}}, [0 \;\; 1 \;\; 0]^{\mathrm{T}}, [\alpha \;\; 0 \;\; -\alpha]^{\mathrm{T}}\}$, where $\alpha = 1/\sqrt{2}$, can be used for AA^*. Hence

$$A = \begin{bmatrix} 1 & 0 \\ 0 & 1 \\ 1 & 0 \end{bmatrix} = \begin{bmatrix} \alpha & 0 & \alpha \\ 0 & 1 & 0 \\ \alpha & 0 & -\alpha \end{bmatrix} \begin{bmatrix} \sqrt{2} & 0 \\ 0 & 1 \\ 0 & 0 \end{bmatrix} \begin{bmatrix} 1 & 0 \\ 0 & 1 \end{bmatrix}$$

is a singular-value decomposition of A. □

The relation (7) gives rise to the general notion of *unitary equivalence* of two $m \times n$ matrices A and B. We say that A and B are *unitarily equivalent* if there exist unitary matrices U and V such that $A = UBV^*$. It is clear that unitary equivalence is an equivalence relation and that Theorem 2 can be interpreted as asserting the existence of a *canonical form* (the matrix D) *with respect to unitary equivalence* in each equivalence class. Now the next result is to be expected.

Proposition 2. *Two $m \times n$ matrices are unitarily equivalent if and only if they have the same singular values.*

PROOF. Assume that the singular values of the matrices $A, B \in \mathscr{F}^{m \times n}$ coincide. Writing a singular-value decomposition of the form (7) for each matrix (with the same D), the unitary equivalence readily follows.

Conversely, let A and B be unitarily equivalent, and let D_1 and D_2 be the matrices of singular values of A and B, respectively, that appear in the singular-value decompositions of the matrices. Without loss of generality, it may be assumed that the singular values on the diagonals of D_1 and D_2 are in nondecreasing order. It is easily found that the equivalence of A and B implies that D_1 and D_2 are also unitarily equivalent: $D_2 = U_1 D_1 V_1^*$. Viewing the last relation as the singular-value decomposition of D_2 and noting that the middle factor D_1 in such a representation is uniquely determined, we conclude that $D_1 = D_2$. ■

Corollary 1. *Two $m \times n$ matrices A and B are unitarily equivalent if and only if the matrices A^*A and B^*B are similar.*

PROOF. If $A = UBV^*$, then

$$A^*A = VB^*U^*UBV^* = VB^*BV^* = VB^*BV^{-1}.$$

Conversely, the similarity of A^*A and B^*B yields, in particular, that they have the same spectrum. Hence A and B have the same singular values and, consequently, are unitarily equivalent. ■

5.8 Idempotent Matrices (Projectors)

By definition, an idempotent matrix P satisfies the condition $P^2 = P$. Another name for such a matrix is a *projector*. The name has geometrical origins that, among other things, will be developed in this section. Observe that the definition can also be applied to transformations $P \in \mathcal{L}(\mathcal{S})$ for any linear space $\mathcal{S}$. It is left as an exercise for the reader to formulate analogues of Theorems 1–5 in this case. We start with the following simple proposition.

Theorem 1. *If P is idempotent, then:*

(a) *$I - P$ is idempotent;*
(b) $\text{Im}\,(I - P) = \text{Ker}\, P$;
(c) $\text{Ker}(I - P) = \text{Im}\, P$.

PROOF. For part (*a*), observe that $(I - P)^2 = I - 2P + P^2 = I - 2P + P = I - P$. Proceeding to part (*b*), we see that if $y \in \text{Im}(I - P)$, then $y = (I - P)x$ for some $x \in \mathcal{F}^n$. Hence $Py = P(I - P)x = (P - P^2)x = 0$ and $y \in \text{Ker}\, P$. Conversely, if $Py = 0$, then $(I - P)y = y$ and therefore $y \in \text{Im}(I - P)$. Thus $\text{Im}(I - P) = \text{Ker}\, P$. A similar argument gives the proof of part (*c*). ■

Note that the idempotent matrix $Q = I - P$ is called the *complementary projector* to P. The importance of the last theorem becomes clearer when we recognize that the kernel and image of P constitute a direct sum decomposition of the space $\mathcal{F}^n$ on which P acts. Note also that, for the complementary projector Q,

$$\text{Im}\, Q = \text{Ker}\, P, \qquad \text{Ker}\, Q = \text{Im}\, P.$$

Theorem 2. *If P is a projector, then* $\text{Ker}\, P \dotplus \text{Im}\, P = \mathcal{F}^n$.

PROOF. For any $x \in \mathcal{F}^n$, we can write $x = x_1 + x_2$, where $x_1 = (I - P)x \in \text{Ker}\, P$ and $x_2 = Px \in \text{Im}\, P$. Hence $\mathcal{F}^n = \text{Ker}\, P + \text{Im}\, P$. Finally, if $x \in \text{Ker}\, P \cap \text{Im}\, P$, it must be the zero element. For, $x \in \text{Ker}\, P$ implies $Px = 0$, while $x \in \text{Im}\, P$ implies by Theorem 1 that $(I - P)x = 0$. These two conclusions mean that $x = 0$ and so the sum $\text{Ker}\, P + \text{Im}\, P$ is direct. ■

Thus, each idempotent matrix P generates two uniquely defined mutually complementary subspaces $\mathscr{S}_1 = \text{Ker } P$ and $\mathscr{S}_2 = \text{Im } P$ such that their direct sum is the entire space. The converse is also true: If $\mathscr{S}_1 \dotplus \mathscr{S}_2 = \mathscr{F}^n$, there exists a unique idempotent P for which $\mathscr{S}_1 = \text{Ker } P$ and $\mathscr{S}_2 = \text{Im } P$. To see this, just define[†] P on $\mathscr{F}^n$ by $P\mathbf{x} = \mathbf{x}_2$, where $\mathbf{x} = \mathbf{x}_1 + \mathbf{x}_2$ and $\mathbf{x}_1 \in \mathscr{S}_1$, $\mathbf{x}_2 \in \mathscr{S}_2$. Then it is clear that $P^2 = P$ and $\mathscr{S}_1 = \text{Ker } P$, $\mathscr{S}_2 = \text{Im } P$. Also, the uniqueness of P follows immediately from the definition. Thus, there is a one-to-one correspondence between idempotent matrices in $\mathscr{F}^{n \times n}$ and pairs of subspaces $\mathscr{S}_1, \mathscr{S}_2$ such that $\mathscr{S}_1 \dotplus \mathscr{S}_2 = \mathscr{F}^n$.

In view of this result, we shall say that the idempotent matrix P *performs the projection of the space* $\mathscr{F}^n$ *on the subspace* $\mathscr{S}_2$ *parallel to* $\mathscr{S}_1$, or *onto* $\mathscr{S}_2$ *along* $\mathscr{S}_1$. Hence the name "projector" as an alternative to "idempotent matrix."

We now go on to study the spectral properties of projectors.

Theorem 3. *An idempotent matrix is simple.*

PROOF. We have already seen in Exercise 4.11.7 that the eigenvalues of an idempotent matrix are all equal to 1 or 0. It follows then that the eigenvectors of P all belong to either $\text{Im } P$ $(\lambda = 1)$ or $\text{Ker } P$ $(\lambda = 0)$. In view of Theorem 2, the eigenvectors of P must span $\mathscr{F}^n$ and hence P is a simple matrix. ■

The spectral theorem (Theorem 4.10.3) can now be applied.

Corollary 1. *An* $n \times n$ *projector* P *admits the representation*

$$P = \sum_{j=1}^{r} \mathbf{x}_j \mathbf{y}_j^*, \tag{1}$$

where $r = \text{rank } P$ *and* $\{\mathbf{x}_1, \ldots, \mathbf{x}_r\}$, $\{\mathbf{y}_1, \ldots, \mathbf{y}_r\}$ *are biorthogonal systems in* $\mathscr{F}_n$.

Note that a direct computation of P^2 shows that, conversely, any matrix of the form (1) is idempotent.

An important case arises when P is both idempotent and Hermitian: $P^2 = P$ and $P^* = P$. In this case the subspaces $\mathscr{S}_1 = \text{Ker } P$ and $\mathscr{S}_2 = \text{Im } P$ are found to be orthogonal. For, if $\mathbf{x} \in \mathscr{S}_1$ and $\mathbf{y} \in \mathscr{S}_2$, then

$$(\mathbf{x}, \mathbf{y}) = (\mathbf{x}, P\mathbf{y}) = (P^*\mathbf{x}, \mathbf{y}) = (P\mathbf{x}, \mathbf{y}) = 0$$

and therefore $\mathscr{S}_1 \oplus \mathscr{S}_2 = \mathscr{F}^n$. A Hermitian idempotent P is called an *orthogonal projector*: it carries out the projection of $\mathscr{F}^n$ onto $\mathscr{S}_2 = \text{Im } P$ along the *orthogonal complement* $\mathscr{S}_1 = \text{Ker } P$.

[†] Strictly speaking, this construction first defines a linear transformation from $\mathscr{F}^n$ to $\mathscr{F}^n$. The matrix P is then its representation in the standard basis.

The effect of an orthogonal projector can be described in terms of its eigenvectors. To see this, we choose an orthonormal set of (right) eigenvectors $\{\mathbf{x}_1, \mathbf{x}_2, \ldots, \mathbf{x}_n\}$ for P. Then the corresponding left eigenvectors are just the adjoints of the (right) eigenvectors and the representation (1) becomes (see Exercise 5.2.5)

$$P = \sum_{j=1}^{r} \mathbf{x}_j \mathbf{x}_j^*,$$

and for any $\mathbf{x} \in \mathscr{F}^n$,

$$P\mathbf{x} = \sum_{j=1}^{r} \mathbf{x}_j \mathbf{x}_j^* \mathbf{x} = \sum_{j=1}^{r} (\mathbf{x}, \mathbf{x}_j)\mathbf{x}_j.$$

Now we see that each term of this summation is an orthogonal projection of $\mathbf{x}$ onto the span of a (right) eigenvector $\mathbf{x}_j$. Hence P is an orthogonal projector onto the space spanned by $\mathbf{x}_1, \mathbf{x}_2, \ldots, \mathbf{x}_r$, that is, onto Im P.

It is instructive to make a more direct approach to orthogonal projectors as follows. Suppose we are given a subspace $\mathscr{S} \subset \mathscr{F}^n$ with a basis $\{\mathbf{u}_1, \mathbf{u}_2, \ldots, \mathbf{u}_k\}$ and we want to find an orthogonal projector onto $\mathscr{S}$. Define the $n \times k$ matrix $A = [\mathbf{u}_1, \mathbf{u}_2, \ldots, \mathbf{u}_k]$, then a typical member of $\mathscr{S}$ is of the form $A\mathbf{x}$ for some $\mathbf{x} \in \mathscr{F}^k$. Given an $\mathbf{a} \in \mathscr{F}^n$, we want to find a vector $A\mathbf{x} \in \mathscr{S}$ that can appropriately be called the projection of $\mathbf{a}$ onto $\mathscr{S}$.

We determine $A\mathbf{x}$ by demanding that $\mathbf{a} - A\mathbf{x}$ should be orthogonal to every member of $\mathscr{S}$. This will be the case if and only if $\mathbf{a} - A\mathbf{x}$ is orthogonal to each of the basis vectors. Thus, in the standard inner product,

$$(\mathbf{a} - A\mathbf{x}, \mathbf{u}_j) = \mathbf{u}_j^*(\mathbf{a} - A\mathbf{x}) = 0, \qquad j = 1, 2, \ldots, k,$$

which is equivalent to the matrix equation

$$A^*(\mathbf{a} - A\mathbf{x}) = \mathbf{0}.$$

Solving for $\mathbf{x}$, we obtain $\mathbf{x} = (A^*A)^{-1}A^*\mathbf{a}$, noting that the inverse exists because A^*A is the Gram matrix of linearly independent vectors (Theorem 3.12.2). Thus if P is to be an orthogonal projector on $\mathscr{S}$, we must have

$$P\mathbf{a} = A\mathbf{x} = A(A^*A)^{-1}A^*\mathbf{a}, \qquad \mathbf{a} \in \mathscr{F}^n,$$

that is,

$$P = A(A^*A)^{-1}A^*. \tag{2}$$

In fact, it is easy to verify that P is idempotent and Hermitian. Thus, we have determined an orthogonal projector on $\mathscr{S}$ in terms of an arbitrary basis for $\mathscr{S}$.

What is of particular interest, however, is the fact that P is *independent* of the choice of basis vectors for $\mathscr{S}$. To see this, observe that if X is the $n \times k$

matrix of another set of basis vectors, (and $k \leq n$) then there is a nonsingular $k \times k$ matrix M such that $A = XM$, whence

$$\begin{aligned} P &= XM(M^*X^*XM)^{-1}M^*X^* \\ &= XM(M^{-1}(X^*X)^{-1}(M^*)^{-1})M^*X^* \\ &= X(X^*X)^{-1}X^*. \end{aligned}$$

Hence, we get the same projector P whatever the choice of basis vectors for $\mathscr{S}$ may be. In particular, if we choose an orthonormal basis for $\mathscr{S}$, then $X^*X = I$ and

$$P = XX^* = \sum_{j=1}^{k} x_j x_j^*,$$

and we have returned to the form for P that we had before.

Exercise 1. If $P \in \mathscr{F}^{n \times n}$ is a projector on $\mathscr{S}_1$ along $\mathscr{S}_2$, show that:

(a) P^* is a projector on $\mathscr{S}_2^{\perp}$ along $\mathscr{S}_1^{\perp}$ (see Section 3.13);

(b) For any nonsingular $H \in \mathscr{F}^{n \times n}$, the matrix $H^{-1}PH$ is a projector. Describe its image and kernel. □

The next results concern properties of invariant subspaces (see Section 4.7), which can be readily described with the help of projectors.

Theorem 4. *Let $A \in \mathscr{F}^{n \times n}$. A subspace $\mathscr{S}$ of $\mathscr{F}^n$ is A-invariant if and only if $PAP = AP$ for any projector P of $\mathscr{F}^n$ onto $\mathscr{S}$.*

PROOF. Let $\mathscr{S}$ be A-invariant, so that $Ax \in \mathscr{S}$ for all $x \in \mathscr{S}$. If P is any projector onto $\mathscr{S}$, then $(I - P)Ax = Ax - Ax = 0$ for all $x \in \mathscr{S}$. Since for any $y \in \mathscr{F}^n$ the vector $x = Py$ is in $\mathscr{S}$, it follows that $(I - P)APy = \mathbf{0}$ for all $y \in \mathscr{F}^n$, and we have $PAP = AP$, as required.

Conversely, if P is a projector onto $\mathscr{S}$ and $PAP = AP$, then for any $x \in \mathscr{S}$, we have $Px = x$ and $Ax = APx = PAPx$. Thus, $Ax \in \operatorname{Im} P = \mathscr{S}$ and, since x is any member of $\mathscr{S}$, the space $\mathscr{S}$ is A-invariant. ■

Note that the sufficient condition for $\mathscr{S}$ to be invariant is reduced, in fact, to the existence of at least one projector P onto $\mathscr{S}$ such that $QAP = (I - P)AP = O$.

A particularly important role is played in the spectral theory of matrices by A-invariant subspaces that have a complementary subspace that is *also* A-invariant. Such a subspace is said to be *A-reducing*. Thus, subspace $\mathscr{S}_1$ of $\mathscr{F}^n$ is A-reducing if $A(\mathscr{S}_1) \subset \mathscr{S}_1$ and there is a subspace $\mathscr{S}_2$ such that $A(\mathscr{S}_2) \subset \mathscr{S}_2$ and $\mathscr{F}^n = \mathscr{S}_1 \dotplus \mathscr{S}_2$. In this case, we may also say that the *pair of subspaces $\mathscr{S}_1$, $\mathscr{S}_2$ is A-reducing*. When a subspace is A-reducing, there is a more symmetric characterization than that of Theorem 4.

Theorem 5. *Let $A \in \mathscr{F}^{n \times n}$. A pair of subspaces $\mathscr{S}_1, \mathscr{S}_2 \subset \mathscr{F}^n$ is A-reducing if and only if $AP = PA$, where P is the projector onto $\mathscr{S}_2$ along $\mathscr{S}_1$.*

PROOF. Let $\mathscr{F}^n = \mathscr{S}_1 \dotplus \mathscr{S}_2$, and $A(\mathscr{S}_1) \subset \mathscr{S}_1$, $A(\mathscr{S}_2) \subset \mathscr{S}_2$. For any $\boldsymbol{x} \in \mathscr{F}^n$ write $\boldsymbol{x} = \boldsymbol{x}_1 + \boldsymbol{x}_2$, where $\boldsymbol{x}_1 \in \mathscr{S}_1$, $\boldsymbol{x}_2 \in \mathscr{S}_2$. Then $AP\boldsymbol{x} = A\boldsymbol{x}_2$ and, since $A\boldsymbol{x}_1 \in \mathscr{S}_1 = \operatorname{Ker} P$,

$$PA\boldsymbol{x} = PA\boldsymbol{x}_1 + PA\boldsymbol{x}_2 = PA\boldsymbol{x}_2.$$

But $A\boldsymbol{x}_2 \in \mathscr{S}_2 = \operatorname{Im} P$, so

$$PA\boldsymbol{x} = PA\boldsymbol{x}_2 = A\boldsymbol{x}_2 = AP\boldsymbol{x}$$

for all $\boldsymbol{x} \in \mathscr{F}^n$. Thus $PA = AP$.

Now let P be a projector on $\mathscr{S}_2$ along $\mathscr{S}_1$ and suppose that $AP = PA$. We are to show that both $\mathscr{S}_1$ and $\mathscr{S}_2$ are A-invariant.

If $\boldsymbol{x}_2 \in \mathscr{S}_2$ then $P\boldsymbol{x}_2 = \boldsymbol{x}_2$ and

$$A\boldsymbol{x}_2 = AP\boldsymbol{x}_2 = PA\boldsymbol{x}_2 \in \mathscr{S}_2,$$

so that $A(\mathscr{S}_2) \subset \mathscr{S}_2$. Next, if $\boldsymbol{x}_1 \in \mathscr{S}_1$,

$$A\boldsymbol{x}_1 = A(I - P)\boldsymbol{x}_1 = (I - P)A\boldsymbol{x}_1 \in \mathscr{S}_1,$$

so that $A(\mathscr{S}_1) \subset \mathscr{S}_1$ and the proof is complete. ■

Let us take this opportunity to bring together several points concerning the image and kernel of a matrix. For the sake of clarity and future use, we consider a matrix $A \in \mathbb{C}^{m \times n}$ to be a linear transformation from $\mathbb{C}^n$ into $\mathbb{C}^m$, although the discussion can also be put into the context of abstract linear transformations. Recall first that $\operatorname{Im} A \subset \mathbb{C}^m$ and $\operatorname{Ker} A \subset \mathbb{C}^n$. Then, by Theorem 4.5.5,

$$\dim(\operatorname{Im} A) + \dim(\operatorname{Ker} A) = n. \tag{3}$$

In particular, if $m = n$ it is *possible* that this statement can be strengthened to yield

$$\operatorname{Im} A \dotplus \operatorname{Ker} A = \mathbb{C}^n. \tag{4}$$

Theorem 2 says that this is true if $A^2 = A$, that is, if A is a projector, and it is easily seen to be true if A is normal (and especially, if A is Hermitian or unitary). In the latter case more is true: when A is normal the direct sum is orthogonal,

$$\operatorname{Im} A \oplus \operatorname{Ker} A = \mathbb{C}^n. \tag{5}$$

However, it is important to bear in mind that Eq. (4) is not *generally* true for square matrices, as the example

$$A = \begin{bmatrix} 0 & 1 \\ 0 & 0 \end{bmatrix}$$

quickly shows.

The general property (3) can be seen as one manifestation of the results in Exercise 5.1.7. These results are widely useful and are presented here as a formal proposition about matrices, including some of the above remarks.

Proposition 1. *If* $A \in \mathbb{C}^{m \times n}$ *then*

$$\mathbb{C}^n = \operatorname{Ker} A \oplus \operatorname{Im} A^*, \qquad \mathbb{C}^m = \operatorname{Ker} A^* \oplus \operatorname{Im} A.$$

Exercise 2. Use Theorem 5.1.2 to prove Proposition 1.

Exercise 3. Verify that the matrix

$$\frac{1}{3}\begin{bmatrix} 2 & 1 & -1 \\ 1 & 2 & 1 \\ -1 & 1 & 2 \end{bmatrix}$$

is an orthogonal projector and find bases for its image and kernel.

Exercise 4. Let P and Q be *commuting* projectors.

(a) Show that PQ is a projector and

$$\operatorname{Im} PQ = (\operatorname{Im} P) \cap (\operatorname{Im} Q), \qquad \operatorname{Ker} PQ = (\operatorname{Ker} P) + (\operatorname{Ker} Q).$$

(b) Show that if we define $P \boxplus Q = P + Q - PQ$ (the "Boolean sum" of P and Q), then $P \boxplus Q$ is a projector and

$$\operatorname{Im}(P \boxplus Q) = (\operatorname{Im} P) + (\operatorname{Im} Q), \qquad \operatorname{Ker}(P \boxplus Q) = (\operatorname{Ker} P) \cap (\operatorname{Ker} Q).$$

Hint. For part (b) observe that $I - (P \boxplus Q) = (I - P)(I - Q)$ and use part (a).

Exercise 5. Let P and Q be projectors. Show that $QP = Q$ if and only if $\operatorname{Ker} P \subset \operatorname{Ker} Q$.

Exercise 6. Let $T \in \mathscr{L}(\mathscr{U}, \mathscr{V})$ (as in Theorem 5.1.4). Show that there is a *unique* solution $\boldsymbol{x}_0 \in \operatorname{Im} T^*$ of $T(\boldsymbol{x}) = \boldsymbol{b}$ if and only if $\boldsymbol{b} \in \operatorname{Im} T$.

Hint. Using Exercise 5.1.7 define $P \in \mathscr{L}(\mathscr{V})$ as the orthogonal projector on $\operatorname{Im} T^*$ along $\operatorname{Ker} T$. If $\boldsymbol{x}$ satisfies $T(\boldsymbol{x}) = \boldsymbol{b}$, define $\boldsymbol{x}_0 = P\boldsymbol{x}$. □

5.9 Matrices over the Field of Real Numbers

Let $A \in \mathbb{R}^{n \times n}$. Clearly, all statements made above for an arbitrary field $\mathcal{F}$ are valid in the particular case $\mathcal{F} = \mathbb{R}$. However, $\mathbb{R}$ fails to be algebraically closed and therefore matrices in $\mathbb{R}^{n \times n}$ may lose some vital properties. For instance, a matrix with real elements may have no (real) eigenvalues. An example of such a matrix is

$$\begin{bmatrix} \mu & \nu \\ -\nu & \mu \end{bmatrix}, \qquad , \nu \in \mathbb{R}, \quad \nu \neq 0. \tag{1}$$

(See also the matrix in Example 4.1.2, in which γ is not an integer multiple of π.) On the other hand, if the spectrum of a real matrix is entirely real, then there is no essential difference between the real and complex cases.

Now we discuss a process known as the *method of complexification*, which reduces the study of real matrices to that of matrices with complex entries.

First we observe the familiar Cartesian decomposition $\mathbf{z} = \mathbf{x} + i\mathbf{y}$, where $\mathbf{z} \in \mathbb{C}^n$ and $\mathbf{x}, \mathbf{y} \in \mathbb{R}^n$. Hence if $\{\mathbf{a}_1, \mathbf{a}_2, \ldots, \mathbf{a}_n\}$ is a basis in $\mathbb{R}^n$,

$$\mathbf{x} = \sum_{j=1}^{n} \alpha_j \mathbf{a}_j \qquad \text{and} \qquad \mathbf{y} = \sum_{j=1}^{n} \beta_j \mathbf{a}_j \qquad (\alpha_j, \beta_j \in \mathbb{R}, j = 1, 2, \ldots, n),$$

then

$$\mathbf{z} = \mathbf{x} + i\mathbf{y} = \sum_{j=1}^{n} (\alpha_j + i\beta_j)\mathbf{a}_j.$$

Thus $\mathbb{C}^n$ can be viewed as the set of all vectors having complex coordinates in this basis, while $\mathbb{R}^n$ is the set of all vectors with real coordinates with respect to the same basis. The last remark becomes obvious if the basis in $\mathbb{R}^n$ is standard. Note that the vectors $\mathbf{x}$ in $\mathbb{R}^n$ can be viewed as vectors of the form $\mathbf{x} + i\mathbf{0}$ in $\mathbb{C}^n$.

Proceeding to a matrix $A \in \mathbb{R}^{n \times n}$, we extend its action to vectors from $\mathbb{C}^n$ by the rule

$$\tilde{A}\mathbf{z} = A\mathbf{x} + iA\mathbf{y} \tag{2}$$

for $\mathbf{z} = \mathbf{x} + i\mathbf{y}, \mathbf{x}, \mathbf{y} \in \mathbb{R}^n$. Note that the matrices A and $\tilde{A}$ are in fact the same and therefore have the same characteristic polynomial with real coefficients. It should also be emphasized that only real numbers are admitted as eigenvalues of A, while only vectors with real entries (i.e., *real vectors*) are admitted as eigenvectors of A.

Let $\lambda_0 \in \sigma(A)$ and $A\mathbf{x}_0 = \lambda_0 \mathbf{x}_0$, where $\mathbf{x}_0 \in \mathbb{R}^n$ and $\mathbf{x}_0 \neq \mathbf{0}$. By Eq. (2), $\tilde{A}\mathbf{x}_0 = A\mathbf{x}_0 = \lambda_0 \mathbf{x}_0$ and the matrix $\tilde{A}$ has the same eigenvalue and eigenvector. If λ_0 is a complex zero of the characteristic polynomial $c(\lambda)$ of $\tilde{A}$ then so is $\bar{\lambda}_0$, and $\lambda_0, \bar{\lambda}_0 \in \sigma(\tilde{A})$. The eigenvectors of $\tilde{A}$ corresponding to λ_0 and $\bar{\lambda}_0$, respectively, are *mutually complex conjugate*, that is, of the form $\boldsymbol{x} + i\boldsymbol{y}$ and $\boldsymbol{x} - i\boldsymbol{y}$. Indeed, if $\tilde{A}\boldsymbol{z} = \lambda_0 \boldsymbol{z}$, where $\boldsymbol{z} = \boldsymbol{x} + i\boldsymbol{y}$, then it is easily checked that $\tilde{A}(\boldsymbol{x} - i\boldsymbol{y}) = \bar{\lambda}_0(\boldsymbol{x} - i\boldsymbol{y})$. Note that since $\lambda_0 \neq \bar{\lambda}_0$, the eigenvectors $\boldsymbol{x} + i\boldsymbol{y}$ and $\boldsymbol{x} - i\boldsymbol{y}$ are linearly independent (Proposition 4.10.4). Consider the subspace $\mathscr{S} = \text{span}\{\boldsymbol{x} + i\boldsymbol{y}, \boldsymbol{x} - i\boldsymbol{y}\}$. It has a real basis $\{\boldsymbol{x}, \boldsymbol{y}\}$. Referring to $\mathscr{S}$ as a subspace of $\mathbb{R}^n$, that is, as the set of all linear combinations of $\boldsymbol{x}, \boldsymbol{y}$ with real coefficients, it is clear from the relations

$$A\boldsymbol{x} = \mu\boldsymbol{x} - \nu\boldsymbol{y}, \qquad A\boldsymbol{y} = \nu\boldsymbol{x} + \mu\boldsymbol{y}, \tag{3}$$

where $\mu + i\nu = \lambda_0$, that $\mathscr{S}$ is A-invariant. However, the matrix A has no (real) eigenvectors in $\mathscr{S}$ (Compare with Exercise 4.9.8.) For example, if A is the coefficient matrix given in (1), the linear system

$$A(\alpha\boldsymbol{x} + \beta\boldsymbol{x}) = \lambda(\alpha\boldsymbol{x} + \beta\boldsymbol{y}), \qquad \lambda \in \mathbb{R},$$

implies that $\alpha\boldsymbol{x} + \beta\boldsymbol{y} = \boldsymbol{0}$.

Thus, if the characteristic polynomial of a real matrix A has a complex zero, then it is associated with a two-dimensional A-invariant subspace $\mathscr{S}$ in $\mathbb{R}^n$, although $A|_{\mathscr{S}}$ has no eigenvalues and, consequently, no eigenvectors in $\mathscr{S}$.

We summarize this discussion.

Theorem 1. *A real matrix $A \in \mathbb{R}^{n \times n}$ has a one-dimensional (two-dimensional) invariant subspace in $\mathbb{R}^n$ associated with each real (complex) zero of the characteristic polynomial.*

In the first case, the invariant subspace is spanned by a (real) eigenvector of A, while in the second case the matrix has no eigenvectors lying in this invariant subspace.

A matrix $U \in \mathbb{R}^{n \times n}$ satisfying the condition $U^{\mathrm{T}}U = UU^{\mathrm{T}} = I$ is an *orthogonal matrix*. The properties of orthogonal matrices viewed as matrices in $\mathbb{C}^{n \times n}$ can be deduced from those of unitary ones. In particular, since a unitary matrix is simple, it is similar to a diagonal matrix of its eigenvalues. Some of the eigenvalues may be complex numbers and therefore are not eigenvalues of U viewed as a matrix in $\mathbb{R}^{n \times n}$. Hence the following result is of interest.

Corollary 1. *Any orthogonal matrix $U \in \mathbb{R}^{n \times n}$ is similar (over $\mathbb{R}$) to a matrix of the form*

$$\text{diag}[1, \ldots, 1, -1, \ldots, -1, D_1, D_2, \ldots, D_k],$$

where $D_i \in \mathbb{R}^{2 \times 2}$ $(i = 1, 2, \ldots, k)$ are matrices of the form (1), in which $\mu = \cos\gamma$, $\nu = -\sin\gamma$, $\gamma \neq \pi t$, $t = 0, \pm 1, \ldots$.

5.10 Bilinear, Quadratic, and Hermitian Forms

Let $A \in \mathscr{F}^{n \times n}$ and $x, y \in \mathscr{F}^n$. Consider the function f_A defined on $\mathscr{F}^n \times \mathscr{F}^n$ by

$$f_A(x, y) \triangleq (Ax, y). \tag{1}$$

Recalling the appropriate properties of the inner product, it is clear that, for a fixed A, the function f_A in Eq. (1) is *additive* with respect to both vector variables (the subscript A is omitted):

$$\begin{aligned} f(x_1 + x_2, y) &= f(x_1, y) + f(x_2, y), \\ f(x, y_1 + y_2) &= f(x, y_1) + f(x, y_2), \end{aligned} \tag{2}$$

and *homogeneous* with respect to the *first* vector variable:

$$f(\alpha x, y) = \alpha f(x, y), \qquad \alpha \in \mathscr{F}. \tag{3}$$

If $\mathscr{F} = \mathbb{R}$, the function is also homogeneous with respect to the second vector variable:

$$f(x, \beta y) = \beta f(x, y), \qquad \beta \in \mathbb{R}, \tag{4}$$

while in the case $\mathscr{F} = \mathbb{C}$, the function $f \equiv f_A$ is *conjugate-homogeneous* with respect to the second argument:

$$f(x, \beta y) = \bar{\beta} f(x, y), \qquad \beta \in \mathbb{C}. \tag{4'}$$

Now we formally define a function $f(x, y)$, defined on pairs of vectors from $\mathscr{F}^n$, to be *bilinear* (respectively, *conjugate bilinear*) if it satisfies the conditions (1) through (4) (respectively, (1) through (3) and (4′)). Consider the conjugate bilinear function $f_A(x, y)$ defined by Eq. (1) on $\mathbb{C}^n \times \mathbb{C}^n$. If $\mathscr{E} = \{x_1, x_2, \ldots, x_n\}$ is a basis in $\mathbb{C}^n$ and $x = \sum_{i=1}^n \alpha_i x_i$, $y = \sum_{j=1}^n \beta_j x_j$, then

$$f_A(x, y) = (Ax, y) = \sum_{i, j=1}^{n} (Ax_i, x_j)\alpha_i \bar{\beta}_j = \sum_{i, j=1}^{n} \gamma_{ij}\alpha_i \bar{\beta}_j, \tag{5}$$

where

$$\gamma_{ij} = (Ax_i, x_j), \qquad i, j = 1, 2, \ldots, n. \tag{6}$$

When expressed in coordinates as in Eq. (5), f_A is generally called a *conjugate bilinear form* or simply a *bilinear form* according as $\mathscr{F} = \mathbb{C}$ or $\mathscr{F} = \mathbb{R}$, respectively. The form is said to be *generated* by the matrix A, and the matrix $\tilde{A} = [\gamma_{ij}]_{i, j=1}^n$ is said to be the *matrix of the form f_A with respect to the basis* $\mathscr{E}$. Our first objective is to investigate the connection between matrices of the bilinear form (1) with respect to different bases and, in particular, between them and the generating matrix A.

We first note that the matrix A is, in fact, the matrix of the form (1) with respect to the standard basis $\{\mathbf{e}_1, \mathbf{e}_2, \ldots, \mathbf{e}_n\}$ in $\mathbb{C}^n$. Furthermore, if P denotes the transition matrix from the standard basis to a basis $\{\boldsymbol{x}_1, \boldsymbol{x}_2, \ldots, \boldsymbol{x}_n\}$ in $\mathbb{C}^n$ then, by Eq. (3.9.3), $\boldsymbol{x}_i = P\boldsymbol{e}_i, i = 1, 2, \ldots, n$. Hence for $i, j = 1, 2, \ldots, n$,

$$\gamma_{ij} = (A\boldsymbol{x}_i, \boldsymbol{x}_j) = (AP\boldsymbol{e}_i, P\boldsymbol{e}_j) = (P^*AP\boldsymbol{e}_i, \boldsymbol{e}_j).$$

Recalling that the standard basis is orthonormal in the standard inner product, we find that the γ_{ij} are just the elements of the matrix P^*AP. Thus $\tilde{A} = P^*AP$ and matrices of the same conjugate bilinear form with respect to different bases must be congruent.

Theorem 1. *Matrices of a conjugate bilinear form with respect to different bases in $\mathbb{C}^n$ generate an equivalence class of congruent matrices.*

Now consider some special kinds of conjugate bilinear forms. A conjugate bilinear form $f_A(\boldsymbol{x}, \boldsymbol{y}) = (A\boldsymbol{x}, \boldsymbol{y})$ is *Hermitian* if $\overline{f_A(\boldsymbol{x}, \boldsymbol{y})} = f_A(\boldsymbol{y}, \boldsymbol{x})$ for all $\boldsymbol{x}, \boldsymbol{y} \in \mathbb{C}^n$.

Exercise 1. Check that a conjugate bilinear form is Hermitian if and only if the generating matrix A is Hermitian. Moreover, all matrices of such a form are Hermitian matrices having the same inertia characteristics.

Hint. Use Theorem 1 and Sylvester's law of inertia. □

Particularly interesting functions generated by conjugate bilinear forms are the *quadratic forms* obtained by substituting $\boldsymbol{y} = \boldsymbol{x}$ in $f_A(\boldsymbol{y}, \boldsymbol{x})$, that is, the forms $f_A(\boldsymbol{x}) = (A\boldsymbol{x}, \boldsymbol{x})$ defined on $\mathscr{F}^n$ with values in $\mathscr{F}$.

Exercise 2. By using the identity

$$f(x, y) = \tfrac{1}{4}(f(x + y, x + y) + if(x + iy, x + iy)$$
$$- if(x - iy, x - iy) - f(x - y, x - y)),$$

show that any conjugate bilinear form is uniquely determined by the corresponding quadratic form.

Exercise 3. Show that the conjugate bilinear form $f_A(\boldsymbol{x}, \boldsymbol{y}) = (A\boldsymbol{x}, \boldsymbol{y})$ is Hermitian if and only if the corresponding quadratic form is a real-valued function. □

The rest of this section is devoted to the study of quadratic forms generated by a Hermitian matrix, therefore called *Hermitian forms*. Clearly, such a form is defined on $\mathscr{F}^n$ but now takes values in $\mathbb{R}$. In this case

$$f_A(\boldsymbol{x}) = (A\boldsymbol{x}, \boldsymbol{x}) = \sum_{i,j=1}^{n} \gamma_{ij}\alpha_i\bar{\alpha}_j, \tag{7}$$

where $\gamma_{ij} = \bar{\gamma}_{ji}$ $(i, j = 1, 2, \ldots, n)$ and the α_i $(i = 1, 2, \ldots, n)$ are the coordinates of x with respect to some basis in $\mathbb{C}^n$. In view of Theorem 1, the Hermitian matrices $\tilde{A} = [\gamma_{ij}]_{i,j=1}^n$ defined by Eq. (7) are congruent. Consequently, Theorem 5.5.1 asserts the existence of a basis in $\mathbb{C}^n$ with respect to which there is a matrix of the form $\tilde{A}_0 = \text{diag}[I_s, -I_{r-s}, 0]$. In this (canonical) basis, the Hermitian form is reduced to an (algebraic) sum of squares:

$$(\tilde{A}_0 x, x) = \sum_{i,j=1}^{n} \gamma_{ij}^{(0)} \alpha_j \bar{\alpha}_j = \sum_{i=1}^{s} |\alpha_i|^2 - \sum_{j=s+1}^{r} |\alpha_j|^2, \tag{8}$$

where $[\alpha_1 \ \ \alpha_2 \ \ \cdots \ \ \alpha_n]^T$ is the representation of x with respect to the canonical basis.

Theorem 2. *For any Hermitian form* (Ax, x), *there exists a basis in* $\mathbb{C}^n$ *with respect to which the form is represented as the sum of squares in Eq.* (8), *where* $s = \pi(A)$ *and* $r = \text{rank } A$.

Corollary 1. (Sylvester's law of inertia).† *The numbers of positive and negative squares in Eq.* (8) *are invariants of the Hermitian form* f_A, *denoted* $s = \pi(A)$, $r - s = \nu(A)$.

Note that the representation (8) is referred to as the *first canonical form* of (Ax, x) in contrast to the *second canonical form*:

$$(Ax, x) = \sum_{i=1}^{r} \lambda_i |\alpha_i|^2, \tag{9}$$

where $\{\lambda_i\}_{i=1}^r$ are the nonzero eigenvalues of A and $[\alpha_1 \ \ \alpha_2 \ \ \cdots \ \ \alpha_n]^T$ is the representation of x with respect to a corresponding orthonormal eigenbasis. The representation (9) readily follows from Theorem 5.2.1. In addition, since $A = UDU^*$, where $D = \text{diag}[\lambda_1, \lambda_2, \ldots, \lambda_r, 0, \ldots, 0]$ and U is unitary, the second canonical form (9) can be obtained by a *unitary congruent* transformation of bases. This is an advantage of this canonical form. (See the next section.)

Consider some particular cases of Hermitian forms. By analogy with matrices, define the Hermitian form $f(x)$ to be *positive* (respectively, *negative*) *definite* if $f(x) > 0$ (respectively, $f(x) < 0$) for all nonzero $x \in \mathbb{C}^n$. This analog extends to the definition of semi-definite Hermitian forms also.

Exercise 4. Check that a Hermitian form is definite or semi-definite according as its generating matrix (or any other matrix of the form) is definite or semi-definite of the same kind.

Hint. See Theorem 5.3.3. □

It should also be noted that a parallel exists between Hermitian matrices and Hermitian forms in that the rank and the inertia characteristics of a

† *Philosophical Mag.* **4** (1852), 138–142.

matrix can be ascribed to the form that it generates. Also, the simultaneous reduction of two Hermitian matrices to diagonal matrices by the same congruent transformation, as discussed in Section 5.5, is equivalent to a simultaneous reduction of the corresponding Hermitian forms to sums of squares.

Exercise 5. Consider a Hermitian form $f(\mathbf{x})$ defined on $\mathbb{R}^3$. The set of points in $\mathbb{R}^3$ (if any) that satisfies the equation $f(x_1, x_2, x_3) = 1$ is said to form a *central quadric*. Show that if the equation of the quadric is referred to suitable new axes, it is reduced to the form

$$\lambda_1 z_1^2 + \lambda_2 z_2^2 + \lambda_3 z_3^2 = 1. \tag{10}$$

The new coordinate axes are called the *principal axes* of the quadric. (Note that the nature of the quadric surface is determined by the relative magnitudes and signs of $\lambda_1, \lambda_2, \lambda_3$.)

Hint. Use $f(x_1, x_2, x_3) = \mathbf{x}^*H\mathbf{x}$ and Eq. (9).

Exercise 6. Verify that $\lambda_1 \leq \lambda_2 \leq \lambda_3$ in Eq. (10) are the eigenvalues of H and that for $H > O$, the quadric surface (10) is an ellipsoid with the point $(1/\sqrt{\lambda_1}, 0, 0)$ farthest from the origin.

Exercise 7. Examine the nature of the central quadric whose equation is

$$3x_2^2 + 3x_3^2 - 2x_2x_3 - 4x_1x_3 - 4x_1x_2 = 1.$$

Hint. Diagonalize the matrix of the Hermitian form on the left using Theorem 5.2.1. □

5.11 Finding the Canonical Forms

We shall describe in this section some techniques for computing the canonical forms of a given Hermitian quadratic form under congruence.

If A is an $n \times n$ Hermitian matrix, we note that the second canonical form of $(A\mathbf{x}, \mathbf{x})$ is easily found if all the eigenvalues $\{\lambda_i\}_{i=1}^n$ and eigenvectors $\{\mathbf{x}_i\}_{i=1}^n$ are known. Indeed, proceeding from $\{\mathbf{x}_i\}_{i=1}^n$ to an orthonormal eigenbasis $\{\tilde{\mathbf{x}}_i\}_{i=1}^n$ for A in $\mathbb{C}^n$ and constructing the unitary matrix

$$U = [\tilde{\mathbf{x}}_1 \quad \tilde{\mathbf{x}}_2 \quad \cdots \quad \tilde{\mathbf{x}}_n],$$

we deduce by use of Theorem 5.2.1 that U^*AU is a diagonal matrix D of the eigenvalues of A. Hence for $\mathbf{x} = \sum_{i=1}^n \alpha_i \mathbf{x}_i$, we obtain the second canonical form

$$(A\mathbf{x}, \mathbf{x}) = \sum_{i=1}^{r} \lambda_i |\alpha_i|^2,$$

where $r = \operatorname{rank} A$ and $\lambda_1, \ldots, \lambda_r$ denote the nonzero eigenvalues of A corresponding (perhaps, after reordering) to the eigenvectors $\tilde{\mathbf{x}}_1, \ldots, \tilde{\mathbf{x}}_r$. However, the difficulties involved in finding the full spectral properties of a matrix encourage us to develop a simpler method for reducing a Hermitian form. A method suggested by Lagrange allows us to obtain the first canonical form, as well as the transforming matrix, by relatively simple calculations.

For convenience we confine our attention to real Hermitian forms. If $\mathbf{x} = [\alpha_1 \quad \alpha_2 \quad \cdots \quad \alpha_n]^T$, we write

$$
\begin{aligned}
f(\mathbf{x}) &= \sum_{i,j=1}^{n} \gamma_{ij}\alpha_i\alpha_j \\
&= \gamma_{11}\alpha_1^2 + \cdots + \gamma_{nn}\alpha_n^2 + 2\gamma_{12}\alpha_1\alpha_2 + \cdots + 2\gamma_{n-1,n}\alpha_{n-1}\alpha_n,
\end{aligned} \tag{1}
$$

where all numbers involved are real and $\gamma_{ij} = \gamma_{ji}$ for $i, j = 1, 2, \ldots, n$. Note that the expression (1) can be viewed as a real function in n real variables $\alpha_1, \alpha_2, \ldots, \alpha_n$.

Case 1. Assume that not all of γ_{ii} are zero. Then assuming, for instance, that $\gamma_{11} \neq 0$, we rewrite Eq. (1) in the form

$$
f(\mathbf{x}) = \gamma_{11}\left(\alpha_1^2 + 2\frac{\gamma_{12}}{\gamma_{11}}\alpha_1\alpha_2 + \cdots + 2\frac{\gamma_{1n}}{\gamma_{11}}\alpha_1\alpha_n\right) + f_1(\mathbf{x}_1),
$$

where $\mathbf{x}_1 = [\alpha_2 \quad \alpha_3 \quad \cdots \quad \alpha_n]^T$. Obviously, $f_1(\mathbf{x}_1)$ is a Hermitian form. Following the process of "completing the square," we obtain

$$
f(\mathbf{x}) = \gamma_{11}\left(\alpha_1 + \frac{\gamma_{12}}{\gamma_{11}}\alpha_2 + \cdots + \frac{\gamma_{1n}}{\gamma_{11}}\alpha_n\right)^2 + f_2(\mathbf{x}_1),
$$

where $f_2(\mathbf{x}_1)$ is another Hermitian form in $\mathbf{x}_1$. Thus, the nonsingular transformation of coordinates determined by

$$
\beta_1 = \alpha_1 + \frac{\gamma_{12}}{\gamma_{11}}\alpha_2 + \cdots + \frac{\gamma_{1n}}{\gamma_{11}}\alpha_n, \qquad \beta_2 = \alpha_2, \ldots, \beta_n = \alpha_n
$$

transforms the given form into the form

$$
\gamma_{11}\beta_1^2 + f_2(\mathbf{x}_1). \tag{2}
$$

and we can write $\mathbf{x}_1 = [\beta_2 \quad \beta_3 \quad \cdots \quad \beta_n]^T$.

Case 2. If $\gamma_{11} = \gamma_{22} = \cdots = \gamma_{nn} = 0$, then we consider a coefficient $\alpha_{ij} \neq 0$, where $i \neq j$, and assume, for instance, that $\alpha_{12} \neq 0$. We now make a congruence transformation of f determined by the coordinate transformation $\beta_1 = \alpha_1$, $\beta_2 = \alpha_2 - \alpha_1$, $\beta_3 = \alpha_3, \ldots,$ and $\beta_n = \alpha_n$. It is easily seen that the form determined by the transformed matrix has a term in β_1^2. We may now apply the technique described in case 1 to reduce f to the form (2).

Thus, we have proved that any real Hermitian form can be reduced to the form (2). Applying the same process to $f_2(x_1)$ and so on, the reduction is completed in at most $n-1$ such steps. ■

Exercise 1. Find the rank and inertia of the form $f(x) = 2\alpha_1\alpha_2 + 2\alpha_2\alpha_3 + \alpha_3^2$ and determine a transformation of the coordinates $\alpha_1, \alpha_2, \alpha_3$ that reduces f to a sum of squares.

SOLUTION. Performing the transformation $\alpha_1 = \beta_3, \alpha_2 = \beta_2, \alpha_3 = \beta_1$, we have

$$f(x) = (\beta_1^2 + 2\beta_1\beta_2) + 2\beta_2\beta_3 = (\beta_1 + \beta_2)^2 - \beta_2^2 + 2\beta_2\beta_3.$$

Now put $\gamma_1 = \beta_1 + \beta_2, \gamma_2 = \beta_2, \gamma_3 = \beta_3$ and obtain

$$f(x) = \gamma_1^2 - \gamma_2^2 + 2\gamma_2\gamma_3 = \delta_1^2 - \delta_2^2 + \delta_3^2$$

where $\delta_1 = \gamma_1, \delta_2 = \gamma_2 - \gamma_3, \delta_3 = \gamma_3$. Thus the rank of the form is $r = 3$, the number of positive squares is $\pi = 2$, and the signature is $\pi - \nu = 2 - 1 = 1$. Note that the transformation P yielding the reduction is obtained by combining the preceding transformations:

$$\begin{bmatrix} \delta_1 \\ \delta_2 \\ \delta_3 \end{bmatrix} = \begin{bmatrix} 0 & 1 & 1 \\ -1 & 1 & 0 \\ 1 & 0 & 0 \end{bmatrix} \begin{bmatrix} \alpha_1 \\ \alpha_2 \\ \alpha_3 \end{bmatrix}.$$

Inverting the transforming matrix it is found that substitutions which reduce $f(\mathbf{x})$ to a sum of squares are given by

$$\begin{bmatrix} \alpha_1 \\ \alpha_2 \\ \alpha_3 \end{bmatrix} = \begin{bmatrix} 0 & 0 & 1 \\ 0 & 1 & 1 \\ 1 & -1 & -1 \end{bmatrix} \begin{bmatrix} \delta_1 \\ \delta_2 \\ \delta_3 \end{bmatrix}.$$

Note also that, by using Lagrange's method, we have avoided the computation of eigenvalues of the matrix

$$A = \begin{bmatrix} 0 & 1 & 0 \\ 1 & 0 & 1 \\ 0 & 1 & 1 \end{bmatrix}$$

that generates the form. Such a computation would require the calculation of all the zeros of the characteristic equation $\lambda^3 - \lambda^2 - 2\lambda + 1 = 0$. As we already know, two of them must be positive and one negative. □

Another method (due to Jacobi) for finding the inertia of A or, equivalently, of the corresponding Hermitian form will be discussed in Section 8.6.

Exercise 2. Find the rank and inertia of the following Hermitian forms and find transformations reducing them to sums of squares:

(a) $2\alpha_1^2 - \alpha_3^2 + \alpha_1\alpha_2 + \alpha_1\alpha_3$;
(b) $\alpha_1^2 + 2(\alpha_1\alpha_2 + \alpha_2\alpha_3 + \alpha_3\alpha_4 + \alpha_4\alpha_5 + \alpha_1\alpha_5)$;
(c) $2\alpha_1\alpha_2 - \alpha_1\alpha_3 + \alpha_1\alpha_4 - \alpha_2\alpha_3 + \alpha_2\alpha_4 - 2\alpha_3\alpha_4$.

Answers. (a) $r = 3$, $\pi = 1$, $\nu = 2$; (b) $r = 5$, $\pi = 3$, $\nu = 2$; (c) $r = 3$, $\pi = 1$, $\nu = 2$. The respective transforming matrices are

$$\begin{bmatrix} 1 & \frac{1}{4} & \frac{1}{4} \\ 0 & 1 & 1 \\ 0 & 0 & 1 \end{bmatrix}, \quad \begin{bmatrix} 1 & 1 & 0 & 0 & 1 \\ 0 & 1 & -1 & 0 & 1 \\ 0 & 0 & 1 & 1 & -1 \\ 0 & 0 & 0 & 1 & -2 \\ 0 & 0 & 0 & 0 & 1 \end{bmatrix}, \quad \begin{bmatrix} \frac{1}{2} & \frac{1}{2} & -\frac{1}{2} & \frac{1}{2} \\ -1 & 1 & 0 & 0 \\ 0 & 0 & 1 & 1 \\ 0 & 0 & 0 & 1 \end{bmatrix}. \quad \square$$

5.12 The Theory of Small Oscillations

In this section we introduce an important physical problem whose solution depends on the problem resolved in Theorem 5.5.2: the simultaneous reduction of two Hermitian matrices (or forms). We will describe the small oscillations of mechanical systems, although the resulting equations and their solution also arise in quite different circumstances; the oscillations of current and voltage in electrical circuits, for example.

Consider the motion of an elastic mechanical system whose displacement from some stable equilibrium position can be specified by n coordinates. Let these coordinates be $p_1, p_2, \ldots, p_n$. If the system performs small oscillations about the equilibrium position, its kinetic energy T is represented by a quadratic form in the time rates of change of the coordinates; $\dot{p}_1, \ldots, \dot{p}_n$, say. Thus, there exists a real symmetric matrix A such that

$$T = \tfrac{1}{2} \sum_{j=1}^{n} \sum_{k=1}^{n} a_{jk} \dot{p}_j \dot{p}_k = \tfrac{1}{2} \dot{\boldsymbol{p}}^{\mathrm{T}} A \dot{\boldsymbol{p}},$$

where $\boldsymbol{p} = [p_1 \quad p_2 \quad \cdots \quad p_n]^{\mathrm{T}}$ and $\dot{\boldsymbol{p}} = [\dot{p}_1 \quad \dot{p}_2 \quad \cdots \quad \dot{p}_n]^{\mathrm{T}}$. Furthermore, for any motion whatever, $\dot{\boldsymbol{p}} \in \mathbb{R}^n$ and for physical reasons, T is positive. Thus, A is a real, positive definite matrix (see Exercise 5.10.4).

When the system is distorted from its equilibrium position, there is generally potential energy stored in the system. It can be shown that if the coordinates are defined so that $\boldsymbol{p} = \boldsymbol{0}$ in the equilibrium position and the potential energy V is agreed to be zero in this position, then for small oscillations,

$$V = \tfrac{1}{2} \sum_{j=1}^{n} \sum_{k=1}^{n} c_{jk} p_j p_k = \tfrac{1}{2} \boldsymbol{p}^{\mathrm{T}} C \boldsymbol{p},$$

where C is a real symmetric nonnegative (or positive) definite matrix, that is, $V \geq 0$ for any $\boldsymbol{p} \in \mathbb{R}^n$.

We define the derivative of a matrix (or a vector) P to be the matrix whose elements are the derivatives of those of P. For any conformable matrices P and Q whose elements are differentiable functions of x, it is easily seen that

$$\frac{d}{dx}(PQ) = \frac{dP}{dx} Q + P \frac{dQ}{dx}. \tag{1}$$

We now view $p_1, p_2, \ldots, p_n, \dot{p}_1, \dot{p}_2, \ldots, p_n$ as a set of $2n$ independent variables and we obtain

$$\begin{aligned} \frac{\partial T}{\partial \dot{p}_i} &= \frac{1}{2}\left(\frac{\partial \dot{\boldsymbol{p}}^{\mathrm{T}}}{\partial \dot{p}_i} A \dot{\boldsymbol{p}} + \dot{\boldsymbol{p}}^{\mathrm{T}} A \frac{\partial \dot{\boldsymbol{p}}}{\partial \dot{p}_i} \right) \\ &= \frac{1}{2}(\boldsymbol{e}_i^{\mathrm{T}} A \dot{\boldsymbol{p}} + \dot{\boldsymbol{p}}^{\mathrm{T}} A \boldsymbol{e}_i) \\ &= \boldsymbol{A}_{i*} \dot{\boldsymbol{p}}, \end{aligned}$$

where $\boldsymbol{A}_{i*}$ denotes the ith row of A. Similarly,

$$\frac{\partial V}{\partial p_i} = \boldsymbol{C}_{i*} \boldsymbol{p}.$$

Equations of motion for the system (with no external forces acting) are obtained by Lagrange's method in the following form:

$$\frac{d}{dt}\left(\frac{\partial T}{\partial \dot{p}_i} \right) - \frac{\partial T}{\partial p_i} = -\frac{\partial V}{\partial p_i}, \qquad i = 1, 2, \ldots, n.$$

From the above expressions for the derivatives we find that the equations of motion are

$$\boldsymbol{A}_{i*} \ddot{\boldsymbol{p}} = -\boldsymbol{C}_{i*} \boldsymbol{p}, \qquad i = 1, 2, \ldots, n,$$

or, in matrix form,

$$A \ddot{\boldsymbol{p}} + C \boldsymbol{p} = \boldsymbol{0}. \tag{2}$$

Following the methods in Section 4.13, and anticipating sinusoidal solutions, we look for a solution $\boldsymbol{p}(t)$ of Eq. (2) of the form $\boldsymbol{p}(t) = \boldsymbol{q} e^{i\omega t}$, where $\boldsymbol{q}$ is independent of t and ω is real. Then $\ddot{\boldsymbol{p}} = -\omega^2 \boldsymbol{q} e^{i\omega t}$. and if we write $\lambda = \omega^2$, Eq. (2) becomes

$$(-\lambda A + C)\boldsymbol{q} = \boldsymbol{0}, \tag{3}$$

or

$$(\lambda I - A^{-1}C)\boldsymbol{q} = \boldsymbol{0}. \tag{4}$$

The admissible values of λ (defining *natural frequencies of vibration*) are the n eigenvalues of the matrix $A^{-1}C$. The *modes of vibration* are determined by corresponding right eigenvectors. However, we need to ensure that the eigenvalues are *real* and *nonnegative* before our solutions have the required form with a real value of ω.

The algebraic problem posed by Eq. (3) is just a special case of Eq. (5.5.5) and, using the argument of that section, the eigenvalues of $A^{-1}C$ are just those of $A^{-1/2}CA^{-1/2}$. But the latter matrix is obviously nonnegative (or positive) definite along with C. Hence, all eigenvalues of Eq. (4) are real and nonnegative (or positive if C is positive definite). Write for the jth eigenvalue $\lambda_j = \omega_j^2$, $(\omega_j \geq 0)$, and set $W = \text{diag}[\omega_1, \ldots, \omega_n]$.

The proof of Theorem 5.5.2 shows that if we define new coordinates in $\mathbb{C}^n$ by $\boldsymbol{q} = A^{-1/2}U\boldsymbol{s}$, where $U^*U = I$ and

$$U^*(A^{-1/2}CA^{-1/2})U = \Lambda = W^2,$$

then A and C can be simultaneously diagonalised. (In fact, U is here a (real) orthogonal matrix.) Reverting to the time-dependent equations, this suggests the definition of new generalised coordinates for the system: $\xi_1, \ldots, \xi_n$, given by $\boldsymbol{p}(t) = A^{-1/2}U\boldsymbol{\xi}(t)$, so that $\ddot{\boldsymbol{p}} = A^{-1/2}U\ddot{\boldsymbol{\xi}}$. Substitution into Eq. (2) and premultiplication by $U^*A^{-1/2}$ leads to the "decoupled" equation:

$$\ddot{\boldsymbol{\xi}} + W^2\boldsymbol{\xi} = \boldsymbol{0},$$

or

$$\ddot{\xi}_i + \omega_i^2\xi_i = 0, \qquad i = 1, 2, \ldots, n.$$

This implies that

$$\xi_i(t) = \alpha_i \cos(\omega_i t + \varepsilon_i) \qquad \text{if} \quad \omega_i \neq 0.$$

The α's and ε's in these equations are arbitrary constants that are usually determined by applying given initial conditions. When ξ_i varies in this fashion while all other ξ's are zero, we say that the system is oscillating in the ith *normal mode*. The coordinates $\xi_1, \xi_2, \ldots, \xi_n$ are *normal coordinates* for the system. Notice that

$$\begin{aligned} T &= \tfrac{1}{2}\dot{\boldsymbol{p}}^{\mathrm{T}}A\dot{\boldsymbol{p}} = \tfrac{1}{2}\dot{\boldsymbol{\xi}}^{\mathrm{T}}\dot{\boldsymbol{\xi}} \\ &= \tfrac{1}{2}(\dot{\xi}_1^2 + \dot{\xi}_2^2 + \cdots + \dot{\xi}_n^2) \end{aligned}$$

and

$$\begin{aligned} V &= \tfrac{1}{2}\boldsymbol{p}^{\mathrm{T}}C\boldsymbol{p} = \tfrac{1}{2}\boldsymbol{\xi}^{\mathrm{T}}W^2\boldsymbol{\xi} \\ &= \tfrac{1}{2}(\omega_1^2\xi_1^2 + \cdots + \omega_n^2\xi_n^2). \end{aligned}$$

Thus, when expressed in terms of the normal coordinates, both the kinetic and potential energies are expressed as sums of squares of the appropriate variables.

We can obtain some information about the normal modes by noting that when the system vibrates in a normal mode of frequency ω_i, we have $\boldsymbol{s} = \boldsymbol{e}_i$ and so the solution for the ith *normal-mode vector* $\boldsymbol{q}_i$ is

$$\boldsymbol{q}_i = (A^{-1/2}U)\boldsymbol{e}_i = i\text{th column of } A^{-1/2}U.$$

Hence the matrix of normal-mode vectors, $Q = [\boldsymbol{q}_1 \quad \boldsymbol{q}_2 \quad \cdots \quad \boldsymbol{q}_n]$ is given by $Q = A^{-1/2}U$. Furthermore,

$$Q^{\mathrm{T}}AQ = I, \qquad Q^{\mathrm{T}}CQ = W^2, \tag{5}$$

which may be described as the biorthogonality properties of the normal-mode vectors.

Now we briefly examine the implications of this analysis for a system to which *external forces* are applied. We suppose that, as in the above analysis, the elastic system makes only small displacements from equilibrium but that, in addition, there are prescribed forces $f_1(t), \ldots, f_n(t)$ applied in the coordinates $p_1, p_2, \ldots, p_n$. Defining $\boldsymbol{f}(t) = [f_1(t) \quad \cdots \quad f_n(t)]^{\mathrm{T}}$, we now have the equations of motion:

$$A\ddot{\boldsymbol{p}} + C\boldsymbol{p} = \boldsymbol{f}(t). \tag{6}$$

Putting $\boldsymbol{p} = Q\boldsymbol{\xi}$, premultiplying by Q^{T}, and using Eq. (5), Eq. (6) is immediately reduced to

$$\ddot{\boldsymbol{\xi}} + W^2\boldsymbol{\xi} = Q^{\mathrm{T}}\boldsymbol{f}(t),$$

or

$$\ddot{\xi}_j + \omega_j^2\xi_j = \boldsymbol{q}_j^{\mathrm{T}}\boldsymbol{f}(t), \qquad j = 1, 2, \ldots, n.$$

If $\omega_j \neq 0$, a particular integral of such an equation can generally be obtained in the form

$$\xi_j(t) = \frac{1}{\omega_j}\int_0^t \boldsymbol{q}_j^{\mathrm{T}}\boldsymbol{f}(\tau)\sin(\omega_j(t-\tau))\,d\tau,$$

and if $\omega_j = 0$, then

$$\xi_j(t) = \int_0^t \boldsymbol{q}_j^{\mathrm{T}}\boldsymbol{f}(\tau)(t-\tau)\,d\tau.$$

In particular, if $\det W \neq 0$, we may write

$$\boldsymbol{\xi}(t) = W^{-1}\int_0^t \sin(W(t-\tau))Q^{\mathrm{T}}\boldsymbol{f}(\tau)\,d\tau,$$

where integration of the vector integrand is understood to mean term-wise integration, and the jth diagonal term of the diagonal matrix $\sin W(t - \tau)$ is $\sin(\omega_j(t - \tau))$.

An important case that arises in practice can be described by writing $f(t) = f_0 e^{i\omega t}$, where f_0 is independent of t. This is the case of a prescribed sinusoidal external (or *exciting*) force. It is left as an exercise to prove that if $\omega \neq \omega_j$ $(j = 1, 2, \ldots, n)$, then

$$p(t) = Q\xi(t) = e^{i\omega t} \sum_{j=1}^{n} q_j \frac{\omega_j(q_j^T f_0)}{\omega_j^2 - \omega^2} \tag{7}$$

is a solution of Eq. (6).

This result can be used to discuss the resonance phenomenon. If we imagine the applied frequency ω to be varied continuously in a neighbourhood of a natural frequency ω_j, Eq. (7) indicates that in general the magnitude of the solution vector (or response) tends to infinity as $\omega \to \omega_j$. However, the singularity may be avoided if $q_j^T f_0 = 0$ for all q_j associated with ω_j.

Exercise 1. Establish Eq. (7).

Exercise 2. Show that if C is singular, that is, C is nonnegative (but not positive) definite, then there may be normal coordinates $\xi_i(t)$ for which $|\xi_i(t)| \to \infty$ as $t \to \infty$.

Exercise 3. Consider variations in the "stiffness" matrix C of the form $C + kI$, where k is a real parameter. What is the effect on the natural frequencies of such a change when $k > 0$ and when $k < 0$? □

5.13 Admissible Pairs of Matrices

When an $r \times p$ matrix X is multiplied on the right by a $p \times p$ matrix T, the resulting matrix XT is again $r \times p$. In fact, all matrices in the sequence $X, XT, XT^2, \ldots$ are of size $r \times p$. Such sequences appear frequently in some modern applications of the theory of matrices and particularly in systems theory. We give a brief introduction to some of their properties in this section.

First, any pair of matrices (X, T) of sizes $r \times p$ and $p \times p$, respectively, with $\operatorname{Ker} X \neq \{0\}$ is said to be an *admissible pair of order* p. Of course, r and p are arbitrary positive integers, although in applications it is frequently the case that $r \leq p$, and if $r < p$, then certainly $\operatorname{Ker} X \neq \{0\}$ (see Proposition

3.8.1). Indeed, the case $r = 1$ is of considerable interest. Let $K_1 = \text{Ker } X$ and let K_2 be the set of all vectors in both Ker X and Ker XT. Obviously, $K_2 \subset K_1$, although we *may* have $K_2 = K_1$. (When $p = 1$ and $T \neq O$, for example.) It is also clear that we may write

$$K_2 = \text{Ker}\begin{bmatrix} X \\ XT \end{bmatrix}.$$

More generally, for $l = 1, 2, \ldots$, define

$$K_l = \text{Ker}\begin{bmatrix} X \\ XT \\ XT^2 \\ \vdots \\ XT^{l-1} \end{bmatrix},$$

and it is seen that we have a "nested" sequence of subspaces of $\mathbb{C}^p$:

$$K_1 \supset K_2 \supset K_3 \supset \cdots.$$

A fundamental property of this sequence is contained in the next lemma.

Lemma 1. *Let (X, T) be an admissible pair of order p and let the sequence $K_1, K_2, \ldots$ of subspaces of $\mathbb{C}^p$ be defined as above. If $K_j = K_{j+1}$, then $K_j = K_{j+1} = K_{j+2} = \cdots$.*

PROOF. The matrices

$$\begin{bmatrix} X \\ XT \\ \vdots \\ XT^{j-1} \end{bmatrix} \quad \text{and} \quad \begin{bmatrix} X \\ XT \\ \vdots \\ XT^{j} \end{bmatrix} \tag{1}$$

have sizes $jr \times p$ and $(j + 1)r \times p$, respectively. If $K_j = K_{j+1}$ then, using Proposition 3.8.1, these two matrices are seen to have the same rank. Thinking in terms of "row rank," this means that the rows of XT^j are linear combinations of the rows of the first matrix in (1). Thus, there is an $r \times jr$ matrix E such that

$$XT^j = E\begin{bmatrix} X \\ XT \\ \vdots \\ XT^{j-1} \end{bmatrix}. \tag{2}$$

Now let $\mathbf{x} \in K_j = K_{j+1}$. Then, using Eq. (2),

$$XT^{j+1}\mathbf{x} = (XT^j)T\mathbf{x} = E\begin{bmatrix} X \\ XT \\ \vdots \\ XT^{j-1} \end{bmatrix} T\mathbf{x} = E\begin{bmatrix} XT \\ XT^2 \\ \vdots \\ XT^j \end{bmatrix}\mathbf{x} = \mathit{0},$$

since $\mathbf{x} \in K_{j+1}$. Consequently, $\mathbf{x} \in \operatorname{Ker}(XT^{j+1})$ and so $\mathbf{x} \in K_{j+2}$. Then it follows inductively that for any $i \geq j$, we have $K_i = K_j$. ■

The lemma now shows that the inclusions in the sequence $K_1 \supset K_2 \supset K_3 \supset \cdots$ are proper until equality first holds, and thereafter all K_j's are equal. The least positive integer s such that $K_s = K_{s+1}$ is a characteristic of the admissible pair (X, T) and is called the *index* (*of stabilization*) of the pair. Thus, if K_j has dimension k_j, then $k_1 > k_2 > \cdots > k_s = k_{s+1} = \cdots$. Since the first $s - 1$ inclusions are proper it is clear that $s \leq p$ (see also Exercise 3 below). The subspace K_s is said to be *residual*.

A distinction is now made between the cases when the residual subspace K_s is just $\{\mathit{0}\}$, so that $k_s = 0$, and when K_s contains nonzero vectors, so that $k_s \geq 1$. If s is the index of stabilization of the admissible pair (X, T) and $k_s = 0$, the pair (X, T) is said to be *observable*. Otherwise, the residual subspace K_s is said to be *unobservable*, a terminology originating in systems theory (see Section 9.11).

Exercise 1. Let (X, T) be an admissible pair of order p, and establish the following special cases.

(a) If $r = 1$, $X \neq \mathit{0}$, and $T = I_p$, then the index of (X, I_p) is one, and $k_1 = k_2 = \cdots = p - 1$.

(b) If $r = 1$, $X \neq 0$, and (X, T) are observable, then the index is p.

(c) If $r = p$ and $\det X \neq \mathit{0}$, then for any $p \times p$ matrix T, the pair (X, T) is observable with index one.

(d) If $\mathbf{x}^{\mathrm{T}}$ is a left eigenvector of T, then the pair $(\mathbf{x}^{\mathrm{T}}, T)$ has index $s = 1$.

Exercise 2. Show that the unobservable subspace of an admissible pair (X, T) is T-invariant.

Exercise 3. If (X, T) is an admissible pair of order p and has index s, and if X has rank $\rho \neq 0$, show that

$$\frac{p - k_s}{\rho} \leq s \leq p + 1 - \rho - k_s.$$

Exercise 4. Let (X, T) be admissible of order p with X of size $r \times p$. Show that for any matrix G of size $p \times r$, the pairs

$$(X, T) \qquad \text{and} \qquad (X, T + GX)$$

have the same index and unobservable subspace.

***Exercise* 5.** Let (X, T) be as in Exercise 4 and let R, S be nonsingular matrices of sizes $r \times r$ and $p \times p$, respectively. Show that the pairs

$$(X, T) \qquad \text{and} \qquad (RXS, S^{-1}TS)$$

have the same index and unobservable subspaces of the same dimension (admitting zero dimension in the case of observable pairs). Observe that, using Theorems 4.3.1 and 4.3.2 this exercise implies that observability can be interpreted as a property of a pair of linear transformations which have X and T as matrix representations. □

The next result is the form in which the observability of (X, T) is usually presented. Note that the value of the index of the pair (X, T) does not appear explicitly in the statement.

Theorem 1. *The admissible pair (X, T) of order p is observable if and only if*

$$\operatorname{rank}\begin{bmatrix} X \\ XT \\ \vdots \\ XT^{p-1} \end{bmatrix} = p. \tag{3}$$

PROOF. If condition (3) holds, then in the above notation, the kernel $K_p = \{\mathbf{0}\}$. Thus, the unobservable subspace is trivial and (X, T) is an observable pair.

Conversely, if (X, T) is observable then, by Exercise 3, the index s cannot exceed p. Hence $K_s = K_p = \{0\}$, and this implies the rank condition (3). ■

The results described so far in this section have a dual formulation. The discussion could begin with consideration of the subspaces of $\mathbb{C}^p$,

$$R_1 = \operatorname{Im} X^*, \qquad R_2 = \operatorname{Im} X^* + \operatorname{Im} T^*X^*,$$

and the trivial observation that $R_1 \subset R_2 \subset \mathbb{C}^p$. But there is a quicker way. Define $R_j = \sum_{k=1}^{j} \operatorname{Im}((T^*)^{k-1}X^*)$ and observe that

$$R_j = \operatorname{Im}[X^* \quad T^*X^* \cdots T^{*j-1}X^*]. \tag{4}$$

Then Proposition 5.8.1 shows that

$$\operatorname{Ker}\begin{bmatrix} X \\ XT \\ \vdots \\ XT^{j-1} \end{bmatrix} \oplus \operatorname{Im}[X^* \quad T^*X^* \cdots T^{*j-1}X^*] = \mathbb{C}^p.$$

Consequently, we have the relation

$$K_j \oplus R_j = \mathbb{C}^p \tag{5}$$

for $j = 1, 2, \ldots$. The properties of the sequence $K_1, K_2, \ldots$ now tell us immediately that $R_1 \subset R_2 \subset \cdots \subset \mathbb{C}^p$ with proper inclusions up to $R_{s-1} \subset R_s$, where s is the index of (X, T), and thereafter $R_s = R_{s+1} = \cdots$. Furthermore, $R_s = \mathbb{C}^p$ if the pair (X, T) is observable.

In this perspective, it is said that the pair (T^*, X^*) is *controllable* when (X, T) is observable, and in this case $K_s = \{0\}$ and $R_s = \mathbb{C}^p$. If $R_s \neq \mathbb{C}^p$, then R_s is called the *controllable subspace* and the pair (T^*, X^*) is *uncontrollable*.

Note that, using Eq. (4) and the fact that $s \leq p$, we may write

$$R_s = \sum_{k=0}^{s-1} \operatorname{Im}((T^*)^k X^*) = \sum_{k=0}^{p-1} \operatorname{Im}((T^*)^k X^*). \tag{6}$$

The criterion of Theorem 1 now gives a criterion for controllability which is included in the following summary of our discussion.

Corollary 1. *Let A and B be matrices of sizes $p \times p$ and $p \times r$, respectively. Then the following statements are equivalent:*

(a) *The pair (A, B) is controllable;*
(b) *The pair (B^*, A^*) is observable;*
(c) $\operatorname{rank}[B \quad AB \quad \cdots \quad A^{p-1}B] = p$;
(d) $\mathbb{C}^p = \sum_{k=0}^{p-1} \operatorname{Im}(A^k B)$.

Exercise 6. Consider a pair $(A, \boldsymbol{b})$, where A is $p \times p$ and $\boldsymbol{b} \in \mathbb{C}^p$ (i.e., $r = 1$ in Corollary 1). Let A be simple with the spectral representation of Theorem 4.10.3, and let $\boldsymbol{b} = \sum_{j=1}^{p} \beta_j \boldsymbol{x}_j$ be the representation of $\boldsymbol{b}$ in the eigenbasis for A (with n replaced by p). Show that if $\beta_j = 0$ for some j, then $(A, \boldsymbol{b})$ is not controllable.

Exercise 7. Use Exercise 4 to show that if A, B, and F are matrices of sizes $p \times p$, $p \times r$, and $r \times p$, respectively, then the pairs

$$(A, B) \qquad \text{and} \qquad (A + BF, B)$$

have the same controllable subspace.

Exercise 8. Let A, B be as in Exercise 7 and let U, V be nonsingular matrices of sizes $p \times p$ and $r \times r$, respectively. Use Exercise 5 to show that the pair (A, B) is controllable if and only if the pair $(U^{-1}AU, U^{-1}BV)$ is controllable.

Exercise 9. Let $A \in \mathbb{C}^{n \times n}$ and $B \in \mathbb{C}^{n \times m}$. Show that the following three statements are equivalent:

(a) The pair (A, B) is controllable;
(b) For any $m \times n$ matrix K the pair $(A + BK, B)$ is controllable;
(c) For any complex number λ, the pair $(A - \lambda I, B)$ is controllable.

Hint. Prove that there are $m \times m$ matrices M_{jk}, $k \geq 2, j = 1, 2, \ldots, k-1$, such that

$$[B \quad (A+BK)B \quad \cdots \quad (A+BK)^{n-1}B] =$$

$$[B \quad AB \quad \cdots \quad A^{n-1}B]\begin{bmatrix} I & M_{12} & \cdots & & M_{1n} \\ 0 & I & & & \vdots \\ \vdots & & \ddots & \ddots & M_{n-1,n} \\ 0 & & \cdots & 0 & I \end{bmatrix}. \quad \square$$

5.14 Miscellaneous Exercises

1. For which real values of α are the following matrices positive definite:

$$\text{(a)} \quad \begin{bmatrix} 1 & \alpha & \alpha \\ \alpha & 1 & \alpha \\ \alpha & \alpha & 1 \end{bmatrix}, \qquad \text{(b)} \quad \begin{bmatrix} 1 & 1 & \alpha \\ 1 & 1 & 1 \\ \alpha & 1 & 1 \end{bmatrix}.$$

Answer. (a) $-\frac{1}{2} < \alpha < 1$; (b) none.

2. Let $A \in \mathbb{C}^{n \times n}$ and suppose $A\mathbf{x} = \lambda\mathbf{x}$ and $A^*\mathbf{y} = \mu\mathbf{y}$ $(\mathbf{x}, \mathbf{y} \neq \mathbf{0})$. Prove that if $\lambda \neq \mu$, then $\mathbf{x} \perp \mathbf{y}$, and that if $\mathbf{x} = \mathbf{y}$, then $\lambda = \bar{\mu}$.

3. Check that the matrix $H \geq 0$ is positive definite if and only if $\det H \neq 0$.

4. Let $A, B \in \mathbb{C}^{n \times n}$ and define

$$R = \begin{bmatrix} A & B \\ B & -A \end{bmatrix}$$

Show that λ is an eigenvalue of R only if $-\lambda$ is an eigenvalue of R.

Hint. Observe that $PR + RP = 0$ where

$$P = \begin{bmatrix} 0 & I \\ -I & 0 \end{bmatrix}.$$

5. Verify that a Hermitian unitary matrix may have only $+1$ and -1 as its eigenvalues.

6. If P is an orthogonal projector and α_1, α_2 are nonzero real numbers, prove that the matrix $\alpha_1^2 P + \alpha_2^2 (I - P)$ is positive definite.

Hint. Use Theorem 5.2.1.

7. Prove that an orthogonal projector that is not a zero-matrix or a unit matrix is positive semi-definite.

8. Prove that if A is Hermitian, then the matrix $I + \varepsilon A$ is positive definite for sufficiently small real numbers ε.

9. Show that a circulant matrix is normal.

10. Prove that an idempotent matrix is normal if and only if it is an orthogonal projector.

11. Prove that A is a normal matrix if and only if each eigenvector of A is an eigenvector of A^*.

Hint. See Exercise 5.2.3 and compare with Exercise 2 above.

12. Prove that if $A^*B = O$, then the subspaces Im A and Im B are orthogonal.

13. Prove that any leading principal submatrix of a positive semi-definite matrix is positive semi-definite.

Hint. See Corollary 5.5.2.

14. If $A, B \in \mathbb{C}^{n \times n}$ are positive semi-definite matrices, prove that

(a) The eigenvalues of AB are real and nonnegative;
(b) $AB = O$ if and only if tr $AB = 0$.

Hint. For part (a) consider the characteristic polynomial of AB, where $A = U \operatorname{diag}[D_0, O]U^*$, $D_0 \geq O$.

15. Prove that the Gram matrix is a positive semi-definite matrix for any system of elements and is positive definite if the system is linearly independent.

16. Let $\mathbf{x}, \mathbf{y}$ be elements of a unitary space $\mathscr{U}$ with a basis $\mathscr{E} = \{\mathbf{x}_1, \mathbf{x}_2, \ldots, \mathbf{x}_n\}$. Prove that, if $\boldsymbol{\alpha}$ and $\boldsymbol{\beta}$ are representations of $\mathbf{x}$ and $\mathbf{y}$, respectively, with respect to $\mathscr{E}$, then

$$(\mathbf{x}, \mathbf{y}) = (\boldsymbol{\alpha}, G\boldsymbol{\beta}), \tag{1}$$

where G is the Gram matrix of $\mathscr{E}$.

Conversely, show that for any $G > O$, the rule in Eq. (1) gives an inner product in $\mathscr{U}$ and that G is the Gram matrix of the system $\mathscr{E}$ with respect to this inner product.

17. Let $A, B, C \in \mathbb{C}^{n \times n}$ be positive definite.

(a) Prove that all $2n$ zeros λ_i of $\det(\lambda^2 A + \lambda B + C)$ have negative real parts.

Hint. Consider the scalar polynomial $(\mathbf{x}, (\lambda_i^2 A + \lambda_i B + C)\mathbf{x})$ for $\mathbf{x} \in \operatorname{Ker}(\lambda_i^2 A + \lambda_i B + C)$ and use Exercise 5.10.4.

(b) Show that all zeros of $\det(\lambda^2 A + \lambda B - C)$ are real.

(c) Prove the conclusion of part (a) under the weaker hypotheses that A and C are positive definite and B has a positive definite real part.

18. Let $A \in \mathbb{C}^{n \times n}$. Prove that if there exists a positive definite $n \times n$ matrix

H such that the matrix $B \triangleq A^*H + HA$ is positive definite, then In $A = \{n, 0, 0\}$.

Hint. Consider the form (Bx, x) evaluated at the eigenvectors of A.

19. Let $A = H_1 + iH_2$ be the Cartesian decomposition of A. Show that if $H_1 < O$, then $\dot{\pi}(A) = 0$.

20. Check that if $\pi(A) = 0$, then $\pi(AH) = 0$ for any $H > 0$.

Hint. Observe that $H^{1/2}(AH)H^{(-1/2)} = H^{1/2}A(H^{1/2})^*$.

21. Let $H_1 = H_1^*$, $H_2 = H_2^*$. We write $H_1 \geq H_2$ if $H_1 - H_2$ is a positive semi-definite matrix. Prove the following statements:

(a) $H_1 \geq H_1$ (*symmetry*);
(b) If $H_1 \geq H_2$ and $H_2 \geq H_3$, then $H_1 \geq H_3$ (*transitivity*);
(c) If $H_1 > O$, $H_2 > O$, and $H_1 \geq H_2$, then $H_2^{-1} \geq H_1^{-1}$.

Hint. For part (c) use Exercise 20.

22. Prove the *Fredholm alternative* for matrices: If $A \in \mathbb{C}^{m \times n}$ then either the system $A\boldsymbol{x} = \boldsymbol{b}$ is solvable for any $\boldsymbol{b} \in \mathbb{C}^m$ or the homogeneous system $A^*\boldsymbol{y} = \boldsymbol{0}$ has nonzero solutions.

23. Prove that if S is real and skew-symmetric, then $I + S$ is nonsingular and the *Cayley transform*,

$$T = (I - S)(I + S)^{-1},$$

is an orthogonal matrix.

24. (a) If A is a real orthogonal matrix and $I + A$ is nonsingular, prove that we can write

$$A = (I - S)(I + S)^{-1},$$

where S is a real skew-symmetric matrix.

(b) If U is a unitary matrix and $I + U$ is nonsingular, show the matrix

$$A = i(I - U)(I + U)^{-1}$$

is Hermitian.

25. Let $B, C \in \mathbb{R}^{n \times n}$, $A = B + iC$, and

$$R = \begin{bmatrix} B & -C \\ C & B \end{bmatrix}.$$

Prove that

(a) If A is normal, then so is the real matrix R;
(b) If A is Hermitian, then R is symmetric;
(c) If A is positive definite, then so is the real matrix R;
(d) If A is unitary, then R is orthogonal.

CHAPTER 6

The Jordan Canonical Form: A Geometric Approach

Let T denote a linear transformation acting on an n-dimensional linear space $\mathscr{S}$. It has been seen in Section 4.3 that a matrix representation of T is determined by each choice of basis for $\mathscr{S}$, that all such matrices are similar to one another, and that they form an equivalence class in the set of all $n \times n$ matrices. The simplest (canonical) matrix in the equivalence class $\mathscr{A}_T^0$ of similar matrices (representations of T in different bases of $\mathscr{S}$) will be found in this chapter. For simple transformations, this problem has already been solved in Section 4.8; its solution was based on the existence of a basis in $\mathscr{S}$ consisting of eigenvectors of the transformation. Furthermore, the canonical form was found to be a diagonal matrix of eigenvalues.

For a general transformation $T \in \mathscr{L}(\mathscr{S})$ there may be no eigenbasis of T in $\mathscr{S}$. Therefore, additional elements from $\mathscr{S}$ must be joined to a set of linearly independent eigenvectors of T to construct a "canonical" basis for $\mathscr{S}$ with respect to which the representation of T is in a canonical form (although no longer diagonal). This procedure will be performed in Section 6.5, after the structure of T-invariant subspaces is better understood. This is called the *geometric* (or *operator*) *approach*, in contrast to that presented in the next chapter and based on algebraic properties of matrices of the form $\lambda I - A$.

6.1 Annihilating Polynomials

We shall first consider the notion of annihilating polynomials for a linear transformation $T: \mathscr{S} \to \mathscr{S}$. The significance of this for our geometric approach mentioned above is not immediately obvious, but it will be seen that the main result of the section, Theorem 4, does indeed present a decomposition of $\mathscr{S}$ as a direct sum of subspaces. This will be important later in the chapter.

Let $\mathscr{S}$ be a linear space over $\mathscr{F}$, let $T \in \mathscr{L}(\mathscr{S})$, and let $p(\lambda) = \sum_{i=0}^{l} p_i \lambda^i$ ($p_l \neq 0$) be a polynomial over $\mathscr{F}$. As with matrices (see Section 1.7), the transformation $p(T)$ is defined by the equality $p(T) \triangleq \sum_{i=0}^{l} p_i T^i$.

Exercise 1. Show that for any two scalar polynomials $p_1(\lambda)$ and $p_2(\lambda)$ over $\mathscr{F}$ and any $T \in \mathscr{L}(\mathscr{S})$, the transformations $p_1(T)$ and $p_2(T)$ commute. □

The result of the first exercise yields the possibility of substituting T instead of λ in any expression consisting of sums and products of polynomials.

Exercise 2. Show that the equality

$$p(\lambda) = \sum_{j=1}^{s} \prod_{i=1}^{k} p_{ij}(\lambda),$$

where $p_{ij}(\lambda)$ are polynomials in λ, implies

$$p(T) = \sum_{j=1}^{s} \prod_{i=1}^{k} p_{ij}(T).$$

Exercise 3. Let $T \in \mathscr{L}(\mathscr{S})$ and let $\mathscr{S}_0$ be a subspace in $\mathscr{S}$. Show that if $\mathscr{S}_0$ is T-invariant, then it is $p(T)$-invariant for any polynomial $p(\lambda)$ over $\mathscr{F}$.

Exercise 4. Let $T \in \mathscr{L}(\mathscr{S})$ with $\mathscr{S}$ a linear space over $\mathscr{F}$. If $\lambda_0 \in \sigma(T)$ ($\lambda_0 \in \mathscr{F}$), and $p(\lambda)$ is any polynomial over $\mathscr{F}$, show that $p(\lambda_0) \in \sigma(p(T))$. Moreover, if $\mathbf{x}$ is an eigenvector of T corresponding to λ_0, then $\mathbf{x}$ is an eigenvector of $p(T)$ associated with $p(\lambda_0)$.

SOLUTION. If $T(\mathbf{x}) = \lambda_0 \mathbf{x}$, then

$$p(T)(\mathbf{x}) = \left(\sum_{i=0}^{l} p_i T^i\right)(\mathbf{x}) = \sum_{i=0}^{l} p_i \lambda_0^i \mathbf{x} = p(\lambda_0)\mathbf{x}. \quad \square$$

Theorem 1. *Let $p_1(\lambda), p_2(\lambda), \ldots, p_k(\lambda)$ be arbitrary nonzero polynomials over $\mathscr{F}$ and let $p(\lambda)$ denote their least common multiple. Then for any $T \in \mathscr{L}(\mathscr{S})$,*

$$\operatorname{Ker} p(T) = \sum_{i=1}^{k} \operatorname{Ker} p_i(T). \tag{1}$$

PROOF. Let $x \in \sum_{i=1}^{k} \operatorname{Ker} p_i(T)$. Then $x = \sum_{i=1}^{k} x_i$, where $p_i(T)(x_i) = 0$, $i = 1, 2, \ldots, k$. Since $p(\lambda)$ is a common multiple of $p_1(\lambda), \ldots, p_k(\lambda)$, there exist polynomials $q_i(\lambda)$ such that $p(\lambda) = q_i(\lambda)p_i(\lambda)$, $i = 1, 2, \ldots, k$. Using Exercise 2, we obtain $p(T) = q_i(T)p_i(T)$. Hence $x_i \in \operatorname{Ker} p_i(T)$ implies $x_i \in \operatorname{Ker} p(T)$ and consequently $x = \sum_{i=1}^{k} x_i \in \operatorname{Ker} p(T)$. Thus, $\sum_{i=1}^{k} \operatorname{Ker} p_i(T) \subset \operatorname{Ker} p(T)$ for any common multiple $p(\lambda)$ of $p_1(\lambda), p_2(\lambda), \ldots, p_k(\lambda)$.

The opposite inclusion can be proved by induction on k. The crux of the argument is contained in the case $k = 2$. So first let $k = 2$ and assume $p(T)(x) = 0$. Then, as in the first part of the proofs, $p_i(T)q_i(T)(x) = 0$ for $i = 1, 2$. If $y_i = q_i(T)(x)$ then, obviously, $y_i \in \operatorname{Ker} p_i(T)$. Furthermore, $p(\lambda)$ is the least common multiple of $p_1(\lambda)$ and $p_2(\lambda)$ and, therefore, the polynomials $q_1(\lambda)$ and $q_2(\lambda)$ are relatively prime. (See Exercise 8 in Appendix 1.) This is equivalent (see Eq. (3) in Appendix 1) to saying that $c_1(\lambda)q_1(\lambda) + c_2(\lambda)q_2(\lambda) = 1$ for some polynomials $c_1(\lambda)$ and $c_2(\lambda)$. Thus there are polynomials $c_i(\lambda)$ $(i = 1, 2)$ such that

$$c_1(T)q_1(T) + c_2(T)q_2(T) = I.$$

Writing $x_i = c_i(T)(y_i)$, and applying this representation of I to x we obtain $x = x_1 + x_2$, where $x_i \in \operatorname{Ker} p_i(T)$ $(i = 1, 2)$. Hence $x \in \operatorname{Ker} p_1(T) + \operatorname{Ker} p_2(T)$, as required.

Now let the result hold for the polynomials $p_1(\lambda), p_2(\lambda), \ldots, p_{n-1}(\lambda)$ $(2 < n \le k)$. Thus if $\tilde{p}(\lambda)$ denotes their least common multiple, then

$$\operatorname{Ker} \tilde{p}(T) = \sum_{i=1}^{n-1} \operatorname{Ker} p_i(T).$$

But the least common multiple $p(\lambda)$ of $p_1(\lambda), \ldots, p_n(\lambda)$ coincides with the least common multiple of $\tilde{p}(\lambda)$ and $p_n(\lambda)$ (Exercise 9 in Appendix 1). Therefore, using the already proved part of the theorem and the inductive hypothesis,

$$\operatorname{Ker} p(T) = \operatorname{Ker} \tilde{p}(T) + \operatorname{Ker} p_n(T) = \sum_{i=1}^{n} \operatorname{Ker} p_i(T).$$

The proof is complete. ■

Theorem 2. *If, in the notation of Theorem 1, each pair of polynomials selected from* $\{p_1(\lambda), p_2(\lambda), \ldots, p_k(\lambda)\}$ *is relatively prime, then*

$$\operatorname{Ker} p(T) = \sum_{i=1}^{k} \cdot \operatorname{Ker} p_i(T). \tag{2}$$

PROOF. Consider the case $k = 2$. To show that the given sum is direct, it suffices to show that $\operatorname{Ker} p_1(T) \cap \operatorname{Ker} p_2(T) = \{0\}$. But if simultaneously

$p_1(T)(\mathbf{x}) = \mathbf{0}$ and $p_2(T)(\mathbf{x}) = \mathbf{0}$ then, since $p_1(\lambda)$ and $p_2(\lambda)$ are relatively prime (see Theorem 3 in Appendix 1)

$$I(\mathbf{x}) = c_1(T)p_1(T)(\mathbf{x}) + c_2(T)p_2(T)(\mathbf{x}) = \mathbf{0},$$

and consequently the required result: $\mathbf{x} = \mathbf{0}$. An induction argument like that of Theorem 1 completes the proof. ■

Note that the hypothesis on the polynomials in Theorem 2 is *stronger* than the condition that the set $\{p_i(\lambda)\}_{i=1}^{k}$ be *relatively prime*. (See Exercise 4 in Appendix 1 and recall that the polynomials being considered are assumed to be monic.)

Now we define a nonzero scalar polynomial $p(\lambda)$ over $\mathcal{F}$ to be an *annihilating polynomial* of the transformation T if $p(T)$ is the zero transformation: $p(T) = O$. It is assumed here, and elsewhere, that $T \in \mathcal{L}(\mathcal{S})$ and $\mathcal{S}$ is over $\mathcal{F}$.

Exercise 5. Show that if $p(\lambda)$ is an annihilating polynomial for T, then so is the polynomial $p(\lambda)q(\lambda)$ for any $q(\lambda)$ over $\mathcal{F}$.

Exercise 6. Check that if $p(\lambda)$ is an annihilating polynomial of T and $\lambda_0 \in \sigma(T)$, $\lambda_0 \in \mathcal{F}$, then $p(\lambda_0) = 0$. (See Exercise 4.) □

In other words, Exercise 6 shows that the spectrum of a transformation T is included in the set of zeros of any annihilating polynomial of T. Note that, in view of Exercise 5, the converse cannot be true.

Theorem 3. *For any $T \in \mathcal{L}(\mathcal{S})$ there exists an annihilating polynomial of degree $l \leq n^2$, where $n = \dim \mathcal{S}$.*

PROOF. We first observe that the existence of a (monic) annihilating polynomial $p(\lambda)$ for T is provided by Exercise 4.1.12 and is equivalent to the condition that T^l be a linear combination of the transformations T^i, $i = 0, 1, \ldots, l - 1$, viewed as elements in $\mathcal{L}(\mathcal{S})$. (We define $T^0 \triangleq I$.) Thus, for some $p_0, p_1, \ldots, p_{l-1} \in \mathcal{F}$,

$$T^l = -p_0 I - p_1 T - \cdots - p_{l-1}T^{l-1}. \tag{3}$$

To verify Eq. (3), we examine the sequence $I, T, T^2, \ldots$. Since the dimension of the linear space $\mathcal{L}(\mathcal{S})$ is n^2 (see Exercise 4.1.8), at most n^2 elements in this sequence are linearly independent. Thus there exists a positive integer l $(1 \leq l \leq n^2)$ such that $I, T, \ldots, T^{l-1}$ viewed as elements in $\mathcal{L}(\mathcal{S})$ are linearly independent, while $I, T, \ldots, T^l$ form a linearly dependent system. Hence T^l is a linear combination of T^i, $i = 0, 1, \ldots, l - 1$. This proves the theorem. ■

The main result follows readily from Theorem 2, bearing in mind that if $p(T) = O$ then $\mathrm{Ker}(p(T)) = \mathcal{S}$.

Theorem 4. *Let $T \in \mathscr{L}(\mathscr{S})$ and let $p(\lambda)$ denote an annihilating polynomial of T. If $p(\lambda) = \prod_{i=1}^{k} p_i(\lambda)$ for some polynomials $p_1(\lambda), \ldots, p_k(\lambda)$ and if each pair of polynomials selected from $\{p_1(\lambda), p_2(\lambda), \ldots, p_k(\lambda)\}$ is relatively prime, then*

$$\mathscr{S} = \sum_{i=1}^{k} \cdot \operatorname{Ker} p_i(T). \tag{4}$$

Theorem 4 is the main conclusion of this section and, as it will be seen, allows us to study the action of T on $\mathscr{S}$ in terms of its action on the (T-invariant) subspaces $p_i(T)$.

Exercise 7. Suppose $\lambda_1 \neq \lambda_2$ and the matrix

$$A = \begin{bmatrix} \lambda_1 & 1 & 0 & 0 \\ 0 & \lambda_1 & 0 & 0 \\ 0 & 0 & \lambda_2 & 1 \\ 0 & 0 & 0 & \lambda_2 \end{bmatrix}$$

is viewed as a linear transformation on $\mathbb{C}^4$. Show that

$$p(\lambda) \triangleq (\lambda - \lambda_1)^2(\lambda - \lambda_2)^2$$

is an annihilating polynomial for A and verify Theorem 4 with $p_1(\lambda) = (\lambda - \lambda_1)^2$ and $p_2(\lambda) = (\lambda - \lambda_2)^2$.

Show that the conclusion does not hold with the choice of factors $q_1(\lambda) = q_2(\lambda) = (\lambda - \lambda_1)(\lambda - \lambda_2)$ for $p(\lambda)$. □

6.2 Minimal Polynomials

It is clear that a fixed linear transformation $T \in \mathscr{L}(\mathscr{S})$ has many annihilating polynomials. It is also clear that there is a lower bound for the degrees of all annihilating polynomials for a fixed T. It is natural to seek a monic annihilating polynomial of the least possible degree, called a *minimal polynomial* for T and written $m_T(\lambda)$, or just $m(\lambda)$. It will be shown next that the set of all annihilating polynomials for T is "well ordered" by the degrees of the polynomials and, hence, there is a *unique* minimal polynomial.

Theorem 1. *A minimal polynomial of $T \in \mathscr{L}(\mathscr{S})$ divides any other annihilating polynomial of T.*

PROOF. Let $p(\lambda)$ be an annihilating polynomial of T and let $m(\lambda)$ denote a minimal polynomial of the transformation. Using Theorem 1 of Appendix 1, we write

$$p(\lambda) = m(\lambda)d(\lambda) + r(\lambda), \tag{1}$$

where either $\deg r(\lambda) < \deg m(\lambda)$ or $r(\lambda) \equiv 0$. Assuming that $m(\lambda)$ does not divide $p(\lambda)$, that is, $r(\lambda) \not\equiv 0$, we arrive at a contradiction. Indeed, applying the result of Exercise 6.1.2 to Eq. (1), we find from $p(T) = m(T) = 0$ that $r(T) = 0$. Hence $r(\lambda)$ is an annihilating polynomial of T of degree less than that of the minimal polynomial $m(\lambda)$. This conflicts with the definition of $m(\lambda)$. ■

Corollary 1. *The minimal polynomial of T is unique.*

PROOF. If $m_1(\lambda)$ and $m_2(\lambda)$ are minimal polynomials of the transformation T, then by Theorem 1 each of them divides the other. Since both of them are monic, Exercise 2 of Appendix 1 yields $m_1(\lambda) = m_2(\lambda)$. ■

Corollary 2. *The set of all annihilating polynomials of T consists of all polynomials divisible by the minimal polynomial of T.*

PROOF. To see that every polynomial divisible by $m(\lambda)$ is an annihilating polynomial of T, it suffices to recall the result of Exercise 6.1.5. ■

Theorem 2. *The set of the distinct zeros of the minimal polynomial of T coincides with the spectrum of T.*

PROOF. In view of Exercise 6.1.6, it suffices to show that any zero of the minimal polynomial $m(\lambda)$ of T belongs to the spectrum $\sigma(T)$ of T.

Let $m(\lambda) = q(\lambda)\prod_{i=1}^{k}(\lambda - \lambda_i)$, where $q(\lambda)$ is irreducible over $\mathscr{F}$. Assume on the contrary that $\lambda_s \notin \sigma(T)$ $(1 \le s \le k)$; in other words, $T - \lambda_s I$ is nonsingular. Since $m(T) = 0$, Exercise 6.1.2 implies

$$q(T)\prod_{i=1, i\neq s}^{k}(T - \lambda_i I) = 0.$$

Hence $\tilde{m}(\lambda) = m(\lambda)/(\lambda - \lambda_s)$ is an annihilating polynomial of T. This contradicts the assumption that $m(\lambda)$ is minimal and shows that $\lambda_1, \lambda_2, \ldots, \lambda_k \in \sigma(T)$ as required. ■

In this chapter the nature of the field $\mathscr{F}$ has not played an important role. Now, and for the rest of the chapter, it is necessary to assume that $\mathscr{F}$ is algebraically closed (see Appendix 1). In fact, we assume for simplicity that $\mathscr{F} = \mathbb{C}$, although any algebraically closed field could take the place of $\mathbb{C}$.

Corollary 1. *Let $\mathscr{F} = \mathbb{C}$, and let $T \in \mathscr{L}(\mathscr{S})$, with $\sigma(T) = \{\lambda_1, \lambda_2, \ldots, \lambda_s\}$. The minimal polynomial $m(\lambda)$ of T is given by*

$$m(\lambda) = \prod_{i=1}^{s} (\lambda - \lambda_i)^{m_i} \tag{2}$$

for some positive integers m_i $(i = 1, 2, \ldots, s)$.

The exponent m_i associated with the eigenvalue $\lambda_i \in \sigma(A)$ is clearly unique (see Corollary 1 of Theorem 1). It plays an important role in this and subsequent chapters and is known as the *index* of λ_i.

Applying Theorem 6.1.4 to $m(\lambda)$ written in the form (2), we arrive at the following important decomposition of a linear transformation.

Theorem 3. *Let $T \in \mathscr{L}(\mathscr{S})$, let $\sigma(T) = \{\lambda_1, \lambda_2, \ldots, \lambda_s\}$; and let λ_i have index m_i for each i. Then*

$$T = \sum_{i=1}^{s} \cdot\, T|_{\mathscr{G}_i}, \tag{3}$$

where, for $i = 1, 2, \ldots, s$,

$$\mathscr{G}_i = \operatorname{Ker}((T - \lambda_i I)^{m_i}) \tag{4}$$

is a T-invariant subspace.

PROOF. By Eq. (6.1.5) we have, on putting $p_i(\lambda) = (\lambda - \lambda_i)^{m_i}$, $i = 1, 2, \ldots, s$,

$$\mathscr{S} = \sum_{i=1}^{s} \cdot\, \mathscr{G}_i. \tag{5}$$

Therefore, recalling the definition of the direct sum of transformations, it suffices to show that each $\mathscr{G}_i$ is T-invariant. Indeed, if $\mathbf{x} \in \mathscr{G}_i$ then $(T - \lambda_i I)^{m_i}(\mathbf{x}) = \mathbf{0}$. Thus, by Exercise 6.1.1,

$$(T - \lambda_i I)^{m_i} T(\mathbf{x}) = T(T - \lambda_i I)^{m_i}(\mathbf{x}) = \mathbf{0}$$

and $T(\mathbf{x}) \in \mathscr{G}_i$ for $1 \le i \le s$, showing that $\mathscr{G}_i$ is T-invariant.

Now it is to be shown that for $j = 1, 2, \ldots, s$, the minimal polynomial of $T_j = T|_{\mathscr{G}_j}$ acting in $\mathscr{G}_j$ is $(\lambda - \lambda_j)^{m_j}$. Indeed, if I_j denotes the identity transformation in $\mathscr{G}_j$, then for any $\mathbf{x} \in \mathscr{G}_j$ we have

$$(T_j - \lambda_j I_j)^{m_j}(\mathbf{x}) = (T - \lambda_j I)^{m_j}(\mathbf{x}) = \mathbf{0}$$

and, therefore, $(T_j - \lambda_k I_j)^{m_j} = 0$ (in $\mathscr{G}_j$). Thus, $(\lambda - \lambda_j)^{m_j}$ is an annihilating polynomial of T_j. Assume that this polynomial is *not* minimal, and let $(\lambda - \lambda_j)^{r_j}$ $(r_j < m_j)$ be the minimal polynomial of T_j. Write

$$m_1(\lambda) = (\lambda - \lambda_j)^{r_j} \prod_{i=1, i \ne j}^{s} (\lambda - \lambda_i)^{m_i}.$$

In view of Eq. (5), we may represent any $\boldsymbol{x} \in \mathscr{S}$ as $\boldsymbol{x} = \sum_{i=1}^{s} \boldsymbol{x}_i$, where $\boldsymbol{x}_i \in \mathscr{G}_i$ for $i = 1, 2, \ldots, s$. Then, using the commutativity of polynomials in T and the conditions $(T_i - \lambda_i I_i)^{m_i}\boldsymbol{x}_i = \boldsymbol{0}$ $(i = 1, 2, \ldots, s, i \neq j)$ and $(T_j - \lambda_j I_j)^{r_j}(\boldsymbol{x}_j) = \boldsymbol{0}$, it is easily found that $m_1(T) = O$. Since $m_1(\lambda)$ is annihilating and divides the minimal polynomial $m(\lambda)$, a contradiction is obtained. ■

Proposition 1. *With the notation of Theorem* 3, $\dim \mathscr{G}_i \geq m_i$.

PROOF. It has been shown above that $(\lambda - \lambda_i)^{m_i}$ is the minimal polynomial of T_i. Hence $(T_i - \lambda_i I)^{m_i} = O$ and $(T_i - \lambda_i I)^{m_i = 1} \neq O$. From Exercise 4.5.14 we see that

$$\operatorname{Ker}(T_i - \lambda_i I) \subset \operatorname{Ker}(T_i - \lambda_i I)^2 \subset \cdots \subset \operatorname{Ker}(T_i - \lambda_i I)^{m_i},$$

each inclusion being strict, and $\dim(\operatorname{Ker}(T_i - \lambda_i I)) \geq 1$. Consequently,

$$\dim \mathscr{G}_i = \dim(\operatorname{Ker}(T_i - \lambda_i I)^{m_i}) \geq m_i. \quad ■$$

Exercise 1. Show that if $\mathscr{S}$ has dimension n and $T \in \mathscr{L}(\mathscr{S})$, then the degree of the minimal polynomial of T does not exceed n. □

Note that the spectrum of each $T_i = T|_{\mathscr{G}_i}$ in Eq. (3) consists of one point only and that $\sigma(T_i) \cap \sigma(T_j) = \phi$ for $i \neq j$. It turns out that the decomposition of a linear transformation into a direct sum of such transformations is unique.

Theorem 4 *The representation* (3) *is unique (up to the order of summands).*

PROOF. Indeed, let $T \sum_{i=1}^{s} \cdot \tilde{T}_i$, where $\sigma(\tilde{T}_i) = \{\lambda_i\}$ $(1 \leq i \leq s)$ and $\lambda_i \neq \lambda_j$ if $i \neq j$. If $(\lambda - \lambda_i)^{r_i}$ is the minimal polynomial of $\tilde{T}_i$ (by Theorem 2, only such polynomials can be minimal for $\tilde{T}_i$), then the product $\prod_{i=1}^{s} (\lambda - \lambda_i)^{r_i}$ $(i = 1, 2, \ldots, s)$ must coincide with the minimal polynomial $m(\lambda)$ of T. This is easily checked by repeating the argument used in the proof of Theorem 3. ■

The fundamental result obtained in Theorem 3 will be developed further in Section 4 after the structure of the subspaces $\mathscr{G}_1, \mathscr{G}_2, \ldots, \mathscr{G}_s$ in Eq. (4) is studied.

Considering the representations of T and the T_i's together with the ideas of Section 4.8, we immediately obtain the following consequences of Theorems 3 and 4.

Corollary 1. *Let $T \in \mathscr{L}(\mathscr{S})$ with $\mathscr{F} = \mathbb{C}$. There exists a basis in $\mathscr{S}$ in which the representation A of T is a complex block-diagonal matrix*

$$A = \operatorname{diag}[A_1, A_2, \ldots, A_s], \tag{6}$$

and the spectrum of A_i $(1 \leq i \leq s)$ consists of one point only. Moreover, these points are distinct and if q_i is the size of A_i, $1 \leq i \leq s$, then $q_1, q_2, \ldots, q_s$ are uniquely determined by T.

It will be shown later (see Proposition 6.6.1) that the numbers $q_1, q_2, \ldots, q_s$ are, in fact, the algebraic multiplicities of the eigenvalues of T (or of A).

Note that each matrix A_i in Eq. (6) can also be described as a representation of the transformation $T_i = T|_{\mathscr{G}_i}$ with respect to an arbitrary basis in $\mathscr{G}_i$. Corollary 1 above can be restated in the following equivalent form.

Corollary 2. *Any complex square matrix B is similar to a matrix A of the form (6), where $A_i \in \mathscr{F}^{q_i \times q_i}$, $\sigma(A_i) = \{\lambda_i\}$, $1 \le i \le s$, and the numbers q_i are uniquely determined by B.*

Note that the positive integer s in Eq. (6) can be equal to 1, in which case the spectrum of T consists of a single point.

Exercise 2. Let λ be an eigenvalue of two similar matrices A and B. Show that the indices of λ with respect to A and to B are equal. □

The final theorem of the section is known as the Cayley–Hamilton theorem and plays an important part in matrix theory. Another proof will appear in Section 7.2 in terms of matrices rather than linear transformations.

Theorem 5. *The characteristic polynomial $c(\lambda)$ of $T \in \mathscr{L}(\mathscr{S})$ is one of its annihilating polynomials: $c(T) = O$.*

PROOF. First recall that the characteristic polynomial of a linear transformation is just that of any of its representations (see Section 4.11). Referring to Eq. (6), we see that $c(T)$ is just the product of the characteristic polynomials of $A_1, A_2, \ldots, A_s$. For $j = 1, 2, \ldots, s$, the size of the square matrix A_j is just $q_j = \dim \mathscr{G}_j$ and $\sigma(A_j) = \{\lambda_j\}$. Consequently, the characteristic polynomial of A_j is $(\lambda - \lambda_j)^{q_j}$. Then, by Proposition 1, $q_j \ge m_j$. Finally, since

$$c(T) = c(A) = \prod_{j=1}^{s} (\lambda - \lambda_j)^{q_j}$$

and

$$m(T) = \sum_{j=1}^{s} (\lambda - \lambda_j)^{m_j},$$

it is clear that $c(T)$ is divisible by $m(T)$. By Corollary 2 to Theorem 1 the conclusion is obtained. ■

Exercise 3. Use the Cayley–Hamilton theorem to check that the matrix A of Exercise 4.11.1 satisfies the equation

$$A^3 = 4A^2 - A + 6I.$$

Exercise 4. Check that $m(\lambda) = (\lambda - 4)^2(\lambda - 2)$ is the minimal polynomial of the matrix

$$A = \begin{bmatrix} 6 & 2 & 2 \\ -2 & 2 & 0 \\ 0 & 0 & 2 \end{bmatrix}. \quad \square$$

6.3 Generalized Eigenspaces

In the previous section it was shown that if $T \in \mathscr{L}(\mathscr{S})$, the action of T on $\mathscr{S}$ can be broken down to the study of the action of the more primitive transformations $T_i = T|_{\mathscr{G}_i}$ on the T-invariant subspaces $\mathscr{G}_i$ of Eq. (6.2.4). To continue with the "decomposition" of T, we now study the structure of these subspaces more closely. Also, a technique will be developed in Section 9.5 that generates a system of projectors onto the subspaces $\mathscr{G}_1, \ldots, \mathscr{G}_s$.

Let $T \in \mathscr{L}(\mathscr{S})$ with $\mathscr{F} = \mathbb{C}$, $\lambda \in \sigma(T)$, and $\mathscr{S}_r = \operatorname{Ker}(T - \lambda I)^r$, where r varies over the set of nonnegative integers. Exercise 4.5.14 asserts that

$$\{\boldsymbol{0}\} = \mathscr{S}_0 \subset \mathscr{S}_1 \subset \mathscr{S}_2 \subset \cdots \subset \mathscr{S}_p = \mathscr{S}_{p+1} = \cdots \subset \mathscr{S} \tag{1}$$

for some positive integer p. Note that $\mathscr{S}_1$ is the eigenspace of T associated with λ; the *strict* inclusion $\{\boldsymbol{0}\} \subset \mathscr{S}_1$ follows from the assumption $\lambda \in \sigma(T)$.

Exercise 1. Check that $\boldsymbol{x} \in \mathscr{S}_r$ if and only if $(T - \lambda I)(\boldsymbol{x}) \in \mathscr{S}_{r-1}$. $\square$

The subspace $\mathscr{S}_r$ $(1 \le r \le p)$ in (1) is referred to as a *generalized eigenspace of T of order r* associated with the eigenvalue λ. Also, a nonzero element $\boldsymbol{x}$ such that $\boldsymbol{x} \in \mathscr{S}_r$ but $\boldsymbol{x} \notin \mathscr{S}_{r-1}$ is said to be a *generalized eigenvector of T of order r* corresponding to the eigenvalue λ, that is,

$$\boldsymbol{x} \in \operatorname{Ker}(T - \lambda I)^r, \qquad \text{but} \qquad \boldsymbol{x} \notin \operatorname{Ker}(T - \lambda I)^{r-1}.$$

In particular, the customary eigenvectors of T can be viewed as generalized eigenvectors of T of order 1 corresponding to the same eigenvalue. The same can be said with regard to eigenspaces and generalized eigenspaces of order 1 of the transformation. Note that throughout this section, all generalized eigenvectors and eigenspaces are assumed (if not indicated otherwise) to be associated with the fixed eigenvalue λ of T.

It is important to note that the subspaces $\mathscr{G}_i$ of Eq. (6.2.4) are generalized eigenspaces, and the basic decomposition of Theorem 6.2.3 depends on the decomposition (6.2.5) of the whole space $\mathscr{S}$ into a direct sum of generalized eigenspaces.

Exercise 2. Show that $\boldsymbol{x}_r$ is a generalized eigenvector of T of order $r \geq 2$ (corresponding to the eigenvalue λ) if and only if the element $\boldsymbol{x}_{r-1} = (T - \lambda I)(\boldsymbol{x}_r)$ is a generalized eigenvector of T of order $r - 1$, or, equivalently,

$$T(\boldsymbol{x}_r) = \lambda \boldsymbol{x}_r + \boldsymbol{x}_{r-1}, \tag{2}$$

where $\boldsymbol{x}_{r-1} \in \mathscr{S}_{r-1}$, $\boldsymbol{x}_{r-1} \notin \mathscr{S}_{r-2}$.

Exercise 3. Check that $\boldsymbol{x}_r$ is a generalized eigenvector of T of order $r \geq 1$ if and only if the vector $(T - \lambda I)^{r-1}(\boldsymbol{x}_r)$ is an eigenvector of T or, equivalently, $(T - \lambda I)^k(\boldsymbol{x}_r) = \boldsymbol{0}$ for $k \geq r$. □

Using Exercise 2, observe that if $\boldsymbol{x}_r$ is a generalized eigenvector of T of order r, then there are vectors $\boldsymbol{x}_{r-1}, \ldots, \boldsymbol{x}_2, \boldsymbol{x}_1$ for which

$$\begin{aligned} T(\boldsymbol{x}_1) &= \lambda \boldsymbol{x}_1 \\ T(\boldsymbol{x}_2) &= \lambda \boldsymbol{x}_2 + \boldsymbol{x}_1, \\ &\ \vdots \\ T(\boldsymbol{x}_r) &= \lambda \boldsymbol{x}_r + \boldsymbol{x}_{r-1}, \end{aligned} \tag{3}$$

where $\boldsymbol{x}_j \in \mathscr{S}_j$ for $j = 1, 2, \ldots, r$. Such a sequence $\boldsymbol{x}_1, \boldsymbol{x}_2, \ldots, \boldsymbol{x}_r$ is called a *Jordan chain of length* r associated with the eigenvalue λ. The chain may also be seen as being associated with $\boldsymbol{x}_1$. From this point of view, an eigenvector $\boldsymbol{x}_1$ is selected and the vectors $\boldsymbol{x}_2, \ldots, \boldsymbol{x}_r$ are generated by successively solving the equations of (3) for as long as there exist solutions to the nonhomogeneous equation $(T - \lambda I)\boldsymbol{x}_j = \boldsymbol{x}_{j-1}$ for $j = 2, 3, \ldots$. Obviously, the length of any Jordan chain of T is finite since $r \leq \dim \mathscr{S}_p \leq \dim \mathscr{S}$ for a chain of length r. Furthermore, the members of a Jordan chain are linearly independent, as the next exercise demonstrates.

Exercise 4. Show that any Jordan chain consists of linearly independent elements.

SOLUTION. Let $\sum_{i=1}^{r} \alpha_i \boldsymbol{x}_i = \boldsymbol{0}$. Applying the transformation $(T - \lambda I)^{r-1}$ to both sides of this equation and noting that $(T - \lambda I)^{r-1}(\boldsymbol{x}_i) = \boldsymbol{0}$ for $1 \leq i \leq r - 1$ (Exercise 3), we obtain $\alpha_r(T - \lambda I)^{r-1}(\boldsymbol{x}_r) = \boldsymbol{0}$. By Exercise 3 the element $(T - \lambda I)^{r-1}(\boldsymbol{x}_r)$ is an eigenvector of T and hence nonzero. Thus, $\alpha_r = 0$. Applying the transformation $(T - \lambda I)^{r-2}$ to the equation $\sum_{i=1}^{r-1} \alpha_i \boldsymbol{x}_i = \boldsymbol{0}$, we similarly derive $\alpha_{r-1} = 0$. Repeating this procedure, the linear independence of the set of generalized eigenvectors of the Jordan chain is established. □

In particular, Exercise 4 shows that the *Jordan subspace* for T,

$$\mathscr{J} = \operatorname{span}\{\boldsymbol{x}_1, \boldsymbol{x}_2, \ldots, \boldsymbol{x}_r\}, \tag{4}$$

generated by the elements of a Jordan chain of T of length r has dimension r.

Exercise 5. Check that any Jordan subspace for T is T-invariant.

Exercise 6. Prove that any Jordan subspace contains only one (linearly independent) eigenvector.

SOLUTION. Suppose the Jordan subspace is generated by the chain $\mathbf{x}_1, \ldots, \mathbf{x}_r$ and has associated eigenvalue λ. If $\mathbf{x} = \sum_{i=1}^{r} \alpha_i \mathbf{x}_i$ is an eigenvector of T associated with λ_0, then using Eq. (2) we have (with $\mathbf{x}_0 = \mathbf{0}$),

$$\lambda_0 \sum_{i=1}^{r} \alpha_i \mathbf{x}_i = T\left(\sum_{i=1}^{r} \alpha_i \mathbf{x}_i\right) = \sum_{i=1}^{r} \alpha_i(\lambda \mathbf{x}_i + \mathbf{x}_{i-1}).$$

Consequently, comparing the coefficients (see Exercise 4),

$$\begin{aligned} \lambda_0 \alpha_i &= \lambda \alpha_i + \alpha_{i+1}, \qquad 1 \le i \le r-1, \\ \lambda_0 \alpha_r &= \lambda \alpha_r, \end{aligned}$$

so that $\lambda = \lambda_0$ and $\alpha_2 = \alpha_3 = \cdots = \alpha_r = 0$. Thus $\mathbf{x} = \alpha_1 \mathbf{x}_1$ with $\alpha_1 \neq 0$, as required. □

Let $T \in \mathscr{L}(\mathscr{S})$ and let the subspace $\mathscr{S}_0 \subset \mathscr{S}$ be T-invariant. If $\mathbf{x} \in \mathscr{S}_0$ and there exists a basis in $\mathscr{S}_0$ of the form

$$\{\mathbf{x}, T(\mathbf{x}), \ldots, T^{k-1}(\mathbf{x})\}, \qquad k = \dim \mathscr{S}_0,$$

then $\mathscr{S}_0$ is said to be a *cyclic subspace* of $\mathscr{S}$ (with respect to T or $T|_{\mathscr{S}_0}$).

Proposition 1. *A Jordan subspace is cyclic.*

PROOF. Let $\mathbf{x}_1, \mathbf{x}_2, \ldots, \mathbf{x}_r$ be a Jordan chain and observe that Eqs. (3) imply

$$\begin{aligned} \mathbf{x}_{r-1} &= (T - \lambda I)\mathbf{x}_r, \\ &\vdots \\ \mathbf{x}_2 &= (T - \lambda I)^{r-2}\mathbf{x}_r, \\ \mathbf{x}_1 &= (T - \lambda I)^{r-1}\mathbf{x}_r. \end{aligned} \tag{5}$$

Using the binomial theorem, it is clear that the Jordan subspace generated by $\mathbf{x}_1, \mathbf{x}_2, \ldots, \mathbf{x}_r$ is also generated by $\mathbf{x}_r, T\mathbf{x}_r, \ldots, T^{r-1}\mathbf{x}_r$. Since the Jordan subspace is T-invariant (Exercise 5) and $\mathbf{x}_r$ is in this subspace, it follows that the subspace is cyclic with respect to T. ■

Observe that Eqs. (5) show that if $\mathbf{x}_r$ is a generalized eigenvector of order r, then $(T - \lambda I)^i \mathbf{x}_r$ is a generalized eigenvector of order $r - i$, for $i = 0, 1, \ldots, r - 1$.

6.4 The Structure of Generalized Eigenspaces

In the preceding section some basic ideas on generalized eigenspaces and Jordan chains have been introduced. In this section we put them to work to prove that each generalized eigenspace $\mathscr{G}_i$ of Eq. (6.2.4) has a basis made up of one or several Jordan chains.

Let $T \in \mathscr{L}(\mathscr{S})$ with $\mathscr{F} = \mathbb{C}$ and let $\mathscr{S}_p = \operatorname{Ker}(T - \lambda I)^p$ be the generalized eigenspace of T of maximal order p associated with the eigenvalue λ, that is, $\mathscr{S}_{p-1} \neq \mathscr{S}_p = \mathscr{S}_{p+1}$. The subspace $\mathscr{S}_p$ satisfying this condition is referred to as the *generalized eigenspace* of T associated with λ. Thus, "the generalized eigenspace" of an eigenvalue, without reference to order, implies that the order is, in fact, maximal.

Theorem 1. *There exists a unique decomposition*

$$\mathscr{S}_p = \sum_{i=1}^{t_p} \cdot \mathscr{J}_p^{(i)} \dotplus \sum_{i=1}^{t_{p-1}} \cdot \mathscr{J}_{p-1}^{i} \dotplus \cdots \dotplus \sum_{i=1}^{t_1} \cdot \mathscr{J}_1^{(i)},$$

where $\mathscr{J}_j^{(i)}$ $(1 \leq j \leq p,\ 1 \leq i \leq t_j)$ is a cyclic Jordan subspace of dimension j.

In this decomposition of $\mathscr{S}_p$, there is the possibility that $t_j = 0$ when the corresponding sum on the right is interpreted as the trivial subspace $\{0\}$.

Theorem 1 implies that there exists a basis in $\operatorname{Ker}(T - \lambda I)^p$ consisting of Jordan chains; each chain is a basis in a cyclic Jordan subspace $\mathscr{J}_j^{(i)}$ $(1 \leq j \leq p,\ 1 \leq i \leq t_j)$. Namely, starting with $j = p$, we have the basis

$$\begin{array}{lllll}
x_1^{(1)}, & x_2^{(1)}, & \ldots, & x_{p-1}^{(1)}, & x_p^{(1)}, \\
\vdots & \vdots & & \vdots & \vdots \\
x_1^{(t_p)}, & x_2^{(t_p)}, & \ldots, & x_{p-1}^{(t_p)}, & x_p^{(t_p)}, \\
x_1^{(t_p+1)}, & x_2^{(t_p+1)}, & \ldots, & x_{p-1}^{(t_p+1)}, & \\
\vdots & \vdots & & \vdots & \\
x_1^{(t_p+t_{p-1})}, & x_2^{(t_p+t_{p-1})}, & \ldots, & x_{p-1}^{(t_p+t_{p-1})}, & \\
\vdots & & & & \\
x_1^{(t-t_1+1)}, & & & & \\
\vdots & & & & \\
x_1^{(t)}, & & & &
\end{array} \tag{1}$$

where $t = \sum_{j=1}^{p} t_j$ and the elements of each row form a Jordan chain of maximal possible length. Also, the elements of column r in (1) are generalized eigenvectors of T of order r associated with λ. The basis (1) of $\operatorname{Ker}(T - \lambda I)^p$ is called a *Jordan basis* of the generalized eigenspace.

PROOF. Bearing in mind the relationships (6.3.1), let $x_p^{(1)}, x_p^{(2)}, \ldots, x_p^{(t_p)}$ denote linearly independent elements in $\mathscr{S}_p$ such that

$$\mathscr{S}_{p-1} \dotplus \operatorname{span}\{x_p^{(1)}, x_p^{(2)}, \ldots, x_p^{(t_p)}\} = \mathscr{S}_p. \tag{2}$$

Obviously, the vectors $x_p^{(1)}, \ldots, x_p^{(t_p)}$ in Eq. (2) are generalized eigenvectors of T of order p. It is claimed that the pt_p vectors of the form

$$(T - \lambda I)^k(x_p^{(1)}), \quad (T - \lambda I)^k(x_p^{(2)}), \quad \ldots, \quad (T - \lambda I)^k(x_p^{(t_p)}), \tag{3}$$

$k = 0, 1, \ldots, p - 1$, are linearly independent. Indeed, if

$$\sum_{k=0}^{p-1} \sum_{i=1}^{t_p} \alpha_{ik}(T - \lambda I)^k(x_p^{(i)}) = 0, \tag{4}$$

then, applying the transformation $(T - \lambda I)^{p-1}$, we deduce by use of Exercise 6.3.3 that

$$(T - \lambda I)^{p-1}\left(\sum_{i=1}^{t_p} \alpha_{i0} x_p^{(i)}\right) = 0.$$

Hence $\sum_{i=1}^{t_p} \alpha_{i0} x_p^{(i)} \in \mathscr{S}_{p-1}$ and, recalling the choice of $\{x_p^{(i)}\}_{i=1}^{t_p}$ in Eq. (2), we obtain $\sum_{i=1}^{t_p} \alpha_{i0} x_p^{(i)} = 0$. Consequently, $\alpha_{i0} = 0$ $(i = 1, 2, \ldots, t_p)$. Now applying the transformation $(T - \lambda I)^{p-2}$ to Eq. (4), it is shown similarly that $\alpha_{i1} = 0$ $(i = 1, 2, \ldots, t_p)$, and so on.

Thus, the cyclic Jordan subspaces $\mathscr{J}_p^{(i)}$ $(i = 1, 2, \ldots, t_p)$ constructed by the rule

$$\mathscr{J}_p^{(i)} = \operatorname{span}\{(T - \lambda I)^k(x_p^{(i)})\}_{k=0}^{p-1}, \qquad 1 \le i \le t_p,$$

satisfy the conditions $\dim(\mathscr{J}_p^{(i)}) = p$ $(1 \le i \le t_p)$ and, if $i \ne j$, then

$$\mathscr{J}_p^{(i)} \cap \mathscr{J}_p^{(j)} = \{0\}.$$

Now we will show that for any fixed k $(0 \le k \le p - 1)$,

$$\mathscr{S}_{p-k-1} \cap \operatorname{span}\{(T - \lambda I)^k(x_p^{(i)});\, 1 \le i \le t_p\} = \{0\}. \tag{5}$$

This is obvious from Eq. (2) when $k = 0$. More generally, suppose that $\sum_{i=1}^{t_p} \alpha_{ik}(T - \lambda I)^k(x_p^{(i)}) \in \mathscr{S}_{p-k-1}$. Then

$$(T - \lambda I)^{p-1}\left(\sum_{i=1}^{t_p} \alpha_{ik} x_p^{(i)}\right) = 0$$

and $\sum_{i=1}^{t_p} \alpha_{ik} x_p^{(i)} \in \mathscr{S}_{p-1}$, which contradicts (2) again unless $\alpha_{1k} = \alpha_{2k} = \cdots = \alpha_{t_pk} = 0$, so that Eq. (5) follows.

Now consider the elements $x_{p-1}^{(i)} = (T - \lambda I)(x_p^{(i)})$, $i = 1, 2, \ldots, t_p$, from (3), which are generalized eigenvectors of T of order $p - 1$ (Exercise 6.3.2). It follows from above that $x_{p-1}^{(i)}$ are linearly independent and that, by Eq. (5),

$$\mathscr{S}_{p-2} \cap \operatorname{span}\{x_{p-1}^{(1)}, x_{p-1}^{(2)}, \ldots, x_{p-1}^{(t_p)}\} = \{0\}.$$

Hence there is a set of linearly independent elements $\{x_{p-1}^{(i)}\}_{i=1}^{t_p+t_{p-1}}$ in $\mathscr{S}_{p-1}$ such that

$$\mathscr{S}_{p-2} \dotplus \operatorname{span}\{x_{p-1}^{(i)}\}_{i=1}^{t_p+t_{p-1}} = \mathscr{S}_{p-1}. \tag{6}$$

Applying the previous argument to Eq. (6) as we did with Eq. (2), we obtain a system of elements of the form (3), where p is replaced by $p-1$, and having similar properties. It remains to denote

$$\mathscr{J}_{p-1}^{(i)} = \operatorname{span}\{(T-\lambda I)^k(x_{p-1}^{(i+t_p)})\}_{k=0}^{p-2}, \qquad 1 \le i \le t_{p-1},$$

to obtain cyclic Jordan subspaces of dimension $p-1$.

Continuing this process, we eventually arrive at the relationship

$$\mathscr{S}_1 \dotplus \operatorname{span}\{x_2^{(i)}\}_{i=1}^{t_p+t_{p-1}+\cdots+t_2} = \mathscr{S}_2, \tag{7}$$

where the vectors $x_2^{(i)}$, with $i = 1, 2, \ldots, \sum_{j=2}^{p} t_j$, denote linearly independent generalized eigenvectors of T of order 2. As before, it can be shown that the elements $x_1^{(i)} = (T-\lambda I)^{(k)}x_2^{(i)}$ $(k = 0, 1)$ are linearly independent members of $\mathscr{S}_1$ and, together with a few other linearly independent elements from $\mathscr{S}_1$, form a basis $\{x_1^{(i)}\}_{i=1}^{t}$ $(t = \sum_{j=1}^{p} t_p)$ in $\mathscr{S}_1$. For $i = 1, 2, \ldots, t$, we write

$$\mathscr{J}_1^{(i)} = \operatorname{span}\{x_1^{(t_p+\cdots+t_2+i)}\}, \qquad 1 \le i \le t_1.$$

The system of elements built up above constitute a basis in $\mathscr{S}_p$. Indeed, $\{x_1^{(i)}\}_{i=1}^{t}$ $(t = \sum_{j=1}^{p} t_j)$ is a basis in $\mathscr{S}_1$, while Eq. (7) shows that the union of $\{x_1^{(i)}\}_{i=1}^{t}$ and $\{x_2^{(i)}\}_{i=1}^{t_p+t_{p-1}+\cdots+t_2}$ generates a basis in $\mathscr{S}_2$, and so on. Note also (see Eqs. (2), (6), and (7)) that for $r = p, p-1, \ldots, 1$,

$$t_p + t_{p-1} + \cdots + t_r = \dim \mathscr{S}_r - \dim \mathscr{S}_{r-1}, \tag{8}$$

and therefore the numbers t_j $(1 \le j \le p)$ are uniquely defined by T and λ. The proof is complete. ■

Exercise 1. Find a Jordan basis in $\mathscr{F}^2$ associated with the transformation T discussed in Exercises 4.7.7 and 4.9.5.

SOLUTION. The transformation T has only one eigenvector, $x_1 = [1 \quad 0]^{\mathrm{T}}$, corresponding to the eigenvalue $\lambda = 1$. Hence there is a generalized eigenvector x_2 such that $\operatorname{span}\{x_1, x_2\} = \mathscr{F}^2$. Indeed, writing $x_2 = [\alpha \quad \beta]^{\mathrm{T}}$, it follows from $T(x_2) = x_2 + x_1$ that

$$\begin{bmatrix} \alpha + \beta \\ \beta \end{bmatrix} = \begin{bmatrix} \alpha \\ \beta \end{bmatrix} + \begin{bmatrix} 1 \\ 0 \end{bmatrix},$$

and consequently $\beta = 1$. Thus, any vector $[\alpha \quad 1]^{\mathrm{T}}$ $(\alpha \in \mathscr{F})$ is a generalized eigenvector of T and a required Jordan basis is, for instance, $\{[1 \quad 0]^{\mathrm{T}}, [0 \quad 1]^{\mathrm{T}}\}$.

Exercise 2. Check that the number k_m of Jordan chains in (1) of order m is

$$k_m = 2l_m - l_{m-1} - l_{m+1}$$

where $l_s = \dim(\operatorname{Ker}(T - \lambda I)^s)$.

Hint. Use Eq. (8).

Exercise 3. Let the subspace $\mathscr{G}_i = \operatorname{Ker}(T - \lambda_i I)^{m_i}$ be defined as in Theorem 6.2.3. Show that for $i = 1, 2, \ldots, s$,

$$\operatorname{Ker}(T - \lambda_i I)^{m_i - 1} \subset \operatorname{Ker}(T - \lambda_i I)^{m_i} = \operatorname{Ker}(T - \lambda_i I)^{m_i + r} \qquad (9)$$

for any positive integer r, and the inclusion is strict.

In other words, for $\lambda = \lambda_i$, the number p of Theorem 1 coincides with m_i and the generalized eigenspace of T associated with the eigenvalue λ_i is just

$$\operatorname{Ker}(T - \lambda_i I)^{m_i},$$

where m_i is the index of λ_i, $1 \leq i \leq s$.

SOLUTION. Let $p_j(\lambda) = (\lambda - \lambda_j)^{m_j}$ so that, as in Eq. (6.2.2),

$$m(\lambda) = p_1(\lambda)p_2(\lambda) \cdots p_s(\lambda)$$

is the minimal polynomial for T and, by Theorem 6.1.4,

$$\mathscr{S} = \sum_{j=1}^{s} \cdot \operatorname{Ker} p_j(T). \qquad (10)$$

Now define $\hat{p}_i(\lambda) = (\lambda - \lambda_i)^r p_i(\lambda)$ and $\hat{p}_j(\lambda) = p_j(\lambda)$ if $j \neq i$. Then by Corollary 2 of Theorem 6.2.1 the polynomial

$$\hat{m}(\lambda) = \hat{p}_1(\lambda)\hat{p}_2(\lambda) \cdots \hat{p}_s(\lambda)$$

annihilates T and Theorem 6.1.4 applies once more to give

$$\mathscr{S} = \sum_{j=1}^{s} \cdot \operatorname{Ker} \hat{p}_j(T)$$

Comparing with (10) it is clear that $\operatorname{Ker} p_i(T) = \operatorname{Ker} \hat{p}_i(T)$, as required.

Clearly $\operatorname{Ker}(T - \lambda_i I)^{m_i - 1} \subset \operatorname{Ker} p_i(T)$ and if equality obtains then the polynomial $q(\lambda) \triangleq m(\lambda)(\lambda - \lambda_i)^{-1}$ annihilates T. This is because

$$\operatorname{Ker} q(T) = \sum_{j=1}^{s} \cdot \operatorname{Ker} p_j(T) = \mathscr{S},$$

which implies $q(T) = O$. But this contradicts the fact that $m(\lambda)$ is the minimal polynomial for T. So the inclusion in (9) must be strict. □

Recalling the result of Exercise 6.3.5, Theorem 1 allows us to develop the result of Theorem 6.2.3 further with the following statement.

Theorem 2. *Using notation of Theorem* 6.2.3 *and defining* $T_i = T|_{\mathscr{G}_i}$,

$$T_i = \sum_{j=1}^{t^{(i)}} {}^{\cdot}\, T_i^{(j)}, \qquad 1 \le i \le s, \tag{11}$$

where $t^{(i)} = \dim(\operatorname{Ker}(T - \lambda_i I))$ *and the transformations* $T_i^{(j)}$ $(1 \le i \le s,$ $j = 1, 2, \ldots, t^{(i)})$ *act in cyclic Jordan subspaces.*

The decomposition (6.2.3) together with Eq. (11) is said to be the *Jordan decomposition* of the transformation T.

6.5 The Jordan Theorem

Consider a linear transformation T acting in a finite-dimensional space $\mathscr{S}$. In view of Corollary 1 to Theorem 6.2.4, the problem of finding the canonical representation of T is reduced to that for a transformation T_i $(1 \le i \le s)$ acting in the typical generalized eigenspace $\operatorname{Ker}(T - \lambda_i I)^{m_i}$ (m_i being the index of λ_i). Then Theorem 6.4.2 states that, for this purpose, it suffices to find a simple representation of the transformations $T_i^{(j)}$ $(1 \le j \le t^{(i)})$ acting on cyclic Jordan subspaces.

Let $\mathscr{J}$ denote a cyclic Jordan subspace for T spanned by

$$(T - \lambda_i I)^{r-1}(\boldsymbol{x}), \quad \ldots, \quad (T - \lambda_i I)(\boldsymbol{x}), \quad \boldsymbol{x}, \tag{1}$$

where $\lambda_i \in \sigma(T)$ and $\boldsymbol{x}$ is a generalized eigenvector of T of order r associated with λ_i. Denote $\boldsymbol{x}_j = (T - \lambda_i I)^{r-j}(\boldsymbol{x})$ $(j = 1, 2, \ldots, r)$ and observe [see Exercise 6.3.2 and Eqs. (6.3.3)] that $T(\boldsymbol{x}_1) = \lambda_i \boldsymbol{x}_1$ and

$$T(\boldsymbol{x}_j) = \lambda_i \boldsymbol{x}_j + \boldsymbol{x}_{j-1}, \qquad j = 2, 3, \ldots, r. \tag{2}$$

Hence, comparing Eq. (2) with the definition of the representation $J_i = [\alpha_{ij}]_{i,j=1}^{r} \in \mathbb{C}^{r \times r}$ of $T|_{\mathscr{J}}$ with respect to the basis (1), that is

$$T(\boldsymbol{x}_j) = \sum_{i=1}^{r} \alpha_{ij} \boldsymbol{x}_i, \qquad j = 1, 2, \ldots, r,$$

we obtain

$$J_i = \begin{bmatrix} \lambda_i & 1 & 0 & \cdots & 0 \\ 0 & \lambda_i & 1 & \ddots & \vdots \\ \vdots & & & \ddots & 0 \\ & & \ddots & \ddots & 1 \\ 0 & & \cdots & 0 & \lambda_i \end{bmatrix}, \tag{3}$$

where all superdiagonal elements are ones and all unmarked elements are zeros. The matrix in (3) is referred to as a *Jordan block* (or *cell*) of order r

corresponding to the eigenvalue λ_i. Thus, the Jordan block (3) is the representation of $T|_{\mathscr{J}}$ with respect to the basis (1), which is a Jordan chain generating $\mathscr{J}$.

Exercise 1. Show that the Jordan block in (3) cannot be similar to a block-diagonal matrix $\text{diag}[A_1, A_2, \ldots, A_k]$, where $k \geq 2$.

SOLUTION. If, for some nonsingular $P \in \mathbb{C}^{r \times r}$,

$$J = P\,\text{diag}[A_1, A_2]P^{-1},$$

then the dimension of the eigenspace for J is greater than or equal to 2 (one for each diagonal block). This contradicts the result of Exercise 6.3.6. □

Thus, a Jordan block cannot be "decomposed" into smaller blocks by any similarity transformation.

Let $T \in \mathscr{L}(\mathscr{S})$ with $\mathscr{F} = \mathbb{C}$, and write $\sigma(T) = \{\lambda_1, \lambda_2, \ldots, \lambda_s\}$. By Theorem 6.2.3, $T = \sum_{i=1}^{s}{}^{\cdot}\, T_i$, where $T_i = T|_{\mathscr{G}_i}$ and $\mathscr{G}_i = \text{Ker}(T - \lambda_i I)^{m_i}$ $(1 \leq i \leq s)$. Choosing in each $\mathscr{G}_i$ a Jordan basis of the form (6.4.1), we obtain a *Jordan basis* for the whole space $\mathscr{S}$ as a union of Jordan bases; one for each generalized eigenspace. This is described as *a T-Jordan basis* for $\mathscr{S}$. We now appeal to Theorem 6.4.1.

Theorem 1. *In the notation of the preceding paragraph, the representation of T with respect to a T-Jordan basis is the matrix*

$$J = \text{diag}[J(\lambda_1), J(\lambda_2), \ldots, J(\lambda_s)], \tag{4}$$

where for $1 \leq i \leq s$,

$$J(\lambda_i) = \text{diag}[J_p^{(i)}, \ldots, J_p^{(i)}, J_{p-1}^{(i)}, \ldots, J_{p-1}^{(i)}, \ldots, J_1^{(i)}, \ldots, J_1^{(i)}], \tag{5}$$

in which the $j \times j$ *matrix* $J_j^{(i)}$ *of the form* (3) *appears* $t_j^{(i)}$ *times in Eq.* (5). *The numbers* $t_j^{(i)}$ $(i = 1, 2, \ldots, s; j = 1, 2, \ldots, p)$ *are uniquely determined by the transformation T.*

It is possible that $t_j^{(i)} = 0$ for some $j < m_i$, in which case no blocks of size j appear in Eq. (5).

Note that the decomposition of J in Eqs. (4) and (5) is the best possible in the sense of Exercise 1. The representation J in Eqs. (4) and (5) is called a *Jordan matrix* (or *Jordan canonical form*)† of the transformation T. We now apply Corollary 2 of Theorem 6.2.4 to obtain the following.

Corollary 1. *Any complex square matrix A is similar to a Jordan matrix that is uniquely determined by A up to the order of the matrices* $J(\lambda_i)$ *in Eq.* (4) *and the Jordan blocks in Eq.* (5)

† Jordan, C., *Traité des substitutions et des équations algébriques*, Paris, 1870 (p. 125).

Example 2. Consider the matrix

$$A = \begin{bmatrix} 6 & 2 & 2 \\ -2 & 2 & 0 \\ 0 & 0 & 2 \end{bmatrix}.$$

Since $\det(A - \lambda I) = (2 - \lambda)(\lambda - 4)^2$, it follows that $\sigma(A) = \{2, 4\}$. It is easily seen that the matrix A has one (linearly independent) eigenvector, say $\boldsymbol{x}_1 = [0 \ \ -1 \ \ 1]^{\mathrm{T}}$, corresponding to the eigenvalue $\lambda_1 = 2$ and one (linearly independent) eigenvector, say $\boldsymbol{x}_2 = [2 \ \ -2 \ \ 0]^{\mathrm{T}}$, associated with $\lambda_2 = 4$. From the equation $(A - 4I)\boldsymbol{x}_3 = \boldsymbol{x}_2$ it is found that $\boldsymbol{x}_3 = [1 \ \ 0 \ \ 0]^{\mathrm{T}}$ is a generalized eigenvector of A of order 2 associated with $\lambda_2 = 4$. Hence a Jordan basis for A is $\{\boldsymbol{x}_1, \boldsymbol{x}_2, \boldsymbol{x}_3\}$ and the matrix

$$P = [\boldsymbol{x}_1 \ \ \boldsymbol{x}_2 \ \ \boldsymbol{x}_3] = \begin{bmatrix} 0 & 2 & 1 \\ -1 & -2 & 0 \\ 1 & 0 & 0 \end{bmatrix}$$

transforms A into the Jordan form:

$$A = P \begin{bmatrix} 2 & 0 & 0 \\ 0 & 4 & 1 \\ 0 & 0 & 4 \end{bmatrix} P^{-1}. \quad \square$$

The problem of determining the parameters of the Jordan matrix for a given $A \in \mathbb{C}^{n \times n}$ without constructing the Jordan basis is considered in the next section. We emphasize again that, in discussing the Jordan form of a matrix $A \in \mathscr{F}^{n \times n}$, the field $\mathscr{F}$ is assumed to be algebraically closed. Hence $n \times n$ real matrices should be considered as members of $\mathbb{C}^{n \times n}$ in order to guarantee the existence of the Jordan form presented here.

Exercise 3. Let $A \in \mathbb{C}^{n \times n}$ and define a *set of Jordan vectors for* A to be a set of linearly independent vectors in $\mathbb{C}^n$ made up of a union of Jordan chains for A. Show that a subspace $\mathscr{S}_0 \subset \mathbb{C}^n$ is A-invariant if and only if $\mathscr{S}_0$ has a basis consisting of a set of Jordan vectors. (In other words, every A-invariant subspace is a direct sum of subspaces that are cyclic with respect to A.)

Hints. Given that $\mathscr{S}_0$ has a set of Jordan vectors as a basis, use Exercise 6.3.5 to show that $\mathscr{S}_0$ is T-invariant. For the converse, let $X = [\boldsymbol{x}_1 \ \ \boldsymbol{x}_2 \ \ \cdots \ \ \boldsymbol{x}_r]$ be an $n \times r$ matrix whose columns form an arbitrary basis for $\mathscr{S}_0$. Observe that $AX = XG$ for some $G \in \mathbb{C}^{r \times r}$. Write $G = SJS^{-1}$, where J is a Jordan matrix, and examine the columns of the matrix XS. $\square$

The following exercise shows that there are, in general, many similarities which will reduce a given matrix to a fixed Jordan normal form and also shows how they are related to one another.

Exercise 4. Show that if $A = SJS^{-1}$ and $A = TJT^{-1}$, then $T = SU$ for some nonsingular U commuting with J.

Conversely, if U is nonsingular and commutes with J, $A = SJS^{-1}$, and if we define $T = SU$, then $A = TJT^{-1}$. □

6.6 Parameters of a Jordan Matrix

Let $A \in \mathbb{C}^{n \times n}$ and let J, given by Eqs. (6.5.4) and (6.5.5) be a Jordan canonical form of A, that is, $A = PJP^{-1}$ for some invertible P. In this section we shall indicate some relations between the multiplicities of an eigenvalue and the parameters of the corresponding submatrices $J(\lambda_i)$ of J.

Proposition 1. *The dimension of the generalized eigenspace of $A \in \mathbb{C}^{n \times n}$ associated with the eigenvalue $\lambda_i \in \sigma(A)$ [that is, the order of the matrix $J(\lambda_i)$ in Eq. (6.5.4)] is equal to the algebraic multiplicity of λ_i.*

PROOF. Let $A = PJP^{-1}$, where J is given by Eq. (6.5.4). The characteristic polynomials of A and J coincide (Section 4.11) and hence

$$c(\lambda) = \det(\lambda I - J) = \prod_{i=1}^{s} \det(\lambda I - J(\lambda_i)).$$

In view of Eqs. (6.5.5) and (6.5.3),

$$c(\lambda) = \prod_{i=1}^{s} (\lambda - \lambda_i)^{q_i}, \tag{1}$$

where q_i is the order of $J(\lambda_i)$ $(1 \le i \le s)$. In remains to recall that q_i in Eq. (1) is, by definition, the algebraic multiplicity of λ_i. ■

Note that the number of Jordan blocks in $J(\lambda_i)$ is equal to the number of rows in (6.4.1), that is, to the geometric multiplicity of the eigenvalue λ_i, so that it readily follows from (6.4.1) and Proposition 1 that the geometric multiplicity of an eigenvalue does not exceed its algebraic multiplicity (the total number of elements in (6.4.1)). However, these two characteristics uniquely determine the Jordan submatrix $J(\lambda_i)$ associated with λ_i only in the cases considered next.

Proposition 2. *The matrix A is simple if and only if the geometric multiplicity of each of its eigenvalues is equal to the algebraic multiplicity. Equivalently, A is simple if and only if the minimal polynomial of A has only simple zeros.*

PROOF. A matrix is simple if and only if it is similar to a diagonal matrix (see Theorem 4.8.2′). Hence, in view of Exercise 6.5.1 and Theorem 6.5.1, the corresponding Jordan matrix must be diagonal and therefore consists of 1×1 Jordan blocks. Thus, recalling Proposition 1 and the foregoing discussion, the multiplicities of the matrix coincide. Proceeding to the second statement, we note that the multiplicity of an eigenvalue as a zero of the minimal polynomial (the index) is equal to the maximal order of Jordan blocks associated with the eigenvalue (remember the meaning of p in (6.4.1). This observation completes the proof.

Note that in this case the matrix has no proper generalized eigenvectors. ■

Exercise 1. Show that if the geometric multiplicity of an eigenvalue λ_i is equal to one, then the Jordan matrix has only one Jordan block associated with λ_i. The order of this block is equal to the algebraic multiplicity of the eigenvalue. □

When Exercise 1 holds for all the eigenvalues, the minimal and the characteristic polynomials coincide. The converse is also true.

Exercise 2. Check that the minimal and the characteristic polynomials of a matrix coincide if and only if its Jordan matrix has only one Jordan block associated with each distinct eigenvalue of the matrix.

Hint. See the proof of Proposition 2. □

A matrix $A \in \mathscr{F}^{n \times n}$ is said to be *nonderogatory* if its characteristic and minimal polynomials coincide; otherwise it is *derogatory*. Thus, Exercise 2 reveals the structure of nonderogatory matrices.

Exercise 3. Check that a companion matrix is nonderogatory.

Exercise 4. Find a Jordan matrix J of the matrix A having the characteristic polynomial $c(\lambda) = (\lambda - 1)^3(\lambda - 2)^4$ if the geometric multiplicities are also known: $\dim(\operatorname{Ker}(A - I)) = 2$ and $\dim(\operatorname{Ker}(A - 2I)) = 3$.

SOLUTION. Consider Jordan blocks of A associated with $\lambda_1 = 1$. The geometric multiplicity 2 means that there are two blocks associated with this eigenvalue, while the algebraic multiplicity 3 implies that one of the blocks must be of order 2. Reasoning in a similar way for the second eigenvalue, we obtain

$$J = \operatorname{diag}\left[\begin{bmatrix} 1 & 1 \\ 0 & 1 \end{bmatrix}, [1], \begin{bmatrix} 2 & 1 \\ 0 & 2 \end{bmatrix}, [2], [2]\right]. \quad \square$$

In cases different from those discussed above, the orders of the Jordan blocks are usually found by use of Eq. (6.4.8) and Exercise 6.4.2, which require the determination of $\dim(\operatorname{Ker}(A - \lambda I)^s)$ for certain positive integers s.

Exercise 5. Determine the Jordan form of the matrix

$$A = \begin{bmatrix} 2 & 0 & 0 & 0 \\ -3 & 2 & 0 & 1 \\ 0 & 0 & 2 & 1 \\ 0 & 0 & 0 & 2 \end{bmatrix}.$$

SOLUTION. Obviously, the characteristic polynomial is $c(\lambda) = (\lambda - 2)^4$ and $\dim(\operatorname{Ker}(A - 2I)) = 2$. Hence the Jordan form consists of two blocks. To determine their orders (2 and 2, or 3 and 1), we calculate that $(A - 2I)^2 = O$ and, therefore, the number k_2 in Exercise 6.4.2 is equal to 2. Hence $k_2 = 2l_2 - l_1 - l_3 = 8 - 2 - 4 = 2$, and consequently there are two blocks of order 2:

$$J = \operatorname{diag}\left[\begin{bmatrix} 2 & 1 \\ 0 & 2 \end{bmatrix}, \begin{bmatrix} 2 & 1 \\ 0 & 2 \end{bmatrix}\right]. \quad \square$$

There is a concept that summarizes the parameters of the Jordan form. Let $A \in \mathbb{C}^{n \times n}$ and let $\lambda_1, \lambda_2, \ldots, \lambda_s$ denote all its distinct eigenvalues of geometric multiplicities $p_1, p_2, \ldots, p_s$, respectively. Thus, there are p_j Jordan blocks in the Jordan form J of A associated with the eigenvalue λ_j $(1 \le j \le s)$. Denote by $\delta_j^{(1)} \ge \delta_j^{(2)} \ge \cdots \ge \delta_j^{(p_j)}$ the orders of these blocks and write

$$\{(\delta_1^{(1)}, \delta_1^{(2)}, \ldots, \delta_1^{(p_1)}), (\delta_2^{(1)}, \delta_2^{(2)}, \ldots, \delta_2^{(p_2)}), \ldots, (\delta_s^{(1)}, \delta_s^{(2)}, \ldots, \delta_s^{(p_s)})\}. \quad (2)$$

The form (2) is referred to as the *Segre characteristic*† of the matrix A. Obviously this characteristic, along with the eigenvalues of the matrix, uniquely defines its Jordan form. Note that the numbers $q_j = \delta_j^{(1)} + \delta_j^{(2)} + \cdots + \delta_j^{(p_j)}$, $j = 1, 2, \ldots, s$, are the algebraic multiplicities of the eigenvalues $\lambda_1, \lambda_2, \ldots, \lambda_s$, respectively.

Exercise 6. Verify that the Segre characteristic of the matrices discussed in Exercises 4 and 5 are, respectively,

$$\{(2, 1), (2, 1, 1)\} \quad \text{and} \quad \{(2), (2)\}. \qquad \square$$

†Segre, C., *Atti Accad. naz. Lincei, Mem. III*, **19** (1884), 127–148.

6.7 The Real Jordan Form

Let $A \in \mathbb{R}^{n \times n}$. If A is viewed as a complex matrix, then it follows from Theorem 6.5.1 that A is similar to a Jordan matrix as defined in Eqs. (6.5.4) and (6.5.5). However, this Jordan matrix J, and the matrix P for which $P^{-1}AP = J$, will generally be complex. In this section we describe a normal form for A that can be achieved by confining attention to *real* transforming matrices P, and hence a real canonical form. To achieve this, we take advantage of the (complex) Jordan form already derived and the fundamental fact (see Section 5.9) that the eigenvalues of A may be real, and if not, that they appear in complex conjugate pairs.

We shall use the construction of a (complex) Jordan basis for A from the preceding sections to obtain a *real* basis. The vectors of this basis will form the columns of our real transforming matrix P.

First, if λ is a real eigenvalue of A, then there is a real Jordan basis for the generalized eigenspace of λ, as described in (6.4.1); these vectors will form a part of our new real basis for $\mathbb{R}^n$. If $\lambda \in \sigma(A)$ and is *not* real, then neither is $\bar{\lambda} \in \sigma(A)$. Construct Jordan chains (necessarily non-real) to span the generalized eigenspace of λ as in (6.4.1). Thus, each chain in this spanning set satisfies relations of the form (6.3.3) (with T replaced by A). Taking complex conjugates in these relations, the complex conjugate vectors form chains spanning the generalized eigenspace of $\bar{\lambda}$.

It will be sufficient to show how the vectors for a single chain of length r, together with those of the conjugate chain, are combined to produce the desired real basis for a $2r$-dimensional real invariant subspace of A. Let λ be the eigenvalue in question and write $\lambda = \mu + i\omega$, where μ, ω are real and $\omega \neq 0$. Let P be the $n \times r$ complex matrix whose columns are the vectors $\mathbf{x}_1, \mathbf{x}_2, \ldots, \mathbf{x}_r$ of a Jordan chain corresponding to λ. If $J(\lambda)$ is the $r \times r$ Jordan block with eigenvalue λ, then $AP = PJ(\lambda)$ and, taking conjugates, $A\bar{P} = \bar{P}J(\bar{\lambda})$. Thus,

$$A[P \quad \bar{P}] = [P \quad \bar{P}]\begin{bmatrix} J(\lambda) & 0 \\ 0 & J(\bar{\lambda}) \end{bmatrix},$$

and for any unitary matrix U of size $2r$,

$$A[P \quad \bar{P}]U = [P \quad \bar{P}]UU^*\begin{bmatrix} J(\lambda) & 0 \\ 0 & J(\bar{\lambda}) \end{bmatrix}U.$$

We claim that there is a unitary matrix U such that the $2r$ columns of $Q \triangleq [P \quad \bar{P}]U$ are real and such that $J_r(\lambda, \bar{\lambda}) \triangleq U^* \operatorname{diag}[J(\lambda), J(\bar{\lambda})]U$ is also real. The matrix $J_r(\lambda, \bar{\lambda})$ then plays the same role in the real Jordan form for A that $\operatorname{diag}[J(\lambda), J(\bar{\lambda})]$ plays in the complex form.

We confirm our claim simply by verification. We confine attention to the case $r = 3$; the modifications for other values of r are obvious. Define the unitary matrix

$$U = \frac{1}{\sqrt{2}} \begin{bmatrix} 1 & -i & 0 & 0 & 0 & 0 \\ 0 & 0 & 1 & -i & 0 & 0 \\ 0 & 0 & 0 & 0 & 1 & -i \\ 1 & i & 0 & 0 & 0 & 0 \\ 0 & 0 & 1 & i & 0 & 0 \\ 0 & 0 & 0 & 0 & 1 & i \end{bmatrix}.$$

Recalling that $\lambda = \mu + i\omega$, it is easily verified that

$$J_r(\lambda, \bar{\lambda}) = U^* \begin{bmatrix} J(\lambda) & 0 \\ 0 & J(\bar{\lambda}) \end{bmatrix} U = \begin{bmatrix} \mu & \omega & 1 & 0 & 0 & 0 \\ -\omega & \mu & 0 & 1 & 0 & 0 \\ 0 & 0 & \mu & \omega & 1 & 0 \\ 0 & 0 & -\omega & \mu & 0 & 1 \\ 0 & 0 & 0 & 0 & \mu & \omega \\ 0 & 0 & 0 & 0 & -\omega & \mu \end{bmatrix}.$$

Furthermore, with the Cartesian decompositions $\mathbf{x}_j = \mathbf{y}_j + i\mathbf{z}_j$ for $j = 1, 2, \ldots, r$, we get

$$Q = [P \quad \bar{P}]U = \sqrt{2}[\mathbf{y}_1 \quad \mathbf{z}_1 \quad \mathbf{y}_2 \quad \mathbf{z}_2 \quad \mathbf{y}_3 \quad \mathbf{z}_3],$$

which is real and of full rank since $[P \quad \bar{P}]$ has full rank and U is unitary. These vectors span a six-dimensional real invariant subspace of A associated with the non-real eigenvalue pair λ, $\bar{\lambda}$ (see Theorem 5.9.1). This statement is clear from the relation $AQ = QJ_r(\lambda, \bar{\lambda})$. Our discussion leads to the following conclusion.

Theorem 1. *If $A \in \mathbb{R}^{n \times n}$, there is a nonsingular matrix $X \in \mathbb{R}^{n \times n}$ such that $X^{-1}AX$ is a real block-diagonal matrix, each block of which has one of two forms:*

(*a*) *the form of* (6.5.3), *in which case the size of the block is equal to the length of a complete Jordan chain of a set* (6.4.1) *corresponding to a real eigenvalue;*

(*b*) *the form $J_r(\lambda, \bar{\lambda})$ indicated above in which case the size of the block is equal to twice the length of a (complex) complete Jordan chain of a set* (6.4.1) *corresponding to a non-real eigenvalue $\lambda = \mu + i\omega$.*

Exercise 1. Describe the real Jordan form for a matrix that is real and simple. ☐

6.8 Miscellaneous Exercises

1. Prove that two simple matrices are similar if and only if they have the same characteristic polynomial.

2. Show that any square matrix is similar to its transpose.

3. Check that the Jordan canonical form of an idempotent matrix (a projector) is a diagonal matrix with zeros and ones in the diagonal positions.

4. Show that

(a) the minimal and the characteristic polynomials of a Jordan block of order r are both $(\lambda - \lambda_0)^r$;

(b) the minimal polynomial of a Jordan matrix is the least common multiple of the minimal polynomials of the Jordan blocks.

5. Show that a Jordan block is similar to the companion matrix associated with its characteristic (or minimal) polynomial.

Hint. For a geometrical solution: Show that the representation of $T|_{\mathscr{J}}$ with respect to the basis $\{\mathbf{x}, T(\mathbf{x}), \ldots, T^{r-1}(\mathbf{x})\}$ for a Jordan subspace $\mathscr{J}$ is a companion matrix.

6. Prove that any matrix $A \in \mathbb{C}^{n \times n}$ can be represented as a sum of a simple matrix and a nilpotent matrix.

Hint. Represent an $r \times r$ Jordan block in the form $\lambda_i I + N$, where $N = [\delta_{j+1,k}]_{j,k=1}^r$, and observe that $N^r = O$.

7. Check that the matrices

$$\begin{bmatrix} a_0 & a_{12} & \cdots & a_{1r} \\ 0 & & \ddots & \vdots \\ \vdots & \ddots & & a_{r-1,r} \\ 0 & \cdots & 0 & a_0 \end{bmatrix} \quad \text{and} \quad \begin{bmatrix} a_0 & 1 & 0 & \cdots & 0 \\ 0 & & & \ddots & \vdots \\ & & & \ddots & 0 \\ \vdots & & \ddots & & 1 \\ 0 & & \cdots & 0 & a_0 \end{bmatrix}$$

are similar, provided $a_{i,i+1} \neq 0$, $i = 1, 2, \ldots, r - 1$.

8. Check that for an $r \times r$ Jordan block J associated with an eigenvalue λ_0, and for any polynomial $p(\lambda)$,

$$p(J) = \begin{bmatrix} p(\lambda_0) & p'(\lambda_0)/1! & \cdots & p^{(r-1)}(\lambda_0)/(r-1)! \\ 0 & & \ddots & \vdots \\ \vdots & \ddots & & p'(\lambda_0)/1! \\ 0 & \cdots & 0 & p(\lambda_0) \end{bmatrix}.$$

9. If $m(\lambda)$ is the minimal polynomial of $A \in \mathbb{C}^{n \times n}$ and $p(\lambda)$ is a polynomial over $\mathbb{C}$, prove that $p(A)$ is nonsingular if and only if $m(\lambda)$ and $p(\lambda)$ are relatively prime.

10. If $\alpha_1, \alpha_2, \alpha_3 \in \mathbb{R}$ and are not all zero, show that the characteristic and minimal polynomials of the matrix

$$\begin{bmatrix} \alpha_0 & \alpha_1 & \alpha_2 & \alpha_3 \\ -\alpha_1 & \alpha_0 & -\alpha_3 & \alpha_2 \\ -\alpha_2 & \alpha_3 & \alpha_0 & -\alpha_1 \\ -\alpha_3 & -\alpha_2 & \alpha_1 & \alpha_0 \end{bmatrix}$$

are

$$(\lambda^2 - 2\alpha_0\lambda + (\alpha_0^2 + \alpha_1^2 + \alpha_2^2 + \alpha_3^2))^j$$

with $j = 2$ and $j = 1$, respectively.

Hint. First reduce to the case $\alpha_0 = 0$ and show that in this case $\lambda_1 = i(\alpha_1^2 + \alpha_2^2 + \alpha_3^2)^{1/2}$ is an eigenvalue with eigenvectors

$$\begin{bmatrix} \lambda_1 \\ \alpha_1 \\ \alpha_2 \\ \alpha_3 \end{bmatrix} \quad \text{and} \quad \begin{bmatrix} \alpha_3 \\ \alpha_2 \\ -\alpha_1 \\ -\lambda_1 \end{bmatrix}.$$

11. Let $A \in \mathbb{C}^{n \times n}$. Show that $A = A^{-1}$ if and only if there is an $m \le n$ and a nonsingular S such that

$$A = S\begin{bmatrix} I_m & 0 \\ 0 & -I_{n-m} \end{bmatrix}S^{-1}.$$

Hint. Use the Jordan form for A.

CHAPTER 7

Matrix Polynomials and Normal Forms

The important theory of invariant polynomials and λ-matrices developed in this chapter allows us to obtain an independent proof of the possibility of reducing a matrix to the Jordan canonical form, as well as a method of constructing that form directly from the elements of the matrix. Also, other normal forms of a matrix are derived. This approach to the Jordan form can be viewed as an algebraic one in contrast to the geometrical approach developed in Chapter 6. It requires less mathematical abstraction than the geometrical approach and may be more accessible for some readers. In addition, the analysis of matrix polynomials is important in its own right. Direct applications to differential and difference equations are developed in this chapter, and foundations are laid for the spectral theory of matrix polynomials to be studied in Chapter 14.

7.1 The Notion of a Matrix Polynomial

Consider an $n \times n$ matrix $A(\lambda)$ whose elements are polynomials in λ:

$$A(\lambda) = \begin{bmatrix} a_{11}(\lambda) & a_{12}(\lambda) & \cdots & a_{1n}(\lambda) \\ a_{21}(\lambda) & a_{22}(\lambda) & \cdots & a_{2n}(\lambda) \\ \vdots & \vdots & \ddots & \vdots \\ a_{n1}(\lambda) & a_{n2}(\lambda) & \cdots & a_{nn}(\lambda) \end{bmatrix}, \tag{1}$$

and suppose that λ and the coefficients of the polynomials $a_{ij}(\lambda)$ are from a field $\mathscr{F}$, so that when the elements of $A(\lambda)$ are evaluated for a particular value of λ, say $\lambda = \lambda_0$, then $A(\lambda_0) \in \mathscr{F}^{n \times n}$. The matrix $A(\lambda)$ is known as a *matrix polynomial* or a *lambda matrix*, or a *λ-matrix*. Where the underlying field needs to be stressed we shall call $A(\lambda)$ a *λ-matrix over $\mathscr{F}$*. Note that an example of a λ-matrix is provided by the matrix $\lambda I - A$ and that the matrices whose elements are scalars (i.e., polynomials of degree zero, or just zeros) can be viewed as a particular case of λ-matrices.

A λ-matrix $A(\lambda)$ is said to be *regular* if the determinant $\det A(\lambda)$ is not identically zero. Note that $\det A(\lambda)$ is itself a polynomial with coefficients in $\mathscr{F}$, so $A(\lambda)$ is *not* regular if and only if each coefficient of $\det A(\lambda)$ is the zero of $\mathscr{F}$. The *degree* of a λ-matrix $A(\lambda)$ is defined to be the greatest degree of the polynomials appearing as entries of $A(\lambda)$ and is denoted by $\deg A(\lambda)$. Thus, $\deg(\lambda I - A) = 1$ and $\deg A = 0$ for any nonzero $A \in \mathscr{F}^{n \times n}$.

Lambda matrices of the same order may be added and multiplied together in the usual way, and in each case the result is another λ-matrix. In particular, a λ-matrix $A(\lambda)$ is said to be *invertible* if there is a λ-matrix $B(\lambda)$ such that $A(\lambda)B(\lambda) = B(\lambda)A(\lambda) = I$. In other words, $A(\lambda)$ is invertible if $B(\lambda) = [A(\lambda)]^{-1}$ exists and is also a λ-matrix.

Proposition 1. *A λ-matrix $A(\lambda)$ is invertible if and only if* $\det A(\lambda)$ *is a nonzero constant.*

PROOF. If $\det A(\lambda) = c \neq 0$, then the entries of the inverse matrix are equal to the minors of $A(\lambda)$ of order $n - 1$ divided by $c \neq 0$ and hence are polynomials in λ. Thus $[A(\lambda)]^{-1}$ is a λ-matrix. Conversely, if $A(\lambda)$ is invertible, then the equality $A(\lambda)B(\lambda) = I$ yields

$$\det A(\lambda) \det B(\lambda) = 1.$$

Thus, the product of the polynomials $\det A(\lambda)$ and $\det B(\lambda)$ is a nonzero constant. This is possible only if they are each nonzero constants. ■

A λ-matrix with a nonzero constant determinant is also referred to as a *unimodular* λ-matrix. This notion permits the restatement of Proposition 1: *A λ-matrix is invertible if and only if it is unimodular.* Note that the degree of a unimodular matrix is arbitrary.

Exercise 1. Check that the following λ-matrices are unimodular:

$$A_1(\lambda) = \begin{bmatrix} 1 & \lambda & -2\lambda^2 \\ 0 & 1 & \lambda^4 \\ 0 & 0 & 1 \end{bmatrix}, \qquad A_2(\lambda) = \begin{bmatrix} (\lambda - 1)^2 & \lambda \\ \lambda - 2 & 1 \end{bmatrix}. \quad \square$$

Unimodular λ-matrices play an important role in reducing an arbitrary λ-matrix to a simpler form. Before going on to develop this idea in the next two sections, we wish to point out a different viewpoint on λ-matrices.

Let $A(\lambda)$ denote an $n \times n$ λ-matrix of degree l. The (i, j)th element of $A(\lambda)$ can be written in the form

$$a_{ij}(\lambda) = a_{ij}^{(0)} + a_{ij}^{(1)}\lambda + \cdots + a_{ij}^{(l)}\lambda^l,$$

and there is at least one element for which $a_{ij}^{(l)} \neq 0$ $(1 \leq i, j \leq n)$. If we let A_r be the matrix whose i, jth element is $a_{ij}^{(r)}$ $(r = 0, 1, \ldots, l$ for each $i, j)$, then we have $A_l \neq O$ and

$$A(\lambda) = A_0 + \lambda A_1 + \cdots + \lambda^{l-1}A_{l-1} + \lambda^l A_l.$$

This formulation motivates the term "matrix polynomial" as an alternative for "λ-matrix."

Exercise 2. Check that

$$\begin{bmatrix} \lambda^2 + \lambda + 1 & \lambda^2 - \lambda + 2 \\ 2\lambda & \lambda^2 - 3\lambda - 1 \end{bmatrix} = \begin{bmatrix} 1 & 2 \\ 0 & -1 \end{bmatrix} + \lambda \begin{bmatrix} 1 & -1 \\ 2 & -3 \end{bmatrix} + \lambda^2 \begin{bmatrix} 1 & 1 \\ 0 & 1 \end{bmatrix}. \quad \square$$

The description of λ-matrices as matrix polynomials suggests problems concerning divisibility, the solution of polynomial equations, and other generalizations of the theory of scalar polynomials. Of course, the lack of commutativity of the coefficients of matrix polynomials causes some difficulties and modifications in this program. However, an important and useful theory of matrix polynomials has recently been developed, and an introduction to this theory is presented in Chapter 14.

In the next section we present some fundamental facts about divisibility of matrix polynomials needed for use in this and subsequent chapters.

7.2 Division of Matrix Polynomials

Let $A(\lambda) = \sum_{i=0}^{l} \lambda^i A_i$ be an $n \times n$ matrix polynomial (i.e., $A_i \in \mathcal{F}^{n \times n}$, $i = 0, 1, \ldots, l$) of degree l. If $A_l = I$, then the matrix polynomial $A(\lambda)$ is said to be *monic*. If $B(\lambda) = \sum_{i=0}^{m} \lambda^i B_i$ denotes an $n \times n$ matrix polynomial of degree m, then obviously

$$A(\lambda) + B(\lambda) = \sum_{i=0}^{k} \lambda^i (A_i + B_i),$$

where $k = \max(l, m)$. Thus, the degree of $A(\lambda) + B(\lambda)$ doex not exceed k.

Furthermore, the degree of the product

$$A(\lambda)B(\lambda) = A_0B_0 + \lambda(A_1B_0 + A_0B_1) + \cdots + \lambda^{l+m}A_lB_m$$

does not exceed $l + m$. Clearly, if either $A(\lambda)$ or $B(\lambda)$ is a matrix polynomial with an invertible leading coefficient, then the degree of the product is exactly $l + m$.

Suppose that $B(\lambda)$ is a matrix polynomial of degree m with an invertible leading coefficient and that there exist matrix polynomials $Q(\lambda)$, $R(\lambda)$, with $R(\lambda) \equiv 0$ or the degree of $R(\lambda)$ less than m, such that

$$A(\lambda) = Q(\lambda)B(\lambda) + R(\lambda).$$

If this is the case we call $Q(\lambda)$ a *right quotient* of $A(\lambda)$ on division by $B(\lambda)$ and $R(\lambda)$ is a *right remainder* of $A(\lambda)$ on division by $B(\lambda)$. Similarly, $\tilde{Q}(\lambda)$, $\tilde{R}(\lambda)$ are a *left quotient* and *left remainder* of $A(\lambda)$ on division by $B(\lambda)$ if

$$A(\lambda) = B(\lambda)\tilde{Q}(\lambda) + \tilde{R}(\lambda)$$

and $\tilde{R}(\lambda) \equiv 0$ or the degree of $R(\lambda)$ is less than m.

If the right remainder of $A(\lambda)$ on division by $B(\lambda)$ is zero, then $Q(\lambda)$ is said to be a *right divisor* of $A(\lambda)$ on division by $B(\lambda)$. A similar definition applies for a *left divisor* of $A(\lambda)$ on division by $B(\lambda)$.

In order to make these formal definitions meaningful, we must show that, given this $A(\lambda)$ and $B(\lambda)$, there do exist quotients and remainders as defined. When we have done this we shall prove that they are also unique. The proof of the next theorem is a generalization of the division algorithm for scalar polynomials.

Theorem 1. *Let $A(\lambda) = \sum_{i=0}^{l} \lambda^i A_i$, $B(\lambda) = \sum_{i=0}^{m} \lambda^i B_i$ be $n \times n$ matrix polynomials of degrees l and m, respectively, with* $\det B_m \neq 0$. *Then there exists a right quotient and right remainder of $A(\lambda)$ on division by $B(\lambda)$ and similarly for a left quotient and left remainder.*

PROOF. If $l < m$, we have only to put $Q(\lambda) = 0$ and $R(\lambda) = A(\lambda)$ to obtain the result.

If $l \geq m$, we first "divide by" the leading term of $B(\lambda)$, namely, $B_m\lambda^m$ Observe that the term of highest degree of the λ-matrix $A_lB_m^{-1}\lambda^{l-m}B(\lambda)$ is just $A_l\lambda^l$. Hence

$$A(\lambda) = A_lB_m^{-1}\lambda^{l-m}B(\lambda) + A^{(1)}(\lambda),$$

where $A^{(1)}(\lambda)$ is a matrix polynomial whose degree, l_1, does not exceed $l - 1$. Writing $A^{(1)}(\lambda)$ in decreasing powers, let

$$A^{(1)}(\lambda) = A_{l_1}^{(1)}\lambda^{l_1} + \cdots + A_0^{(1)}, \qquad A_{l_1}^{(1)} \neq 0, \qquad l_1 < l.$$

If $l_1 \geq m$ we repeat the process, but on $A^{(1)}(\lambda)$ rather than $A(\lambda)$ to obtain

$$A^{(1)}(\lambda) = A_{l_1}^{(1)} B_m^{-1} \lambda^{l_1 - m} B(\lambda) + A^{(2)}(\lambda),$$

where

$$A^{(2)}(\lambda) = A_{l_2}^{(2)} \lambda^{l_2} + \cdots + A_0^{(2)}, \qquad A_{l_2}^{(2)} \neq 0, \quad l_2 < l_1.$$

In this manner we can construct a sequence of matrix polynomials $A(\lambda)$, $A^{(1)}(\lambda)$, $A^{(2)}(\lambda), \ldots$ whose degrees are strictly decreasing, and after a finite number of terms we arrive at a matrix polynomial $A^{(r)}(\lambda)$ of degree $l_r < m$, with $l_{r-1} \geq m$. Then, if we write $A(\lambda) = A^{(0)}(\lambda)$,

$$A^{(s-1)}(\lambda) = A_{l_{s-1}}^{(s-1)} B_m^{-1} \lambda^{l_{s-1} - m} B(\lambda) + A^{(s)}(\lambda), \qquad s = 1, 2, \ldots, r.$$

Combining these equations, we have

$$\begin{aligned} A(\lambda) = (&A_l B_m^{-1} \lambda^{l-m} + A_{l_1}^{(1)} B_m^{-1} \lambda^{l_1 - m} + \cdots \\ &+ A_{l_{r-1}}^{(r-1)} B_m^{-1} \lambda^{l_{r-1} - m}) B(\lambda) + A^{(r)}(\lambda). \end{aligned}$$

The matrix in parentheses can now be identified as a right quotient of $A(\lambda)$ on division by $B(\lambda)$, and $A^{(r)}(\lambda)$ is the right remainder.

It is not difficult to provide the modifications to the proof needed to prove the existence of a left quotient and remainder. ■

Theorem 2. *With the hypotheses of Theorem 1, the right quotient, right remainder, left quotient, and left remainder are each unique.*

Proof. Suppose that there exist matrix polynomials $Q(\lambda)$, $R(\lambda)$ and $Q_1(\lambda)$, $R_1(\lambda)$ such that

$$A(\lambda) = Q(\lambda)B(\lambda) + R(\lambda)$$

and

$$A(\lambda) = Q_1(\lambda)B(\lambda) + R_1(\lambda)$$

where $R(\lambda)$, $R_1(\lambda)$ each have degree less than m. Then

$$(Q(\lambda) - Q_1(\lambda))B(\lambda) = R_1(\lambda) - R(\lambda).$$

If $Q(\lambda) \neq Q_1(\lambda)$, then the left-hand side of this equation is a matrix polynomial whose degree is at least m. However, the right-hand side is a matrix polynomial of degree less than m. Hence $Q_1(\lambda) = Q(\lambda)$, whence $R_1(\lambda) = R(\lambda)$ also.

A similar argument can be used to establish the uniqueness of the left quotient and remainder. ■

Exercise 1. Examine the right and left quotients and remainders of $A(\lambda)$ on division by $B(\lambda)$, where

$$A(\lambda) = \begin{bmatrix} \lambda^4 + \lambda^2 + \lambda - 1 & \lambda^3 + \lambda^2 + \lambda + 2 \\ 2\lambda^3 - \lambda & 2\lambda^2 + 2\lambda \end{bmatrix},$$

$$B(\lambda) = \begin{bmatrix} \lambda^2 + 1 & 1 \\ \lambda & \lambda^2 + \lambda \end{bmatrix}.$$

SOLUTION. Note first of all that $B(\lambda)$ has an invertible leading coefficient. It is found that

$$\begin{aligned} A(\lambda) &= \begin{bmatrix} \lambda^2 - 1 & \lambda - 1 \\ 2\lambda & 2 \end{bmatrix} \begin{bmatrix} \lambda^2 + 1 & 1 \\ \lambda & \lambda^2 + \lambda \end{bmatrix} + \begin{bmatrix} 2\lambda & 2\lambda + 3 \\ -5\lambda & -2\lambda \end{bmatrix} \\ &= Q(\lambda)B(\lambda) + R(\lambda), \end{aligned}$$

and

$$A(\lambda) = \begin{bmatrix} \lambda^2 + 1 & 1 \\ \lambda & \lambda^2 + \lambda \end{bmatrix} \begin{bmatrix} \lambda^2 & \lambda + 1 \\ \lambda - 1 & 1 \end{bmatrix} = B(\lambda)\tilde{Q}(\lambda).$$

Thus, $\tilde{Q}(\lambda)$ is a left divisor of $A(\lambda)$. □

Now we consider the important special case in which a divisor is *linear* (i.e., a matrix polynomial of the first degree). But first note that when discussing a scalar polynomial $p(\lambda)$, we may write

$$p(\lambda) = a_l\lambda^l + a_{l-1}\lambda^{l-1} + \cdots + a_0 = \lambda^l a_l + \lambda^{l-1}a_{l-1} + \cdots + a_0.$$

For a matrix polynomial with a matrix argument, this is not generally possible. If $A(\lambda)$ is a matrix polynomial over $\mathscr{F}$ and $B \in \mathscr{F}^{n \times n}$, we define the *right value* $A(B)$ of $A(\lambda)$ at B by

$$A(B) = A_l B^l + A_{l-1}B^{l-1} + \cdots + A_0$$

and the *left value* $\tilde{A}(B)$ of $A(\lambda)$ at B by

$$\tilde{A}(B) = B^l A_l + B^{l-1}A_{l-1} + \cdots + A_0.$$

The reader should be familiar with the classical *remainder theorem*: On dividing the scalar polynomial $p(\lambda)$ by $\lambda - b$, the remainder is $p(b)$. We now prove a remarkable extension of this result to matrix polynomials. Note first of all that $\lambda I - B$ is monic.

Theorem 3. *The right and left remainders of a matrix polynomial $A(\lambda)$ on division by $\lambda I - B$ are $A(B)$ and $\tilde{A}(B)$, respectively.*

PROOF. The factorization

$$\lambda^j I - B^j = (\lambda^{j-1}I + \lambda^{j-2}B + \cdots + \lambda B^{j-2} + B^{j-1})(\lambda I - B)$$

can be verified by multiplying out the product on the right. Premultiply both sides of this equation by A_j and sum the resulting equations for $j = 1, 2, \ldots, l$. The right-hand side of the equation obtained is of the form $C(\lambda)(\lambda I - B)$, where $C(\lambda)$ is a matrix polynomial.

The left-hand side is

$$\sum_{j=1}^{l} A_j \lambda^j - \sum_{j=1}^{l} A_j B^j = \sum_{j=0}^{l} A_j \lambda^j - \sum_{j=0}^{l} A_j B^j = A(\lambda) - A(B).$$

Thus,

$$A(\lambda) = C(\lambda)(\lambda I - B) + A(B).$$

The result now follows from the uniqueness of the right remainder on division of $A(\lambda)$ by $(\lambda I - B)$.

The result for the left remainder is obtained by reversing the factors in the initial factorization, multiplying on the right by A_j, and summing. ■

An $n \times n$ matrix $X \in \mathscr{F}^{n \times n}$ such that $A(X) = 0$ (respectively, $\tilde{A}(X) = 0$) is referred to as a *right* (respectively, *left*) *solvent* of the matrix polynomial $A(\lambda)$.

Corollary 1. *The matrix polynomial $A(\lambda)$ is divisible on the right (respectively, left) by $\lambda I - B$ with zero remainder if and only if B is a right (respectively, left) solvent of $A(\lambda)$.*

This result provides a second proof of the Cayley–Hamilton theorem (see Theorem 6.2.5).

Theorem 4. *If A is a square matrix with characteristic polynomial $c(\lambda)$, then $c(A) = 0$. In other words, A is a zero of its characteristic polynomial.*

PROOF. Let $A \in \mathscr{F}^{n \times n}$ and let $c(\lambda)$ denote the characteristic polynomial of A. Define the matrix $B(\lambda) = \operatorname{adj}(\lambda I - A)$ and (see Section 2.6) observe that

$B(\lambda)$ is, by definition, an $n \times n$ matrix polynomial over $\mathscr{F}$ of degree $n - 1$ and that

$$(\lambda I - A)B(\lambda) = B(\lambda)(\lambda I - A) = c(\lambda)I. \tag{1}$$

Now, $c(\lambda)I$ is a matrix polynomial of degree n that is divisible on both the left and the right by $\lambda I - A$ with zero remainder. It remains now to apply Corollary 1 above. ■

Corollary 2. *If f is any scalar polynomial over $\mathscr{F}$ and $A \in \mathscr{F}^{n \times n}$, then there is a polynomial p (depending on A) of degree less than n for which $f(A) = p(A)$.*

PROOF. Let $f(\lambda) = q(\lambda)c(\lambda) + r(\lambda)$, where $c(\lambda)$ is the characteristic polynomial of A and $r(\lambda)$ is either zero or a polynomial of degree not exceeding $n - 1$. Then $f(A) = q(A)c(A) + r(A)$ and, by Theorem 4, $f(A) = r(A)$. ■

This result appears again later as a consequence of the definition of $f(A)$ for more general classes of scalar functions f.

7.3 Elementary Operations and Equivalence

In this section our immediate objective is the reduction of a matrix polynomial to a simpler form by means of equivalence transformations, which we will now describe. This analysis is a natural extension of that developed in Sections 2.7 and 2.10 for matrices with entries from a field $\mathscr{F}$ to matrices whose entries are scalar polynomials and so no longer form a field. Afterwards, on confining our attention to the matrix polynomial $\lambda I - A$, the desired normal (canonical) forms for the (constant) matrix A will be derived.

We start by defining the *elementary row* and *column operations* on a matrix polynomial over $\mathscr{F}$ by an appropriate generalization of those (in fact, only of the third one) on a matrix with constant entries as described in Section 2.7. The parentheses refer to the right elementary operations.

(1) Multiply any row (column) by a nonzero $c \in \mathscr{F}$.
(2) Interchange any two rows (columns).
(3) Add to any row (column) any other row (column) multiplied by an arbitrary polynomial $b(\lambda)$ over $\mathscr{F}$.

It is easy to verify (comparing with Section 2.7) that performing an elementary row operation on an $n \times n$ matrix polynomial is equivalent to premultiplication of the matrix polynomial by an $n \times n$ matrix of one of the following three forms, respectively:

$$E^{(1)} = \begin{bmatrix} 1 &&&&&& \\ & \ddots &&&&& \\ && 1 &&&& \\ &&& c &&& \\ &&&& 1 && \\ &&&&& \ddots & \\ &&&&&& 1 \end{bmatrix},$$

$$E^{(2)} = \begin{bmatrix} 1 &&&&&& \\ & \ddots &&&&& \\ && 0 & \cdots & 1 && \\ &&& 1 &&& \\ && \vdots & \ddots & \vdots && \\ &&& 1 &&& \\ && 1 & \cdots & 0 && \\ &&&&& \ddots & \\ &&&&&& 1 \end{bmatrix},$$

$$E_1^{(3)}(\lambda) = \begin{bmatrix} 1 &&&&& \\ & \ddots &&&& \\ && 1 & \cdots & b(\lambda) && \\ &&& \ddots & \vdots && \\ &&&& 1 && \\ &&&&& \ddots & \\ &&&&&& 1 \end{bmatrix},$$

or

$$E_2^{(3)}(\lambda) = \begin{bmatrix} 1 &&&&&& \\ & \ddots &&&&& \\ && 1 &&&& \\ && \vdots & \ddots &&& \\ && b(\lambda) & \cdots & 1 && \\ &&&&& \ddots & \\ &&&&&& 1 \end{bmatrix}.$$

As in Section 2.7, the matrices above are said to be *elementary matrices* of the first, second; or third kind. Observe that the elementary matrices can be generated by performing the desired operation on the unit matrix, and that postmultiplication by an appropriate elementary matrix produces an elementary column operation on the matrix polynomial.

Exercise 1. Prove that elementary matrices are unimodular and that their inverses are also elementary matrices. □

We can now make a formal definition of equivalence transformations: two matrix polynomials $A(\lambda)$ and $B(\lambda)$, over $\mathscr{F}$, are said to be *equivalent*, or to be connected by an *equivalence transformation*, if $B(\lambda)$ can be obtained from $A(\lambda)$ by a finite sequence of elementary operations. It follows from the equivalence of elementary operations and operations with elementary matrices that $A(\lambda)$ and $B(\lambda)$ are equivalent if and only if there are elementary matrices $E_1(\lambda), E_2(\lambda), \ldots, E_k(\lambda), E_{k+1}(\lambda), \ldots, E_s(\lambda)$ such that

$$B(\lambda) = E_k(\lambda) \cdots E_1(\lambda)A(\lambda)E_{k+1}(\lambda) \cdots E_s(\lambda)$$

or, equivalently,

$$B(\lambda) = P(\lambda)A(\lambda)Q(\lambda), \tag{1}$$

where $P(\lambda) = E_k(\lambda) \cdots E_1(\lambda)$ and $Q(\lambda) = E_{k+1}(\lambda) \cdots E_s(\lambda)$ are matrix polynomials over $\mathscr{F}$. Exercise 1 implies that the matrices $P(\lambda)$ and $Q(\lambda)$ are unimodular and hence the following propositions.

Proposition 1. *The matrix polynomials $A(\lambda)$ and $B(\lambda)$ are equivalent if and only if there are unimodular matrices $P(\lambda)$ and $Q(\lambda)$ such that Eq.* (1) *is valid.*

The last fact readily implies that equivalence between matrix polynomials is an equivalence relation.

Exercise 2. Denoting the fact that $A(\lambda)$ and $B(\lambda)$ are equivalent by $A \sim B$, prove that

(a) $A \sim A$;
(b) $A \sim B$ implies $B \sim A$;
(c) $A \sim B$ and $B \sim C$ implies $A \sim C$.

Exercise 3. Prove that any unimodular matrix is equivalent to the identity matrix.

SOLUTION. If $A(\lambda)$ is unimodular, then $A(\lambda)[A(\lambda)]^{-1} = I$ and hence relation (1) holds with $P(\lambda) = I$ and $Q(\lambda) = [A(\lambda)]^{-1}$. □

Exercise 2 shows that the set of all $n \times n$ matrix polynomials splits into subsets of mutually equivalent matrices. Exercise 3 gives one of these classes, namely, the class of unimodular matrices.

Exercise 4. Prove that a matrix polynomial is unimodular if and only if it can be decomposed into a product of elementary matrices.

Hint. For the necessary part, use Exercise 3.

Exercise 5. Show that the matrices

$$A(\lambda) = \begin{bmatrix} \lambda & \lambda + 1 \\ \lambda^2 - \lambda & \lambda^2 - 1 \end{bmatrix} \quad \text{and} \quad B = \begin{bmatrix} 1 & 0 \\ 0 & 0 \end{bmatrix}$$

are equivalent and find the *transforming matrices* $P(\lambda)$ and $Q(\lambda)$.

SOLUTION. To simplify $A(\lambda)$, add $-(\lambda - 1)$ times the first row to the second row, and then add -1 times the first column to the second. Finally, adding $-\lambda$ times the second column of the new matrix to the first and interchanging columns, obtain the matrix B. Translating the operations performed into elementary matrices and finding their product, it is easily found that

$$Q(\lambda) = \begin{bmatrix} -1 & 1 + \lambda \\ 1 & -\lambda \end{bmatrix} \qquad P(\lambda) = \begin{bmatrix} 1 & 0 \\ 1 - \lambda & 1 \end{bmatrix}$$

and that Eq. (1) holds.

Exercise 6. Show that the following pairs of matrices are equivalent:

$$\text{(a)} \quad \begin{bmatrix} 0 & 1 & \lambda \\ \lambda & \lambda & 1 \\ \lambda^2 - \lambda & \lambda^2 - 1 & \lambda^2 - 1 \end{bmatrix} \quad \text{and} \quad \begin{bmatrix} 1 & 0 & 0 \\ 0 & 1 & 0 \\ 0 & 0 & 0 \end{bmatrix};$$

$$\text{(b)} \quad \begin{bmatrix} \lambda^2 & \lambda + 1 \\ \lambda - 1 & \lambda^2 \end{bmatrix} \quad \text{and} \quad \begin{bmatrix} 1 & 0 \\ 0 & \lambda^4 - \lambda^2 + 1 \end{bmatrix}.$$

Exercise 7. Show that if $E(\lambda)$ is a unimodular matrix polynomial, then all minors of size two or more are independent of λ.

SOLUTION. Observe that this is true for all elementary matrices. Express $E(\lambda)$ as a product of elementary matrices and use the Binet–Cauchy formula. □

7.4 A Canonical Form for a Matrix Polynomial

The objective of this section is to find the simplest matrix polynomial in each equivalence class of mutually equivalent matrix polynomials. In more detail, it will be shown that any $n \times n$ matrix polynomial is equivalent to a diagonal matrix polynomial

$$A_0(\lambda) = \operatorname{diag}[a_1(\lambda), a_2(\lambda), \ldots, a_n(\lambda)], \tag{1}$$

in which $a_j(\lambda)$ is zero or a monic polynomial, $j = 1, 2, \ldots, n$, and $a_j(\lambda)$ is divisible by $a_{j-1}(\lambda)$, $j = 2, 3, \ldots, n$. A matrix polynomial with these properties is called a *canonical* matrix polynomial. Note that if there are zeros among the polynomials $a_j(\lambda)$ then they must occupy the last places, since nonzero polynomials are not divisible by the zero polynomial. Furthermore, if some of the polynomials $a_j(\lambda)$ are (nonzero) scalars, then they must be equal to 1 and be placed in the first positions of the canonical matrix, since 1 is divisible only by 1. Thus, the canonical matrix is generally of the form

$$A_0(\lambda) = \operatorname{diag}[1, \ldots, 1, a_1(\lambda), \ldots, a_k(\lambda), 0, \ldots. 0], \tag{2}$$

where $a_j(\lambda)$ is a monic polynomial of degree at least 1 and is divisible (for $j = 2, \ldots, k$) by $a_{j-1}(\lambda)$. It is also possible that the diagonal of the matrix in (2) contains no zeros or ones.

Theorem 1. *Any $n \times n$ matrix polynomial over $\mathcal{F}$ is equivalent to a canonical matrix polynomial.*

PROOF. We may assume that $A(\lambda) \not\equiv 0$, for otherwise there is nothing to prove. The proof is merely a description of a sequence of elementary transformations needed to reduce successively the rows and columns of $A(\lambda)$ to the required form.

Step 1. Let $a_{ij}(\lambda)$ be a nonzero element of $A(\lambda)$ of least degree; by interchanging rows and columns (elementary operations of type 2) bring this element to the 1, 1 position and call it $a_{11}(\lambda)$. For each element of the first row and column of the resulting matrix, we find the quotient and remainder on division by $a_{11}(\lambda)$:

$$\begin{aligned} a_{1j}(\lambda) &= a_{11}(\lambda)q_{1j}(\lambda) + r_{1j}(\lambda), \qquad j = 2, 3, \ldots, n,\\ a_{i1}(\lambda) &= a_{11}(\lambda)q_{i1}(\lambda) + r_{i1}(\lambda), \qquad i = 2, 3, \ldots, n. \end{aligned}$$

Now, for each i and j, subtract $q_{1j}(\lambda)$ times the first column from the jth column and subtract $q_{i1}(\lambda)$ times the first row from the ith row (elementary operations of type 3). Then the elements $a_{1j}(\lambda)$, $a_{i1}(\lambda)$ are replaced by $r_{1j}(\lambda)$ and $r_{i1}(\lambda)$, respectively ($i, j = 2, 3, \ldots, n$), all of which are either the zero polynomial or have degree less than that of $a_{11}(\lambda)$. If the polynomials are not all zero, we use an operation of type 2 to interchange $a_{11}(\lambda)$ with an element $r_{1j}(\lambda)$ or $r_{i1}(\lambda)$ of least degree. Now repeat the process of reducing the degree of the off-diagonal elements of the first row and column to be less than that of the new $a_{11}(\lambda)$. Clearly, since the degree of $a_{11}(\lambda)$ is strictly decreasing at each step, we eventually reduce the λ-matrix to the form

$$\begin{bmatrix} a_{11}(\lambda) & 0 & \cdots & 0 \\ 0 & a_{22}(\lambda) & \cdots & a_{2n}(\lambda) \\ \vdots & \vdots & \ddots & \vdots \\ 0 & a_{n2}(\lambda) & \cdots & a_{nn}(\lambda) \end{bmatrix}. \tag{3}$$

Step 2. In the form (3) there may now be nonzero elements $a_{ij}(\lambda)$, $2 \le i$, $j \le n$, whose degree is less than that of $a_{11}(\lambda)$. If so, we repeat Step 1 again and arrive at another matrix of the form (3) but with the degree of $a_{11}(\lambda)$ further reduced. Thus, by repeating Step 1 a sufficient number of times, we can find a matrix of the form (3) that is equivalent to $A(\lambda)$ and for which $a_{11}(\lambda)$ is a nonzero element of least degree.

Step 3. Having completed Step 2, we now ask whether there are nonzero elements that are not divisible by $a_{11}(\lambda)$. If there is one such, say $a_{ij}(\lambda)$, we add column j to column 1, find remainders and quotients of the new column 1 on division by $a_{11}(\lambda)$, and go on to repeat Steps 1 and 2, winding up with a form (3), again with $a_{11}(\lambda)$ replaced by a polynomial of smaller degree.

Again, this process can continue only for a finite number of steps before we arrive at a matrix of the form

$$\begin{bmatrix} a_1(\lambda) & 0 & \cdots & 0 \\ 0 & b_{22}(\lambda) & \cdots & b_{2n}(\lambda) \\ \vdots & \vdots & \ddots & \vdots \\ 0 & b_{n2}(\lambda) & \cdots & b_{nn}(\lambda) \end{bmatrix},$$

where, after an elementary operation of type 1 (if necessary), $a_1(\lambda)$ is monic and all the nonzero elements $b_{ij}(\lambda)$ are divisible by $a_1(\lambda)$ without remainder.

Step 4. If all $b_{ij}(\lambda)$ are zero, the theorem is proved. If not, the above matrix may be reduced to the form

$$\begin{bmatrix} a_1(\lambda) & 0 & 0 & \cdots & 0 \\ 0 & a_2(\lambda) & 0 & \cdots & 0 \\ 0 & 0 & c_{33}(\lambda) & \cdots & c_{3n}(\lambda) \\ \vdots & \vdots & \vdots & \ddots & \vdots \\ 0 & 0 & c_{n3}(\lambda) & \cdots & c_{nn}(\lambda) \end{bmatrix}$$

where $a_2(\lambda)$ is divisible by $a_1(\lambda)$ and the elements $c_{ij}(\lambda)$, $3 \le i, j \le n$, are divisible by $a_2(\lambda)$. Continuing the process we arrive at the statement of the theorem. ■

Note that a matrix polynomial is regular if and only if all the elements $a_1(\lambda), \ldots, a_n(\lambda)$ in Eq. (1) are nonzero polynomials.

Exercise 1. Prove that, by using left (right) elementary operations only, a matrix polynomial can be reduced to an upper- (lower-) triangular matrix polynomial $B(\lambda)$ with the property that if the degree of $b_{jj}(\lambda)$ is l_j $(j = 1, 2, \ldots, n)$, then

(a) $l_j = 0$ implies that $b_{kj} = 0$ $(b_{jk} = 0)$, $k = 1, 2, \ldots, j - 1$, and
(b) $l_j > 0$ implies that the degree of nonzero elements $b_{kj}(\lambda)$ $(b_{jk}(\lambda))$ is less than l_j, $k = 1, 2, \ldots, j - 1$. ☐

The reduction of matrix polynomials described above takes on a simple form in the important special case in which $A(\lambda)$ does not depend explicitly on λ at all, that is, when $A(\lambda) \equiv A \in \mathscr{F}^{n \times n}$. This observation can be used to show that Theorem 1 implies Theorem 2.7.2 (for square matrices).

The uniqueness of the polynomials $a_1(\lambda), \ldots, a_n(\lambda)$ appearing in Eq. (1) is shown in the next section.

7.5 Invariant Polynomials and the Smith Canonical Form

We start this section by constructing a system of polynomials that is uniquely defined by a given matrix polynomial and that is invariant under equivalence transformations of the matrix. For this purpose, we first define the *rank* of a matrix polynomial to be the order of its largest minor that is not equal to the zero polynomial. For example, the rank of $A(\lambda)$ defined in Exercise 7.3.5 is equal to 1, while the rank of the canonical matrix polynomial $A_0(\lambda) = \operatorname{diag}[a_1(\lambda), \ldots, a_r(\lambda), 0, \ldots, 0]$ is equal to r. Note also that if $A(\lambda)$ is an $n \times n$ matrix polynomial, then it is regular if and only if its rank is n.

Proposition 1. *The rank of a matrix polynomial is invariant under equivalence transformations.*

PROOF. Let the matrix polynomials $A(\lambda)$ and $B(\lambda)$ be equivalent. Then by Proposition 7.3.1 there are unimodular matrices $P(\lambda)$ and $Q(\lambda)$ such that $B(\lambda) = P(\lambda)A(\lambda)Q(\lambda)$. Apply the Binet–Cauchy formula twice to this equation to express a minor $b(\lambda)$ of order j of $B(\lambda)$ in terms of minors $a_s(\lambda)$ of $A(\lambda)$ of the same order as follows (after a reordering):

$$b(\lambda) = \sum_s p_s(\lambda)a_s(\lambda)q_s(\lambda), \tag{1}$$

where $p_s(\lambda)$ and $q_s(\lambda)$ denote the appropriate minors of order j of the matrix polynomials $P(\lambda)$ and $Q(\lambda)$, respectively. If now $b(\lambda)$ is a nonzero minor of $B(\lambda)$ of the greatest order r (that is, the rank of $B(\lambda)$ is r), then it follows from Eq. (1) that at least one minor $a_s(\lambda)$ (of order r) is a nonzero polynomial and hence the rank of $B(\lambda)$ does not exceed the rank of $A(\lambda)$. However, applying the same argument to the equation $A(\lambda) = [P(\lambda)]^{-1}B(\lambda)[Q(\lambda)]^{-1}$, we see that rank $A(\lambda) \leq$ rank $B(\lambda)$. Thus, the ranks of equivalent matrix polynomials coincide. ■

Suppose that the $n \times n$ matrix polynomial $A(\lambda)$ over $\mathcal{F}$ has rank r and let $d_j(\lambda)$ be the greatest common divisor of all minors of $A(\lambda)$ of order j, where $j = 1, 2, \ldots, r$. Clearly, any minor of order $j \geq 2$ may be expressed as a linear combination of minors of order $j - 1$, so that $d_{j-1}(\lambda)$ is necessarily a factor $d_j(\lambda)$. Hence, if we define $d_0(\lambda) \equiv 1$, then in the sequence

$$d_0(\lambda), d_1(\lambda), \ldots, d_r(\lambda),$$

$d_j(\lambda)$ is divisible by $d_{j-1}(\lambda)$, $j = 1, 2, \ldots, r$.

Exercise 1. Show that the polynomials $d_j(\lambda)$ defined in the previous paragraph for the canonical matrix polynomial $A_0(\lambda) = \operatorname{diag}[a_1(\lambda), a_2(\lambda), \ldots, a_r(\lambda), 0, \ldots, 0]$ are, respectively,

$$d_1(\lambda) = a_1(\lambda), \quad d_2(\lambda) = a_1(\lambda)a_2(\lambda), \quad \ldots, \quad d_r(\lambda) = \prod_{j=1}^{r} a_j(\lambda). \quad \square \qquad (2)$$

The polynomials $d_0(\lambda), d_1(\lambda), \ldots, d_r(\lambda)$ have the important property of invariance under equivalence transformations. To see this, let $d_j(\lambda)$ and $\delta_j(\lambda)$ denote the (monic) greatest common divisor of all minors of order j of the matrix polynomials $A(\lambda)$ and $B(\lambda)$, respectively. Note that by Proposition 1 the number of polynomials $d_j(\lambda)$ and $\delta_j(\lambda)$ is the same provided that $A(\lambda)$ and $B(\lambda)$ are equivalent.

Proposition 2. *Let the matrix polynomials $A(\lambda)$ and $B(\lambda)$ of rank r be equivalent. Then, with the notation of the previous paragraph, the polynomials $d_j(\lambda)$ and $\delta_j(\lambda)$ coincide ($j = 1, 2, \ldots, r$).*

PROOF. Preserving the notation used in the proof of Proposition 1, it is easily seen from Eq. (1) that any common divisor of minors $a_s(\lambda)$ of $A(\lambda)$ of order j ($1 \leq j \leq r$) is a divisor of $b(\lambda)$. Hence $\delta_j(\lambda)$ is divisible by $d_j(\lambda)$. But again, the equation $A(\lambda) = [P(\lambda)]^{-1}B(\lambda)[Q(\lambda)]^{-1}$ implies that $d_j(\lambda)$ is divisible by $\delta_j(\lambda)$ and, since both polynomials are assumed to be monic, we obtain

$$\delta_j(\lambda) = d_j(\lambda), \qquad j = 1, 2, \ldots, r. \quad ■$$

Now consider the quotients

$$i_1(\lambda) = \frac{d_1(\lambda)}{d_0(\lambda)}, \qquad i_2(\lambda) = \frac{d_2(\lambda)}{d_1(\lambda)}, \ldots, \qquad i_r = \frac{d_r(\lambda)}{d_{r-1}(\lambda)}.$$

In view of the divisibility of $d_j(\lambda)$ by $d_{j-1}(\lambda)$, the quotients $i_j(\lambda)$ $(j = 1, 2, \ldots, r)$ are polynomials. They are called the *invariant polynomials* of $A(\lambda)$; the invariance refers to equivalence transformations. Note that for $j = 1, 2, \ldots, r$,

$$d_j(\lambda) = i_1(\lambda)i_2(\lambda)\cdots i_j(\lambda), \tag{3}$$

and for $j = 2, 3, \ldots, r$, $i_j(\lambda)$ is divisible by $i_{j-1}(\lambda)$.

***Exercise* 2.** Find (from their definition) the invariant polynomials of the matrix polynomials

$$\text{(a)} \begin{bmatrix} 0 & 1 & \lambda \\ \lambda & \lambda & 1 \\ \lambda^2 - \lambda & \lambda^2 - 1 & \lambda^2 - 1 \end{bmatrix}; \qquad \text{(b)} \begin{bmatrix} \lambda & \lambda^2 & 0 \\ \lambda^3 & \lambda^5 & 0 \\ 0 & 0 & 2\lambda \end{bmatrix}.$$

Answer. (a) $i_1(\lambda) = 1, i_2(\lambda) = 1$; (b) $i_1(\lambda) = \lambda, i_2(\lambda) = \lambda, i_3(\lambda) = \lambda^5 - \lambda^4$. □

Recalling the result of Exercise 1 and observing that the invariant polynomials of the canonical matrix polynomial are merely $a_j(\lambda)$, we now use Proposition 2 to obtain the main result of the section.

Theorem 1. *A matrix polynomial $A(\lambda)$ of rank r is equivalent to the canonical matrix polynomial* $\text{diag}[i_1(\lambda), \ldots, i_r(\lambda), 0, \ldots, 0]$, *where $i_1(\lambda), i_2(\lambda), \ldots, i_r(\lambda)$ are the invariant polynomials of $A(\lambda)$.*

The canonical form

$$\text{diag}[i_1(\lambda), \ldots, i_r(\lambda), 0, \ldots, 0]$$

is known as the *Smith canonical form.* H. J. S. Smith† obtained the form for matrices of integers (which have much the same algebraic structure as matrix polynomials) in 1861. Frobenius‡ obtained the result for matrix polynomials in 1878. The matrix polynomial $A(\lambda)$ and its canonical form are, of course, over the same field $\mathscr{F}$.

Corollary 1. *Two matrix polynomials are equivalent if and only if they have the same invariant polynomials.*

PROOF. The "only if" statement is just Proposition 2. If two matrix polynomials have the same invariant polynomials, then Theorem 1 implies that

† *Philos. Trans. Roy. Soc. London* **151** (1861), 293–326.
‡ *Jour. Reine Angew. Math.* (*Crelle*) **86** (1878), 146–208.

they have the same Smith canonical form. The transitive property of equivalence relations (Exercise 7.3.2) then implies that they are equivalent. ■

Exercise 3. Find the Smith canonical form of the matrices defined in Exercise 2.

Answer.

$$\text{(a)} \quad \begin{bmatrix} 1 & 0 & 0 \\ 0 & 1 & 0 \\ 0 & 0 & 0 \end{bmatrix} \qquad \text{(b)} \quad \begin{bmatrix} \lambda & 0 & 0 \\ 0 & \lambda & 0 \\ 0 & 0 & \lambda^5 - \lambda^4 \end{bmatrix}. \quad \square$$

Exercise 3 illustrates the fact that the Smith canonical form of a matrix polynomial can be found either by performing elementary operations as described in the proof of Theorem 7.4.1 or by finding its invariant polynomials.

Exercise 4. Find the invariant polynomials of $\lambda I - C_a$ where, as in Exercise 2.2.9, C_a denotes the $l \times l$ companion matrix associated with the polynomial $a(\lambda) = \lambda^l + \sum_{i=0}^{l-1} a_i \lambda^i$.

SOLUTION. Observe that there are nonzero minors of orders $1, 2, \ldots, l-1$ of $\lambda I - C_a$ that are independent of λ and therefore, in the previous notation, we have $d_1 = d_2 = \cdots = d_{l-1} = 1$. Since, by definition $d_l(\lambda) = \det(\lambda I - C_a)$, then in view of Exercise 4.11.3, $d_l(\lambda) = a(\lambda)$. Hence $i_1 = \cdots = i_{l-1} = 1$, $i_l(\lambda) = a(\lambda)$.

Exercise 5. Show that if $A \in \mathscr{F}^{n \times n}$, then the sum of the degrees of the invariant polynomials of $\lambda I - A$ is n.

SOLUTION. Since $\lambda I - A$ is a matrix polynomial of rank n, there are (see Theorem 1) unimodular matrix polynomials $P(\lambda)$ and $Q(\lambda)$ such that

$$P(\lambda)(\lambda I - A)Q(\lambda) = \operatorname{diag}[i_1(\lambda), i_2(\lambda), \ldots, i_n(\lambda)].$$

On taking determinants, the left-hand side becomes a scalar multiple of the characteristic polynomial and hence has degree n. The result now follows immediately. □

7.6 Similarity and the First Normal Form

This section is devoted to the development of the connection between equivalence and similarity transformations. The first theorem will establish an important link between the theory of matrix polynomials, as discussed in this chapter, and the notion of similarity of matrices from $\mathscr{F}^{n \times n}$.

Theorem 1. *Matrices $A, B \in \mathscr{F}^{n \times n}$ are similar if and only if the matrix polynomials $\lambda I - A$ and $\lambda I - B$ are equivalent.*

Thus, using Theorem 7.5.1, A and B are similar if and only if $\lambda I - A$ and $\lambda I - B$ have the same invariant polynomials.

PROOF. Suppose first that A and B are similar. Then there is a nonsingular $S \in \mathscr{F}^{n \times n}$ such that $A = SBS^{-1}$; whence

$$\lambda I - A = S(\lambda I - B)S^{-1},$$

which implies that $\lambda I - A$ and $\lambda I - B$ are equivalent. The corollary to Theorem 7.5.1 now implies that $\lambda I - A$ and $\lambda I - B$ have the same invariant polynomials.

Conversely, suppose that $\lambda I - A$ and $\lambda I - B$ have the same invariant polynomials. Then they are equivalent and there exist unimodular matrix polynomials $P(\lambda)$ and $Q(\lambda)$ such that

$$P(\lambda)(\lambda I - A)Q(\lambda) = \lambda I - B.$$

Thus, defining $M(\lambda) = [P(\lambda)]^{-1}$, we observe that $M(\lambda)$ is a matrix polynomial and

$$M(\lambda)(\lambda I - B) = (\lambda I - A)Q(\lambda).$$

Now division of $M(\lambda)$ on the left by $\lambda I - A$ and of $Q(\lambda)$ on the right by $\lambda I - B$ yields

$$M(\lambda) = (\lambda I - A)S(\lambda) + M_0, \tag{1}$$
$$Q(\lambda) = R(\lambda)(\lambda I - B) + Q_0,$$

where M_0, Q_0 are independent of λ. Then, substituting in the previous equation, we obtain

$$((\lambda I - A)S(\lambda) + M_0)(\lambda I - B) = (\lambda I - A)(R(\lambda)(\lambda I - B) + Q_0),$$

whence

$$(\lambda I - A)\{S(\lambda) - R(\lambda)\}(\lambda I - B) = (\lambda I + A)Q_0 - M_0(\lambda I - B).$$

But this is an identity between matrix polynomials, and since the degree of the polynomial on the right is 1, it follows that $S(\lambda) \equiv R(\lambda)$. (Otherwise the degree of the matrix polynomial on the left is at least 2.) Hence,

$$M_0(\lambda I - B) = (\lambda I - A)Q_0, \tag{2}$$

so that

$$M_0 = Q_0, \qquad M_0 B = AQ_0, \qquad \text{and} \qquad M_0 B = AM_0. \tag{3}$$

Now we have only to prove that M_0 is nonsingular and we are finished.

Suppose that division of $P(\lambda)$ on the left by $\lambda I - B$ yields

$$P(\lambda) = (\lambda I - B)U(\lambda) + P_0,$$

where P_0 is independent of λ. Then using Eqs. (1) and (2) we have

$$\begin{aligned} I &= M(\lambda)P(\lambda) \\ &= ((\lambda I - A)S(\lambda) + M_0)((\lambda I - B)U(\lambda) + P_0) \\ &= (\lambda I - A)S(\lambda)(\lambda I - B)U(\lambda) + (\lambda I - A)Q_0 U(\lambda) \\ &\quad + (\lambda I - A)S(\lambda)P_0 + M_0 P_0 \\ &= (\lambda I - A)\{Q(\lambda)U(\lambda) + S(\lambda)P_0\} + M_0 P_0. \end{aligned}$$

Hence $Q(\lambda)U(\lambda) + S(\lambda)P_0 = 0$ and $M_0 P_0 = I$. Thus, $\det M_0 \neq 0$ and the last equation in (3) implies that A and B are similar. ■

Now we are in a position to prove a theorem concerning a simple normal form to which a matrix from $\mathscr{F}^{n \times n}$ can be reduced by similarity transformations. It is important to observe that the normal form obtained is also in $\mathscr{F}^{n \times n}$ and that $\mathscr{F}$ is an *arbitrary* field.

Theorem 2. *If $A \in \mathscr{F}^{n \times n}$ and the invariant polynomials of $\lambda I - A$ of nonzero degree are $i_s(\lambda), i_{s+1}(\lambda), \ldots, i_n(\lambda)$, then the matrix A is similar to the block-diagonal matrix*

$$C_1 = \operatorname{diag}[C_{i_s}, C_{i_{s+1}}, \ldots, C_{i_n}], \tag{4}$$

where C_{i_k} denotes the companion matrix associated with the (invariant) polynomial $i_k(\lambda)$, $(s \leq k \leq n)$.

The matrix C_1 in Eq. (4) is often referred to as the *first natural normal form* of A.

PROOF. First we note that by Exercise 7.5.5 the matrix C_1 in Eq. (4) is $n \times n$, and since the invariant polynomials are over $\mathscr{F}$, we have $C_1 \in \mathscr{F}^{n \times n}$. We now show that $\lambda I - A$ and $\lambda I - C_1$ have the same invariant polynomials and hence the result will follow from Theorem 1.

Indeed, by Theorem 7.5.1 and Exercise 7.5.4, the matrix polynomial $\lambda I - C_{i_k}$ $(s \leq k \leq n)$ is equivalent to the diagonal matrix $D_k(\lambda) = \operatorname{diag}[1, 1, \ldots, 1, i_k(\lambda)]$. Then it follows that $\lambda I - C_1$ is equivalent to the matrix $D(\lambda) = \operatorname{diag}[D_s(\lambda), D_{s+1}(\lambda), \ldots, D_n(\lambda)]$ and thus

$$\lambda I - C_1 = P(\lambda)D(\lambda)Q(\lambda) \tag{5}$$

for some unimodular matrices $P(\lambda), Q(\lambda)$.

Now define

$$\Delta(\lambda) = \operatorname{diag}[1, \ldots, 1, i_s(\lambda), \ldots, i_n(\lambda)]$$

and observe that $\Delta(\lambda)$ is the Smith canonical form of $\lambda I - A$. Since by elementary operations of the second kind the matrix polynomial $D(\lambda)$ can be transformed into $\Delta(\lambda)$, they are equivalent. The relation (5) then implies the equivalence of $\lambda I - C_1$ and $\Delta(\lambda)$; hence $1, \ldots, 1, i_s(\lambda), \ldots, i_n(\lambda)$ are the invariant polynomials of both $\lambda I - C_1$ and $\lambda I - A$. ■

Exercise 1. Check that the first natural normal forms of the matrices

$$A_1 = \begin{bmatrix} 3 & -1 & 0 \\ -1 & 3 & 0 \\ 1 & -1 & 2 \end{bmatrix}, \qquad A_2 = \begin{bmatrix} 6 & 2 & 2 \\ -2 & 2 & 0 \\ 0 & 0 & 2 \end{bmatrix}$$

are, respectively,

$$C_1^{(1)} = \left[\begin{array}{c|cc} 2 & 0 & 0 \\ \hline 0 & 0 & 1 \\ 0 & -8 & 6 \end{array}\right], \qquad C_1^{(2)} = \begin{bmatrix} 0 & 1 & 0 \\ 0 & 0 & 1 \\ 32 & -32 & 10 \end{bmatrix}. \quad \square$$

7.7 Elementary Divisors

The problem of reducing the companion matrices of invariant polynomials appearing in the first natural normal form to matrices of a smaller order depends on the field $\mathcal{F}$ over which these polynomials are defined. Indeed, we obtain the further reduction by considering the factorization of each invariant polynomial into irreducible polynomials over $\mathcal{F}$. For simplicity, and because this is the most important case in applications, we are going to confine our attention to matrices (and hence invariant polynomials) defined over the field of complex numbers, $\mathbb{C}$ (but see also Section 6.7).

We consider the problem of an $n \times n$ matrix polynomial $A(\lambda)$, with elements defined over $\mathbb{C}$, having rank r, $(r \leq n)$ and invariant polynomials $i_1(\lambda), i_2(\lambda), \ldots, i_r(\lambda)$. We define the zeros of $\det A(\lambda)$ to be the *latent roots* of $A(\lambda)$, and since $\det A(\lambda)$ is a polynomial over $\mathbb{C}$ we may write

$$\det A(\lambda) = k_1 \prod_{j=1}^{s} (\lambda - \lambda_j)^{m_j},$$

where $k_1 \neq 0$; $\lambda_1, \lambda_2, \ldots, \lambda_s$ are the distinct latent roots of $A(\lambda)$; and $m_j \geq 1$ for each j. From the Smith canonical form we deduce that

$$\det A(\lambda) = k_2 \prod_{j=1}^{r} i_j(\lambda),$$

where $k_2 \neq 0$. Since the invariant polynomials are monic, it follows that $k_1 = k_2$ and

$$\prod_{j=1}^{r} i_j(\lambda) = \prod_{k=1}^{s} (\lambda - \lambda_k)^{m_k}.$$

Moreover, since $i_j(\lambda)$ is a divisor of $i_{j+1}(\lambda)$ for $j = 1, 2, \ldots, r - 1$, it follows that there are integers α_{jk}, $1 \le j \le r$ and $1 \le k \le s$, such that

$$\begin{aligned} i_1(\lambda) &= (\lambda - \lambda_1)^{\alpha_{11}}(\lambda - \lambda_2)^{\alpha_{12}} \cdots (\lambda - \lambda_s)^{\alpha_{1s}}, \\ i_2(\lambda) &= (\lambda - \lambda_1)^{\alpha_{21}}(\lambda - \lambda_2)^{\alpha_{22}} \cdots (\lambda - \lambda_s)^{\alpha_{2s}}, \\ &\vdots \\ i_r(\lambda) &= (\lambda - \lambda_1)^{\alpha_{r1}}(\lambda - \lambda_2)^{\alpha_{r2}} \cdots (\lambda - \lambda_s)^{\alpha_{rs}}, \end{aligned} \tag{1}$$

and for $k = 1, 2, \ldots, s$,

$$0 \le \alpha_{1k} \le \alpha_{2k} \le \cdots \le \alpha_{rk} \le m_k, \qquad \sum_{j=1}^{r} \alpha_{jk} = m_k.$$

Each factor $(\lambda - \lambda_k)^{\alpha_{jk}}$ appearing in the factorizations (1) with $\alpha_{jk} > 0$ is called an *elementary divisor* of $A(\lambda)$. An elementary divisor for which $\alpha_{jk} = 1$ is said to be *linear*; otherwise it is *nonlinear*. We may also refer to the elementary divisors $(\lambda - \lambda_k)^{\alpha_{jk}}$ as those associated with λ_k, with the obvious meaning. Note that (over $\mathbb{C}$) all elementary divisors are powers of a linear polynomial.

Exercise 1. Find the elementary divisors of the matrix polynomial defined in Exercise 7.5.2(b).

Answer. $\lambda, \lambda, \lambda^4, \lambda - 1$. □

Note that an elementary divisor of a matrix polynomial may appear several times in the set of invariant polynomials of the matrix polynomial. Also, the factorizations (1) show that the system of all elementary divisors (along with the rank and order) of a matrix polynomial completely defines the set of its invariant polynomials and vice versa.

It therefore follows from Corollary 1 of Theorem 7.5.1 that the elementary divisors are invariant under equivalence transformations of the matrix polynomial.

Theorem 1. *Two complex matrix polynomials are equivalent if and only if they have the same elementary divisors.*

Exercise 2. Determine the invariant polynomials of a matrix polynomial of order 5 having rank 4 and elementary divisors $\lambda, \lambda, \lambda^3, \lambda - 2, (\lambda - 2)^2, \lambda + 5$.

SOLUTION. Since the order of the matrix polynomial is 5, there are five invariant polynomials $i_k(\lambda)$, $k = 1, 2, \ldots, 5$. And since the rank is 4, $i_5(\lambda) = d_5(\lambda)/d_4(\lambda) \equiv 0$. Now recall that the invariant polynomial $i_j(\lambda)$ is divisible by $i_{j-1}(\lambda)$ for $j \geq 2$. Hence, $i_4(\lambda)$ must contain elementary divisors associated with all eigenvalues of the matrix polynomial and those elementary divisors must be of the highest degrees. Thus, $i_4(\lambda) = \lambda^3(\lambda - 2)^2(\lambda + 5)$. Applying the same argument to $i_3(\lambda)$ and the remaining elementary divisors, we obtain $i_3(\lambda) = \lambda(\lambda - 2)$ and then $i_2(\lambda) = \lambda$. All elementary divisors are now exhausted and hence $i_1(\lambda) \equiv 1$. □

The possibility of describing an equivalence class of matrix polynomials in terms of the set of elementary divisors permits a reformulation of the statement of Theorem 7.6.1.

Proposition 1. *Two $n \times n$ matrices A and B with elements from $\mathbb{C}$ are similar if and only if $\lambda I - A$ and $\lambda I - B$ have the same elementary divisors.*

Before proceeding to the general reduction of matrices in $\mathbb{C}^{n \times n}$, there is one more detail to be cleared up.

Theorem 2. *If $A(\lambda)$, $B(\lambda)$ are matrix polynomials over $\mathbb{C}$, then the set of elementary divisors of the block-diagonal matrix*

$$C(\lambda) = \begin{bmatrix} A(\lambda) & 0 \\ 0 & B(\lambda) \end{bmatrix}$$

is the union of the sets of elementary divisors of $A(\lambda)$ and $B(\lambda)$.

PROOF. Let $D_1(\lambda)$ and $D_2(\lambda)$ be the Smith forms of $A(\lambda)$ and $B(\lambda)$, respectively. Then clearly

$$C(\lambda) = E(\lambda)\begin{bmatrix} D_1(\lambda) & 0 \\ 0 & D_2(\lambda) \end{bmatrix}F(\lambda)$$

for some matrix polynomials $E(\lambda)$ and $F(\lambda)$ with constant nonzero determinants. Let $(\lambda - \lambda_0)^{\alpha_1}, \ldots, (\lambda - \lambda_0)^{\alpha_p}$ and $(\lambda - \lambda_0)^{\beta_1}, \ldots, (\lambda - \lambda_0)^{\beta_q}$ be the elementary divisors of $D_1(\lambda)$ and $D_2(\lambda)$, respectively, corresponding to the same complex root λ_0. Arrange the set of exponents $\alpha_1, \ldots, \alpha_p, \beta_1, \ldots, \beta_q$, in a nondecreasing order: $\{\alpha_1, \ldots, \alpha_p, \beta_1, \ldots, \beta_q\} = \{\gamma_1, \ldots, \gamma_{p+q}\}$, where $0 < \gamma_1 \leq \cdots \leq \gamma_{p+q}$. From the definition of invariant polynomials (Section 7.5), it is clear that in the Smith form $D = \operatorname{diag}[i_1(\lambda), \ldots, i_r(\lambda), 0, \ldots, 0]$ of $\operatorname{diag}[D_1(\lambda), D_2(\lambda)]$, the invariant polynomial $i_r(\lambda)$ is divisible by $(\lambda - \lambda_0)^{\gamma_{p+q}}$ but not by $(\lambda - \lambda_0)^{\gamma_{p+q}+1}$; and $i_{r-1}(\lambda)$ is divisible by $(\lambda - \lambda_0)^{\gamma_{p+q-1}}$ but not by $(\lambda - \lambda_0)^{\gamma_{p+q-1}+1}$; and so on. It follows that the elementary divisors of

$$\begin{bmatrix} D_1(\lambda) & 0 \\ 0 & D_2(\lambda) \end{bmatrix}$$

(and therefore also those of $C(\lambda)$) corresponding to λ_0) are just $(\lambda - \lambda_0)^{\gamma_1}, \ldots, (\lambda - \lambda_0)^{\gamma_{p+q}}$, and the theorem is proved. ■

Exercise 3. Determine the elementary divisors of the matrix

$$\begin{bmatrix} \lambda - 1 & 0 & 0 \\ 0 & \lambda^2 + 1 & \lambda^2 + 1 \\ 0 & -1 & \lambda^2 - 1 \end{bmatrix}$$

(a) over $\mathbb{C}$; (b) over $\mathbb{R}$.

Answers. (a) $\lambda - 1, \lambda^2, \lambda - i, \lambda + i$; (b) $\lambda - 1, \lambda^2, \lambda^2 + 1$.

Exercise 4. Let J be the Jordan block of order p associated with λ_0:

$$J = \begin{bmatrix} \lambda_0 & 1 & 0 & \cdots & 0 \\ 0 & \lambda_0 & \ddots & \ddots & \vdots \\ \vdots & \ddots & \ddots & \ddots & 0 \\ \vdots & & \ddots & \ddots & 1 \\ 0 & \cdots & & 0 & \lambda_0 \end{bmatrix}.$$

Show that the matrix $\lambda I - J$ has a unique elementary divisor $(\lambda - \lambda_0)^p$.

Exercise 5. Check that unimodular matrices, and only they, have no elementary divisors.

Exercise 6. Let $D \in \mathbb{C}^{n \times n}$ be a diagonal matrix. Check that the elementary divisors of $\lambda I - D$ are linear.

Exercise 7. Verify that $(\lambda - \alpha_0)^r$ is the unique elementary divisor of the matrix polynomial $\lambda I - A$, where

$$A = \begin{bmatrix} \alpha_0 & \alpha_1 & \alpha_2 & \cdots & \alpha_{r-1} \\ 0 & \alpha_0 & \alpha_1 & \ddots & \vdots \\ \vdots & \ddots & \ddots & \ddots & \alpha_2 \\ \vdots & & \ddots & \ddots & \alpha_1 \\ 0 & \cdots & & 0 & \alpha_0 \end{bmatrix} \tag{2}$$

$(A \in \mathbb{C}^{r \times r})$ and $\alpha_1 \neq 0$.

Hint. Observe that the minor of $\lambda I - A$ of order $r - 1$ obtained by striking out the first column and the last row is a nonzero number and, therefore, in the notation of Section 7.5, $d_1 = d_2 = \cdots = d_{r-1} = 1$ and $d_r(\lambda) = (\lambda - \alpha_0)^r$.

Exercise 8. Let $A \in \mathbb{C}^{r \times r}$ be defined by Eq. (2) and let $\alpha_1 = \alpha_2 = \cdots = \alpha_{l-1} = 0$, $\alpha_l \neq 0$ $(1 \leq l \leq r - 1)$. Verify that

(a) $$l_s \triangleq \dim(\operatorname{Ker}(\alpha_0 I - A)^s) = \begin{cases} ls & \text{if } ls \leq r \\ r & \text{if } ls > r \end{cases}$$

or, equivalently,

(b) $$l_1 = l, \quad l_2 = 2l, \quad \ldots, \quad l_p = pl, \quad l_{p+i} = r, \quad i \geq 1,$$

where $p = [r/l]$, the integer part of r/l.

(c) The elementary divisors of $\lambda I - A$ are

$$\underbrace{(\lambda - \alpha_0)^p, \ldots, (\lambda - \alpha_0)^p,}_{l - q \text{ times}} \qquad \underbrace{(\lambda - \alpha_0)^{p+1}, \ldots, (\lambda - \alpha_0)^{p+1},}_{q \text{ times}}$$

where the integers p and q satisfy the condition $r = pl + q$ $(0 \leq q \leq l - 1, p > 0)$.

Hint. For part (c) use part (b) and Exercise 6.4.2 to obtain $k_1 = \cdots = k_{p-1} = 0$, $k_p = l - q$, $k_{p+1} = q$ (in the notation of Exercise 6.4.2). Also, observe that in view of Exercise 4, the number k_m gives the number of elementary divisors $(\lambda - \lambda_0)^m$ of A.

Exercise 9. Let λ be an eigenvalue of $A \in \mathbb{C}^{n \times n}$. Show that the index of λ is equal to the maximal degree of the elementary divisors of $\lambda I - A$ associated with λ. □

7.8 The Second Normal Form and the Jordan Normal Form

Let $A \in \mathbb{C}^{n \times n}$ and let $l_1(\lambda), l_2(\lambda), \ldots, l_p(\lambda)$ denote the elementary divisors of the matrix $\lambda I - A$.

Theorem 1 (The *second* natural normal form). *With the notation of the previous paragraph, the matrix A is similar to the block-diagonal matrix*

$$C_2 = \text{diag}[C_{l_1}, C_{l_2}, \ldots, C_{l_p}], \tag{1}$$

where C_{l_k} $(1 \leq k \leq p)$ denotes the companion matrix associated with the polynomial $l_k(\lambda)$.

Proof. By Exercise 7.5.4 the only nonconstant invariant polynomial of the matrix $\lambda I - C_{l_k}$ is $l_k(\lambda)$. Since $l_k(\lambda)$ is an elementary divisor of $\lambda I - A$ over the field $\mathbb{C}$, it is a power of a linear polynomial in λ, and hence the unique elementary divisor of $\lambda I - C_{l_k}$. Applying Theorem 7.7.2, it follows that the matrix $\lambda I - C_2$, in which C_2 is defined in Eq. (1), has the same elementary divisors as $\lambda I - A$. Proposition 7.7.1 now provides the similarity of A and C_2. ■

Exercise 1. Show that the first and second natural normal forms of the matrix

$$\begin{bmatrix} 3 & -1 & -5 & 1 \\ 1 & 1 & -1 & 0 \\ 0 & 0 & -2 & -1 \\ 0 & 0 & 1 & 0 \end{bmatrix}$$

are, respectively, the matrices

$$\begin{bmatrix} 0 & 1 & 0 & 0 \\ 0 & 0 & 1 & 0 \\ 0 & 0 & 0 & 1 \\ -4 & -4 & 3 & 2 \end{bmatrix} \quad \text{and} \quad \begin{bmatrix} 0 & 1 & 0 & 0 \\ -1 & -2 & 0 & 0 \\ 0 & 0 & 0 & 1 \\ 0 & 0 & -4 & 4 \end{bmatrix}. \quad \square$$

Note that in contrast to the first natural normal form, the second normal form is defined only up to the order of companion matrices associated with the elementary divisors. For instance, the matrix

$$\begin{bmatrix} 0 & 1 & 0 & 0 \\ -4 & 4 & 0 & 0 \\ 0 & 0 & 0 & 1 \\ 0 & 0 & -1 & -2 \end{bmatrix}$$

is also a second natural normal form of the matrix in Exercise 1.

In the statement of Theorem 1, each elementary divisor has the form $l_j(\lambda) = (\lambda - \lambda_j)^{\alpha_j}$. Observe also that there may be repetitions in the eigenvalues $\lambda_1, \lambda_2, \ldots, \lambda_p$, and that $\sum_{i=1}^{p} \alpha_i = n$.

Now we apply Theorem 1 to meet one of the declared objectives of this chapter, namely, the reduction of a matrix to the Jordan normal form.

Theorem 2 (Jordan normal form). *If $A \in \mathbb{C}^{n \times n}$ and $\lambda I - A$ has t elementary divisors $(\lambda - \lambda_i)^{p_i}$, $i = 1, 2, \ldots, t$, then A is similar to the matrix*

$$J = \operatorname{diag}[J_1, J_2, \ldots, J_t] \in \mathbb{C}^{n \times n}$$

where J_i is the $p_i \times p_i$ Jordan block corresponding to $(\lambda - \lambda_i)^{p_i}$ for $i = 1, 2, \ldots, t$;

$$J_i = \begin{bmatrix} \lambda_i & 1 & 0 & \cdots & 0 \\ 0 & \lambda_i & \ddots & \ddots & \vdots \\ \vdots & & \ddots & \ddots & 0 \\ & & & \ddots & 1 \\ 0 & \cdots & & 0 & \lambda_i \end{bmatrix}.$$

PROOF. By Exercise 7.7.4, the only elementary divisor of $\lambda I - J_i$ is $(\lambda - \lambda_i)^{p_i}$. The same is true for the matrix $\lambda I - C_{l_i}$, where l_i refers to $l_i(\lambda) = (\lambda - \lambda_i)^{p_i}$ (see the proof of Theorem 1). Thus, the matrices J_i and C_{l_i} are similar (by Proposition 7.7.1) as are the matrices J and C_2 in Eq. (1) (see Theorem 7.7.1). It remains now to apply Theorem 1 to obtain the desired result. ■

***Exercise* 2.** Show that the Jordan normal forms of the matrix discussed in Exercise 1 are

$$\begin{bmatrix} -1 & 1 & 0 & 0 \\ 0 & -1 & 0 & 0 \\ 0 & 0 & 2 & 1 \\ 0 & 0 & 0 & 2 \end{bmatrix} \quad \text{or} \quad \begin{bmatrix} 2 & 1 & 0 & 0 \\ 0 & 2 & 0 & 0 \\ 0 & 0 & -1 & 1 \\ 0 & 0 & 0 & -1 \end{bmatrix}. \quad \square$$

Corollary 1. *A matrix $A \in \mathbb{C}^{n \times n}$ is simple if and only if all the elementary divisors of $\lambda I - A$ are linear.*

PROOF. First, if A is simple then (Theorem 4.8.2′) $A = PDP^{-1}$, where $\det P \neq 0$ and $D = \operatorname{diag}[\lambda_1, \ldots, \lambda_n]$. Hence $\lambda I - A = P(\lambda I - D)P^{-1}$ and the matrices $\lambda I - A$ and $\lambda I - D$ are equivalent. But the elementary divisors of D are linear (Exercise 7.7.6) and, therefore, so are those of $\lambda I - A$.

Conversely, if all the elementary divisors of $\lambda I - A$ are linear, then all the Jordan blocks in the Jordan normal form are 1×1 matrices. Thus, A is similar to a diagonal matrix and therefore is simple. ■

***Exercise* 3.** Let A be a singular matrix. Show that rank A = rank A^2 if and only if the zero eigenvalue of A has only linear elementary divisors. □

7.9 The Characteristic and Minimal Polynomials

In this section we continue our investigation of the relationships between the characteristic and minimal polynomials. This was begun in Chapter 6.

Let $A \in \mathscr{F}^{n \times n}$ and let $B(\lambda) = \operatorname{adj}(\lambda I - A)$. As noted in Section 7.2, $B(\lambda)$ is a monic polynomial over $\mathscr{F}$ of order n and degree $n - 1$. Let $\delta(\lambda)$ denote the (monic) greatest common divisor of the elements of $B(\lambda)$. If $c(\lambda)$ and $m(\lambda)$ denote, respectively, the characteristic and the minimal polynomials of A, then we already know (see Theorem 6.2.1) that $c(\lambda)$ is divisible by $m(\lambda)$. It turns out that the quotient $c(\lambda)/m(\lambda)$ is simply the polynomial $\delta(\lambda)$ just defined.

Theorem 1. *With the notation of the previous paragraph,*

$$c(\lambda) = \delta(\lambda)m(\lambda). \tag{1}$$

PROOF. First, we introduce the matrix polynomial

$$C(\lambda) = \frac{1}{\delta(\lambda)} B(\lambda),$$

known as the *reduced adjoint* of A, and observe that the elements of $C(\lambda)$ are relatively prime. In view of Eq. (7.2.1) we have $c(\lambda)I = \delta(\lambda)(\lambda I - A)C(\lambda)$. This equation implies that $c(\lambda)$ is divisible by $\delta(\lambda)$ and, since both of them are monic, there exists a monic polynomial $\tilde{m}(\lambda)$ such that $c(\lambda) = \delta(\lambda)\tilde{m}(\lambda)$. Thus

$$\tilde{m}(\lambda)I = (\lambda I - A)C(\lambda). \tag{2}$$

Our aim is to show that $\tilde{m}(\lambda) = m(\lambda)$, that is, that $\tilde{m}(\lambda)$ is the minimal polynomial of A. Indeed, the relation (2) shows that $\tilde{m}(\lambda)I$ is divisible by $\lambda I - A$ and therefore, by the corollary to Theorem 7.2.3, $\tilde{m}(A) = O$. Thus, $\tilde{m}(\lambda)$ is an annihilating polynomial of A and it remains to check that it is an annihilating polynomial for A of the least degree. To see this, suppose that $\tilde{m}(\lambda) = \theta(\lambda)m(\lambda)$. Since $m(A) = 0$, the same corollary yields

$$m(\lambda)I = (\lambda I - A)\tilde{C}(\lambda)$$

for some matrix polynomial $\tilde{C}(\lambda)$, and hence

$$\tilde{m}(\lambda)I = (\lambda I - A)\theta(\lambda)\tilde{C}(\lambda).$$

Comparing this with Eq. (2) and recalling the uniqueness of the left quotient, we deduce that $C(\lambda) = \theta(\lambda)\tilde{C}(\lambda)$. But this would mean (if $\theta(\lambda) \not\equiv 1$) that $\theta(\lambda)$ is a common divisor of the elements of $C(\lambda)$, which contradicts the definition of $\delta(\lambda)$. Thus $\theta(\lambda) \equiv 1$ and $\tilde{m}(\lambda) = m(\lambda)$. ■

This result implies another useful description of the minimal polynomial.

Theorem 2. *The minimal polynomial of a matrix $A \in \mathcal{F}^{n\times n}$ coincides with the invariant polynomial of $\lambda I - A$ of highest degree.*

PROOF. Note that $\lambda I - A$ has rank n and that by Proposition 7.3.1 and Theorem 7.5.1,

$$\det(\lambda I - A) = \alpha i_1(\lambda)i_2(\lambda)\cdots i_n(\lambda), \tag{3}$$

in which none of $i_1(\lambda), \ldots, i_n(\lambda)$ is identically zero.

Also, $\alpha = 1$ since the invariant polynomials, as well as $\det(\lambda I - A)$, are monic. Recalling Eq. (7.5.3), we deduce from Eq. (3) that

$$c(\lambda) = \det(\lambda I - A) = d_{n-1}(\lambda)i_n(\lambda), \tag{4}$$

in which $d_{n-1}(\lambda)$ is the (monic) greatest common divisor of the minors of $\lambda I - A$ of order $n - 1$ or, equivalently, the greatest common divisor of the elements of the adjoint matrix $B(\lambda)$ of $A(\lambda)$. Thus, $d_{n-1}(\lambda) = \delta(\lambda)$, where $\delta(\lambda)$ is defined in the proof of Theorem 1. This yields $i_n(\lambda) = m(\lambda)$ on comparison of Eqs. (4) and (1). ■

Now we can give an independent proof of the result obtained in Proposition 6.6.2.

Corollary 1. *A matrix $A \in \mathbb{C}^{n \times n}$ is simple if and only if its minimal polynomial has only simple zeros.*

PROOF. By Corollary 7.8.1, A is simple if and only if all the elementary divisors of $\lambda I - A$ are linear, that is, if and only if the invariant polynomial $i_n(\lambda) = \prod_{i=1}^{s} (\lambda - \lambda_i)$, where $\lambda_1, \lambda_2, \ldots, \lambda_s$ are the *distinct* eigenvalues of A. The result follows immediately from Theorem 2. ■

Corollary 2. *A matrix $A \in \mathcal{F}^{n \times n}$ is nonderogatory if and only if the $n - 1$ first invariant polynomials of $\lambda I - A$ are of degree zero, that is, they are identically equal to* 1.

PROOF. Let A be nonderogatory and let $i_1(\lambda), i_2(\lambda), \ldots, i_n(\lambda)$ denote the invariant polynomials of $\lambda I - A$. If $c(\lambda)$ is a characteristic polynomial of A then $c(\lambda) = \prod_{k=1}^{n} i_k(\lambda)$, and by assumption $c(\lambda)$ coincides with the minimal polynomial $m(\lambda)$ of A. But by Theorem 2, $m(\lambda) = i_n(\lambda)$ and therefore the comparison gives $i_1(\lambda) = \cdots = i_{n-1}(\lambda) = 1$. Note that there are no zero invariant polynomials of $\lambda I - A$ since it is a regular matrix polynomial. ■

Now we show that the second natural normal form of Theorem 7.8.1 (as well as the Jordan form that is a consequence of it) is the best in the sense that no block C_{l_k}, $k = 1, 2, \ldots, p$, in Eq. (7.8.1) can be split into smaller blocks. More precisely, it will be shown that C_{l_k} cannot be similar to a block-diagonal matrix (of more than one block). To see this, we first define a matrix $A \in \mathcal{F}^{n \times n}$ to be *decomposable* if it is similar to a block-diagonal matrix diag$[A_1, A_2]$, where $A_1 \in \mathbb{C}^{n_1 \times n_1}$, $A_2 \in \mathbb{C}^{n_2 \times n_2}$, and $n_1, n_2 > 0$, $n_1 + n_2 = n$. Otherwise, the matrix is said to be *indecomposable*.

Proposition 1. *A matrix $A \in \mathbb{C}^{n \times n}$ is indecomposable if and only if it is nonderogatory and its characteristic polynomial $c(\lambda)$ is of the form $(\lambda - \lambda_0)^p$ for some $\lambda_0 \in \mathbb{C}$ and positive integer p.*

PROOF. Let A be indecomposable. If it is derogatory, then by Corollary 2 the matrix polynomial $\lambda I - A$ has at least two nonconstant invariant polynomials. Now Theorem 7.6.2 shows that A is decomposable, which is a contradiction. If the characteristic polynomial $c(\lambda)$ has two or more distinct zeros, then $\lambda I - A$ has at least two elementary divisors. Theorem 7.8.1 then states that again A is decomposable.

Conversely, if A is nonderogatory and $c(\lambda) = (\lambda - \lambda_0)^p$, then A must be indecomposable. Indeed, these conditions imply, repeating the previous argument, that $\lambda I - A$ has only one elementary divisor. Since a matrix B is decomposable only if the matrix $\lambda I - B$ has at least two elementary divisors, the result follows. ■

Corollary 1. *The matrices* C_{l_k} $(k = 1, 2, \ldots, s)$ *in* Eq. (7.8.1) *are indecomposable.*

The proof relies on the fact that a companion matrix is nonderogatory (Exercise 6.6.3) and on the form of its characteristic polynomial (see Exercise 4.11.3).

Corollary 2. *Jordan blocks are indecomposable matrices.*

To prove this see Exercises 6.6.2 and 7.7.4. Thus, these forms cannot be reduced further and are the best in this sense.

Exercise 1. The proof of Theorem 1 above [see especially Eq. (2)] shows that the reduced adjoint matrix $C(\lambda)$ satisfies

$$(\lambda I - A)C(\lambda) = m(\lambda)I.$$

Use this relation to show that if λ_0 is an eigenvalue of A, then all nonzero columns of $C(\lambda_0)$ are eigenvectors of A corresponding to λ_0. Obtain an analogous result for the adjoint matrix $B(\lambda_0)$. □

7.10 The Smith Form: Differential and Difference Equations

In this section an important application of the Smith normal form is made to the analysis of differential and difference equations. Consider first a set of n homogeneous scalar differential equations with constant complex coefficients in n scalar variables $x_1(t), \ldots, x_n(t)$. Define $\boldsymbol{x}(t)$ by

$$\boldsymbol{x}(t) = [x_1(t) \quad \cdots \quad x_n(t)]^{\mathrm{T}}$$

and let l be the maximal order of derivatives in the n equations. Then the equations can be written in the form

$$L_l\boldsymbol{x}^{(l)}(t) + L_{l-1}\boldsymbol{x}^{(l-1)}(t) + \cdots + L_1\boldsymbol{x}^{(1)}(t) + L_0\boldsymbol{x}(t) = \boldsymbol{0} \tag{1}$$

for certain matrices $L_0, L_1, \ldots, L_l \in \mathbb{C}^{n\times n}$, where $L_l \neq 0$ and the indices on the vector $\boldsymbol{x}(t)$ denote componentwise derivatives.

Let $\mathscr{C}(\mathbb{C}^n)$ be the set of all continuous vector-valued functions $\boldsymbol{y}(t)$ defined for all real t with values in $\mathbb{C}^n$. Thus $\boldsymbol{y}(t) = [y_1(t) \ \cdots \ y_n(t)]^{\mathrm{T}}$ and $y_1(t), \ldots, y_n(t)$ are continuous for all real t. With the natural definitions of vector addition and scalar multiplication, $\mathscr{C}(\mathbb{C}^n)$ is obviously a (infinite-dimensional) linear space.

Here and in subsequent chapters we will investigate the nature of the solution space of Eq. (1), and our first observation is that every solution is a member of $\mathscr{C}(\mathbb{C}^n)$. Then it is easily verified that the solution set is, in fact, a linear *subspace* of $\mathscr{C}(\mathbb{C}^n)$. In this section, we examine the dimension of this subspace.

The matrix polynomial $L(\lambda) = \sum_{j=0}^{l} \lambda^j L_j$ is associated with Eq. (1), which can be abbreviated to the form $L(d/dt)\boldsymbol{x}(t) = \boldsymbol{0}$. Thus,

$$L\left(\frac{d}{dt}\right)\boldsymbol{x}(t) \triangleq \sum_{j=0}^{l} L_j \frac{d^j\boldsymbol{x}}{dt^j}.$$

To illustrate, if

$$L(\lambda) = \begin{bmatrix} \lambda^2 & 2 \\ \lambda - 1 & \lambda^2 \end{bmatrix} \quad \text{and} \quad \boldsymbol{x}(t) = \begin{bmatrix} x_1(t) \\ x_2(t) \end{bmatrix} \in \mathscr{C}(\mathbb{C}^2),$$

and, in addition, $x_1(t)$ and $x_2(t)$ are twice differentiable, then

$$L\left(\frac{d}{dt}\right)\boldsymbol{x}(t) = \begin{bmatrix} x_1^{(2)}(t) + 2x_2(t) \\ x_1^{(1)}(t) - x_1(t) + x_2^{(2)}(t) \end{bmatrix}.$$

Exercise 1. Let $L_1(\lambda), L_2(\lambda)$ be $n \times n$ matrix polynomials and

$$L(\lambda) = L_1(\lambda)L_2(\lambda).$$

Show that

$$L\left(\frac{d}{dt}\right)\boldsymbol{x}(t) = L_1\left(\frac{d}{dt}\right)\left(L_2\left(\frac{d}{dt}\right)\boldsymbol{x}(t)\right). \quad \square$$

With these conventions and the Smith normal form, we can now establish the dimension of the solution space of Eq. (1) as a subspace of $\mathscr{C}(\mathbb{C}^n)$. This result is sometimes known as Chrystal's theorem. The proof is based on the well-known fact that a *scalar* differential equation of the form (1) with order l has a solution space of dimension l.

Theorem 1 (G. Chrystal[†]). *If* det $L(\lambda) \not\equiv 0$ *then the solution space of Eq.* (1) *has dimension equal to the degree of* det $L(\lambda)$.

PROOF. Since det $L(\lambda) \not\equiv 0$, the Smith canonical form of $L(\lambda)$ (see Section 7.5) is of the form $D(\lambda) = \operatorname{diag}[i_1(\lambda), i_2(\lambda), \ldots, i_n(\lambda)]$, where $i_1(\lambda), i_2(\lambda), \ldots, i_n(\lambda)$ are the nonzero invariant polynomials of $L(\lambda)$. Thus, by Theorem 7.5.1,

$$L(\lambda) = P(\lambda)D(\lambda)Q(\lambda) \tag{2}$$

for some unimodular matrix polynomials $P(\lambda)$ and $Q(\lambda)$. Applying the result of Exercise 1, we may rewrite Eq. (1) in the form

$$P\left(\frac{d}{dt}\right)D\left(\frac{d}{dt}\right)Q\left(\frac{d}{dt}\right)\boldsymbol{x}(t) = \boldsymbol{0}. \tag{3}$$

Let $\boldsymbol{y}(t) = Q(d/dt)\boldsymbol{x}(t)$ and multiply Eq. (3) on the left by $P^{-1}(d/dt)$ (recall that $P^{-1}(\lambda)$ is a polynomial) to obtain the system

$$\operatorname{diag}\left[i_1\left(\frac{d}{dt}\right), i_2\left(\frac{d}{dt}\right), \ldots, i_n\left(\frac{d}{dt}\right)\right]\boldsymbol{y}(t) = \boldsymbol{0}, \tag{4}$$

which is equivalent to Eq. (1). Clearly, the system in (4) splits into n independent scalar equations

$$i_k\left(\frac{d}{dt}\right)y_k(t) = 0, \qquad k = 1, 2, \ldots, n, \tag{5}$$

where $y_1(t), \ldots, y_n(t)$ are the components of $\boldsymbol{y}(t)$. But the scalar differential equation $a(d/dt)y(t) = 0$ has exactly $d = \deg a(\lambda)$ linearly independent solutions. Apply this result to each of the equations in (5) and observe that the number of linearly independent vector-valued functions $\boldsymbol{y}(t)$ satisfying Eq. (4) is equal to the sum of the numbers of linearly independent solutions of the system (5). Thus the dimension of the solution space of Eq. (3) is $d_1 + d_2 + \cdots + d_n$, where $d_k = \deg i_k(\lambda)$, $1 \le k \le n$. On the other hand, Eq. (2) shows that

$$\det L(\lambda) = \alpha \det D(\lambda) = \alpha \prod_{k=1}^{n} i_k(\lambda),$$

where $\alpha = \det P(\lambda)Q(\lambda)$. Since α is a nonzero constant, it follows that

$$\deg(\det L(\lambda)) = \sum_{k=1}^{n} \deg i_k(\lambda) = d_1 + d_2 + + \cdots + d_n.$$

The equivalence of the systems (4) and (1) in the sense that

$$\boldsymbol{x}(t) = Q^{-1}\left(\frac{d}{dt}\right)\boldsymbol{y}(t)$$

now gives the required result. ■

[†] *Trans. Roy. Soc. Edin.* **38** (1895), 163.

The following special cases arise frequently: if $\det L_l \neq 0$ and, in particular, if $L(\lambda)$ is *monic*, then the solution space of Eq. (1) has dimension ln. This follows immediately from the obvious fact that the leading coefficient of $\det L(\lambda)$ is just $\det L_l$. One special case is so important that it justifies a separate formulation:

Corollary 1. *If $A \in \mathbb{C}^{n \times n}$ then the solution space of $\dot{x}(t) = Ax(t)$ has dimension n.*

We turn now to the study of a set of n homogeneous scalar *difference equations* with constant complex coefficients and order l. Such a system can be written in matrix form as

$$L_l x_{j+l} + L_{l-1} x_{j+l-1} + \cdots + L_1 x_{j+1} + L_0 x_j = 0 \tag{6}$$

for $j = 0, 1, 2, \ldots$, where $L_0, L_1, \ldots, L_l \in \mathbb{C}^{n \times n}$ ($L_l \neq O$), and a solution is a *sequence* of vectors $(x_0, x_1, x_2, \ldots)$ for which all the relations (6) hold with $j = 0, 1, 2, \ldots$. In this case, we introduce the linear space $\mathscr{S}(\mathbb{C}^n)$ consisting of all infinite sequences of vectors from $\mathbb{C}^n$, and with the natural component-wise definitions of vector addition and scalar multiplication. Then it is easily verified that the solution set of (6) is a subspace of $\mathscr{S}(\mathbb{C}^n)$.

It will be convenient to introduce a transformation E acting on $\mathscr{S}(\mathbb{C}^n)$ defined by

$$E(u_0, u_1, \ldots) = (u_1, u_2, \ldots)$$

for any $(u_0, u_1, \ldots) \in \mathscr{S}(\mathbb{C}^n)$. It is clear that E is linear. It is called a *shift operator* and will be seen to play a role for difference equations analogous to that played by d/dt for differential equations.

For $r = 1, 2, 3, \ldots$, the powers of E are defined recursively by

$$E^r u = E(E^{r-1}u), \qquad r = 1, 2, \ldots, \qquad u \in \mathscr{S}(\mathbb{C}^n),$$

where E^0 is the identity of $\mathscr{S}(\mathbb{C}^n)$. Consequently,

$$E^r(u_0, u_1, \ldots) = (u_r, u_{r+1}, \ldots).$$

Also, if $A \in \mathbb{C}^{n \times n}$ and $(u_0, u_1, \ldots) \in \mathscr{S}(\mathbb{C}^n)$, we define

$$A(u_0, u_1, \ldots) \triangleq (Au_0, Au_1, \ldots) \in \mathscr{S}(\mathbb{C}^n).$$

Then it is easily verified that, if we write $x = (x_0, x_1, \ldots)$, Eq. (6) can be written in the form

$$L(E)x \triangleq L_l(E^l x) + \cdots + L_1(Ex) + L_0 x = 0, \tag{7}$$

where the zero on the right side of Eq. (7) is the zero element of $\mathscr{S}(\mathbb{C}^n)$.

Another illuminating way to see Eq. (6) is as an equation in infinite vectors and matrices. For simplicity, consider the case $l = 2$:

$$\begin{bmatrix} L_0 & L_1 & L_2 & 0 & \cdots \\ 0 & L_0 & L_1 & L_2 & \cdots \\ 0 & 0 & L_0 & L_1 & \cdots \\ \vdots & \vdots & \vdots & \vdots & \end{bmatrix} \begin{bmatrix} \mathbf{x}_0 \\ \mathbf{x}_1 \\ \mathbf{x}_2 \\ \vdots \end{bmatrix} = \begin{bmatrix} 0 \\ 0 \\ 0 \\ \vdots \end{bmatrix}.$$

Theorem 2. *If* det $L(\lambda) \not\equiv 0$, *then the dimension of the solution space of Eq. (6) is equal to the degree of* det $L(\lambda)$.

PROOF. Using the decomposition (2), reduce Eq. (7) to the equivalent form

$$D(E)\mathbf{y} = \mathbf{0}, \tag{8}$$

where $\mathbf{y} = Q(E)\mathbf{x}$ and $\mathbf{0}, \mathbf{x}, \mathbf{y} \in \mathcal{S}(\mathbb{C}^n)$. Observe that Eq. (8) splits into n independent scalar equations. Since the assertion of the theorem holds true for scalar difference equations, it remains to apply the argument used in the proof of Theorem 1. ■

Corollary 2. *If* $A \in \mathbb{C}^{n \times n}$, *then the solution space of the recurrence relation*

$$\mathbf{x}_{j+1} = A\mathbf{x}_j, \qquad j = 0, 1, 2, \ldots,$$

has dimension n.

7.11 Miscellaneous Exercises

1. Let $A \in \mathbb{C}^{10 \times 10}$ and suppose the invariant polynomials of $\lambda I - A$ are $i_1(\lambda) = \cdots = i_7(\lambda) = 1, i_8(\lambda) = \lambda + 1, i_9(\lambda) = \lambda^3 + 1, i_{10}(\lambda) = (\lambda^3 + 1)^2$. Check that the first and second natural normal forms and the Jordan normal form of A are, respectively,

$$C_1 = \operatorname{diag}\left[-1, \begin{bmatrix} 0 & 1 & 0 \\ 0 & 0 & 1 \\ -1 & 0 & 0 \end{bmatrix}, \begin{bmatrix} 0 & 1 & 0 & 0 & 0 & 0 \\ 0 & 0 & 1 & 0 & 0 & 0 \\ 0 & 0 & 0 & 1 & 0 & 0 \\ 0 & 0 & 0 & 0 & 1 & 0 \\ 0 & 0 & 0 & 0 & 0 & 1 \\ -1 & 0 & 0 & -2 & 0 & 0 \end{bmatrix}\right],$$

$$C_2 = \operatorname{diag}\left[-1, -1, \begin{bmatrix} 0 & 1 \\ -1 & -2 \end{bmatrix}, \mu_2, \begin{bmatrix} 0 & 1 \\ -\mu_2^2 & 2\mu_2 \end{bmatrix}, \mu_3, \begin{bmatrix} 0 & 1 \\ -\mu_3^2 & 2\mu_3 \end{bmatrix}\right],$$

$$J = \operatorname{diag}\left[-1, -1, \begin{bmatrix} -1 & 1 \\ 0 & -1 \end{bmatrix}, \mu_2, \begin{bmatrix} \mu_2 & 1 \\ 0 & \mu_2 \end{bmatrix}, \mu_3, \begin{bmatrix} \mu_3 & 1 \\ 0 & \mu_3 \end{bmatrix}\right],$$

where

$$\mu_2 = \frac{(1 + \sqrt{3}i)}{2}, \qquad \mu_3 = \frac{(1 - \sqrt{3}i)}{2}.$$

2. If $A(\lambda)$, $B(\lambda)$, $D(\lambda)$ are matrix polynomials of size n, and $A(\lambda)$, and $B(\lambda)$ have invertible leading coefficients and degrees l and m, respectively, show that the degree of $A(\lambda)D(\lambda)B(\lambda)$ is not less than $l + m$.

3. Two $m \times n$ matrix polynomials $A(\lambda)$ and $B(\lambda)$ are said to be *equivalent* if there are unimodular matrices $P(\lambda)$ of order m and $Q(\lambda)$ of order n such that $B(\lambda) = P(\lambda)A(\lambda)Q(\lambda)$. Show that any $m \times n$ matrix polynomial is equivalent to a matrix polynomial of the form

$$\begin{bmatrix} a_1(\lambda) & 0 & \cdots & 0 & 0 & \cdots & 0 \\ 0 & a_2(\lambda) & & & & & \\ \vdots & & \ddots & \vdots & \vdots & & \vdots \\ & \ddots & \ddots & 0 & & & \\ 0 & \cdots & 0 & a_m(\lambda) & 0 & \cdots & 0 \end{bmatrix}$$

if $m \le n$, or

$$\begin{bmatrix} a_1(\lambda) & 0 & \cdots & 0 \\ 0 & & \ddots & \vdots \\ & \ddots & & 0 \\ \vdots & & \ddots & a_n(\lambda) \\ & & & 0 \\ & & & \vdots \\ 0 & & \cdots & 0 \end{bmatrix}$$

if $m > n$, where $a_j(\lambda)$ is divisible by $a_{j-1}(\lambda)$, $j \ge 2$.

4. Show that if $A \in \mathbb{C}^{n \times n}$ is idempotent and $A \ne I$, $A \ne O$, then the minimal polynomial of A is $\lambda^2 - \lambda$ and the characteristic polynomial has the form $(\lambda - 1)^r \lambda^s$. Prove that A is similar to the matrix $\operatorname{diag}[I_r, O]$.

5. Prove that

(a) The geometric multiplicity of an eigenvalue is equal to the number of elementary divisors associated with it;

(b) The algebraic multiplicity of an eigenvalue is equal to the sum of degrees of all elementary divisors associated with it.

6. Define the *multiplicity* of a latent root λ_0 of $A(\lambda)$ as the number of times the factor $\lambda - \lambda_0$ appears in the factorization of $\det A(\lambda)$ into linear factors.

Let $A(\lambda)$ be a monic $n \times n$ matrix polynomial of degree l. Show that $A(\lambda)$ has ln latent roots (counted according to their multiplicities), and that if the ln latent roots are distinct, then all the elementary divisors of $A(\lambda)$ are linear.

7. If λ_0 is a latent root of multiplicity α of the $n \times n$ matrix polynomial $A(\lambda)$, prove that the elementary divisors of $A(\lambda)$ associated with λ_0 are all linear if and only if $\dim(\operatorname{Ker} A(\lambda_0)) = \alpha$.

8. If $m(\lambda)$ is the minimal polynomial of $A \in \mathbb{C}^{n \times n}$ and $f(\lambda)$ is a polynomial over $\mathbb{C}$, prove that $f(A)$ is nonsingular if and only if $m(\lambda)$ and $f(\lambda)$ are relatively prime.

9. Let $A(\lambda)$ be a matrix polynomial and define a vector $\boldsymbol{x} \neq \boldsymbol{0}$ to be a *latent vector* of $A(\lambda)$ associated with λ_0 if $A(\lambda_0)\boldsymbol{x} = \boldsymbol{0}$.

If $A(\lambda) = A_0 + \lambda A_1 + \lambda^2 A_2$ ($\det A_2 \neq 0$) is an $n \times n$ matrix polynomial, prove that $\boldsymbol{x}$ is a latent vector of $A(\lambda)$ associated with λ_0 if and only if

$$\left(\lambda_0 \begin{bmatrix} A_2 & 0 \\ A_1 & A_2 \end{bmatrix} + \begin{bmatrix} 0 & -A_2 \\ A_0 & 0 \end{bmatrix}\right)\begin{bmatrix} \mathbf{x}_0 \\ \lambda_0 \mathbf{x}_0 \end{bmatrix} = \boldsymbol{0}.$$

Generalise this result for matrix polynomials with invertible leading coefficients of general degree.

10. Let $A(\lambda)$ be an $n \times n$ matrix polynomial with latent roots $\lambda_1, \lambda_2, \ldots, \lambda_n$ and suppose that there exist linearly independent latent vectors $\boldsymbol{x}_1, \boldsymbol{x}_2, \ldots, \boldsymbol{x}_n$ of $A(\lambda)$ associated with $\lambda_1, \lambda_2, \ldots, \lambda_n$, respectively. Prove that if $X = [\boldsymbol{x}_1 \ \ \boldsymbol{x}_2 \ \ \cdots \ \ \boldsymbol{x}_n]$ and $D = \operatorname{diag}[\lambda_1, \lambda_2, \ldots, \lambda_n]$, then $S = XDX^{-1}$ is a right solvent of $A(\lambda)$.

11. Let the matrix $A \in \mathbb{C}^{n \times n}$ have distinct eigenvalues $\lambda_1, \lambda_2, \ldots, \lambda_s$ and suppose that the maximal degree of the elementary divisors associated with λ_k $(1 \leq k \leq s)$ is m_k (the index of λ_k). Prove that

$$\mathbb{C}^n = \sum_{k=1}^{s} \cdot \operatorname{Ker}(\lambda_k I - A)^{m_k}.$$

12. Prove that if $X \in \mathbb{C}^{n \times n}$ is a solvent of the $n \times n$ matrix polynomial $A(\lambda)$, then each eigenvalue of X is a latent root of $A(\lambda)$.

Hint. Use the Corollary to Theorem 7.2.3.

13. Consider a matrix polynomial $A(\lambda) = \sum_{j=0}^{l} \lambda^j A_j$ with $\det A_l \neq 0$, and define its companion matrix

$$C_A \triangleq \begin{bmatrix} 0 & I_n & 0 & \cdots & 0 \\ 0 & 0 & I_n & \ddots & \vdots \\ \vdots & \vdots & & \ddots & 0 \\ 0 & 0 & \cdots & 0 & I_n \\ -\hat{A}_0 & -\hat{A}_1 & & \cdots & -\hat{A}_{l-1} \end{bmatrix},$$

where $\hat{A}_j = A_l^{-1}A_j$, $j = 0, 1, \ldots, l-1$. Show that λ_0 is a latent root of $A(\lambda)$ if and only if it is an eigenvalue of C_A.

14. Let $X_1, X_2, \ldots, X_l \in \mathbb{C}^{n\times n}$ be right solvents of the $n \times n$ matrix polynomial $A(\lambda)$. Show that, with C_A defined as in Exercise 13,

(a) $C_A V = V \operatorname{diag}[X_1, X_2, \ldots, X_l]$, where V denotes the generalized Vandermonde matrix

$$V = \begin{bmatrix} I_n & I_n & \cdots & I_n \\ X_1 & X_2 & \cdots & X_l \\ \vdots & \vdots & & \vdots \\ X_1^{l-1} & X_2^{l-1} & \cdots & X_l^{l-1} \end{bmatrix},$$

(b) If V is invertible then the *spectrum* of $A(\lambda)$ (i.e., the set of all its latent roots) is the union of the eigenvalues of $X_1, X_2, \ldots, X_l$;

(c) C_A is similar to $\operatorname{diag}[X_1, X_2, \ldots, X_l]$ if and only if the set of all elementary divisors of C_A coincides with the union of the elementary divisors of $X_1, X_2, \ldots, X_l$.

CHAPTER 8

The Variational Method

The technique developed in this chapter can be visualized as the generalization of a geometrical approach to the eigenvalue problem for 3×3 real symmetric matrices. We saw in Exercise 5.10.5 that a quadric surface can be associated with such a matrix H by means of the equation $\boldsymbol{x}^{\mathrm{T}}H\boldsymbol{x} = 1$ or $(H\boldsymbol{x}, \boldsymbol{x}) = 1$, where $\boldsymbol{x} \in \mathbb{R}^3$. In particular (see Exercise 5.10.6), if H is positive definite, then the corresponding surface is an ellipsoid. The vector $\boldsymbol{x}_0 \in \mathbb{R}^3$ from the origin to the farthest point on the ellipsoid can then be described as a vector at which $(\boldsymbol{x}, \boldsymbol{x})$ attains the maximal value subject to the condition that $\boldsymbol{x}$ satisfies $(H\boldsymbol{x}, \boldsymbol{x}) = 1$.

Then the problem can be reformulated and the side condition eliminated by asking for the maximal value of the quotient $(\boldsymbol{x}, \boldsymbol{x})/(H\boldsymbol{x}, \boldsymbol{x})$, where $\boldsymbol{x}$ varies over all nonzero vectors in $\mathbb{R}^3$. Or, what is essentially the same thing, we seek the minimal value of the *Rayleigh quotient*:

$$\frac{(H\boldsymbol{x}, \boldsymbol{x})}{(\boldsymbol{x}, \boldsymbol{x})}, \qquad \boldsymbol{x} \neq \boldsymbol{0}.$$

Now, this quotient makes sense for any $n \times n$ Hermitian matrix H and vector $\boldsymbol{x} \in \mathbb{C}^n$ and is also known in this more general context as the Rayleigh quotient for H. The study of its properties in this context is one of the main subjects of this chapter.

8.1 Field of Values. Extremal Eigenvalues of a Hermitian Matrix

The eigenvalues of a matrix $A \in \mathbb{C}^{n \times n}$ form a set of n (not necessarily distinct) points in the complex plane. Some useful ideas concerning the distribution of these points can be developed from the concept of the *field of values* of A, defined as the set $F(A)$ of complex numbers $(A\boldsymbol{x}, \boldsymbol{x})$, where $\boldsymbol{x}$ ranges over all vectors in $\mathbb{C}^n$ that are normalized so that $(\boldsymbol{x}, \boldsymbol{x}) = \boldsymbol{x}^*\boldsymbol{x} = 1$.†

Observe that the quadratic form

$$f(\boldsymbol{x}) = (A\boldsymbol{x}, \boldsymbol{x}) = \sum_{i,j=1}^{n} a_{ij}x_i\bar{x}_j, \qquad \boldsymbol{x} = [x_1 \quad x_2 \quad \cdots \quad x_n]^{\mathrm{T}},$$

is a continuous function in n variables on the *unit sphere* in $\mathbb{C}^n$, that is, the set of vectors for which

$$(\boldsymbol{x}, \boldsymbol{x}) = \sum_{i=1}^{n} |x_i|^2 = 1.$$

Hence it follows (see Theorem 2 in Appendix 2) that $F(A)$ is a closed and bounded set in the complex plane. It can also be proved that $F(A)$ is a *convex* set (see Appendix 2 for the definition), but we omit this proof.

Theorem 1. *The field of values of a matrix $A \in \mathbb{C}^{n \times n}$ is invariant under unitary similarity transformations. Thus, if $U \in \mathbb{C}^{n \times n}$ is unitary, then*

$$F(A) = F(UAU^*).$$

PROOF. We have

$$(UAU^*\boldsymbol{x}, \boldsymbol{x}) = (AU^*\boldsymbol{x}, U^*\boldsymbol{x}) = (A\boldsymbol{y}, \boldsymbol{y}),$$

where $\boldsymbol{y} = U^*\boldsymbol{x}$. If $(\boldsymbol{x}, \boldsymbol{x}) = 1$, then $(\boldsymbol{y}, \boldsymbol{y}) = (UU^*\boldsymbol{x}, \boldsymbol{x}) = (\boldsymbol{x}, \boldsymbol{x}) = 1$. Hence a number α is equal to $(UAU^*\boldsymbol{x}, \boldsymbol{x})$ for some $\boldsymbol{x} \in \mathbb{C}^n$ with $(\boldsymbol{x}, \boldsymbol{x}) = 1$ if and only if $\alpha = (A\boldsymbol{y}, \boldsymbol{y})$ for some normalized $\boldsymbol{y} \in \mathbb{C}^n$; that is, α belongs to $F(UAU^*)$ if and only if $\alpha \in F(A)$. ■

This result admits an elegant geometrical description of the field of values of a normal matrix in terms of its eigenvalues. First note that all eigenvalues of an arbitrary matrix $A \in \mathbb{C}^{n \times n}$ are in $F(A)$. For, if $\lambda \in \sigma(A)$, then there is an eigenvector $\boldsymbol{x}$ of A corresponding to this eigenvalue for which $A\boldsymbol{x} = \lambda\boldsymbol{x}$ and $(\boldsymbol{x}, \boldsymbol{x}) = 1$. Hence

$$(A\boldsymbol{x}, \boldsymbol{x}) = (\lambda\boldsymbol{x}, \boldsymbol{x}) = \lambda(\boldsymbol{x}, \boldsymbol{x}) = \lambda,$$

and so $\lambda \in F(A)$. Thus, $\sigma(A) \subset F(A)$.

† Throughout this chapter the inner product (,) is assumed to be the standard inner product in $\mathbb{C}^n$.

For normal matrices, the geometry of the field of values is now easily characterized. The notion of the convex hull appearing in the next theorem is defined and discussed in Appendix 2.

Theorem 2. *The field of values of a normal matrix $A \in \mathbb{C}^{n \times n}$ coincides with the convex hull of its eigenvalues.*

PROOF. If A is normal and its eigenvalues are $\lambda_1, \lambda_2, \ldots, \lambda_n$, then by Theorem 5.2.1 there is a unitary $n \times n$ matrix U such that $A = UDU^*$, where $D = \text{diag}[\lambda_1, \lambda_2, \ldots, \lambda_n]$. Theorem 1 now asserts that $F(A) = F(D)$ and so it suffices to show that the field of values of the diagonal matrix D coincides with the convex hull of the entries on its main diagonal.

Indeed, $F(D)$ is the set of numbers

$$\alpha = (D\mathbf{x}, \mathbf{x}) = \sum_{i=1}^{n} \lambda_i |x_i|^2,$$

where $(\mathbf{x}, \mathbf{x}) = \sum_{i=1}^{n} |x_i|^2 = 1$.

On the other hand, the convex hull of the points $\lambda_1, \lambda_2, \ldots, \lambda_n$ is, according to Exercise 2 in Appendix 2 the set

$$\left\{\sum_{i=1}^{n} \theta_i \lambda_i \colon \theta_i \geq 0, \quad \sum_{i=1}^{n} \theta_i = 1\right\}.$$

Putting $\theta_i = |x_i|^2$ $(i = 1, 2, \ldots, n)$ and noting that, as $\mathbf{x}$ runs over all vectors satisfying $(\mathbf{x}, \mathbf{x}) = 1$, the θ_i run over all possible choices such that $\theta_i \geq 0$, $\sum_{i=1}^{n} \theta_i = 1$, we obtain the desired result. ■

For a general $n \times n$ matrix A, the convex hull of the points of the spectrum of a matrix A is a subset of $F(A)$ (see Appendix 2). But for the special case of Hermitian matrices, the geometry simplifies beautifully.

Theorem 3. *The field of values of the matrix $A \in \mathbb{C}^{n \times n}$ is an interval of the real line if and only if A is Hermitian.*

PROOF. First, if A is Hermitian then A is normal and, by Theorem 2, $F(A)$ is the convex hull of the eigenvalues of A, which are all real (Theorem 5.3.1). Since the only convex sets on the real line are intervals, the required result follows.

Conversely, if $A \in \mathbb{C}^{n \times n}$ and $F(A)$ is an interval of the real line, then, writing

$$B = \mathcal{Re}\, A = \frac{1}{2}(A + A^*), \qquad C = \mathcal{Im}\, A = \frac{1}{2i}(A - A^*),$$

we obtain

$$(A\mathbf{x}, \mathbf{x}) = ((B + iC)\mathbf{x}, \mathbf{x}) = (B\mathbf{x}, \mathbf{x}) + i(C\mathbf{x}, \mathbf{x}),$$

where $(B\mathbf{x}, \mathbf{x})$ and $(C\mathbf{x}, \mathbf{x})$ are real since B and C are Hermitian (see Section 5.10). But $F(A)$ consists only of real numbers, therefore $(C\mathbf{x}, \mathbf{x}) = 0$ for all

$\boldsymbol{x}$ such that $(\boldsymbol{x}, \boldsymbol{x}) = 1$. Hence, $C = O$ (see Exercise 3.13.2) and $A = A^*$, that is, A is Hermitian. ■

Corollary 1. *If H is Hermitian with the eigenvalues $\lambda_1 \leq \lambda_2 \leq \cdots \leq \lambda_n$, then*

$$F(H) = [\lambda_1, \lambda_n]. \tag{1}$$

Conversely, if $F(A) = [\lambda_1, \lambda_n]$, then A is Hermitian and λ_1, λ_n are the minimal and maximal eigenvalues of A, respectively.

This result is an immediate consequence of Theorems 2 and 3.

Recalling the definition of $F(H)$, we deduce from Eq. (1) the existence of vectors $\boldsymbol{x}_1$ and $\boldsymbol{x}_n$ on the unit sphere such that $(H\boldsymbol{x}_1, \boldsymbol{x}_1) = \lambda_1$ and $(H\boldsymbol{x}_n, \boldsymbol{x}_n) = \lambda_n$. Obviously,

$$\lambda_1 = (H\boldsymbol{x}_1, \boldsymbol{x}_1) = \min_{(\boldsymbol{x}, \boldsymbol{x}) = 1} (H\boldsymbol{x}, \boldsymbol{x}), \tag{2}$$

and

$$\lambda_n = (H\boldsymbol{x}_n, \boldsymbol{x}_n) = \max_{(\boldsymbol{x}, \boldsymbol{x}) = 1} (H\boldsymbol{x}, \boldsymbol{x}). \tag{3}$$

Moreover, the vectors $\boldsymbol{x}_1$ and $\boldsymbol{x}_n$ are eigenvectors of H corresponding to λ_1 and λ_n, respectively. In fact, it follows from Eq. (2) that if $\boldsymbol{x}$ is any vector with $(\boldsymbol{x}, \boldsymbol{x}) = 1$, then

$$(H\boldsymbol{x}, \boldsymbol{x}) \geq \lambda_1 = \lambda_1(\boldsymbol{x}, \boldsymbol{x}).$$

Hence

$$((H - \lambda_1 I)\boldsymbol{x}, \boldsymbol{x}) \geq 0,$$

and this implies that $H - \lambda_1 I$ is positive semidefinite. But Eq. (2) also implies that

$$((H - \lambda_1 I)\boldsymbol{x}_1, \boldsymbol{x}_1) = 0,$$

and it follows (as in Exercise 5.4.1) that $(H - \lambda_1 I)\boldsymbol{x}_1 = \boldsymbol{0}$. Since $(\boldsymbol{x}_1, \boldsymbol{x}_1) = 1$ it follows that $\boldsymbol{x}_1 \neq \boldsymbol{0}$ and $\boldsymbol{x}_1$ is therefore an eigenvector of H corresponding to λ_1.

Similarly, $\boldsymbol{x}_n$ is an eigenvector of H corresponding to the maximal eigenvalue λ_n. Thus, we have proved the following:

Theorem 4. *If λ_1 and λ_n denote the minimal and maximal eigenvalues of a Hermitian matrix $H \in \mathbb{C}^{n \times n}$, respectively, then*

$$\lambda_1 = \min_{(\boldsymbol{x}, \boldsymbol{x}) = 1} (H\boldsymbol{x}, \boldsymbol{x}), \qquad \lambda_n = \max_{(\boldsymbol{x}, \boldsymbol{x}) = 1} (H\boldsymbol{x}, \boldsymbol{x}).$$

Moreover, the extremal values of $(H\boldsymbol{x}, \boldsymbol{x})$ are attained at corresponding eigenvectors of H.

This result is the first step in the development of a min-max theory of eigenvalues of Hermitian matrices, one of the main topics in this chapter.

Exercise 1. Let $H = [h_{ij}]_{i,j=1}^{n}$ denote a Hermitian matrix. Prove that, preserving the notation of this section,

(a) $\lambda_1 \leq h_{jj} \leq \lambda_n$ for $j = 1, 2, \ldots, n$;
(b) If $\alpha = n^{-1} \sum_{i,j=1}^{n} h_{ij}$, then $\lambda_1 \leq \alpha \leq \lambda_n$.

Hint. For part (a) put $\boldsymbol{x}_j = \boldsymbol{e}_j$, (a unit vector in $\mathbb{C}^n$) in $(H\boldsymbol{x}, \boldsymbol{x})$ and use Theorem 4. For part (b) use the vector $n^{-1/2}[1 \quad 1 \quad \cdots \quad 1]^{\mathrm{T}}$. □

8.2 Courant–Fischer Theory and the Rayleigh Quotient

In this section, the variational description of extremal eigenvalues of a Hermitian matrix achieved in the preceding section will be extended to all of the eigenvalues. For this and other purposes the notion of the Rayleigh quotient for an $n \times n$ Hermitian matrix H is useful. We define it to be

$$R_H(\boldsymbol{x}) \equiv R(\boldsymbol{x}) \triangleq \frac{(H\boldsymbol{x}, \boldsymbol{x})}{(\boldsymbol{x}, \boldsymbol{x})}, \qquad \boldsymbol{x} \neq \boldsymbol{0}.$$

Thus, R is a real-valued function defined by H, whose domain is the nonzero vectors of $\mathbb{C}^n$.

It is clear that, replacing $\boldsymbol{x}$ by $\alpha\boldsymbol{x}$, the Rayleigh quotient remains invariant. Hence, in particular,

$$\min_{(\boldsymbol{x},\boldsymbol{x})=1} (H\boldsymbol{x}, \boldsymbol{x}) = \min_{\boldsymbol{x}\neq \boldsymbol{0}} R(\boldsymbol{x})$$

$$\max_{(\boldsymbol{x},\boldsymbol{x})=1} (H\boldsymbol{x}, \boldsymbol{x}) = \max_{\boldsymbol{x}\neq \boldsymbol{0}} R(\boldsymbol{x}),$$

and the result of Theorem 8.1.4 can be rewritten in the form

$$\lambda_1 = \min_{\boldsymbol{x}\neq \boldsymbol{0}} R(\boldsymbol{x}), \qquad \lambda_n = \max_{\boldsymbol{x}\neq \boldsymbol{0}} R(\boldsymbol{x}).$$

and these extrema are attained on eigenvectors of H.

Now let $\lambda_1 \leq \lambda_2 \leq \cdots \leq \lambda_n$ be the eigenvalues of an $n \times n$ Hermitian matrix H and suppose that $\boldsymbol{x}_1, \boldsymbol{x}_2, \ldots, \boldsymbol{x}_n$ are associated eigenvectors forming an orthonormal basis. Denote by $\mathscr{C}_p$ the subspace of $\mathbb{C}^n$ generated by the vectors $\boldsymbol{x}_p, \boldsymbol{x}_{p+1}, \ldots, \boldsymbol{x}_n$ $(1 \leq p \leq n)$. Obviously, $\mathscr{C}_p$ has dimension $n - p + 1$, we have the chain of inclusions

$$\mathbb{C}^n = \mathscr{C}_1 \supset \mathscr{C}_2 \supset \cdots \supset \mathscr{C}_n,$$

and the subspace $\mathscr{C}_p$ is orthogonal to the space generated by $\boldsymbol{x}_1, \boldsymbol{x}_2, \ldots, \boldsymbol{x}_{p-1}$. It has already been shown that

$$\lambda_1 = \min_{0 \neq \boldsymbol{x} \in \mathscr{C}_1} R(\boldsymbol{x}), \tag{1}$$

and it will now be shown that the other eigenvalues of H can be characterized in a way that generalizes statement (1).

Theorem 1. *Let $R(\boldsymbol{x})$ be the Rayleigh quotient defined by a Hermitian matrix H, and let the subspaces $\mathscr{C}_1, \ldots, \mathscr{C}_n$ (associated with H) be as defined in the previous paragraph. Then, for $i = 1, 2, \ldots, n$, the numbers*

$$\lambda_i = \min_{0 \neq \boldsymbol{x} \in \mathscr{C}_i} R(\boldsymbol{x}) \tag{2}$$

are the eigenvalues of H. Moreover, the minima in Eq. (2) *are attained on eigenvectors $\boldsymbol{x}_i$ of H.*

PROOF. Observe first that the case $i = 1$ is just that of Theorem 8.1.4 written in the form (1). Let $U = [\boldsymbol{x}_1 \quad \boldsymbol{x}_2 \quad \cdots \quad \boldsymbol{x}_n]$ be the matrix formed by the set of orthonormal eigenvectors of H (as in Theorem 5.2.1), and recall that $\mathscr{C}_i$ is spanned by $\boldsymbol{x}_i, \boldsymbol{x}_{i+1}, \ldots, \boldsymbol{x}_n$. Then for any $\boldsymbol{x} \in \mathscr{C}_i$ there are numbers $\alpha_i, \alpha_{i+1}, \ldots, \alpha_n$ such that $\boldsymbol{x} = \sum_{j=i}^{n} \alpha_j \boldsymbol{x}_j$. Introduce the vector $\boldsymbol{\alpha} \in \mathbb{C}^n$ defined by

$$\boldsymbol{\alpha} = [0 \quad \cdots \quad 0 \quad \alpha_i \quad \alpha_{i+1} \quad \cdots \quad \alpha_n]^{\mathrm{T}}, \tag{3}$$

so that $\boldsymbol{x} = U\boldsymbol{\alpha}$, and observe that, using Theorem 5.2.1,

$$(H\boldsymbol{x}, \boldsymbol{x}) = \boldsymbol{x}^*H\boldsymbol{x} = \boldsymbol{\alpha}^*U^*HU\boldsymbol{\alpha} = \boldsymbol{\alpha}^*D\boldsymbol{\alpha} = \sum_{j=i}^{n} \lambda_j|\alpha_j|^2. \tag{4}$$

Also we have $(\boldsymbol{x}, \boldsymbol{x}) = \sum_{j=1}^{n} |\alpha_j|^2$ and so

$$R(\boldsymbol{x}) = \left(\sum_{j=i}^{n} \lambda_j|\alpha_j|^2\right)\Bigg/\left(\sum_{j=1}^{n} |\alpha_j|^2\right). \tag{5}$$

Since $\lambda_i \leq \lambda_{i+1} \leq \cdots \leq \lambda_n$, it is clear that $R(\boldsymbol{x}) \geq \lambda_i$ for all $\alpha_i, \alpha_{i+1}, \ldots, \alpha_n$ (not all zero, and hence for all nonzero $\boldsymbol{x} \in \mathscr{C}_i$) and that $R(\boldsymbol{x}) = \lambda_i$ when $\alpha_i \neq 0$ and $\alpha_{i+1} = \alpha_{i+2} = \cdots = \alpha_n = 0$. Thus relation (2) is established. Note that the minimizing vector of $\mathscr{C}_i$ is just $\alpha_i \boldsymbol{x}_i$, an eigenvector associated with λ_i. ■

The following dual result is proved similarly.

Exercise 1. In the notation of Theorem 1, prove that

$$\lambda_i = \max_{\boldsymbol{0} \neq \boldsymbol{x} \in \mathscr{C}_{n-i+1}} R(\boldsymbol{x}), \qquad i = 1, 2, \ldots, n,$$

where the maxima are attained on eigenvectors of H. □

The result of Exercise 1, like that of Theorem 1, suffers from the disadvantage that λ_j $(j > 1)$ is characterized with the aid of the eigenvectors $\boldsymbol{x}_{n-j+1}, \ldots, \boldsymbol{x}_n$. Thus in employing Theorem 1 we must compute the eigenvalues in nondecreasing order together with orthonormal bases for their eigenspaces. The next theorem is a remarkable result allowing us to characterize each eigenvalue with no reference to other eigenvalues or eigenvectors. This is the famous Courant–Fischer theorem, which originated with E. Fischer[†] in 1905 but was rediscovered by R. Courant[‡] in 1920. It was Courant who saw, and made clear, the great importance of the result for problems in mathematical physics. We start with a preliminary result.

Proposition 1. *Let $\mathscr{S}_j$ $(1 \le j \le n)$ denote an arbitrary $(n - j + 1)$-dimensional subspace of $\mathbb{C}^n$, and let H be a Hermitian matrix with eigenvalues $\lambda_1 \le \lambda_2 \le \cdots \le \lambda_n$ and Rayleigh quotient $R(\boldsymbol{x})$. Then*

$$\min_{\mathbf{0} \neq \boldsymbol{x} \in \mathscr{S}_j} R(\boldsymbol{x}) \le \lambda_j, \qquad \max_{\mathbf{0} \neq \boldsymbol{x} \in \mathscr{S}_j} R(\boldsymbol{x}) \ge \lambda_{n-j+1}. \tag{6}$$

PROOF. Let $\boldsymbol{x}_1, \boldsymbol{x}_2, \ldots, \boldsymbol{x}_n$ be an orthonormal system of eigenvectors for H corresponding to the eigenvalues $\lambda_1 \le \lambda_2 \le \cdots \le \lambda_n$ of H, and let $\hat{\mathscr{S}}_j = \operatorname{span}\{\boldsymbol{x}_1, \boldsymbol{x}_2, \ldots, \boldsymbol{x}_j\}$ for $j = 1, 2, \ldots, n$. Fix j and observe that, since dim $\mathscr{S}_j$ + dim $\hat{\mathscr{S}}_j = (n - j + 1) + j > n$, there exists a nonzero vector $\boldsymbol{x}_0 \in \mathbb{C}^n$ belonging to both of these subspaces (see Exercise 3.14.3). In particular, since $\boldsymbol{x}_0 \in \hat{\mathscr{S}}_j$ it follows that $\boldsymbol{x}_0 = \sum_{k=1}^{j} \alpha_k \boldsymbol{x}_k$ and hence

$$(H\boldsymbol{x}_0, \boldsymbol{x}_0) = \left(\sum_{k=1}^{j} \lambda_k \alpha_k \boldsymbol{x}_k, \sum_{k=1}^{j} \alpha_k \boldsymbol{x}_k \right) = \sum_{k=1}^{j} \lambda_k |\alpha_k|^2.$$

Thus obviously

$$\lambda_1 \sum_{k=1}^{j} |\alpha_k|^2 \le (H\boldsymbol{x}_0, \boldsymbol{x}_0) \le \lambda_j \sum_{k=1}^{j} |\alpha_k|^2$$

and, dividing by $\sum_{k=1}^{j} |\alpha_k|^2 = (\boldsymbol{x}_0, \boldsymbol{x}_0)$, we obtain

$$\lambda_1 \le R(\boldsymbol{x}_0) \le \lambda_j. \tag{7}$$

Recall now that $\boldsymbol{x}_0$ also belongs to $\mathscr{S}_j$ and that our immediate purpose is to establish the first relation in (6). But this follows readily from (7) since

$$\min_{0 \neq \boldsymbol{x} \in \mathscr{S}_j} R(\boldsymbol{x}) \le R(\boldsymbol{x}_0) \le \lambda_j.$$

The second statement of the proposition is proved similarly on replacing $\hat{\mathscr{S}}_j$ by $\mathscr{C}_{n-j+1}$. ■

[†] *Monatshefte für Math. und Phys.* **16** (1905): 234–409.

[‡] *Math. Zeitschrift* **7** (1920): 1–57.

Theorem 2. (Courant–Fischer). *In the previous notation,*

$$\lambda_j = \max_{\mathscr{S}_j} \min_{\mathbf{0} \neq \boldsymbol{x} \in \mathscr{S}_j} R(\boldsymbol{x}) \tag{8}$$

for $j = 1, 2, \ldots, n$, *or, in dual form,*

$$\lambda_{n-j+1} = \min_{\mathscr{S}_j} \max_{\mathbf{0} \neq \boldsymbol{x} \in \mathscr{S}_j} R(\boldsymbol{x}). \tag{9}$$

In other words, Eq. (8) says that by taking the minimum of $R(\boldsymbol{x})$ over all nonzero vectors from an $(n - j + 1)$-dimensional subspace of $\mathbb{C}^n$, and then the maximum of these minima over all subspaces of this dimension in $\mathbb{C}^n$, we obtain the jth eigenvalue of H.

PROOF. To establish Eq. (8) it suffices, in view of the first statement of Eq. (6), to show the existence of an $(n - j + 1)$-dimensional subspace such that $\min R(\boldsymbol{x})$, where $\boldsymbol{x}$ varies over all nonzero vectors of this subspace, is equal to λ_j. Taking $\mathscr{C}_j = \text{span}\{\boldsymbol{x}_j, \boldsymbol{x}_{j+1}, \ldots, \boldsymbol{x}_n\}$, we obtain such a subspace. In fact, by (6),

$$\min_{\mathbf{0} \neq \boldsymbol{x} \in \mathscr{C}_j} R(\boldsymbol{x}) \leq \lambda_j,$$

while by Theorem 1,

$$\min_{\mathbf{0} \neq \boldsymbol{x} \in \mathscr{C}_j} R(\boldsymbol{x}) = \lambda_j.$$

This means that the equality (8) holds. Equation (9) is established similarly. ■

Exercise 2. Check that for $H \in \mathbb{R}^{3 \times 3}$, $H > 0$, and $j = 2$, the geometrical interpretation of Eq. (8) (as developed in Exercises 5.10.5 and 5.10.6) is the following: the length of the major axis of any central cross section of an ellipsoid (which is an ellipse) is not less than the length of the second axis of the ellipsoid, and there is a cross section with the major axis equal to the second axis of the ellipsoid. □

8.3 The Stationary Property of the Rayleigh Quotient

The eigenvalues of a real symmetric matrix can be described in terms of the Rayleigh quotient in another way. Let A denote a real symmetric matrix acting in $\mathbb{R}^n$. We may then view $R(\boldsymbol{x})$ associated with A as a real-valued function of the n independent real variables $x_1, x_2, \ldots, x_n$ that make up the components of $\boldsymbol{x}$. We write $R(\boldsymbol{x}) = R(x_1, x_2, \ldots, x_n)$.

Theorem 1. *If A is a real symmetric matrix with an eigenvalue λ_0 and associated eigenvector $\boldsymbol{x}_0$, then $R(x_1, x_2, \ldots, x_n)$ has a stationary value with respect to $x_1, x_2, \ldots, x_n$ at $\boldsymbol{x}_0 = [x_1^0 \quad x_2^0 \quad \cdots \quad x_n^0]^{\mathrm{T}}$ and $R(\boldsymbol{x}_0) = \lambda_0$.*

PROOF. We have to show that for $k = 1, 2, \ldots, n$, the derivatives $\partial R/\partial x_k$ vanish when evaluated at $\boldsymbol{x}_0$. Indeed, representing $(A\boldsymbol{x}, \boldsymbol{x})$ as a quadratic form, and writing $\boldsymbol{x} = [x_1 \quad x_2 \quad \cdots \quad x_n]^{\mathrm{T}}$, we have

$$(A\boldsymbol{x}, \boldsymbol{x}) = \boldsymbol{x}^{\mathrm{T}} A \boldsymbol{x} = \sum_{i,j=1}^{n} a_{ij} x_i x_j.$$

Since $a_{ij} = a_{ji}$, it follows that

$$\frac{\partial}{\partial x_k}(A\boldsymbol{x}, \boldsymbol{x}) = 2 \sum_{j=1}^{n} a_{kj} x_j$$

or, equivalently,

$$\frac{\partial}{\partial x_k}(A\boldsymbol{x}, \boldsymbol{x}) = 2\boldsymbol{A}_{k*}\boldsymbol{x}, \tag{1}$$

where $\boldsymbol{A}_{k*}$ denotes the kth row of A.

In particular, setting $A = I$ in Eq. (1), we obtain

$$\frac{\partial}{\partial x_k}(\boldsymbol{x}, \boldsymbol{x}) = 2x_k$$

and hence

$$\frac{\partial R}{\partial x_k} = \frac{\partial}{\partial x_k}\frac{(A\boldsymbol{x}, \boldsymbol{x})}{(\boldsymbol{x}, \boldsymbol{x})} = \frac{(\boldsymbol{x}, \boldsymbol{x})(2\boldsymbol{A}_{k*}\boldsymbol{x}) - (A\boldsymbol{x}, \boldsymbol{x})2x_k}{(\boldsymbol{x}, \boldsymbol{x})^2}. \tag{2}$$

Evaluating this at $\boldsymbol{x}_0 = [x_1^0 \quad x_2^0 \quad \cdots \quad x_n^0]^{\mathrm{T}}$, where $A\boldsymbol{x}_0 = \lambda_0 \boldsymbol{x}_0$ and thus $\boldsymbol{A}_{k*}\boldsymbol{x}_0 = \lambda_0 x_k^0$, we deduce that the numerator in Eq. (2) is equal to zero:

$$(\boldsymbol{x}_0, \boldsymbol{x}_0)(2\boldsymbol{A}_{k*}\boldsymbol{x}_0) - (A\boldsymbol{x}_0, \boldsymbol{x}_0)2x_k^0 = (\boldsymbol{x}_0, \boldsymbol{x}_0)2\lambda_0 x_k^0 - \lambda_0(\boldsymbol{x}_0, \boldsymbol{x}_0)2x_k^0 = 0.$$

This completes the proof. ∎

8.4 Problems with Constraints

Long before the formulation and proof of the Courant–Fischer theorem, mathematicians were faced with eigenvalue computations that arose from maximization problems subject to a *constraint*. In the variational approach, this means that we seek the stationary values of $R_H(\boldsymbol{x})$ where $\boldsymbol{x}$ varies over those nonzero members of $\mathbb{C}^n$ that satisfy an equation (or *constraint*) of the form $\boldsymbol{a}^*\boldsymbol{x} = 0$ where $\boldsymbol{a}^* = [\alpha_1 \quad \alpha_2 \quad \cdots \quad \alpha_n]$, or

$$\alpha_1 x_1 + \alpha_2 x_2 + \cdots + \alpha_n x_n = 0,$$

where not all the α_j are zero. Clearly, if $\alpha_j \neq 0$ we can use the constraint to express x_j as a linear combination of the other x's, substitute it into $R(\boldsymbol{x})$, and obtain a new quotient in $n - 1$ variables whose stationary values (and $n - 1$ associated eigenvalues) can be examined directly without reference to constraints. As it happens, this obvious line of attack is not the most profitable and if we already know the properties of the unconstrained system, we can say quite a lot about the constrained one. We shall call eigenvalues and eigenvectors of a problem with constraints *constrained eigenvalues* and *constrained eigenvectors*.

More generally, consider the eigenvalue problem subject to r constraints:

$$(\boldsymbol{x}, \boldsymbol{a}_i) = \boldsymbol{a}_i^*\boldsymbol{x} = 0, \qquad i = 1, 2, \ldots, r, \tag{1}$$

where $\boldsymbol{a}_1, \boldsymbol{a}_2, \ldots, \boldsymbol{a}_r$ are linearly independent vectors in $\mathbb{C}^n$. Given a Hermitian matrix $H \in \mathbb{C}^{n \times n}$, we wish to find the constrained eigenvalues and constrained eigenvectors of H, that is, the scalars $\lambda \in \mathbb{C}$ and vectors $\boldsymbol{x} \in \mathbb{C}^n$ such that

$$H\boldsymbol{x} = \lambda\boldsymbol{x}, \quad \boldsymbol{x} \neq \boldsymbol{0}, \qquad \text{and} \qquad (\boldsymbol{x}, \boldsymbol{a}_i) = \boldsymbol{a}_i^*\boldsymbol{x} = 0, \quad i = 1, 2, \ldots, r. \tag{2}$$

Let $\mathscr{A}_r$ be the subspace generated by $\boldsymbol{a}_1, \boldsymbol{a}_2, \ldots, \boldsymbol{a}_r$ appearing in Eq. (1) and let $\mathscr{C}_{n-r}$ denote the orthogonal complement to $\mathscr{A}_r$ in $\mathbb{C}^n$:

$$\mathscr{A}_r \oplus \mathscr{C}_{n-r} = \mathbb{C}^n.$$

If $\boldsymbol{b}_1, \boldsymbol{b}_2, \ldots, \boldsymbol{b}_{n-r}$ is an orthonormal basis in $\mathscr{C}_{n-r}$ and

$$B = [\boldsymbol{b}_1 \quad \boldsymbol{b}_2 \quad \cdots \quad \boldsymbol{b}_{n-r}],$$

then, obviously, $B \in \mathbb{C}^{n \times (n-r)}$, $B^*B = I_{n-r}$, and BB^* is an orthogonal projector onto $\mathscr{C}_{n-r}$ (see Section 5.8).

The vector $\boldsymbol{x} \in \mathbb{C}^n$ satisfies the constraints of Eq. (1) if and only if $\boldsymbol{x} \in \mathscr{C}_{n-r}$. That is, $BB^*\boldsymbol{x} = \boldsymbol{x}$ and the vector $\boldsymbol{y} = B^*\boldsymbol{x} \in \mathbb{C}^{n-r}$ satisfies $\boldsymbol{x} = B\boldsymbol{y}$. So, for $\boldsymbol{x} \neq \boldsymbol{0}$,

$$R_H(\boldsymbol{x}) = \frac{(H\boldsymbol{x}, \boldsymbol{x})}{(\boldsymbol{x}, \boldsymbol{x})} = \frac{(HB\boldsymbol{y}, B\boldsymbol{y})}{(B\boldsymbol{y}, B\boldsymbol{y})} = \frac{(B^*HB\boldsymbol{y}, \boldsymbol{y})}{(B^*B\boldsymbol{y}, \boldsymbol{y})}.$$

Since $B^*B = I_{n-r}$, we obtain

$$R(\boldsymbol{x}) \equiv R_H(\boldsymbol{x}) = R_{B^*HB}(\boldsymbol{y}) \tag{3}$$

and the problem in Eq. (2) is reduced to the standard form of finding the stationary values of the Rayleigh quotient for the $(n - r) \times (n - r)$ Hermitian matrix B^*HB (Section 8.3). We have seen that these occur only at the eigenvalues of B^*HB. Thus λ is a constrained eigenvalue of H with associated eigenvector $\mathbf{x}$ if and only if λ is an eigenvalue of B^*HB with eigenvector $\mathbf{y}$, where $\mathbf{x} = B\mathbf{y}$.

Let the constrained eigenvalues be $\lambda_1 \le \lambda_2 \le \cdots \le \lambda_{n-r}$ and let

$$\mathbf{y}_1, \mathbf{y}_2, \ldots, \mathbf{y}_{n-r}$$

be associated orthonormal eigenvectors of B^*HB. Then $\mathbf{x}_1 = B\mathbf{y}_1, \ldots, \mathbf{x}_{n-r} = B\mathbf{y}_{n-r}$ are the corresponding constrained eigenvectors. They are also orthonormal since, for $i, j = 1, 2, \ldots, n - r$,

$$\delta_{ij} = (\mathbf{x}_i, \mathbf{x}_j) = (B\mathbf{y}_i, B\mathbf{y}_j) = (B^*B\mathbf{y}_i, \mathbf{y}_j) = (\mathbf{y}_i, \mathbf{y}_j).$$

For $i = 1, 2, \ldots, n - r$, let X_i be the subspace in $\mathbb{C}^n$ generated by the vectors $\mathbf{x}_i, \mathbf{x}_{i+1}, \ldots, \mathbf{x}_{n-r}$. By first applying Theorem 8.2.1 to B^*HB in $\mathbb{C}^{n-r}$ and then transforming the result to $\mathbb{C}^n$ and using Eq. (3), we easily derive the following result.

Proposition 1. *If $\lambda_1 \le \lambda_2 \le \cdots \le \lambda_{n-r}$ are the eigenvalues of the constrained eigenvalue problem* (2) *with associated orthonormal eigenvectors $\mathbf{x}_1, \mathbf{x}_2, \ldots, \mathbf{x}_{n-r}$, then for $i = 1, 2, \ldots, n - r$,*

$$\lambda_i = \min_{0 \ne \mathbf{x} \in X_i} R(\mathbf{x}),$$

where X_i is the subspace generated by $\mathbf{x}_i, \mathbf{x}_{i+1}, \ldots, \mathbf{x}_{n-r}$.

Similarly, using Exercise 8.2.1, we have another result.

Proposition 2. *With the notation of Proposition* 1,

$$\lambda_i = \max_{0 \ne \mathbf{x} \in X_{n-r-i+1}} R(\mathbf{x}), \qquad i = 1, 2, \ldots, n - r,$$

where $X_{n-r-i+1}$ is the space generated by $\mathbf{x}_{n-r-i+1}, \ldots, \mathbf{x}_{n-r}$.

We consider now the relationships between the eigenvalues of a constrained eigenvalue problem and those of the initial unconstrained problem. The main result in this direction is sometimes known as Rayleigh's theorem,† although Rayleigh himself attributes it to Routh and it had been used even earlier by Cauchy. It is easily proved with the aid of the Courant–Fischer theorem.

Theorem 1. *If $\lambda_1 \le \lambda_2 \le \cdots \le \lambda_{n-r}$ are the eigenvalues of the eigenvalue problem* (2) *for a Hermitian $H \in \mathbb{C}^{n \times n}$ with r constraints, then*

$$\mu_i \le \lambda_i \le \mu_{i+r}, \qquad i = 1, 2, \ldots, n - r,$$

where $\mu_1 \le \mu_2 \le \cdots \le \mu_n$ are the eigenvalues of the corresponding unconstrained problem $H\mathbf{x} = \mu\mathbf{x}$, $\mathbf{x} \ne \mathbf{0}$.

† Rayleigh, J. W. S., *The Theory of Sound* 2nd Ed., 1894. Revised, Dover, New York, 1945. Vol. 1, section 92(a).

PROOF. First recall the definition of the space X_i in Proposition 1 and observe that $\dim X_i = n - r - i + 1$. It follows from Theorem 8.2.2 that

$$\mu_{i+r} = \max_{\mathscr{S}_{i+r}} \min_{0 \neq x \in \mathscr{S}_{i+r}} R(x) \geq \min_{0 \neq x \in X_i} R(x).$$

By virtue of Proposition 1,

$$\mu_{i+r} \geq \min_{0 \neq x \in X_i} R(x) = \lambda_i,$$

and the first inequality is established.

For the other inequality we consider the eigenvalues $\nu_1 \leq \nu_2 \leq \cdots \leq \nu_n$ of the matrix $-H$. Obviously, $\nu_i = -\mu_{n+1-i}$. Similarly, let the eigenvalues of $-H$ under the given r constraints in (1) be $\omega_1 \leq \omega_2 \leq \cdots \leq \omega_{n-r}$, where

$$\omega_i = -\lambda_{n-r+1-i} \qquad \text{for} \quad i = 1, 2, \ldots, n - r.$$

Applying the result just proved for H to the matrix $-H$, we obtain

$$\nu_{i+r} \geq \omega_i,$$

for $i = 1, 2, \ldots, n - r$, and hence

$$-\mu_{n+1-i-r} \geq -\lambda_{n-r+1-i}.$$

Writing $j = n + 1 - r - i$, we finally have $\mu_j \leq \lambda_j$ for $j = 1, 2, \ldots, n - r$, and the theorem is proved. ■

Note the following particular case of this theorem in which the "interlacing" of the constrained and unconstrained eigenvalues is most obvious.

Corollary 1. *If $\lambda_1 \leq \lambda_2 \leq \cdots \leq \lambda_{n-1}$ are the eigenvalues of H with one constraint and $\mu_1 \leq \mu_2 \leq \cdots \leq \mu_n$ are eigenvalues of H, then*

$$\mu_1 \leq \lambda_1 \leq \mu_2 \leq \cdots \leq \mu_{n-1} \leq \lambda_{n-1} \leq \mu_n.$$

Note also that the eigenvalue problem for H with linear constraints (1) can be interpreted as an unconstrained problem for a leading principal submatrix of a matrix congruent to H.

More precisely, let $H \in \mathbb{C}^{n \times n}$ be a Hermitian matrix generating the Rayleigh quotient $R(x)$ and let B denote the matrix introduced at the beginning of the section, that is, $B = [b_1 \quad b_2 \quad \cdots \quad b_{n-r}]$, where $\{b_j\}_{j=1}^{n-r}$ forms an orthonormal basis in $\mathscr{C}_{n-r}$, the orthogonal complement of $\mathscr{A}_r$ in $\mathbb{C}^n$. Consider the matrix $C = [B \quad A]$, where the columns of A form an orthonormal basis in $\mathscr{A}_r$, and note that C is unitary and, therefore, the matrix C^*HC has the same eigenvalues as H. Since

$$C^*HC = \begin{bmatrix} B^* \\ A^* \end{bmatrix} H [B \quad A] = \begin{bmatrix} B^*HB & B^*HA \\ A^*HB & A^*HA \end{bmatrix},$$

B^*HB is the $(n - r) \times (n - r)$ leading principal submatrix of C^*HC. Thus, the constrained eigenvalue problem for H described above is reduced to finding the eigenvalues of a leading principal submatrix of a matrix congruent to H. This last problem is interesting in its own right and will be considered in the next section.

Exercise 1. Consider the matrix A of Exercise 5.2.8. Examine the constrained eigenvalues under one constraint for each of the following constraint equations and verify Corollary 1 in each case:

(a) $x_1 = 0$;
(b) $x_1 - x_2 = 0$;
(c) $x_1 - x_3 = 0$. □

8.5 The Rayleigh Theorem and Definite Matrices

Let H be an $n \times n$ Hermitian matrix and let H_{n-r} be an $(n - r) \times (n - r)$ principal submatrix of H, and $1 \le r < n$. The following result can be viewed as a special case of Theorem 8.4.1.

Theorem 1 (Rayleigh). *If $\lambda_1 \le \lambda_2 \le \cdots \le \lambda_{n-r}$ are the eigenvalues of H_{n-r} and $\mu_1 \le \mu_2 \le \cdots \le \mu_n$ are the eigenvalues of H, then for $i = 1, 2, \ldots, n - r$,*

$$\mu_i \le \lambda_i \le \mu_{i+r}. \tag{1}$$

PROOF. If H_{n-r} is not the $(n - r) \times (n - r)$ leading principal submatrix of H we apply a permutation matrix P so that it is in this position in the transformed matrix P^THP. Since $P^{-1} = P^T$, the matrix P^THP has the same spectrum as H, so without loss of generality we may assume that H_{n-r} is the $(n - r) \times (n - r)$ leading principal submatrix of H.

In view of the concluding remarks of the previous section, the eigenvalues of H_{n-r} are those of an eigenvalue problem for CHC^* with r constraints, where C is some unitary matrix. This also follows from the easily checked relation $B^*(CHC^*)B = H_{n-r}$. Appealing to Theorem 8.4.1, we obtain

$$\tilde{\mu}_i \le \lambda_i \le \tilde{\mu}_{i+r}, \qquad i = 1, 2, \ldots, n - r,$$

where $\tilde{\mu}_1, \tilde{\mu}_2, \ldots, \tilde{\mu}_n$ are the eigenvalues of CHC^*. It remains now to observe that, because C is unitary, $\tilde{\mu}_j = \mu_j$ for $j = 1, 2, \ldots, n$ and this completes the proof. ■

Exercise 1. All principal minors of a positive definite matrix are positive.

SOLUTION. For the eigenvalues $\lambda_1 \le \lambda_2 \le \cdots \le \lambda_{n-r}$ of a principal submatrix H_{n-r} of H we have, using Eq. (1),

$$0 < \mu_1 \le \lambda_1 \le \lambda_2 \le \cdots \le \lambda_{n-r},$$

where μ_1 denotes the smallest eigenvalue of the positive definite matrix H. But then $\det H_{n-r} = \lambda_1 \lambda_2 \cdots \lambda_{n-r}$ is positive and the assertion follows. □

We can now develop a different approach to definite matrices, one in which they are characterized by determinantal properties.

Theorem 2 (G. Frobenius†). *An $n \times n$ Hermitian matrix H is positive definite if and only if all of its leading principal minors,*

$$H\begin{pmatrix}1\\1\end{pmatrix}, \quad H\begin{pmatrix}1 & 2\\1 & 2\end{pmatrix}, \quad \ldots, \quad H\begin{pmatrix}1 & 2 & \cdots & n\\1 & 2 & \cdots & n\end{pmatrix}, \tag{2}$$

are positive.

PROOF. If $H > 0$ then, by Exercise 1, all of its principal minors and, in particular, the leading principal minors of H are positive.

Conversely, let the minors (2) be positive. Denote by H_k $(1 \le k \le n)$ the $k \times k$ leading principal submatrix of H; we prove by induction on k that the matrix $H_n = H$ is positive definite.

First, for $k = 1$ the 1×1 matrix H_1 is obviously positive definite. Then suppose that the matrix H_k $(1 \le k \le n-1)$ is positive definite and let us show that H_{k+1} is also positive definite. Note that the assumption $H_k > 0$ means the positivity of the eigenvalues $\lambda_1^{(k)} \le \lambda_2^{(k)} \le \cdots \le \lambda_k^{(k)}$ of H_k. If now $\lambda_1^{(k+1)} \le \lambda_2^{(k+1)} \le \cdots \le \lambda_{k+1}^{(k+1)}$ are the eigenvalues of H_{k+1}, then Theorem 1 yields

$$\lambda_1^{(k+1)} \le \lambda_1^{(k)} \le \lambda_2^{(k+1)} \le \cdots \le \lambda_k^{(k+1)} \le \lambda_k^{(k)} \le \lambda_{k+1}^{(k+1)},$$

and the positivity of the eigenvalues $\lambda_i^{(k)}$ $(i = 1, 2, \ldots, k)$ implies positivity of the numbers $\lambda_2^{(k+1)}, \lambda_3^{(k+1)}, \ldots, \lambda_{k+1}^{(k+1)}$. But also $\lambda_1^{(k+1)}$ is positive since the positivity of the minors in (2) implies

$$\det H_{k+1} = \lambda_1^{(k+1)} \lambda_2^{(k+1)} \cdots \lambda_{k+1}^{(k+1)} > 0.$$

Thus, all the eigenvalues of H_{k+1} are positive, that is, H_{k+1} is positive definite. ■

Note that, in fact, all the leading principal submatrices of a positive definite matrix are positive definite.

† *Sitzungsber der preuss. Akad. Wiss.* (1894), 241–256 März, 407–431 Mai.

Exercise 2. Confirm the following criterion for negative definiteness of an $n \times n$ Hermitian matrix: $H < 0$ if and only if the leading principal minors $\det H_k$ $(k = 1, 2, \ldots, n)$ of H satisfy the conditions $\det H_1 < 0$ and

$$(\det H_k)(\det H_{k+1}) < 0$$

for $k = 1, 2, \ldots, n - 1$. □

Thus, a Hermitian matrix $H = [h_{ij}]_{i,j=1}^{n}$ is negative definite if and only if $h_{11} < 0$ and the signs of its leading principal minors alternate.

The tests for positive and negative definiteness presented in this section are often used in practice since they are readily implemented from a computational point of view. Some generalizations of these results will be considered in the next section.

8.6 The Jacobi–Gundelfinger–Frobenius Method

The following result is a generalization of the Sylvester theorem. For a general Hermitian matrix H, it gives a criterion for finding the *inertia* of H, as defined in Section 5.5.

Theorem 1 (G. Jacobi†). *Let $H_1, H_2, \ldots, H_n$ be the leading principal submatrices of a Hermitian matrix $H \in \mathbb{C}^{n \times n}$ of rank r, and assume that $d_k = \det H_k \neq 0$, $k = 1, 2, \ldots, r$. Then the number of negative (respectively, positive) eigenvalues of H is equal to the number of alterations (respectively, constancies) of sign in the sequence*

$$1, d_1, d_2, \ldots, d_r. \tag{1}$$

PROOF. We prove this result by induction. If $k = 1$, then d_1 is the (unique) eigenvalue of H_1 and obviously d_1 is negative or positive according as there is an alteration or constancy of sign in the sequence $1, d_1$.

Suppose that the statement holds for the leading principal submatrix H_k, that is, the number of negative (respectively, positive) eigenvalues of H_k is equal to the number of alterations (respectively, constancies) of sign in the sequence $1, d_1, d_2, \ldots, d_k$. The transition to H_{k+1} adds to this sequence the nonzero number $d_{k+1} = \det H_{k+1}$. If $d_{k+1} > 0$, then (see Eq. (4.11.3)) the product of all the eigenvalues

$$\lambda_1^{(k+1)} \leq \lambda_2^{(k+1)} \leq \cdots \leq \lambda_{k+1}^{(k+1)}$$

† *Journal für die Reine und Angew. Math.* **53** (1857), 265–270.

of H_{k+1} is positive. Furthermore, $d_k = \lambda_1^{(k)}\lambda_2^{(k)} \cdots \lambda_k^{(k)}$. Consider the case $d_k > 0$. Then in the sequence $1, d_1, \ldots, d_k, d_{k+1}$ the sign is changed the same number of times as in the sequence $1, d_1, \ldots, d_k$, and hence our aim is to show that H_{k+1} has the same number of negative eigenvalues and one more positive eigenvalue than H_k.

Indeed, let H_k have m $(1 \le m \le k)$ negative and $k - m$ positive eigenvalues:

$$\lambda_1^{(k)} \le \cdots \le \lambda_m^{(k)} < 0 < \lambda_{m+1}^{(k)} \le \cdots \le \lambda_k^{(k)}.$$

By Theorem 8.5.1, $\lambda_m^{(k)} \le \lambda_{m+1}^{(k+1)} \le \lambda_{m+1}^{(k)}$ and two cases can occur:

$$\lambda_{m+1}^{(k+1)} < 0 \qquad \text{or} \qquad \lambda_{m+1}^{(k+1)} > 0.$$

The first inequality is impossible in the case $d_k > 0$ and $d_{k+1} > 0$ presently being considered. In fact, if

$$\lambda_1^{(k+1)} \le \lambda_1^{(k)} \le \cdots \le \lambda_m^{(k)} \le \lambda_{m+1}^{(k+1)} < 0 < \lambda_{m+1}^{(k)} \le \cdots \le \lambda_{k+1}^{(k+1)},$$

we have

$$\operatorname{sign} d_{k+1} = \operatorname{sign}(\lambda_1^{(k+1)}\lambda_2^{(k+1)} \cdots \lambda_{k+1}^{(k+1)}) = (-1)^{m+1},$$

$$\operatorname{sign} d_k = \operatorname{sign}(\lambda_1^{(k)}\lambda_2^{(k)} \cdots \lambda_k^{(k)}) = (-1)^m,$$

a contradiction with the assumption $\operatorname{sign} d_{k+1} = \operatorname{sign} d_k = 1$. Thus, $\lambda_{m+1}^{(k+1)} > 0$ and H_{k+1} has one more positive eigenvalue than H_k.

Other cases are proved similarly to complete an inductive proof of the theorem. ■

Exercise 1. Consider the matrix

$$H = \begin{bmatrix} 2 & \frac{1}{2} & \frac{1}{2} \\ \frac{1}{2} & 0 & 0 \\ \frac{1}{2} & 0 & -1 \end{bmatrix}$$

associated with the real Hermitian form discussed in Exercise 5.11.2(a). Since

$$d_1 = 2, \qquad d_2 = \det \begin{bmatrix} 2 & \frac{1}{2} \\ \frac{1}{2} & 0 \end{bmatrix} = -\tfrac{1}{4}, \qquad d_3 = \det H = \tfrac{1}{4},$$

the number of alterations of sign in the sequence $1, 2, -\frac{1}{4}, \frac{1}{4}$ or, what is equivalent, in the sequence of signs $+, +, -, +$ is equal to 2, while the number of constancies of sign is 1. Thus, $\pi(H) = 1$, $\nu(H) = 2$, and the first canonical form of the corresponding Hermitian form contains one positive and two negative squares. □

It is clear that this method of finding the inertia (if it works) is simpler than the Lagrange method described in Section 5.11. However, the latter also supplies the transforming matrix.

A disadvantage of the Jacobi method is the strong hypothesis that the leading principal minors be nonzero. This condition is relaxed in the generalizations due to Gundelfinger and Frobenius.

Theorem 2 (S. Gundelfinger†). *With the notation of Theorem* 1, *suppose that the sequence* (1) *contains no two successive zeros. If* $\det H_r \neq 0$, *then the assertion of Jacobi's theorem remains valid if any sign is assigned to each (isolated) zero* $d_i = 0$ *in* (1).

In other words, for computing the number $\nu(H)$, single zeros in the sequence (1) can be omitted.

PROOF. Assume inductively that the hypothesis holds for $H_1, H_2, \ldots, H_k$. If $\det H_{k+1} \neq 0$, then the appropriate part of the proof of Jacobi's theorem can be repeated to show that the transition from $1, d_1, \ldots, d_k$ (with signs instead of zeros) to $1, d_1, \ldots, d_{k+1}$ adds an alteration or a constancy of sign (between d_k and d_{k+1}) according as H_{k+1} has one more negative or positive eigenvalue than H_k.

Suppose that $d_{k+1} = \det H_{k+1} = 0$. Then at least one eigenvalue $\lambda_i^{(k+1)}$ $(1 \leq i \leq k+1)$ must be zero. Since $d_k \neq 0$ and $d_{k+2} \neq 0$ there is exactly one zero eigenvalue of H_{k+1}. Indeed, if $\lambda_i^{(k+1)} = \lambda_s^{(k+1)} = 0$ $(i < s)$, then by applying Theorem 8.5.1 there are zero eigenvalues of H_k and H_{k+2}, which yields $d_k = d_{k+2} = 0$.

Let $\lambda_{m+1}^{(k+1)} = 0$ $(1 \leq m \leq k-1)$. Then it follows from

$$\lambda_m^{(k)} < \lambda_{m+1}^{(k+1)} = 0 < \lambda_{m+1}^{(k)}$$

and

$$\lambda_{m+1}^{(k+2)} < \lambda_{m+1}^{(k+1)} = 0 < \lambda_{m+2}^{(k+2)} \tag{2}$$

that the number of negative eigenvalues of H_k is m and the number of negative eigenvalues of H_{k+2} is $m+1$. Hence $\operatorname{sign} d_k = (-1)^m$, $\operatorname{sign} d_{k+2} = (-1)^{m+1}$, and therefore $d_k d_{k+2} < 0$.

If $\lambda_1^{(k+1)} = 0$, then all eigenvalues of H_k are positive and so $\operatorname{sign} d_k = 1$. Also, we may put $m = 0$ in (2) and find that $\operatorname{sign} d_{k+2} = -1$. Thus $d_k d_{k+2} < 0$ once more. A similar argument gives the same inequality when $\lambda_{k+1}^{(k+1} = 0$.

It is now clear that, by assigning any sign to $d_{k+1} = 0$ or omitting it, we obtain one alteration of sign from d_k to d_{k+2}. This proves the theorem. ■

***Example* 2.** Consider the matrix

$$H = \begin{bmatrix} 0 & 1 & 0 \\ 1 & 0 & 1 \\ 0 & 1 & 1 \end{bmatrix}$$

† *Journal für die Reine und Angew. Math.* **91** (1881), 221–237.

associated with the quadratic form in Exercise 5.11.1. Compute the number of alterations and constancies of sign in the sequence

$$1, \quad d_1 = 0, \quad d_2 = -1, \quad d_3 = -1,$$

or equivalently, in the sign sequences

$$+, +, -, - \qquad \text{or} \qquad +, -, -, -.$$

Using the Gundelfinger rule, we deduce $\nu(H) = 1$, $\pi(H) = 2$. □

Theorem 3 (G. Frobenius[†]). *If, in the notation of Theorem* 1, $\det H_r \neq 0$ *and the sequence* (1) *contains no three successive zeros, then Jacobi's method can be applied by assigning the same sign to the zeros* $d_{i+1} = d_{i+2} = 0$ *if* $d_i d_{i+3} < 0$, *and different signs if* $d_i d_{i+3} > 0$.

PROOF. We only sketch the proof, because it relies on the same argument as that of the previous theorem.

Let $d_{k+1} = d_{k+2} = 0$, $d_k \neq 0$, $d_{k+3} \neq 0$. As before, there is exactly one zero eigenvalue of H_{k+1}, say, $\lambda_{m+1}^{(k+1)}$. Similarly, it follows from $\lambda_i^{(k+3)} \neq 0$ $(i = 1, 2, \ldots, k+3)$ and Theorem 8.5.1 that there exists exactly one zero, $\lambda_{s+1}^{(k+2)} = 0$ $(0 \leq s \leq k+1)$, of H_{k+2}. Using Theorem 8.5.1 to compare the eigenvalues of H_{k+1} and H_{k+2}, it is easily seen that either $s = m$ or $s = m + 1$.

The distribution of eigenvalues in the case $s = m$ looks as follows:

$$\lambda_m^{(k)} < \lambda_{m+1}^{(k+1)} = 0, \qquad \lambda_m^{(k+1)} < \lambda_{m+1}^{(k+2)} = 0 = \lambda_{m+1}^{(k+1)},$$
$$\lambda_m^{(k+2)} \leq \lambda_{m+1}^{(k+3)} < \lambda_{m+1}^{(k+2)} = 0 < \lambda_{m+2}^{(k+3)}.$$

Hence the matrices H_k, H_{k+1}, H_{k+2}, and H_{k+3} have, respectively, m, m, m, and $m + 1$ negative eigenvalues.

Thus, for $s = m$ we have $d_k d_{k+3} < 0$ and, in this case, H_{k+3} has one more negative eigenvalue than H_k. Assigning to the zeros d_{k+1}, d_{k+2} the same sign, we obtain only one alteration of sign in the sequence d_k, d_{k+1}, d_{k+2}, d_{k+3}, which proves the theorem in this case.

It is similarly found that if $s = m + 1$, then $d_k d_{k+3} > 0$, and the assertion of the theorem is valid. ■

It should be emphasized that, as illustrated in the next exercise, a further improvement of Jacobi's theorem is impossible. In other words, if the sequence (1) contains three or more successive zeros, then the signs of the leading principal minors do not uniquely determine the inertia characteristics of the matrix.

[†] *Sitzungsber. der preuss. Akad. Wiss.* (1894), 241–256 März, 407–431 Mai.

Example 3. The leading principal minors of the Hermitian matrix

$$H = \begin{bmatrix} 0 & 0 & 0 & b \\ 0 & a & 0 & 0 \\ 0 & 0 & a & 0 \\ b & 0 & 0 & 0 \end{bmatrix}, \qquad a, b \neq 0,$$

are

$$d_1 = d_2 = d_3 = 0, \qquad d_4 = a^2b^2 < 0.$$

On the other hand, the characteristic polynomial of H is $(\lambda^2 - b^2)(\lambda - a)^2$ and hence $\sigma(H) = \{b, -b, a, a\}$. Thus for $a > 0$ we have $\pi(H) = 3$, $\nu(H) = 1$, while for $a < 0$ we have $\pi(H) = 1$, $\nu(H) = 3$, and the inertia of H is not uniquely determined by the signs of its leading principal minors. □

Note also that if the rank r of H fails to be attained on the leading principal submatrix H_r of H that is, $d_r = 0$, then the Jacobi theorem cannot be applied.

Exercise 4. Observe that the Hermitian matrix

$$H = \begin{bmatrix} 1 & 1 & 1 \\ 1 & 1 & 1 \\ 1 & 1 & a \end{bmatrix}$$

with $a \neq 1$ has rank 2, while $d_2 = \det H_2 = 0$. Furthermore, the leading principal minors $d_1 = 1, d_2 = d_3 = 0$ do not depend on the values assigned to a, whereas the inertia of H does:

$$\det(\lambda I - H) = \lambda[\lambda^2 - (a + 2)\lambda + (2a - 2)].$$

Since for the nonzero eigenvalues λ_1, λ_2 of H we have $\lambda_1\lambda_2 = 2a - 2$ and $\lambda_1 + \lambda_2 = a + 2$, it follows that $\pi(H) = 2$ if $a > 1$ and $\pi(H) = 1$ for $a < 1$. □

8.7 An Application of the Courant–Fischer Theory

Consider two $n \times n$ Hermitian matrices H_1 and H_2. We say that H_2 is *not greater* than H_1, written $H_1 \geq H_2$ or $H_2 \leq H_1$, if the difference $H_1 - H_2$ is a positive semi-definite matrix. Similarly, we write $H_1 > H_2$ or $H_2 < H_1$ if $H_1 - H_2 > 0$. These relations preserve some properties of the natural ordering of real numbers (see Exercise 5.14.21). However, there are properties of numerical inequalities that fail to be true for Hermitian matrices. For instance, from $H_1 \geq H_2 \geq 0$ the inequality $H_1^2 \geq H_2^2$ does not follow in general.

Exercise 1. Check that $H_1 \geq H_2 \geq 0$ if

$$H_1 = \begin{bmatrix} 2 & 1 \\ 1 & 2 \end{bmatrix}, \qquad H_2 = \begin{bmatrix} 2 & 1 \\ 1 & 1 \end{bmatrix},$$

while $H_1^2 - H_2^2$ fails to be a positive semi-definite matrix. □

Applying the Courant–Fischer theory, a relation between the eigenvalues of matrices H_1 and H_2 satisfying the condition $H_1 \geq H_2$ can be derived. This can also be viewed as a *perturbation* result: What can be said about the perturbations of eigenvalues of H_2 under addition of the positive semi-definite matrix $H_1 - H_2$?

Theorem 1. *Let $H_1, H_2 \in \mathscr{F}^{n \times n}$ be Hermitian matrices for which $H_1 \geq H_2$. If $\mu_1^{(k)} \leq \mu_2^{(k)} \leq \cdots \leq \mu_n^{(k)}$ denote the eigenvalues of H_k $(k = 1, 2)$ and the rank of the matrix $H = H_1 - H_2$ is r, then*

(*a*) $\mu_i^{(1)} \geq \mu_i^{(2)}$ *for* $i = 1, 2, \ldots, n$;
(*b*) $\mu_j^{(1)} \leq \mu_{j+r}^{(2)}$ *for* $j = 1, 2, \ldots, n - r$.

PROOF. Observe that the condition $H \geq 0$ implies, by Theorem 5.3.3, that

$$(H_1 x, x) = ((H_2 + H)x, x) = (H_2 x, x) + (Hx, x) \geq (H_2 x, x)$$

for any $x \in \mathscr{F}^n$. Thus, if $\mathscr{S}$ is any subspace in $\mathscr{F}^n$, then

$$\min(H_1 x, x) \geq \min(H_2 x, x), \tag{1}$$

where the minima are over all vectors $x \in \mathscr{S}$ such that $(x, x) = 1$. In particular, considering all the subspaces $\mathscr{S}_i$ of dimension $n - i + 1$, where $1 \leq i \leq n$, we deduce from (1) and Theorem 8.2.2 that

$$\mu_i^{(1)} = \max_{\mathscr{S}_i} \min_{\substack{x \in \mathscr{S}_i \\ (x, x) = 1}} (H_1 x, x) \geq \max_{\mathscr{S}_i} \min_{\substack{x \in \mathscr{S}_i \\ (x, x) = 1}} (H_2 x, x) = \mu_i^{(2)}$$

for $i = 1, 2, \ldots, n$. This proves part (*a*) of the theorem.

For the second part, assume that H has positive eigenvalues $\nu_1, \ldots, \nu_r$ and eigenvalues $\nu_{r+1} = \cdots = \nu_n = 0$, together with an associated orthonormal basis of eigenvectors $x_1, x_2, \ldots, x_n$. For any $x \in \mathbb{C}^n$ write $x = \sum_{j=1}^n \alpha_j x_j$. Then we have [see also Eq. (5.10.9)]

$$(Hx, x) = \sum_{j=1}^r \nu_j |\alpha_j|^2.$$

Write $y_j = \nu_j^{1/2} x_j$ for $j = 1, 2, \ldots, r$ and use the fact that $\alpha_j = (x, x_j)$ (Exercise 3.12.3) to obtain

$$\sum_{j=1}^r \nu_j |\alpha_j|^2 = \sum_{j=1}^r |(x, y_j)|^2,$$

and note that $\boldsymbol{y}_1, \ldots, \boldsymbol{y}_r$ are obviously linearly independent. Since $H_1 = H_2 + H$, we now have

$$(H_1\boldsymbol{x}, \boldsymbol{x}) = (H_2\boldsymbol{x}, \boldsymbol{x}) + \sum_{i=1}^{r} |(\boldsymbol{x}, \boldsymbol{y}_i)|^2, \tag{2}$$

and we consider the eigenvalue problems for H_1 and H_2 both with constraints $(\boldsymbol{x}, \boldsymbol{y}_i) = 0$, $i = 1, 2, \ldots, r$. Obviously, in view of Eq. (2), the two problems coincide and have the same eigenvalues $\lambda_1 \leq \lambda_2 \leq \cdots \leq \lambda_{n-r}$. Applying Theorem 8.4.1 to both problems, we obtain

$$\mu_j^{(k)} \leq \lambda_j \leq \mu_{j+r}^{(k)}, \qquad k = 1, 2; \quad j = 1, 2, \ldots, n - r,$$

which implies the desired result:

$$\mu_j^{(1)} \leq \lambda_j \leq \mu_{j+r}^{(2)}. \quad \blacksquare$$

8.8 Applications to the Theory of Small Vibrations

We return now to the problem discussed in Section 5.12. In that section we reduced a physical vibration problem to the problem of solving the matrix differential equation $A\ddot{\boldsymbol{p}} + C\boldsymbol{p} = \boldsymbol{0}$, where A is positive definite and C is either positive definite or semi-definite. This then reduced to the purely algebraic problem of using a congruence transformation to reduce A and C simultaneously to diagonal matrices. This, in turn, was seen to depend on the reduction of the matrix $B = A^{-1/2}CA^{-1/2}$ to diagonal form by means of an orthogonal congruence transformation.

The natural frequencies of vibration of the system are given by the non-negative square roots of the zeros of $\det(A\lambda - C)$, which are also the eigenvalues of B. We know now, therefore, that the largest natural frequency, $\omega_n = \lambda_n^{1/2}$, is given by $\lambda_n = \max R_1(\boldsymbol{x})$, where R_1 is the Rayleigh quotient for B and $\boldsymbol{x} \neq \boldsymbol{0}$ ranges over $\mathbb{R}^n$ (see Exercise 8.2.1). Theorems 8.2.1 and 8.2.2 imply that all the natural frequencies can be defined in terms of extreme values of R_1. However, observe that if we write $\boldsymbol{x} = A^{1/2}\boldsymbol{q}$, then

$$R_1(\boldsymbol{x}) = \frac{\boldsymbol{x}^{\mathrm{T}}A^{-1/2}CA^{-1/2}\boldsymbol{x}}{\boldsymbol{x}^{\mathrm{T}}\boldsymbol{x}} = \frac{\boldsymbol{q}^{\mathrm{T}}C\boldsymbol{q}}{\boldsymbol{q}^{\mathrm{T}}A\boldsymbol{q}}.$$

The Rayleigh quotient appropriate to this problem is therefore chosen to be

$$R(\boldsymbol{q}) = \frac{\boldsymbol{q}^{\mathrm{T}}C\boldsymbol{q}}{\boldsymbol{q}^{\mathrm{T}}A\boldsymbol{q}}.$$

Notice also that if $\boldsymbol{x}_j$ is an eigenvector of B, then $\boldsymbol{q}_j = A^{-1/2}\boldsymbol{x}_j$ is a normal-mode vector for the vibration problem and, as in Eqs. (5.12.5),

$$\boldsymbol{q}_j^{\mathrm{T}} A \boldsymbol{q}_k = \delta_{jk}, \qquad \boldsymbol{q}_j^{\mathrm{T}} C \boldsymbol{q}_k = \omega_j^2 \delta_{jk}.$$

On applying each theorem of this chapter to B, it is easily seen that a generalization is obtained that is applicable to the zeros of $\det(A\lambda - C)$. As an example, we merely state the generalized form of the Courant–Fischer theorem of Section 8.2.

Theorem 1. *Given any linearly independent vectors* $\boldsymbol{a}_1, \boldsymbol{a}_2, \ldots, \boldsymbol{a}_{i-1}$ *in* $\mathbb{R}^n$, *let* $\mathscr{S}_i$ *be the subspace consisting of all vectors in* $\mathbb{R}^n$ *that are orthogonal to* $\boldsymbol{a}_1, \ldots, \boldsymbol{a}_{i-1}$. *Then*

$$\lambda_i = \max_{\mathscr{S}_i} \min_{\substack{\boldsymbol{x} \in \mathscr{S}_i \\ \boldsymbol{x} \neq \boldsymbol{0}}} \left(\frac{\boldsymbol{x}^{\mathrm{T}} C \boldsymbol{x}}{\boldsymbol{x}^{\mathrm{T}} A \boldsymbol{x}} \right).$$

The minimum is attained when $\boldsymbol{a}_j = \boldsymbol{q}_j, j = 1, 2, \ldots, i - 1$.

The physical interpretations of Theorems 8.4.1 and 8.7.1 are more obvious in this setting. The addition of a constraint may correspond to holding one point of the system so that there is no vibration at this point. The perturbation considered in Theorem 8.7.1 corresponds to an increase in the stiffness of the system. The theorem tells us that this can only result in an increase in the natural frequencies.

CHAPTER 9

Functions of Matrices

In this chapter we consider matrices A from $C^{n \times n}$ (including those of $\mathbb{R}^{n \times n}$ as a special case) and the possibility of giving a meaning to $f(A)$, where $f(\lambda)$ is a complex-valued function of a complex variable. We would like the definition of $f(A)$ to hold for as wide a class of functions $f(\lambda)$ as possible. We have seen that the question is easily resolved if $f(\lambda) = p(\lambda)$ is a polynomial over the complex numbers; thus,

$$p(A) \triangleq \sum_{i=1}^{l} p_i A^i \qquad \text{if} \qquad p(\lambda) = \sum_{i=0}^{l} p_i \lambda^i.$$

Moreover, if $m(\lambda)$ is the minimal polynomial for A, then for a polynomial $p(\lambda)$ there are polynomials $q(\lambda)$ and $r(\lambda)$ such that

$$p(\lambda) = m(\lambda)q(\lambda) + r(\lambda),$$

where $r(\lambda)$ is the zero polynomial or has degree less than that of $m(\lambda)$. Thus, since $m(A) = O$, we have $p(A) = r(A)$.

The more general functions $f(\lambda)$ that we shall consider will retain this property; that is to say, given $f(\lambda)$ and A, there is a polynomial $r(\lambda)$ (with degree less than that of the minimal polynomial of A) such that $f(A) = r(A)$. It is surprising but true that this constraint still leaves considerable freedom in the nature of the functions $f(\lambda)$ that can be accommodated.

9.1 Functions Defined on the Spectrum of a Matrix

Let $A \in \mathbb{C}^{n \times n}$ and suppose that $\lambda_1, \lambda_2, \ldots, \lambda_s$ are the distinct eigenvalues of A, so that

$$m(\lambda) = (\lambda - \lambda_1)^{m_1}(\lambda - \lambda_2)^{m_2} \cdots (\lambda - \lambda_s)^{m_s} \tag{1}$$

is the minimal polynomial of A with degree $m = m_1 + m_2 + \cdots + m_s$. Recall that the multiplicity m_k of λ_k as a zero of the minimal polynomial is referred to as the index of the eigenvalue λ_k (Section 6.2) and is equal to the maximal degree of the elementary divisors associated with λ_k $(1 \le k \le s)$.

Given a function $f(\lambda)$, we say that it is *defined on the spectrum of A* if the numbers

$$f(\lambda_k), f'(\lambda_k), \ldots, f^{(m_k - 1)}(\lambda_k), \qquad k = 1, 2, \ldots, s, \tag{2}$$

called *the values of $f(\lambda)$ on the spectrum of A*, exist. Clearly, every polynomial over $\mathbb{C}$ is defined on the spectrum of any matrix from $\mathbb{C}^{n \times n}$. More can be said about minimal polynomials.

Exercise 1. Check that the values of the minimal polynomial of A on the spectrum of A are all zeros. □

The next result provides a criterion for two scalar polynomials $p_1(\lambda)$ and $p_2(\lambda)$ to produce the same matrix $p_1(A) = p_2(A)$. These matrices are defined as described in Section 1.7.

Proposition 1. *If $p_1(\lambda)$ and $p_2(\lambda)$ are polynomials over $\mathbb{C}$ and $A \in \mathbb{C}^{n \times n}$, then $p_1(A) = p_2(A)$ if and only if the polynomials $p_1(\lambda)$ and $p_2(\lambda)$ have the same values on the spectrum of A.*

PROOF. If $p_1(A) = p_2(A)$ and $p_0(\lambda) = p_1(\lambda) - p_2(\lambda)$, then obviously, $p_0(A) = 0$ and hence $p_0(\lambda)$ is an annihilating polynomial for A. Thus (Theorem 6.2.1) $p_0(\lambda)$ is divisible by the minimal polynomial $m(\lambda)$ of A given by (1) and there exists a polynomial $q(\lambda)$ such that $p_0(\lambda) = q(\lambda)m(\lambda)$. Computing the values of $p_0(\lambda)$ on the spectrum of A and using Exercise 1, it is easily seen that

$$p_1^{(j)}(\lambda_k) - p_2^{(j)}(\lambda_k) = p_0^{(j)}(\lambda_k) = 0 \tag{3}$$

for $j = 0, 1, \ldots, m_k - 1$, $1 \le k \le s$. Thus, the two polynomials $p_1(\lambda)$ and $p_2(\lambda)$ have the same values on the spectrum of A provided $p_1(A) = p_2(A)$.

Conversely, if (3) holds, then $p_0(\lambda)$ has a zero of multiplicity m_k at the point λ_k for each $k = 1, 2, \ldots, s$. Hence $p_0(\lambda)$ must be divisible by $m(\lambda)$ in (1) and it follows that $p_0(A) = 0$ (see Corollary 1 of Theorem 6.2.2) or, equivalently, $p_1(A) = p_2(A)$. ■

It is this property of polynomials with matrix arguments that we now use to extend the definition of $f(A)$ to more general functions $f(\lambda)$. Thus, we shall demand that all functions that are defined on the spectrum of A and that take on the same values there should yield the same matrix $f(A)$. In particular, for any $f(\lambda)$ defined on the spectrum of A we shall be able to write $f(A) = p(A)$, where $p(\lambda)$ is a polynomial with the same values on the spectrum of A. The existence of such a polynomial $p(\lambda)$ with the prescribed properties follows from the solution of the problem of general Hermite interpolation, which we must now consider.

9.2 Interpolatory Polynomials

Proposition 1. *Given distinct numbers $\lambda_1, \lambda_2, \ldots, \lambda_s$, positive integers $m_1, m_2, \ldots, m_s$ with $m = \sum_{k=1}^{s} m_k$, and a set of numbers*

$$f_{k,0}, \quad f_{k,1}, \ldots, \quad f_{k,m_k-1}, \qquad k = 1, 2, \ldots, s,$$

there exists a polynomial $p(\lambda)$ of degree less than m such that

$$p(\lambda_k) = f_{k,0}, \qquad p^{(1)}(\lambda_k) = f_{k,1}, \ldots, \qquad p^{(m_k-1)}(\lambda_k) = f_{k,m_k-1} \tag{1}$$

for $k = 1, 2, \ldots, s$.

PROOF. It is easily seen that the polynomial $p_k(\lambda) = \alpha_k(\lambda)\psi_k(\lambda)$,† where $1 \le k \le s$ and

$$\alpha_k(\lambda) = \alpha_{k,0} + \alpha_{k,1}(\lambda - \lambda_k) + \cdots + \alpha_{k,m_k-1}(\lambda - \lambda_k)^{m_k-1},$$

$$\psi_k(\lambda) = \prod_{j=1,\, j\neq k}^{s} (\lambda - \lambda_j)^{m_j},$$

has degree less than m and satisfies the conditions

$$p_k(\lambda_i) = p_k^{(1)}(\lambda_i) = \cdots = p_k^{(m_i-1)}(\lambda_i) = 0$$

for $i \neq k$ and arbitrary $\alpha_{k,0}, \alpha_{k,1}, \ldots \alpha_{k,m_k-1}$. Hence the polynomial

$$p(\lambda) = p_1(\lambda) + p_2(\lambda) + \cdots + p_s(\lambda) \tag{2}$$

satisfies conditions (1) if and only if

$$p_k(\lambda_k) = f_{k,0}, \qquad p_k^{(1)}(\lambda_k) = f_{k,1}, \ldots, \qquad p_k^{(m_k-1)}(\lambda_k) = f_{k,m_k-1} \tag{3}$$

† If $s = 1$, then by definition $\psi_1(\lambda) \equiv 1$.

for each $1 \le k \le s$. By differentiation,

$$p_k^{(j)}(\lambda_k) = \sum_{i=0}^{j} \binom{j}{i} \alpha_k^{(i)}(\lambda_k)\psi_k^{(j-i)}(\lambda_k)$$

for $1 \le k \le s$, $0 \le j \le m_k - 1$. Using Eqs. (3) and recalling the definition of $\alpha_k(\lambda)$, we have for $k = 1, 2, \dots, s$, $j = 0, 1, \dots, m_k - 1$,

$$f_{k,j} = \sum_{i=0}^{j} \binom{j}{i} i!\,\alpha_{k,i}\psi_k^{(j-i)}(\lambda_k). \tag{4}$$

Since $\psi_k(\lambda_k) \ne 0$ for each fixed k, Eqs. (4) can now be solved successively (beginning with $j = 0$) to find the coefficients $\alpha_{k,0}, \dots, \alpha_{k,m_k-1}$ for which (3) holds. Thus, a polynomial $p(\lambda)$ of the form given in (2) satisfies the required conditions. ■

Proposition 2. *The polynomial $p(\lambda)$ of Proposition 1 is unique.*

The unique polynomial of degree less than m that satisfies conditions (1) is known as the *Hermite interpolating polynomial.*

This is an important step in our argument but the proof is cumbersome and is therefore omitted. It can be made to depend on the nonsingularity of a generalized Vandermonde matrix (see Exercise 2.11.22).

Exercise 1. If $m_1 = m_2 = \cdots = m_s = 1$, then conditions (1) reduce to the simple "point wise" conditions of interpolation: $p(\lambda_k) = f_{k,0}$ for $k = 1, 2, \dots, s$. Show that these conditions are satisfied by the *Lagrange interpolating polynomial*:

$$p(\lambda) = \sum_{k=1}^{s} f_{k,0} \frac{(\lambda - \lambda_1)\cdots(\lambda - \lambda_{k-1})(\lambda - \lambda_{k+1})\cdots(\lambda - \lambda_s)}{(\lambda_k - \lambda_1)\cdots(\lambda_k - \lambda_{k-1})(\lambda_k - \lambda_{k+1})\cdots(\lambda_k - \lambda_s)}. \tag{5}$$

Exercise 2. Prove the result of Proposition 2 in the case described in Exercise 1, that is, when $m_1 = m_2 = \cdots = m_s = 1$.

Hint. First show that a polynomial of degree less than m that vanishes at m distinct points is necessarily the zero polynomial. This can be done using the nonsingularity of a certain Vandermonde matrix (see Exercise 2.3.6). □

A useful role is played by those polynomials for which numbers $f_{k,i}$ in (1) are all zero except one, and the nonzero number, say $f_{k,j}$ is equal to 1. Thus Propositions 1 and 2 guarantee the existence of a unique polynomial $\varphi_{kj}(\lambda)$ of degree less than m (with $1 \le k \le s$ and $0 \le j \le m_k - 1$) such that

$$\varphi_{kj}^{(r)}(\lambda_k) = \delta_{jr}, \qquad r = 0, 1, \dots, m_k - 1; \tag{6}$$

and when $i \ne k$,

$$\varphi_{kj}^{(r)}(\lambda_i) = 0, \qquad r = 0, 1, \dots, m_i - 1. \tag{7}$$

These m polynomials are called the *fundamental* or *cardinal polynomials* of the interpolation problem.

Once these polynomials are known, the solution to every problem with the general conditions (1) can be expressed as a linear combination of them. For, if $p(\lambda)$ satisfies (1) then, as is easily verified,

$$p(\lambda) = \sum_{k=1}^{s} \sum_{j=0}^{m_k - 1} f_{k,j} \varphi_{kj}(\lambda). \tag{8}$$

Exercise 3. Verify that the m polynomials $\varphi_{kj}(\lambda)$ $(1 \le k \le s$ and $0 \le j \le m_k - 1)$ just defined are linearly independent and hence form a basis for the space $\mathscr{P}_{m-1}$ of all polynomials of degree less than m.

Exercise 4. Show that, with the hypothesis of Exercise 1, the fundamental interpolatory polynomials are given by

$$\varphi_{k0}(\lambda) \equiv \varphi_k(\lambda) = \prod_{j=1,\, j\neq k}^{s} (\lambda - \lambda_j) \Bigg/ \prod_{j=1,\, j\neq k}^{s} (\lambda_k - \lambda_j)$$

for $k = 1, 2, \ldots, s$. □

9.3 Definition of a Function of a Matrix

Using the concepts and results of the two preceding sections, a general definition of the matrix $f(A)$ in terms of matrix A and the function f can be made, provided $f(\lambda)$ is defined on the spectrum of A. If $p(\lambda)$ is any polynomial that assumes the same values as $f(\lambda)$ on the spectrum of A, we simply define $f(A) \triangleq p(A)$. Proposition 9.2.1 assures us that such a polynomial exists, while Proposition 9.1.1 shows that, for the purpose of the formal definition, the choice of $p(\lambda)$ is not important. Moreover, Eq. (9.2.8) indicates how $p(\lambda)$ might be chosen with the least possible degree.

Exercise 1. If A is a simple matrix with distinct eigenvalues $\lambda_1, \ldots, \lambda_s$ and $f(\lambda)$ is any function that is well defined at the eigenvalues of A, show that

$$f(A) = \sum_{k=1}^{s} f(\lambda_k) \left[\prod_{j=1,\, j\neq k}^{s} (A - \lambda_j I) \Bigg/ \prod_{j=1,\, j\neq k}^{s} (\lambda_k - \lambda_j) \right].$$

Exercise 2. Calculate $f(A)$, where $f(\lambda) = e^{2\lambda}$ and

$$A = \begin{bmatrix} 6 & -1 \\ 3 & 2 \end{bmatrix}.$$

SOLUTION. The eigenvalues of A are found to be $\lambda_1 = 3$, $\lambda_2 = 5$, so that the minimal polynomial $m(\lambda)$ of A is $(\lambda - 3)(\lambda - 5)$. The Lagrange interpolatory polynomial is

$$\begin{aligned} p(\lambda) &= f(\lambda_1)\frac{\lambda - \lambda_2}{\lambda_1 - \lambda_2} + f(\lambda_2)\frac{\lambda - \lambda_1}{\lambda_2 - \lambda_1} \\ &= -\tfrac{1}{2}e^6(\lambda - 5) + \tfrac{1}{2}e^{10}(\lambda - 3). \end{aligned}$$

Hence, by definition,

$$\begin{aligned} e^{2A} &= -\tfrac{1}{2}e^6(A - 5I) + \tfrac{1}{2}e^{10}(A - 3I) \\ &= \tfrac{1}{2}\begin{bmatrix} 3e^{10} - e^6 & -e^{10} + e^6 \\ 3e^{10} - 3e^6 & -e^{10} + 3e^6 \end{bmatrix}. \end{aligned}$$

Exercise 3. If

$$A = \frac{\pi}{2}\begin{bmatrix} 2 & 0 & 0 \\ 0 & 1 & 1 \\ 0 & 0 & 1 \end{bmatrix},$$

prove that

$$\sin A = \frac{4}{\pi}A - \frac{4}{\pi^2}A^2 = \begin{bmatrix} 0 & 0 & 0 \\ 0 & 1 & 0 \\ 0 & 0 & 1 \end{bmatrix}.$$

Exercise 4. If $f(\lambda)$ and $g(\lambda)$ are defined on the spectrum of A and $h(\lambda) = \alpha f(\lambda) + \beta g(\lambda)$, where $\alpha, \beta \in \mathbb{C}$, and $k(\lambda) = f(\lambda)g(\lambda)$, show that $h(\lambda)$ and $k(\lambda)$ are defined on the spectrum of A, and

$$\begin{aligned} h(A) &= \alpha f(A) + \beta g(A), \\ k(A) &= f(A)g(A) = g(A)f(A). \end{aligned}$$

Exercise 5. If $f_1(\lambda) = \lambda^{-1}$ and $\det A \neq 0$, show that $f_1(A) = A^{-1}$. Show also that if $f_2(\lambda) = (a - \lambda)^{-1}$ and $a \notin \sigma(A)$, then $f_2(A) = (aI - A)^{-1}$.

Hint: For the first part observe that if $p(\lambda)$ is an interpolatory polynomial for $f_1(\lambda)$, then the values of $\lambda p(\lambda)$ and $\lambda f_1(\lambda) = 1$ coincide on the spectrum of A and hence $Ap(A) = Af_1(A) = I$.

Exercise 6. Use the ideas of this section to determine a positive definite square root for a positive definite matrix (see Theorem 5.4.1). Observe that there are, in general, many square roots. ☐

We conclude this section with the important observation that if $f(\lambda)$ is a polynomial, then the matrix $f(A)$ determined by the definition of Section 1.7 coincides with that obtained from the more general definition introduced in this section.

9.4 Properties of Functions of Matrices

In this section we obtain some important properties of $f(A)$, which, in particular, allow us to compute $f(A)$ very easily, provided the Jordan decomposition of A is known. These properties will also establish important relationships between Jordan forms for A and $f(A)$.

Theorem 1. *If $A \in C^{n \times n}$ is a block-diagonal matrix,*

$$A = \mathrm{diag}[A_1, A_2, \ldots, A_t],$$

and the function $f(\lambda)$ is defined on the spectrum of A, then

$$f(A) = \mathrm{diag}[f(A_1), f(A_2), \ldots, f(A_t)]. \tag{1}$$

PROOF. First, it is clear that for any polynomial $q(\lambda)$,

$$q(A) = \mathrm{diag}[q(A_1), q(A_2), \ldots, q(A_t)].$$

Hence, if $p(\lambda)$ is the interpolatory polynomial for $f(\lambda)$ on the spectrum of A, we have

$$f(A) = p(A) = \mathrm{diag}[p(A_1), p(A_2), \ldots, p(A_t)].$$

Second, since the spectrum of A_j $(1 \leq j \leq t)$ is obviously a subset of the spectrum of A, the function $f(\lambda)$ is defined on the spectrum of A_j for each $j = 1, 2, \ldots, t$. (Note also that the index of an eigenvalue of A_j cannot exceed the index of the same eigenvalue of A.) Furthermore, since $f(\lambda)$ and $p(\lambda)$ assume the same values on the spectrum of A, they must also have the same values on the spectrum of A_j $(j = 1, 2, \ldots, t)$. Hence $f(A_j) = p(A_j)$ and we obtain Eq. (1). ■

The next result provides a relation between $f(A)$ and $f(B)$ when A and B are similar matrices.

Theorem 2. *If $A, B, P \in \mathbb{C}^{n \times n}$, where $B = PAP^{-1}$, and $f(\lambda)$ is defined on the spectrum of A, then*

$$f(B) = Pf(A)P^{-1}. \tag{2}$$

PROOF. Since A and B are similar, they have the same minimal polynomial (Section 6.2). Thus, if $p(\lambda)$ is the interpolatory polynomial for $f(\lambda)$ on the spectrum of A, then it is also the interpolatory polynomial for $f(\lambda)$ on the spectrum of B. Thus we have $f(A) = p(A)$, $f(B) = p(B)$, $p(B) = Pp(A)P^{-1}$, and so the relation (2) follows. ■

In view of the Jordan theorem, Theorems 1 and 2 imply a related theorem about functions of matrices.

Theorem 3. *Let $A \in \mathbb{C}^{n \times n}$ and let $J = \text{diag}[J_j]_{j=1}^{t}$ be the Jordan canonical form of A, where $A = PJP^{-1}$ and J_j is the jth Jordan block of J. Then*

$$f(A) = P\,\text{diag}[f(J_1), f(J_2), \ldots, f(J_t)]P^{-1}. \tag{3}$$

The last step in computing $f(A)$ by use of the Jordan form of A consists, therefore, of the following formula.

Theorem 4. *Let J_0 be a Jordan block of size l associated with λ_0;*

$$J_0 = \begin{bmatrix} \lambda_0 & 1 & & \\ & \lambda_0 & \ddots & \\ & & \ddots & 1 \\ & & & \lambda_0 \end{bmatrix}.$$

If $f(\lambda)$ is an $(l-1)$-times differentiable function in a neighborhood of λ_0, then

$$f(J_0) = \begin{bmatrix} f(\lambda_0) & \frac{1}{1!}f'(\lambda_0) & \cdots & \frac{1}{(l-1)!}f^{(l-1)}(\lambda_0) \\ 0 & f(\lambda_0) & \ddots & \vdots \\ \vdots & \ddots & \ddots & \frac{1}{1!}f'(\lambda_0) \\ 0 & \cdots & 0 & f(\lambda_0) \end{bmatrix}. \tag{4}$$

PROOF. The minimal polynomial of J_0 is $(\lambda - \lambda_0)^l$ (see Exercise 7.7.4) and the values of $f(\lambda)$ on the spectrum of J_0 are therefore $f(\lambda_0), f'(\lambda_0), \ldots, f^{(l-1)}(\lambda_0)$. The interpolatory polynomial $p(\lambda)$, defined by the values of $f(\lambda)$ on the spectrum $\{\lambda_0\}$ of J_0, is found by putting $s = 1, m_k = l, \lambda_1 = \lambda_0$, and $\psi_1(\lambda) \equiv 1$ in Eqs. (9.2.2) through (9.2.4). One obtains

$$p(\lambda) = \sum_{i=0}^{l-1} \frac{1}{i!} f^{(i)}(\lambda_0)(\lambda - \lambda_0)^i.$$

The fact that the polynomial $p(\lambda)$ solves the interpolation problem $p^{(j)}(\lambda_0) = f^{(j)}(\lambda_0)$, $1 \le j \le l-1$, can also be easily checked by a straightforward calculation.

We then have $f(J_0) = p(J_0)$ and hence

$$f(J_0) = \sum_{i=0}^{l-1} \frac{1}{i!} f^{(i)}(\lambda_0)(J_0 - \lambda_0 I)^i.$$

Computing the powers of $J_0 - \lambda_0 I$, we obtain

$$(J_0 - \lambda_0 I)^i = \begin{bmatrix} 0 & 1 & & \\ & 0 & \ddots & \\ & & \ddots & 1 \\ & & & 0 \end{bmatrix}^i = \begin{bmatrix} \overbrace{0 & \cdots & 0}^{i} & 1 & 0 & \cdots & 0 \\ & & & & \ddots & \ddots & \vdots \\ & & & & & \ddots & 0 \\ & & & & & \ddots & 1 \\ & & & & & & 0 \\ & & & & & & \vdots \\ & & & & & & 0 \end{bmatrix}$$

and Eq. (4) follows. ■

Thus, given a Jordan decomposition of the matrix A, the matrix $f(A)$ is easily found by combining Theorems 3 and 4.

Theorem 5. *With the notation of Theorem* 3,

$$f(A) = P \operatorname{diag}[f(J_1), f(J_2), \ldots, f(J_t)]P^{-1},$$

where $f(J_i)$ $(i = 1, 2, \ldots, t)$ *are upper-triangular matrices of the form given in Eq.* (4).

Exercise 1. Verify (using the real natural logarithm) that

$$\ln\left(\begin{bmatrix} 6 & 2 & 2 \\ -2 & 2 & 0 \\ 0 & 0 & 2 \end{bmatrix}\right) = \begin{bmatrix} 0 & 2 & 1 \\ -1 & -2 & 0 \\ 1 & 0 & 0 \end{bmatrix} \begin{bmatrix} \ln 2 & 0 & 0 \\ 0 & \ln 4 & \frac{1}{4} \\ 0 & 0 & \ln 4 \end{bmatrix} \begin{bmatrix} 0 & 2 & 1 \\ -1 & -2 & 0 \\ 1 & 0 & 0 \end{bmatrix}^{-1}$$

Hint. Use Example 6.5.2. □

We observe that if J_i is associated with the eigenvalue λ_i, then the diagonal elements of $f(J_i)$ are all equal to $f(\lambda_i)$, and since the eigenvalues of a triangular matrix are just its diagonal elements, it follows that the eigenvalues of $f(A)$, possibly multiple, are $f(\lambda_1), f(\lambda_2), \ldots, f(\lambda_s)$. Compare the next result with Theorem 4.11.3.

Theorem 6. *If* $\lambda_1, \lambda_2, \ldots, \lambda_n$ *are the eigenvalues of the matrix* $A \in \mathbb{C}^{n \times n}$ *and* $f(\lambda)$ *is defined on the spectrum of* A, *then the eigenvalues of* $f(A)$ *are* $f(\lambda_1), f(\lambda_2), \ldots, f(\lambda_n)$.

Exercise 2. Check that e^A is nonsingular for any $A \in \mathbb{C}^{n \times n}$.

Exercise 3. Show that, if all the eigenvalues of the matrix $A \in \mathbb{C}^{n \times n}$ lie in the open unit disk, then $A + I$ is nonsingular.

Exercise 4. If the eigenvalues of $A \in \mathbb{C}^{n \times n}$ lie in the open unit disk, show that Re $\lambda < 0$ for any $\lambda \in \sigma(C)$, where $C = (A + I)^{-1}(A - I)$. □

The elementary divisors associated with a particular eigenvalue are not necessarily preserved under the transformation determined by $f(\lambda)$, in that an elementary divisor $(\lambda - \lambda_i)^r$ of $\lambda I - A$ is not necessarily carried over to an elementary divisor $(\lambda - f(\lambda_i))^r$ for $\lambda I - f(A)$. Exercise 9.3.3 illustrates this fact by showing that the nonlinear elementary divisor associated with the eigenvalue $\pi/2$ of A splits into two linear elementary divisors of the eigenvalue $f(\pi/2) = 1$ of $f(A)$. A description of the elementary divisors of $\lambda I - f(A)$ in terms of those of $\lambda I - A$ is given in the following theorem.

Theorem 7. *Let $\lambda_1, \lambda_2, \ldots, \lambda_s$ be the distinct eigenvalues of $A \in \mathbb{C}^{n \times n}$, and let the function $f(\lambda)$ be defined on the spectrum of A.*

(a) If $f'(\lambda_k) \neq 0$, the elementary divisors of $\lambda I - f(A)$ associated with the eigenvalue $f(\lambda_k)$ are obtained by replacing each elementary divisor $(\lambda - \lambda_k)^r$ of $\lambda I - A$ associated with λ_k by $(\lambda - f(\lambda_k))^r$.

(b) If $f'(\lambda_k) = f''(\lambda_k) = \cdots = f^{(l-1)}(\lambda_k) = 0$, but $f^{(l)}(\lambda_k) \neq 0$, where $l \geq r$, then an elementary divisor $(\lambda - \lambda_k)^r$ of $\lambda I - A$ splits into r linear elementary divisors of $\lambda I - f(A)$ of the form $\lambda - f(\lambda_k)$.

(c) If $f'(\lambda_k) = f''(\lambda_k) = \cdots = f^{(l-1)}(\lambda_k) = 0$ but $f^{(l)}(\lambda_k) \neq 0$, where $2 \leq l \leq r - 1$, the elementary divisors of $\lambda I - f(A)$ associated with $f(\lambda_k)$ are obtained by replacing each elementary divisor $(\lambda - \lambda_k)^r$ of $\lambda I - A$ associated with λ_k by l elementary divisors $(\lambda - \lambda_k)^p$ (taken $l - q$ times) and $(\lambda - \lambda_k)^{p+1}$ (taken q times), where the integers p and q satisfy the condition $r = lp + q$ $(0 \leq q \leq l - 1, p > 0)$.

Note that the cases (a) and (b) can be viewed as extreme cases of (c).

PROOF. In view of Theorems 5 and 7.7.1, since each elementary divisor $(\lambda - \lambda_k)^r$ of $\lambda I - A$ generates a Jordan block J_k of size r associated with λ_k, it suffices to determine the elementary divisors of $\lambda I - f(J_k)$.

If $f'(\lambda_k) \neq 0$ in Eq. (4), then by Exercise 7.7.7 the unique elementary divisor of $f(J_k)$ is $(\lambda - f(\lambda_k))^r$, which proves part (a). Part (b) is checked by observing that $f(J_k)$ is in this case a diagonal matrix and applying Exercise 7.7.6. The result of Exercise 7.7.8 readily implies the required statement in (c). ■

Exercise 5. If B is positive definite, show that there is a unique Hermitian matrix H such that $B = e^H$. (This matrix is, of course, a natural choice for ln B.) Give an example of a non-Hermitian matrix A such that e^A is positive definite. □

9.5 Spectral Resolution of *f*(*A*)

The following decomposition, or *spectral resolution*, of $f(A)$ is important and is obviously closely related to the spectral theorem for simple matrices, which we have already discussed in Theorem 4.10.3 and Exercise 4.14.3. It gives a useful explicit representation of $f(A)$ in terms of fundamental properties of A.

Theorem 1. *If* $A \in \mathbb{C}^{n \times n}$, $f(\lambda)$ *is defined on the spectrum* $\{\lambda_1, \lambda_2, \ldots, \lambda_s\}$ *of* A, $f_{k,j}$ *is the value of the jth derivative of* $f(\lambda)$ *at the eigenvalue* λ_k ($k = 1, 2, \ldots, s, j = 0, 1, \ldots, m_k - 1$), *and* m_k *is the index of* λ_k, *then there exist matrices* Z_{kj} *independent of* $f(\lambda)$ *such that*

$$f(A) = \sum_{k=1}^{s} \sum_{j=0}^{m_k - 1} f_{k,j} Z_{kj}. \tag{1}$$

Moreover, the matrices Z_{kj} *are linearly independent members of* $\mathbb{C}^{n \times n}$ *and commute with* A *and with each other.*

PROOF. Recall that the function $f(\lambda)$ and the polynomial $p(\lambda)$ defined by Eq. (9.2.8) take the same values on the spectrum of A. Hence $f(A) = p(A)$ and Eq. (1) is obtained on putting $Z_{kj} = \varphi_{kj}(A)$. It also follows immediately that A and all of the Z_{kj} are mutually commutative.

It remains only to prove that these matrices are linearly independent. Suppose that

$$\sum_{k=1}^{s} \sum_{j=0}^{m_k - 1} c_{kj} Z_{kj} = O$$

and define $h(\lambda) = \sum_{k,j} c_{kj} \varphi_{kj}(\lambda)$; then $h(\lambda)$ is a polynomial of the space $\mathscr{P}_{m-1}$; that is, the degree of $h(\lambda)$ does not exceed $m - 1$. Then $Z_{kj} = \varphi_{kj}(A)$ implies that $h(A) = O$. But m is the degree of the minimal polynomial of A and, since h annihilates A, it follows from Theorem 6.2.1 that h is the zero polynomial. Hence, since the φ_{kj}'s are independent by Exercise 9.2.3, we have $c_{kj} = 0$ for each k and j. Thus, the matrices Z_{kj} are linearly independent. ■

We shall call the matrices Z_{kj} the *components* of A; note that, because they are linearly independent, none of them can be a zero-matrix.

Let us investigate the case of simple matrices more closely. We now have $m_1 = m_2 = \cdots = m_s = 1$ in (1) and so, for a simple matrix A,

$$f(A) = \sum_{k=1}^{s} f_k Z_{k0}, \qquad f_k = f_{k,0}, \quad k = 1, 2, \ldots, s.$$

Recall that, by definition, $Z_{k0} = \varphi_{k0}(A)$. If we apply the result of Exercise 4.14.3 to the polynomials $\varphi_{k0}(\lambda)$, $k = 1, 2, \ldots, s$, we see that Z_{k0} is the sum of the constituent matrices associated with the eigenvalue λ_k. It is then easily seen that Z_{k0} is idempotent, and the range of Z_{k0} is the (right) eigenspace associated with λ_k (that is, the kernel of $\lambda_k I - A$). Thus, the result of Exercise 4.14.3 is seen to be a special case of Theorem 1. We deduce from it that for a simple matrix A

$$\sum_{k=1}^{s} Z_{k0} = I \qquad \text{and} \qquad Z_{k0} Z_{j0} = \delta_{kj} Z_{k0},$$

although the first of these results is easily seen to be true for all matrices $A \in \mathbb{C}^{n \times n}$ (see Exercise 1). We shall show in Theorem 2 that the second result is also generally true.

Finally, we note that, in view of Exercise 9.2.4,

$$Z_{k0} = \varphi_{k0}(A) = \prod_{j=1,\, j \neq k}^{s} (A - \lambda_j I) \Big/ \prod_{j=1,\, j \neq k}^{s} (\lambda_k - \lambda_j).$$

Exercise 1. If $A \in \mathbb{C}^{n \times n}$ prove that, in the previous notation,

$$\text{(a) } I = \sum_{k=1}^{s} Z_{k0} \qquad \text{and} \qquad \text{(b) } A = \sum_{k=1}^{s} (\lambda_k Z_{k0} + Z_{k1}).$$

Hint. Put $f(\lambda) = 1$ in Eq. (1) for part (a) and $f(\lambda) = \lambda$ for part (b).

Exercise 2. If $A \in \mathbb{C}^{n \times n}$ and $\lambda \notin \sigma(A)$, show that

$$R_\lambda(A) \triangleq (\lambda I - A)^{-1} = \sum_{k=1}^{s} \sum_{j=0}^{m_k - 1} \frac{j!}{(\lambda - \lambda_k)^{j+1}} Z_{kj}.$$

Hint. Use Exercise 9.3.5. □

Let us examine more closely the properties of the components of a general matrix $A \in \mathbb{C}^{n \times n}$. Note particularly that part (c) of the next theorem shows that the components $Z_{10}, \ldots, Z_{s0}$ of *any* matrix are projectors and, by part (a), their sum is the identity. A formula like that in part (a) is sometimes known as a *resolution of the identity*.

Theorem 2. *The components Z_{kj} ($j = 0, 1, \ldots, m_k - 1$; $k = 1, 2, \ldots, s$) of the matrix $A \in \mathbb{C}^{n \times n}$ satisfy the conditions*

(*a*) $\sum_{k=1}^{s} Z_{k0} = I$;

(*b*) $Z_{kp} Z_{l} = 0$ *if* $k \neq l$;

(*c*) $Z_{kj}^2 = Z_{kj}$ *if and only if* $j = 0$;

(*d*) $Z_{k0} Z_{kr} = Z_{kr}$, $r = 0, 1, \ldots, m_k - 1$.

PROOF. The relation (a) was established in Exercise 1. However, this and the other properties readily follow from the structure of the components Z_{kj}, which we discuss next.

First observe that if $A = PJP^{-1}$, where $J = \text{diag}[J_1, J_2, \ldots, J_t]$ is a Jordan form for A, then, by Theorem 9.4.5,

$$Z_{kj} = \varphi_{kj}(A) = P\,\text{diag}[\varphi_{kj}(J_1), \ldots, \varphi_{kj}(J_t)]P^{-1}, \tag{2}$$

where $\varphi_{kj}(J_i)$ $(1 \le i \le t,\ 1 \le k \le s,\ 0 \le j \le m_k - 1)$ is a matrix of the form (9.4.4). The structure of $\varphi_{kj}(J_i)$ is now studied by considering two possibilities.

Case 1. The Jordan block J_i (of size n_i) is associated with the eigenvalue λ_k. Since (see Eq. (9.2.6)) $\varphi_{kj}^{(r)}(\lambda_k) = \delta_{jr}$ $(r, j = 0, 1, \ldots, m_k - 1)$, it follows from Eq. (9.4.4) that

$$\varphi_{kj}(J_i) = \frac{1}{j!} N_{n_i}^j, \tag{3}$$

where N_{n_i} is the $n_i \times n_i$ Jordan block associated with the zero eigenvalue, and it is assumed that for $j = 0$ the expression on the right in Eq. (3) is the $n_i \times n_i$ identity matrix. Note also that $n_i \le m_k$ (see Exercises 7.7.4 and 7.7.9).

Case 2. The Jordan block J_i (of order n_i) is associated with an eigenvalue $\lambda_l \ne \lambda_k$. Then again $n_i \le m_l$ and [as in Eq. (9.2.7)] we observe that $\varphi_{kj}(\lambda)$, along with its first $m_l - 1$ derivatives, is zero at $\lambda = \lambda_l$. Using (9.4.4) again, we see that

$$\varphi_{kj}(J_i) = 0. \tag{4}$$

To simplify the notation, consider the case $k = 1$. Substituting Eqs. (3) and (4) into (2) it is easily seen that

$$Z_{1j} = \frac{1}{j!} P\,\text{diag}[N_{n_1}^j, N_{n_2}^j, \ldots, N_{n_q}^j, 0, \ldots, 0]P^{-1} \tag{5}$$

for $j = 0, 1, \ldots, m_1 - 1$, provided $J_1, J_2, \ldots, J_q$ are the only Jordan blocks of J associated with λ_1. More specifically, Eq. (5) implies

$$Z_{10} = P\,\text{diag}[I_r, 0]P^{-1}, \tag{6}$$

where $r = n_1 + n_2 + \cdots + n_q$.

The nature of the results for general λ_k should now be quite clear, and the required statements follow. ■

The results of Theorems 1 and 2 should be regarded as generalizations of Exercise 4.14.3. They may help in computing $f(A)$ and especially in simultaneously finding several functions of the same matrix. In the last case each function can be represented as in Eq. (1); only the coefficients $f_{k,j}$ differ from one function to another.

When faced with the problem of finding the matrices Z_{kj}, the obvious technique is first to find the interpolatory polynomials $\varphi_{kj}(\lambda)$ and then to put $Z_{kj} = \varphi_{kj}(A)$. However, the computation of the polynomials $\varphi_{kj}(\lambda)$ may be a very laborious task, and the following exercise suggests how this can be avoided.

Example 3 (see Exercises 6.2.4, 6.5.2, and 9.4.1). We want to find the component matrices of the matrix

$$A = \begin{bmatrix} 6 & 2 & 2 \\ -2 & 2 & 0 \\ 0 & 0 & 2 \end{bmatrix}.$$

We also want to find ln A using Theorem 1. We have seen that the minimal polynomial of A is $(\lambda - 2)(\lambda - 4)^2$ and so, for any f defined on the spectrum of A, Theorem 1 gives

$$f(A) = f(2)Z_{10} + f(4)Z_{20} + f^{(1)}(4)Z_{21}. \tag{7}$$

We substitute for $f(\lambda)$ three judiciously chosen linearly independent polynomials and solve the resulting simultaneous equations for Z_{10}, Z_{20}, and Z_{21}. Thus, putting $f(\lambda) = 1$, $f(\lambda) = \lambda - 4$, and $f(\lambda) = (\lambda - 4)^2$ in turn, we obtain

$$Z_{10} + Z_{20} = I,$$

$$-2Z_{10} + Z_{21} = A - 4I = \begin{bmatrix} 2 & 2 & 2 \\ -2 & -2 & 0 \\ 0 & 0 & -2 \end{bmatrix},$$

$$4Z_{10} = (A - 4I)^2 = \begin{bmatrix} 0 & 0 & 0 \\ 0 & 0 & -4 \\ 0 & 0 & 4 \end{bmatrix}.$$

These equations are easily solved to give

$$Z_{10} = \begin{bmatrix} 0 & 0 & 0 \\ 0 & 0 & -1 \\ 0 & 0 & 1 \end{bmatrix}, \qquad Z_{20} = \begin{bmatrix} 1 & 0 & 0 \\ 0 & 1 & 1 \\ 0 & 0 & 0 \end{bmatrix}, \qquad Z_{21} = \begin{bmatrix} 2 & 2 & 2 \\ -2 & -2 & -2 \\ 0 & 0 & 0 \end{bmatrix}.$$

By comparison, the straightforward calculation of a component matrix, say Z_{20}, using its definition, $Z_{20} = \varphi_{20}(A)$, is more complicated. Seeking a polynomial $\varphi_{20}(\lambda)$ of degree not exceeding 2 and such that

$$\varphi_{20}(2) = 0, \qquad \varphi_{20}(4) = 1, \qquad \varphi_{20}'(4) = 0, \tag{8}$$

we write (see the proof of Proposition 9.2.1)

$$\varphi_{20}(\lambda) = [\alpha_0 + \alpha_1(\lambda - 4)](\lambda - 2)$$

and observe that the conditions (8) give $\alpha_0 = \frac{1}{2}$, $\alpha_1 = -\frac{1}{4}$. Hence

$$Z_{20} = -\tfrac{1}{4}A^2 + 2A - 3I = \begin{bmatrix} 1 & 0 & 0 \\ 0 & 1 & 1 \\ 0 & 0 & 0 \end{bmatrix},$$

as required.

Note that formula (7) for $f(\lambda) = \ln \lambda$ provides another method for finding $\ln A$ (compare with that used in Exercise 9.4.1):

$$\ln A = (\ln 2)Z_{10} + (\ln 4)Z_{20} + \tfrac{1}{4}Z_{21} = \begin{bmatrix} \ln 4 + \frac{1}{2} & \frac{1}{2} & \frac{1}{2} \\ -\frac{1}{2} & \ln 4 - \frac{1}{2} & \ln 2 - \frac{1}{2} \\ 0 & 0 & \ln 2 \end{bmatrix},$$

which coincides with the result of Exercise 9.4.1.

Exercise 4. Check formulas (3) through (6) and the statement of Theorem 2 for the matrix A considered in Exercise 3. □

Note that, in general, the results of Theorem 2 may help in solving the equations for the component matrices.

Exercise 5. Show that the component matrices of

$$A = \begin{bmatrix} -1 & 1 & 2 & 1 \\ 0 & -1 & 0 & -3 \\ 0 & 0 & 1 & 1 \\ 0 & 0 & 0 & 0 \end{bmatrix}$$

are (taking $\lambda_1 = 0, \lambda_2 = 1, \lambda_3 = -1$)

$$Z_{10} = \begin{bmatrix} 0 & 0 & 0 & -4 \\ 0 & 0 & 0 & -3 \\ 0 & 0 & 0 & -1 \\ 0 & 0 & 0 & 1 \end{bmatrix}, \qquad Z_{20} = \begin{bmatrix} 0 & 0 & 1 & 1 \\ 0 & 0 & 0 & 0 \\ 0 & 0 & 1 & 1 \\ 0 & 0 & 0 & 0 \end{bmatrix},$$

$$Z_{30} = \begin{bmatrix} 1 & 0 & -1 & 3 \\ 0 & 1 & 0 & 3 \\ 0 & 0 & 0 & 0 \\ 0 & 0 & 0 & 0 \end{bmatrix}, \qquad Z_{31} = \begin{bmatrix} 0 & 1 & 0 & 3 \\ 0 & 0 & 0 & 0 \\ 0 & 0 & 0 & 0 \\ 0 & 0 & 0 & 0 \end{bmatrix}. \quad □$$

The next result allows us to find all component matrices very easily once the projectors Z_{k0}, $k = 1, 2, \ldots, s$, are known.

Theorem 3. *For $k = 1, 2, \ldots, s$ and $j = 0, 1, \ldots, m_k - 1$,*

$$Z_{kj} = \frac{1}{j!}(A - \lambda_k I)^j Z_{k0}. \tag{9}$$

PROOF. Observe first that the result is trivial for $j = 0$. Without loss of generality, we will again suppose $k = 1$ and use the associated notations introduced above. Since $A - \lambda_1 I = P(J - \lambda_1 I)P^{-1}$, it follows that

$$J - \lambda_1 I = P^{-1}(A - \lambda_1 I)P = \text{diag}[N_{n_1}, \ldots, N_{n_q}, J_{q+1} - \lambda_1 I, \ldots, J_t - \lambda_1 I], \tag{10}$$

where $J_1, J_2, \ldots, J_q$ are the only Jordan blocks of J associated with λ_1. Multiplying Eq. (10) on the right by $P^{-1}Z_{1j}P$ and using Eq. (5), we obtain

$$j!P^{-1}(A - \lambda_1 I)Z_{1j}P = \text{diag}[N_{n_1}^{j+1}, \ldots, N_{n_q}^{j+1}, 0, \ldots, 0].$$

Using Eq. (5) again, we thus have

$$j!P^{-1}(A - \lambda_1 I)Z_{1j}P = (j + 1)!P^{-1}Z_{1,j+1}P$$

and

$$Z_{1,j+1} = \frac{1}{j+1}(A - \lambda_1 I)Z_{1j}$$

for $j = 0, 1, \ldots, m_k - 1$. Applying this result repeatedly, we obtain Eq. (9) for $k = 1$. ■

Theorem 3 should now be complemented with some description of the projectors Z_{k0}, $k = 1, 2, \ldots, s$ (see also Section 9.6).

Theorem 4. *With the previous notation,*

$$\text{Im } Z_{k0} = \text{Ker}(A - \lambda_k I)^{m_k}.$$

PROOF. We will again consider only the case $k = 1$. If J is the Jordan form of A used in the above discussion, then (see Eq. (10))

$$(J - \lambda_1 I)^{m_1} = \text{diag}[0, \ldots, 0, (J_{q+1} - \lambda_1 I)^{m_1}, \ldots, (J_t - \lambda_1 I)^{m_1}],$$

since $n_1, n_2, \ldots, n_q \leq m_1$. The blocks $(J_r - \lambda_1 I)^{m_1}$ are easily seen to be nonsingular for $r = q + 1, \ldots, t$, and it follows that the kernel of $(J - \lambda_1 I)^{m_1}$ is spanned by the unit vectors $\mathbf{e}_1, \mathbf{e}_2, \ldots, \mathbf{e}_r$, where $r = n_1 + n_2 + \cdots + n_q$. Let $\mathscr{S}$ denote this space.

We also have $(A - \lambda_1 I)^{m_1} = P(J - \lambda_1 I)^{m_1}P^{-1}$, and it follows immediately that $\mathbf{x} \in \text{Ker}(A - \lambda_1 I)^{m_1}$ if and only if $P^{-1}\mathbf{x} \in \mathscr{S}$.

Using Eq. (6) we see that $\mathbf{x} \in \text{Im } Z_{10}$ if and only if

$$P^{-1}\mathbf{x} \in \text{Im}[I_r \quad 0]P^{-1} = \mathscr{S},$$

and hence $\text{Ker}(A - \lambda_1 I)^{m_1} = \text{Im } Z_{10}$. ■

It turns out that the result of the theorem cannot be extended to give an equally simple description of Im Z_{kj} for $j \geq 1$. However, a partial result for Im Z_{k,m_k-1} is given in Exercise 7.

Exercise 6. Prove that $(A - \lambda_k I)^{m_k} Z_{k0} = O$.

Exercise 7. Prove that Im $Z_{k0} \supset$ Im $Z_{k1} \supset \cdots \supset$ Im Z_{k,m_k-1} and that Im Z_{k,m_k-1} is a subspace $\mathscr{S}_k$ of Ker$(A - \lambda_k I)$. Show that the dimension of $\mathscr{S}_k$ is equal to the number of Jordan blocks associated with λ_k having order m_k.

Exercise 8. If $\mathbf{x}, \mathbf{y}$ are, respectively, right and left eigenvectors (see Section 4.10) associated with λ_1, prove that $Z_{10}\mathbf{x} = \mathbf{x}$ and $\mathbf{y}^{\mathrm{T}} Z_{10} = \mathbf{y}^{\mathrm{T}}$.

Exercise 9. If A is Hermitian, prove that Z_{k0} is an orthogonal projector onto the (right) eigenspace of λ_k.

Exercise 10. Consider the $n \times n$ matrix A with all diagonal elements equal to a number a and all off-diagonal elements equal to 1. Show that the eigenvalues of A are $\lambda_1 = n + a - 1$ and $\lambda_2 = a - 1$ with λ_2 repeated $n - 1$ times. Show also that

$$Z_{10} = \frac{1}{n}\begin{bmatrix} 1 & 1 & \cdots & 1 \\ 1 & 1 & \cdots & 1 \\ \vdots & \vdots & \ddots & \vdots \\ 1 & 1 & \cdots & 1 \end{bmatrix}, \quad Z_{20} = \frac{1}{n}\begin{bmatrix} n-1 & -1 & -1 & \cdots & -1 \\ -1 & n-1 & -1 & \cdots & -1 \\ \vdots & \vdots & & & \vdots \\ -1 & -1 & \cdots & & n-1 \end{bmatrix}.$$

□

9.6 Component Matrices and Invariant Subspaces

It was shown in Chapter 6 (see especially Theorem 6.2.3 and Section 6.3) that, given an arbitrary matrix $A \in \mathbb{C}^{n \times n}$, the space $\mathbb{C}^n$ is a direct sum of the generalized eigenspaces of A. Now, Theorem 9.5.4 states that each component matrix Z_{k0} is a projector onto the generalized eigenspace associated with eigenvalue λ_k, $k = 1, 2, \ldots, s$. Furthermore, the result of Theorem 9.5.2(a), the resolution of the identity, shows that the complementary projector $I - Z_{k0}$ can be expressed in the form

$$I - Z_{k0} = \sum_{j=0,\, j \neq k}^{s} Z_{j0}$$

and, since $\text{Im } Z_{i0} \cap \text{Im } Z_{j0} = \{0\}$ when $i \neq j$,

$$\text{Ker } Z_{k0} = \text{Im}(I - Z_{k0}) = \sum_{j=0,\, j \neq k}^{s} \cdot \text{Im } Z_{j0}. \tag{1}$$

Thus, both $\text{Im } Z_{k0}$ and $\text{Ker } Z_{k0}$ are A-invariant and the following result holds.

Theorem 1. *The component matrix Z_{k0} associated with an eigenvalue λ_k of $A \in \mathbb{C}^{n \times n}$ is a projector onto the generalized eigenspace of λ_k and along the sum of generalized eigenspaces associated with all eigenvalues of A different from λ_k.*

The discussion of Section 4.8 shows that if we choose bases in $\text{Im } Z_{k0}$ and $\text{Ker } Z_{k0}$, they will determine a basis for $\mathbb{C}^n$ in which A has a representation $\text{diag}[A_1, A_2]$, where the sizes of A_1 and A_2 are just the dimensions of $\text{Im } Z_{k0}$ and $\text{Ker } Z_{k0}$, respectively. In particular, if these bases are made up of Jordan chains for A, as in Section 6.4, then A_1 and A_2 both have Jordan form.

For any $A \in \mathbb{C}^{n \times n}$ and any $z \notin \sigma(A)$, the function $f(\lambda) = (z - \lambda)^{-1}$ is defined on $\sigma(A)$; hence the *resolvent* of A, $(zI - A)^{-1}$, exists. Holding A fixed, the resolvent can be considered to be a function of z on any domain in $\mathbb{C}$ not containing points of $\sigma(A)$, and this point of view plays an important role in matrix analysis. Indeed, we shall develop it further in Section 9.9 and make substantial use of it in Chapter 11, in our introduction to perturbation theory.

A representation for the resolvent was obtained in Exercise 9.5.2:

$$(zI - A)^{-1} = \sum_{k=1}^{s} \sum_{j=0}^{m_k - 1} \frac{j!}{(z - \lambda_k)^{j+1}} Z_{kj}.$$

Using the orthogonality properties of the matrices Z_{kj}, it is easily deduced that

$$(zI - A)^{-1}(I - Z_{10}) = \sum_{k=2}^{s} \sum_{j=0}^{m_k - 1} \frac{j!}{(z - \lambda_k)^{j+1}} Z_{kj}$$

and, using Eq. (1) and Exercise 9.5.6, the resolvent has $\text{Im}(I - Z_{10})$ as its range. Now as $z \to \lambda_1$, elements of $(zI - A)^{-1}$ will become unbounded. But, since the limit on the right of the equation exists, we see that multiplication of the resolvent by $I - Z_{10}$ has the effect of eliminating the singular part of $(zI - A)^{-1}$ at $z = \lambda_1$. In other words, the restriction of the resolvent to the subspace

$$\text{Im}(I - Z_{10}) = \sum_{j=2}^{s} \text{Im } Z_{j0}$$

has no singularity at $z = \lambda_1$.

Define the matrix

$$E_1 = \lim_{z \to \lambda_1} (zI - A)^{-1}(I - Z_{10}) = \sum_{k=2}^{s} \sum_{j=0}^{m_k - 1} \frac{j!}{(\lambda_1 - \lambda_k)^{j+1}} Z_{kj}. \tag{2}$$

The next proposition expresses $I - Z_{10} = \sum_{k=2}^{s} Z_{k0}$ in terms of E_1 and will be used in Chapter 11.

Theorem 2. *In the notation of the previous paragraphs,*

$$I - Z_{10} = E_1(\lambda_1 I - A).$$

PROOF. The simple statement $\lambda_1 I - A = (\lambda_1 - z)I + (zI - A)$ implies

$$(zI - A)^{-1}(\lambda_1 I - A) = (\lambda_1 - z)(zI - A)^{-1} + I.$$

Multiply on the left by $I - Z_{10}$, take the limit as $z \to \lambda_1$, and take advantage of Eq. (2) to obtain the result. ∎

Exercise 1. Let $A \in \mathbb{C}^{n \times n}$ and $\sigma(A) = \{\lambda_1, \lambda_2, \ldots, \lambda_s\}$. Define a function $f_k(\lambda)$ on the spectrum of A by setting $f_k(\lambda) \equiv 1$ in an arbitrarily small neighborhood of λ_k and $f_k(\lambda) \equiv 0$ in an arbitrarily small neighborhood of each $\lambda_j \in \sigma(A)$ with $\lambda_j \neq \lambda_k$. Show that $f_k(A) = Z_{k0}$ and use Exercise 9.3.4 to show that Z_{k0} is idempotent.

Exercise 2. If $\lambda = \lambda_1$ is an eigenvalue of A and $\lambda_1, \ldots, \lambda_s$ are the distinct eigenvalues of A, show that, for any $z \notin \sigma(A)$,

$$(zI - A)^{-1} = \sum_{j=0}^{m_1 - 1} \frac{j! Z_{1j}}{(z - \lambda)^{j+1}} + E(z),$$

where $E(z)$ is a matrix that is analytic on a neighborhood of λ and $Z_{11}E(z) = E(z)Z_{11} = 0$. Write $E = E(\lambda)$ and use this representation to obtain Theorem 2. □

9.7 Further Properties of Functions of Matrices

In this section we obtain some additional properties of functions of matrices, which, in particular, allow us to extend some familiar identities for scalar functions to their matrix-valued counterparts. For example, we ask whether the identity $\sin^2 \alpha + \cos^2 \alpha = 1$ implies that $\sin^2 A + \cos^2 A = I$, and does $e^{\alpha}e^{-\alpha} = 1$ imply $e^{A}e^{-A} = I$? We answer these questions (and a whole class of such questions) in the following theorems.

Theorem 1. *Let $P(u_1, u_2, \ldots, u_t)$ be a scalar polynomial in $u_1, u_2, \ldots, u_t$ and let $f_1(\lambda), f_2(\lambda), \ldots, f_t(\lambda)$ be functions defined on the spectrum of $A \in \mathbb{C}^{n \times n}$. If the function $f(\lambda) = P(f_1(\lambda), f_2(\lambda), \ldots, f_t(\lambda))$ assumes zero values on the spectrum of A, then*

$$f(A) = P(f_1(A), f_2(A), \ldots, f_t(A)) = 0.$$

PROOF. Let $p_1(\lambda), p_2(\lambda), \ldots, p_t(\lambda)$ be interpolatory polynomials on the spectrum of A for $f_1(\lambda), f_2(\lambda), \ldots, f_t(\lambda)$, respectively. Then by definition of the p_i, we have $f_i(A) = p_i(A)$ for $i = 1, 2, \ldots, t$. Denote the polynomial $P(p_1(\lambda), p_2(\lambda), \ldots, p_t(\lambda))$ by $p(\lambda)$ and observe that since $p_i(\lambda)$ and $f_i(\lambda)$ have the same values on the spectrum of A $(1 \le i \le t)$, so do $p(\lambda)$ and $f(\lambda)$. Hence our assumption that $f(\lambda)$ is zero on the spectrum of A implies the same for $P(\lambda)$ and therefore $f(A) = p(A) = 0$. ■

For the particular cases mentioned above, we first choose $f_1(\lambda) = \sin \lambda$, $f_2(\lambda) = \cos \lambda$, and $P(u_1, u_2) = u_1^2 + u_2^2 - 1$ to prove that, for every $A \in \mathbb{C}^{n \times n}$, it is true that $\sin^2 A + \cos^2 A = I$. Next, taking $f_1(\lambda) = e^{\lambda}$, $f_2(\lambda) = e^{-\lambda}$, and $P(u_1, u_2) = u_1 u_2 - 1$, we obtain $e^A e^{-A} = I$. Thus, $e^{-A} = (e^A)^{-1}$.

Exercise 1. If $A \in \mathbb{C}^{n \times n}$ and the functions $f_1(\lambda)$ and $f_2(\lambda)$ are defined on the spectrum of A, show that the functions $F_1(\lambda) = f_1(\lambda) + f_2(\lambda)$, $F_2(\lambda) = f_1(\lambda) f_2(\lambda)$ are also defined on the spectrum of A, and $F_1(A) = f_1(A) + f_2(A)$, $F_2(A) = f_1(A) f_2(A)$.

Exercise 2. If $p(\lambda)$ and $q(\lambda)$ are relatively prime polynomials, $\det q(A) \neq 0$, and $r(\lambda)$ denotes the rational function $p(\lambda)/q(\lambda)$, prove that

$$r(A) = p(A)[q(A)]^{-1} = [q(A)]^{-1} p(A).$$

Exercise 3. Prove that for any $A \in \mathbb{C}^{n \times n}$,

$$e^{iA} = \cos A + i \sin A.$$

(Note that, in general, the cosine and sine functions must be considered as defined on $\mathbb{C}$.) □

We can also discuss the possibility of taking pth roots of a matrix with the aid of Theorem 1. Suppose that p is a positive integer and let $f_1(\lambda) = \lambda^{1/p}$, $f_2(\lambda) = \lambda$, and $P(u_1, u_2) = u_1^p - u_2$. We would normally choose a single-valued branch of the function $\lambda^{1/p}$ for $f_1(\lambda)$, but this is not essential. If, however, the matrix A has a zero eigenvalue with index greater than 1, then, since derivatives of $f_1(\lambda)$ at the origin are needed, $f_1(\lambda)$ would not be defined on the spectrum of A. Thus, if A is nonsingular, or is singular with a zero eigenvalue of index one, we define $f_1(\lambda) = \lambda^{1/p}$ (with a known convention concerning the choice of branches at the eigenvalues), and

$$A^{1/p} = f_1(A) = \sum_{k=1}^{s} \sum_{j=0}^{m_k - 1} f_1^{(j)}(\lambda_k) Z_{kj}.$$

Then, by Theorem 1 applied to the functions $f_1(\lambda)$, $f_2(\lambda)$, and $P(u_1, u_2)$ defined above,

$$(A^{1/p})^p = A.$$

This discussion should be compared with the definition of the square root of a positive definite matrix made in Section 5.4.

It should be noted that our definition of $A^{1/2}$, for example, does not include all possible matrices B for which $B^2 = A$. To illustrate this point, any real symmetric orthogonal matrix B has the property that $B^2 = I$, but B will not necessarily be generated by applying our definitions to find $I^{1/2}$. A particular example is the set of matrices defined by

$$B(\theta) = \begin{bmatrix} \cos\theta & \sin\theta \\ \sin\theta & -\cos\theta \end{bmatrix}$$

(see Exercise 2.6.16).

Theorem 2. *Let $h(\lambda) = g(f(\lambda))$, where $f(\lambda)$ is defined on the spectrum of $A \in \mathbb{C}^{n\times n}$ and $g(\mu)$ is defined on the spectrum of $B = f(A)$. Then $h(\lambda)$ is defined on the spectrum of A and $h(A) = g(f(A))$.*

PROOF. Let $\lambda_1, \lambda_2, \ldots, \lambda_s$ be the distinct eigenvalues of A with indices $m_1, m_2, \ldots, m_s$, respectively. Denote $f(\lambda_k) = \mu_k$ for $k = 1, 2, \ldots, s$. Since

$$\begin{aligned} h(\lambda_k) &= g(\mu_k), \\ h'(\lambda_k) &= g'(\mu_k) f'(\lambda_k), \\ &\ \vdots \\ h^{(m_k - 1)}(\lambda_k) &= g^{(m_k - 1)}(\mu_k)[f'(\lambda_k)]^{m_k - 1} + \cdots + g'(\mu_k) f^{(m_k - 1)}(\lambda_k), \end{aligned} \tag{1}$$

it follows that $h(\lambda)$ is defined on the spectrum of A and assigns values given by the formulas in (1). Let $r(\mu)$ denote any polynomial that has the same values as $g(\mu)$ on the spectrum of B and consider the function $r(f(\lambda))$. If the values of $h(\lambda)$ and $r(f(\lambda))$ coincide on the spectrum of A, then $h(\lambda) - r(f(\lambda))$ assigns zero values on the spectrum of A and, by Theorem 1,

$$h(A) = r(f(A)) = r(B) = g(B) = g(f(A)),$$

and the result is established.

Now observe that in view of (1) the values of $h(\lambda)$ and $r(f(\lambda))$ coincide on the spectrum of A if

$$r^{(i)}(\mu_k) = g^{(i)}(\mu_k) \tag{2}$$

for $i = 0, 1, \ldots, m_k - 1$, $k = 1, 2, \ldots, s$.

Let us show that $\{g(\mu_k), g'(\mu_k), \ldots, g^{(m_k-1)}(\mu_k)\}$ contains all the values of $g(\mu)$ on the spectrum of B. This fact is equivalent to saying that the index n_k of the eigenvalue μ_k of B does not exceed m_k or that the polynomial $p(\mu) = (\mu - \mu_1)^{m_1}(\mu - \mu_2)^{m_2} \cdots (\mu - \mu_s)^{m_k}$ annihilates B $(1 \le k \le s)$. To see this, observe that $q(\lambda) = p(f(\lambda)) = (f(\lambda) - f(\lambda_1))^{m_1} \cdots (f(\lambda) - f(\lambda_s))^{m_s}$ has zeros λ_k with multiplicity at least m_k $(k = 1, 2, \ldots, s)$ and therefore $q(\lambda)$ assigns zero values on the spectrum of A. Thus, by Theorem 1 with $t = 1$, $p(B) = p(f(A)) = 0$ and $p(\lambda)$ is an annihilating polynomial for B. Now defining the polynomial $r(\mu)$ as satisfying the conditions (2), we obtain a polynomial with the same values as $g(\mu)$ on the spectrum of B and the proof is completed. Note that $r(\mu)$ is not necessarily the interpolatory polynomial for $g(\mu)$ on the spectrum of B of least possible degree. ■

Exercise 4. Prove that for any $A \in \mathbb{C}^{n \times n}$,

$$\sin 2A = 2 \sin A \cos A. \quad \square$$

9.8 Sequences and Series of Matrices

In this section we show how other concepts of analysis can be applied to matrices and particularly to functions of matrices as we have defined them. We first consider sets of matrices defined on the nonnegative integers, that is, sequences of matrices.

Let $A_1, A_2, \ldots$ be a sequence of matrices belonging to $\mathbb{C}^{m \times n}$ and let $a_{ij}^{(p)}$ be the i, jth element of A_p, $p = 1, 2, \ldots$. The sequence $A_1, A_2, \ldots$ is said to *converge* to the matrix $A \in \mathbb{C}^{m \times n}$ if there exist numbers a_{ij} (the elements of A) such that $a_{ij}^{(p)} \to a_{ij}$ as $p \to \infty$ for each pair of subscripts i, j. A sequence that does not converge is said to *diverge*. Thus the convergence of sequences of matrices is defined as elementwise convergence. The definition also includes the convergence of column and row vectors as special cases.

Exercise 1. Let $\{A_p\}_{p=1}^{\infty}$, $\{B_p\}_{p=1}^{\infty}$ be convergent sequences of matrices from $\mathbb{C}^{m \times n}$ and assume that $A_p \to A$, $B_p \to B$ as $p \to \infty$. Prove that as $p \to \infty$,

(a) $A_p + B_p \to A + B$;
(b) $A_p B_p \to AB$ and, in particular, $\alpha_p B_p \to \alpha B$ for any sequence $\alpha_p \to \alpha$ in $\mathbb{C}$;
(c) $QA_pQ^{-1} \to QAQ^{-1}$;
(d) $\text{diag}[A_p, B_p] \to \text{diag}[A, B]$.

Hint. These results follow from fundamental properties of sequences of complex numbers. For part (b), if $\alpha_p \to \alpha$ and $\beta_p \to \beta$ as $p \to \infty$, write

$$\alpha_p\beta_p - \alpha\beta = (\alpha_p - \alpha)\beta + (\beta_p - \beta)\alpha + (\alpha_p - \alpha)(\beta_p - \beta). \quad \square$$

We now consider the case in which the members of a sequence of matrices are defined as functions of $A \in \mathbb{C}^{n \times n}$. Thus, let the sequence of functions $f_1(\lambda), f_2(\lambda), \ldots$ be defined on the spectrum of A, and define $A_p \triangleq f_p(A)$, $p = 1, 2, \ldots$. For each term of the sequence $A_1, A_2, \ldots$, Theorem 9.5.1 yields

$$f_p(A) = \sum_{k=1}^{s} \sum_{j=0}^{m_k - 1} f_p^{(j)}(\lambda_k) Z_{kj}. \tag{1}$$

Now the component matrices Z_{kj} depend only on A and not on p, so that each term of the sequence is defined by the m numbers $f_p^{(j)}(\lambda_k)$ for $k = 1, 2, \ldots, s$ and $j = 0, 1, \ldots, m_k - 1$. We might therefore expect that the convergence of $A_1, A_2, \ldots$ can be made to depend on the convergence of the m scalar sequences $f_1^{(j)}(\lambda_k), f_2^{(j)}(\lambda_k), \ldots$ rather than the n^2 scalar sequences of the elements of $A_1, A_2, \ldots$.

Theorem 1. *Let the functions $f_1, f_2, \ldots$ be defined on the spectrum of $A \in \mathbb{C}^{n \times n}$ and let $A_p = f_p(A)$, $p = 1, 2, \ldots$. Then the sequence $A_1, A_2, \ldots, A_p, \ldots$ converges as $p \to \infty$ if and only if the m scalar sequences $f_1^{(j)}(\lambda_k), f_2^{(j)}(\lambda_k), \ldots, f_p^{(j)}(\lambda_k), \ldots$ ($k = 1, 2, \ldots, s$, $j = 0, 1, \ldots, m_k - 1$, $m = m_1 + m_2 + \cdots + m_s$) converge as $p \to \infty$.*

Moreover, if $f_p^{(j)}(\lambda_k) \to f^{(j)}(\lambda_k)$ as $p \to \infty$ for each j and k and for some function $f(\lambda)$, then $A_p \to f(A)$ as $p \to \infty$. Conversely, if $\lim A_p$ exists as $p \to \infty$, then there is a function $f(\lambda)$ defined on the spectrum of A such that $\lim A_p = f(A)$ as $p \to \infty$.

PROOF. If $f_p^{(j)}(\lambda_k) \to f^{(j)}(\lambda_k)$ as $p \to \infty$ for each j and k and some $f(\lambda)$, then Eq. (1) and Exercise 1 imply that

$$\lim_{p \to \infty} A_p = \lim_{p \to \infty} f_p(A) = \sum_{k=1}^{s} \sum_{j=0}^{m_k - 1} \left(\lim_{p \to \infty} f_p^{(j)}(\lambda_k) \right) Z_{kj} \tag{2}$$

and

$$\lim_{p \to \infty} A_p = \sum_{k=1}^{s} \sum_{j=0}^{m_k - 1} f^{(j)}(\lambda_k) Z_{kj} = f(A). \tag{3}$$

Conversely, suppose that $\lim A_p$ exists as $p \to \infty$ and denote the elements of $A_p = f_p(A)$ by $a_{ir}^{(p)}$, $i, r = 1, 2, \ldots, n$, $p = 1, 2, \ldots$. Viewing Eq. (1) as a set of n^2 equations for the $m = m_1 + m_2 + \cdots + m_s$ unknowns $f_p^{(j)}(\lambda_k)$, where p is fixed, and recalling the linear independence (Theorem 9.5.1) of the component matrices Z_{kj} as members of $\mathbb{C}^{n \times n}$, we deduce that the $n^2 \times m$ coefficient matrix of this system has rank m. Since the degree m of the minimal polynomial of A does not exceed the degree of the characteristic polynomial for A, it follows that $m \le n \le n^2$ and therefore (see Exercise 3.10.7) the system in (1) has a unique solution of the form

$$f_p^{(j)}(\lambda_k) = \sum_{i,r=1}^{n} \gamma_{ir} a_{ir}^{(p)}, \tag{4}$$

where the γ_{ir} $(i, r = 1, 2, \ldots, n)$ depend on j and k but not on p. Now if $\lim A_p$ exists as $p \to \infty$, then so does $\lim a_{ir}^{(p)}$ and it follows from Eq. (4) that $\lim f_p^{(j)}(\lambda_k)$ also exists as $p \to \infty$ for each j and k. This completes the first part of the theorem.

If $f_p^{(j)}(\lambda_k) \to f^{(j)}(\lambda_k)$ as $p \to \infty$ for each j and k and for some $f(\lambda)$, then Eq. (3) shows that $A_p \to f(A)$ as $p \to \infty$. Conversely, the existence of $\lim A_p$ as $p \to \infty$ provides the existence of $\lim f_p^{(j)}(\lambda_k)$ as $p \to \infty$ and hence the equality (2). Thus, we may choose for $f(\lambda)$ the interpolatory polynomial whose values on the spectrum of A are the numbers $\lim_{p\to\infty} f_p^{(j)}(\lambda_k)$, $k = 1, 2, \ldots, s$, $j = 0, 1, \ldots, m_k - 1$. The proof is complete. ■

We now wish to develop a similar theorem for series of matrices. First, the *series* of matrices $\sum_{p=0}^{\infty} A_p$ is said to *converge* to the matrix A (the *sum* of the series) if the sequence of partial sums $B_\nu = \sum_{p=1}^{\nu} A_p$ converges to A as $\nu \to \infty$. A series that does not converge is said to *diverge*.

The following theorem is immediately obtained by applying Theorem 1 to a sequence of partial sums.

Theorem 2. *If the functions $u_1, u_2, \ldots$ are defined on the spectrum of $A \in \mathbb{C}^{n\times n}$, then $\sum_{p=1}^{\infty} u_p(A)$ converges if and only if the series $\sum_{p=1}^{\infty} u_p^{(j)}(\lambda_k)$ converge for $k = 1, 2, \ldots, s$ and $j = 0, 1, \ldots, m_k - 1$.*

Moreover, if

$$\sum_{p=1}^{\infty} u_p^{(j)}(\lambda_k) = f^{(j)}(\lambda_k)$$

for each j and k, then

$$\sum_{p=1}^{\infty} u_p(A) = f(A),$$

and vice versa.

We can now apply this theorem to obtain an important result that helps to justify our development of the theory of functions of matrices. The next theorem shows the generality of the definition made in Section 9.3, which may have seemed rather arbitrary at first reading. Indeed, the properties described in the next theorem are often used as a (more restrictive) definition of a function of a matrix.

Theorem 3. *Let the function $f(\lambda)$ have a Taylor series about $\lambda_0 \in \mathbb{C}$:*

$$f(\lambda) = \sum_{p=0}^{\infty} \alpha_p(\lambda - \lambda_0)^p, \tag{5}$$

with radius of convergence r. If $A \in \mathbb{C}^{n \times n}$, then $f(A)$ is defined and is given by

$$f(A) = \sum_{p=0}^{\infty} \alpha_p(A - \lambda_0 I)^p \tag{6}$$

if and only if each of the distinct eigenvalues $\lambda_1, \lambda_2, \ldots, \lambda_s$ of A satisfies one of the following conditions:

(*a*) $|\lambda_k - \lambda_0| < r$;
(*b*) $|\lambda_k - \lambda_0| = r$ *and the series for* $f^{(m_k - 1)}(\lambda)$ (*where* m_k *is the index of* λ_k) *is convergent at the point* $\lambda = \lambda_k$, $1 \le k \le s$.

PROOF. First we recall that $f(A)$ exists if and only if $f^{(j)}(\lambda_k)$ exists for $j = 0, 1, \ldots, m_k - 1$ and $k = 1, 2, \ldots, s$. Furthermore, the function $f(\lambda)$ given by Eq. (5) has derivatives of all orders at any point inside the circle of convergence and these derivatives can be obtained by differentiating the series for $f(\lambda)$ term by term. Hence $f(\lambda)$ is obviously defined on the spectrum of A if $|\lambda_k - \lambda_0| < r$ for $k = 1, 2, \ldots, s$. Also, if $|\lambda_k - \lambda_0| = r$ for some k and the series obtained by $m_k - 1$ differentiations of the series in (5) is convergent at the point $\lambda = \lambda_k$, then $f(\lambda_k), f'(\lambda_k), \ldots, f^{(m_k - 1)}(\lambda_k)$ exist and, consequently, $f(A)$ is defined. Note that, starting with the existence of $f(A)$, the above argument can be reversed.

If we now write $u_p(\lambda) = \alpha_p(\lambda - \lambda_0)^p$ for $p = 0, 1, 2, \ldots$, then $u_p(\lambda)$ is certainly defined on the spectrum of A and under the hypotheses (a) or (b), the series $\sum_{p=1}^{\infty} u_p^{(j)}(\lambda_k)$, $k = 1, 2, \ldots, s$, $j = 0, 1, \ldots, m_k - 1$, all converge. It now follows from Theorem 2 that $\sum_{p=0}^{\infty} u_p(A)$ converges and has sum $f(A)$ and vice versa. ■

Note that for functions of the form (5), a function $f(A)$ of A can be defined by formula (6). Clearly, this definition cannot be applied for functions having only a finite number of derivatives at the points $\lambda_k \in \sigma(A)$, as assumed in Section 9.3.

If the Taylor series for $f(\lambda)$ about the origin converges for all points in the complex plane (the radius of convergence is infinite), then $f(\lambda)$ is said to be an *entire* function and the power series representation for $f(A)$ will converge for all $A \in \mathbb{C}^{n \times n}$. Among the most important functions of this kind are the trigonometric and exponential functions. Thus, from the corresponding results for (complex) scalar functions, we deduce that for all $A \in \mathbb{C}^{n \times n}$,

$$\sin A = \sum_{p=0}^{\infty} \frac{(-1)^p}{(2p+1)!} A^{2p+1}, \qquad \cos A = \sum_{p=0}^{\infty} \frac{(-1)^p}{(2p)!} A^{2p},$$

$$e^A = \sum_{p=0}^{\infty} \frac{A^p}{p!}, \qquad \sinh A = \sum_{p=0}^{\infty} \frac{A^{2p+1}}{(2p+1)!}, \qquad \cosh A = \sum_{p=0}^{\infty} \frac{A^{2p}}{(2p)!}.$$

If the eigenvalues $\lambda_1, \lambda_2, \ldots, \lambda_n$ of A satisfy the conditions $|\lambda_j| < 1$ for $j = 1, 2, \ldots, n$, then we deduce from the binomial theorem and the logarithmic series that

$$(I - A)^{-1} = \sum_{p=0}^{\infty} A^p, \qquad \log(I + A) = \sum_{p=1}^{\infty} \frac{(-1)^{p-1}}{p} A^p. \tag{7}$$

Exercise 2. Prove that if $A \in \mathbb{C}^{n \times n}$, then $e^{A(s+t)} = e^{As}e^{At}$ for any complex numbers s and t.

Exercise 3. Prove that if A and B commute, then $e^{A+B} = e^A e^B$, and if $e^{(A+B)t} = e^{At}e^{Bt}$ for all t, then A and B commute.

Exercise 4. Show that if

$$A = \begin{bmatrix} 0 & 1 \\ 1 & 0 \end{bmatrix}, \qquad B = \begin{bmatrix} 0 & 1 \\ 0 & \varepsilon \end{bmatrix}, \qquad C = \begin{bmatrix} 0 & 1 \\ -1 & 0 \end{bmatrix},$$

then

$$e^{At} = \begin{bmatrix} e^t - t & t \\ t & e^t - t \end{bmatrix}, \qquad e^{Bt} = \begin{bmatrix} 1 & \varepsilon^{-1}(e^{\varepsilon t} - 1) \\ 0 & e^{\varepsilon t} \end{bmatrix},$$

$$e^{Ct} = \begin{bmatrix} \cos t & \sin t \\ -\sin t & \cos t \end{bmatrix}.$$

Exercise 5. Show that the sequence $A, A^2, A^3, \ldots$ converges to the zero-matrix if and only if $|\lambda| < 1$ for every $\lambda \in \sigma(A)$. ☐

9.9 The Resolvent and the Cauchy Theorem for Matrices

The theory of (scalar) functions of a complex variable provides a third approach to the definition of $f(A)$ that is applicable when $f(\lambda)$ is an analytic function. This approach is also consistent with the general definition of Section 9.3.

Consider a matrix $A(\lambda)$ whose elements are functions of a complex variable λ and define the *derivative* (or *integral*) of the matrix to be the matrix obtained by differentiating (or integrating) each element of $A(\lambda)$. For $r = 0, 1, 2, \ldots,$ we write

$$\frac{d^r}{d\lambda^r} A(\lambda) \qquad \text{or} \qquad A^{(r)}(\lambda)$$

for the derivative of $A(\lambda)$ of order r.

Exercise 1. Prove that

$$\frac{d}{dt}(e^{At}) = Ae^{At} = e^{At}A.$$

Hint. Use the series representation of e^{At}.

Exercise 2. Prove that (with the variable t omitted from the right-hand expressions)

$$\frac{d}{dt}(A(t))^2 = A^{(1)}A + AA^{(1)},$$

and construct an example to show that, in general,

$$\frac{d}{dt}(A(t))^2 \neq 2AA^{(1)}.$$

Exercise 3. Prove that, when p is a positive integer and the matrices exist,

$$\frac{d}{dt}(A(t))^p = \sum_{j=1}^{p} A^{j-1}A^{(1)}A^{p-j},$$

$$\frac{d}{dt}(A(t))^{-p} = -A^{-p}\frac{dA^p}{dt}A^{-p}. \quad \square$$

The notation

$$\int_L A(\lambda)\, d\lambda$$

will be used for the integral of $A(\lambda)$ taken in the positive direction along a path L in the complex plane, which, for our purposes, will always be a simple piecewise smooth closed contour, or a finite system of such contours, without points of intersection.

Let $A \in \mathbb{C}^{n \times n}$. The matrix $(\lambda I - A)^{-1}$ is known as the *resolvent* of A and it has already played an important part in our analysis. Viewed as a function of a complex variable λ, the resolvent is defined only at those points

λ that do not belong to the spectrum of A. Since the matrix A is fixed in this section, we use the notation $R_\lambda = (\lambda I - A)^{-1}$ for the resolvent of A.

Exercise 4. Verify that for $\lambda \notin \sigma(A)$,

$$\frac{d}{d\lambda} R_\lambda = -R_\lambda^2,$$

and, more generally,

$$\frac{d^j}{d\lambda^j} R_\lambda = (-1)^j j! R^{j+1} \qquad \text{for} \qquad j = 1, 2, \ldots.$$

Hint. Use Exercise 4.14.4(a). □

Theorem 1. *The resolvent R_λ of $A \in \mathbb{C}^{n \times n}$ is a rational function of λ with poles at the points of the spectrum of A and $R_\infty = O$. Moreover, each $\lambda_k \in \sigma(A)$ is a pole of R_λ of order m_k, where m_k is the index of the eigenvalue λ_k.*

PROOF. Let $m(\lambda) = \sum_{i=0}^{m} \alpha_i \lambda^i$ be the minimal polynomial of A. It is easily seen that

$$\frac{m(\lambda) - m(\mu)}{\lambda - \mu} = \sum_{i,j=0}^{m-1} \gamma_{ij} \lambda^i \mu^j, \qquad \lambda \neq \mu,$$

for some γ_{ij} $(1 \le i, j \le m - 1)$ and therefore, substituting A for μ, we have for $\lambda \notin \sigma(A)$

$$(m(\lambda)I - m(A))R_\lambda = \sum_{j=0}^{m-1} \left(\sum_{i=0}^{m-1} \gamma_{ij} \lambda^i \right) A^j.$$

Since $m(A) = 0$, it follows that, for the values of λ for which $m(\lambda) \neq 0$,

$$R_\lambda = \frac{1}{m(\lambda)} \sum_{j=0}^{m-1} m_j(\lambda) A^j, \tag{1}$$

where $m_j(\lambda)$ $(0 \le j \le m - 1)$ is a polynomial of degree not exceeding $m - 1$. The required results now follow from the representation (1). ■

Thus, using the terminology of complex analysis, the spectrum of a matrix can be described in terms of its resolvent. This suggests the application of results from complex analysis to the theory of matrices and vice versa. In the next theorem we use the classical Cauchy integral theorem to obtain its generalization for a matrix case, and this turns out to be an important and useful result in matrix theory.

First, let us agree to call a function $f(\lambda)$ of the complex variable λ *analytic* on an open set of points D of the complex plane if $f(\lambda)$ is continuously differentiable at every point of D. Or, what is equivalent, $f(\lambda)$ is analytic if it has a convergent Taylor series expansion about each point of D. If $f(\lambda)$ is analytic within an open domain D bounded by a contour L and continuous on the closure $\overline{D}$, then the Cauchy theorem asserts that

$$f^{(j)}(\lambda_0) = \frac{j!}{2\pi i} \int_L \frac{f(\lambda_0)}{(\lambda - \lambda_0)^{j+1}} \, d\lambda \tag{2}$$

for any $\lambda_0 \in D$ and $j = 0, 1, \ldots$.

Theorem 2. *If $A \in \mathbb{C}^{n \times n}$ has distinct eigenvalues $\lambda_1, \ldots, \lambda_s$, the path L is a closed contour having $\lambda_1, \ldots, \lambda_s$ in its interior, and $f(\lambda)$ is continuous in and on L and analytic within L, then*

$$f(A) = \frac{1}{2\pi i} \int_L f(\lambda)(\lambda I - A)^{-1} \, d\lambda = \frac{1}{2\pi i} \int_L f(\lambda) R_\lambda \, d\lambda. \tag{3}$$

PROOF. Use the representation of Exercise 9.5.2 for the resolvent and multiply both sides by $f(\lambda)(2\pi i)^{-1}$. Applying formula (2) with $\lambda_0 = \lambda_k$ $(1 \le k \le s)$ and integrating along L, we obtain

$$\frac{1}{2\pi i} \int_L f(\lambda)(\lambda I - A)^{-1} \, d\lambda = \sum_{k=1}^{s} \sum_{j=0}^{m_k - 1} f^{(j)}(\lambda_k) Z_{kj},$$

and the result follows from the spectral resolution of Theorem 9.5.1. ■

Note that the relation (3) is an exact analog of (2) in the case $j = 0$; $(\lambda - \lambda_0)^{-1}$ is simply replaced by $(\lambda I - A)^{-1}$. It should also be remarked that (3) is frequently used as a definition of $f(A)$ for analytic functions $f(\lambda)$.

Corollary 1. *With the notation of Theorem 2,*

$$I = \frac{1}{2\pi i} \int_L R_\lambda \, d\lambda, \qquad A = \frac{1}{2\pi i} \int_L \lambda R_\lambda \, d\lambda.$$

The component matrices Z_{kj} of A can also be described in terms of the resolvent of A.

Theorem 3. *Preserving the previous notation, let L_k be a simple piecewise smooth closed contour that contains $\lambda_k \in \sigma(A)$ in its interior, and assume that no other points of the spectrum of A are inside or on L_k. Then*

$$Z_{kj} = \frac{1}{j!\,2\pi i} \int_{L_k} (\lambda - \lambda_k)^j R_\lambda \, d\lambda \tag{4}$$

for $j = 0, 1, \ldots, m_k - 1$; $k = 1, 2, \ldots, s$.

PROOF. Multiplying both sides of the equation in Exercise 9.5.2 by $(\lambda - \lambda_p)^r$ $(1 \le p \le s, 0 \le r \le m_p - 1)$ and integrating around L_p, we obtain

$$\int_{L_p} (\lambda - \lambda_p)^r R_\lambda \, d\lambda = \sum_{k=1}^{s} \sum_{j=0}^{m_k - 1} j!\, Z_{kj} \int_{L_p} \frac{(\lambda - \lambda_p)^r}{(\lambda - \lambda_k)^{j+1}} \, d\lambda. \tag{5}$$

The functions $(\lambda - \lambda_p)^r(\lambda - \lambda_k)^{-j-1}$ with $k \neq p$ or with $k = p$ and $j \leq r - 1$ are analytic in the domain bounded by L_p. Therefore the integrals of these functions around L_p are equal to zero. Now observe that for the remaining cases we have, in view of Eq. (2) for $f(\lambda) \equiv 1$,

$$\int_{L_p} (\lambda - \lambda_p)^{r-j-1}\, d\lambda = \begin{cases} 2\pi i & \text{if } j = r, \\ 0 & \text{if } j > r. \end{cases}$$

Hence Eq. (5) implies

$$\int_{L_p} (\lambda - \lambda_p)^r R_\lambda\, d\lambda = r!\, Z_{pr}(2\pi i),$$

which is equivalent to (4). ■

We remark that, for those familiar with Laurent expansions, Eq. (4) expresses Z_{kj} as a coefficient of the Laurent expansion for the resolvent R_λ about the point λ_k.

Using the results of Theorems 9.5.2 and 9.5.4, and that of Theorem 3 for $j = 0$, the following useful result is obtained (see also Theorem 9.6.1).

Theorem 4. *With the previous notation, the integral*

$$P_k = Z_{k0} = \frac{1}{2\pi i} \int_{L_k} R_\lambda\, d\lambda, \qquad 1 \leq k \leq s,$$

called the Riesz projector, is a projector. The system $\{P_k\}_{k=1}^s$ is an orthogonal system of projectors in the sense that $P_k P_j = 0$ for $k \neq j$, giving a spectral resolution of the identity matrix (with respect to A):

$$I = \sum_{k=1}^{s} P_k \qquad \textit{and} \qquad \operatorname{Im} P_k = \operatorname{Ker}(A - \lambda_k I)^{m_k},$$

where m_k is the index of the eigenvalue λ_k of A.

Exercise 5. Let $A \in \mathbb{C}^{n \times n}$ and L_0 be a simple piecewise smooth closed contour. Show that

$$\int_{L_0} R_\lambda\, d\lambda = \sum_{k=1}^{r} Z_{k0}$$

if L_0 contains in its interior only the points $\lambda_1, \lambda_2, \ldots, \lambda_r$ from the spectrum of A, and that the integral is the zero-matrix if there are no points of $\sigma(A)$ inside L_0.

Exercise 6. Prove that, with the notation of Theorems 3 and 4,

$$\lambda_k Z_{k0} + Z_{k1} = A Z_{k0} = \frac{1}{2\pi i} \int_{L_k} \lambda R_\lambda\, d\lambda. \quad \square$$

Exercise 7 shows that, when λ_k is an eigenvalue of A of index 1, the projector Z_{k0} can easily be described in terms of the reduced adjoint $C(\lambda)$ of A and the minimal polynomial $m(\lambda)$ of A.

Exercise 7. Show that, if λ_k is an eigenvalue of A of index 1, then

$$Z_{k0} = \frac{1}{m^{(1)}(\lambda_k)} C(\lambda_k).$$

SOLUTION. Use Theorem 3 to write

$$Z_{k0} = \frac{1}{2\pi i} \int_{L_k} (\lambda I - A)^{-1}\, d\lambda.$$

Then use Exercise 7.9.1 to obtain

$$Z_{k0} = \frac{1}{2\pi i} \int_{L_k} \frac{1}{m(\lambda)} C(\lambda)\, d\lambda.$$

Since the index of λ_k is 1, we have $m(\lambda) = (\lambda - \lambda_j)m_1(\lambda)$, where $m_1(\lambda_j) = m^{(1)}(\lambda_j) \neq 0$. Substitute this into the last displayed equation and use the theorem of residues to obtain the result.

Exercise 8. Let J be a matrix in Jordan normal form. Use Theorem 4 to show that the projectors Z_{k0} for J have the form $\operatorname{diag}[d_{k1}, d_{k2}, \ldots, d_{kn}]$ where d_{ki} is equal to 0 or 1 for $1 \le k \le s$ and $1 \le i \le n$. ☐

9.10 Applications to Differential Equations

Consider again the systems of differential equations with constant coefficients introduced in Section 4.13. We wish to find functions of time, $\boldsymbol{x}(t)$, with values in $\mathbb{C}^n$ satisfying the homogeneous equation

$$\dot{\boldsymbol{x}}(t) = A\boldsymbol{x}(t) \tag{1}$$

or the nonhomogeneous equation

$$\dot{\boldsymbol{x}}(t) = A\boldsymbol{x}(t) + \boldsymbol{f}(t), \tag{2}$$

where $A \in \mathbb{C}^{n \times n}$ and $\boldsymbol{f}(t)$ is a prescribed function of time. We will now analyse these problems with no hypotheses on $A \in \mathbb{C}^{n \times n}$.

It was seen in Section 4.13 that primitive solutions of the form $\boldsymbol{x}_0 e^{\lambda_0 t}$, where λ_0 is an eigenvalue and $\boldsymbol{x}_0$ an associated eigenvector, play an important part.

More generally, we look for a solution of Eq. (1) in the form

$$x(t) = x_0(t)e^{\lambda_0 t}, \qquad \lambda_0 \in \mathbb{C}, \tag{3}$$

where $x_0(t)$ is a vector of order n whose coordinates are polynomials in t. For convenience we write $x_0(t)$ in the form

$$x_0(t) = \frac{t^{k-1}}{(k-1)!}x_1 + \cdots + \frac{t}{1!}x_{k-1} + x_k, \tag{4}$$

where $x_i \in \mathbb{C}^n$, $i = 1, 2, \ldots, k$, and x_1 is assumed to be nonzero.

It is easily seen that $x(t)$ in the form (3) is a solution of (1) if and only if

$$\dot{x}_0(t) = (A - \lambda_0 I)x_0(t),$$

where $x_0(t)$ is given by Eq. (4). By differentiation, we then have

$$x_0^{(i)}(t) = (A - \lambda_0 I)^i x_0(t), \qquad i = 1, 2, \ldots, \tag{5}$$

and since $x_0(t)$ is a vector-valued polynomial of degree $k - 1$, it follows that $0 = (A - \lambda_0 I)^k x_0(t)$.

Now we find from Eq. (4) that $x_0^{(k-1)}(t) = x_1$, and comparing this with Eq. (5), $x_1 = (A - \lambda_0 I)^{k-1}x_0(t)$. Multiplying on the left by $(A - \lambda_0 I)$ we obtain $(A - \lambda_0 I)x_1 = 0$, and since $x_1 \neq 0$ by hypothesis, x_1 is an eigenvector of A corresponding to the eigenvalue λ_0.

Now compare the $(k - 2)$nd derivative of $x_0(t)$ calculated from Eqs. (4) and (5) to get

$$tx_1 + x_2 = (A - \lambda_0 I)^{k-2}x_0(t).$$

Multiply on the left by $(A - \lambda_0 I)$, and use the results of the preceding paragraph to get $(A - \lambda_0 I)x_2 = x_1$. Thus, (see Eqs. (6.3.3)) x_1, x_2 form a Jordan chain of length 2.

Proceeding in the same way, it is found that $x_1, x_2, \ldots, x_k$ form a Jordan chain for A associated with λ_0. A similar argument shows that if $x_1, x_2, \ldots, x_k$ generate a Jordan chain for A corresponding to the eigenvalue λ_0, then Eq. (3) along with Eq. (4) gives a solution of Eq. (1). Thus, we have proved the following.

Proposition 1. *The function $x(t)$ defined by Eqs. (3) and (4) is a solution of Eq. (1) if and only if $x_1, x_2, \ldots, x_k$ constitute a Jordan chain for A associated with the eigenvalue λ_0 of A.*

Now recall that, by Theorem 7.10.1, the dimension of the solution space $\mathcal{S}$ of Eq. (1) is n and by the Jordan theorem, there is a basis in $\mathbb{C}^n$ consisting entirely of eigenvectors and generalized eigenvectors of A. Hence the following result is clear.

Theorem 1. *Let $A \in \mathbb{C}^{n \times n}$ and let $\lambda_1, \lambda_2, \ldots, \lambda_s$ be the distinct eigenvalues of A of algebraic multiplicity $\alpha_1, \alpha_2, \ldots, \alpha_s$, respectively. If the Jordan chains for λ_k $(1 \leq k \leq s)$ are*

$$\begin{array}{llll} \mathbf{x}_{11}^{(k)}, & \mathbf{x}_{12}^{(k)}, & \ldots, & \mathbf{x}_{1\nu_{k1}}^{(k)}, \\ \mathbf{x}_{21}^{(k)}, & \mathbf{x}_{22}^{(k)}, & \ldots, & \mathbf{x}_{2\nu_{k2}}^{(k)}, \\ \vdots & & & \\ \mathbf{x}_{r_k,1}^{(k)} & \mathbf{x}_{r_k,2}^{(k)} & \ldots, & \mathbf{x}_{r_k,\nu_{k,r_k}}^{(k)} \end{array}$$

where $\sum_{i=1}^{r_k} \nu_{ki} = \alpha_k$, then any solution $\mathbf{x}(t)$ of (1) is a linear combination of the n vector-valued functions

$$\left[\frac{t^{j-1}}{(j-1)!}\mathbf{x}_{i1}^{(k)} + \cdots + \frac{t}{1!}\mathbf{x}_{i,j-1}^{(k)} + \mathbf{x}_{ij}^{(k)}\right]e^{\lambda_k t}, \tag{6}$$

where $j = 1, 2, \ldots, \nu_{ki}$, $i = 1, 2, \ldots r_k$, $k = 1, 2, \ldots, s$.

In other words, the expressions in (6) form a basis in the solution space $\mathscr{S}$ of (1) and therefore are referred to as the *fundamental solutions* of (1). Note also that λ_k has geometric multiplicity r_k and $\mathbf{x}_{11}^{(k)}, \mathbf{x}_{21}^{(k)}, \ldots, \mathbf{x}_{r_k 1}^{(k)}$ span $\operatorname{Ker}(A - \lambda_k I)$.

Exercise 1. Find the general form of the solutions (the *general solution*) of the system

$$\dot{\mathbf{x}}(t) = \begin{bmatrix} \dot{x}_1 \\ \dot{x}_2 \\ \dot{x}_3 \end{bmatrix} = \begin{bmatrix} 6 & 2 & 2 \\ -2 & 2 & 0 \\ 0 & 0 & 2 \end{bmatrix} \begin{bmatrix} x_1 \\ x_2 \\ x_3 \end{bmatrix} = A\mathbf{x}(t).$$

SOLUTION. It was found in Exercise 6.5.2 that the matrix A has an eigenvalue $\lambda_1 = 2$ with an associated eigenvector $[0 \ \ -1 \ \ 1]^{\mathrm{T}}$ and an eigenvalue $\lambda_2 = 4$ associated with the eigenvector $[2 \ \ -2 \ \ 0]^{\mathrm{T}}$ and generalised eigenvector $[1 \ \ 0 \ \ 0]^{\mathrm{T}}$. Hence the general solution, according to (6), is given by

$$\mathbf{x}(t) = \beta_1 \begin{bmatrix} 0 \\ -1 \\ 1 \end{bmatrix} e^{2t} + \beta_2 \begin{bmatrix} 2 \\ -2 \\ 0 \end{bmatrix} e^{4t} + \beta_3 \left(\begin{bmatrix} 2 \\ -2 \\ 0 \end{bmatrix} t + \begin{bmatrix} 1 \\ 0 \\ 0 \end{bmatrix} \right) e^{4t}. \quad \square \tag{7}$$

It is frequently the case that a particular solution of (1) is required that is specified by additional conditions. The *initial-value problems* are of this kind, in which a vector $\mathbf{x}_0 \in \mathbb{C}^n$ is given and a solution $\mathbf{x}(t)$ of (1) is to be found, where $\mathbf{x}(t)$ takes the value $\mathbf{x}_0$ at $t = 0$.

Theorem 2. *Let $A \in \mathbb{C}^{n \times n}$. There is a unique $\mathbf{x}(t)$ such that*

$$\dot{\mathbf{x}}(t) = A\mathbf{x}(t) \qquad \text{and} \qquad \mathbf{x}(0) = \mathbf{x}_0 \tag{8}$$

for any given vector $\mathbf{x}_0 \in \mathbb{C}^n$.

PROOF. By Theorem 1 any solution $\boldsymbol{x}(t)$ of (1) is a linear combination of the n expressions $\boldsymbol{x}_p(t)$ $(p = 1, 2, \ldots, n)$ of the form (6);

$$\boldsymbol{x}(t) = \sum_{p=1}^{n} \beta_p \boldsymbol{x}_p(t).$$

Observe that $\boldsymbol{x}_p(0)$, $p = 1, 2, \ldots, n$, gives the vectors of the Jordan basis of A in $\mathbb{C}^n$, and therefore the matrix $B = [\boldsymbol{x}_1(0) \quad \boldsymbol{x}_2(0) \quad \cdots \quad \boldsymbol{x}_n(0)]$ is nonsingular. Thus, the system

$$\sum_{p=1}^{n} \beta_p \boldsymbol{x}_p(0) = \boldsymbol{x}_0$$

or in matrix form $B\boldsymbol{\beta} = \boldsymbol{x}_0$ with $\boldsymbol{\beta} = [\beta_1 \quad \beta_2 \quad \cdots \quad \beta_n]^{\mathrm{T}}$, has a unique solution for every $\boldsymbol{x}_0 \in \mathbb{C}^n$. ■

Exercise 2. Find the solution of the system considered in Exercise 1 that satisfies the condition $\boldsymbol{x}(0) = [0 \quad 1 \quad 1]^{\mathrm{T}}$.

SOLUTION. It is easily seen that by putting $t = 0$ in Eq. (7) we obtain a linear system

$$\begin{bmatrix} 0 & 2 & 1 \\ -1 & -2 & 0 \\ 1 & 0 & 0 \end{bmatrix} \begin{bmatrix} \beta_1 \\ \beta_2 \\ \beta_3 \end{bmatrix} = \begin{bmatrix} 0 \\ 1 \\ 1 \end{bmatrix}$$

with an invertible coefficient matrix. The vector $[\beta_1 \quad \beta_2 \quad \beta_3]^{\mathrm{T}}$ is found to be $[1 \quad -1 \quad 2]^{\mathrm{T}}$ and therefore the required solution is

$$\boldsymbol{x}(t) = \begin{bmatrix} 0 \\ -1 \\ 1 \end{bmatrix} e^{2t} + \left(\begin{bmatrix} 0 \\ 2 \\ 0 \end{bmatrix} + \begin{bmatrix} 4 \\ -4 \\ 0 \end{bmatrix} t \right) e^{4t}. \quad \square$$

Now we show how the solution of (8) can be found by use of the notion of a function of A. Start by using the idea in the beginning of the section but applying it directly to the matrix differential equation $\dot{\boldsymbol{x}} = A\boldsymbol{x}$. We look for solutions in the form $e^{At}\boldsymbol{z}$ with $\boldsymbol{z} \in \mathbb{C}^n$. Exercise 9.9.1 clearly shows that the vector-valued function $\boldsymbol{x}(t) = e^{At}\boldsymbol{z}$ is a solution of (1) for any $\boldsymbol{z} \in \mathbb{C}^n$:

$$\dot{\boldsymbol{x}}(t) = \frac{d}{dt} e^{At}\boldsymbol{z} = Ae^{At}\boldsymbol{z} = A\boldsymbol{x}(t).$$

Obviously $\boldsymbol{x}(t) = e^{At}\boldsymbol{z}$ satisfies the initial condition $\boldsymbol{x}(0) = \boldsymbol{x}_0$ if and only if $\boldsymbol{z} = \boldsymbol{x}_0$. Thus, using Theorem 2, we find the following.

Theorem 3. *The unique solution of the problem* (8) *can be written in the form*

$$\boldsymbol{x}(t) = e^{At}\boldsymbol{x}_0. \tag{9}$$

Let $\mathbf{x}(t)$ be a solution of (1). Computing $\mathbf{x}(0)$, we see from Theorem 3 that $\mathbf{x}(t) = e^{At}\mathbf{x}(0)$. Hence we have a further refinement.

Theorem 4. *Any solution $\mathbf{x}(t)$ of (1) can be represented in the form $\mathbf{x}(t) = e^{At}\mathbf{z}$ for some $\mathbf{z} \in \mathbb{C}^n$.*

It is a remarkable and noteworthy fact that all the complications of Theorem 1 can be streamlined into the simple statement of Theorem 4. This demonstrates the power of the "functional calculus" developed in this chapter.

Another representation for solutions of (1) is derived on putting $f(\lambda) = e^{\lambda}$ in Theorem 9.5.1 and using the spectral resolution for e^{At}. Hence, denoting $\mathbf{x}(0) = \mathbf{x}_0$, we use Theorem 3 to derive

$$\mathbf{x}(t) = \sum_{k=1}^{s} \sum_{j=0}^{m_k - 1} Z_{kj} t^j e^{\lambda_k t} \mathbf{x}_0 = \sum_{k=1}^{s} (\mathbf{z}_{k0} + \mathbf{z}_{k1} t + \cdots + \mathbf{z}_{k, m_k - 1} t^{m_k - 1}) e^{\lambda_k t},$$

where $\mathbf{z}_{kj} = Z_{kj}\mathbf{x}_0$ and is independent of t. Thus, we see again that the solution is a combination of polynomials in t with exponential functions. If A is a simple matrix, the solution reduces to

$$\mathbf{x}(t) = \sum_{k=1}^{s} \mathbf{z}_{k0}\, e^{\lambda_k t},$$

and we see that each element of $x(t)$ is a linear combination of exponential functions with no polynomial contributions (see Section 4.13).

Exercise 3. By means of component matrices, find the solution of the initial-value problem discussed in Exercise 2. ☐

Finally, consider the inhomogeneous Eq. (2) in which the elements of f are prescribed piecewise-continuous functions $f_1(t), \ldots, f_n(t)$ of t that are not all identically zero. Let $\mathcal{S}$ be the space of all solutions of the associated homogeneous problem $\dot{\mathbf{x}} = A\mathbf{x}$ and let $\mathcal{T}$ be the set of all solutions of $\dot{\mathbf{x}} = A\mathbf{x} + f$. If we are given just one solution $\mathbf{y} \in \mathcal{T}$, then (compare with Theorem 3.10.2) $\mathbf{z} \in \mathcal{T}$ if and only if $\mathbf{z} = \mathbf{y} + \mathbf{w}$ for some $\mathbf{w} \in \mathcal{S}$. The verification of this result is left as an exercise. Thus, to obtain $\mathcal{T}$ we need to know just one solution of the inhomogeneous equation, known as a *particular integral*, and add to it all the solutions of the homogeneous equation, each of which we express in the form $e^{At}\mathbf{x}$ for some $\mathbf{x} \in C^n$.

We look for a solution of $\dot{\mathbf{x}} = A\mathbf{x} + f$ in the form

$$\mathbf{x}(t) = e^{At}\mathbf{z}(t).$$

If there is such a solution, then $\dot{\mathbf{x}} = A\mathbf{x} + e^{At}\dot{\mathbf{z}}$ and so $e^{At}\dot{\mathbf{z}} = f$, hence

$$\mathbf{z}(t) = e^{-At_0}\mathbf{x}_0 + \int_{t_0}^{t} e^{-A\tau} \mathbf{f}(\tau)\, d\tau,$$

where $x_0 = e^{At_0}z(t_0) = x(t_0)$. This suggests that

$$x(t) = e^{At}\left(e^{-At_0}x_0 + \int_{t_0}^{t} e^{-A\tau}f(\tau)\,d\tau\right)$$

$$= e^{A(t-t_0)}x_0 + \int_{t_0}^{t} e^{A(t-\tau)}f(\tau)\,d\tau$$

is the solution of $\dot{x} = Ax + f$, where $x(t_0) = x_0$. This is easily verified.

Theorem 5. *If $f(t)$ is a piecewise-continuous function of t with values in $\mathbb{C}^n$, $t_0 \in R$, and $A \in \mathbb{C}^{n\times n}$, then every solution of the differential equation*

$$\dot{x}(t) = Ax(t) + f(t)$$

has the form

$$x(t) = e^{A(t-t_0)}x_0 + \int_{t_0}^{t} e^{A(t-\tau)}f(\tau)\,d\tau, \tag{10}$$

and $x_0 = x(t_0)$.

Exercise 4. Find the solution of $\dot{x} = Ax + f$, with $x(0) = x_0$, if A and x_0 are as defined in Exercises 1 and 2 and

$$f = \begin{bmatrix} ae^{4t} \\ 0 \\ 0 \end{bmatrix} = ae^{4t}e_1.$$

SOLUTION. The vector $e^{At}x_0$ is known from Exercise 1. Using Eq. (9.5.1) to find the particular integral, we have

$$y(t) = \int_0^t e^{A(t-\tau)}f(\tau)\,d\tau = \sum_{k=1}^{s}\sum_{j=0}^{m_k-1} Z_{kj}\int_0^t (t-\tau)^{j-1}e^{\lambda_k(t-\tau)}f(\tau)\,d\tau.$$

The integrals

$$a\int_0^t e^{4(t-\tau)}e^{4\tau}\,d\tau, \qquad a\int_0^t (t-\tau)e^{4(t-\tau)}e^{4\tau}\,d\tau, \qquad a\int_0^t e^{2(t-\tau)}e^{4\tau}\,d\tau$$

are easily found to be ate^{4t}, $\frac{1}{2}at^2e^{4t}$, and $\frac{1}{2}ae^{2t}(e^{2t} - 1)$, respectively. Hence

$$y(t) = ate^{4t}Z_{10}e_1 + \tfrac{1}{2}at^2e^{4t}Z_{11}e_1 + \tfrac{1}{2}ae^{2t}(e^{2t} - 1)Z_{20}e_1$$

$$= \tfrac{1}{2}ate^{4t}\left(2\begin{bmatrix} 1 \\ 0 \\ 0 \end{bmatrix} + t\begin{bmatrix} 2 \\ -2 \\ 0 \end{bmatrix}\right) = \begin{bmatrix} a(1+t)te^{4t} \\ -at^2e^{4t} \\ 0 \end{bmatrix}.$$

The solution is then given by $x(t) = e^{At}x_0 + y(t)$.

Exercise 5. Let A be nonsingular and $\boldsymbol{\beta} = A^{-1}\boldsymbol{b}$, where $\boldsymbol{b} \in \mathbb{C}^n$. Show that if $\boldsymbol{x}(t)$ is a solution of $\dot{\boldsymbol{x}}(t) = A\boldsymbol{x}(t) - \boldsymbol{b}$, then

$$\boldsymbol{x}(t) - \boldsymbol{\beta} = e^{At}(\boldsymbol{x}(0) - \boldsymbol{\beta}).$$

(Here $\boldsymbol{\beta}$ is the *steady-state solution* of the differential equation.)

Exercise 6. Let $A \in \mathbb{C}^{n \times n}$ and let a sequence of vectors $\boldsymbol{f}_0, \boldsymbol{f}_1, \ldots$ in $\mathbb{C}^n$ be given. Show that any solution of the difference equation

$$\boldsymbol{x}_{r+1} = A\boldsymbol{x}_r + \boldsymbol{f}_r, \qquad r = 0, 1, 2, \ldots,$$

can be expressed in the form $\boldsymbol{x}_0 = \boldsymbol{z}$ and, for $j = 1, 2, \ldots$,

$$\boldsymbol{x}_{j+1} = A^j\boldsymbol{z} + \sum_{k=0}^{j-1} A^{j-k-1}\boldsymbol{f}_k,$$

for some $\boldsymbol{z} \in \mathbb{C}^n$. □

9.11 Observable and Controllable Systems

In this section we bring together several important results developed in this and earlier chapters. This will allow us to present some interesting ideas and results from the theory of systems and control.

Let $A \in \mathbb{C}^{n \times n}$ and $C \in \mathbb{C}^{r \times n}$ and consider the system of equations

$$\begin{aligned} \dot{\boldsymbol{x}}(t) &= A\boldsymbol{x}(t), \qquad \boldsymbol{x}(0) = \boldsymbol{x}_0, \\ \boldsymbol{y}(t) &= C\boldsymbol{x}(t). \end{aligned} \tag{1}$$

Thus $\boldsymbol{x}(t)$ has values in $\mathbb{C}^n$ determined by the initial-value problem (supposing $\boldsymbol{x}_0 \in \mathbb{C}^n$ is given) and is called the *state vector*. Then $\boldsymbol{y}(t) \in \mathbb{C}^r$ for each $t \geq 0$ and is the *observed* or *output vector*. It represents the information about the state of the system that can be recorded after "filtering" through the matrix C. The system (1) is said to be *observable* if $\boldsymbol{y}(t) \equiv \boldsymbol{0}$ occurs only when $\boldsymbol{x}_0 = \boldsymbol{0}$. Thus, for the system to be observable *all* nonzero initial vectors must give rise to a solution which can be observed, in the sense that the output vector is not identically zero.

The connection with observable pairs, as introduced in Section 5.13, is established by the first theorem. For convenience, write

$$Q = Q(C, A) = \begin{bmatrix} C \\ CA \\ \vdots \\ CA^{n-1} \end{bmatrix},$$

a matrix of size $nr \times n$.

Theorem 1. *The system* (1) *is observable if and only if* Ker $Q(C, A) = \{0\}$, *that is, if and only if the pair* (C, A) *is observable.*

Before proving the theorem we establish a technical lemma that will simplify subsequent arguments involving limiting processes. An *entire function* is one that is analytic in the whole complex plane; the only possible singularity is at infinity.

Lemma 1. *For any matrix* $A \in \mathbb{C}^{n \times n}$ *there exist entire functions* $\psi_1(t), \ldots, \psi_n(t)$ *such that*

$$e^{At} = \sum_{j=1}^{n} \psi_j(t) A^{j-1}. \tag{2}$$

PROOF. Let $f(\lambda) = e^{\lambda t}$; then f is an entire function of λ that is defined on the spectrum of any $n \times n$ matrix A. Since $f^{(j)}(\lambda) = t^j e^{\lambda t}$, it follows from Theorem 9.5.1 that

$$f(A) = e^{At} = \sum_{k=1}^{s} \sum_{j=0}^{m_k - 1} t^j e^{\lambda_k t} Z_{kj}. \tag{3}$$

Since $Z_{kj} = \varphi_{kj}(A)$ for each k, j and φ_{kj} is a polynomial whose degree does not exceed n [see the discussion preceding Eqs. (9.2.6) and (9.2.7)], and because the φ_{kj} depend only on A, the right-hand term in Eq. (3) can be rearranged as in Eq. (2). Furthermore, each coefficient $\psi_j(t)$ is just a linear combination of functions of the form $t^j e^{\lambda_k t}$ and is therefore entire. ■

PROOF OF THEOREM 1. If Ker $Q(C, A) \neq \{0\}$ then there is a nonzero $\boldsymbol{x}_0$ such that $CA^j \boldsymbol{x}_0 = \boldsymbol{0}$ for $j = 0, 1, \ldots, n - 1$. It follows from Lemma 1 that $Ce^{At}\boldsymbol{x}_0 = \boldsymbol{0}$ for any real t. But then Theorem 9.10.3 and Eqs. (1) yield

$$\boldsymbol{y}(t) = Ce^{At}\boldsymbol{x}_0 \equiv \boldsymbol{0}$$

and the system is not observable. Thus, if the system is observable, the pair (C, A) must be observable.

Conversely, if the system is not observable, there is an $\boldsymbol{x}_0 \neq \boldsymbol{0}$ such that $\boldsymbol{y}(t) = Ce^{At}\boldsymbol{x}_0 \equiv \boldsymbol{0}$. Differentiating repeatedly with respect to t, it is found that

$$Q(C, A)e^{At}\boldsymbol{x}_0 \equiv \boldsymbol{0}$$

and Ker $Q(C, A) \neq \{0\}$. Consequently, observability of the pair (C, A) must imply observability of the system. ■

It has been seen in Section 5.13 that there is a notion of "controllability" of matrix pairs that is dual to that of observability. We now develop this notion in the context of systems theory. Let $A \in \mathbb{C}^{n \times n}$ and $B \in \mathbb{C}^{n \times m}$ and consider a differential equation with initial condition

$$\dot{\boldsymbol{x}}(t) = A\boldsymbol{x}(t) + B\boldsymbol{u}(t), \qquad \boldsymbol{x}(0) = \boldsymbol{x}_0. \tag{4}$$

Consistency demands that $\boldsymbol{x}(t) \in \mathbb{C}^n$ and $\boldsymbol{u}(t) \in \mathbb{C}^m$ for each t. We adopt the point of view that $\boldsymbol{u}(t)$ is a vector function, called the *control*, that can be used to guide, or control, the solution $\boldsymbol{x}(t)$ of the resulting initial-value problem to meet some objective. The objective that we consider is that of controlling the solution $\boldsymbol{x}(t)$ in such a way that after some time t, the function $\boldsymbol{x}(t)$ takes some preassigned vector value in $\mathbb{C}^n$.

To make these ideas precise we first define a class of admissible control functions $\boldsymbol{u}(t)$. We define $\mathscr{U}_m$ to be the linear space of all $\mathbb{C}^m$-valued functions that are defined and piecewise continuous on the domain $t \geq 0$, and we assume that $\boldsymbol{u}(t) \in \mathscr{U}_m$. Then observe that, using Eq. (9.10.10), when $\boldsymbol{u}(t) \in \mathscr{U}_m$ is chosen, the solution of our initial-value problem can be written explicitly in the form

$$\boldsymbol{x}(t; \boldsymbol{x}_0, \boldsymbol{u}) = e^{At}\boldsymbol{x}_0 + \int_0^t e^{(t-s)A}B\boldsymbol{u}(s)\, ds. \tag{5}$$

The purpose of the complicated notation on the left is to emphasize the dependence of the solution $\boldsymbol{x}$ on the control function $\boldsymbol{u}$ and the initial value $\boldsymbol{x}_0$, as well as on the independent variable t.

A vector $\boldsymbol{y} \in \mathbb{C}^n$ is said to be *reachable from the initial vector* $\boldsymbol{x}_0$ if there is *a* $\boldsymbol{u}(t) \in \mathscr{U}_m$ such that $\boldsymbol{x}(t; \boldsymbol{x}_0, \boldsymbol{u}) = \boldsymbol{y}$ for some positive time t. Under the same condition we may say that $\boldsymbol{y}$ is *reachable from* $\boldsymbol{x}_0$ *in time* t.

The *system* of Eq. (4) is said to be *controllable* (in contrast to controllability of a matrix pair) if every vector in $\mathbb{C}^n$ is reachable from every other vector in $\mathbb{C}^n$. Our next major objective is to show that the system (4) is controllable if and only if the pair (A, B) is controllable, and this will follow easily from the next result.

Recall first that the controllable subspace of (A, B), which we will denote $\mathscr{C}_{A,B}$, is given by

$$\mathscr{C}_{A,B} = \sum_{r=0}^{n-1} \operatorname{Im}(A^r B). \tag{6}$$

(See Corollary 1(d) of Theorem 5.13.1.)

Theorem 2. *The vector* $\boldsymbol{z} \in \mathbb{C}^n$ *is reachable from* $\boldsymbol{y} \in \mathbb{C}^n$ *in time* t *if and only if*

$$\boldsymbol{z} - e^{At}\boldsymbol{y} \in \mathscr{C}_{A,B}.$$

The proof will be broken down with the help of another lemma, which provides a useful description of the controllable subspace of (A, B).

Lemma 2. *The matrix-valued function $W(t)$ defined by*

$$W(t) = \int_0^t e^{As}BB^*e^{A^*s}\,ds \tag{7}$$

has the property that, for any $t > 0$, $\operatorname{Im} W(t) = \mathscr{C}_{A,B}$.

PROOF. Clearly, $W(t)$ is Hermitian for each $t > 0$ and hence [see Eq. (5.8.5)]

$$\operatorname{Im} W(t) = (\operatorname{Ker} W(t))^{\perp}.$$

Also (see Eq. 5.13.5), if $\mathscr{K}_{A,B}$ denotes the unobservable subspace of (B^*, A^*) then

$$\mathscr{C}_{A,B} = (\mathscr{K}_{A,B})^{\perp}.$$

Thus, the conclusion of the lemma is equivalent to the statement $\operatorname{Ker} W(t) = \mathscr{K}_{A,B}$ and we shall prove it in this form.

If $\boldsymbol{x} \in \operatorname{Ker} W(t)$ then

$$\boldsymbol{x}^*W(t)\boldsymbol{x} = \int_0^t \|B^*e^{A^*s}\boldsymbol{x}\|^2\,ds = 0;$$

here the Euclidean vector norm is used. Hence, for all $s \in [0, t]$,

$$B^*e^{A^*s}\boldsymbol{x} = \boldsymbol{0}.$$

Differentiate this relation with respect to s repeatedly and take the limit as $s \to 0_+$ to obtain

$$B^*(A^*)^{j-1}\boldsymbol{x} = \boldsymbol{0} \qquad \text{for} \quad j = 1, 2, \ldots, n.$$

In other words,

$$\boldsymbol{x} \in \bigcap_{j=1}^{n} \operatorname{Ker}(B^*(A^*)^{j-1}) = \mathscr{K}_{A,B}.$$

Thus $\operatorname{Ker} W(t) \subset \mathscr{K}_{A,B}$. It is easy to verify that, reversing the argument and using Lemma 1, the opposite inclusion is obtained. So $\operatorname{Ker} W(t) = \mathscr{K}_{A,B}$ and the lemma is proved. ■

This lemma yields another criterion for controllability of the pair (A, B).

Corollary 1. *The pair (A, B) is controllable if and only if the matrix $W(t)$ of Eq. (7) is positive definite for all $t > 0$.*

PROOF. It is clear from definition (7) that $W(t)$ is at least positive semi-definite. However, for (A, B) to be controllable means that $\mathscr{C}_{A,B} = \mathbb{C}^n$, and so Lemma 2 implies that $\operatorname{Im} W(t) = \mathbb{C}^n$. Hence $W(t)$ has rank n and is therefore positive definite. ■

PROOF OF THEOREM 2. If z is reachable from y in time t, then, using Eq. (5),

$$z - e^{At}y = \int_0^t e^{(t-s)A}Bu(s)\,ds$$

for some $u \in \mathscr{U}_m$. Use Lemma 1 to write

$$z - e^{At}y = \sum_{j=1}^{n} A^{j-1}B \int_0^t \psi_j(t-s)u(s)\,ds,$$

and, using definition (6), it is seen immediately that $z - e^{At}y \in \mathscr{C}_{A,B}$.

For the converse, let $z - e^{At}y \in \mathscr{C}_{A,B}$. We are to construct an input, or control, $u(t)$ that will "steer" the system from y to z. Use Lemma 2 to write

$$z - e^{At}y = W(t)w = \int_0^t e^{As}BB^*e^{A^*s}w\,ds$$

for some $w \in \mathbb{C}^n$. Put $s = t - \tau$ to obtain

$$z - e^{At}y = \int_0^t e^{A(t-\tau)}BB^*e^{A^*(t-\tau)}w\,d\tau = \int_0^t e^{A(t-\tau)}Bu(\tau)\,d\tau,$$

where we define $u(\tau) = B^*e^{A^*(t-\tau)}w \in \mathscr{U}_m$. Comparing this with Eq. (5), we see that z is reachable from y in time t. ■

Corollary 1. *The pair (A, B) is controllable if and only if the system $\dot{x} = Ax + Bu$, $x(0) = x_0$, is controllable.*

PROOF. If the pair (A, B) is controllable then $\mathscr{C}_{A,B} = \mathbb{C}^n$. Thus, for any $y, z \in \mathbb{C}^n$, it follows trivially that $z - e^{At}y \in \mathscr{C}_{A,B}$. Then Theorem 2 asserts that the system is controllable.

Conversely, if the system is controllable then, in particular, every vector $z \in \mathscr{C}^n$ is reachable from the zero vector and, by Theorem 2, $\mathscr{C}_{A,B} = \mathbb{C}^n$. Thus, the pair (A, B) is controllable. ■

When the system is controllable, it turns out that any $z \in \mathbb{C}^n$ is reachable from any $y \in \mathbb{C}^n$ at *any time* $t > 0$. To see this, we construct an explicit control function $u(s)$ that will take the solution from y to z in any prescribed time t. Thus, in the controllable case, $W(t)$ of Lemma 2 is positive definite and we may define a function

$$u(s) = B^*e^{A^*(t-s)}W(t)^{-1}(z - e^{At}y). \tag{8}$$

Then Eq. (5) gives

$$x(t; y, u) = e^{At}y + \left(\int_0^t e^{A(t-s)}BB^*e^{A^*(t-s)}\,ds\right)W(t)^{-1}(z - e^{At}y).$$

Using definition (7) of $W(t)$, this equation reduces to

$$\mathbf{x}(t;\mathbf{y},\mathbf{u}) = e^{At}\mathbf{y} + \mathbf{z} - e^{At}\mathbf{y} = \mathbf{z},$$

as required.

Exercise 1. Show that if $\mathbf{z}$ is reachable from $\mathbf{y}$ and $\mathbf{y} \in \mathscr{C}_{A,B}$, then $\mathbf{z} \in \mathscr{C}_{A,B}$.

Exercise 2. The system dual to that of Eqs. (4) is defined to be a system of the type in Eqs. (1), given in terms of A and B by

$$\dot{\mathbf{y}}(t) = -A^*\mathbf{y}(t),$$
$$\mathbf{z}(t) = B^*\mathbf{y}(t).$$

Show that the system (4) is controllable if and only if its dual system is observable.

Exercise 3. Let $\mathscr{C}_{A,B}$ have dimension k. Use the ideas of Section 4.8 to form a matrix P such that, if $\hat{A} = P^{-1}AP$ and $\hat{B} = P^{-1}B$, then

$$\hat{A} = \begin{bmatrix} A_1 & A_3 \\ 0 & A_2 \end{bmatrix} \quad \text{and} \quad \hat{B} = \begin{bmatrix} B_1 \\ 0 \end{bmatrix},$$

where A_1 is $k \times k$, B_1 is $k \times m$, and the pair (A_1, B_1) is controllable.

Hint. Recall that $\mathscr{C}_{A,B}$ is A-invariant and $\operatorname{Im} B \subset \mathscr{C}_{A,B}$. Use the transformation of variables $\mathbf{x} = P\mathbf{z}$ to transform $\dot{\mathbf{x}}(t) = A\mathbf{x}(t) + B(t)$ to the form $\dot{\mathbf{z}}(t) = \hat{A}\mathbf{z}(t) + \hat{B}\mathbf{u}(t)$ and show that if

$$\mathbf{z}(t) = \begin{bmatrix} \mathbf{z}_1(t) \\ \mathbf{z}_2(t) \end{bmatrix},$$

where $\mathbf{z}_1(t)$ takes values in $\mathbb{C}^k$, then $\dot{\mathbf{z}}_1(t) = A_1\mathbf{z}_1(t) + B_1\mathbf{u}(t)$ is a controllable system.

Exercise 4. Establish the following analogue of Lemma 1. If $A \in \mathbb{C}^{n\times n}$, then there are functions $\varphi_1(\lambda), \ldots, \varphi_n(\lambda)$ analytic in any neighborhood not containing points of $\sigma(A)$ such that

$$(\lambda I - A)^{-1} = \sum_{j=1}^{n} \varphi_j(\lambda)A^{j-1}. \quad \square$$

9.12 Miscellaneous Exercises

1. If $f(\lambda)$ is defined on the spectrum of $A \in \mathbb{C}^{n\times n}$, prove that $f(A^{\mathrm{T}}) = [f(A)]^{\mathrm{T}}$.

   Hint. Observe that the interpolatory polynomials for A and A^{T} coincide.

2. If $A \in \mathbb{C}^{n\times n}$, show that $A^2 = \sum_{k=1}^{s} (\lambda_k^2 Z_{k0} + 2\lambda_k Z_{k1} + 2Z_{k2})$ by use of Theorems 9.5.1 and 9.5.2 and Exercise 9.5.1(b).

3. Prove that the series $I + A + A^2 + \cdots$ converges if and only if all the eigenvalues of A have modulus less than 1.

Hint. See Exercise 9.8.5.

4. Show that a unitary matrix U can be expressed in the form $U = e^{iH}$, where H is Hermitian. Observe the analogy between this and the representation $u = e^{i\alpha}$ (α real) of any point on the unit circle.

Hint. Show that $U = V \operatorname{diag}[e^{i\alpha_1}, e^{i\alpha_2}, \ldots, e^{i\alpha_n}]V^*$ for some unitary V and real scalars $\alpha_1, \alpha_2, \ldots, \alpha_n$ and define $H = V \operatorname{diag}[\alpha_1, \alpha_2, \ldots, \alpha_n]V^*$.

5. If $A \in \mathbb{C}^{n\times n}$, prove that there is a real skew-symmetric matrix S such that $A = e^S$ if and only if A is a real orthogonal matrix with $\det A = 1$.

6. Determine $A^n(\lambda)$, where

$$A(\lambda) = \begin{bmatrix} 1 - \lambda + \lambda^2 & 1 - \lambda \\ \lambda - \lambda^2 & \lambda \end{bmatrix},$$

for all integers n and all λ by use of Eq. (9.5.1).

7. If the elements of the square matrix $D(\lambda)$ are differentiable and $\Delta(\lambda) = \det D(\lambda)$, show that

$$\Delta^{(1)}(\lambda) = \operatorname{tr}((\operatorname{adj} D(\lambda))D^{(1)}(\lambda))$$

and if $\Delta(\lambda) \neq 0$,

$$\frac{\Delta^{(1)}(\lambda)}{\Delta(\lambda)} = \operatorname{tr}(D^{-1}(\lambda)D^{(1)}(\lambda)).$$

8. Suppose that λ is a real variable. Let $D(\lambda)$, defined in Exercise 7, be a solution of the matrix differential equation $\dot{X}(\lambda) = X(\lambda)A(\lambda)$, where $A(\lambda)$ exists for all λ. Prove that if $\Delta(\lambda) = \det D(\lambda)$, and if there is a λ_0 for which $\Delta(\lambda_0) \neq 0$, then

$$\Delta(\lambda) = \Delta(\lambda_0) \exp \int_{\lambda_0}^{\lambda} \operatorname{tr}(A(\mu))\, d\mu,$$

and hence $\Delta(\lambda) \neq 0$ for all λ. (This relation is variously known as the *Jacobi identity*, *Liouville's formula*, or *Abel's formula*.)

9. If all the eigenvalues of A have negative real parts, prove that

$$A^{-1} = -\int_0^{\infty} e^{At}\, dt,$$

and hence that if all eigenvalues of A are in the half-plane given by $\{\lambda \in \mathbb{C} : \mathcal{Re}\, \lambda < \mathcal{Re}\, \lambda_0\}$, then

$$(I\lambda_0 - A)^{-1} = \int_0^{\infty} e^{-(I\lambda_0 - A)t}\, dt.$$

Hint. Apply the result of Exercise 9.5.2.

10. Prove that for any $A \in \mathbb{C}^{n \times n}$, the component matrices Z_{kj} ($j = 0, 1, \ldots, m_k - 1, k = 1, 2, \ldots, s$) satisfy

$$Z_{kj} Z_{kt} = \binom{j + t}{j} Z_{k, j+t}, \qquad 0 \le j, t \le m_k - 1,$$

with the convention that $Z_{kl} = O$ if $l \ge m_k$.

Hint. Use Eq. (9.5.9).

11. Let $L(\lambda) = \sum_{i=0}^{l} \lambda^i L_i$ be an $n \times n$ λ-matrix over $\mathbb{C}$, let $A \in \mathbb{C}^{n \times n}$, and let $L(A) = \sum_{i=0}^{l} L_i A^i$. Prove that

(a) $$L(A) = \sum_{k=1}^{s} \sum_{j=0}^{m_k - 1} L^{(j)}(\lambda_k) Z_{kj},$$

where Z_{kj} ($1 \le k \le s$, $0 \le j \le m_k - 1$) are the component matrices for A.

(b) $$L(A) = \frac{1}{2\pi i} \int_L L(\lambda) R_\lambda \, d\lambda,$$

where $R_\lambda = (\lambda I - A)^{-1}$, $\lambda \notin \sigma(A)$, and L is a contour like that described in Theorem 9.9.2.

Hint. For part (a) first find A^i ($i = 0, 1, \ldots, l$) by Eq. (9.5.1); for part (b) use Eq. (9.8.4) and part (a).

12. With the notation of Exercise 11, prove that

$$L(A) Z_{kt} = \sum_{j=0}^{m_k - 1} L^{(j)}(\lambda_k) \binom{j + t}{j} Z_{k, j+t}$$

for $k = 1, 2, \ldots, s$ and $t = 0, 1, \ldots, m_k - 1$, where $Z_{k, j+t} = O$ for $j + t \ge m_k$.

Hint. Use Exercises 10 and 11 and Theorem 9.5.2(b).

13. Prove that $\operatorname{Ker} Z_{k0} \subset \operatorname{Im}(A - \lambda_k I)$ and that $m_k = 1$ implies $\operatorname{Ker} Z_{k0} = \operatorname{Im}(A - \lambda_k I)$ for $k = 1, 2, \ldots, s$.

Hint. Represent $\varphi_{k0}(\lambda) - 1 = (\lambda - \lambda_k)q(\lambda)$, where $\varphi_{kj}(\lambda)$ are defined in Section 9.2, and substitute A for λ. For the second part use Exercise 9.5.1(b).

14. If $A, B \in \mathbb{C}^{n\times n}$ and $X = X(t)$ is an $n \times n$ matrix dependent on t, check that a solution of

$$\dot{X} = AX + XB, \qquad X(0) = C,$$

is given by $X = e^{At}Ce^{Bt}$.

15. Consider the matrix differential equations

$$\text{(a) } \dot{X} = AX, \qquad \text{(b)} \dot{X} = AX + XB,$$

in which $A, B \in \mathbb{C}^{n\times n}$ are skew-Hermitian matrices and $X = X(t)$. Show that in each case there is a solution matrix $X(t)$ that is unitary for each t.

Hint. Use Exercise 14 with $C = I$ for part (b).

16. Let $A(t), B(t) \in \mathbb{C}^{n\times n}$ for each $t \in [t_1, t_2]$ and suppose that any solution of $\dot{X} + AX + XBX = 0$ is nonsingular for each $t \in [t_1, t_2]$. Show the following.

(a) The matrix-valued function $Y = X^{-1}$ satisfies the linear equation

$$\dot{Y} - YA = B. \tag{1}$$

(b) If A is independent of t, then the solution of Eq. (1) with $Y(t_0) = C^{-1}$ $(t_1 \le t_0 \le t_2)$ is given by

$$Y(t) = C^{-1}e^{A(t-t_0)} + \int_{t_0}^{t} B(\tau)e^{A(t-\tau)}\, d\tau.$$

17. Find the solution of $\ddot{\boldsymbol{x}} = A\boldsymbol{x} + \boldsymbol{f}$, $\boldsymbol{x}(0) = \boldsymbol{x}_0$, $\dot{\boldsymbol{x}}(0) = \dot{\boldsymbol{x}}_0$.

18. Verify that if A is nonsingular, then

$$\boldsymbol{x}(t) = (\cos A^{1/2}t)\boldsymbol{x}_0 + (A^{1/2})^{-1}(\sin A^{1/2}t)\dot{\boldsymbol{x}}_0$$

is the solution of $\ddot{\boldsymbol{x}} = -A\boldsymbol{x}$, $\boldsymbol{x}(0) = \boldsymbol{x}_0$, $\dot{\boldsymbol{x}}(0) = \dot{\boldsymbol{x}}_0$. How can the result be extended to include a singular matrix A?

19. If $A \in \mathbb{C}^{n\times n}$ and $\dot{\boldsymbol{y}}(t) = A\boldsymbol{y}(t)$ has a nonzero solution $\boldsymbol{y}(t)$ for which $\boldsymbol{y}(t + \tau) = \lambda\boldsymbol{y}(t)$ for all t and some fixed $\tau > 0$, prove that $\lambda = e^{\mu\tau}$ for some $\mu \in \sigma(A)$. Hence show that a periodic solution exists only if A is singular.

20. Show that the controllable subspace of the pair (A, B) is the smallest A-invariant subspace containing Im B.

Hint. Note Exercise 9.11.3.

21. Let $f(\lambda)$ be a function defined on the spectrum of a Hermitian matrix H with the property $|f(\lambda)| = 1$ whenever λ is real. Show that $f(H)$ is unitary.

22. (a) Let λ_0 be an eigenvalue of $A \in \mathbb{C}^{n \times n}$ with algebraic multiplicity μ. Show that

$$\mu = \lim_{\lambda \to \lambda_0} [(\lambda - \lambda_0)\operatorname{tr}(\lambda I - A)^{-1}].$$

Hint. Use Ex. 7 above with $D(\lambda) = \lambda I - A$.

(b) If A has component matrices $Z_0, Z_1, \ldots, Z_{m-1}$ corresponding to λ_0, show that

$$\operatorname{tr}(Z_j) = \begin{cases} \mu & \text{for } j = 0 \\ 0 & \text{for } j = 1, 2, \ldots, m - 1. \end{cases}$$

(c) Check the above properties in Exs. 9.5.3 and 9.5.5.

CHAPTER 10

Norms and Bounds for Eigenvalues

In many situations it is useful to be able to assign a single nonnegative number to a vector or a matrix as a measure of its magnitude. We already have such a measure of magnitude for a complex number, namely, its modulus; and for a vector $\boldsymbol{x}$ of $\mathbb{C}^n$, the magnitude can be measured by the length $(\boldsymbol{x}, \boldsymbol{x})^{1/2}$, introduced in Section 3.11. We will take an axiomatic approach to the formulation of measures of magnitude or *norms* of elements of an arbitrary linear space and begin by setting down the properties we demand of such a measure. We shall see that there are many possible choices for norms.

A study of norms of matrices and vectors is the first topic of this chapter, while different results concerning bounds for eigenvalues (some of which are expressed in terms of a norm of a matrix) are studied in the second part of the chapter.

10.1 The Notion of a Norm

Consider a linear space $\mathscr{S}$ over the field of complex or real numbers. A real-valued function defined on all elements $\boldsymbol{x}$ from $\mathscr{S}$ is called a *norm* (on $\mathscr{S}$); usually written $\| \ \|$, if it satisfies the following axioms:

(1) $\|\boldsymbol{x}\| \geq 0$ for all $\boldsymbol{x} \in \mathscr{S}$ and $\|\boldsymbol{x}\| = 0$ if and only if $\boldsymbol{x} = \boldsymbol{0}$;
(2) $\|\alpha\boldsymbol{x}\| = |\alpha| \, \|\boldsymbol{x}\|$ for all $\boldsymbol{x} \in \mathscr{S}$ and $\alpha \in \mathbb{C}$;
(3) $\|\boldsymbol{x} + \boldsymbol{y}\| \leq \|\boldsymbol{x}\| + \|\boldsymbol{y}\|$ for all $\boldsymbol{x}, \boldsymbol{y} \in \mathscr{S}$.

It is helpful to bear in mind the analogy of a norm with the modulus of a complex number or with the familiar length of a vector in $\mathbb{R}_3$. The axioms 1–3 are immediate generalizations for these ideas and therefore, in particular, axiom 3 is called the *triangle inequality*.

Note first of all that in a unitary space $\mathscr{U}$ (see Section 3.11) the inner product generates a norm. Thus, one defines $\|\boldsymbol{x}\| = (\boldsymbol{x}, \boldsymbol{x})^{1/2}$ for any $\boldsymbol{x} \in \mathscr{U}$, and the norm axioms above follow readily from the inner product axioms. It can be shown, however, that not all norm functions are generated by an inner product and, since several such norms are widely useful, much of the analysis of this chapter is completed without reference to inner products.

The next set of exercises is important. The first points out some immediate and frequently used consequences of the norm axioms. The second introduces the most important norms on the spaces $\mathbb{R}^n$ and $\mathbb{C}^n$ of n-tuples. The third exercise shows, in particular, that norms can always be defined on an abstract finite-dimensional linear space. A linear space together with a norm defined on it is called a *normed linear space*.

Exercise 1. (a) For any $\boldsymbol{x}, \boldsymbol{y}$ in a normed linear space $\mathscr{S}$, prove that

$$\big|\,\|\boldsymbol{x}\| - \|\boldsymbol{y}\|\,\big| \le \|\boldsymbol{x} + \boldsymbol{y}\| \le \|\boldsymbol{x}\| + \|\boldsymbol{y}\|.$$

(b) For any $\boldsymbol{x} \in \mathscr{S}$, show that $\|-\boldsymbol{x}\| = \|\boldsymbol{x}\|$.

(c) For any $\boldsymbol{x} \in \mathscr{S}$ with $\boldsymbol{x} \ne \boldsymbol{0}$, show that there is a member of $\mathscr{S}$ of the form $\alpha\boldsymbol{x}$ with norm equal to 1. (For example, take $\alpha = \|\boldsymbol{x}\|^{-1}$.)

Exercise 2. Let $\boldsymbol{x}$ be a typical vector of $\mathbb{C}^n$ (or $\mathbb{R}^n$), $\boldsymbol{x} = [x_1 \quad x_2 \quad \cdots \quad x_n]^{\mathrm{T}}$. Show that the functions $\|\ \|_\infty$, $\|\ \|_1$, $\|\ \|_2$, $\|\ \|_p$ defined on $\mathbb{C}^n$ (or $\mathbb{R}^n$) as follows are all norms on $\mathbb{C}^n$ (or $\mathbb{R}^n$):

(a) $\|\boldsymbol{x}\|_\infty = \max\limits_{1 \le j \le n} |x_j|$ (the *infinity* norm);

(b) $\|\boldsymbol{x}\|_1 = \sum\limits_{j=1}^{n} |x_j|$;

(c) $\|\boldsymbol{x}\|_2 = \left(\sum\limits_{j=1}^{n} |x_j|^2\right)^{1/2} = (\boldsymbol{x}^*\boldsymbol{x})^{1/2}$ (the *euclidean* norm);

(d) $\|\boldsymbol{x}\|_p = \left(\sum\limits_{j=1}^{n} |x_j|^p\right)^{1/p}$, $p \ge 1$ (the *Hölder* norms).

Hint. For part (d) the Minkowski inequality should be assumed, that is, for any $p \ge 1$ and any complex n-tuples $(x_1, x_2, \ldots, x_n)$ and $(y_1, y_2, \ldots, y_n)$,

$$\left(\sum_{j=1}^{n} |x_j + y_j|^p\right)^{1/p} \le \left(\sum_{j=1}^{n} |x_j|^p\right)^{1/p} + \left(\sum_{j=1}^{n} |y_j|^p\right)^{1/p}.$$

Exercise 3. Let $\{\boldsymbol{b}_1, \boldsymbol{b}_2, \ldots, \boldsymbol{b}_n\}$ be a fixed basis for a linear space $\mathscr{S}$, and for any $\boldsymbol{x} \in \mathscr{S}$, write $\boldsymbol{x} = \sum_{j=1}^{n} \alpha_j \boldsymbol{b}_j$.

(a) Show that $\|\boldsymbol{x}\|_\infty = \max_{1 \le j \le n} |\alpha_j|$ defines a norm on $\mathscr{S}$.
(b) If $p \ge 1$, show that $\|\boldsymbol{x}\|_p = (\sum_{j=1}^{n} |\alpha_j|^p)^{1/p}$ defines a norm on $\mathscr{S}$. □

The euclidean norm of Exercise 2(c) is often denoted by $\| \ \|_{\mathrm{E}}$ and is just one of the Hölder norms of part (d). It is clear from these exercises that many different norms can be defined on a given finite-dimensional linear space, and a linear space combined with one norm produces a different "normed linear space" from the same linear space combined with a different norm. For example, $\mathbb{C}^n$ with $\| \ \|_\infty$ is not the same normed linear space as $\mathbb{C}^n$ with $\| \ \|_2$. For this reason it is sometimes useful to denote a normed linear space by a pair, $\mathscr{S}$, $\| \ \|$.

It is also instructive to consider the examples of Exercise 2 as special cases of Exercise 3. This is done by introducing the basis of unit coordinate vectors into $\mathbb{C}^n$, or $\mathbb{R}^n$, in the role of the general basis $\{\boldsymbol{b}_1, \ldots, \boldsymbol{b}_n\}$ of Exercise 3.

The notation $\| \ \|_\infty$ used for the norm in Exercise 3(a) suggests that it might be interpreted as a norm of the type introduced in Exercise 2(d) with $p = \infty$. This can be justified in a limiting sense in the following way: If $\boldsymbol{x} = \sum_{j=1}^{n} \alpha_j \boldsymbol{b}_j$ and there are ν components α_i $(1 \le i \le n)$ of absolute value $\|\boldsymbol{x}\|_\infty = \rho = \max_{1 \le i \le n} |\alpha_i|$, then

$$\|\boldsymbol{x}\|_p = (|\alpha_1|^p + |\alpha_2|^p + \cdots + |\alpha_n|^p)^{1/p} = \rho(\nu + \varepsilon_1^p + \cdots + \varepsilon_{n-\nu}^p)^{1/p},$$

where for some k depending on j, $\varepsilon_j = |\alpha_k|/\rho < 1$ $(j = 1, 2, \ldots, n - \nu)$. Now it is easily seen that $\|\boldsymbol{x}\|_\infty = \rho = \lim_{p \to \infty} \|\boldsymbol{x}\|_p$.

Exercise 4. If $\| \ \|$ denotes a norm on a linear space $\mathscr{S}$ and $k > 0$, show that $k\| \ \|$ is also a norm on $\mathscr{S}$.

Exercise 5. Let $\mathscr{U}$ be a unitary space with inner product (,), as defined in Section 3.11. Show that $\|\boldsymbol{x}\| = (\boldsymbol{x}, \boldsymbol{x})^{1/2}$ defines a norm on $\mathscr{U}$. (Thus, the natural definition of a norm on a unitary space makes it a normed linear space.)

Exercise 6. Show that the euclidean vector norm on $\mathbb{C}^n$ is invariant under unitary transformations. In other words, if $\boldsymbol{x} \in \mathbb{C}^n$ and $U \in \mathbb{C}^{n \times n}$ is a unitary matrix, then $\|U\boldsymbol{x}\|_2 = \|\boldsymbol{x}\|_2$. Formulate an abstract generalization of this result concerning an arbitrary unitary space $\mathscr{U}$.

Exercise 7. If H is a positive definite matrix, show that

$$\|\boldsymbol{x}\| = (H\boldsymbol{x}, \boldsymbol{x})^{1/2} = (\boldsymbol{x}^* H \boldsymbol{x})^{1/2}$$

is a norm on $\mathbb{C}^n$. What is the generalization appropriate for an abstract unitary space? □

The remainder of this section is devoted to two important results concerning norms in general. The first tells us that a norm is necessarily a "smooth" function of the coordinates of a vector $\mathbf{x} \in \mathscr{S}$ with respect to a fixed basis in $\mathscr{S}$. Thus, "small" variations in the coordinates of $\mathbf{x}$ give rise to small variations in the size of the norm. Observe first that in a normed linear space $\|\mathbf{x}\|$ depends continuously on $\mathbf{x}$ itself. This follows immediately from Exercise 1(a), which implies that, in any normed linear space $\mathscr{S}$,

$$| \|\mathbf{x}\| - \|\mathbf{y}\| | \leq \|\mathbf{x} - \mathbf{y}\|.$$

for all $\mathbf{x}, \mathbf{y} \in \mathscr{S}$.

Theorem 1. *Let $\{\mathbf{b}_1, \mathbf{b}_2, \ldots, \mathbf{b}_n\}$ be a fixed basis in a linear space $\mathscr{S}$. Then any norm defined on $\mathscr{S}$ depends continuously on the coordinates of members of $\mathscr{S}$ with respect to $\mathbf{b}_1, \ldots, \mathbf{b}_n$. In other words, if $\mathbf{x} = \sum_{j=1}^n x_j \mathbf{b}_j$, $\mathbf{y} = \sum_{j=1}^n y_j \mathbf{b}_j$ denote arbitrary members of $\mathscr{S}$ and $\| \;\|$ is a norm on $\mathscr{S}$, then given $\varepsilon > 0$ there is a $\delta > 0$ depending on ε such that*

$$| \|\mathbf{x}\| - \|\mathbf{y}\| | < \varepsilon$$

whenever $|x_j - y_j| < \delta$ for $j = 1, 2, \ldots, n$.

PROOF. Let $M = \max_{1 \leq j \leq n} \|\mathbf{b}_j\|$. By the first norm axiom, $M > 0$, so we may choose a δ satisfying $0 < \delta \leq \varepsilon/Mn$.

Then for any $\mathbf{x}, \mathbf{y} \in \mathscr{S}$ for which $|x_j - y_j| < \delta$ for each j we have, using the second and third norm axioms,

$$\|\mathbf{x} - \mathbf{y}\| \leq \sum_{j=1}^{n} |x_j - y_j| \, \|\mathbf{b}_j\| \leq M \sum_{j=1}^{n} |x_j - y_j| < Mn\delta \leq \varepsilon. \quad \blacksquare$$

The next result is a comparison theorem between different norms defined on the same space. If any two norms, $\| \;\|_1$ and $\| \;\|_2$, on $\mathscr{S}$ satisfy $\|\mathbf{x}\|_1 \leq \|\mathbf{x}\|_2$ for all $\mathbf{x} \in \mathscr{S}$, we say that $\| \;\|_2$ *dominates* $\| \;\|_1$. Now, we have noted in Exercise 4 that a positive multiple of a norm is a norm. The next result says that all norms are equivalent in the sense that for any two norms $\| \;\|_1$ and $\| \;\|_2$ on $\mathscr{S}$, $\| \;\|_1$ is dominated by a positive multiple of $\| \;\|_2$ and, at the same time, $\| \;\|_2$ is dominated by a positive multiple of $\| \;\|_1$. It should be noted that the proof depends on the fact that $\mathscr{S}$ is finite-dimensional.

First we make a formal definition: two norms $\| \;\|_1$ and $\| \;\|_2$ defined on a linear space $\mathscr{S}$ are said to be *equivalent* if there exist positive numbers r_1 and r_2 depending only on the choice of norms such that, for any nonzero $\mathbf{x} \in \mathscr{S}$,

$$r_1 \leq \|\mathbf{x}\|_1 / \|\mathbf{x}\|_2 \leq r_2. \tag{1}$$

Theorem 2. *Any two norms on a finite-dimensional linear space are equivalent.*

PROOF. Let $\| \;\|_1$ and $\| \;\|_2$ be any two norms defined on $\mathscr{S}$. We have observed that the functions f_1 and f_2 defined on $\mathscr{S}$ by $f_1(\mathbf{x}) = \|\mathbf{x}\|_1$ and $f_2(\mathbf{x}) = \|\mathbf{x}\|_2$ are continuous. Then it is easily verified that the sets

$$C_1 = \{\mathbf{x} \in \mathscr{S} : \|\mathbf{x}\|_1 = 1\}, \qquad C_2 = \{\mathbf{x} \in \mathscr{S} : \|\mathbf{x}\|_2 = 1\}$$

are bounded and closed in $\mathscr{S}$ (see Appendix 2). It follows that the greatest lower bounds

$$\gamma_1 = \inf_{x \in C_1} \|x\|_2, \qquad \gamma_2 = \inf_{x \in C_2} \|x\|_1 \tag{2}$$

are attained on members of C_1 and C_2, respectively. Hence $\gamma_1 = \|x_0\|_2$ for some $x_0 \in C_1$ and so $\gamma_1 > 0$. Similarly, $\gamma_2 > 0$ and for any nonzero $x \in \mathscr{S}$, it follows from (2) and the second norm axiom that

$$\gamma_1 \le \left\| \frac{x}{\|x\|_1} \right\|_2 = \frac{\|x\|_2}{\|x\|_1}$$

and

$$\gamma_2 \le \left\| \frac{x}{\|x\|_2} \right\|_1 = \frac{\|x\|_1}{\|x\|_2}.$$

Taking $r_1 = \gamma_2$ and $r_2 = 1/\gamma_1$, inequalities (1) are obtained and the theorem is proved. ■

Exercise 8. Verify that equivalence of norms on $\mathscr{S}$ is an equivalence relation on the set of all norms on $\mathscr{S}$. □

Theorem 2 can be useful since it can be applied to reduce the study of properties of a general norm to those of a relatively simple norm. The strong candidates for the choice of a "relatively simple" norm are generally $\|\ \|_\infty$, $\|\ \|_1$, or most frequently, $\|\ \|_2$ of Exercises 2 and 3.

10.2 A Vector Norm as a Metric: Convergence

In Section 10.1 the emphasis placed on a norm was its usefulness as a measure of the size of members of a normed linear space. It is also very important as a measure of *distance*. Thus, if $\mathscr{S}$, $\|\ \|$ is a normed linear space then for any $x, y \in \mathscr{S}$, the quantity $\|x - y\|$ is thought of as a measure of the distance between x and y. The special case of $\mathbb{R}_3$ (or $\mathbb{R}_2$) with the euclidean norm produces the most familiar physical measure of distance.

The following properties are easily verified from the norm axioms for any x, y, z in a normed linear space:

(a) $\|x - y\| \ge 0$ with equality if and only if $y = x$;
(b) $\|x - y\| = \|y - x\|$;
(c) $\|x - y\| \le \|x - z\| + \|z - y\|$.

They correspond to the axioms for a *metric*, that is, a function defined on pairs of elements of the set $\mathscr{S}$ that is a mathematically sensible measure of distance.

The set of all vectors in a normed linear space whose distance from a fixed point $\boldsymbol{x}_0$ is $r > 0$ is known as the *sphere* with center $\boldsymbol{x}_0$ and radius r. Thus, written explicitly in terms of the norm, this sphere is

$$\{\boldsymbol{x} \in \mathscr{S} : \|\boldsymbol{x} - \boldsymbol{x}_0\| = r\}.$$

The set of $\boldsymbol{x} \in \mathscr{S}$ for which $\|\boldsymbol{x} - \boldsymbol{x}_0\| \leq r$ is the *ball* with centre $\boldsymbol{x}_0$ and radius r. The sphere with centre $\boldsymbol{0}$ and radius $r = 1$ is called the *unit sphere* and plays an important part in the analysis of normed linear spaces, as it did in the proof of Theorem 10.1.2. The *unit ball* is, of course, $\{\boldsymbol{x} \in \mathscr{S} : \|\boldsymbol{x}\| \leq 1\}$.

It is instructive to examine the unit sphere in $\mathbb{R}_2$ with the norms $\| \ \|_\infty$, $\| \ \|_2$, and $\| \ \|_1$ introduced in Exercise 10.1.2. It is easy to verify that the portions of the unit spheres in the first quadrant for these norms (described by $p = \infty, 2, 1$, respectively) are as sketched in Fig. 10.1. It is also instructive to examine the nature of the unit spheres for Hölder norms with intermediate values of p and to try to visualize corresponding unit spheres in $\mathbb{R}_3$.

Exercise 1. Show that the function $\| \ \|$ defined on $\mathbb{C}^n$ by

$$\|\boldsymbol{x}\| = |x_1| + \cdots + |x_{n-1}| + n^{-1}|x_n|,$$

where $\boldsymbol{x} = [x_1 \ \ x_2 \ \ \cdots \ \ x_n]^T$, is a norm on $\mathbb{C}^n$ (or $\mathbb{R}^n$). Sketch the unit spheres in $\mathbb{R}^2$ and $\mathbb{R}^3$. □

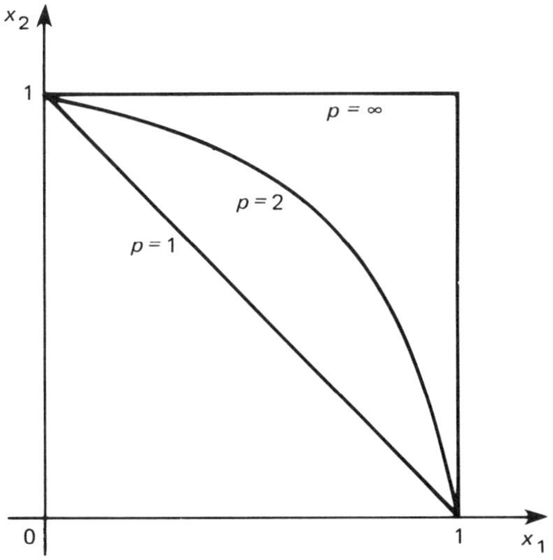

Fig. 10.1 Unit spheres in R_2 for p-norms.

Norms can be defined on a linear space in a great variety of ways and give rise to unit balls with varied geometries. A property that all unit balls have in common, however, is that of *convexity*. (See Appendix 2 for a general discussion of convexity.)

Theorem 1. *The unit ball in a normed linear space is a convex set.*

PROOF. Let $\mathcal{S}$, $\| \ \|$ be a normed linear space with unit ball B. If $\boldsymbol{x}, \boldsymbol{y} \in B$, then for any $t \in [0, 1]$, the vector $\boldsymbol{z} = t\boldsymbol{x} + (1 - t)\boldsymbol{y}$ is a member of the line segment joining $\boldsymbol{x}$ and $\boldsymbol{y}$. We have only to prove that $\boldsymbol{z} \in B$. But the second and third norm axioms immediately give

$$\|\boldsymbol{z}\| \leq t\|\boldsymbol{x}\| + (1 - t)\|\boldsymbol{y}\|,$$

and since $\|\boldsymbol{x}\| \leq 1$ and $\|\boldsymbol{y}\| \leq 1$ we get $\|\boldsymbol{z}\| \leq 1$, so that $\boldsymbol{z} \in B$. ■

Exercise 2. Show that any ball in a normed linear space is a convex set. □

A primitive open set in a normed linear space is a *neighborhood*, which is simply the interior of a ball. More formally, if $\mathcal{S}$, $\| \ \|$ is a normed linear space, the neighborhood with centre $\boldsymbol{x}_0$ and radius r is the set

$$N(\boldsymbol{x}_0, r) = \{\boldsymbol{x} \in \mathcal{S} : \|\boldsymbol{x} - \boldsymbol{x}_0\| < r\}. \tag{1}$$

Neighborhoods are frequently useful in analysis. For example, if $\mathcal{S}$, $\| \ \|_1$ and $\mathcal{R}$, $\| \ \|_2$ are two normed linear spaces and f is a function defined on $\mathcal{S}$ with values in $\mathcal{R}$, then f is *continuous at* $\boldsymbol{x}_0 \in \mathcal{S}$ if given $\varepsilon > 0$ there is a $\delta > 0$, depending on ε, such that $\boldsymbol{x} \in N(\boldsymbol{x}_0, \delta)$ implies $f(\boldsymbol{x}) \in N(f(\boldsymbol{x}_0), \varepsilon)$. Note that $N(\boldsymbol{x}_0, \delta)$ is a neighborhood in $\mathcal{S}$ defined via the norm $\| \ \|_1$ and $N(f(\boldsymbol{x}_0), \varepsilon)$ is a neighborhood in $\mathcal{R}$ defined using the norm $\| \ \|_2$.

A basic concept of analysis is convergence, and we have already considered the convergence of sequences and series in $\mathbb{C}^n$ and $\mathbb{R}^n$ in Section 9.8. For a general finite-dimensional linear space $\mathcal{S}$, the approach adopted there can be generalized by first introducing a basis into $\mathcal{S}$, say $\{\boldsymbol{b}_1, \boldsymbol{b}_2, \ldots, \boldsymbol{b}_n\}$. If $\{\boldsymbol{x}_k\}_{k=1}^{\infty}$ is a sequence in $\mathcal{S}$ then each member has a representation

$$\boldsymbol{x}_k = \sum_{j=1}^{n} \alpha_j^{(k)} \boldsymbol{b}_j, \qquad k = 1, 2, \ldots, \tag{2}$$

and the sequence is said to *converge coordinatewise* to an element $\boldsymbol{x}_0 = \sum_{j=1}^{n} \alpha_j^{(0)} \boldsymbol{b}_j$ if $\alpha_j^{(k)} \to \alpha_j^{(0)}$ as $k \to \infty$ for each of the n scalar sequences obtained by putting $j = 1, 2, \ldots, n$. Thus, the convergence of sequences of vectors considered in section 9.8 is just coordinatewise convergence in the linear space $\mathbb{C}^n$ with respect to its standard basis.

However, this process is cumbersome and appears to depend on the choice of basis for $\mathscr{S}$ (although, in fact, it does not). Furthermore it turns out to be equivalent to a notion of convergence defined by a norm and, indeed, *any* norm, as we shall soon verify.

A sequence $\{x_k\}_{k=1}^{\infty}$ in a normed linear space $\mathscr{S}, \|\ \|$ is said to be *convergent in norm* (or simply *convergent*) to $x_0 \in \mathscr{S}$ if $\|x_k - x_0\| \to 0$ as $k \to \infty$. We will write, informally, $x_k \to x_0$ as $k \to \infty$.

Exercise 3. Show that, if $x_k \to x_0, y_k \to y_0$, and $\alpha_k \to \alpha_0$ as $k \to \infty$, where α_0 and the α_k are scalars, then $\|x_k - x_0\| \to 0$ and $\alpha_k x_k + y_k \to \alpha_0 x_0 + y_0$ as $k \to \infty$.

Exercise 4. Show that if $\{x_k\}_{k=1}^{\infty}$ converges in one norm to x_0, then it converges to x_0 in any other norm.

Hint. Use Theorem 10.1.2. □

The next theorem shows that coordinatewise convergence and convergence in norm for infinite sequences (and hence series) in a finite-dimensional space are equivalent. However, the result is not true if the space is not of finite dimension. Indeed, the possibility of different forms of convergence is a major issue in the study of infinite-dimensional linear spaces.

Theorem 2. *Let $\{x_k\}_{k=1}^{\infty}$ be a sequence in a finite-dimensional linear space $\mathscr{S}$. The sequence converges coordinatewise to $x_0 \in \mathscr{S}$ if and only if it converges to x_0 in (any) norm.*

PROOF. Let $\{b_1, b_2, \ldots, b_n\}$ be a basis in the linear space $\mathscr{S}$ and write $x_k = \sum_{j=1}^{n} \alpha_j^{(k)} b_j$ for $k = 0, 1, 2, \ldots$. Given coordinatewise convergence to $x_0 \in \mathscr{S}$ we have, for any norm $\|\ \|$ defined on $\mathscr{S}$,

$$\|x_k - x_0\| = \left\| \sum_{j=1}^{n} (\alpha_j^{(k)} - \alpha_j^{(0)}) b_j \right\| \le \sum_{j=1}^{n} |\alpha_j^{(k)} - \alpha_j^{(0)}|\, \|b_j\| \to 0$$

as $k \to \infty$, since n and the basis are fixed. Hence, we have convergence in norm.

Conversely, given that $x_k \to x_0$ in any norm we can, using the equivalence of norms (as in Exercise 4), replace the norm by $\|\ \|_\infty$ of Exercise 10.1.3(a). Then

$$\|x_k - x_0\|_\infty = \left\| \sum_{j=1}^{n} (\alpha_j^{(k)} - \alpha_j^{(0)}) b_j \right\|_\infty = \max_{1 \le j \le n} |\alpha_j^{(k)} - \alpha_j^{(0)}|.$$

Thus, $\|x_k - x_0\| \to 0$ implies $\|x_k - x_0\|_\infty \to 0$ and hence

$$\max_{1 \le j \le n} |\alpha_j^{(k)} - \alpha_j^{(0)}| \to 0$$

as k $\to \infty$. This clearly implies coordinatewise convergence. ■

10.3 Matrix Norms

We start this section by confining our attention to norms on the linear space $\mathbb{C}^{n\times n}$, that is, to norms of square matrices. Obviously, all properties of a general norm discussed in Sections 1 and 2 remain valid for this particular case. For instance, we have the following.

Exercise 1. Verify that any norm on $\mathbb{C}^{n\times n}$ depends continuously on the matrix elements.

Hint. Consider the standard basis $\{E_{ij}\}_{i,j=1}^{n}$ for $\mathbb{C}^{n\times n}$ (see Exercise 3.5.1) and use Theorem 10.1.1. □

The possibility of multiplying any two matrices $A, B \in \mathbb{C}^{n\times n}$ gives rise to an important question about the relation between the norms $\|A\|$, $\|B\|$ and the norm $\|AB\|$ of their product. We say that the norm $\|\ \|$ on $\mathbb{C}^{n\times n}$ is a *matrix norm* or is *submultiplicative* (in contrast to a norm of a matrix as a member of the normed space $\mathbb{C}^{n\times n}$) if

$$\|AB\| \leq \|A\|\,\|B\| \tag{1}$$

for all $A, B \in \mathbb{C}^{n\times n}$. Before examining some matrix norms note first that not all norms on $\mathbb{C}^{n\times n}$ are matrix norms.

Exercise 2. Verify that the real-valued function $\max_{1\leq i,j\leq n}|a_{ij}|$, where a_{ij} is the i, jth element of $A \in \mathbb{C}^{n\times n}$, defines a norm on $\mathbb{C}^{n\times n}$ but is not a matrix norm.

Exercise 3. (The euclidean or Frobenius norm.) Prove that

$$\|A\|_{\mathrm{E}} = \left(\sum_{i,j=1}^{n} |a_{ij}|^2\right)^{1/2}, \qquad A = [a_{ij}]_{i,j=1}^{n},$$

defines a matrix norm on $\mathbb{C}^{n\times n}$.

Hint. For the norm axioms the argument is that of Exercise 10.1.2(c), which requires the use of the Cauchy–Schwarz inequality (Exercise 2.5.4). For property (1) use the Cauchy–Schwarz inequality again.

Exercise 4. This generalizes Exercise 3. Show that the Hölder norm,

$$\|A\|_p = \left(\sum_{i,j=1}^{n} |a_{ij}|^p\right)^{1/p}, \qquad p \geq 1, \tag{2}$$

is a matrix norm on $\mathbb{C}^{n\times n}$ if and only if $1 \leq p \leq 2$.

Hint. Use the Hölder inequality:

$$\sum_{j=1}^{n} |x_j y_j| \le \left(\sum_{j=1}^{n} |x_j|^p\right)^{1/p} \left(\sum_{j=1}^{n} |y_j|^q\right)^{1/q}, \tag{3}$$

where $p \ge 1$, $q \ge 1$, and $p^{-1} + q^{-1} = 1$.

Exercise 5. Prove that, for any matrix norm $\| \ \|$,

$$\|I\| \ge 1, \qquad \|A^n\| \le \|A\|^n, \qquad \|A^{-1}\| \ge \frac{1}{\|A\|} \quad (\det A \ne 0).$$

Exercise 6. If S is nonsingular and $\| \ \|$ is a matrix norm, then the real-valued function f defined by $f(A) = \|SAS^{-1}\|$ is a matrix norm.

Exercise 7. Check that the following are true.

(a) The function $\|A\| = n \max_{1 \le i, j \le n} |a_{ij}|$ is a matrix norm. (Compare with Exercises 2 and 10.1.4.)

(b) If $\| \ \|_E$ denotes the euclidean matrix norm (see Exercise 3) and $\| \ \|$ is defined as in part (a), then $\|A\|_E \le \|A\|$.

(c) $\|A\|_E^2 = \operatorname{tr}(A^*A)$.

Exercise 8. If $p \ge 2$ and $(1/p) + (1/q) = 1$, then in the notation of Eq. (2),

$$\|AB\|_p \le \min(\|A\|_p \|B\|_q, \|A\|_q \|B\|_p).$$

Hint. Use Eq. (3). □

The most useful matrix norms are often those that are easily defined in terms of the elements of the matrix argument. It is conceivable that a measure of magnitude for matrices could be based on the magnitudes of the eigenvalues, although this may turn out to be of no great practical utility, since the eigenvalues are often hard to compute. However, if $\lambda_1, \ldots, \lambda_n$ are the eigenvalues of A, then $\mu_A = \max_{1 \le j \le n} |\lambda_j|$ is such a measure of magnitude and is known as the *spectral radius* of A.

Consider the matrix N of Exercise 6.8.6. It is easily seen that all of its eigenvalues are zero and hence the spectral radius of N is zero. Since N is not the zero-matrix, we see immediately that the spectral radius does not satisfy the first norm axiom. However, although the spectral radius is not a norm, it is intimately connected with the magnitude of matrix norms. Indeed, matrix norms are often used to obtain bounds for the spectral radius. The next theorem is a result of this kind.

Theorem 1. *If $A \in \mathbb{C}^{n \times n}$ and μ_A is the spectral radius of A then, for any matrix norm,*

$$\|A\| \ge \mu_A.$$

PROOF. Let λ be an eigenvalue of A such that $\mu_A = |\lambda|$. Then there is a vector $\boldsymbol{x} \neq \boldsymbol{0}$ such that $A\boldsymbol{x} = \lambda\boldsymbol{x}$. Define the $n \times n$ matrix

$$A_x = [\boldsymbol{x} \quad 0 \quad 0 \quad \cdots \quad 0]$$

and observe that

$$AA_x = \lambda A_x.$$

Using the second norm axiom and property (1) of the matrix norm, we deduce that

$$|\lambda| \, \|A_x\| \leq \|A\| \, \|A_x\|,$$

Since $A_x \neq 0$, it follows that $\|A_x\| \neq 0$ and hence

$$|\lambda| = \mu_A \leq \|A\|. \quad \blacksquare$$

We are very often faced with the problem of finding the norm of a vector that is given in the form $A\boldsymbol{x}$, where $A \in \mathbb{C}^{n \times n}$ and $\boldsymbol{x} \in \mathbb{C}^n$. Given our experience with property (1) for a matrix norm, we might expect that there will be matrix norms $\| \;\|$ and vector norms $\| \;\|_v$ (that is, norms on $\mathbb{C}^n$) for which

$$\|A\boldsymbol{x}\|_v \leq \|A\| \, \|\boldsymbol{x}\|_v. \tag{4}$$

When Eq. (4) holds for all $\boldsymbol{x} \in \mathbb{C}^n$ and all $A \in \mathbb{C}^{n \times n}$, then the vector norm $\| \;\|_v$ and the matrix norm $\| \;\|$ are said to be *compatible.*

Exercise 9. Prove that the euclidean matrix and vector norms are compatible.

SOLUTION. For arbitrary $\boldsymbol{x} = [\alpha_1 \quad \alpha_2 \quad \cdots \quad \alpha_n]^T \in \mathbb{C}^n$ and

$$A = [a_{ij}]_{i,j=1}^n \in \mathbb{C}^{n \times n},$$

compute $A\boldsymbol{x} = [\beta_1 \quad \beta_2 \quad \cdots \quad \beta_n]^T$, where $\beta_i = \sum_{j=1}^n a_{ij}\alpha_j$, $i = 1, 2, \ldots, n$. Hence by Eq. (3) for $p = 2$, $q = 2$,

$$\|A\boldsymbol{x}\|_v^2 = \sum_{i=1}^n |\beta_i|^2 = \sum_{i=1}^n \left| \sum_{j=1}^n a_{ij}\alpha_j \right|^2 \leq \sum_{i=1}^n \left(\sum_{j=1}^n |a_{ij}|^2 \right) \left(\sum_{j=1}^n |\alpha_j|^2 \right).$$

Finally, since $\|A\|_E^2 = \sum_{i,j=1}^n |a_{ij}|^2$ and $\|\boldsymbol{x}\|_v^2 = \sum_{j=1}^n |\alpha_j|^2$, we have

$$\|A\boldsymbol{x}\|^2 \leq \|A\|_E^2 \|\boldsymbol{x}\|^2,$$

which is equivalent to Eq. (4).

Exercise 10. Prove that the matrix norm defined in Exercise 7(a) is compatible with the Hölder norms on $\mathbb{C}^n$ for $p = 1, 2, \infty$.

Exercise 11. Let $\| \;\|_v$ be a norm on $\mathbb{C}^n$ and let $S \in \mathbb{C}^{n \times n}$ be nonsingular. Show that $\|\boldsymbol{x}\|_S = \|S\boldsymbol{x}\|_v$ is also a norm on $\mathbb{C}^n$. Then show that $\| \;\|_S$ and the matrix norm of Exercise 6 are compatible if and only if $\| \ \|_v$ and $\| \ \|$ are compatible. □

As with column vectors and their norms, norms applied to matrices provide a useful vehicle for the study of convergence of sequences and series of matrices. Indeed, comparing the discussions of sections 9.8 and 10.2 it is easily seen that the following proposition is just a special case of Theorem 10.2.2.

Proposition 1. *Let $\{A_k\}_{k=1}^{\infty}$ be a sequence in $\mathbb{C}^{m\times n}$. The sequence converges coordinatewise to A if and only if $\|A_k - A\| \to 0$ as $k \to \infty$ in any norm.*

Here, the norm needs to have only the three basic properties of a vector norm, and the sub-multiplicative property plays no part.

The next exercises illustrate the use of norms in the analysis of an important numerical procedure.

Exercise 12. Suppose that $A \in \mathbb{C}^{n\times n}$ has an eigenvalue λ_1 with the properties that $|\lambda_1|$ exceeds the absolute value of all other eigenvalues of A and the index of λ_1 is one. (Then λ_1 is called the *dominant* eigenvalue of A.) Show that, as $r \to \infty$,

$$\left(\frac{1}{\lambda_1} A\right)^r \to Z_{10}. \tag{5}$$

Hint. Let $f(z) = z^r$ and use Theorem 9.5.1. (The hypothesis that λ_1 has index one is made for convenience. Otherwise, the limit on the right of (5) is a little more complicated.)

Exercise 13. (The *power method* for calculating eigenvalues). Let $A \in \mathbb{C}^{n\times n}$ and have a dominant eigenvalue λ_1 of index one (as in Exercise 12). Let $\boldsymbol{x}_0$ be any vector for which $\boldsymbol{x}_0 \notin \operatorname{Ker} A^n$, $\|\boldsymbol{x}_0\| = 1$ in the euclidean norm, and $Z_{10}\boldsymbol{x}_0 \neq 0$. Define a sequence of vectors $\boldsymbol{x}_0, \boldsymbol{x}_1, \boldsymbol{x}_2, \ldots$ of norm one by

$$\boldsymbol{x}_r = \frac{1}{k_r} A\boldsymbol{x}_{r-1}, \qquad r = 1, 2, \ldots,$$

where $k_r = \|A\boldsymbol{x}_{r-1}\|$. Show that:

(a) The sequence is well-defined (i.e., $k_r \neq 0$ for any r).

(b) As $r \to \infty$, $k_r \to |\lambda_1|$.

(c) There is a sequence $\{\beta_r\}_{r=1}^{\infty}$ of complex numbers such that as $r \to \infty$, $\beta_{r+1}/\beta_r \to |\lambda_1|/\lambda_1$ and

$$\beta_r \boldsymbol{x}_r \to \boldsymbol{p}_1, \tag{6}$$

where $\boldsymbol{p}_1$ is an eigenvector of A of norm one determined by the choice of $\boldsymbol{x}_0$.

(Note that by normalizing the vectors $\boldsymbol{x}_r$ to have euclidean norm equal to one, they are not defined to within a unimodular scalar multiple. Thus, the presence of β_r in (6) is to be expected. In numerical practice a different normalization is likely to be used, with $\| \ \|_\infty$ for example. But as in so many cases, the euclidean norm is the most convenient for analysis.)

OUTLINE OF THE SOLUTION. Assume inductively that $k_1, k_2, \ldots, k_{r-1}$ are nonzero. Then show that

$$A\boldsymbol{x}_{r-1} = \frac{1}{k_{r-1} \cdots k_2 k_1} A^r \boldsymbol{x}_0.$$

Since Ker A^r = Ker A^n for all $r \geq n$ (ref. Ex. 4.5.14), $\boldsymbol{x}_0 \notin$ Ker A^r, and so $A\boldsymbol{x}_{r-1} \neq \boldsymbol{0}$. Hence $k_r \neq 0$ and (a) follows by induction.
Then observe

$$k_r = \|A^r \boldsymbol{x}_0\| / \|A^{r-1} \boldsymbol{x}_0\|. \tag{7}$$

Define $\boldsymbol{p}_1 = \alpha^{-1} Z_{10} \boldsymbol{x}_0$, where $\alpha = \|Z_{10} \boldsymbol{x}_0\| \neq 0$, and use the preceding exercise to show that $|\lambda_1|^{-r} \|A^r \boldsymbol{x}_0\| \to \alpha$. Combine with (7) to obtain part (b).

Use the fact that

$$\boldsymbol{x}_r = \frac{1}{k_r \cdots k_2 k_1} A^r \boldsymbol{x}_0$$

together with Theorem 9.5.1 to show that

$$\frac{k_r \cdots k_2 k_1}{\lambda_1^r} \boldsymbol{x}_r \to \alpha \boldsymbol{p}_1,$$

and define $\beta_r = (k_r \cdots k_2 k_1)/(\alpha \lambda_1^r)$ to obtain (c). □

10.4 Induced Matrix Norms

We now investigate a method for obtaining a matrix norm from any given vector norm. If $\| \ \|_v$ is a vector norm in $\mathbb{C}^n$ and $A \in \mathbb{C}^{n \times n}$, we consider the quotient $\|A\boldsymbol{x}\|_v / \|\boldsymbol{x}\|_v$ defined for each nonzero $\boldsymbol{x} \in \mathbb{C}^n$. The quotient is obviously nonnegative; we consider the set of all numbers obtained when $\boldsymbol{x}$ takes all possible nonzero values in $\mathbb{C}^n$ and define

$$f(A) = \sup_{\boldsymbol{x} \in \mathbb{C}^n, \boldsymbol{x} \neq \boldsymbol{0}} \frac{\|A\boldsymbol{x}\|_v}{\|\boldsymbol{x}\|_v}. \tag{1}$$

We shall prove that, for a fixed matrix A, this is a well-defined nonnegative number, and the resulting function $f(A)$ is a matrix norm. We shall call it the *matrix norm induced by the vector norm* $\| \ \|_v$. Note that there is a close analogy with the Rayleigh quotient (for Hermitian matrices), as discussed in Chapter 8. Also, observe that the definition can be applied to the definition of a norm for any linear transformation acting on a normed linear space.

To begin our investigation of the quotient in (1), we first note some important properties:

Exercise 1. Show that, for the function $f(A)$ defined on $\mathbb{C}^{n \times n}$ by Eq. (1),

(a) $f(A) = \sup_{\boldsymbol{x} \in \mathbb{C}^n, \|\boldsymbol{x}\|_v = 1} \|A\boldsymbol{x}\|_v$;

(b) There is a vector $\boldsymbol{x}_0 \in \mathbb{C}^n$, depending on A, such that $\|\boldsymbol{x}_0\|_v = 1$ and $f(A) = \|A\boldsymbol{x}_0\|_v$. Thus

$$f(A) = \max_{\|\boldsymbol{x}\|_v = 1} \|A\boldsymbol{x}\|_v, \qquad \boldsymbol{x} \in \mathbb{C}^n. \tag{2}$$

Hint. Observe first that replacing $\boldsymbol{x}$ by $\alpha\boldsymbol{x}$ does not alter the quotient in Eq. (1). For part (b) take advantage of the technique used in the proof of Theorem 10.1.2. □

The importance of the induced norms is partly explained by part (c) of the following theorem.

Theorem 1. *Let* $\| \ \|_v$ *denote any vector norm on* $\mathbb{C}^n$.

(*a*) *The function* f *defined on* $\mathbb{C}^{n \times n}$ *by*

$$f(A) = \sup_{\boldsymbol{x} \in \mathbb{C}^n, \boldsymbol{x} \neq \boldsymbol{0}} \frac{\|A\boldsymbol{x}\|_v}{\|\boldsymbol{x}\|_v} = \max_{\boldsymbol{x} \in \mathbb{C}^n, \|\boldsymbol{x}\|_v = 1} \|A\boldsymbol{x}\|_v$$

is a matrix norm and is denoted by $\| \ \|$.

(*b*) *The norms* $\| \ \|$ *and* $\| \ \|_v$ *are compatible.*

(*c*) *If* $\| \ \|_1$ *is any matrix norm compatible with* $\| \ \|_v$ *then, for all* $A \in \mathbb{C}^{n \times n}$,

$$\|A\| \leq \|A\|_1.$$

PROOF. We set out to establish the axioms for a matrix norm. For axiom 1 it is clear that $f(A) \geq 0$, and that $f(A) = 0$ if $A = 0$. If $A \neq 0$, then there is an $\boldsymbol{x} \in \mathbb{C}^n$ such that $A\boldsymbol{x} \neq \boldsymbol{0}$ and $\|\boldsymbol{x}\|_v = 1$. Hence $f(A) > 0$. Axiom 1 is therefore satisfied.

Using axiom 2 for the vector norm and Eq. (2), we obtain

$$f(\lambda A) = \max_{\|\boldsymbol{x}\|_v = 1} \|\lambda A\boldsymbol{x}\|_v = |\lambda| \max_{\|\boldsymbol{x}\|_v = 1} \|A\boldsymbol{x}\|_v = |\lambda| f(A),$$

and so axiom 2 for the matrix norm is also satisfied. Furthermore,

$$\|(A + B)\mathbf{x}\|_v = \|A\mathbf{x} + B\mathbf{x}\|_v \le \|A\mathbf{x}\|_v + \|B\mathbf{x}\|_v,$$

hence

$$\max_{\|\mathbf{x}\|_v=1} \|(A + B)\mathbf{x}\|_v \le \max_{\|\mathbf{x}\|_v=1} \|A\mathbf{x}\|_v + \max_{\|\mathbf{x}\|_v=1} \|B\mathbf{x}\|_v.$$

Thus, $f(A + B) \le f(A) + f(B)$ and we have axiom 3 for the matrix norm.

Proceeding finally to the submultiplicative property (10.3.1) of a matrix norm, we infer from Exercise 1(b) the existence of an $\mathbf{x}_0 \in \mathbb{C}^n$ with $\|\mathbf{x}_0\|_v = 1$ such that $f(AB) = \|AB\mathbf{x}_0\|_v$ for given $A, B \in \mathbb{C}^{n\times n}$. Then

$$f(AB) = \|AB\mathbf{x}_0\|_v = \|A(B\mathbf{x}_0)\|_v \le f(A)\|B\mathbf{x}_0\|_v,$$

because it follows from Eq. (1) that

$$\|A\mathbf{x}\|_v \le f(A)\|\mathbf{x}\|_v \tag{3}$$

for all nonzero $\mathbf{x} \in \mathbb{C}^n$ and any $A \in \mathbb{C}^{n\times n}$. Using Eq. (3) once more (with respect to B), we have

$$f(AB) \le f(A)\|B\mathbf{x}_0\|_v \le f(A)f(B)\|\mathbf{x}_0\|_v = f(A)f(B),$$

and the property (10.3.1) holds. Part (*a*) of the theorem is thus proved and we may write $f(A) = \|A\|$.

Observe that part (*b*) of the theorem is just the relation (3). For part (*c*) we have, with an element $\mathbf{x}_0$ defined as in Exercise 1(b),

$$\|A\| = f(A) = \|A\mathbf{x}_0\|_v \le \|A\|_1\|\mathbf{x}_0\|_v = \|A\|_1,$$

for any matrix norm compatible with $\|\ \|_v$. ■

Exercise 2. Verify that $\|I\| = 1$ for any induced matrix norm.

Exercise 3. Let $N_n \in \mathbb{C}^{n\times n}$ denote a matrix with ones in the i, $i + 1$st positions $(1 \le i \le n - 1)$ and zeros elsewhere. Check that there are induced matrix norms such that

(a) $\|N_n\| = n$; (b) $\|N_n\| = 1$; (c) $\|N_n\| < 1$.

Hint. Consider the vector norm of Exercise 10.2.1 for part (a) and the norms

$$\|\mathbf{x}\|_v = \left(\frac{1}{k}\right)^{n-1}|x_1| + \left(\frac{1}{k}\right)^{n-2}|x_2| + \cdots + \frac{1}{k}|x_{n-1}| + |x_n|$$

with $k = 1$ and $k > 1$ for parts (b) and (c), respectively. □

Some of the most important and useful matrix norms are the induced norms discussed in the next exercises. In particular, the matrix norm induced by the euclidean vector norm is known as the *spectral norm*, and it is probably the most widely used norm in matrix analysis.

Exercise 4. If A is a unitary matrix, show that the spectral norm of A is 1.

Exercise 5. Let $A \in \mathbb{C}^{n\times n}$ and let λ_A be the spectral radius of A^*A. If $\| \ \|_s$ denotes the spectral norm, show that $\|A\|_s = \lambda_A^{1/2}$. (In other words, $\|A\|_s$ is just the largest singular value of A.)

SOLUTION. The matrix A^*A is positive semi-definite and therefore its eigenvalues are nonnegative numbers. Hence λ_A is actually an eigenvalue of A^*A.

Let $\boldsymbol{x}_1, \boldsymbol{x}_2, \ldots, \boldsymbol{x}_n$ be a set of orthonormal eigenvectors of A^*A with associated eigenvalues $0 \le \lambda_1 \le \lambda_2 \le \cdots \le \lambda_n$. Note that $\lambda_A = \lambda_n$ and if $\boldsymbol{x} = \sum_{i=1}^n \alpha_i \boldsymbol{x}_i$, then $A^*A\boldsymbol{x} = \sum_{i=1}^n \alpha_i \lambda_i \boldsymbol{x}_i$. Let $\| \ \|_E$ denote the euclidean vector norm on $\mathbb{C}^n$. Then

$$\|A\boldsymbol{x}\|_E^2 = (A\boldsymbol{x}, A\boldsymbol{x}) = (\boldsymbol{x}, A^* \ A\boldsymbol{x}) = \sum_{i=1}^{n} |\alpha_i|^2 \lambda_i.$$

For $\boldsymbol{x} \in \mathbb{C}^n$ such that $\|\boldsymbol{x}\|_E = 1$, that is, $\sum_{i=1}^n |\alpha_i|^2 = 1$, the expression above does not exceed λ_n and attains this value if $\alpha_1 = \alpha_2 = \cdots = \alpha_{n-1} = 0$ and $\alpha_n = 1$. Thus

$$\|A\|_s = \max_{\|\boldsymbol{x}\|_E = 1} \|A\boldsymbol{x}\|_E = \lambda_n^{1/2} = \lambda_A^{1/2}.$$

Exercise 6. If $A = \mathrm{diag}[d_1, d_2, \ldots, d_n]$ and $\| \ \|$ is a matrix norm induced by one of the Hölder norms of Exercise 10.1.2(d), show that

$$\|A\| = \max_{1 \le j \le n} |d_j|.$$

Exercise 7. If $A, U \in \mathbb{C}^{n\times n}$ and U is unitary, prove that, for both the euclidean matrix and spectral norms, $\|A\| = \|UA\| = \|AU\|$ and hence that, for these two norms, $\|A\|$ is invariant under unitary similarity transformations.

Hint. For the euclidean norm use Exercise 10.3.7(c); for the spectral norm use Exercises 4 and 10.1.6.

Exercise 8. If A is a normal matrix in $\mathbb{C}^{n\times n}$ and $\| \ \|_s$ denotes the spectral norm, prove that $\|A\|_s = \mu_A$, the spectral radius of A. (Compare this result with Theorem 10.3.1.) If f is a complex-valued function defined on the spectrum of A, prove that $\|f(A)\|_s$ is equal to the spectral radius of $f(A)$.

Hint. Use Theorem 5.2.1.

Exercise 9. Show that the matrix norm $\| \ \|_\rho$ induced by the vector norm $\| \ \|_\infty$ of Exercise 10.1.2 is given by

$$\|A\|_\rho = \max_{1 \le i \le n} \sum_{j=1}^{n} |a_{ij}|.$$

(Note that we form the *row sums* of the *absolute values* of the matrix elements, and then take the greatest of these sums.)

SOLUTION. Let $\mathbf{x} = [x_1 \ \ x_2 \ \ \cdots \ \ x_n]^{\mathrm{T}}$ be a vector in $\mathbb{C}^n$ with $\|\mathbf{x}\|_\infty = 1$. Then

$$\|A\mathbf{x}\|_\infty = \max_{1 \le i \le n} \left| \sum_{j=1}^{n} a_{ij} x_j \right| \le \max_{1 \le i \le n} \sum_{j=1}^{n} |a_{ij}||x_j| \le \|\mathbf{x}\|_\infty \|A\|_\rho = \|A\|_\rho .$$

Thus,

$$\max_{\|\mathbf{x}\|_\infty = 1} \|A\mathbf{x}\|_\infty \le \|A\|_\rho ,$$

and so the norm induced by $\| \ \ \|_\infty$ cannot exceed $\|A\|_\rho$. To prove that they are equal we have only to show that there is an $\mathbf{x}_0 \in \mathbb{C}^n$ with $\|\mathbf{x}_0\|_\infty = 1$ and $\|A\mathbf{x}_0\|_\infty \ge \|A\|_\rho$. It is easy to see that if $\|A\|_\rho = \sum_{j=1}^{n} |a_{kj}|$, then the vector $\mathbf{x}_0 = [\alpha_1 \ \ \alpha_2 \ \ \cdots \ \ \alpha_n]^{\mathrm{T}}$, where

$$\alpha_j = \begin{cases} \dfrac{|a_{kj}|}{a_{kj}} & \text{if} \quad a_{kj} \ne 0, \\ 0 & \text{if} \quad a_{kj} = 0, \end{cases}$$

$j = 1, 2, \ldots, n$, satisfies the above conditions.

Exercise 10. Show that the matrix norm induced by the vector norm $\| \ \ \|_1$ of Exercise 10.1.2 is given by

$$\|A\|_\gamma = \max_{1 \le j \le n} \sum_{i=1}^{n} |a_{ij}|.$$

(Note that we form the *column sums* of the *absolute values* of the matrix elements, and then take the greatest of these sums.)

Hint. Use the fact that $\|A\|_\gamma = \|A^*\|_\rho$.

Exercise 11. If $\| \ \ \|_s$ denotes the spectral norm and $\| \ \ \|_\rho$ and $\| \ \ \|_\gamma$ are as defined in Exercises 9 and 10, prove that

$$\|A\|_s^2 \le \|A\|_\rho \|A\|_\gamma .$$

SOLUTION. Using the result of Exercise 5, we have $\|A\|_s = \lambda_A^{1/2}$, where λ_A is the maximum eigenvalue of A^*A. Since A^*A is positive semidefinite, λ_A is just the spectral radius of A^*A. By Theorem 10.3.1, $\lambda_A \le \|A^*A\|_\gamma$. Thus

$$\|A\|_s^2 \le \|A^*A\|_\gamma \le \|A^*\|_\gamma \|A\|_\gamma = \|A\|_\rho \|A\|_\gamma . \quad \square$$

Before leaving the idea of inducing matrix norms from vector norms, we ask the converse question: Given a matrix norm $\| \ \|$, can we use it to induce a compatible vector norm? This can be done by defining $\| \ \|_v$ by

$$\|x\|_v = \|xa^T\| \tag{4}$$

for some fixed nonzero vector a.

Exercise 12. Let

$$A = \begin{bmatrix} 3/2 & -1/2 \\ -1/2 & 1 \end{bmatrix}.$$

Use Exercise 11 to show that $\|A\|_s \le 2$ and Exercise 8 to show that $\|A\|_s = (5 + \sqrt{5})/4 \simeq 1.8$.

Exercise 13. Consider an $n \times n$ matrix with block diagonal form, $A = \mathrm{diag}[A_1 \quad A_2]$, where A_1 and A_2 are square matrices. Show that, in the spectral norm,

$$\|A\| = \max(\|A_1\|, \|A_2\|).$$

Exercise 14. Let $A \in \mathbb{C}^{n \times n}$ and consider a Hermitian matrix of the form

$$H = \begin{bmatrix} 0 & iA \\ -iA^* & 0 \end{bmatrix}.$$

Show that $\|H\| = \|A\|$, in the spectral norm. □

10.5 Absolute Vector Norms and Lower Bounds of a Matrix

Almost all the vector norms we have introduced in the exercises depend only on the absolute values of the elements of the vector argument. Such norms are called *absolute* vector norms. The general theory of norms that we have developed up to now does not assume that this is necessarily the case. We shall devote the first part of this section to proving two properties of absolute vector norms.

One of the properties of an absolute vector norm that we shall establish is called the *monotonic property*. Let us denote by $|x|$ the vector whose elements are the absolute values of the components of x, and if $x_1, x_2 \in \mathbb{R}^n$, we shall say that $x_1 \le x_2$ (or $x_2 \ge x_1$) if this relation holds for corresponding components of x_1 and x_2. A vector norm $\| \ \|_v$ on $\mathbb{C}^n$ is said to be *monotonic* if $|x_1| \le |x_2|$ implies $\|x_1\|_v \le \|x_2\|_v$ for any $x_1, x_2 \in \mathbb{C}^n$.

The second property we will develop concerns the induced matrix norm. When a matrix norm is evaluated at a diagonal matrix

$$D = \operatorname{diag}[d_1, d_2, \dots, d_n],$$

we think highly of the norm if it has the property $\|D\| = \max_j |d_j|$. We will show that $\| \;\|$ has this property if and only if it is a norm induced by an absolute vector norm.

Theorem 1 (F. L. Bauer, J. Stoer, and C. Witzgall†). *If $\| \;\|_v$ is a vector norm in $\mathbb{C}^n$ and $\| \;\|$ denotes the matrix norm induced by $\| \;\|_v$, then the following conditions are equivalent:*

(*a*) $\| \;\|_v$ *is an absolute vector norm;*
(*b*) $\| \;\|_v$ *is monotonic;*
(*c*) $\|D\| = \max_{1 \le j \le n} |d_j|$ *for any diagonal matrix*

$$D = \operatorname{diag}[d_1, d_2, \dots, d_n]$$

from $\mathbb{C}^{n \times n}$.

PROOF. (*a*) ⇒ (*b*) We shall prove that if $\| \;\|_v$ is an absolute vector norm and $\|\boldsymbol{x}\|_v = F(x_1, x_2, \dots, x_n)$, where $\boldsymbol{x} = [x_1 \quad x_2 \quad \cdots \quad x_n]^T$, then F is a nondecreasing function of the absolute values of $x_1, x_2, \dots, x_n$. Let us focus attention on F as a function of the first component, x_1. The same argument will apply to each of the n variables. Thus, we suppose that $x_2, x_3, \dots, x_n$ are fixed, define

$$f(x) = F(x, x_2, x_3, \dots, x_n)$$

and consider the behaviour of $f(x)$ as x varies through the real numbers.

If the statement in part (*b*) is false, there exist nonnegative numbers p, q for which $p < q$ and $f(p) > f(q)$. Now since $f(x) = f(|x|)$, we have $f(-q) = f(q)$, and so the vectors $\boldsymbol{x}_1 = [-q \quad x_2 \quad \cdots \quad x_n]^T$ and $\boldsymbol{x}_2 = [q \quad x_2 \quad \cdots \quad x_n]^T$ belong to the ball consisting of vectors $\boldsymbol{x}$ for which $\|\boldsymbol{x}\|_v \le f(q)$. This ball is convex (Exercise 10.2.2) and since the vector $\boldsymbol{x}_0 = [p \quad x_2 \quad \cdots \quad x_n]^T$ can be written in the form $\boldsymbol{x}_0 = t\boldsymbol{x}_1 + (1 - t)\boldsymbol{x}_2$ for some $t \in [0, 1]$, then $\boldsymbol{x}_0$ also belongs to the ball. Thus $\|\boldsymbol{x}_0\|_v \le f(q)$. But $\|\boldsymbol{x}_0\|_v = f(p) > f(q)$, so we arrive at a contradiction and the implication (*a*) ⇒ (*b*) is proved.

(*b*) ⇒ (*c*) Suppose without loss of generality that $D \ne O$ and define $d = \max_{1 \le j \le n} |d_j|$. Clearly, $|D\boldsymbol{x}| \le |d\boldsymbol{x}|$ for all $\boldsymbol{x} \in \mathbb{C}^n$ and so condition (*b*) implies that $\|D\boldsymbol{x}\|_v \le \|d\boldsymbol{x}\|_v = d\|\boldsymbol{x}\|_v$. Thus

$$\|D\| = \sup_{\boldsymbol{x} \ne \boldsymbol{0}} \frac{\|D\boldsymbol{x}\|_v}{\|\boldsymbol{x}\|_v} \le d.$$

† *Numerische Math.* **3** (1961), 257–264.

If we can show that there is an $x_0 \in \mathbb{C}^n$ for which $\|Dx_0\|_v/\|x_0\|_v = d$, the proof will be complete. Suppose that m is an integer for which $d = |d_m|$. Then

$$\frac{\|De_m\|_v}{\|e_m\|_v} = \frac{\|d_m e_m\|_v}{\|e_m\|_v} = |d_m| = d,$$

where e_m is the mth unit vector in $\mathbb{C}^n$. Hence

$$\|D\| = \max_{x \neq 0} \frac{\|Dx\|_v}{\|x\|_v} = d,$$

and (c) follows from (b).

$(c) \Rightarrow (a)$ Suppose that the condition (c) holds; let $x = [x_1 \quad x_2 \quad \cdots \quad x_n]^T$. Choosing $d_j = 1$ if $x_j = 0$ and $d_j = |x_j|/x_j$ if $x_j \neq 0$, we see that for any $x \in \mathbb{C}^n$ there is a diagonal matrix $D = \operatorname{diag}[d_1, d_2, \ldots, d_n]$ such that $|x| = Dx$. Hence property (c) yields $\|D\| = \|D^{-1}\| = 1$. We then have

$$\|\,|x|\,\|_v = \|Dx\|_v \leq \|D\|\,\|x\|_v = \|x\|_v$$

and since $x = D^{-1}|x|$, it follows that

$$\|x\|_v \leq \|D^{-1}\|\,\|\,|x|\,\|_v = \|\,|x|\,\|_v$$

Thus $\|x\|_v = \|\,|x|\,\|_v$ for any $x \in \mathbb{C}^n$, and so (c) implies (a). This completes the proof. ■

Exercise 1. If $\|\ \|$ is a matrix norm induced by an absolute vector norm and $|A|$ is the matrix with elements $|a_{ij}|$, show that $\|A\| < \|\,|A|\,\|$.

Hint. Use Eq. (10.4.2). □

Proceeding to the second topic of this section, recall that our discussion of norms was started by looking for a measure of magnitude or a measure of the departure of an element from the zero element. Confining our attention to matrix norms, a measure of departure of a matrix from singularity (rather than the zero-matrix) would also be useful. Thus we seek a nonnegative function that is defined on all square matrices A and that is zero if and only if $\det A = 0$. At first sight, it is surprising that such a measure should be obtained from a definition complementary to that of the induced norm given in Eq. (10.4.1). At second sight it is less surprising, because we know that there is an $x \neq 0$ with $Ax = 0$ if and only if $\det A = 0$. We define the *lower bound* of A, written $l(A)$, with respect to the vector norm $\|\ \|_v$, by

$$l(A) = \inf_{x \in \mathbb{C}^n, x \neq 0} \frac{\|Ax\|_v}{\|x\|_v}. \tag{1}$$

If we accept the idea that $\|A^{-1}\|$ increases indefinitely in magnitude as A approaches a singular matrix, then the next theorem justifies our claim that the lower bound is a measure of departure from singularity.

Theorem 2. *If $\| \ \|$ is the matrix norm induced by the vector norm $\| \ \|_v$ and l is the corresponding lower bound, then*

$$l(A) = \begin{cases} 1/\|A^{-1}\| & \text{if } \det A \neq 0, \\ 0 & \text{if } \det A = 0. \end{cases}$$

PROOF. If $\det A \neq 0$, define a vector $\boldsymbol{y}$ in terms of a vector $\boldsymbol{x}$ by $\boldsymbol{y} = A\boldsymbol{x}$. Then for any $\boldsymbol{x} \neq \boldsymbol{0}$, $\boldsymbol{y} \neq \boldsymbol{0}$,

$$l(A) = \inf \frac{\|A\boldsymbol{x}\|_v}{\|\boldsymbol{x}\|_v} = \inf \frac{\|\boldsymbol{y}\|_v}{\|A^{-1}\boldsymbol{y}\|_v} = \left(\sup \frac{\|A^{-1}\boldsymbol{y}\|_v}{\|\boldsymbol{y}\|_v}\right)^{-1} = \frac{1}{\|A^{-1}\|}.$$

If $\det A = 0$, there is an $\boldsymbol{x}_0 \in \mathbb{C}^n$ such that $A\boldsymbol{x}_0 = \boldsymbol{0}$ and $\|\boldsymbol{x}_0\|_v = 1$. Then obviously $l(A) = 0$ by the definition of $l(A)$. ■

Corollary 1. *If $A, B \in \mathbb{C}^{n \times n}$, then*

$$l(AB) \geq l(A)l(B).$$

Corollary 2. *If $\| \ \|_v$ is an absolute norm and*

$$D = \operatorname{diag}[d_1, d_2, \ldots, d_n] \in \mathbb{C}^{n \times n},$$

then

$$l(D) = \min_{1 \leq j \leq n} |d_j|.$$

PROOF. If D is nonsingular, then $d_j \neq 0$ for each j and, using Theorems 2 and 1,

$$l(D) = \frac{1}{\|D\|^{-1}} = \frac{1}{\max\limits_{1 \leq j \leq n} (1/|d_j|)} = \min_{1 \leq j \leq n} |d_j|.$$

If $\det D = 0$, then $l(D) = 0$ and obviously $\min_{1 \leq j \leq n} |d_j| = 0$ also. ■

Exercise 2. Let $\lambda_1, \lambda_2, \ldots, \lambda_n$ be the eigenvalues of $A \in \mathbb{C}^{n \times n}$. Prove that, with respect to any vector norm $\| \ \|_v$,

$$l(A) \leq \min_{1 \leq j \leq n} |\lambda_j|.$$

Prove also that equality obtains if A is normal and $\| \ \|_v$ is the euclidean vector norm. □

Note that by combining the result of Exercise 2 with Theorem 10.3.1, we can define an annular region of the complex plane within which the eigenvalues of A must lie.

Exercise 3. Show that if A is nonsingular then, in the notation of Theorem 2,

$$\|A\boldsymbol{x}\|_v \geq l(A)\|\boldsymbol{x}\|_v. \quad \square$$

10.6 The Geršgorin Theorem

A description of regions of the complex plane containing the eigenvalues $\lambda_1, \lambda_2, \ldots, \lambda_n$ of a matrix $A \in \mathbb{C}^{n \times n}$ is presented in this section. Recall that a first result of this kind was obtained in Theorem 10.3.1 (see also Exercise 10.5.2).

Before entering into further details, let us point out the essential fact that *the eigenvalues of matrix A depend continuously on the elements of A*. Since the eigenvalues are just the zeros of the characteristic polynomial, this result follows immediately once it is known that the zeros of a polynomial depend continuously on its coefficients. This is a result of the elementary theory of algebraic functions and will not be pursued here.

We now prove one of the most useful and easily applied theorems that give bounds for eigenvalues. This is known as Geršgorin's theorem and was first published as recently as 1931†.

Theorem 1. *If $A \in \mathbb{C}^{n \times n}$ and a_{jk} denotes the elements of $A, j, k = 1, \ldots, n$, and*

$$\rho_j = \sum_k{}' |a_{jk}|,$$

where $\sum_k'$ denotes the sum from $k = 1$ to n, $k \neq j$, then every eigenvalue of A lies in at least one of the disks

$$\{z : |z - a_{jj}| \leq \rho_j\}, \qquad j = 1, 2, \ldots, n, \tag{1}$$

in the complex z-plane.

Furthermore, a set of m disks having no point in common with the remaining $n - m$ disks contains m and only m eigenvalues of A.

PROOF. Let λ be an eigenvalue of A with the associated eigenvector $\mathbf{x}$. Then $A\mathbf{x} = \lambda\mathbf{x}$ or, writing this relation out as n scalar equations,

$$\sum_{k=1}^{n} a_{jk} x_k = \lambda x_j, \qquad j = 1, 2, \ldots, n.$$

Let $|x_p| = \max_j |x_j|$; then the pth equation gives

$$|\lambda - a_{pp}||x_p| = \left| \sum_{k=1}^{n}{}' a_{pk} x_k \right| \leq \sum_{k=1}^{n}{}' |a_{pk}||x_k| \leq |x_p| \sum_{k=1}^{n}{}' |a_{pk}|.$$

Now since $\mathbf{x} \neq \mathbf{0}$ it must be that $|x_p| \neq 0$, and so we have $|\lambda - a_{pp}| \leq \rho_p$. The first result is proved.

† *Izv. Akad. Nauk. SSSR Ser. Fiz.-Mat.* **6** (1931), 749–754.

Suppose now that $A = D + C$, where $D = \text{diag}[a_{11}, \ldots, a_{nn}]$, and let $B(t) = D + tC$. Then $B(0) = D$ and $B(1) = A$. We consider the behaviour of the eigenvalues of $B(t)$ as t varies in the interval $[0, 1]$ and use the continuity of the eigenvalues as functions of t. Thus, for any t in this interval, the eigenvalues of $B(t)$ lie in the disks with centers a_{jj} and radii $t\rho_j, j = 1, 2, \ldots, n$.

Now suppose that the jth disk of $A = B(1)$ has no point in common with the remaining $n - 1$ disks. Then it is obviously true that the jth disk of $B(t)$ is isolated from the rest for all t in $[0, 1]$. Now when $t = 0$, the eigenvalues of $B(0)$ are $a_{11}, \ldots, a_{nn}$ and, of these, a_{jj} is the only one in the jth (degenerate) disk. Since the eigenvalues of $B(t)$ are continuous functions of t, and the jth disk is always isolated from the rest, it follows that there is one and only one eigenvalue of $B(t)$ in the jth disk for all $t \in [0, 1]$. In particular, this is the case when $t = 1$ and $B(1) = A$.

This proves the second part of the theorem for $m = 1$. The completion of the proof for any $m \le n$ is left as an exercise. ∎

Note that by applying the theorem to A^T, a similar result is obtained using *column* sums of A, rather than *row* sums, to define the radii of the new disks. The n disks in (1) are often referred to as the *Geršgorin disks.*

Exercise 1. Sketch the Geršgorin disks for the following matrix and for its transpose:

$$A = \begin{bmatrix} 0 & 1 & 0 & i \\ 1 & 6 & 1 & 1 \\ \frac{1}{2}i & i & 5i & 0 \\ 0 & \frac{1}{2} & \frac{1}{2} & -2 \end{bmatrix}.$$

If $S = \text{diag}[1, 1, 1, 4]$, compute SAS^{-1} and deduce that A has an eigenvalue in the disk $|z + 2| \le \frac{1}{2}$.

Exercise 2. If $A \in \mathbb{C}^{n \times n}$, show that the eigenvalues of A lie in the disk $\{z \in \mathbb{C} : |z| \le \min(\rho, \nu)\}$, where

$$\rho = \max_{1 \le j \le n} \sum_{k=1}^{n} |a_{jk}|, \qquad \nu = \max_{1 \le k \le n} \sum_{j=1}^{n} |a_{jk}|. \tag{2}$$

SOLUTION. For any eigenvalue λ_i of A $(1 \le i \le n)$,

$$|\lambda_i| - |a_{jj}| \le |\lambda - a_{jj}| \le \rho_j.$$

Therefore

$$|\lambda_i| \le \rho_j + |a_{jj}| = \sum_{k=1}^{n} |a_{jk}| \le \rho.$$

A similar proof can be performed for column sums, and the result follows. □

Note that ρ in Eq. (2) can be viewed as the matrix norm induced by the vector norm $\|\mathbf{x}\|_\infty$ (Exercise 10.4.9) and similarly for ν (Exercise 10.4.10). This shows that the result of Exercise 2 is a particular case of Theorem 10.3.1 (for a special norm).

Exercise 3. With the notation of Eq. (2), prove that

$$|\det A| \leq \min(\rho^n, \nu^n)$$

for any $A \in \mathbb{C}^{n \times n}$. □

A matrix $A = [a_{jk}]_{j,k=1}^n \in \mathbb{C}^{n \times n}$ is said to be *diagonally dominant* if

$$|a_{jj}| > \sum_{k=1}^{n}{}' |a_{jk}| = \rho_j$$

for all $j = 1, 2, \ldots, n$. In other words, if

$$d_j = |a_{jj}| - \rho_j, \qquad j = 1, 2, \ldots, n,$$
$$d = \min_{1 \leq j \leq n} d_j, \tag{3}$$

then A is diagonally dominant if and only if $d > 0$.

Exercise 4. Let A be diagonally dominant and λ_i be an eigenvalue of A. With the previous notation, show that $|\lambda_i| \geq d, i = 1, 2, \ldots, n$, and hence that $|\det A| \geq d^n$. In particular, confirm the nonsingularity of any diagonally dominant matrix, and illustrate with the matrix of Exercise 4.11.10. (The nonsingularity of a diagonally dominant matrix is sometimes referred to as the Lévy–Desplanques–Hadamard theorem.)

Hint. Consider the matrix $B = (1/d)A$ and apply Theorem 1 to show that $|\mu_j| \geq 1$ for any $\mu_j \in \sigma(B)$.

Exercise 5. Prove that an Hermitian matrix that is diagonally dominant and has all positive diagonal elements is positive definite.

Exercise 6. If $A = B + C$, where $B = \text{diag}[1, 2, 2]$ and $|c_{jk}| \leq \varepsilon < \frac{1}{6}$ for $j, k = 1, 2, 3$, prove that there is an eigenvalue of A in the disk

$$|z - 1 - c_{11}| \leq 12\varepsilon^2.$$

Exercise 7. This is a generalization of Exercise 6. If $A = B + C$, where $B = \text{diag}[b_1, \ldots, b_n]$, $\rho_j = \min_{k \neq j} |b_j - b_k| > 0$, and $|c_{jk}| \leq \varepsilon < \beta_j/2n$ for $j, k = 1, 2, \ldots, n$, prove that there is an eigenvalue of A in the disk

$$|z - b_j - c_{jj}| \leq 2n(n - 1)\varepsilon^2/\beta_j.$$

Exercise 8. If $f(z) = a_0 + a_1 z + \cdots + a_l z^l$, $a_l \neq 0$, and z_0 is a zero of f, prove that

$$|z_0| \leq 1 + \max_k |a_k/a_l|. \quad \square$$

The localization of eigenvalues of a simple matrix will be discussed in Section 11.2.

10.7 Geršgorin Disks and Irreducible Matrices

In this section two useful results are presented that can be seen as refinements of Geršgorin's theorem. However, they both involve the further hypothesis that the matrix in question be "irreducible." This is a property that is easily seen to hold in many practical problems. We make a formal definition here and return for further discussion of the concept in Chapter 15.

If $n \geq 2$, the matrix $A \in \mathbb{C}^{n \times n}$ is said to be *reducible* if there is an $n \times n$ permutation matrix P such that

$$P^{\mathrm{T}}AP = \begin{bmatrix} A_{11} & A_{12} \\ 0 & A_{22} \end{bmatrix}, \tag{1}$$

where A_{11} and A_{22} are square matrices of order less than n. If no such P exists then A is *irreducible.*

If A is considered to be a representation of a linear transformation T on an n-dimensional space $\mathscr{S}$, the fact that A is irreducible simply means that, among the basis vectors for $\mathscr{S}$ chosen for this representation, there is no proper subset that spans an invariant subspace of A (see Section 4.8 and Eq. (4.8.3) in particular).

Exercise 1. If $A \in \mathbb{R}^{n \times n}$ is reducible, prove that for any positive integer p, the matrix A^p is reducible. Also, show that any polynomial in A is a reducible matrix.

Exercise 2. Check that the square matrices A and A^{T} are both reducible or both irreducible. $\square$

The first result concerns a test for a matrix $A \in \mathbb{C}^{n \times n}$ to be nonsingular. Clearly, Theorem 10.6.1 provides a test (or class of tests) for nonsingularity (see Exercise 10.6.4): A is certainly nonsingular if it is diagonally dominant. In other words, if $|a_{jj}| > \sum_{k \neq j} |a_{jk}|$ for $j = 1, 2, \ldots, n$, then the origin of the complex plane is excluded from the union of the Geršgorin disks in which $\sigma(A)$ must lie. These inequalities can be weakened at the expense of the irreducibility hypothesis.

Theorem 1 (O. Taussky[†]). *If* $A = [a_{jk}]_{j,k=1}^{n}$ *is irreducible and*

$$|a_{jj}| \geq \sum_{k=1,\,k\neq j}^{n} |a_{jk}|, \qquad j = 1, 2, \ldots, n, \tag{2}$$

with strict inequality for at least one j, then A is nonsingular.

PROOF. Suppose, on the contrary, that $A\boldsymbol{x} = \boldsymbol{0}$ for some

$$\boldsymbol{x} = [x_1 \quad x_2 \quad \cdots \quad x_n]^{\mathrm{T}} \neq \boldsymbol{0}.$$

Clearly, there is a permutation matrix $P \in \mathbb{R}^{n \times n}$ such that $P\boldsymbol{x} = \tilde{\boldsymbol{x}}$, where $\tilde{\boldsymbol{x}} = [\tilde{x}_1 \quad \tilde{x}_2 \quad \cdots \quad \tilde{x}_n]^{\mathrm{T}}$ and either

$$|\tilde{x}_1| = |\tilde{x}_2| = \cdots = |\tilde{x}_n| \tag{3}$$

or

$$|\tilde{x}_1| = \cdots = |\tilde{x}_r| > |\tilde{x}_{r+1}| \geq \cdots \geq |\tilde{x}_n|. \tag{4}$$

Denote $\tilde{A} = PAP^{\mathrm{T}} = [\tilde{a}_{jk}]_{j,k=1}^{n}$ and observe that this simultaneous permutation of rows and columns of A allows us to rewrite Eq. (2) in the form

$$|\tilde{a}_{jj}| \geq \sum_{k=1,\,k\neq j}^{n} |\tilde{a}_{jk}|, \qquad j = 1, 2, \ldots, n. \tag{5}$$

We also have $\tilde{A}\tilde{\boldsymbol{x}} = \boldsymbol{0}$ and so $\sum_{k=1}^{n} \tilde{a}_{jk}\tilde{x}_k = 0$ for $j = 1, 2, \ldots, n$. Therefore we obtain, in the case of Eq. (3),

$$|\tilde{a}_{jj}|\,\|\boldsymbol{x}\|_\infty = |\tilde{a}_{jj}|\,|\tilde{x}_j| = \left|\sum_{k=1,\,k\neq j}^{n} \tilde{a}_{jk}\tilde{x}_k\right| \leq \left(\sum_{k=1,\,k\neq j}^{n} |\tilde{a}_{jk}|\right)\|\boldsymbol{x}\|_\infty, \qquad 1 \leq j \leq n.$$

Dividing by $\|\boldsymbol{x}\|_\infty \neq 0$, we then have

$$|\tilde{a}_{jj}| \leq \sum_{k=1,\,k\neq j}^{n} |\tilde{a}_{jk}|, \qquad 1 \leq j \leq n,$$

and comparison with (5) gives equalities throughout (5), a contradiction that proves the theorem in this case.

In the case of Eq. (4) we have, for $1 \leq j \leq r$,

$$\begin{aligned}
|\tilde{a}_{jj}|\,\|\boldsymbol{x}\|_\infty &= |\tilde{a}_{jj}|\,|\tilde{\boldsymbol{x}}_j| \\
&= \left|\sum_{k=1,\,k\neq j}^{n} \tilde{a}_{jk}\tilde{x}_k\right| \\
&\leq \left(\sum_{k=1,\,k\neq j}^{r} |\tilde{a}_{jk}|\right)\|\boldsymbol{x}\|_\infty + \sum_{k=r+1}^{n} |\tilde{a}_{jk}|\,|\tilde{\boldsymbol{x}}_k| \\
&\leq \left(\sum_{k=1,\,k\neq j}^{n} |\tilde{a}_{jk}|\right)\|\boldsymbol{x}\|_\infty.
\end{aligned} \tag{6}$$

[†] *Duke Math. J.* **15** (1948), 1043–1044.

Dividing by $\|\boldsymbol{x}\|_\infty \neq 0$, we see that

$$|\tilde{a}_{jj}| \leq \sum_{k=1, k\neq j}^{n} |\tilde{a}_{jk}|, \qquad 1 \leq j \leq r,$$

and comparing with Eq. (5), we obtain equalities in (5) for $j = 1, 2, \ldots, r$. Hence it follows from Eq. (6) that

$$\sum_{k=r+1}^{n} |\tilde{a}_{jk}| = 0, \qquad j = 1, 2, \ldots, r,$$

or, what is equivalent, the matrix $\tilde{A}$ is lower block-triangular. Thus, $A = P^{\mathrm{T}}\tilde{A}P$ is reducible and a contradiction is obtained. This proves the theorem. ■

Exercise 3. Check that the $n \times n$ matrices

$$\begin{bmatrix} n & 1 & 1 & \cdots & 1 \\ 1 & 2 & 0 & \cdots & 0 \\ 1 & 0 & 3 & \ddots & \vdots \\ \vdots & \vdots & \ddots & \ddots & 0 \\ 1 & 0 & \cdots & 0 & n \end{bmatrix}, \qquad \begin{bmatrix} 2 & -1 & 0 & \cdots & 0 \\ -1 & 2 & -1 & \ddots & \vdots \\ 0 & \ddots & \ddots & \ddots & 0 \\ \vdots & \ddots & & & -1 \\ 0 & \cdots & 0 & -1 & 2 \end{bmatrix}$$

are nonsingular. □

The next result shows that either all eigenvalues of an irreducible matrix A lie in the interior of the union of Geršgorin's disks or an eigenvalue of A is a common boundary point of all n disks.

Theorem 2 (O. Taussky). *Let $A = [a_{jk}]_{j,k=1}^{n}$ be a complex irreducible matrix and let λ be an eigenvalue of A lying on the boundary of the union of the Geršgorin disks. Then λ is on the boundary of each of the n Geršgorin disks.*

PROOF. Given the hypotheses of the theorem, let $|\lambda - a_{ii}| = \rho_i$ for some i, $1 \leq i \leq n$. If λ lies outside the remaining $n - 1$ Geršgorin disks, then $|\lambda - a_{jj}| > \rho_j$ for all $j \neq i$. Therefore by Theorem 1 the matrix $\lambda I - A$ must be nonsingular. This contradicts the assumption that λ is an eigenvalue of A. Thus, since λ must lie on the boundary of the union of the disks, it must be a point of intersection of m circumferences where $2 \leq m \leq n$. Thus there are m equalities in the system

$$|\lambda - a_{jj}| \geq \sum_{k=1, k\neq j}^{n} |a_{jk}|, \qquad 1 \leq j \leq n,$$

and $n - m$ strict inequalities. In view of Theorem 1, this conflicts again with the definition of λ unless $m = n$. Hence the assertion of the theorem. ■

In the study of iterative processes it is often useful to have criteria (sharper than that of Theorem 10.3.1) for deciding when the spectral radius of a matrix is less than 1 (see Exercise 9.8.5). Theorem 2 yields such a criterion as a corollary.

Corollary 1. *Let $A = [a_{ij}]_{i,j=1}^n$ be irreducible and satisfy*

$$\sum_{j=1}^{n} |a_{ij}| \leq 1 \qquad \textit{for} \quad i = 1, 2, \ldots, n, \tag{5}$$

with strict inequality for at least one i. Then $\mu_A < 1$.

PROOF. Note first that the hypothesis (5) implies at once that all Geršgorin disks are in the unit disk. So Theorem 10.6.1 immediately gives $\mu_A \leq 1$.

The fact that there is strict inequality in (5) for at least one i means that there is at least one Geršgorin disk that is entirely in the interior of the unit disk. Since points on the unit circle can only be on the circumference of the Geršgorin disks, it follows from Theorem 2 that an eigenvalue of A on the unit circle would have to be a common boundary point of all n disks, and this is impossible. Hence the result. ■

10.8 The Schur Theorem

We present next a rather different result that originates with I. Schur.[†] Let us recall the decomposition of a matrix $A \in \mathbb{R}^{n \times n}$ into the sum of a Hermitian matrix B and a skew-Hermitian matrix C. We write $B = \frac{1}{2}(A + A^*)$, $C = \frac{1}{2}(A - A^*)$, and observe that $A = B + C$, $B^* = B$, and $C^* = -C$. Now if $A^* = A$, then $C = O$ and all the eigenvalues of A are real. This suggests that a norm of C may give an upper bound for the magnitudes of the *imaginary parts* of the eigenvalues of A. Similarly, if $A^* = -A$, then $B = O$ and all the eigenvalues of A have zero real parts, so a norm of B might be expected to give an upper bound for the magnitudes of the *real parts* of the eigenvalues of A. Schur's theorem contains results of this kind. In the following statement $\mathcal{Re}\,\lambda$ and $\mathcal{Im}\,\lambda$ will denote the real and imaginary parts of the complex number λ.

[†] *Math. Ann.* **65** (1909), 488–510.

Theorem 1. *If $A \in \mathbb{C}^{n \times n}$, $\| \ \|$ denotes the euclidean matrix norm, and $\lambda_1, \lambda_2, \ldots, \lambda_n$ are the eigenvalues of A, then*

$$\sum_{i=1}^{n} |\lambda_i|^2 \le \|A\|^2,$$

$$\sum_{i=1}^{n} |\mathcal{R}e\, \lambda_i|^2 \le \|B\|^2,$$

$$\sum_{i=1}^{n} |\mathcal{I}m\, \lambda_i|^2 \le \|C\|^2,$$

where $B = \frac{1}{2}(A + A^*)$ and $C = \frac{1}{2}(A - A^*)$. Equality in any one of these relations implies equality in all three and occurs if and only if A is normal.

PROOF. Theorem 5.2.2 informs us that there is a unitary matrix U and an upper-triangular matrix Δ such that $A = U\Delta U^*$, and the eigenvalues of A are the diagonal elements of Δ. Now the euclidean norm is invariant under unitary similarity transformations (Exercise 10.4.7) and so $\|A\|^2 = \|\Delta\|^2$. If the elements of Δ are written d_{rs} $(r, s = 1, 2, \ldots, n)$, then

$$\|A\|^2 = \sum_{i=1}^{n} |\lambda_i|^2 + \sum_{r<s} |d_{rs}|^2 \ge \sum_{i=1}^{n} |\lambda_i|^2,$$

and the first inequality is proved.

We also have $B = \frac{1}{2}U(\Delta + \Delta^*)U^*$ so that, as before,

$$\|B\|^2 = \sum_{i=1}^{n} \left| \frac{\lambda_i + \bar{\lambda}_i}{2} \right|^2 + \sum_{r \ne s} \left| \frac{d_{rs} + \bar{d}_{sr}}{2} \right|^2$$

$$= \sum_{i=1}^{n} |\mathcal{R}e\, \lambda_i|^2 + \sum_{r \le s} \frac{|d_{rs}|^2}{2} \ge \sum_{i=1}^{n} |\mathcal{R}e\, \lambda_i|^2.$$

A similar proof also yields the third inequality.

It is clear that equality obtains in each of the three relations if and only if $d_{rs} = 0$ for all $r \ne s$, and $r, s = 1, 2, \ldots, n$; that is, if and only if A is unitarily similar to a diagonal matrix of its eigenvalues. The result now follows from Theorem 5.2.1. ■

Corollary 1 (A. Hirsch[†]). *If we define*

$$\rho = \max_{r,s} |a_{rs}|, \qquad \sigma = \max_{r,s} |b_{rs}|, \qquad \tau = \max_{r,s} |c_{rs}|,$$

and if λ is any eigenvalue of A, then

$$|\lambda| \le n\rho, \qquad |\mathcal{R}e\, \lambda| \le n\sigma, \qquad |\mathcal{I}m\, \lambda| \le n\tau.$$

[†] *Acta Math.* **25** (1902), 367–370.

PROOF. Clearly, $\|A\|_E^2 \le \sum_{r,s} \rho^2 = n^2\rho^2$. The first inequality of Theorem 1 then gives $|\lambda| \le (\sum_i |\lambda_i|^2)^{1/2} \le \|A\| \le n\rho$, which is the first result of the corollary. The two remaining results follow similarly. ■

Note that the particular case $\tau = 0$ provides another (high-powered) proof that the eigenvalues of a Hermitian matrix are real. Similarly the case $\sigma = 0$ yields the result that the eigenvalues of a skew-Hermitian matrix have their real parts zero.

Finally, we should note that $n\rho$ is the norm defined in Exercise 10.3.7a. The result $|\lambda| \le n\rho$ is then a special case of Theorem 10.3.1.

Corollary 2 (I. Bendixson†). *If* $A \in \mathbb{R}^{n \times n}$ *and* $\tau = \frac{1}{2}\max_{r,s}|a_{rs} - a_{sr}|$, *then for any eigenvalue* λ *of* A,

$$|\mathcal{I}m\, \lambda| \le \tau\sqrt{n(n-1)/2}.$$

PROOF. When A is real, the third inequality of Theorem 1 implies that

$$\sum_{i=1}^{n} |\mathcal{I}m\, \lambda_i|^2 \le \sum_{r,s=1,\, r\neq s}^{n} \left|\frac{a_{rs} - a_{sr}}{2}\right|^2 \le \tau^2 n(n-1).$$

Since A is real, all of its complex eigenvalues arise in complex conjugate pairs, so that every nonzero term in the sum on the left appears at least twice. Hence $2|\mathcal{I}m\, \lambda_i|^2 \le \tau^2 n(n-1)$ and the result follows. ■

Before leaving these results, we should remark that the bounds obtained in the corollaries are frequently less sharp than the corresponding bounds obtained from Geršgorin's theorem. This is well illustrated on applying Corollary 1 to Exercise 10.6.1.

Exercise 1. Prove that $|\det A| \le n^{n/2}\rho^n$, where $\rho = \max_{r,s}|a_{rs}|$.

SOLUTION. First of all note the *inequality of the arithmetic and geometric means*: namely, for any nonnegative numbers $a_1, a_2, \ldots, a_n$,

$$(a_1 a_2 \cdots a_n)^{1/n} \le \frac{a_1 + a_2 + \cdots + a_n}{n}.$$

If $\lambda_1, \lambda_2, \ldots, \lambda_n$ are the eigenvalues of A, we have, by Theorem 1,

$$\begin{aligned}|\det A|^2 &= |\lambda_1|^2\,|\lambda_2|^2 \cdots |\lambda_n|^2 \\ &\le \left(\frac{|\lambda_1|^2 + \cdots + |\lambda_n|^2}{n}\right)^n \\ &\le \left(\frac{1}{n}\|A\|_E^2\right)^n \le \left(\frac{1}{n} n^2\rho^2\right)^n,\end{aligned}$$

and the result follows on taking square roots.

† *Acta Math.* **25** (1902), 359–365.

Exercise 2. If $\lambda_1, \lambda_2, \ldots, \lambda_n$ are the eigenvalues of A and $\mu_1, \mu_2, \ldots, \mu_n$ are those of A^*A, show that

$$\sum_{i=1}^{n} |\lambda_i|^2 \le \sum_{i=1}^{n} \mu_i,$$

with equality if and only if A is normal.

Exercise 3. With the notation of Exercise 2, show that A is normal if and only if, for some ordering of the eigenvalues, $\mu_i = |\lambda_i|^2$, $i = 1, 2, \ldots, n$. □

10.9 Miscellaneous Exercises

1. Prove that for any $A \in \mathbb{C}^{n \times n}$,

$$\|A\|_E^2 = s_1^2 + s_2^2 + \cdots + s_n^2,$$

where $s_1, s_2, \ldots, s_n$ are the singular values of A.

Hint. Use Theorem 5.7.2 and Exercise 10.4.7.

2. Let $P \in \mathbb{C}^{n \times n}$ be an orthogonal projector on $\mathscr{S} \subset \mathbb{C}^n$. Prove that

$$\|\boldsymbol{x} - P\boldsymbol{x}\|_E \le \|\boldsymbol{x} - Q\boldsymbol{x}\|_E$$

for any projector Q on $\mathscr{S}$ and any $\boldsymbol{x} \in \mathbb{C}^n$. Interpret the result geometrically (in terms of a distance) in the case of $\mathbb{R}^3$ and $\dim \mathscr{S} = 1$.

3. Let $A = H_1 + iH_2$, $H_1^* = H_1$, $H_2^* = H_2$.

(a) show that $\|A\|_E^2 = \|H_1\|_E^2 + \|H_2\|_E^2$;
(b) For any Hermitian matrix $H \in \mathbb{C}^{n \times n}$, show that

$$\|A - H_1\|_E \le \|A - H\|_E$$

and

$$\|A - iH_2\|_E \le \|A - iH\|_E.$$

Interpret the result geometrically.

4. If $A = HU$ is the polar decomposition of A, prove that $\|A\|_E = \|H\|_E$. Indicate the corresponding property of complex numbers.

5. Define a vector norm $\|\ \|_v$ by Eq. (10.4.4) and prove that $\|\boldsymbol{x}\|_v \ge |\boldsymbol{a}^T\boldsymbol{x}|$. Also, if $\|\ \|_1$ is the matrix norm induced by $\|\ \|_v$, then $\|A\|_1 \le \|A\|$ for all $A \in \mathbb{C}^{n \times n}$, where $\|\ \|$ is the matrix norm in Eq. (10.4.4).

Hint. Start with $\|(\boldsymbol{y}\boldsymbol{a}^T)(\boldsymbol{x}\boldsymbol{a}^T)\|$ $(\boldsymbol{y} \ne \boldsymbol{0})$ to obtain the first result.

6. If $\| \ \|_v$ is the vector norm of Exercise 10.1.7, and $\| \ \|$ and $\| \ \|_s$ denote the matrix norms induced by $\| \ \|_v$ and $\| \ \|_E$, respectively, prove that for any $A \in \mathbb{C}^{n\times n}$,

$$\|A\| = \|H^{1/2}AH^{1/2}\|_s,$$

where $H^{1/2}$ denotes the positive definite square root of H. □

If $\| \ \|_v$ is a vector norm in $\mathbb{C}^n$, then the function $\| \ \|_d$ defined by

$$\|y\|_d \triangleq \sup_{x\neq 0} \frac{|y^*x|}{\|x\|_v}, \qquad y \in \mathbb{C}^n, \tag{1}$$

is called the *dual norm* of $\| \ \|_v$.

7. Prove that

(a) $\| \ \|_d$ is a vector norm;

(b) $|y^*x| \le \|x\|_v\|y\|_d$ for all $x, y \in \mathbb{C}^n$ (the generalized Cauchy–Schwarz inequality);

(c) If $\|x\|_p = (\sum_{i=1}^n |x_i|^p)^{1/p}$ is a Hölder vector norm, then the corresponding dual norm of x is

$$\|x\|_q = \left(\sum_{i=1}^n |x_i|^q\right)^{1/q},$$

where $(1/q) + (1/p) = 1$. By use of part (b) for these norms, obtain the Hölder inequality.

8. If $A = ab^*$ and $\| \ \|$ is the matrix norm induced by a vector norm $\| \ \|_v$, prove that $\|A\| = \|a\|_v\|b\|_d$, where $\| \ \|_d$ is defined in Eq. (1).

9. If $x \neq 0$, prove that there is a $y \in \mathbb{C}^n$ such that $y^*x = \|x\|_v\|y\|_d = 1$.

10. Prove that if $\| \ \|_{dd}$ is the dual norm of $\| \ \|_d$, then $\|x\|_{dd} = \|x\|_v$ for all $x \in \mathbb{C}^n$. The norm $\| \ \|_d$ is defined in Eq. (1).

11. Let $\| \ \|$ be a norm in $\mathbb{C}^{n\times n}$ (not necessarily a matrix norm). Define

$$N(A) = \sup_{0\neq X\in\mathbb{C}^{n\times n}} \frac{\|AX\|}{\|X\|}$$

and prove that

(a) N is a matrix norm;

(b) If $\| \ \|$ is a matrix norm, then $N(A) \le \|A\|$ for all $A \in \mathbb{C}^{n\times n}$;

(c) $N(A) = \|A\|$ for all $A \in \mathbb{C}^{n\times n}$ if and only if $\| \ \|$ is a matrix norm and $\|I\| = 1$.

12. If $\| \ \|$ is a norm in $\mathbb{C}^{n \times n}$ (not necessarily a matrix norm), then there exists a number $c > 0$ such that the function

$$M(A) = c\|A\|, \qquad A \in \mathbb{C}^{n \times n},$$

is a matrix norm.

13. Use Exercise 1.8.5 to show that if, for any matrix norm $\|X\| \le M$ and $\|Y\| \le M$ (where $X, Y \in \mathbb{C}^{n \times n}$), then for $r = 1, 2, \ldots,$

$$\|X^r - Y^r\| \le rM^{r-1}\|X - Y\|.$$

14. (a) Let $\hat{J}$ be a Jordan block of size p with eigenvalue λ. Let $\mu = \mathcal{Re}(\lambda)$ and $f(t) = t^{p-1}e^{\mu t}$. Using Theorem 9.4.4 show that there are positive constants k_1, k_2, and t_0 such that

$$k_1 f(t) \le \|e^{\hat{J}t}\|_s \le k_2 f(t), \qquad t \ge t_0.$$

(b) Let $A \in \mathbb{C}^{n \times n}$, $\mu = \max_{\lambda \in \sigma(A)} \mathcal{Re}(\lambda)$ and let p be the largest index of any eigenvalue for which $\mu = \mathcal{Re}(\lambda)$. Show that there are positive constants K_1, K_2, and t_0 such that

$$K_1 f(t) \le \|e^{At}\|_s \le K_2 f(t), \qquad t \ge t_0.$$

(c) Use the result of part (b) to show that

(i) $\|e^{At}\|_s \to 0$ as $t \to \infty$ if and only if $\mu < 0$.

(ii) If $\mu \le 0$ and $p = 1$, then $\|e^{At}\|_s \le K_2$ for all $t \ge t_0$.

(iii) If $\mu < 0$ there are positive constants K and α such that $\|e^{At}\|_s \le Ke^{-\alpha t}$ for all $t \ge 0$.

15. Let $F \in \mathbb{C}^{n \times n}$ and

$$A = \begin{bmatrix} I & F \\ F^* & I \end{bmatrix}.$$

Prove that A is positive definite if and only if $\|F\|_s < 1$.

Hint. Use the factorization

$$\begin{bmatrix} I & F \\ F^* & I \end{bmatrix} = \begin{bmatrix} I & 0 \\ F^* & I \end{bmatrix} \begin{bmatrix} I & 0 \\ 0 & I - F^*F \end{bmatrix} \begin{bmatrix} I & F \\ 0 & I \end{bmatrix}$$

together with Exercise 10.4.5.

16. Let $A \in \mathbb{C}^{n \times n}$, $B \in \mathbb{C}^{r \times r}$ and $U \in \mathbb{C}^{n \times r}$ with orthonormal columns, i.e. $U^*U = I_r$. If $A = UBU^*$ show that $\|A\|_s = \|B\|_s$.

CHAPTER 11

Perturbation Theory

The greater part of this chapter is devoted to investigations of the behavior of matrix eigenvalues under perturbations of the elements of the matrix. Before starting on this subject it is instructive (and particularly important for the numerical analyst) to examine the effects of perturbations of A and $\boldsymbol{b}$ on a solution $\boldsymbol{x}$ of the linear equation $A\boldsymbol{x} = \boldsymbol{b}$.

11.1 Perturbations in the Solution of Linear Equations

Given $A \in \mathbb{C}^{n \times n}$, consider the equation $A\boldsymbol{x} = \boldsymbol{b}$ ($\boldsymbol{x}, \boldsymbol{b} \in \mathbb{C}^n$) and assume that $\det A \neq 0$, so that $\boldsymbol{x}$ is the unique solution of the equation.

In general terms, a problem is said to be *stable*, or *well conditioned*, if "small" causes (perturbations in A or $\boldsymbol{b}$) give rise to "small" effects (perturbations in $\boldsymbol{x}$). The degree of smallness of the quantities in question is usually measured in relation to their magnitudes in the unperturbed state. Thus if $\boldsymbol{x} + \boldsymbol{y}$ is the solution of the problem after the perturbation, we measure the magnitude of the perturbation in the solution vector by $\|\boldsymbol{y}\|_v / \|\boldsymbol{x}\|_v$ for some vector norm $\| \; \|_v$. We call this quotient the *relative perturbation* of $\boldsymbol{x}$.

Before investigating the problem with a general coefficient matrix A, it is very useful to examine the prototype problem of variations in the coefficients of the equation $\boldsymbol{x} = \boldsymbol{b}$. Thus, we are led to investigate solutions of $(I + M)\boldsymbol{\xi} = \boldsymbol{b}$, where M is "small" compared to I. The departure of $\boldsymbol{\xi}$ from $\boldsymbol{x}$ will obviously depend on the nature of $(I + M)^{-1}$, if this inverse exists. In this connection we have one of the most useful results of linear analysis:

Theorem 1. *If* $\| \ \|$ *denotes any matrix norm for which* $\|I\| = 1$ *and if* $\|M\| < 1$, *then* $(I + M)^{-1}$ *exists,*

$$(I + M)^{-1} = I - M + M^2 - \cdots$$

and

$$\|(I + M)^{-1}\| \le \frac{1}{1 - \|M\|}.$$

Note that in view of Theorem 10.3.1, the hypotheses of the theorem can be replaced by a condition on the spectral radius: $\mu_M < 1$.

PROOF. Theorem 10.3.1 implies that for any eigenvalue λ_0 of M, $|\lambda_0| \le \|M\| < 1$. Moreover, the function f defined by $f(\lambda) = (1 + \lambda)^{-1}$ has a power series expansion about $\lambda = 0$ with radius of convergence equal to 1. Hence (see Theorem 9.8.3), $f(M) = (I + M)^{-1}$ exists and

$$(I + M)^{-1} = I - M + M^2 - \cdots.$$

If we write $S_p = I - M + M^2 - \cdots \pm M^{p-1}$ for the pth partial sum and note that $\|M^r\| \le \|M\|^r$, we obtain

$$\|S_p\| \le \sum_{k=1}^{p-1} \|M^k\| \le \sum_{k=1}^{p-1} \|M\|^k = \frac{1 - \|M\|^p}{1 - \|M\|} \le \frac{1}{1 - \|M\|}.$$

As this bound is independent of p, the result follows on taking the limit as $p \to \infty$. ■

Let us return now to our original problem. We have

$$A\boldsymbol{x} = \boldsymbol{b},$$

with $\det A \ne 0$, and we suppose that $\boldsymbol{b}$ is changed to $\boldsymbol{b} + \boldsymbol{k}$, that A is changed to $A + F$, and that $\boldsymbol{x} + \boldsymbol{y}$ satisfies

$$(A + F)(\boldsymbol{x} + \boldsymbol{y}) = \boldsymbol{b} + \boldsymbol{k}. \tag{1}$$

We will compute an upper bound for the relative perturbation $\|\boldsymbol{y}\|_v / \|\boldsymbol{x}\|_v$.

Subtracting the first equation from the second, we have

$$(A + F)\boldsymbol{y} + F\boldsymbol{x} = \boldsymbol{k},$$

or

$$A(I + A^{-1}F)\boldsymbol{y} = \boldsymbol{k} - F\boldsymbol{x}.$$

We assume now that $\delta = \|A^{-1}\| \, \|F\| < 1$, and that $\|I\| = 1$. Then, since $\|A^{-1}F\| \leq \|A^{-1}\| \, \|F\| < 1$, Theorem 1 implies that $(I + A^{-1}F)^{-1}$ exists, and that $\|(I + A^{-1}F)^{-1}\| < (1 - \delta)^{-1}$. Thus

$$\boldsymbol{y} = (I + A^{-1}F)^{-1}A^{-1}\boldsymbol{k} - (I + A^{-1}F)^{-1}A^{-1}F\boldsymbol{x},$$

and, if $\| \;\|_v$ is any vector norm compatible with $\| \;\|$,

$$\|\boldsymbol{y}\|_v \leq \frac{\|A^{-1}\|}{1 - \delta} \|\boldsymbol{k}\|_v + \frac{\delta}{1 - \delta} \|\boldsymbol{x}\|_v.$$

To obtain a convenient expression for our measure, $\|\boldsymbol{y}\|_v / \|\boldsymbol{x}\|_v$ of the magnitude of the perturbation, we use the fact that $A\boldsymbol{x} = \boldsymbol{b}$ implies $\|\boldsymbol{b}_v\| \leq \|A\| \, \|\boldsymbol{x}\|_v$, hence $1/\|\boldsymbol{x}\|_v \leq \|A\|/\|\boldsymbol{b}\|_v$, $\boldsymbol{b} \neq \boldsymbol{0}$. Thus

$$\frac{\|\boldsymbol{y}\|_v}{\|\boldsymbol{x}\|_v} \leq \frac{\|A\| \, \|A^{-1}\|}{1 - \delta} \frac{\|\boldsymbol{k}\|_v}{\|\boldsymbol{b}\|_v} + \frac{\delta}{1 - \delta}.$$

The first term on the right contains the relative perturbation of $\boldsymbol{b}$, and the last term is independent of the right-hand side of the original equation. We now define $\kappa(A) = \|A\| \, \|A^{-1}\|$ to be the *condition number of A* (with respect to $\| \;\|$) and note that

$$\delta = \|A^{-1}\| \, \|F\| = \frac{\kappa(A) \, \|F\|}{\|A\|}.$$

Thus, we obtain finally

$$\frac{\|\boldsymbol{y}\|_v}{\|\boldsymbol{x}\|_v} \leq \frac{\kappa(A)}{1 - (\kappa(A)\|F\|/\|A\|)} \left(\frac{\|\boldsymbol{k}\|_v}{\|\boldsymbol{b}\|_v} + \frac{\|F\|}{\|A\|} \right). \tag{2}$$

This gives us an upper bound for the relative perturbation of $\boldsymbol{x}$ in terms of the relative perturbations of $\boldsymbol{b}$ and A and the condition number $\kappa(A)$. In particular, the decisive part played by the condition number should be noted. This is the case whether perturbations occur in $\boldsymbol{b}$ alone, in A alone, or in $\boldsymbol{b}$ and A simultaneously. The condition number will also play an important part in the investigation of perturbations of eigenvalues (Section 2).

Exercise 1. Prove that if $\|A^{-1}\| \, \|F\| < 1$, then $\|F\| < \|A\|$.

Exercise 2. If $A = \text{diag}[1, 2, 3, \ldots, 10]$ and $\|\boldsymbol{b}\|_\infty = 1$, and it is known only that the elements of F and $\boldsymbol{k}$ are bounded in absolute value by ε, where $\varepsilon < 0.1$, then

$$\frac{\|\boldsymbol{y}\|_\infty}{\|\boldsymbol{x}\|_\infty} \leq \frac{20\varepsilon}{1 - 10\varepsilon}.$$

If, in addition,

$$b = \begin{bmatrix} 0.1 \\ 0.2 \\ \vdots \\ 1.0 \end{bmatrix}, \qquad k = \begin{bmatrix} -\varepsilon \\ 0 \\ \vdots \\ 0 \end{bmatrix}, \qquad F = \begin{bmatrix} \varepsilon & \varepsilon & \cdots & \varepsilon \\ 0 & 0 & \cdots & 0 \\ \vdots & \vdots & & \vdots \\ 0 & 0 & \cdots & 0 \end{bmatrix},$$

show that

$$\frac{\|\mathbf{y}\|_\infty}{\|\mathbf{x}\|_\infty} = \frac{20\varepsilon}{1+\varepsilon}.$$

Exercise 3. Prove that for any $A \in \mathbb{C}^{n \times n}$ and any matrix norm, the condition number $\kappa(A)$ satisfies the inequality $\kappa(A) \geq 1$.

Exercise 4. If $\kappa(A)$ is the *spectral condition number* of A, that is, the condition number obtained using the spectral norm, and s_1 and s_n are the largest and smallest singular values of A, respectively, prove that $\kappa(A) = s_1/s_n$.

Exercise 5. If $\kappa(A)$ is the spectral condition number of a normal nonsingular matrix $A \in \mathbb{C}^{n \times n}$, and λ_1 and λ_n are (in absolute value) the largest and smallest eigenvalues of A, respectively, prove that $\kappa(A) = |\lambda_1|/|\lambda_n|$.

Exercise 6. Prove that the spectral condition number $\kappa(A)$ is equal to 1 if and only if αA is a unitary matrix for some $\alpha \neq 0$.

Exercise 7. If the linear equation $A\mathbf{x} = \mathbf{b}$, $\det A \neq 0$, is replaced by the equivalent equation $B\mathbf{x} = A^*\mathbf{b}$, where $B = A^*A$, prove that this may be done at the expense of the condition number; that is, that $\kappa(B) \geq \kappa(A)$.

Exercise 8. Use Theorem 1 to prove that $M^p \to 0$ as $p \to \infty$. More precisely, prove that $M^p \to 0$ as $p \to \infty$ if and only if the spectral radius of M is less than 1.

Exercise 9. If $A \in \mathbb{C}^{n \times n}$ satisfies the condition $\|A\|^p < k$ for some matrix norm $\|\ \|$, all nonnegative integers p, and some constant k independent of p, show that A also satisfies the *resolvent condition*: for all complex z with $|z| > 1$, the matrix $zI - A$ is nonsingular and

$$\|(zI - A)^{-1}\| \leq \frac{k}{|z| - 1}.$$

Exercise 10. Let $A \in \mathbb{C}^{n \times n}$ be an invertible normal matrix with eigenvalues $\lambda_{\max}$ and $\lambda_{\min}$ of largest and smallest absolute values, respectively. Let $A\mathbf{v} = \lambda_{\max}\mathbf{v}$, $A\mathbf{u} = \lambda_{\min}\mathbf{u}$ with $\|\mathbf{v}\| = \|\mathbf{u}\| = 1$ in the euclidean norm. Write $\mathbf{b} = \mathbf{v}$, $\mathbf{k} = \varepsilon\mathbf{u}$, with $\varepsilon > 0$. Show that if $A\mathbf{x} = \mathbf{b}$ and $A(\mathbf{x} + \mathbf{y}) = \mathbf{b} + \mathbf{k}$ [cf. Eq. (1)], then

$$\frac{\|\mathbf{y}\|}{\|\mathbf{x}\|} = \varepsilon\kappa(A).$$

(This is a "worst case" in which the general error bound of (2) is actually attained. Note the result of Ex. 10.4.8) □

11.2 Perturbations of the Eigenvalues of a Simple Matrix

We now take up the question of the dependence of the eigenvalues of a simple matrix A on the elements of A. The results of this section are essentially due to F. L. Bauer and A. S. Householder.[†]

Let A be a simple matrix with eigenvalues $\lambda_1, \lambda_2, \ldots, \lambda_n$. Then if $D = \text{diag}[\lambda_1, \ldots, \lambda_n]$, there is a nonsingular matrix P such that $A = PDP^{-1}$. In Theorem 10.5.1 we have been able to characterize a class of matrix norms with favorable properties when applied to diagonal matrices, and so we are led to the idea of using $\|D\|$ to estimate $\|A\|$ for such a norm. Thus, if $\| \ \|$ is a matrix norm induced by an absolute vector norm, the equation $A = PDP^{-1}$ and Theorem 10.5.1 imply that

$$\|A\| \le \kappa(P) \max_{1 \le j \le n} |\lambda_j|,$$

where $\kappa(P) = \|P\| \, \|P^{-1}\|$ is the condition number of P with respect to $\| \ \|$, as defined in Section 1. However, as Exercise 6.5.4 shows, the matrix P is not uniquely defined by the equation $A = PDP^{-1}$. In particular, $A = QDQ^{-1}$ for any matrix $Q = PD_1$, where D_1 is a nonsingular diagonal matrix. We therefore obtain the best possible bound,

$$\|A\| \le \nu(A) \max_{1 \le j \le n} |\lambda_j|,$$

if we define $\nu(A) = \inf_P \kappa(P)$, the infimum being taken with respect to all matrices P for which $P^{-1}AP$ is diagonal. This result may be viewed as a lower bound for the spectral radius of A; coupled with Theorem 10.3.1, it yields the first half of the following result.

Theorem 1. *If A is a simple matrix, $\| \ \|$ is a matrix norm induced by an absolute vector norm, and $\nu(A) = \inf_P \kappa(P)$, where $P^{-1}AP$ is diagonal, then*

$$\frac{\|A\|}{\nu(A)} \le \max_{1 \le j \le n} |\lambda_j| \le \|A\|,$$

and (1)

$$l(A) \le \min_{1 \le j \le n} |\lambda_j| \le \nu(A) l(A),$$

where $l(A)$ is defined in Eq. (10.5.1).

[†] *Numerische Math.* **3** (1961), 241–246.

PROOF. The first inequality in (1) is just the result of Exercise 10.5.2. For the second inequality, the equation $A = PDP^{-1}$ and Corollaries 10.5.1 and 10.5.2 imply that

$$l(A) \geq l(P)l(P^{-1}) \min_{1 \leq j \leq n} |\lambda_j|.$$

Theorem 10.5.2 then gives

$$l(A) \geq (\kappa(P))^{-1} \min_{1 \leq j \leq n} |\lambda_j|,$$

and the second inequality in (1) follows. ■

For a general simple matrix A this theorem is of doubtful computational value because the quantity $\nu(A)$ is so difficult to estimate. However, we should observe that $PP^{-1} = I$ implies that $\kappa(P) = \|P\| \, \|P^{-1}\| \geq 1$ (see also Exercise 11.1.3), and hence that $\nu(A) \geq 1$.

If A is a normal matrix then we know that P may be chosen to be a unitary matrix. If, in addition, we use the euclidean vector norm, and hence the spectral matrix norm $\| \;\|$, then $\|P\| = \|P^{-1}\| = 1$. Thus, for a *normal matrix* and *spectral norm* we have $\nu(A) = 1$ and the theorem reduces to the known cases:

$$\|A\| = \max_{1 \leq j \leq n} |\lambda_j|, \qquad l(A) = \min_{1 \leq j \leq n} |\lambda_j|.$$

Our next theorem demonstrates the significance of $\nu(A)$ as a condition number for A reflecting the stability of the eigenvalues under perturbation of the matrix elements.

Theorem 2. *Let $A, B \in \mathbb{C}^{n \times n}$, with A simple. If A has eigenvalues $\lambda_1, \lambda_2, \ldots, \lambda_n$ and μ is an eigenvalue of $A + B$, and if for a matrix norm induced by an absolute vector norm $\| \;\|_v$ we have*

$$r = \|B\| \nu(A),$$

then μ lies in at least one of the disks

$$\{z : |z - \lambda_j| \leq r\}, \qquad j = 1, 2, \ldots, n,$$

of the complex z-plane.

PROOF. Since μ is an eigenvalue of $A + B$, there is a $\mathbf{y} \neq \mathbf{0}$ for which

$$(A + B)\mathbf{y} = \mu\mathbf{y}.$$

If $A = PDP^{-1}$, where $D = \operatorname{diag}[\lambda_1, \ldots, \lambda_n]$, then

$$(D + P^{-1}BP)\mathbf{x} = \mu\mathbf{x},$$

where $x = P^{-1}y \neq 0$. Thus,

$$(\mu I - D)x = P^{-1}BPx \tag{2}$$

and hence

$$l(\mu I - D) \leq \frac{\|(\mu I - D)x\|_v}{\|x\|_v} = \frac{\|P^{-1}BPx\|_v}{\|x\|_v} \leq \|P^{-1}BP\| \leq \kappa(P)\,\|B\|.$$

Therefore we have

$$\min_{1 \leq j \leq n} |\mu - \lambda_j| \leq \kappa(P)\,\|B\|.$$

Since this is true for every P diagonalizing A, we have

$$\min_{1 \leq j \leq n} |\mu - \lambda_j| \leq \nu(A)\,\|B\| = r.$$

Hence, μ lies in at least one of the disks $|z - \lambda_j| \leq r, j = 1, 2, \ldots, n$. ■

Exercise 1. Show that the eigenvalues in Exercise 10.6.1 lie in the disks with radius 2 and centers at $3 \pm \sqrt{10}$, $5i$, and -2.

Hint. Use Exercise 10.4.11 to estimate $\|B\|$. □

If the perturbation matrix B of Theorem 2 is also known to be simple, then the radius of the disks obtained in that theorem can be reformulated, as indicated in the next theorem.

Theorem 3. *If, in addition to the hypotheses of Theorem 2, the matrix B is simple with eigenvalues $\mu_1, \ldots, \mu_n$, then μ lies in at least one of the disks*

$$\{z : |z - \lambda_j| \leq \nu(A, B) \max_j |\mu_j|\}, \qquad j = 1, 2, \ldots, n,$$

where

$$\nu(A, B) = \inf_{P, Q} \kappa(P^{-1}Q)$$

and P and Q diagonalize A and B, respectively.

PROOF. We may now write Eq. (2) in the form

$$(\mu I - D)x = (P^{-1}Q)D_1(P^{-1}Q)^{-1}x,$$

where $D_1 = \operatorname{diag}[\mu_1, \ldots, \mu_n]$. Hence

$$l(\mu I - D) \leq \kappa(P^{-1}Q)\,\|D_1\|,$$

and the result follows from Theorem 10.5.1. ■

We can easily obtain a stronger form of this theorem. Observe that P will diagonalize A if and only if P will diagonalize $\alpha I + A$ for any complex number α, and similarly for Q, B, and $-\alpha I + B$. Since

$$(\alpha I + A) + (-\alpha I + B) = A + B,$$

we can now apply Theorem 3 to $\alpha I + A$ and $-\alpha I + B$.

Corollary 1. *Under the hypotheses of Theorem 3, the eigenvalue μ lies in at least one of the disks*

$$\{z : |z - (\alpha + \lambda_j)| \le \nu(A, B) \max_{1 \le j \le n} |\mu_j - \alpha|\}, \qquad j = 1, 2, \ldots, n,$$

for any fixed complex number α.

The advantage of this result lies in the fact that α may now be chosen to minimize $\max_j |\mu_j - \alpha|$ and hence minimize the area of the disks containing the eigenvalues.

The case in which A and B are normal is important enough to warrant a separate statement.

Corollary 2. *If, in the hypotheses of Theorem 3, the matrices A and B are normal, then the eigenvalues of $A + B$ are contained in the disks*

$$\{z : |z - (\alpha + \lambda_j)| \le \max_j |\mu_j - \alpha|\}, \qquad j = 1, 2, \ldots, n,$$

for any fixed complex number α.

PROOF. Observe that P and Q may be supposed to be unitary. This implies that $P^{-1}Q$ is unitary and hence, if we use the spectral matrix norm, $\nu(A, B) = \inf_{P,Q} \kappa(P^{-1}Q) = 1$. The result now follows from the first corollary. ■

Exercise 2. If A is a normal matrix with eigenvalues $\lambda_1, \lambda_2, \ldots, \lambda_n$, and B has all its elements equal to ε, prove that the eigenvalues of $A + B$ lie in the disks

$$\left\{z : \left| z - \left(\frac{n\varepsilon}{2} + \lambda_j\right)\right| \le \frac{n}{2}|\varepsilon|\right\}, \qquad j = 1, 2, \ldots, n.$$

Exercise 3. If A is a normal matrix and $B = \operatorname{diag}[b_1, \ldots, b_n]$ is real, prove that the eigenvalues of $A + B$ lie in the disks

$$\{z : |z - (\alpha + \lambda_j)| \le \beta\}, \qquad j = 1, 2, \ldots, n,$$

where $\lambda_1, \lambda_2, \ldots, \lambda_n$ are the eigenvalues of A and

$$2\alpha = \max_{1 \le j \le n} b_j + \min_{1 \le j \le n} b_j \qquad \text{and} \qquad 2\beta = \max_{1 \le j \le n} b_j - \min_{1 \le j \le n} b_j.$$

Exercise 4. Let $A \in \mathbb{R}^{n \times n}$, and let $\mu_1, \ldots, \mu_n$ be the eigenvalues of $\frac{1}{2}(A + A^T)$. If k is the spectral radius of $\frac{1}{2}(A - A^T)$ and

$$\min_{j \neq k} |\mu_j - \mu_k| \geq 2k,$$

prove that the eigenvalues of A are real.

Exercise 5. Suppose that $A, B \in \mathbb{R}^{n \times n}$, $A^T = A$, and $|b_{jk}| \leq b$ for $j, k = 1, 2, \ldots, n$. If $\mu_1, \ldots, \mu_n$ are the eigenvalues of A and

$$\min_{j \neq k} |\mu_j - \mu_k| > 2nb,$$

prove that the eigenvalues of $A + B$ are real.

Exercise 6. Let A and $A + E$ be $n \times n$ matrices with eigenvalues $\lambda_1, \lambda_2, \ldots, \lambda_n$ and $\mu_1, \mu_2, \ldots, \mu_n$, respectively. Show that the eigenvalues can be ordered in such a way that

$$|\lambda_j - \mu_j| \leq \|E\|_s,$$

for $j = 1, 2, \ldots, n$. □

11.3 Analytic Perturbations

We now take quite a different approach to the perturbation problem. Suppose that $A \in \mathbb{C}^{n \times n}$ is the unperturbed matrix and has eigenvalues $\lambda_1, \ldots, \lambda_n$. We then suppose that $A(\zeta)$ is an $n \times n$ matrix whose elements are analytic functions of ζ in a neighborhood of ζ_0 and that $A(\zeta_0) = A$. For convenience, and without loss of generality, we may suppose that $\zeta_0 = 0$, so $A(0) = A$. We write $\lambda_1(\zeta), \ldots, \lambda_n(\zeta)$ for the eigenvalues of $A(\zeta)$ and then (Section 10.6), since the eigenvalues of $A(\zeta)$ depend continuously on the elements of $A(\zeta)$, we may suppose that $\lambda_j(\zeta) \to \lambda_j$ as $\zeta \to 0$ for $j = 1, 2, \ldots, n$.

The following exercise illustrates the fact that, though the eigenvalues depend continuously on the matrix elements, they may nevertheless vary quite rapidly in pathological circumstances.

Exercise 1. Consider the 10×10 matrix $A(\zeta)$ defined as a function of the real variable ζ by

$$A(\zeta) = \begin{bmatrix} 0 & 1 & 0 & \cdots & 0 \\ 0 & 0 & 1 & \cdots & 0 \\ \vdots & \vdots & \ddots & \ddots & \vdots \\ 0 & 0 & \cdots & 0 & 1 \\ \zeta & 0 & \cdots & 0 & 0 \end{bmatrix}.$$

Prove that the eigenvalues at $\zeta = 0$ and those at $\zeta = 10^{-10}$ differ in absolute value by 0.1.

Note that the difference between $A(0)$ and $A(10^{-10})$ is so small that they may not be distinguishable on a computer with too short a word length. □

What can we say about the nature of the functions $\lambda_j(\zeta)$ for sufficiently small $|\zeta|$? We shall see that in certain important special cases, the functions $\lambda_j(\zeta)$ are also analytic in a neighborhood of $\zeta = 0$. That this is not generally the case is illustrated by the next example.

Example 2. The eigenvalues of the matrix

$$A(\zeta) = \begin{bmatrix} 0 & \zeta \\ \zeta^2 & 0 \end{bmatrix}$$

are the zeros of the polynomial

$$\det(\lambda I - A(\zeta)) = \lambda^2 - \zeta^3,$$

that is, $\lambda_1(\zeta) = \zeta^{3/2}$ (defined by a particular branch of the function $\zeta^{1/2}$) and $\lambda_2(\zeta) = -\lambda_1(\zeta)$. Thus, we see that the eigenvalues are certainly not analytic in a neighbourhood of $\zeta = 0$. □

It should be noted that $\det(\lambda I - A(\zeta))$ is a polynomial in λ whose coefficients are analytic functions of ζ. It then follows from the theory of algebraic functions that

(a) If λ_j is an unrepeated eigenvalue of A, then $\lambda_j(\zeta)$ is analytic in a neighbourhood of $\zeta = 0$;

(b) If λ_j has algebraic multiplicity m and $\lambda_k(\zeta) \to \lambda_j$ for $k = \alpha_1, \ldots, \alpha_m$, then $\lambda_k(\zeta)$ is an analytic function of $\zeta^{1/l}$ in a neighborhood of $\zeta = 0$, where $l \le m$ and $\zeta^{1/l}$ is one of the l branches of the function $\zeta^{1/l}$. Thus $\lambda_k(\zeta)$ has a power-series expansion in $\zeta^{1/l}$.

We shall prove result (a) together with a result concerning the associated eigenvectors. Note that, in result (b), we *may* have $l = 1$ and $m > 1$, in which case $\lambda_k(\zeta)$ could be analytic. As might be expected it is case (b) with $m > 1$ that gives rise to difficulties in the analysis. The expansions in fractional powers of ζ described in (b) are known as *Puiseux series*, and their validity will be assumed in the discussions of Section 11.7.

It is when we try to follow the behavior of eigenvectors that the situation becomes more complicated. The matrix

$$A(\zeta) = \begin{bmatrix} 0 & \zeta \\ 0 & 0 \end{bmatrix}$$

serves to show that even though the eigenvalues may be analytic, the eigenvectors need not be continuous at $\zeta = 0$.

Nevertheless, we shall approach the perturbation problem by analysis of the spaces generated by eigenvectors, insofar as these are determined by the component matrices of $A(\zeta)$ (see Section 9.5). The reason for this is that, first, the resolvent $R_z(\zeta) = (zI - A(\zeta))^{-1}$ is expressed directly in terms of $A(\zeta)$ and, second, by means of Eq. (9.9.4) we can then examine the component matrices of $A(\zeta)$. In particular, we shall be able to infer properties of vectors in the range of a component matrix.

11.4 Perturbation of the Component Matrices

Suppose that an unperturbed eigenvalue λ "splits" into distinct eigenvalues $\lambda_1(\zeta), \ldots, \lambda_p(\zeta)$ under the perturbation determined by $A(\zeta)$. We can prove that the *sum* of the first component matrices for each of these eigenvalues is analytic near $\zeta = 0$. This theorem is the peg on which we hang all our subsequent analysis of the perturbation problem.

Theorem 1. *Let $A(\zeta)$ be analytic in a neighborhood of $\zeta = 0$ and $A(0) = A$. Suppose that $\lambda_1(\zeta), \ldots, \lambda_p(\zeta)$ are the distinct eigenvalues of $A(\zeta)$ for which $\lambda_k(\zeta) \to \lambda$ as $\zeta \to 0$, $k = 1, 2, \ldots, p$, and let $Z_{k0}(\zeta)$ be the idempotent component matrix of $A(\zeta)$ associated with $\lambda_k(\zeta)$. If Z is the idempotent component matrix of A associated with λ and we define*

$$Y(\zeta) = \sum_{k=1}^{p} Z_{k0}(\zeta),$$

then there is a neighborhood of $\zeta = 0$ in which the following developments hold:

$$Y(\zeta) = Z + \zeta C_1 + \zeta^2 C_2 + \cdots,$$
$$A(\zeta)Y(\zeta) = AZ + \zeta B_1 + \zeta^2 B_2 + \cdots,$$

for matrices C_r, B_r independent of ζ, $r = 1, 2, \ldots$.

PROOF. Consider a fixed $\zeta \neq 0$ for which $|\zeta|$ is small enough to admit the construction of a simple closed contour L that encircles $\lambda, \lambda_1(\zeta), \ldots, \lambda_p(\zeta)$ and no other eigenvalues of A or of $A(\zeta)$. Then writing $R_z = (zI - A)^{-1}$ and $R_z(\zeta) = (zI - A(\zeta))^{-1}$, we have

$$Z = \frac{1}{2\pi i}\int_L R_z\, dz, \qquad Y(\zeta) = \frac{1}{2\pi i}\int_L R_z(\zeta)\, dz. \qquad (1)$$

We note that the elements of $R_z(\zeta)$ are rational expressions in functions of ζ that are analytic in a neighborhood of $\zeta = 0$ for z on the contour L. Thus, there exist matrices $R_z^{(r)}$, $r = 1, 2, \ldots$, independent of ζ such that

$$R_z(\zeta) = R_z + \zeta R_z^{(1)} + \zeta^2 R_z^{(2)} + \cdots . \tag{2}$$

Using (1) we have only to integrate this series term by term around the contour to obtain the first result of the theorem. This is permissible provided only that each coefficient $R_z^{(r)}$ $(r = 0, 1, 2, \ldots)$ is integrable on L. Using Exercise 9.9.3 we see that, for a fixed z,

$$R_z^{(1)}(\zeta) = R_z(\zeta)A^{(1)}(\zeta)R_z(\zeta), \tag{3}$$

and hence that, for each r, the matrix $R_z^{(r)}$ is a linear combination of matrices $R_z(A^{(p)}R_z)^s$ for certain integers p, s with $p \le r$. Now, $A^{(p)}(0)$ is independent of z and R_z is certainly continuous in z on L. Hence $R_z^{(r)}$ is continuous on L. Integrating the series for $R_z(\zeta)$ around L, we now obtain

$$Y(\zeta) = Z + \zeta C_1 + \zeta^2 C_2 + \cdots ,$$

where

$$C_r = \frac{1}{2\pi i}\int_L R_z^{(r)}\, dz, \qquad r = 1, 2, \ldots .$$

Now both $Y(\zeta)$ and $A(\zeta)$ are analytic on a neighborhood of $\zeta = 0$. It follows that their product $A(\zeta)Y(\zeta)$ is also analytic there and has a Taylor expansion given by the product of the expansions for $A(\zeta)$ and $Y(\zeta)$, so the theorem is proved.

A similar theorem can be proved showing that the *continuity* of $A(\zeta)$ at $\zeta = 0$ implies that of $Y(\zeta)$ at $\zeta = 0$. ■

Now we show that the *individual* first component matrices do not necessarily have the property of being analytic near $\zeta = 0$.

Exercise 1. Find the matrices Z, $Z_{10}(\zeta)$, $Z_{20}(\zeta)$, and $Y(\zeta)$ for the matrix

$$A(\zeta) = \begin{bmatrix} 0 & \zeta \\ \zeta^2 & 0 \end{bmatrix}.$$

SOLUTION. At $\zeta = 0$, the matrix A is the zero-matrix and it is easily seen that $Z = I$. The eigenvalues of $A(\zeta)$ are $\mu_1(\zeta) = \zeta^{3/2}$ and $\mu_2(\zeta) = -\mu_1(\zeta)$. We then have

$$R_z(\zeta) = \begin{bmatrix} z & -\zeta \\ -\zeta^2 & z \end{bmatrix}^{-1} = \frac{1}{z^2 - \zeta^3}\begin{bmatrix} z & \zeta \\ \zeta^2 & z \end{bmatrix}.$$

Partial fraction expansions now give

$$R_z(\zeta) = \frac{1}{2}\begin{bmatrix} \dfrac{1}{z-\zeta^{3/2}} + \dfrac{1}{z+\zeta^{3/2}} & \dfrac{\zeta^{-1/2}}{z-\zeta^{3/2}} - \dfrac{\zeta^{-1/2}}{z+\zeta^{3/2}} \\ \dfrac{\zeta^{1/2}}{z-\zeta^{3/2}} - \dfrac{\zeta^{1/2}}{z+\zeta^{3/2}} & \dfrac{1}{z-\zeta^{3/2}} + \dfrac{1}{z+\zeta^{3/2}} \end{bmatrix}.$$

Using Eq. (9.9.4) we obtain

$$Z_{10}(\zeta) = \tfrac{1}{2}\begin{bmatrix} 1 & \zeta^{-1/2} \\ \zeta^{1/2} & 1 \end{bmatrix} \quad \text{and} \quad Z_{20}(\zeta) = \tfrac{1}{2}\begin{bmatrix} 1 & -\zeta^{-1/2} \\ -\zeta^{1/2} & 1 \end{bmatrix}.$$

Hence $Y(\zeta) = Z_{10}(\zeta) + Z_{20}(\zeta) = I$. This example emphasizes the fact that although $Y(\zeta)$ may be analytic in ζ, the component matrices (of which it is the sum) need not be analytic. Indeed, in this example the component matrices have a singularity at $\zeta = 0$. It should also be noted that although the eigenvalues are not analytic in a neighborhood containing $\zeta = 0$, yet they are "better behaved" than the component matrices.

Exercise 2. If the matrix

$$A(\zeta) = \begin{bmatrix} 0 & \zeta & 0 \\ 0 & 0 & \zeta \\ \zeta & 0 & 1 \end{bmatrix}$$

has eigenvalues $\lambda_1(\zeta)$, $\lambda_2(\zeta)$, and $\lambda_3(\zeta)$ and $\lambda_j(\zeta) \to 0$ as $\zeta \to 0$, $j = 1, 2$, show that we may write

$$\lambda_j(\zeta) = (-1)^{j+1} i\zeta^{3/2} - \tfrac{1}{2}\zeta^3 + \cdots, \qquad j = 1, 2.$$

If $\lambda = 0$ is the unperturbed eigenvalue under investigation, find Z, $Z_{10}(\zeta)$, $Z_{20}(\zeta)$, and $Y(\zeta)$, and verify that $Y(\zeta)$ is analytic in a neighbourhood of $\zeta = 0$.

Exercise 3. Consider eigenvectors $\boldsymbol{e}_1$, $\boldsymbol{e}_2$, $\boldsymbol{e}_3$ (unit coordinate vectors) of $A(0)$ in Exercise 2. Consider the possibility of extending these, as eigenvectors of $A(\zeta)$, into a neighbourhood of $\zeta = 0$. Show that $\boldsymbol{e}_3$ can be extended analytically, $\boldsymbol{e}_1$ has two continuous extensions, and $\boldsymbol{e}_2$ has no such extension. □

11.5 Perturbation of an Unrepeated Eigenvalue

We are now in a position to prove a perturbation theorem of the greatest practical importance. The notation is that developed in the two preceding sections.

Theorem 1. *If λ is an eigenvalue of A of multiplicity 1, then, for sufficiently small $|\zeta|$, there is an eigenvalue $\lambda(\zeta)$ of $A(\zeta)$ such that*

$$\lambda(\zeta) = \lambda + \zeta\lambda^{(1)} + \zeta^2\lambda^{(2)} + \cdots. \tag{1}$$

Also, there are right and left eigenvectors $\mathbf{x}(\zeta)$ and $\mathbf{y}(\zeta)$, respectively, associated with $\lambda(\zeta)$ for which

$$\mathbf{y}^{\mathrm{T}}(\zeta)\mathbf{x}(\zeta) = 1$$

and

$$\mathbf{x}(\zeta) = \mathbf{x} + \zeta\mathbf{x}^{(1)} + \zeta^2\mathbf{x}^{(2)} + \cdots, \tag{2}$$

$$\mathbf{y}(\zeta) = \mathbf{y} + \zeta\mathbf{y}^{(1)} + \zeta^2\mathbf{y}^{(2)} + \cdots. \tag{3}$$

PROOF. We suppose throughout that $|\zeta|$ is so small that $\lambda(\zeta)$ has multiplicity 1. It then follows that $Y(\zeta) = Z_{10}(\zeta)$, a component matrix. Furthermore (see Theorem 4.10.3) we may write

$$Z = \mathbf{x}\mathbf{y}^{\mathrm{T}}, \qquad Y(\zeta) = \mathbf{x}(\zeta)\mathbf{y}^{\mathrm{T}}(\zeta), \tag{4}$$

where

$$\mathbf{y}^{\mathrm{T}}\mathbf{x} = \mathbf{y}^{\mathrm{T}}(\zeta)\mathbf{x}(\zeta) = 1.$$

In this case we obviously have

$$A(\zeta)Y(\zeta) = \lambda(\zeta)Y(\zeta),$$

so premultiplying by $\mathbf{y}^{\mathrm{T}}$ and postmultiplying by $\mathbf{x}$ we obtain

$$\lambda(\zeta) = \frac{\mathbf{y}^{\mathrm{T}}A(\zeta)Y(\zeta)\mathbf{x}}{\mathbf{y}^{\mathrm{T}}Y(\zeta)\mathbf{x}} = \frac{\lambda + \zeta b_1 + \zeta^2 b_2 + \cdots}{1 + \zeta c_1 + \zeta^2 c_2 + \cdots}$$

after making use of Theorem 11.4.1. The coefficients on the right are given by

$$b_r = \mathbf{y}^{\mathrm{T}}B_r\mathbf{x}, \qquad c_r = \mathbf{y}^{\mathrm{T}}C_r\mathbf{x}, \qquad r = 1, 2, \ldots.$$

It follows that $\lambda(\zeta)$ is analytic in a neighbourhood of $\zeta = 0$, which is the first part of the theorem.

It is clear that, in the statement of the theorem, $\mathbf{x}(\zeta)$ is not defined to within a nonzero multiplicative constant (which may vary with ζ). We define $\mathbf{x}(\zeta)$ more precisely by noting first of all that there is no ambiguity in the definition of $Y(\zeta)$. We then choose the eigenvectors $\mathbf{x}$, $\mathbf{y}$ with $\mathbf{y}^{\mathrm{T}}\mathbf{x} = 1$, and hence $\mathbf{x}\mathbf{y}^{\mathrm{T}} = Z$.

Then $Y(\zeta) = \mathbf{x}(\zeta)\mathbf{y}^{\mathrm{T}}(\zeta)$ implies

$$\mathbf{y}^{\mathrm{T}}Y(\zeta)\mathbf{x} = (\mathbf{y}^{\mathrm{T}}\mathbf{x}(\zeta))(\mathbf{y}^{\mathrm{T}}(\zeta)\mathbf{x}),$$

and since $Y(\zeta) \to \boldsymbol{x}\boldsymbol{y}^{\mathrm{T}}$ as $\zeta \to 0$, it follows that $(\boldsymbol{y}^{\mathrm{T}}\boldsymbol{x}(\zeta))(\boldsymbol{y}^{\mathrm{T}}(\zeta)\boldsymbol{x}) \to 1$ as $\zeta \to 0$. Thus we may suppose that $\boldsymbol{y}^{\mathrm{T}}\boldsymbol{x}(\zeta)$ and $\boldsymbol{y}^{\mathrm{T}}(\zeta)$ are nonzero for small enough $|\zeta|$. Now we can define $\boldsymbol{x}(\zeta)$ by normalizing in such a way that

$$\boldsymbol{y}^{\mathrm{T}}\boldsymbol{x}(\zeta) = 1. \tag{5}$$

Since $Y(\zeta)\boldsymbol{x} = \boldsymbol{x}(\zeta)(\boldsymbol{y}^{\mathrm{T}}(\zeta)\boldsymbol{x})$, we have $\boldsymbol{y}^{\mathrm{T}}Y(\zeta)\boldsymbol{x} = \boldsymbol{y}^{\mathrm{T}}(\zeta)\boldsymbol{x}$. Hence $Y(\zeta) = \boldsymbol{x}(\zeta)\boldsymbol{y}^{\mathrm{T}}(\zeta)$ implies

$$\boldsymbol{x}(\zeta) = \frac{1}{\boldsymbol{y}^{\mathrm{T}}Y(\zeta)\boldsymbol{x}}\, Y(\zeta)\boldsymbol{x}.$$

Since $Y(\zeta)$ is analytic in a neighborhood of $\zeta = 0$ and $\boldsymbol{y}^{\mathrm{T}}Y(0)\boldsymbol{x} = 1$, it follows that $\boldsymbol{x}(\zeta)$ is analytic in a neighborhood of $\zeta = 0$.

Finally, Eq. (5) also implies $\boldsymbol{x}^{\mathrm{T}}(\zeta)\boldsymbol{y} = 1$, so the relation $\boldsymbol{y}(\zeta)\boldsymbol{x}^{\mathrm{T}}(\zeta) = Y^{\mathrm{T}}(\zeta)$ implies $\boldsymbol{y}(\zeta) = Y^{\mathrm{T}}(\zeta)\boldsymbol{y}$, which is also analytic in a neighborhood of $\zeta = 0$. ■

We have now proved the existence of power series expansions in ζ for an unrepeated eigenvalue and its eigenvectors. The next step in the analysis is the discussion of methods for computing the coefficients of these expansions.

11.6 Evaluation of the Perturbation Coefficients

We shall first discuss a particular problem that will arise repeatedly later on. If λ is an eigenvalue of the matrix $A \in \mathbb{C}^{n \times n}$ and $\boldsymbol{b} \neq 0$, we want to solve the equation

$$(\lambda I - A)\boldsymbol{x} = \boldsymbol{b}. \tag{1}$$

Given that a solution exists (i.e., $\boldsymbol{b} \in \operatorname{Im}(\lambda I - A)$), we aim to express a solution $\boldsymbol{x}$ of (1) in terms of the spectral properties of A. If $\lambda_1, \lambda_2, \ldots, \lambda_s$ are the distinct eigenvalues of A and $\lambda = \lambda_1$ we first define the matrix

$$E = \sum_{k=2}^{s} \sum_{j=0}^{m_k - 1} \frac{j!}{(\lambda - \lambda_k)^{j+1}} Z_{kj}, \tag{2}$$

just as in Eq. (9.6.2) but replacing λ_1 and E_1 by λ and E. Multiply Eq. (1) on the left by E and use Theorem 9.6.2 to obtain

$$(I - Z)\boldsymbol{x} = E\boldsymbol{b}. \tag{3}$$

(With the convention of Sections 4 and 5 we write Z for Z_{10}.) Since $ZE = O$ it is easily seen that the vector $\boldsymbol{x} = E\boldsymbol{b}$ satisfies Eq. (3), and we would like to assert that it is also a solution of Eq. (1). However, we need an additional hypothesis to guarantee this.

Proposition 1. *If λ is an eigenvalue of A of index 1, then Eqs.* (*1*) *and* (*3*) *have the same solution set $\mathscr{S}$. Furthermore, $\mathbf{x} = E\mathbf{b}$ is a common solution* (*when $\mathscr{S} \neq \varnothing$*).

PROOF. In view of the argument above, we have only to show that the solution sets coincide. But $I - Z = E(\lambda I - A)$ and so, using Exercise 3.10.8, we need to check that $\operatorname{rank}(I - Z) = \operatorname{rank}(\lambda I - A)$. To do this, observe that $I - Z$ and $\lambda I - A$ have the same kernel, namely, the eigenspace of A associated with λ. ■

Suppose now that the hypotheses of Theorem 11.5.1 are fulfilled. Then λ has multiplicity 1 and for small enough $|\zeta|$, we use the results of that theorem to write

$$(A + \zeta A^{(1)} + \zeta^2 A^{(2)} + \cdots)(\mathbf{x} + \zeta \mathbf{x}^{(1)} + \zeta^2 \mathbf{x}^{(2)} + \cdots)$$
$$= (\lambda + \zeta\lambda^{(1)} + \cdots)(\mathbf{x} + \zeta \mathbf{x}^{(1)} + \cdots). \tag{4}$$

We can equate the coefficients of ζ^n obtained by multiplying these series for $n = 0, 1, 2, \ldots$. The coefficient of ζ^0 gives the equation for the unperturbed eigenvalue: $A\mathbf{x} = \lambda\mathbf{x}$. The coefficient of ζ gives

$$(\lambda I - A)\mathbf{x}^{(1)} = (A^{(1)} - \lambda^{(1)} I)\mathbf{x}. \tag{5}$$

Since $\mathbf{y}$ (of Eq. 11.5.4) is a left eigenvector of λ, we deduce that $\mathbf{y}^{\mathrm{T}}(A^{(1)} - \lambda^{(1)} I)\mathbf{x} = 0$. Since $\mathbf{y}^{\mathrm{T}}\mathbf{x} = 1$, this condition yields

$$\lambda^{(1)} = \mathbf{y}^{\mathrm{T}} A^{(1)} \mathbf{x}. \tag{6}$$

We also know that, with our definition of $\mathbf{x}(\zeta)$, the relation (11.5.5) gives us

$$\mathbf{y}^{\mathrm{T}}(\mathbf{x} + \zeta \mathbf{x}^{(1)} + \zeta^2 \mathbf{x}^{(2)} + \cdots) = 1,$$

and this implies that

$$\mathbf{y}^{\mathrm{T}}\mathbf{x}^{(r)} = 0, \qquad r = 1, 2, \ldots. \tag{7}$$

To obtain an expression for $\mathbf{x}^{(1)}$ from Eq. (5), we use Proposition 1. Thus, we know that $\mathbf{x}^{(1)} = E(A^{(1)} - \lambda^{(1)} I)\mathbf{x}$ is a solution. Since $Z = \mathbf{x}\mathbf{y}^{\mathrm{T}}$ and $EZ = O$, it follows that $E\mathbf{x} = EZ\mathbf{x} = \mathbf{0}$, so the expression for $\mathbf{x}^{(1)}$ simplifies to give $\mathbf{x}^{(1)} = EA^{(1)}\mathbf{x}$.

We should confirm that this choice of solution is consistent with (7), that is, that $\mathbf{y}^{\mathrm{T}}\mathbf{x}^{(1)} = 0$. Since $\mathbf{y}^{\mathrm{T}}Z = \mathbf{y}^{\mathrm{T}}$ we have

$$\mathbf{y}^{\mathrm{T}}\mathbf{x}^{(1)} = \mathbf{y}^{\mathrm{T}} ZEA^{(1)}\mathbf{x}.$$

But then $ZE = EZ = 0$, so $\mathbf{y}^{\mathrm{T}}\mathbf{x}^{(1)} = 0$, as required.

Thus, we complete the solution for the first-order perturbation coefficients as follows:

$$\lambda^{(1)} = \mathbf{y}^{\mathrm{T}} A^{(1)} \mathbf{x}, \qquad \mathbf{x}^{(1)} = E A^{(1)} \mathbf{x} \tag{8}$$

The argument can easily be extended to admit the recursive calculation of $\lambda^{(r)}$ and $\mathbf{x}^{(r)}$ for any r. The coefficients of ζ^r in Eq. (4) yield

$$A\mathbf{x}^{(r)} + A^{(1)}\mathbf{x}^{(r-1)} + \cdots + A^{(r)}\mathbf{x} = \lambda\mathbf{x}^{(r)} + \lambda^{(1)}\mathbf{x}^{(r-1)} + \cdots + \lambda^{(r)}\mathbf{x},$$

or

$$(\lambda I - A)\mathbf{x}^{(r)} = (A^{(1)} - \lambda^{(1)} I)\mathbf{x}^{(r-1)} + \cdots + (A^{(r)} - \lambda^{(r)} I)\mathbf{x}.$$

If we now assume that $\lambda^{(1)}, \ldots, \lambda^{(r-1)}$ and $\mathbf{x}^{(1)}, \ldots, \mathbf{x}^{(r-1)}$ are known, then multiplying on the left by $\mathbf{y}^{\mathrm{T}}$ gives

$$\lambda^{(r)} = \mathbf{y}^{\mathrm{T}} A^{(r)} \mathbf{x} + \mathbf{y}^{\mathrm{T}} A^{(r-1)} \mathbf{x}^{(1)} + \cdots + \mathbf{y}^{\mathrm{T}} A^{(1)} \mathbf{x}^{(r-1)}, \tag{9}$$

for $r = 1, 2, \ldots,$ and $\mathbf{x}^{(0)} = \mathbf{x}$. Proposition 1 then yields the result

$$\mathbf{x}^{(r)} = E[(A^{(1)} - \lambda^{(1)} I)\mathbf{x}^{(r-1)} + \cdots + (A^{(r-1)} - \lambda^{(r-1)} I)\mathbf{x}^{(1)} + A^{(r)}\mathbf{x}]. \tag{10}$$

In many applications the perturbations of A are *linear* in ζ. Thus we have $A(\zeta) = A + \zeta B$ for a matrix B independent of ζ. In this case we have $A^{(1)} = B$ and $A^{(2)} = A^{(3)} = \cdots = O$. Equations (9) and (10) now reduce to

$$\lambda^{(r)} = \mathbf{y}^{\mathrm{T}} B \mathbf{x}^{(r-1)}, \qquad \mathbf{x}^{(r)} = E\left(B\mathbf{x}^{(r-1)} - \sum_{j=1}^{r-1} \lambda^{(j)} \mathbf{x}^{(r-j)}\right),$$

for $r = 1, 2, \ldots$. Recall that in these formulas the superscripts do not denote derivatives but are defined by the expansions (11.5.1) and (11.5.2).

Exercise 1. Show that Eq. (9) in the case $r = 2$ may be written

$$\lambda^{(2)} = \operatorname{tr}(A^{(2)} Z_{10}) + \operatorname{tr}(A^{(1)} E A^{(1)} Z_{10}). \quad \square$$

11.7 Perturbation of a Multiple Eigenvalue

We can now consider the problem of analytic perturbations of an unrepeated eigenvalue to be solved. In particular, the result (a) of Section 11.3 has been established. If λ is an iegenvalue of A of multiplicity $m > 1$, then we must return to the Puiseux series representation of a perturbed eigenvalue $\lambda(\zeta)$ that tends to λ as $\zeta \to 0$ (result (b) of Section 11.3). In Section 11.4 we gained some further insight into the behavior of the component matrices of the perturbed eigenvalues and saw that although their sum $Y(\zeta)$ is analytic for small enough $|\zeta|$ (Theorem 11.4.1), the separate component matrices may have singularities at $\zeta = 0$ (Exercise 11.4.1).

Suppose now that we impose the further condition that $A(\zeta)$ be *Hermitian* and analytic on a neighborhood $\mathcal{N}$ of $\zeta = \zeta_0$ in the complex plane. Let $a_{jk}(\zeta)$ be an element of $A(\zeta)$ with $j \neq k$, and let $z = x + iy$ with x and y real. Then define $a_{jk}(x + iy) = u(x, y) + iv(x, y)$, where u and v are real-valued functions.

With these definitions, the Cauchy–Riemann equations imply that

$$\frac{\partial u}{\partial x} = \frac{\partial v}{\partial y} \quad \text{and} \quad \frac{\partial u}{\partial y} = -\frac{\partial v}{\partial x}.$$

But $a_{kj}(\zeta) = \overline{a_{jk}(\zeta)} = u(x, y) - iv(x, y)$, which is also analytic on $\mathcal{N}$. Hence

$$\frac{\partial u}{\partial x} = -\frac{\partial v}{\partial y} \quad \text{and} \quad \frac{\partial u}{\partial y} = \frac{\partial v}{\partial x}.$$

These equations together imply that $a_{jk}(\zeta)$ is a constant on $\mathcal{N}$, and the parameter dependence has been lost. This means that analytic dependence in the sense of complex variables is too strong a hypothesis, and for the analysis of matrices that are both analytic and Hermitian we must confine attention to a *real* parameter ζ and assume that for each j and k the real and imaginary parts of a_{jk} are real analytic functions of the real variable ζ.

For this reason the final theorem of this chapter (concerning normal matrices) must be stated in terms of a real parameter ζ. When applied to Hermitian matrices this means that the perturbed eigenvalues remain real and the component matrices (as studied in Theorem 11.4.1) will be Hermitian. It can also be shown that in this case the perturbed eigenvalues are analytic in ζ, whatever the multiplicity of the unperturbed eigenvalue may be. The perturbed component matrices are also analytic and hence the eigenvectors of $A(\zeta)$ can be supposed to be analytic and orthonormal throughout a neighborhood of $\zeta = 0$. These important results were first proved by F. Rellich in the early 1930s. Rellich's first proof depends primarily on the fact that the perturbed eigenvalues are real. We shall prove a result (Theorem 2) that contains these results by means of a more recent technique due to Kato. (See the references in Appendix 3.)

With this preamble, we now return to a perturbation problem, making no assumption with regard to symmetry of the matrix $A(\zeta)$. Until we make some hypothesis concerning the symmetry of $A(\zeta)$, we retain the assumption of dependence on a complex parameter ζ. The formulation of corresponding results for the case of a real variable ζ is left as an exercise. We first prove that if λ is an eigenvalue of A of index 1 and $\lambda_j(\zeta)$ is an eigenvalue of $A(\zeta)$ for which $\lambda(\zeta) \to \lambda$ as $\zeta \to 0$, then, as $\zeta \to 0$,

$$\lambda_j(\zeta) = \lambda + \alpha_j \zeta + O(|\zeta|^{1+(1/l)}).$$

Thus, in the Puiseux expansion for $\lambda_j(\zeta)$ described in result (b) of Section 11.3, the coefficients of $\zeta^{j/l}$ are zero for $j = 1, 2, \ldots, l - 1$.

As in Section 11.4, the eigenvalue λ of $A = A(0)$ is supposed to "split" into eigenvalues $\lambda_1(\zeta), \ldots, \lambda_p(\zeta)$ for $|\zeta|$ sufficiently small. Also $A(\zeta) = \sum_{n=0}^{\infty} \zeta^n A^{(n)}$, the matrix Z is the component matrix of λ, and $Y(\zeta)$ is the sum of component matrices $Z_{k0}(\zeta)$ of $A(\zeta)$ for $1 \le k \le p$. Thus, $Y(\zeta) \to Z$ as $|\zeta| \to 0$.

Lemma 1. *With the conventions and hypotheses of the previous paragraphs, the function*

$$B(\zeta) \triangleq \zeta^{-1}(\lambda I - A(\zeta))Y(\zeta) \tag{1}$$

has a Taylor expansion about $\zeta = 0$ of the form

$$B(\zeta) = -ZA^{(1)}Z + \sum_{n=1}^{\infty} \zeta^n B^{(n)}.$$

PROOF. Using Eq. (11.4.1) and Exercise 9.9.6, we may write

$$Y(\zeta) = \frac{1}{2\pi i}\int_L R_z(\zeta)\,dz, \qquad A(\zeta)Y(\zeta) = \frac{1}{2\pi i}\int_L zR_z(\zeta)\,dz,$$

where, as in the proof of Theorem 11.4.1, the contour L is simple and closed and contains $\lambda, \lambda_1(\zeta), \ldots, \lambda_p(\zeta)$ and no other eigenvalues of A or $A(\zeta)$. Here, ζ is fixed and $|\zeta|$ is assumed to be small enough to complete this (and subsequent) constructions. Using Eq. (11.4.2) it follows that

$$\begin{aligned}(\lambda I - A(\zeta))Y(\zeta) &= -\frac{1}{2\pi i}\int_L (z - \lambda)R_z(\zeta)\,dz \\ &= -\frac{1}{2\pi i}\int_L (z - \lambda)\sum_{r=0}^{\infty} \zeta^r R_z^{(r)}\,dz.\end{aligned}$$

Now, $Y(0) = Z$ and, since λ has index 1, $(\lambda I - A)Z = O$. Hence the coefficient of ζ^0 on both sides of the last equation must vanish and we find that

$$B(\zeta) = \zeta^{-1}(\lambda I - A(\zeta))Y(\zeta) = \sum_{r=0}^{\infty} \zeta^r B^{(r)},$$

where, for $r = 0, 1, 2, \ldots,$

$$B^{(r)} = -\frac{1}{2\pi i}\int_L (z - \lambda)R_z^{(r+1)}\,dz$$

and is independent of ζ. This confirms the existence of the Taylor expansion for $B(\zeta)$.

To examine the leading coefficient we first take advantage of Eq. (11.4.3) and Exercise 9.5.2 to write

$$R_z^{(1)} = (zI - A)^{-1}A^{(1)}(zI - A)^{-1}$$
$$= \left(\frac{Z}{z-\lambda} + E(z)\right)A^{(1)}\left(\frac{Z}{z-\lambda} + E(z)\right).$$

Hence, using Cauchy's theorem of residues,

$$B^{(0)} = -\frac{1}{2\pi i}\int_L \left\{\frac{ZA^{(1)}Z}{z-\lambda} + (\text{function analytic near } z = \lambda)\right\} dz.$$
$$= -ZA^{(1)}Z. \quad \blacksquare$$

Let $\mathbf{x}_j(\zeta)$ be a right eigenvector of $A(\zeta)$ associated with one of the eigenvalues $\lambda_j(\zeta)$, $1 \le j \le p$. Then $Y(\zeta)\mathbf{x}_j(\zeta) = \mathbf{x}_j(\zeta)$ and we see immediately from Eq. (1) that

$$B(\zeta)\mathbf{x}_j(\zeta) = \zeta^{-1}(\lambda - \lambda_j(\zeta))\mathbf{x}_j(\zeta). \tag{2}$$

Thus, $\zeta^{-1}(\lambda - \lambda_j(\zeta))$ is an eigenvalue of $B(\zeta)$, say $\beta(\zeta)$. But, since $B(\zeta)$ has a Taylor expansion at $\zeta = 0$, $\beta(\zeta)$ also has an expansion in fractional powers of ζ. Consequently, for small enough $|\zeta|$ and some integer l,

$$\zeta^{-1}(\lambda - \lambda_j(\zeta)) = \beta_0 + \beta_1\zeta^{1/l} + \beta_2\zeta^{2/l} + \cdots,$$

and β_0 is an eigenvalue of the unperturbed matrix $-ZA^{(1)}Z$. This implies

$$\lambda_j(\zeta) = \lambda - \beta_0\zeta - \beta_1\zeta^{1+(1/l)} - \beta_2\zeta^{1+(2/l)} - \cdots. \tag{3}$$

In other words,

$$\lambda_j(\zeta) = \lambda + a_j\zeta + O(|\zeta|^{1+(1/l)})$$

as $|\zeta| \to 0$, and a_j is an eigenvalue of $ZA^{(1)}Z$.

We have proved the first part of the following theorem.

Theorem 1. *Let $A(\zeta)$ be a matrix that is analytic in ζ on a neighborhood of $\zeta = 0$, and suppose $A(0) = A$. Let λ be an eigenvalue of A of index 1 and multiplicity m, and let $\lambda_j(\zeta)$ be an eigenvalue of $A(\zeta)$ for which $\lambda_j(0) = \lambda$. Then there is a number a_j and a positive integer $l \le m$ such that*

$$\lambda_j(\zeta) = \lambda + a_j\zeta + 0(|\zeta|^{1+(1/l)}) \tag{4}$$

as $|\zeta| \to 0$, and a_j is an eigenvalue of $ZA^{(1)}Z$.

Moreover, to each $\lambda_j(\zeta)$ corresponds at least one eigenvector $\mathbf{x}_j(\zeta)$ such that, for small enough $|\zeta|$

$$\mathbf{x}_j(\zeta) = \mathbf{x} + \zeta^{1/l}\mathbf{x}_1 + \zeta^{2/l}\mathbf{x}_2 + \cdots, \tag{5}$$

where $\mathbf{x}, \mathbf{x}_1, \ldots, \mathbf{x}_{l-1} \in \operatorname{Ker}(A - \lambda I)$ and $a_j\mathbf{x} = ZA^{(1)}Z\mathbf{x}$.

PROOF. It remains to prove only the statements concerning eigenvectors. But these follow immediately on writing down $A(\zeta)\mathbf{x}_j(\zeta) = \lambda_j(\zeta)\mathbf{x}_j(\zeta)$ and comparing coefficients of powers of $\zeta^{1/l}$. ■

Now it is important to observe that if a_j is an eigenvalue of $ZA^{(1)}Z$ with index 1, then the process can be repeated, that is, we can apply the above analysis to $B(\zeta)$ instead of $A(\zeta)$ and find that in expansions (3), the coefficients $\beta_1, \beta_2, \ldots, \beta_{l-1}$ all vanish. This observation will be exploited in Exercise 3 below.

Exercise 1. Find $ZA^{(1)}Z$ for the matrix $A(\zeta)$ given in Exercise 11.4.2 and show that the result of Theorem 1 is consistent with the expansions for $\lambda_1(\zeta)$ and $\lambda_2(\zeta)$ given in that example.

Exercise 2. Verify the results of Theorem 1 for the matrix $A(\zeta)$ of Exercise 11.4.1 and for the matrix

$$\begin{bmatrix} 0 & \zeta - 1 \\ \zeta(\zeta - 1) & 0 \end{bmatrix}.$$

Exercise 3. Prove that if a_j is an eigenvalue of $ZA^{(1)}Z$ of index 1, then there is a number b_j such that

$$\lambda_j(\zeta) = \lambda + a_j\zeta + b_j\zeta^2 + O(|\zeta|^{2+(1/l)}), \tag{6}$$

and that b_j is an eigenvalue of $Z(A^{(1)}EA^{(1)} + A^{(2)})Z$ with right eigenvector $\mathbf{x}$. □

Consider the matrix function $B(\zeta)$ of Lemma 1 once more. Since $Y(\zeta)$ is just a function of $A(\zeta)$ (for a fixed ζ), $A(\zeta)$ and $Y(\zeta)$ commute. Also $Y(\zeta)^2 = Y(\zeta)$ so we can write, instead of Eq. (1),

$$B(\zeta) = \zeta^{-1}Y(\zeta)(\lambda I - A(\zeta))Y(\zeta), \tag{7}$$

and this representation has a clear symmetry. In particular, if A and $A^{(1)}$ happen to be Hermitian, then $Y(0) = Z$ and hence $B(0)$ is also Hermitian. Thus, $B(0)$ has all eigenvalues of index one and repetition of the process described above is legitimate, thus guaranteeing dependence of the perturbed eigenvalues on ζ like that described by Eq. (6). Furthermore, it looks as though we can repeat the process indefinitely, if all the derivatives of $A(\zeta)$ at $\zeta = 0$ are Hermitian, and we conclude that $\lambda_j(\zeta)$ has a Taylor expansion about $\zeta = 0$.

But once we bring symmetry into the problem, the considerations of the preamble to this section come into play. Thus, it will now be assumed that *ζ is a real parameter*, and that $A(\zeta)$ is normal for every real ζ in some open interval containing $\zeta = 0$. Now it follows that whenever $A(\zeta)$ is normal, $Y(\zeta)$ and Z are Hermitian (i.e., they are *orthogonal* projectors), and $B(\zeta)$

is also normal. In particular, $B(0) = -ZA^{(1)}Z$ is normal and all eigenvalues of $B(0)$ have index 1. So a second step of the reduction process can be taken and, in fact, repeated indefinitely. This leads to the following conclusion.

Theorem 2. *Let $A(\zeta)$ be analytic and normal throughout a real neighborhood of $\zeta = 0$; then the eigenvalues of $A(\zeta)$ are analytic in some real neighborhood of $\zeta = 0$.*

Moreover, to each eigenvalue $\lambda_j(\zeta)$ analytic on a real neighborhood of $\zeta = 0$ corresponds at least one eigenvector $\mathbf{x}_j(\zeta)$ that is analytic on the same neighborhood.

Exercise 4. Show that if ζ is real and $A(\zeta)$ is normal, then $B(\zeta)$ is normal.

Exercise 5. Let $A(\zeta)$ be analytic and Hermitian throughout a real neighborhood of $\zeta = 0$, and suppose that all eigenvalues $\lambda_1(0), \ldots, \lambda_n(0)$ of $A(0)$ are distinct. Show that there is a real neighborhood $\mathcal{N}$ of $\zeta = 0$, real functions $\lambda_1(\zeta), \ldots, \lambda_n(\zeta)$ that are analytic on $\mathcal{N}$, and a matrix-valued function $U(\zeta)$ that is analytic on $\mathcal{N}$ such that, for $\zeta \in \mathcal{N}$,

$$A(\zeta) = U(\zeta)^{-1}\mathrm{diag}[\lambda_1(\zeta), \ldots, \lambda_n(\zeta)]U(\zeta)$$

and

$$U(\zeta)U(\zeta)^* = I.$$

(This result also holds without the hypothesis of distinct unperturbed eigenvalues.) □

There are expansions of the form (5) for eigenvectors even when the unperturbed eigenvalue has index greater than 1. Indeed, Exercise 6 demonstrates the existence of a continuous eigenvector. However, even this cannot be guaranteed if the hypothesis of *analytic* dependence on the parameter is removed. This is illustrated in Exercise 7, due to F. Rellich, in which the matrix is infinitely differentiable at $\zeta = 0$ but not analytic, and it has *no* eigenvector at $\zeta = 0$ that can be extended in a continuous way to eigenvectors for nonzero values of ζ.

Exercise 6. Adopt the notation and hypotheses of Theorem 1, except that the eigenvalue λ of A may have any index m. Show that there is an eigenvector $\mathbf{x}_j(\zeta)$ of $A(\zeta)$ such that the expansion (5) holds for $|\zeta|$ small enough.

SOLUTION. Using statement (b) of Section 11.3, we may assume that there is an eigenvalue $\lambda_j(\zeta)$ of $A(\zeta)$ that has a power series expansion in powers of $\zeta^{1/l}$ for $|\zeta|$ small enough. Then the matrix $D(\zeta) = \lambda_j(\zeta)I - A(\zeta)$ has elements with power series expansions of the same kind.

Suppose that $D(\zeta)$ has rank r in a deleted neighborhood $\mathcal{N}$ of $\zeta = 0$ (i.e., excluding the point $\zeta = 0$ itself). For simplicity, and without loss of generality, assume that the rth-order minor

$$D\begin{pmatrix} 1 & 2 & \cdots & r \\ 1 & 2 & \cdots & r \end{pmatrix}$$

is nonzero in $\mathcal{N}$. Then expand the minor

$$D\begin{pmatrix} 1 & 2 & \cdots & r & r+1 \\ 1 & 2 & \cdots & r & r+1 \end{pmatrix}$$

by its last row, and write $l_1, \ldots, l_{r+1}$ for the cofactors of $d_{r+1,1}, \ldots, d_{r+1,r+1}$. Thus

$$D\begin{pmatrix} 1 & 2 & \cdots & r+1 \\ 1 & 2 & \cdots & r+1 \end{pmatrix} = \sum_{j=1}^{r+1} d_{r+1,j} l_j = 0.$$

Define $l_{r+2} = \cdots = l_n = 0$ (for any ζ in $\mathcal{N}$) and then write the vector $\boldsymbol{l} = [l_1 \;\; l_2 \;\; \cdots \;\; l_n]^{\mathrm{T}}$.

Obviously, $\boldsymbol{l}$ is a function of ζ with a power series expansion in $\zeta^{1/l}$ for $|\zeta|$ small enough. Furthermore, $D(\zeta)\boldsymbol{l}(\zeta) = \boldsymbol{0}$, so that $\boldsymbol{l}(\zeta)$ is an eigenvector with the required property. To see this, let D_{j*} be the jth row of $D(\zeta)$. Then

$$D_{j*}\boldsymbol{l} = \sum_{k=1}^{n} d_{jk} l_k = D\begin{pmatrix} 1 & 2 & \cdots & r & j \\ 1 & 2 & \cdots & r & r+1 \end{pmatrix}.$$

This is zero for each j because if $1 \le j \le r$ we have a determinant with two equal rows, and if $j \ge r + 1$ we have a minor of D of order $r + 1$.

Exercise 7 (Rellich). Verify that the matrix $A(x)$ defined for all real x by

$$A(0) = 0, \qquad A(x) = e^{-1/x^2}\begin{bmatrix} \cos(2/x) & \sin(2/x) \\ \sin(2/x) & -\cos(2/x) \end{bmatrix}$$

is continuous (even infinitely differentiable) at $x = 0$ and that there is no eigenvector $\boldsymbol{v}(x)$ continuous near $x = 0$ with $\boldsymbol{v}(0) \neq \boldsymbol{0}$. □

CHAPTER 12

Linear Matrix Equations and Generalized Inverses

In this chapter we consider the general linear matrix equation

$$A_1XB_1 + A_2XB_2 + \cdots + A_pXB_p = C,$$

for the unknown matrix X; we will also consider its particular cases, especially the equation of great theoretical and practical importance:

$$AX + XB = C.$$

Our investigation of these matrix equations will depend on the idea of the "Kronecker product" of two matrices, a useful notion that has several important applications.

In the second half of the chapter some generalizations of the notion of the inverse of a square matrix are developed. Beginning with "one-sided" inverses, we go on to a widely useful definition of a generalized inverse that is applicable to any matrix, whether square or rectangular. The Moore–Penrose inverse appears as a special case of this analysis, and the chapter concludes with the application of the Moore–Penrose inverse to the solution of $A\mathbf{x} = \mathbf{b}$, where A and $\mathbf{b}$ are an arbitrary (rectangular) matrix and vector, respectively.

12.1 The Notion of a Kronecker Product

In this section we introduce and examine a new operation that, in full generality, is a binary operation from $\mathscr{F}^{m \times l} \times \mathscr{F}^{n \times k}$ to $\mathscr{F}^{mn \times lk}$, although we shall confine our attention to the case $l = m$, $k = n$. This is the most useful case in applications and yields beautiful results concerning the eigenvalues of the product matrix in terms of the factors. The idea of the Kronecker product arises naturally in group theory and, as a result, has important applications in particle physics.

If $A = [a_{ij}]_{i,j=1}^{m} \in \mathscr{F}^{m \times m}$, $B = [b_{ij}]_{i,j=1}^{n} \in \mathscr{F}^{n \times n}$, then the *right Kronecker* (or *direct*, or *tensor*) *product* of A and B, written $A \otimes B$, is defined to be the partitioned matrix

$$A \otimes B = \begin{bmatrix} a_{11}B & a_{12}B & \cdots & a_{1m}B \\ a_{21}B & a_{22}B & \cdots & a_{2m}B \\ \vdots & \vdots & & \vdots \\ a_{m1}B & a_{m2}B & \cdots & a_{mm}B \end{bmatrix} = [a_{ij}B]_{i,j=1}^{m} \in \mathscr{F}^{mn \times mn}.$$

For example, if

$$A = \begin{bmatrix} a_{11} & a_{12} \\ a_{21} & a_{22} \end{bmatrix}, \qquad B = \begin{bmatrix} b_{11} & b_{12} \\ b_{21} & b_{22} \end{bmatrix},$$

then

$$A \otimes B = \begin{bmatrix} a_{11}b_{11} & a_{11}b_{12} & a_{12}b_{11} & a_{12}b_{12} \\ a_{11}b_{21} & a_{11}b_{22} & a_{12}b_{21} & a_{12}b_{22} \\ a_{21}b_{11} & a_{21}b_{12} & a_{22}b_{11} & a_{22}b_{12} \\ a_{21}b_{21} & a_{21}b_{22} & a_{22}b_{21} & a_{22}b_{22} \end{bmatrix}.$$

It should be noted that the *left Kronecker product* of A and B, defined to be the matrix

$$\begin{bmatrix} Ab_{11} & \cdots & Ab_{1n} \\ \vdots & & \vdots \\ Ab_{n1} & \cdots & Ab_{nn} \end{bmatrix},$$

has similar properties to those of $A \otimes B$, but to avoid unnecessary duplication, only the right Kronecker product, called merely the Kronecker product, will be discussed.

Exercise 1. Verify that for any square matrix $A = [a_{ij}]_{i,j=1}^{m}$,

(a) $I_n \otimes A = \operatorname{diag}[A, A, \ldots, A]$,

(b) $A \otimes I_n = \begin{bmatrix} a_{11}I_n & a_{12}I_n & \cdots & a_{1m}I_n \\ \vdots & \vdots & & \vdots \\ a_{m1}I_n & a_{m2}I_n & \cdots & a_{mm}I_n \end{bmatrix}$;

(c) $I_m \otimes I_n = I_{mn}$. □

The properties of the Kronecker product incorporated in Proposition 1 follow immediately from the definition.

Proposition 1. *If the orders of the matrices involved are such that all the operations below are defined, then*

(a) *If* $\mu \in \mathcal{F}$, $(\mu A) \otimes B = A \otimes (\mu B) = \mu(A \otimes B)$;
(b) $(A + B) \otimes C = (A \otimes C) + (B \otimes C)$;
(c) $A \otimes (B + C) = (A \otimes B) + (A \otimes C)$;
(d) $A \otimes (B \otimes C) = (A \otimes B) \otimes C$;
(e) $(A \otimes B)^{\mathrm{T}} = A^{\mathrm{T}} \otimes B^{\mathrm{T}}$.

We need a further property that is less obvious and will be found very useful.

Proposition 2. *If* $A, C \in \mathcal{F}^{m \times m}$ *and* $B, D \in \mathcal{F}^{n \times n}$, *then*

$$(A \otimes B)(C \otimes D) = AC \otimes BD. \tag{1}$$

PROOF. Write $A \otimes B = [a_{ij}B]_{i,j=1}^{m}$, $C \otimes D = [c_{ij}D]_{i,j=1}^{m}$. The i, jth block of $(A \otimes B)(C \otimes D)$ is

$$F_{ij} = \sum_{k=1}^{m} (a_{ik}B)(c_{kj}D) = \sum_{k=1}^{m} a_{ik}c_{kj}BD, \qquad \text{for} \quad 1 \le i, j \le m.$$

On the other hand, by the definition of the Kronecker product, $AC \otimes BD = [\gamma_{ij}BD]_{i,j=1}^{m}$, where γ_{ij} is the i, jth element of AC and is given by the formula $\gamma_{ij} = \sum_{k=1}^{m} a_{ik}c_{kj}$. Thus $F_{ij} = \gamma_{ij}BD$ for each possible pair (i, j), and the result follows. ■

Proposition 2 has some important consequences, which we list in the following corollaries.

Corollary 1. *If* $A \in \mathcal{F}^{m \times m}$, $B \in \mathcal{F}^{n \times n}$, *then*

(a) $A \otimes B = (A \otimes I_n)(I_m \otimes B) = (I_m \otimes B)(A \otimes I_n)$;
(b) $(A \otimes B)^{-1} = A^{-1} \otimes B^{-1}$, *provided that* A^{-1} *and* B^{-1} *exist.*

Note that the assertion of Corollary 1(a) states the commutativity of $A \otimes I_n$ and $I_m \otimes B$.

Corollary 2. *If* $A_1, A_2, \ldots, A_p \in \mathcal{F}^{m \times m}$ *and* $B_1, B_2, \ldots, B_p \in \mathcal{F}^{n \times n}$, *then*

$$(A_1 \otimes B_1)(A_2 \otimes B_2) \cdots (A_p \otimes B_p) = (A_1 A_2 \cdots A_p) \otimes (B_1 B_2 \cdots B_p).$$

The Kronecker product is, in general, not commutative. However, the next Proposition shows that $A \otimes B$ and $B \otimes A$ are intimately connected.

Proposition 3. *If* $A \in \mathcal{F}^{m \times m}$, $B \in \mathcal{F}^{n \times n}$, *then there exists a permutation matrix* $P \in \mathbb{R}^{mn \times mn}$ *such that*

$$P^{\mathrm{T}}(A \otimes B)P = B \otimes A.$$

PROOF. First observe that by a simultaneous interchange of rows and columns with the same number, the matrix $A \otimes I_n$ can be reduced to the matrix $I_n \otimes A$. Thus, there exists a permutation matrix P such that $P^T(A \otimes I_n)P = I_n \otimes A$. In fact, if the permutation of $1, 2, 3, \ldots, mn$, which is equal to

$$1, n+1, 2n+1, \ldots, (m-1)n+1, 2, n+2, 2n+2, \ldots, (m-1)n+2, \ldots, n, 2n, 3n, \ldots, (m-1)n, mn,$$

is denoted by $i_1, i_2, \ldots, i_{mn}$, then the permutation P has the unit coordinate vectors $e_{i_1}, e_{i_2}, \ldots, e_{i_{mn}}$ for its columns. Now observe that the *same* matrix P performs the transformation $P^T(I_m \otimes B)P = B \otimes I_m$. Since $PP^T = I$, then by a repeated use of Corollary 1(a), it is found that

$$\begin{aligned} P^T(A \otimes B)P &= P^T(A \otimes I_n)PP^T(I_m \otimes B)P \\ &= (I_n \otimes A)(B \otimes I_m) = B \otimes A, \end{aligned}$$

and the proposition is established. ■

Exercise 2. If $A \in \mathscr{F}^{m \times m}$, $B \in \mathscr{F}^{n \times n}$, prove that

$$\det(A \otimes B) = (\det A)^n (\det B)^m.$$

SOLUTION. In view of Corollary 1(a), it is sufficient to find the determinants of the matrices $A \otimes I_n$ and $I_m \otimes B$. By Exercise 1(a), $\det(I_m \otimes B) = (\det B)^m$. Using Proposition 3 and the property $\det P = \pm 1$, we have

$$\det(A \otimes I_n) = (\det P^T)\det(I_n \otimes A)(\det P) = (\det A)^n,$$

and the result follows. □

Now we define a vector-valued function associated with a matrix and closely related to the Kronecker product. For a matrix $A \in \mathscr{F}^{m \times n}$ write $A = [A_{*1} \quad A_{*2} \quad \cdots \quad A_{*n}]$, where $A_{*j} \in \mathscr{F}^m, j = 1, 2, \ldots, n$. Then the vector

$$\begin{bmatrix} A_{*1} \\ A_{*2} \\ \vdots \\ A_{*n} \end{bmatrix} \in \mathscr{F}^{mn}$$

is said to be the *vec-function* of A and is written vec A. It is the vector formed by "stacking" the columns of A into one long vector. Note that the vec-function is linear:

$$\text{vec}(\alpha A + \beta B) = \alpha \text{ vec } A + \beta \text{ vec } B,$$

for any $A, B \in \mathscr{F}^{m \times n}$ and $\alpha, \beta \in \mathscr{F}$. Also, the matrices $A_1, A_2, \ldots, A_k$ from $\mathbb{C}^{m \times n}$ are linearly independent as members of $\mathscr{F}^{m \times n}$ if and only if vec A_1, vec $A_2, \ldots,$ vec A_k are linearly independent members of $\mathscr{F}^{mn}$. The next result indicates a close relationship between the vec-function and the Kronecker product.

Proposition 4. *If $A \in \mathscr{F}^{m \times m}$, $B \in \mathscr{F}^{n \times n}$, $X \in \mathscr{F}^{m \times n}$, then*

$$\operatorname{vec}(AXB) = (B^{\mathrm{T}} \otimes A) \operatorname{vec} X.$$

PROOF. For $j = 1, 2, \ldots, n$, the j-column $(AXB)_{*j}$ can be expressed

$$(AXB)_{*j} = AXB_{*j} = \sum_{k=1}^{n} b_{kj}(AX)_{*k} = \sum_{k=1}^{m} (b_{kj}A)X_{*k},$$

where b_{kj} is the element of B in the k, jth position. Consequently,

$$(AXB)_{*j} = [b_{1j}A \quad b_{2j}A \quad \cdots \quad b_{nj}A] \operatorname{vec} X,$$

and it remains to observe that the expression in brackets is the jth block-row in $B^{\mathrm{T}} \otimes A$. ■

Corollary 1. *With the previous notation,*

(a) $\operatorname{vec}(AX) = (I_n \otimes A) \operatorname{vec} X;$
(b) $\operatorname{vec}(XB) = (B^{\mathrm{T}} \otimes I_m) \operatorname{vec} X;$
(c) $\operatorname{vec}(AX + XB) = ((I_n \otimes A) + (B^{\mathrm{T}} \otimes I_m)) \operatorname{vec} X.$

Important applications of these results will be considered in Section 3. We conclude with a few more properties of the Kronecker product.

Exercise 3. If $A \in \mathscr{F}^{m \times m}$, $B \in \mathscr{F}^{n \times n}$, prove that

(a) $\operatorname{tr}(A \otimes B) = (\operatorname{tr} A)(\operatorname{tr} B);$
(b) $\operatorname{rank}(A \otimes B) = (\operatorname{rank} A)(\operatorname{rank} B).$

Hint. For part (a) use the representation $A \otimes B = [a_{ij}B]_{i,j=1}^{m}$. Use Proposition 2 for part (b).

Exercise 4. Write $A^{[1]} = A$ and define $A^{[k+1]} = A \otimes A^{[k]}$, $k = 1, 2, \ldots$. Prove that

(a) $A^{[k+l]} = A^{[k]}A^{[l]};$
(b) If $A, B \in \mathscr{F}^{n \times n}$, then $(AB)^{[k]} = A^{[k]}B^{[k]}$.

Note the contrast with $(AB)^k$.

Exercise 5. Verify that

(a) The Kronecker product of diagonal matrices is a diagonal matrix;
(b) The Kronecker product of upper- (respectively, lower-) triangular matrices is an upper- (respectively, lower-) triangular matrix.

Exercise 6. Prove that, if A_1, B_1 are similar to the matrices A_2, B_2, respectively, then $A_1 \otimes B_1$ is similar to $A_2 \otimes B_2$. In particular, if A_1 and B_1 are simple, then so is $A_1 \otimes B_1$.

Exercise 7. If $A \in \mathscr{F}^{m \times m}$, prove that, for any function $f(\lambda)$ defined on the spectrum of A,

(a) $f(I_n \otimes A) = I_n \otimes f(A)$;
(b) $f(A \otimes I_n) = f(A) \otimes I_n$. □

12.2 Eigenvalues of Kronecker Products and Composite Matrices

One of the main reasons for interest in the Kronecker product is a beautifully simple connection between the eigenvalues of the matrices A and B and $A \otimes B$. We obtain this relation from a more general result concerning composite matrices of Kronecker products.

Consider a polynomial p in two variables, x and y, with complex coefficients. Thus, for certain complex numbers c_{ij} and a positive integer l,

$$p(x, y) = \sum_{i,j=0}^{l} c_{ij} x^i y^j.$$

If $A \in \mathbb{C}^{m \times m}$ and $B \in \mathbb{C}^{n \times n}$, we consider matrices in $\mathbb{C}^{mn \times mn}$ of the form

$$p(A; B) = \sum_{i,j=0}^{l} c_{ij} A^i \otimes B^j.$$

For example, if $p(x, y) = 2x + xy^3$, we write $2x + xy^3 = 2x^1y^0 + x^1y^3$ and

$$p(A; B) = 2A \otimes I_n + A \otimes B^3.$$

The next theorem gives a connection between the eigenvalues of A and B and those of $p(A; B)$.

Theorem 1 (C. Stephanos[†]). *If $\lambda_1, \ldots, \lambda_m$ are the eigenvalues of $A \in \mathbb{C}^{m \times m}$ and $\mu_1, \ldots, \mu_n$ are the eigenvalues of $B \in \mathbb{C}^{n \times n}$, then the eigenvalues of $p(A; B)$ are the mn numbers $p(\lambda_r, \mu_s)$, where $r = 1, 2, \ldots, m$ and $s = 1, 2, \ldots, n$.*

PROOF. Let matrices $P \in \mathbb{C}^{m \times m}$ and $Q \in \mathbb{C}^{n \times n}$ be defined so that

$$PAP^{-1} = J_1, \qquad QBQ^{-1} = J_2,$$

where J_1 and J_2 are matrices in Jordan normal form (Section 6.5). It is clear that J_1^i is an upper-triangular matrix having $\lambda_1^i, \ldots, \lambda_m^i$ as its main diagonal

[†] *Jour. Math. pures appl.* *V* **6** (1900), 73–128.

elements and similarly, $\mu_1^j, \ldots, \mu_n^j$ are the main diagonal elements of the upper-triangular matrix J_2^j.

Furthermore, from Exercise 12.1.5 it follows that $J_1^i \otimes J_2^j$ is also an upper-triangular matrix. It is easily checked that its diagonal elements are $\lambda_r^i \mu_s^j$ for $r = 1, 2, \ldots, m$, $s = 1, 2, \ldots, n$. Hence the matrix $p(J_1; J_2)$ is upper-triangular and has diagonal elements $p(\lambda_r, \mu_s)$. But the diagonal elements of an upper-triangular matrix are its eigenvalues; hence $p(J_1; J_2)$ has eigenvalues $p(\lambda_r, \mu_s)$ for $r = 1, 2, \ldots, m$, $s = 1, 2, \ldots, n$.

Now we have only to show that $p(J_1; J_2)$ and $p(A; B)$ have the same eigenvalues. On applying Corollary 2 to Proposition 12.1.2, we find that

$$\begin{aligned} J_1^i \otimes J_2^j &= PA^iP^{-1} \otimes QB^jQ^{-1} \\ &= (P \otimes Q)(A^i \otimes B^j)(P^{-1} \otimes Q^{-1}), \end{aligned}$$

and by Corollary 1 to Proposition 12.1.2, $P^{-1} \otimes Q^{-1} = (P \otimes Q)^{-1}$. Thus,

$$J_1^i \otimes J_2^j = (P \otimes Q)(A^i \otimes B^j)(P \otimes Q)^{-1},$$

and hence

$$p(J_1; J_2) = (P \otimes Q)p(A; B)(P \otimes Q)^{-1}.$$

This shows that $p(J_1; J_2)$ and $p(A; B)$ are similar and so they must have the same eigenvalues (Section 4.10). ■

The two following special cases (for $p(x, y) = xy$ and $p(x, y) = x + y$) are probably the most frequently used.

Corollary 1. *The eigenvalues of* $A \otimes B$ *are the mn numbers* $\lambda_r \mu_s$, $r = 1, 2, \ldots, m$, $s = 1, 2, \ldots, n$.

Corollary 2. *The eigenvalues of* $(I_n \otimes A) + (B \otimes I_m)$ *(or, what is equivalent in view of Exercise 4.11.9, of* $(I_n \otimes A^T) + (B^T \otimes I_m)$*) are the mn numbers* $\lambda_r + \mu_s$, $r = 1, 2, \ldots, m$, $s = 1, 2, \ldots, n$.

The matrix $(I_n \otimes A) + (B \otimes I_m)$ is often called the *Kronecker sum* of A and B.

Exercise 1. If $C \in \mathscr{F}^{mn \times mn}$ is the Kronecker sum of $A \in \mathscr{F}^{m \times m}$ and $B \in \mathscr{F}^{n \times n}$, prove that $e^C = e^A \otimes e^B$ and indicate a corresponding result in numbers.

Hint. Use Exercise 12.1.7 for $f(\lambda) = e^\lambda$.

Exercise 2. Let $A \in \mathbb{C}^{n \times n}$ and let $B \in \mathbb{C}^{mn \times mn}$ be the partitioned matrix with each of its m^2 blocks equal to A. If A has eigenvalues $\mu_1, \ldots, \mu_n$ (not necessarily distinct), prove that the eigenvalues of B are $m\mu_1, \ldots, m\mu_n$ and $m(n - 1)$ zeros.

Hint. Express B as a Kronecker product.

Exercise 3. If $\boldsymbol{x} = [\alpha_1 \quad \alpha_2 \quad \cdots \quad \alpha_m]^{\mathrm{T}} \in \mathbb{C}^m$ is an eigenvector of $A \in \mathbb{C}^{m \times m}$ corresponding to the eigenvalue λ, and $\boldsymbol{y} \in \mathbb{C}^n$ is an eigenvector of $B \in \mathbb{C}^{n \times n}$ corresponding to the eigenvalue μ, show that an eigenvector $\boldsymbol{z}$ of $A \otimes B$ associated with $\lambda\mu$ is

$$\boldsymbol{z} = [\alpha_1 \boldsymbol{y}^{\mathrm{T}} \quad \alpha_2 \boldsymbol{y}^{\mathrm{T}} \quad \cdots \quad \alpha_m \boldsymbol{y}^{\mathrm{T}}]^{\mathrm{T}}. \tag{1}$$

Exercise 4. Show that if A and B are positive definite matrices, then $A \otimes B$ is positive definite. □

Define the (right) Kronecker product of two *rectangular* matrices $A = [a_{ij}]_{i,j=1}^{m,l} \in \mathscr{F}^{m \times l}$ and $B \in \mathscr{F}^{n \times k}$ to be the matrix

$$A \otimes B \triangleq [a_{ij} B]_{i,j=1}^{m,l} \in \mathscr{F}^{mn \times lk}. \tag{2}$$

Exercise 5. Establish the assertions of Propositions 12.1.1 and 12.1.2 for rectangular matrices.

Exercise 6. Let $\boldsymbol{x}, \boldsymbol{y} \in \mathbb{C}^m$ and $\boldsymbol{u}, \boldsymbol{v} \in \mathbb{C}^n$. Show that the equation $\boldsymbol{x} \otimes \boldsymbol{u} = \boldsymbol{y} \otimes \boldsymbol{v}$ implies that $\boldsymbol{y} = \lambda \boldsymbol{x}$ and $\boldsymbol{v} = \lambda \boldsymbol{u}$ for some $\lambda \in \mathbb{C}$. □

Observe that, using the general definition of a Kronecker product of rectangular matrices (see Eq. (2)), we may write $\boldsymbol{z}$ in Eq. (1) as $\boldsymbol{z} = \boldsymbol{x} \otimes \boldsymbol{y}$ and use the generalized Proposition 12.1.1 (see Exercise 5) for proving Exercise 3. Furthermore, if A and B are simple, so is $A \otimes B$ (Exercise 12.1.6) and hence there is an eigenbasis for $\mathbb{C}^{mn \times mn}$ with respect to $A \otimes B$ consisting entirely of the Kronecker products $\boldsymbol{x}_i \otimes \boldsymbol{y}_j$, where $\{\boldsymbol{x}_i\}_{i=1}^m$, $\{\boldsymbol{y}_j\}_{j=1}^n$ are the eigenbases in $\mathbb{C}^m$ and $\mathbb{C}^n$ for A and B, respectively. In the general case the situation is more complicated.

12.3 Applications of the Kronecker Product to Matrix Equations

We start by studying the general linear matrix equation

$$A_1 X B_1 + A_2 X B_2 + \cdots + A_p X B_p = C, \tag{1}$$

where $A_j \in \mathbb{C}^{m \times m}$, $B_j \in \mathbb{C}^{n \times n}$ $(j = 1, 2, \ldots, p)$, $X, C \in \mathbb{C}^{m \times n}$, and its particular cases. The method we use relies on the assertion of Proposition 12.1.4 and reduces Eq. (1) to a matrix–vector equation of the form $G\boldsymbol{x} = \boldsymbol{c}$, where $G \in \mathbb{C}^{mn \times mn}$ and $\boldsymbol{x}, \boldsymbol{c} \in \mathbb{C}^{mn}$.

Theorem 1. *A matrix $X \in \mathbb{C}^{m \times n}$ is a solution of Eq. (1) if and only if the vector $\mathbf{x} = \operatorname{vec} X$ defined in Section 12.1 is a solution of the equation*

$$G\mathbf{x} = \mathbf{c}, \tag{2}$$

with $G = \sum_{j=1}^{p} (B_j^{\mathrm{T}} \otimes A_j)$ and $\mathbf{c} = \operatorname{vec} C$.

PROOF. By Proposition 12.1.4, $\operatorname{vec}(A_j X B_j) = (B_j^{\mathrm{T}} \otimes A_j) \operatorname{vec} X$ for each $j = 1, 2, \ldots, p$. Since the function vec A is linear, then

$$\operatorname{vec} C = \sum_{j=1}^{p} (B_j^{\mathrm{T}} \otimes A_j) \operatorname{vec} X$$

and the result follows. ■

Having transformed the linear matrix equation to an equation of the form $G\mathbf{x} = \mathbf{c}$, we may now apply the results developed in Section 3.10 to determine criteria for the existence and uniqueness of a solution (see also Theorem 12.6.7 to follow).

Corollary 1. *Equation* (1) *has a solution X if and only if* $\operatorname{rank}[G \quad \mathbf{c}] = \operatorname{rank} G$.

Corollary 2. *Equation* (1) *has a unique solution if and only if the matrix G in Eq.* (2) *is nonsingular.*

We now consider an important particular case of Eq. (1); the matrix equation

$$AX + XB = C. \tag{3}$$

Theorem 2. *Equation* (3) *has a unique solution if and only if the matrices A and $-B$ have no eigenvalues in common.*

PROOF. In the case being considered, the matrix G in Eq. (2) is $(I_n \otimes A) + (B^{\mathrm{T}} \otimes I_m)$ and its eigenvalues are, by Corollary 12.2.2, the numbers $\lambda_r + \mu_s$, $r = 1, 2, \ldots, m$, $s = 1, 2, \ldots, n$. It remains now to apply Corollary 2 and recall that a matrix is nonsingular if and only if all its eigenvalues are nonzero. ■

There is a special case in which we can obtain an explicit form for the solution matrix X of Eq. (3).

Theorem 3. *If all eigenvalues of the matrices $A \in C^{m \times m}$ and $B \in C^{n \times n}$ have negative real parts* (*that is, A and B are* stable), *then the unique solution X of Eq.* (3) *is given by*

$$X = -\int_0^{\infty} e^{At} C e^{Bt}\, dt. \tag{4}$$

PROOF. First observe that the conditions of the theorem imply that A and $-B$ have no eigenvalue in common, and hence there is a unique solution X of Eq. (3).

Consider a matrix-valued function $Z(t)$ defined as the solution of the initial-value problem

$$dZ/dt = AZ + ZB, \qquad Z(0) = C. \tag{5}$$

As indicated in Exercise 9.12.14, the solution of Eq. (5) is

$$Z(t) = e^{At}Ce^{Bt}.$$

Now observe that, integrating the differential equation in (5) from $t = 0$ to $t = \infty$ under the hypothesis that the matrix (4) exists, we obtain

$$Z(\infty) - Z(0) = A \int_0^\infty Z(t)\,dt + \left(\int_0^\infty Z(t)\,dt\right)B.$$

Hence assuming, in addition, that

$$Z(\infty) = \lim_{t \to \infty} e^{At}Ce^{Bt} = 0, \tag{6}$$

it is found that the matrix given by Eq. (4) is the (unique) solution of Eq. (3). Thus it suffices to check that, for stable A and B, the assumptions we have had to make are valid. To this end, write the spectral resolution (Eq. (9.5.1)) for the function e^{At} $(0 < t < \infty)$:

$$e^{At} = \sum_{k=1}^{s} \sum_{j=0}^{m_k - 1} t^j e^{\lambda_k t} Z_{kj}, \tag{7}$$

where $\lambda_1, \lambda_2, \ldots, \lambda_s$ are the distinct eigenvalues of A with indices $m_1, m_2, \ldots, m_s$, respectively. Consider the Cartesian decomposition of the eigenvalues, $\lambda_k = \alpha_k + i\beta_k$, and write

$$e^{\lambda_k t} = e^{\alpha_k t}e^{i\beta_k t} = e^{\alpha_k t}(\cos \beta_k t + i \sin \beta_k t),$$

for each $k = 1, 2, \ldots, s$. Since $\alpha_k < 0$ $(k = 1, 2, \ldots, s)$, it follows that, for each k, $\exp(\lambda_k t) \to 0$ as $t \to \infty$. Hence (see Eq. (7)) $\lim_{t\to\infty} e^{At} = O$ and, similarly, $\lim_{t\to\infty} e^{Bt} = O$. Thus, the relation (6) as well as the existence of the integral in (4) follow. ■

Now consider the case in which A and $-B$ have eigenvalues in common. Then the matrix $G = (I_n \otimes A) + (B^{\mathrm{T}} \otimes I_m)$ is singular and solutions exist, provided $\operatorname{rank}[G \;\; \mathbf{c}] = \operatorname{rank} G$ (Corollary 1). The number of linearly independent solutions is then determined by the dimension of the kernel of G. The direct determination of $\dim(\operatorname{Ker} G)$ poses a relatively difficult question and can be found in the standard reference book by MacDuffee (1946) where the Frobenius proof is presented; this proof dates back to 1910.

However, we prefer at this moment to abandon the idea of using the Kronecker product in investigating the solvability and the number of linearly independent solutions of Eq. (3). Our line of attack will restart with a study of the particular cases $AX = XA$ in Section 12.4 and then $AX + XB = O$ in Section 12.5, where the problems mentioned regarding the general equation $AX + XB = C$ will be resolved.

Exercise 1. Let A be a stable matrix and suppose $W \geq 0$. Show that the Lyapunov equation $AX + XA^* = -W$ has a unique solution X and that $X \geq O$. (See also Section 13.1.)

Exercise 2. Let $A \in \mathbb{C}^{m \times m}$, $B \in \mathbb{C}^{n \times n}$, and $C \in \mathbb{C}^{m \times n}$. If A and B have spectral radii μ_A and μ_B, respectively, and $\mu_A \mu_B < 1$, show that the equation

$$X = AXB + C \tag{8}$$

has a unique solution X and that this solution is given by

$$X = \sum_{j=0}^{\infty} A^j C B^j.$$

Exercise 3. Show that Eq. (8) has a unique solution if $\lambda\mu \neq 1$ for all $\lambda \in \sigma(A)$ and $\mu \in \sigma(B)$.

Exercise 4. Let $A, C \in \mathbb{C}^{n \times n}$ and $C^* = C$. If the equation

$$X = A^*XA + C$$

has a unique solution X, show that X is Hermitian. Also, if $\mu_A < 1$, show that $(X - C) \geq 0$. In particular, if $C \geq 0$, then $X \geq 0$.

Exercise 5. If $A > 0$, $C > 0$, and $X = AXA + C$, show that $\mu_A < 1$. □

12.4 Commuting Matrices

An important special case of Eq. (12.3.3) arises when $C = O$ and $B = -A$:

$$AX = XA, \qquad A, X \in \mathbb{C}^{n \times n}. \tag{1}$$

Clearly, finding all the solutions X of Eq. (1) is equivalent to determining all matrices that commute with the given matrix A. To this end, we first observe that if $A = PJP^{-1}$, where J is a Jordan normal form for A, then Eq. (1) is equivalent to the equation

$$JY = YJ, \tag{2}$$

where $Y = P^{-1}XP$.

Proceeding to the analysis of the reduced Eq. (2) we write $J = \text{diag}[J_1, J_2, \ldots, J_p]$, where $J_s = \lambda_s I_{k_s} + N_{k_s}$ is a $k_s \times k_s$ Jordan block associated with an eigenvalue λ_s $(1 \le s \le p)$. Hence, performing a corresponding partition of Y in Eq. (2),

$$Y = [Y_{st}]_{s,t=1}^{p}, \qquad Y_{st} \in \mathbb{C}^{k_s \times k_t},$$

Eq. (2) is reduced to the following system:

$$J_s Y_{st} = Y_{st} J_t, \qquad s, t = 1, 2, \ldots, p,$$

or, exploiting the structure of Jordan blocks,

$$(\lambda_s - \lambda_t) Y_{st} = Y_{st} N_{k_t} - N_{k_s} Y_{st}, \qquad s, t = 1, 2, \ldots, p. \tag{3}$$

We now consider two possible cases.

Case (1) $\lambda_s \neq \lambda_t$. Multiplying Eq. (3) by $\lambda_s - \lambda_t$ and using Eq. (3) again, we obtain

$$(\lambda_s - \lambda_t)^2 Y_{st} = (Y_{st} N_{k_t} - N_{k_s} Y_{st}) N_{k_t} - N_{k_s}(Y_{st} N_{k_t} - N_{k_s} Y_{st}).$$

Simplifying the right-hand expression and multiplying successively by $\lambda_s - \lambda_t$, it is found that, for $r = 1, 2, \ldots,$

$$(\lambda_s - \lambda_r)^r Y_{st} = \sum_{j=0}^{r} (-1)^j \binom{r}{j} N_{k_s}^j Y_{st} N_{k_t}^{r-j}, \tag{4}$$

where it is assumed that $N_{k_s}^0 = I_{k_s}$, $N_{k_t}^0 = I_{k_t}$.

Now recall that the matrices N_{k_s} are nilpotent and therefore $N_{k_s}^m = O$ for sufficiently large m. Hence if r in Eq. (4) is large enough, then $(\lambda_s - \lambda_t)^r Y_{st} = O$ and the assumption $\lambda_s \neq \lambda_t$ yields $Y_{st} = 0$.

Case (2) $\lambda_s = \lambda_t$. In this case Eq. (3) becomes

$$Y_{st} N_{k_s} = N_{k_t} Y_{st}, \qquad 1 \le s, t \le p, \tag{5}$$

and if $Y_{st} = [\gamma_{ij}]_{i,j=1}^{k_s, k_t}$, comparing the corresponding elements of the matrices on the right and left in Eq. (5), it is not difficult to see that

$$\gamma_{i+1,j} = \gamma_{i,j-1}, \qquad i = 1, 2, \ldots, k_s, \quad j = 1, 2, \ldots, k_t, \tag{6}$$

where it is supposed that $\gamma_{i0} = \gamma_{k_s+1,j} = 0$.

The equalities (6) show that the $k_s \times k_t$ matrix Y_{st} $(1 \le s, t \le p)$ is in one of the following forms, depending on the relation between k_s and k_t:

(a) if $k_s = k_t$, then Y_{st} is an upper-triangular Toeplitz matrix;

(b) if $k_s < k_t$, then $Y_{st} = [0 \quad Y_{k_s}]$;

(c) if $k_s > k_t$, then $Y_{st} = \begin{bmatrix} Y_{k_t} \\ 0 \end{bmatrix}$,

in which Y_{k_s} and Y_{k_t} denote (square) upper-triangular Toeplitz matrices of orders k_s and k_t, respectively.

Thus, we may state the following.

Theorem 1. *If $A \in \mathbb{C}^{n \times n}$ and $A = PJP^{-1}$, where J is a Jordan canonical form of A, then a matrix X commutes with A if and only if it is of the form $X = PYP^{-1}$, where $Y = [Y_{st}]_{s,t=1}^{p}$ is the partition of Y consistent with the partition of J into Jordan blocks, and where $Y_{st} = 0$ for $\lambda_s \neq \lambda_t$ and Y_{st} is in one of the forms (a), (b), or (c) if $\lambda_s = \lambda_t$.* ■

Exercise 1. Check that the set of matrices commuting with the matrix

$$A = P\left(\begin{bmatrix} 2 & 1 \\ 0 & 2 \end{bmatrix} \dotplus \begin{bmatrix} 2 & 1 & 0 \\ 0 & 2 & 1 \\ 0 & 0 & 2 \end{bmatrix} \dotplus \begin{bmatrix} 1 & 1 \\ 0 & 1 \end{bmatrix}\right)P^{-1}$$

consists entirely of the matrices

$$X = P\left(\left[\begin{array}{cc:ccc} \gamma_0 & \gamma_1 & 0 & \alpha_0 & \alpha_1 \\ 0 & \gamma_0 & 0 & 0 & \alpha_0 \\ \hdashline \beta_0 & \beta_1 & \delta_0 & \delta_1 & \delta_2 \\ 0 & \beta_0 & 0 & \delta_0 & \delta_1 \\ 0 & 0 & 0 & 0 & \delta_0 \end{array}\right] \dotplus \begin{bmatrix} \varepsilon_0 & \varepsilon_1 \\ 0 & \varepsilon_0 \end{bmatrix}\right)P^{-1},$$

where the letters denote arbitrary complex numbers. □

The number of arbitrary parameters in the general solution X of Eq. (1) is obviously greater than or equal to n.

Exercise 2. If α_{st} denotes the degree of the greatest common divisor of the elementary divisors $(\lambda - \lambda_s)^{k_s}$ and $(\lambda - \lambda_t)^{k_t}$ $(1 \leq s, t \leq p)$, prove that X in Eq. (1) has

$$\alpha = \sum_{s,t=1}^{p} \alpha_{st} \tag{7}$$

undefined elements.

Hint. Observe that if $\lambda_s = \lambda_t$, then $\alpha_{st} = \min(k_s, k_t)$. □

Rephrasing the result of Exercise 2, the dimension of the linear space of solutions X of $AX = XA$ is the number α given by Eq. (7). To see this, denote the α parameters of X by $\gamma_1, \gamma_2, \ldots, \gamma_\alpha$, and denote by X_i the solution of Eq. (1) obtained by putting $\gamma_i = 1, \gamma_j = 0$ $(j = 1, 2, \ldots, \alpha, j \neq i)$. Clearly, the matrices X_i $(i = 1, 2, \ldots, \alpha)$ are linearly independent and any solution X of Eq. (1) can be written

$$X = \sum_{i=1}^{\alpha} \gamma_i X_i. \tag{8}$$

Thus, $X_1, X_2, \ldots, X_\alpha$ constitute a basis in the solution space of $AX = XA$.

We have seen that, quite obviously, any polynomial $p(A)$ in $A \in \mathbb{C}^{n \times n}$ commutes with A; we now ask about conditions on A such that, conversely, each matrix commuting with A is a polynomial in A.

Proposition 1. *Every matrix commuting with $A \in \mathbb{C}^{n \times n}$ is a polynomial in A if and only if no two Jordan blocks in the Jordan canonical form of A are associated with the same eigenvalue. Or, equivalently, the elementary divisors of A are pairwise relatively prime.*

PROOF. First recall that in view of the Cayley–Hamilton theorem the number of linearly independent powers of A does not exceed n. Thus, if $AX = XA$ implies that X is a polynomial of A, then X can be rewritten as $X = p(A)$, where $l = \deg p(\lambda) \le n$ and all powers $I, A, A^2, \ldots, A^{l-1}$ are linearly independent. Hence the dimension α of the solution set of Eq. (1) is in this case $\alpha = l \le n$. But Eq. (7) shows that $\alpha \ge n$, therefore $\alpha = n$ and the result follows from Eq. (7).

Conversely, if $\lambda_1, \lambda_2, \ldots, \lambda_s$ are the distinct eigenvalues of A with indices $m_1, m_2, \ldots, m_s$, respectively, then the Jordan form J of A consists of exactly s blocks J_i. As we have seen, the solution X of $AX = XA$ is then similar to a direct sum of upper-triangular Toeplitz matrices:

$$Y_i = \begin{bmatrix} \gamma_0^{(i)} & \gamma_1^{(i)} & \cdots & \gamma_{m_i-1}^{(i)} \\ & \gamma_0^{(i)} & \ddots & \vdots \\ & & \ddots & \gamma_1^{(i)} \\ & & & \gamma_0^{(i)} \end{bmatrix},$$

for $i = 1, 2, \ldots, s$. In view of Eq. (9.2.4) it is clear that a polynomial $p(\lambda)$ satisfying the conditions

$$p(\lambda_i) = \gamma_0^{(i)}, \qquad \frac{1}{1!}p'(\lambda_i) = \gamma_1^{(i)}, \ldots, \quad \frac{1}{(m_i - 1)!}p^{(m_i-1)}(\lambda_i) = \gamma_{m_i-1}^{(i)},$$

$i = 1, 2, \ldots, s$, gives the required result:

$$\begin{aligned} X &= P \operatorname{diag}[Y_1, Y_2, \ldots, Y_s]P^{-1} \\ &= P \operatorname{diag}[p(J_1), \ldots, p(J_s)]P^{-1} \\ &= Pp(J)P^{-1} = p(PJP^{-1}) = p(A). \end{aligned}$$

Note that $p(\lambda)$ can be chosen with degree not exceeding $n - 1$. ■

We now confine our attention to matrices commuting with a simple matrix. First, observe that Theorem 1 implies the result of Exercise 3.

Exercise 3. If A is simple and $\lambda_1, \lambda_2, \ldots, \lambda_s$ are the distinct eigenvalues of A, show that the linear space of solutions of $AX = XA$ has dimension $\beta_1^2 + \beta_2^2 + \cdots + \beta_s^2$, where β_i $(1 \le i \le s)$ is the algebraic multiplicity of λ_i.

Exercise 4. Show that if $A \in \mathbb{C}^{n \times n}$ has all its eigenvalues distinct, then every matrix that commutes with A is simple. □

Proposition 2. *Simple matrices $A, X \in \mathbb{C}^{n \times n}$ commute if and only if they have a set of n linearly independent eigenvectors in common.*

In other words, if A and X are simple and $AX = XA$, then there exists a basis in $\mathbb{C}^n$ consisting entirely of common eigenvectors of A and X.

PROOF. Let $\lambda_1, \lambda_2, \ldots, \lambda_s$ be the distinct eigenvalues of A with indices $m_1, m_2, \ldots, m_s$, respectively, and let $A = PJP^{-1}$, where the diagonal matrix J is a Jordan canonical form of A. By Theorem 1, the general form of matrices commuting with A is

$$X = PYP^{-1} = P \operatorname{diag}[Y_1, Y_2, \ldots, Y_s]P^{-1}$$

where $Y_j \in \mathbb{C}^{m_j \times m_j}$, $j = 1, 2, \ldots, s$. Reducing each Y_j to a Jordan canonical form, $Y_j = Q_j J_j Q_j^{-1}$, we obtain

$$X = (PQ) \operatorname{diag}[J_1, J_2, \ldots, J_s](PQ)^{-1},$$

where $Q = \operatorname{diag}[Q_1, Q_2, \ldots, Q_s]$.

Since X is simple so is $\operatorname{diag}[J_1, J_2, \ldots, J_s]$, and therefore the columns of PQ give n linearly independent eigenvectors of X. The same is true for A, because

$$(PQ)J(PQ)^{-1} = P \operatorname{diag}[\lambda_1 Q_1 Q_1^{-1}, \ldots, \lambda_s Q_s Q_s^{-1}]P^{-1} = PJP^{-1} = A.$$

Conversely, if A and X have a set of n linearly independent eigenvectors in common, then A and X are obviously simple, and the eigenvectors determine a matrix P for which $A = PD_1P^{-1}$, $X = PD_2P^{-1}$, and D_1, D_2 are diagonal matrices. Then A and X commute, for

$$AX = PD_1D_2P^{-1} = PD_2D_1P^{-1} = XA. \quad \blacksquare$$

Exercise 5. If A is a normal matrix with all its eigenvalues distinct, prove that every matrix that commutes with A is normal.

Exercise 6. Prove that normal matrices commute if and only if they have a common set of orthonormal eigenvectors.

Exercise 7. If A is a simple matrix with s distinct eigenvalues, prove that every matrix that commutes with A has at least s linearly independent eigenvectors in common with A. □

12.5 Solutions of $AX + XB = C$

In the previous section we described the form of solutions of the equation $AX = XA$. Now observe that a matrix $X \in \mathbb{C}^{m\times n}$ is a solution of the equation $AX + XB = 0$, where $A \in \mathbb{C}^{m\times m}$, $B \in \mathbb{C}^{n\times n}$, if and only if the $(m + n) \times (m + n)$ matrices

$$\begin{bmatrix} I_m & X \\ 0 & I_n \end{bmatrix}, \quad \begin{bmatrix} A & 0 \\ 0 & -B \end{bmatrix} \tag{1}$$

commute. Thus, applying Theorem 12.4.1 to these two matrices, the next result readily follows.

Theorem 1. *Let $A \in \mathbb{C}^{m\times m}$, $B \in \mathbb{C}^{n\times n}$ and let $A = PJ_1P^{-1}$, $B = QJ_2Q^{-1}$, where*

$$J_1 = \operatorname{diag}[\lambda_1 I_{k_1} + N_{k_1}, \ldots, \lambda_p I_{k_p} + N_{k_p}]$$

and

$$J_2 = \operatorname{diag}[\mu_1 I_{r_1} + N_{r_1}, \ldots, \mu_q I_{r_q} + N_{r_q}]$$

are Jordan canonical forms of A and B, respectively. Then any solution X of

$$AX + XB = 0 \tag{2}$$

is of the form $X = PYQ^{-1}$, where $Y = [Y_{st}]_{s,t=1}^{p,q}$ is the general solution of $J_1Y + YJ_2 = 0$, with $Y_{st} \in \mathbb{C}^{k_s\times r_t}$, $s = 1, 2, \ldots, p$, $t = 1, 2, \ldots, q$. Therefore Y has the properties $Y_{st} = 0$ if $\lambda_s \neq -\mu_t$, and Y_{st} is a matrix of the form described in Theorem 12.4.1 for $\lambda_s = -\mu_t$.

The argument used in Section 12.4 can also be exploited here.

Corollary 1. *The general form of solutions of Eq. (2) is given by the formula*

$$X = \sum_{i=1}^{\alpha} \gamma_i X_i, \tag{3}$$

where $X_1, X_2, \ldots, X_\alpha$ are linearly independent solutions of Eq. (2) and

$$\alpha = \sum_{s=1}^{p} \sum_{t=1}^{q} \alpha_{st}, \tag{4}$$

where α_{st} $(1 \le s \le p, 1 \le t \le q)$ is the degree of the greatest common divisor of the elementary divisors $(\lambda - \lambda_s)^{k_s}$ and $(\lambda + \mu_t)^{r_t}$ of A and B, respectively.

Note that the number α in Eq. (4) is equal to dim(Ker G), where $G = I_n \otimes A + B^{\mathrm{T}} \otimes I_m$.

The next exercise indicates a second approach to equation (2) based on the observation that Im X is A-invariant.

Exercise 1. Show that the equation $AX = XB$ has a solution X of rank r if and only if A and B have r eigenvalues in common with corresponding partial multiplicities in A and B.

OUTLINE OF SOLUTION. Observe first that Im X is A-invariant. Then use Ex. 6.5.3 to show that X has a factorization $X = EX_0$ where E is $m \times r$, X_0 is $r \times n$, both have full rank, and the columns of E are made up of Jordan chains of A. Thus $AE = EJ$ and J is an $r \times r$ matrix in Jordan normal form. Then show that $X_0 B = JX_0$, and conclude that the eigenvalues of J (and their partial multiplicities) are common to both A and B.

Conversely, define E and X_0 of full rank so that

$$AE = EJ, \qquad X_0 B = JX_0,$$

and show that $X = EX_0$ satisfies $AX = XB$. □

Proceeding to the general equation

$$AX + XB = C, \tag{5}$$

where $A \in \mathbb{C}^{m \times m}$, $B \in \mathbb{C}^{n \times n}$, $X, C \in \mathbb{C}^{m \times n}$, it is easily seen, after rewriting this as a system of mn linear (scalar) equations, that Eq. (2) (also rewritten as such a system) is the corresponding homogeneous equation. Thus we can combine the results of Section 3.10 with the foregoing discussion.

Theorem 2. *If Eq. (5) is solvable, then either it has a unique solution (when $\sigma(A) \cap \sigma(-B) = \emptyset$) or it has infinitely many solutions given by the formula*

$$X = X_0 + \tilde{X},$$

where X_0 is a fixed particular solution of Eq. (5) and $\tilde{X}$, being the general solution of the homogeneous equation, Eq. (2), is of the form (3).

Thus, to complete an investigation of Eq. (5), it remains to obtain a criterion for its solvability in terms of the matrices A, B, and C. This criterion is due to W. E. Roth† and the proof presented here is that of H. Flanders and H. K. Wimmer‡.

Theorem 3. *Equation (5) has a solution if and only if the matrices*

$$\begin{bmatrix} A & 0 \\ 0 & -B \end{bmatrix} \quad \text{and} \quad \begin{bmatrix} A & C \\ 0 & -B \end{bmatrix} \tag{6}$$

are similar.

† *Proc. Amer. Math. Soc.* **3** (1952), 392–396.

‡ *SIAM Jour. Appl. Math.* **32** (1977), 707–710.

PROOF. If Eq. (5) has a solution X, it is easily checked that the matrices in (6) are similar, with the transforming matrix

$$\begin{bmatrix} I & X \\ 0 & I \end{bmatrix}.$$

Conversely, let

$$\begin{bmatrix} A & 0 \\ 0 & -B \end{bmatrix} = P^{-1}\begin{bmatrix} A & C \\ 0 & -B \end{bmatrix}P \tag{7}$$

for some nonsingular matrix $P \in \mathbb{C}^{(m+n)\times(m+n)}$. Define linear transformations $T_1, T_2 \in \mathscr{L}(\mathbb{C}^{(m+n)\times(m+n)})$ by the rule

$$T_1(Z) = \begin{bmatrix} A & 0 \\ 0 & -B \end{bmatrix}Z - Z\begin{bmatrix} A & 0 \\ 0 & -B \end{bmatrix},$$

$$T_2(Z) = \begin{bmatrix} A & C \\ 0 & -B \end{bmatrix}Z - Z\begin{bmatrix} A & 0 \\ 0 & -B \end{bmatrix},$$

where $Z \in \mathbb{C}^{(m+n)\times(m+n)}$, and observe that it suffices to find a matrix in Ker T_2 of the form

$$\begin{bmatrix} R & U \\ 0 & -I \end{bmatrix}. \tag{8}$$

Indeed, in this case a straightforward calculation shows that the matrix U in (8) is a solution of (5).

Eq. (7) implies that

$$T_1(Z) = P^{-1}T_2(PZ),$$

and hence

$$\text{Ker } T_2 = \{PZ : Z \in \mathbb{C}^{(m+n)\times(m+n)} \quad \text{and} \quad Z \in \text{Ker } T_1\}.$$

Consequently,

$$\dim(\text{Ker } T_1) = \dim(\text{Ker } T_2). \tag{9}$$

Let us consider the set $\mathscr{S}$ consisting of all pairs (V, W), where $V \in \mathbb{C}^{n\times m}$, $W \in \mathbb{C}^{n\times n}$, and such that $BV + VA = 0$ and $BW = WB$. Then $\mathscr{S}$ is a linear space with respect to the natural operations

$$\alpha(V, W) = (\alpha V, \alpha W), \qquad \alpha \in \mathbb{C},$$

$$(V_1, W_1) + (V_2, W_2) = (V_1 + V_2, W_1 + W_2).$$

Define the transformations $\varphi_i \in \mathscr{L}(\text{Ker } T_i, \mathscr{S})$, $i = 1, 2$, by the formula

$$\varphi_i\left(\begin{bmatrix} R & U \\ V & W \end{bmatrix}\right) = (V, W)$$

and observe that

$$\operatorname{Ker} \varphi_1 = \operatorname{Ker} \varphi_2 = \left\{ \begin{bmatrix} R & U \\ 0 & 0 \end{bmatrix} : AR = RA, \quad AU + UB = 0 \right\}. \tag{10}$$

Furthermore, $\operatorname{Im} \varphi_1 = \mathscr{S}$. This is because $(V, W) \in \mathscr{S}$ implies $BV + VA = 0$ and $BW = WB$. Hence

$$\begin{bmatrix} 0 & 0 \\ V & W \end{bmatrix} \in \operatorname{Ker} T_1, \qquad \text{and} \qquad \varphi_1\left(\begin{bmatrix} 0 & 0 \\ V & W \end{bmatrix}\right) = (V, W).$$

Thus, $\mathscr{S} \subset \operatorname{Im} \varphi_1$. Since the reverse inclusion is obvious, $\operatorname{Im} \varphi_1 = \mathscr{S}$. Now we have $\operatorname{Im} \varphi_2 \subset \mathscr{S} = \operatorname{Im} \varphi_1$. But, on the other hand (see Eq. (4.5.2)),

$$\dim(\operatorname{Ker} \varphi_i) + \dim(\operatorname{Im} \varphi_i) = \dim(\operatorname{Ker} T_i), \qquad i = 1, 2,$$

and therefore Eqs. (9) and (10) imply that $\dim(\operatorname{Im} \varphi_1) = \dim(\operatorname{Im} \varphi_2)$ and, consequently, that

$$\operatorname{Im} \varphi_1 = \operatorname{Im} \varphi_2 = \mathscr{S}. \tag{11}$$

Now observe that

$$\begin{bmatrix} I & 0 \\ 0 & -I \end{bmatrix} \in \operatorname{Ker} T_1 \qquad \text{and} \qquad \varphi_1\left(\begin{bmatrix} I & 0 \\ 0 & -I \end{bmatrix}\right) = (0, -I).$$

Thus, in view of Eq. (11), there is a matrix

$$\begin{bmatrix} R_0 & U_0 \\ V_0 & W_0 \end{bmatrix} \in \operatorname{Ker} T_2$$

such that

$$(0, -I) = \varphi_2\left(\begin{bmatrix} R_0 & U_0 \\ V_0 & W_0 \end{bmatrix}\right).$$

This means that $V_0 = 0$ and $W_0 = -I$. Thus we have found a matrix in $\operatorname{Ker} T_2$ of the desired form (8). This completes the proof. ■

12.6 One-Sided Inverses

In this and the following sections we shall introduce some generalizations of the inverse matrix. These generalized inverses retain some of the important properties of the inverse but are defined not only for nonsingular (square) matrices but for singular square and rectangular matrices, as well. We devote this section to a study of the notion of one-sided invertibility of matrices, but first recall the corresponding ideas for linear transformations studied in Section 4.6.

A matrix $A \in \mathscr{F}^{m \times n}$ is said to be *left* (respectively, *right*) *invertible* if there exists a matrix A_L^{-1} (respectively, A_R^{-1}) from $\mathscr{F}^{n \times m}$ such that

$$A_L^{-1}A = I_n \qquad \text{(respectively,} \quad AA_R^{-1} = I_m). \tag{1}$$

A matrix A_L^{-1} (respectively, A_R^{-1}) satisfying Eq. (1) is called a *left* (respectively, *right*) *inverse* of A. Clearly, if $m = n$ and A is nonsingular, then the one-sided inverses of A coincide with the regular inverse A^{-1}.

The first results are obtained directly from Theorems 4.6.1 and 4.6.2 on interpreting members of $\mathscr{F}^{m \times n}$ as linear transformations from $\mathbb{C}^n$ to $\mathbb{C}^m$.

Theorem 1. *The following statements are equivalent for any matrix $A \in \mathscr{F}^{m \times n}$:*

(a) *the matrix A is left invertible;*
(b) *$m \geq n$ and* rank $A = n$;
(c) *The columns of A are linearly independent as members of $\mathscr{F}^m$;*
(d) Ker $A = \{0\}$.

Theorem 2. *The following statements are equivalent for any matrix $A \in \mathscr{F}^{m \times n}$:*

(a) *the matrix A is right invertible;*
(b) *$m \leq n$ and* rank $A = m$;
(c) *The rows of A are linearly independent as members of $\mathscr{F}^n$;*
(d) Im $A = \mathscr{F}^m$.

Thus, $m \times n$ matrices of full rank (see Section 3.8) and only they are one-sided invertible. They are invertible from the left if $m \geq n$ and from the right if $m \leq n$. In the case $m = n$ a matrix of full rank is (both-sided) invertible.

Exercise 1. Let

$$A = \begin{bmatrix} 1 & 2 & -1 \\ 0 & 1 & 0 \end{bmatrix}, \qquad B = \begin{bmatrix} 1 & -2 \\ 0 & 1 \\ 0 & 0 \end{bmatrix}.$$

Verify that $B = A_R^{-1}$ and that $A = B_L^{-1}$. Find other right inverses of A and left inverses of B. $\square$

Let $A \in \mathscr{F}^{m \times n}$, $m \geq n$, and rank $A = n$. Then A is left invertible (Theorem 1) and hence the equation

$$XA = I_n \tag{2}$$

is solvable. To describe all solutions of Eq. (2), that is, the set of all left inverses of A, we first apply a permutation matrix $P \in \mathcal{F}^{m \times m}$ such that the $n \times n$ nonsingular submatrix A_1 in PA occupies the top n rows, that is,

$$PA = \begin{bmatrix} A_1 \\ A_2 \end{bmatrix}, \qquad A_1 \in \mathcal{F}^{n \times n}, \quad \det A_1 \neq 0.$$

It is easily verified that matrices of the form

$$[A_1^{-1} - BA_2 A_1^{-1} \quad B]P \tag{3}$$

for any $B \in \mathcal{F}^{n \times (m-n)}$ satisfy Eq. (2), that is, are left inverses of A. Conversely, if A_L^{-1} is a left inverse of A, then, representing $A_L^{-1} = [C \quad B]P$, where $C \in \mathcal{F}^{n \times n}$, $B \in \mathcal{F}^{n \times (m-n)}$, and substituting it in Eq. (2), we obtain $CA_1 + BA_2 = I_n$. Hence $C = A_1^{-1} - BA_2 A_1^{-1}$ and the following result is proved.

Theorem 3. *If $A \in \mathcal{F}^{m \times n}$ is left invertible, then all left inverses of A are given by formula* (3), *where B is an arbitrary $n \times (m - n)$ matrix.*

A similar result is easily proved for right inverses.

Theorem 4. *Let $A \in \mathcal{F}^{m \times n}$ be right invertible. Then all right inverses of A are given by the formula*

$$A_R^{-1} = Q\begin{bmatrix} A_3^{-1} - A_3^{-1}A_4 C \\ C \end{bmatrix}, \tag{4}$$

for any $(n - m) \times m$ matrix C. Here $Q \in \mathcal{F}^{n \times n}$ is a permutation matrix such that $AQ = [A_3 \quad A_4]$ and the $m \times m$ matrix A_3 is nonsingular.

It should also be noted that the matrices AA_L^{-1} and $A_R^{-1}A$ are idempotents:

$$(AA_L^{-1})(AA_L^{-1}) = A(A_L^{-1}A)A_L^{-1} = AA_L^{-1}.$$

Similarly, $A_R^{-1}A$ is also idempotent.

***Exercise* 2.** If A is left invertible, prove that

(a) AA_L^{-1} is a projector onto Im A;
(b) $\dim(\operatorname{Ker} A_L^{-1}) = m - \dim(\operatorname{Im} A)$.

***Exercise* 3.** If A is right invertible, prove that

(a) $I - A_R^{-1}A$ is a projector onto Ker A;
(b) $\dim(\operatorname{Im} A_R^{-1}) = n - \dim(\operatorname{Ker} A)$. □

We now consider the equation

$$A\mathbf{x} = \mathbf{b}, \qquad A \in \mathcal{F}^{m \times n}, \quad \mathbf{x} \in \mathcal{F}^n, \quad \mathbf{b} \in \mathcal{F}^m, \tag{5}$$

with a one-sided invertible matrix A.

Theorem 5. *If A is left invertible, then the equation $A\mathbf{x} = \mathbf{b}$ is solvable if and only if*

$$(I_m - AA_L^{-1})\mathbf{b} = \mathbf{0}. \tag{6}$$

Furthermore, if Eq. (6) holds, then the solution of Eq. (5) is unique and is given by the formula

$$\mathbf{x} = A_L^{-1}\mathbf{b}.$$

PROOF. Let $\mathbf{x}$ be a solution of Eq. (5). Then the multiplication of Eq. (5) by AA_L^{-1} from the left gives

$$AA_L^{-1}\mathbf{b} = A(A_L^{-1}A)\mathbf{x} = A\mathbf{x} = \mathbf{b},$$

and the condition (6) follows. Conversely, if Eq. (6) holds, then the vector $\mathbf{x} = A_L^{-1}\mathbf{b}$ is a solution of Eq. (5). If $\mathbf{x}_1$ is another solution of (5), then $A(\mathbf{x} - \mathbf{x}_1) = \mathbf{0}$ and the assertion (d) of Theorem 1 yields the required result, $\mathbf{x} = \mathbf{x}_1$. ∎

Theorem 6. *If A is right invertible, then Eq. (5) is solvable for any $\mathbf{b} \in \mathcal{F}^m$, and if $\mathbf{b} \neq \mathbf{0}$, every solution $\mathbf{x}$ is of the form*

$$\mathbf{x} = A_R^{-1}\mathbf{b}, \tag{7}$$

for some right inverse A_R^{-1} of A.

PROOF. The solvability of Eq. (5) for any vector $\mathbf{b} \in \mathcal{F}^m$ is obvious since $A(A_R^{-1}\mathbf{b}) = \mathbf{b}$, and $\mathbf{x}$ in Eq. (7) is a solution of Eq. (5). Suppose now that $\mathbf{x}$ is a solution of Eq. (5). Decompose A as in Theorem 4: $A = [A_3 \quad A_4]Q^T$, where $\det A_3 \neq 0$ and Q is a permutation matrix. Perform a consistent decomposition of $\mathbf{x}$: $\mathbf{x} = Q[\mathbf{x}_1^T \quad \mathbf{x}_2^T]^T$. Then Eq. (5) can be rewritten $A_3\mathbf{x}_1 + A_4\mathbf{x}_2 = \mathbf{b}$, hence $\mathbf{x}_1 = A_3^{-1}\mathbf{b} - A_3^{-1}A_4\mathbf{x}_2$. Now compute $A_R^{-1}\mathbf{b}$ using Eq. (4) and use the fact that (since $\mathbf{b} \neq \mathbf{0}$) a matrix $C \in \mathcal{F}^{(n-m)\times m}$ exists such that $C\mathbf{b} = \mathbf{x}_2$:

$$A_R^{-1}\mathbf{b} = Q\begin{bmatrix} A_3^{-1}\mathbf{b} - A_3^{-1}A_4C\mathbf{b} \\ C\mathbf{b} \end{bmatrix} = Q\begin{bmatrix} \mathbf{x}_1 \\ \mathbf{x}_2 \end{bmatrix} = \mathbf{x},$$

and the assertion follows. ∎

Theorems 5 and 6 are easily included in a single more general statement that will also serve to introduce an idea developed in the next section. Comparison should also be made with Theorem 3.10.2. Let $A \in \mathcal{F}^{m\times n}$, and let $X \in \mathcal{F}^{n\times m}$ be any matrix for which

$$AXA = A. \tag{8}$$

If A is left (or right) invertible, we may choose for X a left (or right) inverse of A, of course, but as we shall see in the next section, such matrices X exist even if X has no one-sided inverse.

Theorem 7. *Let $A \in \mathscr{F}^{m \times n}$, $\boldsymbol{b} \in \mathscr{F}^m$, and let X be a solution of Eq. (8). Then the equation $A\boldsymbol{x} = \boldsymbol{b}$ has a solution if and only if $AX\boldsymbol{b} = \boldsymbol{b}$, and then any solution has the form*

$$\boldsymbol{x} = X\boldsymbol{b} + (I - XA)\boldsymbol{y},$$

for some $\boldsymbol{y} \in \mathscr{F}^n$.

Exercise 4. With the assumptions of Theorem 7, prove that $A\boldsymbol{x} = \boldsymbol{0}$ has the general solution $\boldsymbol{x} = (I - XA)\boldsymbol{y}$ for any $\boldsymbol{y} \in \mathscr{F}^n$. In other words, if $\boldsymbol{x}$ satisfies (8) then $I_m(I - XA) = \text{Ker}A$. Then prove Theorem 7. □

12.7 Generalized Inverses

In the preceding section we introduced the notion of a one-sided inverse for $A \in \mathscr{F}^{m \times n}$, and we showed that such an inverse exists if and only if the matrix A is of full rank. Our purpose is now to define a generalization of the concept of a one-sided inverse that will apply to *any* $m \times n$ matrix A. To this end, we first observe that a one-sided inverse X of A (if it exists) satisfies the equations

$$AXA = A, \tag{1}$$

$$XAX = X. \tag{2}$$

Therefore we define a *generalized inverse of* $A \in \mathscr{F}^{m \times n}$ to be a matrix $X \in \mathscr{F}^{n \times m}$ for which Eq. (1) and (2) hold, and we write $X = A^I$ for such a matrix.

Proposition 1. *A generalized inverse A^{I} exists for any matrix $A \in \mathscr{F}^{m \times n}$.*

PROOF. Observe first that the $m \times n$ zero matrix has the $n \times m$ zero matrix as its generalized inverse. Then let $r \neq 0$ be the rank of A. Then (see Section 2.7)

$$A = R\begin{bmatrix} I_r & 0 \\ 0 & 0 \end{bmatrix}V, \tag{3}$$

for some nonsingular matrices $R \in \mathscr{F}^{m \times m}$, $V \in \mathscr{F}^{n \times n}$. It is easily checked that matrices of the form

$$A^{\mathrm{I}} = V^{-1}\begin{bmatrix} I_r & B_1 \\ B_2 & B_2 B_1 \end{bmatrix}R^{-1}, \tag{4}$$

for any $B_1 \in \mathscr{F}^{r \times (n-r)}$, $B_2 \in \mathscr{F}^{(m-r) \times r}$, satisfy the conditions (1) and (2). ■

Thus, a generalized inverse always exists and is, generally, not unique. Bearing in mind the decomposition (3), it is easy to deduce the following criterion for uniqueness of the generalized inverse.

Exercise 1. Prove that the generalized inverse of A is unique if and only if $A = 0$ or A is a square and nonsingular matrix. □

The following properties of generalized inverses follow readily from the defining relations (1) and (2).

Proposition 2. *If $A \in \mathscr{F}^{m\times n}$ and A^{I} denotes one of its generalized inverses, then*

(a) *$A^{\mathrm{I}}A$ and AA^{I} are idempotent matrices.*
(b) rank A^{I} = rank A.
(c) *the matrix $(A^{\mathrm{I}})^*$ is one of the generalized inverses of A^*.*

PROOF. Parts (a) and (c) follow immediately from the defining properties (1) and (2) as applied to A^{I}. Part (b) follows on applying the result of Exercise 3.8.5 to the same relations. ■

If $A \in \mathscr{F}^{n\times n}$ is a nonsingular matrix, so is A^{-1} and obviously

$$\operatorname{Im} A \dotplus \operatorname{Ker} A^{-1} = \mathscr{F}^n \dotplus \{0\} = \mathscr{F}^n,$$

$$\operatorname{Ker} A \dotplus \operatorname{Im} A^{-1} = \{0\} \dotplus \mathscr{F}^n = \mathscr{F}^n.$$

The next result generalizes these statements for an arbitrary matrix A and its generalized inverses.

Proposition 3. *If $A \in \mathscr{F}^{m\times n}$ and A^{I} denotes one of the generalized inverses of A, then*

$$\operatorname{Im} A \dotplus \operatorname{Ker} A^{\mathrm{I}} = \mathscr{F}^m \tag{5}$$

$$\operatorname{Ker} A \dotplus \operatorname{Im} A^{\mathrm{I}} = \mathscr{F}^n. \tag{6}$$

PROOF. It has been noted in Proposition 2(a) that AA^{I} is a projector and it is clear that $\operatorname{Im}(AA^{\mathrm{I}}) \subset \operatorname{Im} A$. We show that, in fact, AA^{I} projects *onto* $\operatorname{Im} A$. If $\mathbf{y} = A\mathbf{x} \in \operatorname{Im} A$, then since $A = AA^{\mathrm{I}}A$, $\mathbf{y} = (AA^{\mathrm{I}})(A\mathbf{x}) \in \operatorname{Im}(AA^{\mathrm{I}})$. So $\operatorname{Im} A \subset \operatorname{Im}(AA^{\mathrm{I}})$ and hence $\operatorname{Im}(AA^{\mathrm{I}}) = \operatorname{Im} A$.

It is deduced from this, and Proposition 2(b), that

$$\operatorname{rank} A = \operatorname{rank}(AA^{\mathrm{I}}) = \operatorname{rank} A^{\mathrm{I}},$$

and then that $\operatorname{Ker} A^{\mathrm{I}} = \operatorname{Ker}(AA^{\mathrm{I}})$. For $\operatorname{Ker} A^{\mathrm{I}} \subset \operatorname{Ker}(AA^{\mathrm{I}})$, obviously. But also, by Theorem 4.5.5

$$\dim(\operatorname{Ker} A^{\mathrm{I}}) = m - \operatorname{rank} A^{\mathrm{I}} = m - \operatorname{rank}(AA^{\mathrm{I}}) = \dim \operatorname{Ker}(AA^{\mathrm{I}}).$$

Consequently, $\operatorname{Ker} A^{\mathrm{I}} = \operatorname{Ker}(AA^{\mathrm{I}})$.

Finally, combining these results with the fact that AA^{I} is a projector it is found that

$$\mathcal{F}^m = \mathrm{Im}(AA^{\mathrm{I}}) \dotplus \mathrm{Ker}(AA^{\mathrm{I}}) = \mathrm{Im}\, A \dotplus \mathrm{Ker}\, A^{\mathrm{I}},$$

which is equation (5).

Since the defining equations for A^{I} are symmetric in A, A^{I} we can replace A, A^{I} in (5) by A^{I}, A, respectively, and obtain (6). ■

Corollary 1. *If A^{I} is a generalized inverse of A, then AA^{I} is a projector onto* $\mathrm{Im}\, A$ *along* $\mathrm{Ker}\, A^{\mathrm{I}}$, *and $A^{\mathrm{I}}A$ is a projector onto* $\mathrm{Im}\, A^{\mathrm{I}}$ *along* $\mathrm{Ker}\, A$.

Note also the following property of the generalized inverse of A that follows immediately from the Corollary, and develops further the relationship with properties of the ordinary inverse.

Exercise 2. If $A \in \mathcal{F}^{m \times n}$, prove that

$$A^{\mathrm{I}}A\mathbf{x} = \mathbf{x} \qquad \text{and} \qquad AA^{\mathrm{I}}\mathbf{y} = \mathbf{y}$$

for any $\mathbf{x} \in \mathrm{Im}\, A^{\mathrm{I}}$ and $\mathbf{y} \in \mathrm{Im}\, A$. In other words, considered as linear transformations, A and A^{I} are mutual inverses on the pairs of subspaces $\mathrm{Im}\, A^{\mathrm{I}}$ and $\mathrm{Im}\, A$, respectively. □

Proposition 3 shows that a generalized inverse for an $m \times n$ matrix determines a complementary subspace for $\mathrm{Im}\, A$ in $\mathcal{F}^m$ and another complementary subspace for $\mathrm{Ker}\, A$ in $\mathcal{F}^n$. The next results shows that a converse statement also holds.

Proposition 4. *Given a matrix $A \in \mathcal{F}^{m \times n}$, let $\mathcal{S}_1$, $\mathcal{S}_2$ be subspaces of $\mathcal{F}^m$ and $\mathcal{F}^n$, respectively, such that*

$$\mathrm{Im}\, A \dotplus \mathcal{S}_1 = \mathcal{F}^m \qquad \textit{and} \qquad \mathrm{Ker}\, A \dotplus \mathcal{S}_2 = \mathcal{F}^n. \tag{7}$$

Then there is a generalized inverse A^{I} of A such that $\mathrm{Ker}\, A^{\mathrm{I}} = \mathcal{S}_1$ *and* $\mathrm{Im}\, A^{\mathrm{I}} = \mathcal{S}_2$.

PROOF. Let A have rank r and write A in the form (3). It is easily seen that

$$V(\mathrm{Ker}\, A) = \mathrm{Ker}\begin{bmatrix} I_r & 0 \\ 0 & 0 \end{bmatrix}, \qquad R^{-1}(\mathrm{Im}\, A) = \mathrm{Im}\begin{bmatrix} I_r & 0 \\ 0 & 0 \end{bmatrix}$$

and then, using (7), that

$$\mathrm{Im}\begin{bmatrix} I_r & 0 \\ 0 & 0 \end{bmatrix} \dotplus R^{-1}(\mathcal{S}_1) = \mathcal{F}^m, \qquad \mathrm{Ker}\begin{bmatrix} I_r & 0 \\ 0 & 0 \end{bmatrix} \dotplus V(\mathcal{S}_2) = \mathcal{F}^n.$$

Consequently, for some $r \times (m - r)$ matrix X and $(n - r) \times r$ matrix Y,

$$\mathscr{S}_1 = R \operatorname{Im}\begin{bmatrix} X \\ I_{m-r} \end{bmatrix}, \qquad \mathscr{S}_2 = V^{-1} \operatorname{Im}\begin{bmatrix} I_r \\ Y \end{bmatrix}.$$

Now define A^{I} by

$$A^{\mathrm{I}} = V^{-1}\begin{bmatrix} I_r & -X \\ Y & -YX \end{bmatrix} R^{-1},$$

and it is easily verified that A^{I} is a generalized inverse of A with the required properties. ■

This section is to be concluded with a result showing how any generalized inverse can be used to construct the *orthogonal* projector onto Im A; this should be compared with Corollary 1 above in which AA^{I} is shown to be a projector onto Im A, but not necessarily an orthogonal projector, and also with a construction of Section 5.8. The following exercise will be useful in the proof. It concerns situations in which left, or right, cancellation of matrices is permissible.

Exercise 3. Prove that whenever the matrix products exist,

(a) $A^*AB = A^*AC$ implies $AB = AC$
(b) $BA^*A = CA^*A$ implies $BA^* = CA^*$.

Hint. Use Exercise 5.3.6 to establish part (a) and then obtain part (b) from (a). □

Proposition 5. *If* $A \in \mathscr{F}^{m \times n}$ *and* $(A^*A)^{\mathrm{I}}$ *is any generalized inverse of* A^*A, *then* $P = A(A^*A)^{\mathrm{I}}A^*$ *is the orthogonal projector of* $\mathscr{F}^m$ *onto* Im A.

PROOF. Equation (1) yields

$$A^*A(A^*A)^{\mathrm{I}}A^*A = A^*A,$$

and the results of Exercise 3 above show that

$$A(A^*A)^{\mathrm{I}}A^*A = A, \qquad A^*A(A^*A)^{\mathrm{I}}A^* = A^*. \tag{8}$$

Using either equation it is seen immediately that $P^2 = P$.

Using Proposition 2(c) one may write

$$P^* = A((A^*A)^{\mathrm{I}})^*A^* = A(A^*A)^{J}A^*,$$

where $(A^*A)^{J}$ is (possibly) another generalized inverse of A^*A. It can be shown that $P^* = P$ by proving $(P - P^*)(P - P^*)^* = 0$. Write

$$(P - P^*)(P - P^*)^* = PP^* + P^*P - P^2 - (P^*)^2,$$

and use equation (8) (for either inverse) to show that the sum on the right reduces to zero.

Obviously, P is a projector of $\mathscr{F}^m$ into Im A. To show that the projection is *onto* Im A, let $\mathbf{y} \in \text{Im } A$. Then $\mathbf{y} = A\mathbf{x}$ for some $\mathbf{x} \in \mathscr{F}^m$ and, using the first equality in (8), we obtain

$$\mathbf{y} = A(A^*A)^{\text{I}}A^*(A\mathbf{x}) = P\mathbf{z},$$

where $\mathbf{z} = A\mathbf{x}$, and the result follows. ■

12.8 The Moore–Penrose Inverse

It has been shown in Propositions 12.7.3 and 12.7.4 that a generalized inverse of an $m \times n$ matrix determines and is determined by a pair of direct complements to Im A in $\mathscr{F}^m$ and Ker A in $\mathscr{F}^n$. An important case arises when *orthogonal* complements are chosen. To ensure this, it suffices to require the projectors AA^{I} and $A^{\text{I}}A$ to be Hermitian (see Section 5.8). Indeed, in this case AA^{I} is an orthogonal projector onto Im A along Ker A^{I} (Corollary 12.7.1) and then

$$\text{Im } A \oplus \text{Ker } A^{\text{I}} = \mathscr{F}^m, \tag{1}$$

and similarly,

$$\text{Ker } A \oplus \text{Im } A^{\text{I}} = \mathscr{F}^n. \tag{2}$$

Thus, we define the *Moore–Penrose inverse*† of $A \in \mathscr{F}^{m \times n}$, written A^+, as a generalized inverse of A satisfying the additional conditions

$$(AA^{\text{I}})^* = AA^{\text{I}}$$

$$(A^{\text{I}}A)^* = A^{\text{I}}A.$$

Obviously, the existence of an orthogonal complement to a subspace and Proposition 12.7.4 together guarantee the existence of A^+ for any matrix A. It turns out, moreover, that the Moore–Penrose inverse is unique.

Proposition 1. *Let $A \in \mathscr{F}^{m \times n}$. There exists a unique matrix $X \in \mathscr{F}^{n \times m}$ for which*

$$AXA = A, \tag{3}$$

$$XAX = X, \tag{4}$$

$$(AX)^* = AX, \tag{5}$$

$$(XA)^* = XA. \tag{6}$$

† E. H. Moore, *Bull. Amer. Math. Soc.* **26** (1920), 394–395, and R. Penrose, *Proc. Cambridge Phil. Soc.* **51** (1955), 406–413.

PROOF. Let X_1 and X_2 be two matrices satisfying the Eqs. (3) through (6). By Eqs. (4) and (5), we have

$$X_i^* = X_i^* A^* X_i^* = AX_i X_i^*, \qquad i = 1, 2,$$

and hence $X_1^* - X_2^* = A(X_1 X_1^* - X_2 X_2^*)$. Thus

$$\operatorname{Im}(X_1^* - X_2^*) \subset \operatorname{Im} A. \tag{7}$$

On the other hand, using Eqs. (3) and (6), it is found that

$$A^* = A^* X_i^* A^* = X_i AA^*, \qquad i = 1, 2,$$

which implies $(X_1 - X_2)AA^* = 0$ or $AA^*(X_1^* - X_2^*) = 0$. Hence

$$\operatorname{Im}(X_1^* - X_2^*) \subset \operatorname{Ker} AA^* = \operatorname{Ker} A^*$$

(by Exercise 5.3.6). But Ker A^* is orthogonal to Im A (see Exercise 5.1.7) and therefore it follows from Eq. (7) that $X_1^* - X_2^* = 0$. ■

It was pointed out in Exercise 5.1.7 that for any $A \in \mathscr{F}^{m \times n}$,

$$\operatorname{Im} A \oplus \operatorname{Ker} A^* = \mathscr{F}^m,$$

$$\operatorname{Ker} A \oplus \operatorname{Im} A^* = \mathscr{F}^n.$$

Comparing this with Eqs. (1) and (2) the following proposition is obtained immediately.

Proposition 2. *For any matrix A*

$$\operatorname{Ker} A^+ = \operatorname{Ker} A^*, \qquad \operatorname{Im} A^+ = \operatorname{Im} A^*.$$

A straightforward computation can be applied to determine A^+ from A. Let $r = \operatorname{rank} A$ and let

$$A = FR^*, \qquad F \in \mathscr{F}^{m \times r}, \qquad R^* \in \mathscr{F}^{r \times n} \tag{8}$$

be a rank decomposition of A, that is, rank F = rank $R^* = r$. Noting that the $r \times r$ matrices F^*F and R^*R are of full rank, and therefore invertible, we define the $n \times n$ matrix

$$X = R(R^*R)^{-1}(F^*F)^{-1}F^*. \tag{9}$$

It is easy to check that this is the Moore–Penrose inverse of A, that $X = A^+$. Note that the linearly independent columns of A can be chosen to be the columns of the matrix F in Eq. (8). This has the useful consequences of the next exercise.

Exercise 1. Let A be $m \times n$ and have full rank. Show that if $m \le n$, then $A^+ = A^*(AA^*)^{-1}$ and if $m \ge n$, then $A^+ = (A^*A)^{-1}A^*$. □

Representation (9) can also be used in proving the following properties of the Moore–Penrose inverse.

Proposition 3. *Let A^+ be the Moore–Penrose inverse of A. Then*

(a) $(A^*)^+ = (A^+)^*$;
(b) $(\alpha A)^+ = \alpha^{-1}A^+$ *for any* $\alpha \in \mathscr{F}, \alpha \neq 0$;
(c) *If A is normal (Hermitian, positive semi-definite), then so is* A^+;
(d) *For unitary U, V,*

$$(UAV)^+ = V^*A^+U^*.$$

PROOF. Let us illustrate the method used here on the proof of the assertion (d).

If $A = FR^*$ is a rank decomposition of A, then $(UF)(R^*V)$ is a rank decomposition for UAV. Hence, using Eq. (9) and the unitary properties of U and V, we have

$$(UAV)^+ = (V^*R)(R^*VV^*R)^{-1}(F^*UU^*F)^{-1}(F^*U^*) = V^*A^+U^*. \quad ■$$

Additional properties of A^+ are listed in the next exercises.

Exercise 2. Prove that

(a) $0^+ = 0^{\mathrm{T}}$;
(b) $(A^+)^+ = A$;
(c) $(AA^*)^+ = (A^+)^*A^+$ and $(A^*A)^+ = A^+(A^+)^*$;
(d) $(A^k)^+ = (A^+)^k$ for any positive integer k and a normal matrix A.

Exercise 3. Show that

$$A^+A\mathbf{x} = \mathbf{x}, \qquad AA^+\mathbf{y} = \mathbf{y},$$

for any $\mathbf{x} \in \operatorname{Im} A^*$, $\mathbf{y} \in \operatorname{Im} A$.

Hint. Use Exercise 12.7.2 and Proposition 2. □

Now recall that in the proofs of Theorems 5.7.1 and 5.7.2, we introduced a pair of orthonormal eigenbases $\{\mathbf{x}_1, \mathbf{x}_2, \ldots, \mathbf{x}_n\}$ and $\{\mathbf{y}_1, \mathbf{y}_2, \ldots, \mathbf{y}_m\}$ of the matrices A^*A and AA^*, respectively, such that

$$A\mathbf{x}_i = \begin{cases} s_i\mathbf{y}_i & \text{if } i = 1, 2, \ldots, r, \\ \mathbf{0} & \text{if } i = r + 1, \ldots, n, \end{cases} \tag{10}$$

where it is assumed that the nonzero singular values s_i $(1 \leq i \leq r)$ of A are associated with the eigenvectors $\mathbf{x}_1, \mathbf{x}_2, \ldots, \mathbf{x}_r$. Note that since $A^*A\mathbf{x}_i = s_i^2\mathbf{x}_i$ for $i = 1, 2, \ldots, n$, the multiplication of (10) by A^* on the left yields

$$A^*\mathbf{y}_i = \begin{cases} s_i\mathbf{x}_i & \text{if } i = 1, 2, \ldots, r, \\ \mathbf{0} & \text{if } i = r + 1, \ldots, m. \end{cases} \tag{11}$$

The bases constructed in this way are called the *singular bases* of A. It turns out that the singular bases allow us to gain deeper insight into the structure of Moore–Penrose inverses.

Proposition 4. *Let $A \in \mathcal{F}^{m \times n}$ and let $s_1 \geq s_2 \geq \cdots \geq s_r > 0 = s_{r+1} = \cdots = s_n$ be the singular values of A. Then $s_1^{-1}, s_2^{-1}, \ldots, s_r^{-1}$ are the nonzero singular values of A^+. Moreover, if $\{x_i\}_{i=1}^n$ and $\{y_i\}_{i=1}^m$ are singular bases of A, then $\{y_i\}_{i=1}^m$ and $\{x_i\}_{i=1}^n$ are singular bases of A^+, that is,*

$$A^+ y_i = \begin{cases} s_i^{-1} x_i & \text{if } i = 1, 2, \ldots, r, \\ 0 & \text{if } i = r+1, \ldots, m, \end{cases} \tag{12}$$

$$(A^+)^* x_i = \begin{cases} s_i^{-1} y_i & \text{if } i = 1, 2, \ldots, r, \\ 0 & \text{if } i = r+1, \ldots, n. \end{cases} \tag{13}$$

PROOF. Consider the matrix $(A^+)^* A^+$ and observe that, by Exercise 2(c) and Proposition 2,

$$\operatorname{Ker}((A^+)^* A^+) = \operatorname{Ker}((AA^*)^+) = \operatorname{Ker} AA^*,$$

and each subspace has dimension $m - r$. Thus A^+ has $m - r$ zero singular values and the bases $\{y_j\}_{j=1}^m$ and $\{x_i\}_{i=1}^n$ can be used as singular bases of A^+. Furthermore, applying A^+ to Eqs. (10) and using the first result of Exercise 3, we obtain, for $x_i \in \operatorname{Im} A^*$, $i = 1, 2, \ldots, n$,

$$x_i = A^+ A x_i = \begin{cases} s_i A^+ y_i & \text{if } i = 1, 2, \ldots, r, \\ 0 & \text{if } i = r+1, \ldots, n, \end{cases}$$

which implies (12). Clearly, s_i^{-1} $(i = 1, 2, \ldots, r)$ are the nonzero singular values of A^+.

Now rewrite the first relation in Exercise 3 for the matrix A^*:

$$(A^*)^+ A^* z = z, \tag{14}$$

for any $z \in \operatorname{Im} A$, and apply $(A^+)^* = (A^*)^+$ to (11). Then, since $y_1, \ldots, y_r$ belong to $\operatorname{Im} A$, we have

$$y_i = (A^+)^* A^* y_i = \begin{cases} s_i (A^+)^* x_i & \text{if } i = 1, 2, \ldots, r, \\ 0 & \text{if } i = r+1, \ldots, n, \end{cases}$$

and the result in (13) follows. ■

Exercise 4. If $A = UDV^*$ as in Theorem 5.7.2, and $\det D \neq 0$, show that $A^+ = VD^{-1}U^*$. □

12.9 The Best Approximate Solution of the Equation $Ax = b$

The Moore–Penrose inverse has a beautiful application in the study of the equation

$$Ax = b, \tag{1}$$

where $A \in \mathscr{F}^{m \times n}$ and $\boldsymbol{b} \in \mathscr{F}^m$, and we wish to solve for $\boldsymbol{x} \in \mathscr{F}^n$. We have seen that there is a solution of Eq. (1) if and only if $\boldsymbol{b} \in \operatorname{Im} A$. In general, the solution set is a "manifold" of vectors obtained by a "shift" of Ker A, as described in Theorem 3.10.2. Using Exercise 12.8.3, we see that the vector $\boldsymbol{x}_0 = A^+\boldsymbol{b}$ is one of the solutions in this manifold:

$$A\boldsymbol{x}_0 = AA^+\boldsymbol{b} = \boldsymbol{b},$$

and it is the *only* solution of Eq. (1) if and only if A is nonsingular, and then $A^+ = A^{-1}$.

Also, a vector $\boldsymbol{x} \in \mathscr{F}^n$ is a solution of Eq. (1) if and only if $\|A\boldsymbol{x} - \boldsymbol{b}\| = 0$ for any vector norm $\| \ \|$ in $\mathscr{F}^m$.

Consider the case $\boldsymbol{b} \notin \operatorname{Im} A$. Now there are no solutions of Eq. (1) but, keeping in mind the preceding remark, we can define an *approximate solution* of $A\boldsymbol{x} = \boldsymbol{b}$ to be a vector $\boldsymbol{x}_0$ for which $\|A\boldsymbol{x} - \boldsymbol{b}\|$ is minimized:

$$\|A\boldsymbol{x}_0 - \boldsymbol{b}\| = \min_{\boldsymbol{x} \in \mathscr{F}^n} \|A\boldsymbol{x} - \boldsymbol{b}\| \tag{2}$$

in the Euclidean vector norm $\| \ \|$. (Then $\boldsymbol{x}_0$ is also known as a *least squares solution* of the inconsistent, or overdetermined, system $A\boldsymbol{x} = \boldsymbol{b}$. Note that if $\boldsymbol{x}_0$ is such a vector, so is any vector of the form $\boldsymbol{x}_0 + \boldsymbol{u}$ where $\boldsymbol{u} \in \operatorname{Ker} A$.)

Now for *any* $A \in \mathscr{F}^{m \times n}$ and *any* $\boldsymbol{b} \in \mathscr{F}^m$, the *best approximate solution* of Eq. (1) is a vector $\boldsymbol{x}_0 \in \mathscr{F}^n$ of the least euclidean length satisfying Eq. (2). We are going to show that $\boldsymbol{x}_0 = A^+\boldsymbol{b}$ is such a vector. Thus, if $\boldsymbol{b} \in \operatorname{Im} A$, then $A^+\boldsymbol{b}$ will be the vector of least norm in the manifold of solutions of Eq. (1) (or that equation's unique solution), and if $\boldsymbol{b} \notin \operatorname{Im} A$, then $A^+\boldsymbol{b}$ is the unique vector which attains the minimum of (2), and also has minimum norm.

Theorem 1. *The vector* $\boldsymbol{x}_0 = A^+\boldsymbol{b}$ *is the best approximate solution of* $A\boldsymbol{x} = \boldsymbol{b}$.

PROOF. For any $\boldsymbol{x} \in \mathscr{F}^n$ write

$$A\boldsymbol{x} - \boldsymbol{b} = A(\boldsymbol{x} - A^+\boldsymbol{b}) + (I - AA^+)(-\boldsymbol{b}),$$

and observe that, because $I - AA^+$ is the orthogonal projector onto $(\operatorname{Im} A)^\perp$ (ref. Corollary 12.7.1), the summands on the right are mutually orthogonal vectors:

$$A(\boldsymbol{x} - A^+\boldsymbol{b}) \in \operatorname{Im} A, \qquad (I - AA^+)(-b) \in (\operatorname{Im} A)^\perp.$$

Thus, the extension of the Pythagorean theorem implies

$$\begin{aligned} \|A\boldsymbol{x} - \boldsymbol{b}\|^2 &= \|A(\boldsymbol{x} - A^+\boldsymbol{b})\|^2 + \|(I - AA^+)(-\boldsymbol{b})\|^2 \\ &= \|A(\boldsymbol{x} - \boldsymbol{x}_0)\|^2 + \|A\boldsymbol{x}_0 - \boldsymbol{b}\|^2 \end{aligned} \tag{3}$$

and, consequently, the relation (2).

It remains only to show that $\|\boldsymbol{x}_1\| \geq \|\boldsymbol{x}_0\|$ for any vector $\boldsymbol{x}_1 \in \mathscr{F}^n$ minimizing $\|A\boldsymbol{x} - \boldsymbol{b}\|$. But then, substituting $\boldsymbol{x}_1$ instead of $\boldsymbol{x}$ in Eq. (3), $\|A\boldsymbol{x}_1 - \boldsymbol{b}\| = \|A\boldsymbol{x}_0 - \boldsymbol{b}\|$ yields $A(\boldsymbol{x}_1 - \boldsymbol{x}_0) = \boldsymbol{0}$, and hence $\boldsymbol{x}_1 - \boldsymbol{x}_0 \in \operatorname{Ker} A$. According to Proposition 12.8.2, $\boldsymbol{x}_0 = A^+\boldsymbol{b} \in \operatorname{Im} A^*$ and, since $\operatorname{Ker} A \perp \operatorname{Im} A^*$, it follows that $\boldsymbol{x}_0 \perp (\boldsymbol{x}_1 - \boldsymbol{x}_0)$. Applying the Pythagorean theorem again, we finally obtain

$$\|\boldsymbol{x}_1\|^2 = \|\boldsymbol{x}_0\|^2 + \|\boldsymbol{x}_1 - \boldsymbol{x}_0\|^2 \geq \|\boldsymbol{x}_0\|^2.$$

The proof of the theorem is complete. ■

Note that the uniqueness of the Moore–Penrose inverse yields the uniqueness of the best approximate solution of Eq. (1).

Exercise 1. Check that the best approximate solution of the equation $A\boldsymbol{x} = \boldsymbol{b}$, where

$$A = \begin{bmatrix} 1 & 1 \\ 2 & 0 \\ -1 & 3 \end{bmatrix}, \qquad \boldsymbol{b} = \begin{bmatrix} 1 \\ 0 \\ 2 \end{bmatrix},$$

is given by the vector $\boldsymbol{x}_0 = \frac{1}{14}[1 \quad 10]^{\mathrm{T}}$. Find the best approximate solution of $B\boldsymbol{x} = \boldsymbol{c}$ if $B = A^{\mathrm{T}}$ and $\boldsymbol{c} = [1 \quad 1]^{\mathrm{T}}$.

Exercise 2. Generalize the idea of the "best approximate solution" for the matrix equation $AX = B$, where $A \in \mathbb{C}^{m \times n}$, $X \in \mathbb{C}^{n \times k}$, and $B \in \mathbb{C}^{m \times k}$, and obtain the formula $X_0 = A^+B$ for this solution.

Exercise 3. Show that an approximate solution of $A\boldsymbol{x} = \boldsymbol{b}$ is the best approximate solution of the equation if and only if it belongs to the image of A^*. □

If the singular basis $\{\boldsymbol{x}_1, \boldsymbol{x}_2, \ldots, \boldsymbol{x}_n\}$ of $A \in \mathscr{F}^{m \times n}$ in the linear space $\mathscr{F}^n$ and the nonzero singular values $s_1, s_2, \ldots, s_r$ of A are known, then an explicit form of the best approximate solution of Eq. (1) can be given.

Theorem 2. *With the notation of the preceding paragraph, the best approximate solution of $A\boldsymbol{x} = \boldsymbol{b}$ is given by the formula*

$$\boldsymbol{x}_0 = \alpha_1\boldsymbol{x}_1 + \alpha_2\boldsymbol{x}_2 + \cdots + \alpha_r\boldsymbol{x}_r, \tag{4}$$

where

$$\alpha_i = \frac{(A^*\boldsymbol{b}, \boldsymbol{x}_i)}{s_i^2}, \qquad i = 1, 2, \ldots, r. \tag{5}$$

PROOF. In Eq. (5) and in this proof we use the notation (,) for the standard inner product on $\mathscr{F}^m$, or on $\mathscr{F}^n$. Thus, $(\boldsymbol{u}, \boldsymbol{v}) = \boldsymbol{v}^*\boldsymbol{u}$. Let $\boldsymbol{y}_1, \boldsymbol{y}_2, \ldots, \boldsymbol{y}_m$ be an orthonormal eigenbasis of AA^* in $\mathscr{F}^m$ (the second singular basis of A). Write the representation of $\boldsymbol{b}$ with respect to this basis:

$$\boldsymbol{b} = \beta_1 \boldsymbol{y}_1 + \beta_2 \boldsymbol{y}_2 + \cdots + \beta_m \boldsymbol{y}_m,$$

and note that $\beta_i = (\boldsymbol{b}, \boldsymbol{y}_i)$, $i = 1, 2, \ldots, m$. Then, using Eq. (12.8.12), we find that

$$\boldsymbol{x}_0 = A^+\boldsymbol{b} = \sum_{i=1}^{m} \beta_i A^+ \boldsymbol{y}_i = \sum_{i=1}^{r} \beta_i s_i^{-1} \boldsymbol{x}_i.$$

In view of Eq. (12.8.10),

$$\boldsymbol{y}_i = \frac{1}{s_i} A\boldsymbol{x}_i, \qquad i = 1, 2, \ldots, r,$$

and therefore

$$\beta_i s_i^{-1} = \frac{1}{s_i^2} (\boldsymbol{b}, A\boldsymbol{x}_i) = \frac{1}{s_i^2} (A^*\boldsymbol{b}, \boldsymbol{x}_i),$$

for $i = 1, 2, \ldots, r$. This completes the proof. ■

12.10 Miscellaneous Exercises

1. Let $A \in \mathscr{F}^{m \times l}$ and $B \in \mathscr{F}^{n \times k}$.

(a) Prove Proposition 12.1.4 for rectangular matrices of appropriate sizes.

(b) Show that there are permutation matrices P_1 and P_2 such that $P_1(A \otimes B)P_2 = B \otimes A$. (This is a generalization of Proposition 12.1.3.)

2. *If* $\{\boldsymbol{x}_1, \boldsymbol{x}_2, \ldots, \boldsymbol{x}_m\}$ is a basis in $\mathscr{F}^m$ and $\{\boldsymbol{y}_1, \boldsymbol{y}_2, \ldots, \boldsymbol{y}_n\}$ is a basis in $\mathscr{F}^n$, prove that the Kronecker products $\boldsymbol{y}_j^{\mathrm{T}} \otimes \boldsymbol{x}_i$, $1 \le j \le n$, $1 \le i \le m$ constitute a basis in $\mathscr{F}^{m \times n}$.

3. Prove that the Kronecker product of any two square unitary (Hermitian, positive definite) matrices is a unitary (Hermitian, positive definite) matrix.

4. Let $A \in \mathscr{F}^{m \times m}$, $B \in \mathscr{F}^{n \times n}$ be fixed and consider the transformations T_1, $T_2 \in \mathscr{L}(\mathscr{F}^{m \times n})$ defined by the rules

$$T_1(X) = AXB, \qquad T_2(X) = AX + XB,$$

for any $X \in \mathscr{F}^{m \times n}$. Prove that the matrix representations of T_1 and T_2 with respect to the standard basis $[E_{ij}]_{i,j=1}^{m,n}$ for $\mathscr{F}^{m \times n}$ (see Exercise 3.5.1(c)) are the matrices

$$B^{\mathrm{T}} \otimes A \qquad \text{and} \qquad (I_n \otimes A) + (B^{\mathrm{T}} \otimes I_m),$$

respectively.

5. If $A \in \mathscr{F}^{m \times m}$, $B \in \mathscr{F}^{n \times n}$, and $\| \;\|$ denotes either the spectral norm or the norm of Exercise 10.4.9, prove that

$$\|A \otimes B\| = \|A\| \, \|B\|.$$

Hint. For the spectral norm use Exercise 10.4.5. Compute row sums of the absolute values of the elements to prove the equality for the second norm.

6. Let $A \in \mathscr{F}^{m \times m}$ and $C \in \mathscr{F}^{n \times n}$ be invertible matrices. Prove that if $B \in \mathscr{F}^{m \times n}$ is left (respectively, right) invertible, so is the matrix ABC.

7. Let $A \in \mathscr{F}^{m \times n}$ be a left (or right) invertible matrix. Prove that there exists $\varepsilon > 0$ such that every matrix $B \in \mathscr{F}^{m \times n}$ satisfying the condition $\|A - B\| < \varepsilon$ also is left (or right) invertible.

In other words, one-sided invertibility is "stable" under small perturbations.

Hint. Pick $\varepsilon < \|A_{\mathrm{L}}^{-1}\|^{-1}$ for the left invertibility case and show that $(I + C)^{-1} A_{\mathrm{L}}^{-1} B = I$, where $C = A_{\mathrm{L}}^{-1}(B - A)$.

8. Let $A \in \mathscr{F}^{m \times r}$, $B \in \mathscr{F}^{r \times n}$ ($r \leq \min(m, n)$) be matrices of full rank. Prove that

$$(AB)^{+} = B^{+} A^{+}.$$

Find an example of two matrices A and B such that this relation does not hold.

Hint. Use Eq. (12.8.9) with $A = F$, $B = R^*$.

9. If $A \in \mathscr{F}^{n \times n}$ and $A = HU = U_1 H_1$ are the polar decompositions of A, show that $A^{+} = U^* H^{+} = H_1^{+} U_1^*$ are those of A^{+}.

Hint. Use Proposition 12.8.3d.

10. Prove that a matrix X is the Moore–Penrose inverse of A if and only if

$$XAA^* = A^* \qquad \text{and} \qquad X = BA^*$$

for some matrix B.

11. Verify that the matrices

$$\begin{bmatrix} 1 & 0 & 0 \\ 0 & 0 & 0 \end{bmatrix} \quad \text{and} \quad \begin{bmatrix} 1 & 0 \\ 0 & 0 \\ 0 & 0 \end{bmatrix}$$

are mutual Moore–Penrose inverses, each of the other.

12. Consider the permutation matrix P of Proposition 12.1.3. Show that when $m = n$, P is a symmetric matrix. In this case, show also that $A \in \mathscr{F}^{n \times n}$ is a symmetric matrix if and only if vec $A = P(\text{vec } A)$.

13. (a) Generalize Exercise 12.1.3(b) to admit rectangular A and B.
(b) Let $A \in \mathbb{C}^{m \times n}$ and rank $A = r$. Note that the equation $AXA = A$ (Eq. (12.7.1)) can be written in the form $(A^{\mathrm{T}} \otimes A)$ vec $X =$ vec A. Find the rank of $A^{\mathrm{T}} \otimes A$. What can be said about the number of linearly independent solutions of $AXA = A$?

14. Show by example that, in general, $(AB)^+ \neq B^+A^+$.

15. If B^+ is the Moore-Penrose inverse of B and $A = \begin{bmatrix} B \\ B \end{bmatrix}$, show that $A^+ = \frac{1}{2}[B^+ B^+]$.

CHAPTER 13

Stability Problems

In Section 12.3 we defined a stable matrix. Matrices of this kind are of particular significance in the study of differential equations, as demonstrated by the following proposition: *The matrix A is stable if and only if, for every solution vector $\mathbf{x}(t)$ of $\dot{\mathbf{x}} = A\mathbf{x}$, we have $\mathbf{x}(t) \to \mathbf{0}$ as $t \to \infty$.* This result is evident when the solutions $\mathbf{x}(t) = e^{At}\mathbf{x}_0$ of the equation $\dot{\mathbf{x}} = A\mathbf{x}$ (Theorem 9.10.3) are examined, as in the proof of Theorem 12.3.3 (see also Exercise 10.9.14).

In terms of inertia, a matrix $A \in \mathbb{C}^{n \times n}$ is stable if and only if its inertia (see Section 5.5) is $(0, n, 0)$. The Lyapunov method, presented in this chapter, expresses the inertia of any given matrix in terms of the inertia of a related *Hermitian* matrix and is of great theoretical importance.

In particular, the Lyapunov theory provides a criterion for a matrix to be stable that avoids a direct calculation of the characteristic polynomial of A and its zeros. Moreover, given a scalar polynomial with real coefficients, the distribution of its zeros with respect to the imaginary axis (*the Routh-Hurwitz problem*) will be obtained by use of the method of Lyapunov. Also, a solution of the *Schur–Cohn problem* of zero location with respect to the unit circle, as well as other problems concerning the location of zeros of a real polynomial, will be discussed.

13.1 The Lyapunov Stability Theory and Its Extensions

To illustrate Lyapunov's approach to stability theory, we shall be concerned in the beginning of this section with quadratic forms. Denote by v the quadratic form whose matrix is a real and symmetric matrix V (see Section

5.10), and similarly for w and W. The positive definiteness of V (or of W) will be expressed by writing $v > 0$ (or $w > 0$) and similarly for positive semi-definite forms.

Let $A \in \mathbb{R}^{n \times n}$ and consider the differential equation

$$\dot{\boldsymbol{x}}(t) = A\boldsymbol{x}(t). \tag{1}$$

Evaluate a quadratic form v at a solution $\boldsymbol{x}(t)$ of Eq. (1), where $\boldsymbol{x}(t) \in \mathbb{R}^n$ for each t. We have $v(\boldsymbol{x}) = \boldsymbol{x}^T V\boldsymbol{x}$ and hence

$$\dot{v}(\boldsymbol{x}) = \dot{\boldsymbol{x}}^T V\boldsymbol{x} + \boldsymbol{x}^T V\dot{\boldsymbol{x}} = \boldsymbol{x}^T(A^T V + VA)\boldsymbol{x}.$$

If we write

$$A^T V + VA = -W, \tag{2}$$

then clearly W is real and symmetric and $\dot{v}(\boldsymbol{x}) = -w(\boldsymbol{x})$, where $w(\boldsymbol{x}) = \boldsymbol{x}^T W\boldsymbol{x}$.

Lyapunov noted that, given a positive definite form w, the stability of A can be characterized by the existence of a positive *definite* solution matrix V for Eq. (2), for then $v(\boldsymbol{x})$ is a viable measure of the size of $\boldsymbol{x}$. Thus the equation $\dot{\boldsymbol{x}} = A\boldsymbol{x}$ has the property that $\lim_{t\to\infty} \boldsymbol{x}(t) = \boldsymbol{0}$ for every solution vector $\boldsymbol{x}(t)$ if and only if we can find positive definite forms w and v such that $\dot{v}(\boldsymbol{x}) = -w(\boldsymbol{x})$. Furthermore, the positive definite matrix W of Eq. (2) can be chosen arbitrarily, so a natural choice is often $W = I$.

To see the significance of this heuristically, let A be 2×2 and consider a solution

$$\boldsymbol{x}(t) = \begin{bmatrix} x_1(t) \\ x_2(t) \end{bmatrix},$$

in which $x_1(t)$ and $x_2(t)$ are real-valued functions of t. The level curves for a positive definite form v (see Exercise 5.10.6) are ellipses in the x_1, x_2 plane (Fig. 1), and if $\boldsymbol{x}_0 = \boldsymbol{x}(t_0)$ we locate a point on the ellipse $v(\boldsymbol{x}) = v(\boldsymbol{x}_0)$ corresponding to $\boldsymbol{x}(t_0)$. Now consider the continuous path (trajectory) of the solution $\boldsymbol{x}(t)$ in a neighbourhood of $(x_1(t_0), x_2(t_0))$. If $w > 0$ then $\dot{v} < 0$, and the path must go *inside* the ellipse $v(\boldsymbol{x}) = v(\boldsymbol{x}_0)$ as t increases. Furthermore, the path must come within an arbitrarily small distance of the origin for sufficiently large t.

Returning to a more precise and general discussion, let us now state an algebraic version of the classical Lyapunov theorem. More general inertia theorems are to be proved later in this section, as well as a more general stability theorem (Theorem 5).

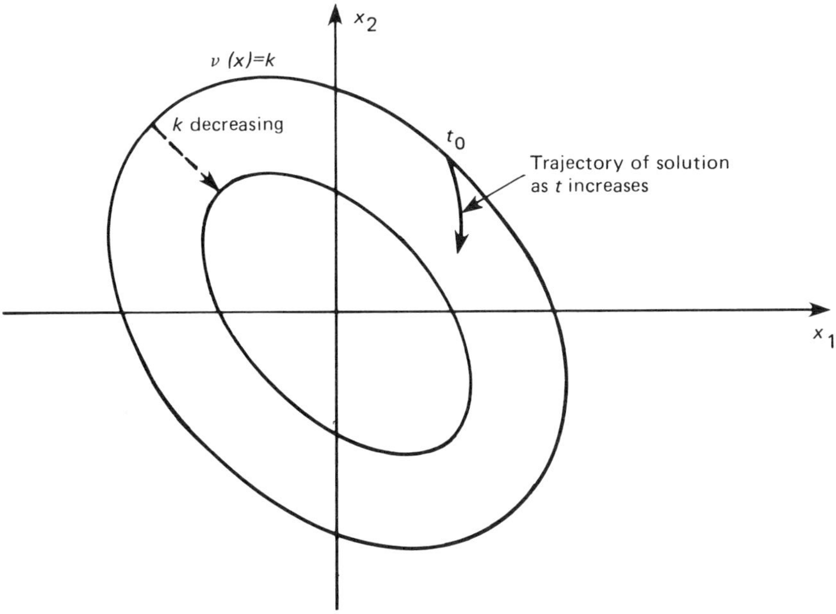

Fig. 13.1 The trajectory for a solution when A is stable.

Theorem 1 (A. M. Lyapunov†). *Let A, $W \in \mathbb{C}^{n\times n}$ and let W be positive definite.*

(*a*) *If A is stable then the equation*

$$AH + HA^* = W \tag{3}$$

has a unique solution matrix H and H is negative definite.

(*b*) *If there is a negative definite matrix H satisfying Eq. (3), then A is stable.*

PROOF. (a) Let A be stable. Then A and $-A^*$ obviously have no eigenvalues in common and therefore Theorem 12.3.2 implies the solvability and uniqueness of the solution of Eq. (3). We may use Theorem 12.3.3 to write the solution H of Eq. (1) explicitly in the form

$$H = -\int_0^\infty e^{At}We^{A^*t}\,dt.$$

† *Problème général de la stabilité du mouvement*, *Ann. Math. Studies*, no. 17, 1947, Princeton Univ. (Translation of an 1897 original.)

To show that $H < O$, compute

$$x^*Hx = -\int_0^\infty (x^*e^{At})W(x^*e^{At})^*\,dt, \tag{4}$$

and observe that the positive definiteness of W and the nonsingularity of e^{At} for all t yield the positivity of the integrand in Eq. (4). Hence $x^*Hx < 0$ for all nonzero $x \in \mathbb{C}^n$ and thus $H < O$.

(b) Conversely, let λ be an eigenvalue of A^*, where $A^*x = \lambda x$ with $x \neq 0$. Then we also have $x^*A = \bar{\lambda}x^*$. Multiplying Eq. (3) on the left and right by x^* and x, respectively, we obtain

$$(\bar{\lambda} + \lambda)x^*Hx = x^*Wx. \tag{5}$$

Our hypothesis implies that $x^*Hx < 0$ and also $x^*Wx > 0$, so we have $\bar{\lambda} + \lambda < 0$, that is, $\mathcal{Re}\,\lambda < 0$. Since this is true for each eigenvalue λ, it follows that A is stable. ■

Note that, in fact, a slightly stronger statement than part (b) of the theorem has been proved. If H is negative definite and satisfies the *Lyapunov equation* $AH + HA^* = W$, then the stability of A can be deduced if $x^*Wx < 0$ for just one eigenvector x associated with each eigenvalue of A^*. Thus, if A has multiple eigenvalues, then W need not be definite on the whole space $\mathbb{C}^n$.

Exercise 1. Verify that if $a > 0$ then both eigenvalues of

$$A = \begin{bmatrix} 0 & -1 \\ 1 & a \end{bmatrix}$$

have positive real parts. Show that the matrix

$$H = \begin{bmatrix} 4 + a^2 & -2a \\ -2a & 4 + a^2 \end{bmatrix}$$

is positive definite, and so is $AH + HA^T$.

Exercise 2. Check that the Matrix

$$A = P(S - Q),$$

where P and Q are positive definite, is stable for any skew-Hermitian matrix S.

Hint. Compute $A^*P^{-1} + P^{-1}A$ and use Theorem 1.

Exercise 3. Show that if $\mathcal{Re}\,A = \frac{1}{2}(A + A^*)$ is negative definite then A is stable. Show that the converse does not hold by considering the matrix

$$A = \begin{bmatrix} -\varepsilon & 1 \\ 0 & -\varepsilon \end{bmatrix},$$

where $\varepsilon > 0$ is small. □

In numerical practice, we can decide whether a given real matrix A is stable by solving for H in the equation $AH + HA^* = I$ (or $AH + HA^* = W$, for some other choice of W) and then applying the result of Exercise 8.5.2 to decide whether H is negative definite. Thus, A is stable if and only if the leading principal minors $d_1, d_2, \ldots, d_n$ of H satisfy the conditions

$$d_1 < 0 \qquad \text{and} \qquad d_i d_{i+1} < 0, \quad i = 1, 2, \ldots, n-1.$$

If the equation $AH + HA^* = I$ can be solved to give the coefficients of H explicitly in terms of the elements of A, then we obtain a direct determinantal characterization of stable matrices. For general matrices this is not a practical proposition, but for relatively simple classes of matrices, explicit stability criteria may be obtained in this way. In subsequent sections we develop this possibility in the particularly important case in which A is defined directly by its characteristic polynomial, for example, when A is a companion matrix. Here we keep the hypothesis that W is positive definite in Eq. (3), and we seek some connection between the inertia of A and that of H (where $H^* = H$), which should reduce to the hypothesis of Theorem 1 when A is stable.

We now consider a generalization of Lyapunov's theorem due to M. G. Krein[†] in the USSR and, independently, to A. Ostrowski and H. Schneider[‡] in the West.

Theorem 2. *Let $A \in \mathbb{C}^{n \times n}$. If H is a Hermitian matrix such that*

$$AH + HA^* = W, \qquad W > 0, \tag{6}$$

then H is nonsingular and

$$\pi(A) = \pi(H), \qquad \nu(A) = \nu(H), \qquad \delta(A) = \delta(H) = 0. \tag{7}$$

Conversely, if $\delta(A) = 0$, then there exists a Hermitian matrix H such that Eqs. (6) and (7) hold.

PROOF. Let Eq. (6) hold. Proceeding as in the proof of Theorem 1(b), we obtain Eq. (5), which shows that $W > 0$ yields $\lambda + \bar{\lambda} \neq 0$. Thus A has no pure imaginary eigenvalues and therefore $\text{In } A = \{p, n-p, 0\}$ for some integer $p \leq n$. A Jordan form J of A can then be written in the form

$$P^{-1}AP = J = \begin{bmatrix} J_1 & 0 \\ 0 & J_2 \end{bmatrix}, \tag{8}$$

[†] *Stability of solutions of differential equations in Banach space*, by Ju. L. Daleckii and M. G. Krein, Amer. Math. Soc., Providence, 1974. Theorem 2 seems to have been included in the lectures of M. G. Krein for some years before first appearing in print in a restricted edition (in Russian) of an early (1964) version of the 1974 edition.

[‡] *J. Math. Anal. Appl.* **4** (1962), 72–84. See also O. Taussky, *J. Soc. Indust. Appl. Math.* **9** (1961), 640–643.

where $J_1 \in \mathbb{C}^{p \times p}$, $J_2 \in \mathbb{C}^{(n-p) \times (n-p)}$, and $\operatorname{In} J_1 = \{p, 0, 0\}$, $\operatorname{In} J_2 = \{0, n-p, 0\}$. Define $R = P^{-1}H(P^*)^{-1}$ and rewrite Eq. (6) in the form

$$JR + RJ^* = W_0,$$

where $W_0 = P^{-1}W(P^*)^{-1} > 0$. The partitions of R and W_0 corresponding to that of J in Eq. (8) give

$$\begin{bmatrix} J_1 & 0 \\ 0 & J_2 \end{bmatrix}\begin{bmatrix} R_1 & R_2 \\ R_2^* & R_3 \end{bmatrix} + \begin{bmatrix} R_1 & R_2 \\ R_2^* & R_3 \end{bmatrix}\begin{bmatrix} J_1^* & 0 \\ 0 & J_2^* \end{bmatrix} = \begin{bmatrix} W_1 & W_2 \\ W_2^* & W_3 \end{bmatrix},$$

and hence

$$J_1R_1 + R_1J_1^* = W_1, \qquad J_2R_3 + R_3J_2^* = W_3,$$

where $W_1 > 0$, $W_3 > 0$. By Theorem 1 the matrices $R_1 \in \mathbb{C}^{p \times p}$ and $-R_3 \in \mathbb{C}^{(n-p) \times (n-p)}$ are positive definite. It remains to show that, in this case,

$$\operatorname{In} H = \operatorname{In} \begin{bmatrix} R_1 & R_2 \\ R_2^* & R_3 \end{bmatrix} = \{p, n-p, 0\} = \operatorname{In} A.$$

This follows from the easily verified congruence relation

$$Q^*RQ = \begin{bmatrix} R_1 & 0 \\ 0 & -R_2^*R_1^{-1}R_2 + R_3 \end{bmatrix}, \tag{9}$$

where

$$Q = \begin{bmatrix} I & -R_1^{-1}R_2 \\ 0 & I \end{bmatrix}.$$

Conversely, suppose that $\delta(A) = 0$, so that J has a Jordan form as described in Eq. (8). A Hermitian matrix H is to be constructed so that Eqs. (6) and (7) are valid. To this end, apply Theorem 1, which ensures the existence of negative definite matrices $H_1 \in \mathbb{C}^{p \times p}$, $H_2 \in \mathbb{C}^{(n-p) \times (n-p)}$ such that

$$-J_1H_1 - H_1J_1^* > 0,$$
$$J_2H_2 + H_2J_2^* > 0.$$

Now it is easily checked by use of Sylvester's law of inertia that the matrix

$$H = P\begin{bmatrix} -H_1 & 0 \\ 0 & H_2 \end{bmatrix}P^*$$

is Hermitian and satisfies condition (6). It remains to recall that $H_1 < 0$, $H_2 < 0$, and hence

$$\operatorname{In} H = \operatorname{In}(-H_1) + \operatorname{In} H_2 = \{p, n-p, 0\} = \operatorname{In} A. \quad \blacksquare$$

In contrast to Theorem 2 the next results are concerned with the case when $AH + HA^* = W$ is merely a positive *semi-definite* matrix. First, let us agree to write $\text{In } A \le \text{In } B$ for any two $n \times n$ matrices A and B such that

$$\pi(A) \le \pi(B), \qquad \nu(A) \le \nu(B).$$

Clearly, in this case $\delta(A) \ge \delta(B)$.

Proposition 1. *Let $A \in \mathbb{C}^{n \times n}$ and $\delta(A) = 0$. If H is Hermitian and*

$$AH + HA^* = W \ge 0, \tag{10}$$

then $\text{In } H \le \text{In } A$.

PROOF. By Theorem 2, there exists a Hermitian matrix H_0 such that

$$AH_0 + H_0 A^* > 0 \qquad \text{and} \qquad \text{In } A = \text{In } H_0.$$

Putting $H_\varepsilon = H + \varepsilon H_0$, we observe that

$$AH_\varepsilon + H_\varepsilon A^* = (AH + HA^*) + \varepsilon(AH_0 + H_0 A^*) > 0,$$

for all $\varepsilon > 0$. Thus, using Theorem 2 again, we find that for any $\varepsilon > 0$,

$$\text{In } A = \text{In } H_\varepsilon. \tag{11}$$

In particular, the last equality implies $\delta(A) = \delta(H_\varepsilon) = 0$, and hence the matrices A and H_ε have no eigenvalues on the imaginary axis. In particular, they are nonsingular. The continuity of the (real) eigenvalues of H_ε as functions of ε (see Section 10.5) then implies that none of them can, in the limit, "jump" through the origin. Hence the only possibility is that, as $\varepsilon \to 0$, one or more eigenvalues of H_ε become zero as H_ε approaches H. The result now follows from Eq. (11):

$$\text{In } A = \text{In } H_\varepsilon \ge \text{In } H. \quad \blacksquare$$

Note that if A is stable, then certainly $\delta(A) = 0$ and the proposition shows that H is semidefinite, a result that has already been seen in Exercise 12.3.1.

Proposition 2. *If $A \in \mathbb{C}^{n \times n}$ then, for any nonsingular Hermitian matrix $H \in \mathbb{C}^{n \times n}$ such that Eq.* (10) *holds,*

$$\text{In } A \le \text{In } H. \tag{12}$$

PROOF. Denote $A_\varepsilon = A + \varepsilon H^{-1}$ for ε real. Then the matrix

$$A_\varepsilon H + HA_\varepsilon^* = (AH + HA^*) + 2\varepsilon I$$

is positive definite for each $\varepsilon > 0$. Hence Theorem 2 implies $\text{In } A_\varepsilon = \text{In } H$ and, in particular, $\delta(A_\varepsilon) = 0$ for all $\varepsilon > 0$. The continuity of the eigenvalues of A_ε as functions of ε now implies that the inertia of A can differ from that of A_ε only by the number of eigenvalues that become pure imaginary in the limit as $\varepsilon \to 0$. Thus,

$$\text{In } H = \text{In } A_\varepsilon \geq \text{In } A,$$

and the proposition is established. ∎

Combining the results of Propositions 1 and 2, we obtain a generalization of Theorem 2 in which W is semi-definite.

Theorem 3 (D. Carlson and H. Schneider[†]). *If* $A \in \mathbb{C}^{n \times n}$, $\delta(A) = 0$, *and the Hermitian nonsingular matrix* $H \in \mathbb{C}^{n \times n}$ *satisfies the condition* $AH + HA^* = W \geq 0$, *then* $\text{In } A = \text{In } H$.

When A, H, and W are related as in Eq. (10), with $H = H^*$ and W positive semi-definite, the equality of In A and In H is achieved at the expense of the further hypotheses that $\delta(A) = 0$ and H is nonsingular. A useful sufficient condition for these latter hypotheses, and hence for equality of the inertias, involves the idea of controllability of a matrix pair (see Section 5.13). Recall that an $n \times n$ matrix A and an $n \times m$ matrix B together are controllable if and only if

$$\text{rank}[B \quad AB \quad \cdots \quad A^{n-1}B] = n. \tag{13}$$

Theorem 4. *If* $AH + HA^* = W \geq BB^*$, *where* $H^* = H$ *and* (A, B) *is controllable, then* $\delta(A) = \delta(H) = 0$ *and* $\text{In } A = \text{In } H$.

PROOF. We shall show that the controllability of (A, W) yields $\delta(A) = 0$ and $\det H \neq 0$, so that the required assertion follows from Theorem 3.

Suppose on the contrary that $\delta(A) \neq 0$; hence there is a nonzero vector $\mathbf{x} \in \mathbb{C}^n$ such that $\mathbf{x}^*A = i\alpha\mathbf{x}^*$ and α is real. Then $A^*\mathbf{x} = -i\alpha\mathbf{x}$ and

$$\mathbf{x}^*W\mathbf{x} = \mathbf{x}^*(AH + HA^*)\mathbf{x} = (-i\alpha + i\alpha)\mathbf{x}^*H\mathbf{x} = 0.$$

Since $W \geq BB^* \geq 0$, it follows that $0 = \mathbf{x}^*W\mathbf{x} \geq \|B^*\mathbf{x}\|^2$, that is, $\mathbf{x}^*B = \mathbf{0}^T$. But then the controllability condition is violated, since

$$\mathbf{x}^*[B \quad AB \quad \cdots \quad A^{n-1}B] = [\mathbf{x}^*B \quad i\alpha\mathbf{x}^*B \quad \cdots \quad (i\alpha)^{n-1}\mathbf{x}^*B] = \mathbf{0}^T,$$

for $\mathbf{x} \neq \mathbf{0}$, and hence $\text{rank}[B \quad AB \quad \cdots \quad A^{n-1}B] < n$. So we must have $\delta(A) = 0$.

Assume now that H is singular. Then there is a vector $\mathbf{x} \neq \mathbf{0}$ such that $H\mathbf{x} = \mathbf{0}$ and

$$\mathbf{x}^*W\mathbf{x} = \mathbf{x}^*(AH + HA^*)\mathbf{x} = 0.$$

[†] *J. Math. Anal. Appl.* **6** (1963), 430–446.

It follows that $x^*W = 0^T$ and, as noted before, this implies that $x^*B = 0^T$. Hence

$$x^*AH = x^*W - x^*HA^* = 0^T. \tag{14}$$

Now we have

$$x^*ABB^*A^*x \leq x^*AWA^*x = x^*(A^2HA^* + AH(A^*)^2)x = 0,$$

using Eq. (14) twice. Thus, $x^*AB = 0^T$. Repeating this procedure, it is found that $x^*A^kB = 0^T$ for $k = 0, 1, \ldots, n - 1$. Therefore we have found an $x \neq 0$ such that

$$x^*[B \quad AB \quad \cdots \quad A^{n-1}B] = 0^T,$$

and the pair (A, B) fails to be controllable. This shows that, actually, $\det H \neq 0$ and hence the requirements of Theorem 3 are fulfilled. ■

It is clear that if (A, W) is controllable, then we may take $B = W^{1/2}$ (the unique positive semi-definite square root of W) and obtain the next conclusion.

Corollary 1 (C. T. Chen[†]; H. K. Wimmer[‡]). *If* $AH + HA^* = W$, $H = H^*$, $W \geq 0$, *and* (A, W) *is controllable, then* $\delta(A) = \delta(H) = 0$ *and* $\text{In } A = \text{In } H$.

Now a useful generalization of the classical stability theorem is readily proved.

Theorem 5. *Let* $A \in \mathbb{C}^{m \times n}$ *with* (A, B) *controllable, and let* W *be positive semi-definite, with*

$$W \geq BB^* \geq 0.$$

Then the conclusions (a) and (b) of Theorem 1 hold.

PROOF. For conclusion (a) follow the proof of Theorem 1 and use Lemma 9.11.2 together with the controllability of (A, B) to establish the definite property of H.

Conclusion (b) is a special case of Theorem 4. ■

We conclude by noting that the inertia we have been using is often referred to as the *inertia of a matrix with respect to the imaginary axis*. In contrast, the triple $\{\pi'(A), \nu'(A), \delta'(A)\}$, where $\pi'(A)$ (respectively, $\nu'(A)$ and $\delta'(A)$)

[†] *SIAM J. Appl. Math.* **25** (1973), 158–161.

[‡] *Linear Alg. Appl.* **8** (1974), 337–343.

denotes the number of eigenvalues (counted with their multiplicities) with positive (respectively, negative and zero) imaginary parts, is said to be the *inertia of A with respect to the real axis*. Note that obviously

$$\{\pi(iA), \nu(iA), \delta(iA)\} = \{\nu'(A), \pi'(A), \delta'(A)\}.$$

Exercise 4. Check that the $n \times n$ *Schwarz matrix*

$$S = \begin{bmatrix} 0 & 1 & 0 & \cdots & 0 \\ -c_n & 0 & 1 & \ddots & \vdots \\ 0 & -c_{n-1} & & \ddots & 0 \\ \vdots & \ddots & \ddots & 0 & 1 \\ 0 & \cdots & 0 & -c_2 & -c_1 \end{bmatrix} \tag{15}$$

satisfies the Lyapunov equation

$$S^{\mathrm{T}}H + HS = W,$$

where $H = \operatorname{diag}[c_1c_2 \cdots c_n, c_2c_3 \cdots c_n, \ldots, c_{n-1}c_n, c_n]$ and

$$W = \operatorname{diag}[0, \ldots 0, -2c_n^2].$$

Exercise 5. If $c_1, c_2, \ldots, c_n$ in (15) are nonzero real numbers, show that the pair of matrices (W, S^{T}) is controllable and hence

$$\pi(S) = n - k, \qquad \nu(S) = k, \qquad \delta(S) = 0,$$

where k is the number of positive terms in the sequence

$$c_1, \quad c_1c_2, \quad c_1c_2c_3, \ldots, \quad c_1c_2 \cdots c_n.$$

Exercise 6. (Sylvester). If A is an $n \times n$ positive definite matrix and $H \in \mathbb{C}^{n \times n}$ is Hermitian, show that In AH = In H.

Exercise 7. (A generalization of Exercise 6 by H. Wielandt). If $A \in \mathbb{C}^{n \times n}$, $\mathcal{R}e\, A$ is positive definite, and $H \in \mathbb{C}^{n \times n}$ is Hermitian, show that In AH = In H.

Exercise 8. (W. Hahn[†]). Show that if $\mathcal{R}e\, A \le 0$ and $(A, \mathcal{R}e\, A)$ is a controllable pair, then A is stable (see Exercise 3).

Exercise 9. Show that if $W_1 \ge W_2 \ge 0$, then Ker $W_1 \subset$ Ker W_2. □

[†] *Monatshefte für Math.* **75** (1971), 118–122.

13.2 Stability with Respect to the Unit Circle

We have seen that the stability problem for a linear differential equation with constant coefficients is closely related to the location of the eigenvalues of a matrix with respect to the imaginary axis. An analogous role for linear difference equations is played by the location of eigenvalues relative to the unit circle.

To illustrate these ideas, let $A \in \mathbb{C}^{n \times n}$ and a nonzero $\mathbf{x}_0 \in \mathbb{C}^n$ be given, and consider the simplest difference equation:

$$\mathbf{x}_j = A\mathbf{x}_{j-1}, \qquad \mathbf{x}_j \in \mathbb{C}^n, \qquad j = 1, 2, \ldots. \tag{1}$$

This equation is a difference analogue of the differential equation (13.1.1), but its solution is much more simple: $\mathbf{x}_j = A^j\mathbf{x}_0$ for $j = 0, 1, \ldots$. We say that the matrix A in Eq. (1) is *stable (with respect to the unit circle)* if the solution sequence $\mathbf{x}_0, \mathbf{x}_1, \ldots, \mathbf{x}_j, \ldots$ of (1) converges to zero for *any* choice of the initial vector $\mathbf{x}_0$. It is easily seen that this occurs if and only if $A^j \to 0$ as $j \to \infty$, and this occurs (see Exercise 11.1.8) if and only if the spectral radius of A is less than 1. Thus, *a matrix A is stable (with respect to the unit circle) if and only if all its eigenvalues $\lambda_1, \lambda_2, \ldots, \lambda_n$ lie inside the unit circle, that is, $|\lambda_i| < 1$ for $i = 1, 2, \ldots, n$.*

In this section we develop a method analogous to that of Lyapunov for determining the number of eigenvalues of a matrix (counted with their multiplicities) lying inside and outside the unit circle and on its circumference. It turns out that this problem is closely related to the existence of positive definite solutions H of a *Stein equation* $H - A^*HA = V$ for some positive definite V.

Theorem 1. *Let $A, V \in \mathbb{C}^{n \times n}$ and let V be positive definite.*

(*a*) *If A is stable with respect to the unit circle, then the equation*

$$H - A^*HA = V \tag{2}$$

has a unique solution H, and H is positive definite.

(*b*) *If there is a positive definite matrix H satisfying Eq. (2), then A is stable with respect to the unit circle.*

PROOF. First suppose that A is stable and let $\lambda_1, \lambda_2, \ldots, \lambda_n$ denote the eigenvalues of A, so that $|\lambda_i| < 1$ for $i = 1, 2, \ldots, n$. Then (see Exercise 9.4.3) $A^* + I$ is nonsingular and therefore we may define the matrix

$$C = (A^* + I)^{-1}(A^* - I). \tag{3}$$

Observe that, by Exercise 9.4.4, the matrix C is stable (with respect to the imaginary axis). A little calculation shows that for any $H \in \mathbb{C}^{n \times n}$

$$\begin{aligned} CH + HC^* &= (A^* + I)^{-1}[(A^* - I)H(A + I) + (A^* + I)H(A - I)] \\ &\quad \times (A + I)^{-1} \\ &= 2(A^* + I)^{-1}[A^*HA - H](A + I)^{-1}. \end{aligned}$$

and hence

$$C(-H) + (-H)C^* = 2(A^* + I)^{-1}[H - A^*HA](A + I)^{-1}. \tag{4}$$

Now apply Theorem 13.1.1(a) to Eq. (4) to see that, if $H - A^*HA$ is positive definite, so that the matrix on the right of Eq. 4 is positive definite, then H is unique and is positive definite.

The converse is proved with the line of argument used in proving Theorem 13.1.1(b). Suppose that Eq. (2) holds with $H > O$. If λ is an eigenvalue of A corresponding to the eigenvector $\mathbf{x}$, then $A\mathbf{x} = \lambda\mathbf{x}$ and $\mathbf{x}^*A^* = \bar{\lambda}\mathbf{x}^*$, so that Eq. (2) implies

$$\mathbf{x}^*V\mathbf{x} = (1 - |\lambda|^2)\mathbf{x}^*H\mathbf{x}. \tag{5}$$

The positive definiteness of the matrices V and H now clearly implies $|\lambda| < 1$ for each eigenvalue λ of A. ■

Following the lead set in the previous section, we now seek a generalization of Theorem 1 of the following kind: retain the hypotheses $V > O$ and $H^* = H$, $\det H \neq 0$, for matrices of Eq. (2) and look for a connection between the inertia of H (relative to the imaginary axis) and that of A (relative to the unit circle). Theorem 2 provides such a generalization and is due to Taussky, Hill, and Wimmer.†

Theorem 2. *Let $A \in \mathbb{C}^{n \times n}$. If H is a Hermitian matrix satisfying the Stein equation* (2) *with $V > 0$, then H is nonsingular and the matrix A has $\pi(H)$ (respectively, $\nu(H)$) eigenvalues lying inside (respectively, outside) the (open) unit circle and no eigenvalues of modulus* 1.

Conversely, if A has no eigenvalues of modulus 1, *then there exists a Hermitian matrix H such that Eq.* (2) *holds and its inertia gives the distribution of eigenvalues of A with respect to the unit circle, as described above.*

PROOF. If Eq. (2) is valid, then the relation (5) shows that A has no eigenvalues of modulus 1. Proceeding as in the proof of Theorem 13.1.2, let the Jordan form of A be

$$J = \begin{bmatrix} J_1 & 0 \\ 0 & J_2 \end{bmatrix} = S^{-1}AS, \tag{6}$$

† See, respectively, *J. Algebra* **1** (1964), 5–10; *Linear Alg. Appl.* **2** (1969), 131–142; *J. Math. Anal. Appl.* **14** (1973), 164–169.

where $J_1 \in \mathbb{C}^{p \times p}$, $J_2 \in \mathbb{C}^{(n-p)\times(n-p)}$, and the eigenvalues of J_1 and J_2 lie inside and outside the unit circle, respectively. Preserving the notation of the proof of Theorem 13.1.2, Eq. (2) yields the equation

$$R - J^*RJ = V_0, \qquad V_0 > 0,$$

and, partitioning R and V_0 compatibly with J, we obtain

$$R_1 - J_1^*R_1J_1 = V_1, \qquad R_3 - J_2^*R_3J_2 = V_3,$$

where $V_1 > 0$, $V_3 > 0$. Now Theorem 1 yields $R_1 > 0$ and $R_3 < 0$ and therefore, using Eq. (13.1.9), it is found that

$$\operatorname{In} H = \operatorname{In} R = (p, n - p, 0),$$

as required.

Conversely, suppose that $|\lambda| \neq 1$ for any eigenvalue λ of A. Then A can be transformed to the Jordan form (6). By Theorem 1 there are positive definite matrices $H_1 \in \mathbb{C}^{p \times p}$, $H_2 \in \mathbb{C}^{(n-p)\times(n-p)}$ such that

$$H_1 - J_1^*H_1J_1 = V_1, \qquad V_1 > 0,$$
$$H_2 - (J_2^*)^{-1}H_2J_2^{-1} = V_2, \qquad V_2 > 0.$$

Hence the positive definite matrix

$$H = (S^*)^{-1}\begin{bmatrix} H_1 & 0 \\ 0 & -(J_2^*)^{-1}H_2J_2^{-1} \end{bmatrix} S^{-1}$$

satisfies the condition $H - A^*HA > 0$. Obviously, the number of eigenvalues of A lying inside (respectively, outside) the unit circle is equal to $p = \pi(H)$ (respectively, $n - p = \nu(H)$), as required. ■

The following results are analogues of Theorems 13.1.3 and 13.1.4 and can be established in the same fashion. They are stated here without proof.

Theorem 3. *Let $A \in \mathbb{C}^{n \times n}$ and let A have no eigenvalues of modulus 1. If there exists an $n \times n$ nonsingular Hermitian matrix H such that*

$$H - A^*HA = V, \qquad V \geq 0, \tag{7}$$

then A has $\pi(H)$ and $\nu(H)$ eigenvalues of modulus less than 1 and greater than 1, respectively.

Theorem 4. *If Eq. (7) holds and the pair (A^*, V) is controllable, then A has no eigenvalues of modulus 1, $\det H \neq 0$, and the assertion of Theorem 3 is valid.*

13.3 The Bezoutian and the Resultant

The proofs of the stability results given in the following sections depend on two broad areas of development. The first is the development of inertia theory for matrices, which we have already presented, and the second is a more recent development of properties of Bezout matrices for a pair of scalar polynomials, which will now be discussed.

Let $a(\lambda) = \sum_{j=0}^{l} a_j\lambda^j$ and $b(\lambda) = \sum_{j=0}^{m} b_j\lambda^j$ $(a_l \neq 0, b_m \neq 0, l \geq m)$ be arbitrary complex polynomials. Since $b(\lambda)$ can be viewed as $\sum_{j=0}^{l} b_j\lambda^j$, where $b_{m+1} = \cdots = b_l = 0$, we may assume without loss of generality that the polynomials are of the form $\sum_{j=0}^{l} a_j\lambda^j$, $a_l \neq 0$, and $\sum_{j=0}^{l} b_j\lambda^j$. The expression

$$\frac{a(\lambda)b(\mu) - b(\lambda)a(\mu)}{\lambda - \mu}$$

is easily verified to be a polynomial in the two variables λ and μ:

$$\frac{a(\lambda)b(\mu) - b(\lambda)a(\mu)}{\lambda - \mu} = \sum_{i,j=0}^{l-1} \gamma_{ij}\lambda^i\mu^j. \tag{1}$$

Then the matrix $B = [\gamma_{ij}]_{i,j=0}^{l-1}$ is called the *Bezout matrix* (or *Bezoutian*) associated with the polynomials $a(\lambda)$ and $b(\lambda)$. To emphasize this, we also write $B = \text{Bez}(a, b)$.

The main objectives of this section are to obtain formulae expressing $\text{Bez}(a, b)$ in terms of the coefficients of a and b (Eqs. (7), (17), and (20), in particular), to make the connection with the *resultant* of $a(\lambda)$ and $b(\lambda)$ (Eq. (24)), and to make the first deductions from these concerning the greatest common divisor of $a(\lambda)$ and $b(\lambda)$ in the form of Theorems 1 and 2.

Exercise 1. Check that

$$\text{Bez}(a, b) = -\text{Bez}(b, a),$$

for any two polynomials $a(\lambda)$ and $b(\lambda)$ and that $\text{Bez}(a, b)$ is symmetric.

Exercise 2. Show that the matrix

$$B = \begin{bmatrix} 0 & -1 & 4 \\ -1 & 4 & 0 \\ 4 & 0 & -2 \end{bmatrix}$$

is the Bezout matrix associated with the polynomials $a(\lambda) = 1 - \lambda^2 + 2\lambda^3$ and $b(\lambda) = 2 - \lambda^2 = 2 - \lambda^2 + 0\lambda^3$. □

If $a(\lambda) = a_0 + a_1\lambda + \cdots + a_l\lambda^l$, the triangular *Hankel matrix of coefficients*,

$$S(a) = \begin{bmatrix} a_1 & a_2 & \cdots & a_l \\ a_2 & \cdots & a_l & 0 \\ \vdots & \iddots & \iddots & \vdots \\ a_l & 0 & \cdots & 0 \end{bmatrix}, \tag{2}$$

arises very frequently. For a reason that will be clear in a moment (see Exercises 3 and 4), we call $S(a)$ the *symmetrizer* of $a(\lambda)$. Note that $S(a)$ does not depend on a_0.

Exercise 3. If C_a denotes the companion matrix of $a(\lambda)$,

$$C_a = \begin{bmatrix} 0 & 1 & 0 & \cdots & 0 \\ 0 & 0 & 1 & \ddots & \vdots \\ \vdots & & \ddots & \ddots & 0 \\ 0 & \cdots & & 0 & 1 \\ -\tilde{a}_0 & -\tilde{a}_1 & & \cdots & -\tilde{a}_{l-1} \end{bmatrix}, \qquad C_a \in \mathbb{C}^{l \times l},$$

where $\tilde{a}_j = a_l^{-1}a_j, j = 0, 1, \ldots, l-1$, check that

$$S(a)C_a = \begin{bmatrix} -a_0 & 0 & & \cdots & 0 \\ 0 & a_2 & & \cdots & a_l \\ \vdots & \vdots & & \iddots & 0 \\ & & \iddots & & \vdots \\ 0 & a_l & 0 & \cdots & 0 \end{bmatrix}.$$

and note that this matrix is symmetric. □

Define the polynomial $\hat{a}(\lambda) = \lambda^l a(\lambda^{-1}) = a_0\lambda^l + \cdots + a_{l-1}\lambda + a_l$; *its* symmetrizer is

$$S(\hat{a}) = \begin{bmatrix} a_{l-1} & a_{l-2} & \cdots & a_0 \\ a_{l-2} & \cdots & a_0 & 0 \\ \vdots & \iddots & \iddots & \vdots \\ a_0 & 0 & \cdots & 0 \end{bmatrix}. \tag{3}$$

Another device will also be useful. Define the permutation matrix P (the *rotation matrix*) by

$$P = \begin{bmatrix} 0 & 0 & \cdots & 0 & 1 \\ 0 & \cdots & 0 & 1 & 0 \\ \vdots & \iddots & \iddots & & \vdots \\ 0 & 1 & & & \\ 1 & 0 & \cdots & 0 & 0 \end{bmatrix}, \tag{4}$$

and observe that $P^2 = I$ and that $S(a)P$ and $PS(a)$ are Toeplitz matrices:

$$S(a)P = \begin{bmatrix} a_l & \cdots & a_2 & a_1 \\ 0 & & & a_2 \\ \vdots & \ddots & \ddots & \vdots \\ 0 & \cdots & 0 & a_l \end{bmatrix}, \qquad PS(a) = \begin{bmatrix} a_l & 0 & \cdots & 0 \\ \vdots & \ddots & \ddots & \vdots \\ a_2 & & \ddots & 0 \\ a_1 & a_2 & \cdots & a_l \end{bmatrix}.$$

Exercise 4. Generalize the result of Exercise 3. Namely, show that, for any $1 \le k \le l - 1$,

$$S(a)C_a^k = \operatorname{diag}[-PS_1P, S_2], \qquad P, S_1 \in \mathbb{C}^{k \times k}, \quad S_2 \in \mathbb{C}^{(l-k)\times(l-k)}, \tag{5}$$

where S_1 and S_2 stand for the symmetrizers of the polynomials $a_1(\lambda) = a_k + a_{k-1}\lambda + \cdots + a_0\lambda^k$ and $a_2(\lambda) = a_k + a_{k+1}\lambda + \cdots + a_l\lambda^{l-k}$, respectively.

Also, show that

$$S(a)C_a^l = -PS(\hat{a})P. \quad \square \tag{6}$$

The formula derived in the following proposition will be useful in the sequel.

Proposition 1. *With the previous notation, the Bezoutian B of scalar polynomials $a(\lambda)$ and $b(\lambda)$ is given by*

$$B = S(a)S(\hat{b})P - S(b)S(\hat{a})P. \tag{7}$$

PROOF. Rewrite Eq. (1) in the form

$$(\lambda - \mu)\sum_{i=0}^{l-1} \gamma_i(\mu)\lambda^i = a(\lambda)b(\mu) - b(\lambda)a(\mu),$$

where $\gamma_i(\mu) = \sum_{j=0}^{l-1} \gamma_{ij}\mu^j$ for $i = 0, 1, \ldots, l-1$. Comparing the coefficients of λ^i on both sides of the equation we obtain

$$\begin{aligned} \gamma_0(\mu) - \mu\gamma_1(\mu) &= a_1 b(\mu) - b_1 a(\mu), \\ \gamma_1(\mu) - \mu\gamma_2(\mu) &= a_2 b(\mu) - b_2 a(\mu), \\ &\vdots \\ \gamma_{l-2}(\mu) - \mu\gamma_{l-1}(\mu) &= a_{l-1} b(\mu) - b_{l-1} a(\mu), \\ \gamma_{l-1}(\mu) &= a_l b(\mu) - b_l a(\mu). \end{aligned}$$

From these equations it is easily deduced that for $i = 0, 1, \ldots, l-1$,

$$\begin{aligned} \gamma_i(\mu) = {} & (\mu^{l-i-1}a_l + \mu^{l-i-2}a_{l-1} + \cdots + \mu a_{i+2} + a_{i+1})b(\mu) \\ & - (\mu^{l-i-1}b_l + \mu^{l-i-2}b_{l-1} + \cdots + \mu b_{i+2} + b_{i+1})a(\mu). \end{aligned}$$

Note that $\deg(\gamma_i(\mu)) \le l - 1$ for $0 \le i \le l - 1$. Defining

$$\gamma_i^{\mathrm{T}} = [\gamma_{i0} \quad \gamma_{i1} \quad \cdots \quad \gamma_{i,l-1}]$$

for $i = 0, 1, \ldots, l - 1$ we see that γ_i^T forms row $(i + 1)$ in the Bezout matrix B and that the last equation implies

$$\gamma_i^T \begin{bmatrix} 1 \\ \mu \\ \vdots \\ \mu^{l-1} \end{bmatrix} = [a_{i+1} \quad a_{i+2} \quad \cdots \quad a_l \quad 0 \quad \cdots \quad 0] \begin{bmatrix} b(\mu) \\ b(\mu)\mu \\ \vdots \\ b(\mu)\mu^{l-1} \end{bmatrix}$$
$$- [b_{i+1} \quad b_{i+2} \quad \cdots \quad b_l \quad 0 \quad \cdots \quad 0] \begin{bmatrix} a(\mu) \\ a(\mu)\mu \\ \vdots \\ a(\mu)\mu^{l-1} \end{bmatrix}. \qquad (8)$$

Now a little calculation will verify that for any polynomial $c(\mu)$ with degree at most l,

$$\operatorname{col}(c(\mu)\mu^{i-1})_{i=1}^{l} = S(\hat{c})P \operatorname{col}(\mu^{i-1})_{i=1}^{l} + PS(c) \operatorname{col}(\mu^{i-1})_{i=1}^{l}\mu^{l}, \qquad (9)$$

where by definition

$$\operatorname{col}(R_i)_{i=0}^{l-1} \triangleq \begin{bmatrix} R_0 \\ R_1 \\ \vdots \\ R_{l-1} \end{bmatrix}, \qquad (10)$$

and, in general, $R_0, R_1, \ldots, R_{l-1}$ may be scalars or matrices of the same size.

Using Eq. (9) for $a(\lambda)$ and $b(\lambda)$, we find from Eq. (8) that

$$B \operatorname{col}(\mu^{i-1})_{i=1}^{l} = S(a)(S(\hat{b})P \operatorname{col}(\mu^{i-1})_{i=1}^{l} + PS(b) \operatorname{col}(\mu^{i-1})_{i=1}^{l}\mu^{l})$$
$$- S(b)(S(\hat{a})P \operatorname{col}(\mu^{i-1})_{i=1}^{l} + PS(a) \operatorname{col}(\mu^{i-1})_{i=1}^{l}\mu^{l}).$$

Note that for any $a(\lambda)$ and $b(\lambda)$,

$$S(a)PS(b) = S(b)PS(a), \qquad S(\hat{a})PS(\hat{b}) = S(\hat{b})PS(\hat{a}). \qquad (12)$$

Thus, using the first equality in (12), we obtain

$$B \operatorname{col}(\mu^{i-1})_{i=1}^{l} = S(a)S(\hat{b})P \operatorname{col}(\mu^{i-1})_{i=1}^{l} - S(b)S(\hat{a})P \operatorname{col}(\mu^{i-1})_{i=1}^{l}.$$

Since the last relation holds for any $\mu \in \mathbb{C}$, and the Vandermonde matrix for distinct $\mu_1, \mu_2, \ldots, \mu_l$ is nonsingular (see Exercise 2.3.6), the required formula (7) is established. ■

A dual expression for the Bezoutian $B = \operatorname{Bez}(a, b)$ is presented next.

Exercise 5. Check that

$$-B = PS(\hat{a})S(b) - PS(\hat{b})S(a). \qquad (13)$$

Hint. Rewrite Eq. (9) in terms of rows. □

It is easily seen that, if the polynomial a is fixed then the matrix $\text{Bez}(a, b)$ is a linear function of the polynomial b. Indeed, if $b(\lambda)$ and $c(\lambda)$ are polynomials with degree not exceeding l and $\beta, \gamma \in \mathbb{C}$, we see that in the definition of Eq. (1),

$$\begin{aligned} &a(\lambda)\{\beta b(\mu) + \gamma c(\mu)\} - \{\beta b(\lambda) + \gamma c(\lambda)\}a(\mu) \\ &\quad = \beta\{a(\lambda)b(\mu) - b(\lambda)a(\mu)\} + \gamma\{a(\lambda)c(\mu) - c(\lambda)a(\mu)\}. \end{aligned}$$

This immediately implies that

$$\text{Bez}(a, \beta b + \gamma c) = \beta \text{ Bez}(a, b) + \gamma \text{ Bez}(a, c). \tag{14}$$

In particular, writing $b(\lambda) = \sum_{j=0}^{m} b_j \lambda^j$ we have

$$\text{Bez}(a, b) = \sum_{j=0}^{m} b_j \text{ Bez}(a, \lambda^j). \tag{15}$$

Now examine the "primitive" Bezout matrices $\text{Bez}(a, \lambda^j)$ for $j = 0, 1, 2, \ldots, m$. Calculating directly with the definition of Eq. (1) we find first of all that $\text{Bez}(a, 1) = S(a)$. More generally, to study $\text{Bez}(a, \lambda^k)$ we note that

$$\begin{aligned} \frac{a(\lambda)\mu^k - \lambda^k a(\mu)}{\lambda - \mu} &= \sum_{j=0}^{l} a_j \frac{\lambda^j \mu^k - \lambda^k \mu^i}{\lambda - \mu} \\ &= -\sum_{j=0}^{k-1} a_j(\lambda^k \mu^j + \cdots + \lambda^j \mu^{k-1}) \\ &\quad + \sum_{j=k+1}^{l} a_j(\lambda^{j-1}\mu^k + \cdots + \lambda^k \mu^{j-1}). \end{aligned}$$

Checking the form of the matrices in Eq. (5) we find that

$$\text{Bez}(a, \lambda^k) = \text{diag}[-PS_1P, S_2] = S(a)C_a^k. \tag{16}$$

Combining this with Eq. (15) we get the first part of the following proposition, known as the *Barnett factorization* of $\text{Bez}(a, b)$.

Proposition 2.

$$\text{Bez}(a, b) = S(a)b(C_a) = -S(b)a(C_b). \tag{17}$$

PROOF. For the second statement, we first use Exercise 1 and then apply the first statement. Thus,

$$\text{Bez}(a, b) = -\text{Bez}(b, a) = -S(b)a(C_b),$$

as required. ■

It will be useful to have at hand another representation for the Barnett factorization. Observe first that by the Cayley–Hamilton theorem, $a(C_a) = 0$, and therefore

$$b(C_a) = b(C_a) - a_l^{-1} b_l a(C_a) = \sum_{j=0}^{l} \tilde{b}_j C_a^j,$$

where $\tilde{b}_j = b_j - a_l^{-1} b_l a_j$, for $j = 0, 1, \ldots, l - 1$. Define

$$\boldsymbol{\beta}^T = [\tilde{b}_0 \quad \tilde{b}_1 \quad \cdots \quad \tilde{b}_{l-1}] \tag{18}$$

and denote the ith row of $b(C_a)$ by $\boldsymbol{d}_i^T$ for $i = 1, 2, \ldots, l$. Since

$$\boldsymbol{e}_1^T C_a^j = [\underbrace{0 \quad \cdots \quad 0}_{j} \quad 1 \quad 0 \quad \cdots \quad 0]$$

we obtain

$$\boldsymbol{d}_1^T = \boldsymbol{e}_1^T b(C_a) = [\tilde{b}_0 \quad \tilde{b}_1 \quad \cdots \quad \tilde{b}_{l-1}] = \boldsymbol{\beta}^T. \tag{19}$$

Furthermore, for $j = 2, 3, \ldots, l$,

$$\boldsymbol{d}_j^T = \boldsymbol{e}_j^T b(C_a) = \boldsymbol{e}_{j-1}^T C_a b(C_a) = \boldsymbol{e}_{j-1}^T b(C_a) C_a = \boldsymbol{d}_{j-1}^T C_a = \boldsymbol{d}_1^T C_a^{j-1} = \boldsymbol{\beta}^T C_a^{j-1}.$$

Thus,

$$b(C_a) = \begin{bmatrix} \boldsymbol{\beta}^T \\ \boldsymbol{\beta}^T C_a \\ \vdots \\ \boldsymbol{\beta}^T C_a^{l-1} \end{bmatrix},$$

and combining with the first equation of (17) we have

$$\operatorname{Bez}(a, b) = S(a) \begin{bmatrix} \boldsymbol{\beta}^T \\ \boldsymbol{\beta}^T C_a \\ \vdots \\ \boldsymbol{\beta}^T C_a^{l-1} \end{bmatrix} \tag{20}$$

Theorem 1. *The polynomials $a(\lambda)$ and $b(\lambda)$ have no zeros in common if and only if the associated Bezout matrix* Bez(a, b) *is nonsingular.*

PROOF. For definiteness, let $l = \deg a(\lambda) \geq \deg b(\lambda)$. Write the corresponding Bezout matrix $B = \operatorname{Bez}(a, b)$ in the first form of (17) and observe that $\det B \neq 0$ if and only if $b(C_a)$ is nonsingular. Since the eigenvalues of the latter matrix are of the form $b(\lambda_i)$, where $\lambda_1, \lambda_2, \ldots, \lambda_l$ are the zeros of $a(\lambda)$, the result follows. ■

Let $a(\lambda) = \sum_{i=0}^{l} a_i\lambda^i$, $a_l \neq 0$, and $b(\lambda) = \sum_{i=0}^{m} b_i\lambda^i$, $b_m \neq 0$, where $l \geq m$. As in the definition of the Bezout matrix, we may assume without loss of generality that $b(\lambda)$ is written in the form $\sum_{i=0}^{l} b_i\lambda^i$, where the coefficients of the higher powers are, perhaps, zeros. With this assumption, we define the *resultant* (or *Sylvester*) *matrix* $R(a, b)$ associated with the polynomials $a(\lambda)$ and $b(\lambda)$ as the $2l \times 2l$ matrix

$$R(a, b) = \begin{bmatrix} a_0 & a_1 & \cdots & a_{l-1} & a_l & & & & \\ & \ddots & \ddots & & & \ddots & \ddots & & \\ & & & a_0 & a_1 & \cdots & & a_{l-1} & a_l \\ b_0 & b_1 & \cdots & b_m & 0 & \cdots & 0 & & \\ & \ddots & \ddots & & \ddots & & \ddots & & \\ & & & b_0 & b_1 & \cdots & b_m & 0 \cdots & 0 \end{bmatrix} \begin{matrix} \}l \\ \\ \}l \end{matrix}.$$

Partitioning $R(a, b)$ into $l \times l$ matrices and using the notation introduced above we may also write

$$R(a, b) = \begin{bmatrix} S(\hat{a})P & PS(a) \\ S(\hat{b})P & PS(b) \end{bmatrix}.$$

Using Eqs. (7) and (12) it follows that

$$\begin{bmatrix} P & 0 \\ -S(b) & S(a) \end{bmatrix} R(a, b) = \begin{bmatrix} PS(\hat{a})P & S(a) \\ \mathrm{Bez}(a, b) & 0 \end{bmatrix}$$

$$= \begin{bmatrix} I & 0 \\ 0 & \mathrm{Bez}(a, b) \end{bmatrix} \begin{bmatrix} PS(\hat{a})P & S(a) \\ I & 0 \end{bmatrix}. \tag{21}$$

This equation demonstrates an intimate connection between the resultant of $a(\lambda)$ and $b(\lambda)$ and their Bezout matrix. In particular, it is clear from equation (21) that $R(a, b)$ and $\mathrm{Bez}(a, b)$ have the same rank. Also, the resultant matrix (as well as the Bezoutian) can be used in solving the problem of whether the polynomials $a(\lambda)$ and $b(\lambda)$ have zeros in common.

Theorem 2. *The polynomials $a(\lambda)$ and $b(\lambda)$ defined above have no common zeros if and only if the resultant matrix $R(a, b)$ is nonsingular.*

PROOF. Since $S(a)$ is nonsingular, it follows immediately from Eq. (21) that $R(a, b)$ is nonsingular if and only if $\mathrm{Bez}(a, b)$ is nonsingular. So the result follows from Theorem 1. ■

There is an interesting and more general result that we will not discuss. Namely, that the degree of the greatest common divisor of $a(\lambda)$ and $b(\lambda)$ is just $l - \mathrm{rank}\, R(a, b)$.

Note that the use of the resultant matrix may be more convenient than the use of the Bezoutian in the search for common divisors since, despite its double size, the elements of the resultant are just the coefficients of the given polynomials (or zero).

Exercise 6. Verify that the Bezout matrix associated with the polynomials $a(\lambda) = \sum_{i=0}^{l} a_i\lambda^i$ and $\bar{a}(\lambda) = \sum_{i=0}^{l} \bar{a}_i\lambda^i$ is skew-Hermitian. Combine this with Exercise 1 to show that the elements of Bez$(a, \bar{a})$ are all pure imaginary.

Exercise 7. Check that the Bezout matrix B associated with the polynomials $a(\lambda) = \sum_{i=0}^{l} a_i\lambda^i$, $a_l \neq 0$, and $b(\lambda) = \sum_{i=0}^{m} b_i\lambda^i$, $l \geq m$, satisfies the equation

$$\hat{C}_a B = BC_a, \tag{22}$$

where C_a and $\hat{C}_a$ denote the first and second companion matrices associated with $a(\lambda)$, respectively. (The second companion matrix is defined by $\hat{C}_a = C_a^{\mathrm{T}}$, and C_a is defined in Exercise 3.)

Hint. Apply Eq. (17).

Exercise 8. Let Bez(a, b) denote the Bezout matrix associated with the polynomials $a(\lambda)$ and $b(\lambda)$ of degrees l and m $(l \geq m)$, respectively. Show that

$$\operatorname{Bez}(\hat{a}, \hat{b}) = -P \operatorname{Bez}(a, b)P,$$

where P is defined in Eq. (4).

Exercise 9. Let $a(\lambda)$ and $b(\lambda)$ be two polynomials of degrees l and m $(l \geq m)$, respectively. Denote $a_\alpha = a_\alpha(\lambda) = a(\lambda + \alpha)$, $b_\alpha = b_\alpha(\lambda) = b(\lambda + \alpha)$, and check that

$$\operatorname{Bez}(a_\alpha, b_\alpha) = V_\alpha^{(l)} \operatorname{Bez}(a, b)(V_\alpha^{(l)})^{\mathrm{T}},$$

where $V_\alpha^{(l)} = \left[\binom{j}{i}\alpha^{j-i}\right]_{i,j=0}^{l-1}$ and it is assumed that $\binom{j}{i} = 0$ for $j < i$.

Exercise 10. Prove that, for any polynomials $a(\lambda)$ and $b(\lambda)$ of degree not exceeding l,

$$R^{\mathrm{T}}(-b, a)\begin{bmatrix} P & 0 \\ 0 & P \end{bmatrix} R(a, b) = \begin{bmatrix} 0 & -B \\ B & 0 \end{bmatrix}, \qquad B = \operatorname{Bez}(a, b). \quad \square$$

13.4 The Hermite and the Routh–Hurwitz Theorems

The question of stability of eigenvalues can always be posed in terms of the zeros of a polynomial rather than the eigenvalues of a matrix. We need only replace the eigenvalue equation, $A\boldsymbol{x} = \lambda\boldsymbol{x}$, by the characteristic equation, $\det(\lambda I - A) = 0$. One advantage of the Lyapunov theorem of Section 13.1 is that it enables us to avoid computation of the coefficients of this polynomial and any consequent loss of accuracy. Nevertheless, it is important that we develop necessary and sufficient conditions that all the zeros of a polynomial lie in the left half of the complex plane. Consider a polynomial

$$a(\lambda) = a_0 + a_1\lambda + \cdots + a_l\lambda^l, \qquad a_l \neq 0,$$

with *complex* coefficients. The *inertia* In(a) of the polynomial $a(\lambda)$ is defined as the triple of nonnegative integers $\{\pi(a), \nu(a), \delta(a)\}$, where $\pi(a)$ (respectively, $\nu(a)$ and $\delta(a)$) denotes the number of zeros of $a(\lambda)$, counting multiplicities, with positive (respectively, negative and zero) real parts. Like matrices, the inertia of a polynomial $a(\lambda)$ is also called the *inertia of* $a(\lambda)$ *with respect to the imaginary axis*. In particular, a polynomial $a(\lambda)$ with $\pi(a) = \delta(a) = 0$ is said to be a *stable* polynomial (with respect to the imaginary axis).

In contrast, there is the triple $In'(a) = \{\pi'(a), \nu'(a), \delta'(a)\}$, where $\pi'(a)$, $\nu'(a)$, $\delta'(a)$ denote the number of zeros of $a(\lambda)$ lying in the open upper half-plane, the open lower half-plane, and on the real axis, respectively. This triple is referred to as the *inertia of* $a(\lambda)$ *with respect to the real axis*. Clearly,

$$\{\pi'(\tilde{a}), \nu'(\tilde{a}), \delta'(\tilde{a})\} = \{\pi(a), \nu(a), \delta(a)\},$$

where $\tilde{a} = \tilde{a}(\lambda) = a(-i\lambda)$. If $a(\lambda)$ is a polynomial with real coefficients, then its complex zeros occur in conjugate pairs and hence $\pi'(a) = \nu'(a)$.

The main result of this section is due to C. Hermite[†] and is concerned with the distribution of zeros of a complex polynomial with respect to the real axis. The formulation in terms of the Bezoutian is due to M. Fujiwara[‡] in 1926. Indeed, it was Fujiwara who first used Bezout matrices to develop a systematic approach to stability criteria, including those of Sections 13.4 and 13.5.

Consider a polynomial $a(\lambda) = \sum_{j=0}^{l} a_j \lambda^j$ $(a_l \neq 0)$ and let $\bar{a}(\lambda) = \sum_{j=0}^{l} \bar{a}_j \lambda^j$. If $B = \text{Bez}(a, \bar{a})$ denotes the Bezout matrix associated with $a(\lambda)$ and $\bar{a}(\lambda)$, then it is skew-Hermitian (see Exercise 13.3.6) and it is also nonsingular if and only if $a(\lambda)$ and $\bar{a}(\lambda)$ have no zeros in common (Theorem 13.3.1). Thus, det $B \neq 0$ if and only if $a(\lambda)$ has no real zeros and no zeros occuring in conjugate pairs.

Theorem 1. *With the notation of the previous paragraph, assume that* det $B \neq 0$. *Then*

$$\pi'(a) = \pi\left(\frac{1}{i} B\right),$$

$$\nu'(a) = \nu\left(\frac{1}{i} B\right),$$

$$\delta'(a) = \delta\left(\frac{1}{i} B\right) = 0.$$

[†] *C. R. Acad. Sci. Paris* **35** (1852), 52–54; **36** (1853), 294–297. See also *J. für die Reine u. Angew. Math.* **52** (1856), 39–51.

[‡] *Math. Zeit.* **24** (1926), 161–169.

PROOF. In applying formula (13.3.22), observe that

$$\hat{C}_a = C_a^* - [0 \quad \cdots \quad 0 \quad \boldsymbol{v}],$$

where $\boldsymbol{v} = [a_0 a_l^{-1} - \bar{a}_0 \bar{a}_l^{-1}, \ldots, a_{l-1} a_l^{-1} - \bar{a}_{l-1} \bar{a}_l^{-1}]^{\mathrm{T}}$.

Thus

$$C_a^* B - BC_a = [0 \quad \cdots \quad 0 \quad \boldsymbol{v}]B = W, \tag{1}$$

and we are now to show that $W \geq 0$. Using the first factorization of (13.3.17), and noting that

$$[0 \quad \cdots \quad 0 \quad \boldsymbol{v}]S(a) = a_l[\boldsymbol{v} \quad 0 \quad \cdots \quad 0],$$

we have

$$W = [0 \quad \cdots \quad 0 \quad \boldsymbol{v}]B = a_l[\boldsymbol{v} \quad 0 \quad \cdots \quad 0]\bar{a}(C_a) = a_l \boldsymbol{v} \boldsymbol{e}_1^T \bar{a}(C_a).$$

But

$$\boldsymbol{e}_1^T \bar{a}(C_a) = \sum_{j=0}^{l} \bar{a}_j \boldsymbol{e}_1^T C_a^j = \sum_{j=0}^{l-1} \bar{a}_j \boldsymbol{e}_{j+1}^T + \bar{a}_l \boldsymbol{e}_l^T C_a = -\bar{a}_l \boldsymbol{v},$$

and consequently, $W = -|a_l|^2 \boldsymbol{v}\boldsymbol{v}^T = |a_l|^2 \boldsymbol{v}\boldsymbol{v}^* \geq 0$.

Write Eq. (1) in the form

$$(-iC_a)^*\left(\frac{1}{i} B\right) + \left(\frac{1}{i} B\right)(-iC_a) = W \geq 0, \tag{2}$$

and recall that the assumption of nonsingularity of B implies that $\delta(iC_a) = 0$. Hence Theorem 13.1.3 yields

$$\operatorname{In}(-iC_a^*) = \operatorname{In}\left(\frac{1}{i} B\right),$$

and the assertion follows in view of the relations

$$\pi(-iC_a)^* = \pi(-iC_a) = \nu(iC_a) = \pi'(C_a) = \pi'(a),$$

and similarly for $\nu(-iC_a)^*$. ■

Corollary 1. *All the zeros of the polynomial $a(\lambda)$ lie in the open upper half-plane if and only if $(1/i)B > O$, where $B = \operatorname{Bez}(a, \bar{a})$.*

Example 1. The polynomial

$$a(\lambda) = \lambda^4 + 4i\lambda^3 - 4\lambda^2 - 1 = (\lambda + i)^2(\lambda^2 + 2i\lambda + 1)$$

clearly has zeros $-i$, $-i$, $(-1 \pm \sqrt{2})i$, We have

$$\begin{aligned} B &= \mathrm{Bez}(a, \tilde{a}) \\ &= \begin{bmatrix} 0 & -4 & 4i & 1 \\ -4 & 4i & 1 & 0 \\ 4i & 1 & 0 & 0 \\ 1 & 0 & 0 & 0 \end{bmatrix} \begin{bmatrix} -1 & 0 & -4 & -4i \\ 0 & -1 & 0 & -4 \\ 0 & 0 & -1 & 0 \\ 0 & 0 & 0 & -1 \end{bmatrix} \\ &\quad - \begin{bmatrix} 0 & -4 & -4i & 1 \\ -4 & -4i & 1 & 0 \\ -4i & 1 & 0 & 0 \\ 1 & 0 & 0 & 0 \end{bmatrix} \begin{bmatrix} -1 & 0 & -4 & 4i \\ 0 & -1 & 0 & -4 \\ 0 & 0 & -1 & 0 \\ 0 & 0 & 0 & -1 \end{bmatrix} \\ &= i \begin{bmatrix} 0 & 0 & -8 & 0 \\ 0 & -8 & 0 & 0 \\ -8 & 0 & -32 & 0 \\ 0 & 0 & 0 & -8 \end{bmatrix}. \end{aligned}$$

Hence, using the Jacobi theorem, Theorem 8.6.1,

$$\mathrm{In}\left(\frac{1}{i} B\right) = \mathrm{In} \begin{bmatrix} -8 & 0 & 0 & 0 \\ 0 & -32 & 0 & -8 \\ 0 & 0 & -8 & 0 \\ 0 & -8 & 0 & 0 \end{bmatrix} = \{1, 3, 0\}.$$

Thus, $\pi'(a) = 1$ and $\nu'(a) = 3$, as required. □

Note that Theorem 1 gives no information if $a(\lambda)$ is a real polynomial since in this case $B = 0$. Methods for evaluating the inertia of a real polynomial with respect to the real axis, as well as a condition for a polynomial to have no real zeros, are discussed in Section 6.

As soon as the zero-location problem is solved with respect to the upper and lower half-planes, the results can be reformulated for any other pair of mutually complementary half-planes in the complex plane. The problem of determining the distribution of zeros relative to the imaginary axis is of particular interest in view of the application to differential equations (see Section 13.1); this is known as the *Routh–Hurwitz problem*.

Let $a(\lambda) = \sum_{j=0}^{l} a_j \lambda^j$ $(a_l \neq 0)$ and let $\hat{B}$ stand for the Bezout matrix associated with $a(\lambda)$ and $\bar{a}(-\lambda)$. Define $D = \mathrm{diag}[1, -1, \ldots, (-1)^{l-1}]$ and note that the matrix $\tilde{B} = \hat{B}D$ is nonsingular if and only if $a(\lambda)$ has no pure imaginary zeros or zeros symmetric relative to the origin.

Theorem 2. *With the notation of the previous paragraph, the matrix $\tilde{B} = \hat{B}D$ is Hermitian. If $\tilde{B}$ is also nonsingular, then*

$$\pi(a) = \nu(\tilde{B}),$$

$$\nu(a) = \pi(\tilde{B}),$$

$$\delta(a) = \delta(\tilde{B}) = 0.$$

PROOF. The assertion of the theorem follows from Theorem 1 and the easily verified relation

$$F\left(\frac{1}{i}B\right)F^* = -\tilde{B},$$

where $F = \mathrm{diag}[i^{l-1}, i^{l-2}, \ldots, i, 1]$ and B is the Bezoutian associated with $a_1(\lambda) = a(-i\lambda)$ and $\bar{a}_1(\lambda) = \bar{a}(i\lambda)$. ■

Corollary 1. *A polynomial $a(\lambda)$ is stable if and only if the matrix $\tilde{B}$ is positive definite.*

***Example* 2.** Consider the polynomials $a(\lambda) = \lambda^2 - 2\lambda + 5$ and $a(-\lambda) = \lambda^2 + 2\lambda + 5$. The corresponding Bezoutian is

$$\hat{B} = \begin{bmatrix} -2 & 1 \\ 1 & 0 \end{bmatrix}\begin{bmatrix} 5 & 2 \\ 0 & 5 \end{bmatrix} - \begin{bmatrix} 2 & 1 \\ 1 & 0 \end{bmatrix}\begin{bmatrix} 5 & -2 \\ 0 & 5 \end{bmatrix} = \begin{bmatrix} -20 & 0 \\ 0 & 4 \end{bmatrix},$$

and hence the matrix

$$\tilde{B} = \begin{bmatrix} -20 & 0 \\ 0 & -4 \end{bmatrix}$$

is negative definite. Thus both zeros, λ_1 and λ_2, of the polynomial lie in the open right half-plane, as required, since $\lambda_1 = 1 + 2i$ and $\lambda_2 = 1 - 2i$. □

If $a(\lambda)$ in Theorem 2 is a real polynomial, then those i, jth elements of the matrix B with $i + j$ odd are zeros. This is illustrated in the following exercise.

***Exercise* 3.** Check that the matrix $\tilde{B}$ defined in Theorem 2 and associated with the real polynomial $a(\lambda) = a_0 + a_1\lambda + a_2\lambda^2 + a_3\lambda^3 + a_4\lambda^4$ is

$$\tilde{B} = 2\begin{bmatrix} a_0 a_1 & 0 & a_0 a_3 & 0 \\ 0 & a_1 a_2 - a_0 a_3 & 0 & a_1 a_4 \\ a_0 a_3 & 0 & a_2 a_3 - a_1 a_4 & 0 \\ 0 & a_1 a_4 & 0 & a_3 a_4 \end{bmatrix}. \quad \square$$

The existence of symmetrical patterns of zeros in the matrix allows us to compute the inertia of the matrix

$$Q^{\mathrm{T}}\tilde{B}Q = \begin{bmatrix} B_1 & 0 \\ 0 & B_2 \end{bmatrix}$$

instead of In $\tilde{B}$. Here

$$Q = [Q_1 \quad Q_2] \tag{3}$$

is a permutation matrix defined by

$$Q_1 = [\boldsymbol{e}_1 \quad \boldsymbol{e}_3 \quad \cdots \quad \boldsymbol{e}_k], \qquad Q_2 = [\boldsymbol{e}_2 \quad \boldsymbol{e}_4 \quad \cdots \quad \boldsymbol{e}_j],$$

where $\boldsymbol{e}_1, \boldsymbol{e}_2, \ldots, \boldsymbol{e}_l$ are the unit vectors in $\mathbb{C}^l$, $k = l - 1$ and $j = l$ if l is even, $k = l$ and $j = l - 1$ if l is odd.

Exercise 4. Show that the inertia of the matrix $\tilde{B}$ discussed in Exercise 3 is equal to In B_1 + In B_2, where

$$B_1 = \begin{bmatrix} a_0 a_1 & a_0 a_3 \\ a_0 a_3 & a_2 a_3 - a_1 a_4 \end{bmatrix}, \qquad B_2 = \begin{bmatrix} a_1 a_2 - a_0 a_3 & a_1 a_4 \\ a_1 a_4 & a_3 a_4 \end{bmatrix}. \quad \square$$

A different form of the Routh–Hurwitz theorem for real polynomials will be obtained in Section 13.9.

13.5 The Schur–Cohn Theorem

The determination of the number of zeros of a polynomial lying inside the unit circle is known as the *Schur–Cohn problem* and, as noted in Section 13.2, it is closely related to the stability problem for linear difference equations. A solution of this problem in terms of a corresponding Bezout matrix is presented in this section.

Consider a complex polynomial $a(\lambda) = a_0 + a_1\lambda + \cdots + a_l\lambda^l$ $(a_l \neq 0)$ and define the *inertia of* $a(\lambda)$ *with respect to the unit circle* to be the triple of numbers $\alpha(a) = \{\alpha_+(a), \alpha_-(a), \alpha_0(a)\}$, where $\alpha_+(a)$ (respectively, $\alpha_-(a)$ and $\alpha_0(a)$) denotes the number of zeros of $a(\lambda)$, counted with their multiplicities, lying inside the open unit circle (respectively, outside and on the unit circumference).

Theorem 1. *Let* $a(\lambda) = \sum_{j=0}^{l} a_j\lambda^j$ $(a_l \neq 0)$ *and let* B *stand for the Bezoutian associated with* $\bar{a}(\lambda)$ *and* $\hat{a}(\lambda) \triangleq \lambda^l\bar{a}(\lambda^{-1})$. *If* P *is a rotation matrix then the* *matrix* $\breve{B} = PB$, *is Hermitian and, if it is nonsingular,*

$$\alpha_+(a) = \pi(\breve{B}),$$

$$\alpha_-(a) = \nu(\breve{B}),$$

$$\alpha_0(a) = \delta(\breve{B}) = 0.$$

PROOF. The proof of the theorem consists in proving the relation

$$\breve{B} - C_a^{*l}\breve{B}C_a^l = (\hat{a}(C_a))^* \hat{a}(C_a) > 0 \tag{1}$$

and in applying Theorem 13.2.2. Note that the equality $\breve{B} = \breve{B}^*$ is easily established by transposing the expression for B in Eq. (13.3.7) and reducing it by use of the relation $P^2 = I$ to the expression in Eq. (13.3.13).

First use the identity (13.3.6) to deduce that

$$S(\hat{\bar{a}})PS(a)C_a^l = -S(\hat{\bar{a}})S(\hat{a})P,$$

and then (13.3.12) implies

$$S(\hat{\bar{a}})S(\hat{a})P = -S(a)PS(\hat{\bar{a}})C_a^l.$$

Use this in the representation for B given by Proposition 13.3.1:

$$\begin{aligned} B &= S(a)S(\bar{a})P - S(\hat{\bar{a}})S(\hat{a})P \\ &= S(a)S(\bar{a})P + S(a)PS(\hat{\bar{a}})C_a^l. \end{aligned}$$

But Proposition 13.3.2 also gives $B = S(a)\hat{\bar{a}}(C_a)$, and so

$$\hat{\bar{a}}(C_a) = S(\bar{a})P + PS(\hat{\bar{a}})C_a^l.$$

Consider the conjugate transposed relation and postmultiply by $\hat{\bar{a}}(C_a) = S(a)^{-1}B$ to obtain

$$\hat{\bar{a}}(C_a)^*\hat{\bar{a}}(C_a) = PB + C_a^{*l}S(\hat{a})P\hat{\bar{a}}(C_a). \tag{2}$$

But using (13.3.6) again,

$$\begin{aligned} S(\hat{a})P\hat{\bar{a}}(C_a) &= -PS(a)C_a^l\hat{\bar{a}}(C_a) = -PS(a)\hat{\bar{a}}(C_a)C_a^l \\ &= -PBC_a^l. \end{aligned}$$

Thus, Eq. (2) takes the form

$$\hat{\bar{a}}(C_a)^*\hat{\bar{a}}(C_a) = \breve{B} - C_a^{*l}\breve{B}C_a^l,$$

which is Eq. (1).

Hence by Theorem 13.2.2, the matrix C_a^l has $\pi(\breve{B})$ eigenvalues inside the unit circle and $\nu(\breve{B})$ eigenvalues outside it. Obviously, the same can be said about the eigenvalues of C_a and, consequently, about the zeros of the polynomial $a(\lambda)$. ■

Corollary 1. *A polynomial $a(\lambda)$ is stable with respect to the unit circle if and only if the matrix $\breve{B}$ is positive definite.*

13.6 Perturbations of a Real Polynomial

Let $a(\lambda)$ be a polynomial of degree l with *real* coefficients. In this section we are concerned with the effect of perturbing the coefficients of $a(\lambda)$ with purely imaginary numbers. In particular, we will study the relation between the inertias (with respect to the real axis) of the original and the perturbed polynomials. These results will be applied in subsequent sections to obtain criteria for a real polynomial to be stable.

The main result of this section is presented first and then is used to provide a unified approach to some classical stability criteria.

Theorem 1 (P. Lancaster and M. Tismenetsky[†]). *Let $a(\lambda) = \sum_{i=0}^{l} a_i\lambda^i$ ($a_l \neq 0$) and $b(\lambda) = \sum_{i=0}^{l-1} b_i\lambda^i$ be polynomials with real coefficients. Write $\tilde{a}(\lambda) = a(\lambda) + ib(\lambda)$. If the Bezout matrix $B = \mathrm{Bez}(a, b)$ is nonsingular, that is, if $a(\lambda)$ and $b(\lambda)$ are relatively prime, then*

$$\delta'(\tilde{a}) = 0 \tag{1}$$

and

$$\pi'(a) \le \pi'(\tilde{a}) = \nu(B), \qquad \nu'(a) \le \nu'(\tilde{a}) = \pi(B). \tag{2}$$

Furthermore, equalities hold throughout Eq. (2) if $\delta'(a) = 0$.

This theorem means that the perturbation $b(\lambda)$ has the effect of shifting all real zeros of $a(\lambda)$ off the real axis, and the numbers $\pi'(\tilde{a})$ and $\nu'(\tilde{a})$ of eigenvalues of the perturbed polynomial above and below the real axis, respectively, can be counted using the inertia properties of B.

PROOF. Suppose that $\tilde{a}(\lambda_0) = 0$ and λ_0 is real. Then $a(\lambda_0)$, $b(\lambda_0)$ are real and so $\tilde{a}(\lambda_0) = a(\lambda_0) + ib(\lambda_0) = 0$ implies that $a(\lambda_0) = b(\lambda_0) = 0$. Thus $a(\lambda)$ and $b(\lambda)$ have a common zero, contradicting the hypothesis that B is nonsingular. Hence the relation (1).

As noted earlier, the Bezout matrix B satisfies Eq. (13.3.22) which, when $\tilde{a}(\lambda) = a(\lambda)$, can be written in the form $C_a^*B = BC_a$, where C_a is the companion matrix associated with $a(\lambda)$, or, what is equivalent,

$$(iC_a)^*B + B(iC_a) = 0. \tag{3}$$

We also know (see Exercise 13.3.1) that B is real and symmetric.

[†] *Linear Alg. Appl.* **52** (1983), 479–496.

The companion matrix $C_{\tilde{a}}$ is related to C_a by

$$C_a = C_{\tilde{a}} + FD, \tag{4}$$

where $D = \text{diag}[b_0, b_1, \ldots, b_{l-1}]$ and

$$F = \begin{bmatrix} 0 & \cdots & 0 \\ \vdots & & \vdots \\ 0 & \cdots & 0 \\ ia_l^{-1} & \cdots & ia_l^{-1} \end{bmatrix}.$$

Substituting from Eq. (4) into Eq. (3), it is found that

$$(iC_{\tilde{a}})^*B + B(iC_{\tilde{a}}) = iDF^*B - iBFD. \tag{5}$$

Now use the representation (13.3.20) for B to obtain

$$iDF^*B = D\begin{bmatrix} 1 & 0 & \cdots & 0 \\ 1 & 0 & \cdots & 0 \\ \vdots & \vdots & & \vdots \\ 1 & 0 & \cdots & 0 \end{bmatrix}\begin{bmatrix} \boldsymbol{b}^{\mathrm{T}} \\ \boldsymbol{b}^{\mathrm{T}}C_a \\ \vdots \\ \boldsymbol{b}^{\mathrm{T}}C_a^{l-1} \end{bmatrix} = DAD,$$

where $\boldsymbol{b}^{\mathrm{T}} = [b_0 \quad b_1 \quad \cdots \quad b_{l-1}]$ and A is the $l \times l$ matrix with all elements equal to 1. Furthermore, $-iBFD = (iDF^*B)^* = DAD$ and therefore Eq. (5) implies

$$(iC_{\tilde{a}})^*B + B(iC_{\tilde{a}}) = W, \tag{6}$$

where $W = 2DAD \geq 0$.

Since $\delta'(\tilde{a}) = 0$, then $\delta(iC_{\tilde{a}}) = 0$ and, by Theorem 13.1.3, $\text{In}(iC_{\tilde{a}}) = \text{In } B$. When B is nonsingular, Proposition 13.1.2 can be applied to Eq. (3) to yield the inequalities in (2).

The additional condition $\delta'(a) = \delta(iC_a) = 0$ together with $\det B \neq 0$ yields (in view of Theorem 13.1.3 and Eq. (3)) equality throughout (2). This completes the proof. ∎

Corollary 1. *If, in the previous notation, $B < 0$ or $B > 0$, then the zeros of the polynomial $\tilde{a}(\lambda)$ lie in the open upper half-plane or open lower half-plane, respectively.*

Example 1. The polynomial $a(\lambda) = \lambda^2 - 1$ has two real zeros. The perturbation $ib(\lambda) = 2i\lambda$ satisfies the hypothesis of Theorem 1 and

$$B = \begin{bmatrix} 0 & 1 \\ 1 & 0 \end{bmatrix}\begin{bmatrix} 0 & 2 \\ 2 & 0 \end{bmatrix} = 2I > 0.$$

Hence both zeros of $\lambda^2 + 2i\lambda - 1$ are in the open lower half-plane.

Exercise 2. By use of Theorem 1, check that the polynomial $\tilde{a}(\lambda) = \lambda^4 + 4i\lambda^3 - 4\lambda^2 - 1$ has three zeros with negative imaginary parts and a zero with a positive imaginary part. □

Another hypothesis leading to the main conclusion (Eq. (2)) of Theorem 1 is expressed in terms of controllability.

Theorem 2. *With the notation of Theorem* 1 (*and its proof*), *let the pair* $(C_{\tilde{a}}^*, W)$ *be controllable. Then the Bezout matrix B associated with* $a(\lambda)$ *and* $b(\lambda)$ *is nonsingular,* $\delta(\tilde{a}) = 0$, *and the relations* (2) *hold.*

The proof of Theorem 2 is easily obtained by applying Theorem 13.1.4 to Eq. (6).

Exercise 3. Let $a(\lambda) = \sum_{i=0}^{l} a_i\lambda^i$, $a_l \neq 0$, $a_i = \bar{a}_i$, $i = 0, 1, \ldots, l$. Prove that the polynomial $\tilde{a}(\lambda) = a(\lambda) + ib_0$ $(b_0 > 0)$ has no real spectrum and that

$$\nu'(a) - \pi'(a) = \begin{cases} 0 & \text{if } l \text{ is even,} \\ \operatorname{sgn} a_l & \text{if } l \text{ is odd.} \end{cases}$$

Hint. Apply Theorem 2. □

13.7 The Liénard–Chipart Criterion

The names of A. Liénard and M. Chipart are strongly associated with a stability criterion for the zeros of real polynomials. Using the results of the previous section, we can now present a modified version of the original Liénard–Chipart[†] stability criterion in terms of a Bezoutian.

Consider a polynomial $a(\lambda) = \sum_{j=0}^{l} a_j\lambda^j$, with $a_l \neq 0$ and real coefficients. If l is even, define

$$h(\lambda) = \sum_{j=0}^{l/2} a_{2j}\lambda^j, \qquad g(\lambda) = \sum_{j=0}^{(l/2)-1} a_{2j+1}\lambda^j.$$

If l is odd, then we put

$$h(\lambda) = \sum_{j=0}^{1/2(l-1)} a_{2j}\lambda^j, \qquad g(\lambda) = \sum_{j=0}^{1/2(l-1)} a_{2j+1}\lambda^j.$$

For brevity, we refer to $h(\lambda)$ and $g(\lambda)$, as well as to the obvious representation

$$a(\lambda) = h(\lambda^2) + \lambda g(\lambda^2), \tag{1}$$

as the *L–C splitting* of $a(\lambda)$.

[†] *J. Math. Pures Appl.* **10** (1914), 291–346.

Theorem 1. *Let Eq.* (1) *be the L–C splitting of the real polynomial* $a(\lambda)$. *Denote*

$$a_1(\lambda) = \begin{cases} h(-\lambda^2) & \text{if } l \text{ is even,} \\ \lambda g(-\lambda^2) & \text{if } l \text{ is odd;} \end{cases}$$

$$a_2(\lambda) = \begin{cases} \lambda g(-\lambda^2) & \text{if } l \text{ is even,} \\ -h(-\lambda^2) & \text{if } l \text{ is odd.} \end{cases}$$

If the Bezout matrix B *associated with* $a_1(\lambda)$ *and* $a_2(\lambda)$ *is nonsingular, then* $a(\lambda)$ *has no pure imaginary zeros and*

$$\pi(a) = \pi(B), \qquad \nu(a) = \nu(B). \tag{2}$$

PROOF. Let l be even. Define the polynomial $\tilde{a}(\lambda)$ by $\tilde{a}(\lambda) = a(i\lambda) = h(-\lambda^2) + i\lambda g(-\lambda^2) = a_1(\lambda) + ia_2(\lambda)$ and observe that $\tilde{a}(\lambda)$ has no real zeros. Indeed, if there were a real λ_0 such that $a_1(\lambda_0) + ia_2(\lambda_0) = 0$ then, taking complex conjugates, we would obtain $a_1(\lambda_0) - ia_2(\lambda_0) = 0$. Thus, it would follow by comparison that $a_2(\lambda_0) = 0$ and hence $a_1(\lambda_0) = 0$. This conflicts with the assumed relative primeness of $a_1(\lambda)$ and $a_2(\lambda)$ (implied by $\det B \neq 0$).

Now applying Theorem 13.6.1, we obtain

$$\nu(B) = \pi'(\tilde{a}) = \nu(a),$$

$$\pi(B) = \nu'(\tilde{a}) = \pi(a),$$

$$\delta(B) = \delta'(\tilde{a}) = \delta(a),$$

and the theorem is established for the case of an even l.

If l is odd, then write

$$-i\tilde{a}(\lambda) = \lambda g(-\lambda^2) - ih(-\lambda^2) = a_1(\lambda) + ia_2(\lambda)$$

and observe that the zeros of $\tilde{a}(\lambda)$ and $-i\tilde{a}(\lambda)$ coincide. Now apply Theorem 13.6.1 again. ■

Corollary 1. *The polynomial* $a(\lambda)$ *is stable if and only if* $B = \text{Bez}(a_1, a_2)$ *is negative definite.*

PROOF. If $a(\lambda)$ is stable, then the polynomials $a_1(\lambda)$ and $a_2(\lambda)$ defined above are relatively prime. Indeed, if $a_1(\lambda_0) = a_2(\lambda_0) = 0$ then also $a_1(\bar{\lambda}_0) = a_2(\bar{\lambda}_0) = 0$ and $\tilde{a}(\lambda_0) = \tilde{a}(\bar{\lambda}_0) = 0$. Hence $a(\lambda)$ has complex zeros μ and $-\bar{\mu}$ $(\mu = i\lambda_0)$ or a pure imaginary zero $i\lambda_0$ $(\lambda_0 = \bar{\lambda}_0)$. In both cases we contradict the assumption that all zeros of $a(\lambda)$ lie in the open left half-plane. Thus, B is nonsingular and, by the theorem, must be negative definite. ■

Example 1. Consider the polynomial $a(\lambda) = (\lambda - 1)^2(\lambda + 2)$ and the representation

$$-i\tilde{a}(\lambda) = -ia(i\lambda) = \lambda(-3 - \lambda^2) - 2i.$$

Here $a_1(\lambda) = -\lambda(3 + \lambda^2)$, $a_2(\lambda) = -2$, and the corresponding Bezoutian is

$$B = \begin{bmatrix} 6 & 0 & 2 \\ 0 & 2 & 0 \\ 2 & 0 & 0 \end{bmatrix},$$

with In $B = \{2, 1, 0\}$. (To see this, use the Jacobi method.) Thus, by Eqs. (2), also $\text{In}(a) = \{2, 1, 0\}$. □

The following result is a consequence of Theorem 1 and permits the description of the inertia of a real polynomial in terms of that of two Bezoutians of smaller size.

Theorem 2. *If $a(\lambda)$ is a real polynomial and $h(\lambda)$ and $g(\lambda)$ form the L–C splitting of $a(\lambda)$, where the Bezoutians $B_1 = \text{Bez}(h(\lambda), g(\lambda))$ and $B_2 = \text{Bez}(h(\lambda), \lambda g(\lambda))$ are nonsingular, then $a(\lambda)$ has no pure imaginary zeros and*

$$\pi(a) = \nu(B_1) + \pi(B_2),$$

$$\nu(a) = \pi(B_1) + \nu(B_2).$$

In particular, $\alpha(\lambda)$ is stable if and only if $B_1 > 0$ and $B_2 < 0$.

PROOF. We give the proof for the case of l even, say $l = 2k$.

Let B denote the Bezout matrix associated with the polynomials $a_1(\lambda)$ and $a_2(\lambda)$ defined in Theorem 1. Applying Proposition 13.3.2 and Eq. (13.3.5), it is found that

$$B = \sum_{j=1}^{k} (-1)^{j-1} a_{2j-1} \operatorname{diag}[A_{2j-1}^{(1)}, A_{2j-1}^{(2)}], \tag{4}$$

where

$$A_{2j-1}^{(1)} = \begin{bmatrix} 0 & \cdots & & 0 & -a_0 \\ & & \cdot^{\cdot^{\cdot}} & -a_0 & 0 \\ \vdots & & & & a_2 \\ & \cdot^{\cdot^{\cdot}} & & \cdot^{\cdot^{\cdot}} & 0 \\ & & & \cdot^{\cdot^{\cdot}} & \vdots \\ 0 & & & \cdot^{\cdot^{\cdot}} & \\ -a_0 & 0 \quad a_2 & 0 & \cdots & (-1)^j a_{2j-2} \end{bmatrix},$$

$$A_{2j-1}^{(2)} = \begin{bmatrix} (-1)^{j+1} a_{2j} & 0 & \cdots & (-1)^k a_l \\ 0 & & & 0 \\ \vdots & & \cdot^{\cdot^{\cdot}} \quad \cdot^{\cdot^{\cdot}} & \vdots \\ (-1)^k a_l & 0 & \cdots & 0 \end{bmatrix}.$$

Consider the matrix $Q = [Q_1 \quad Q_2]$ defined by Eq. (13.4.3). For $j = 1, 2, \ldots, k$, we have

$$Q^T \operatorname{diag}[A_{2j-1}^{(1)}, A_{2j-1}^{(2)}]Q = \operatorname{diag}[A_{2j+1}^{[1]}, A_{2j+1}^{[2]}, A_{2j-1}^{[1]}, A_{2j-1}^{[2]}], \tag{5}$$

where

$$A_{2s-1}^{[1]} = \begin{bmatrix} 0 & \cdots & 0 & -a_0 \\ \vdots & \cdot^{\cdot^{\cdot}} & \cdot^{\cdot^{\cdot}} & a_2 \\ 0 & & \cdot^{\cdot^{\cdot}} & \vdots \\ -a_0 & a_2 & \cdots & (-1)^{s-1}a_{2s-4} \end{bmatrix},$$

$$A_{2s-1}^{[2]} = \begin{bmatrix} (-1)^s a_{2s} & \cdots & (-1)^k a_l \\ \vdots & \cdot^{\cdot^{\cdot}} \quad \cdot^{\cdot^{\cdot}} & 0 \\ & & \vdots \\ (-1)^k a_l & 0 \quad \cdots & 0 \end{bmatrix}.$$

Thus, by use of Eq. (5), it follows from Eq. (4) that

$$Q^{\mathrm{T}}BQ = \operatorname{diag}[\tilde{B}_2, \tilde{B}_1], \tag{6}$$

where, appealing to Eq. (13.3.5) and writing C for the companion matrix of $g(-\lambda)$,

$$\begin{aligned} \tilde{B}_1 &= \sum_{j=1}^{k} (-1)^{j-1} a_{2j-1} \operatorname{diag}[A_{2j-1}^{[1]}, A_{2j-1}^{[2]}] \\ &= \begin{bmatrix} -a_2 & a_4 & \cdots & (-1)^k a_l \\ a_4 & & & 0 \\ \vdots & & \cdot^{\cdot^{\cdot}} \quad \cdot^{\cdot^{\cdot}} & \vdots \\ (-1)^k a_l & 0 & \cdots & 0 \end{bmatrix} (a_1 I - a_3 C + \cdots + (-1)^{k-1} a_{l-1} C^{k-1}) \end{aligned}$$

$$\begin{aligned} \tilde{B}_2 &= \sum_{j=1}^{k} (-1)^{j-1} a_{2j-1} \operatorname{diag}[A_{2j+1}^{[1]}, A_{2j+1}^{[2]}] \\ &= \begin{bmatrix} -a_2 & a_4 & \cdots & (-1)^k a_l \\ a_4 & & & 0 \\ \vdots & & \cdot^{\cdot^{\cdot}} \quad \cdot^{\cdot^{\cdot}} & \vdots \\ (-1)^k a_l & 0 & \cdots & 0 \end{bmatrix} (a_1 C - a_3 C^2 + \cdots + (-1)^{k-1} a_{l-1} C^k). \end{aligned}$$

Now use Proposition 13.3.2 again to obtain

$$\tilde{B}_1 = B(h(-\lambda), g(-\lambda)), \qquad \tilde{B}_2 = B(h(-\lambda), \lambda g(-\lambda)).$$

Using Eq. (13.3.7), we easily check that

$$P\tilde{B}_1 P = -B_1, \qquad P\tilde{B}_2 P = -B(h(\lambda), -\lambda g(\lambda)) = B_2, \tag{7}$$

where $P = \operatorname{diag}[1, -1, \ldots, (-1)^{k-1}] \in \mathbb{C}^k$. Now Sylvester's law of inertia applied to Eqs. (6) and (7) yields, in view of Theorem 1, the required result provided l is even. The case of l odd is similarly established. ■

***Example* 2.** Let $a(\lambda)$ be the polynomial considered in Example 1. We have $h(\lambda) = 2$, $g(\lambda) = -3 + \lambda$, and

$$B_1 = -2, \qquad B_2 = \begin{bmatrix} 6 & -2 \\ -2 & 0 \end{bmatrix}.$$

Thus In $B_1 = \{0, 1, 0\}$, In $B_2 = \{1, 1, 0\}$, and $a(\lambda)$ has two zeros in the open right half-plane and one zero in the open left half-plane, as required.

***Example* 3.** Let $a(\lambda) = (\lambda - 1)^3(\lambda + 2)$. Then $h(\lambda) = -2 - 3\lambda + \lambda^2$, $g(\lambda) = 5 - \lambda$, and

$$B_1 = \begin{bmatrix} -17 & 5 \\ 5 & -1 \end{bmatrix}, \qquad B_2 = \begin{bmatrix} 10 & -2 \\ -2 & 2 \end{bmatrix}.$$

Since In $B_1 = \{1, 1, 0\}$, In $B_2 = \{2, 0, 0\}$, then, by Eq. (3), $\operatorname{In}(a) = \{3, 1, 0\}$, as required. □

A simplified version of the criteria of Theorem 2 will be presented in Section 13.10.

13.8 The Markov Criterion

Let $a(\lambda) = a_0 + a_1\lambda + \cdots + a_l\lambda^l$ be a polynomial with real coefficients and $a_l \neq 0$, and let $a_1(\lambda)$, $a_2(\lambda)$ be defined as in Theorem 13.7.1. Note that generally $\deg a_2(\lambda) \leq \deg a_1(\lambda)$. Hence we may define a rational function $r(\lambda) = a_2(\lambda)/a_1(\lambda)$ with the property $\lim_{\lambda \to \infty} |r(\lambda)| < \infty$.

For sufficiently large $|\lambda|$, we may represent $r(\lambda)$ by a *Laurent series*,

$$r(\lambda) = \sum_{j=0}^{\infty} h_j \lambda^{-j},$$

and construct the corresponding (truncated) Hankel matrix H of the *Markov parameters* h_j: $H = [h_{i+j-1}]_{i,j=1}^{l}$. It turns out that the results obtained in Sections 13.3 and 13.4 can be reformulated in terms of the inertia of a Hankel matrix of Markov parameters. This is possible in view of the relationship between Hankel and Bezout matrices that is established in the next proposition.

Proposition 1. *Let $b(\lambda) = \sum_{j=0}^{l} b_j \lambda^j$ and $c(\lambda) = \sum_{j=0}^{l} c_j \lambda^j$ be two real polynomials in which $c_l \neq 0$. If $B = \text{Bez}(c, b)$ denotes the Bezout matrix associated with $c(\lambda)$ and $b(\lambda)$, then*

$$B = S(c)HS(c), \tag{1}$$

where H denotes the Hankel matrix of Markov parameters of the rational function $R(\lambda) = b(\lambda)/c(\lambda)$.

PROOF. Let $H = [h_{i+j-1}]_{i,j=1}^{l}$, where $h_1, h_2, \ldots, h_{2l-1}$ are defined by

$$b(\lambda) = (h_0 + h_1\lambda^{-1} + h_2\lambda^{-2} + \cdots)c(\lambda). \tag{2}$$

It is easily found, by comparing the corresponding coefficients on both sides of Eq. (2), that

$$\boldsymbol{b}^{\mathrm{T}} = \boldsymbol{h}_1^{\mathrm{T}} S(c), \tag{3}$$

where $\boldsymbol{b}^{\mathrm{T}} = [\tilde{b}_0 \quad \tilde{b}_1 \quad \cdots \quad \tilde{b}_{l-1}]$, $\tilde{b}_j = b_j - b_l c_j c_l^{-1}$, $j = 0, 1, \ldots, l-1$, and $\boldsymbol{h}_1^{\mathrm{T}} = [h_1 \quad h_2 \quad \cdots \quad h_l]$.

Furthermore, using Eq. (13.3.5), it is found that

$$\boldsymbol{b}^{\mathrm{T}} C_c^j = \boldsymbol{h}_1^{\mathrm{T}} \operatorname{diag}\left[-\begin{bmatrix} & & & c_0 \\ & & & c_1 \\ & & \cdot^{\cdot^{\cdot}} & \vdots \\ c_0 & c_1 & \cdots & c_{j-1} \end{bmatrix}, \begin{bmatrix} c_{j+1} & \cdots & c_l \\ \vdots & \cdot^{\cdot^{\cdot}} & \\ c_l & & \end{bmatrix}\right] = \boldsymbol{h}_{j+1}^{\mathrm{T}} S(c),$$

where $\boldsymbol{h}_{j+1}^{\mathrm{T}} = [h_{j+1} \quad h_{j+2} \quad \cdots \quad h_{j+l}]$ and $j = 1, 2, \ldots, l-1$. Combining this with Eq. (3) we obtain

$$\begin{bmatrix} \boldsymbol{b}^{\mathrm{T}} \\ \boldsymbol{b}^{\mathrm{T}} C_c \\ \vdots \\ \boldsymbol{b}^{\mathrm{T}} C_c^{l-1} \end{bmatrix} = HS(c).$$

It remains only to use Eq. (13.3.20) to obtain Eq. (1). ■

Now we are in a position to state a modified Markov stability test.

Theorem 1. *Let H be the Hankel matrix of the Markov parameters of the polynomial $a(\lambda)$ and assume that* $\det H \neq 0$. *Then*

$$\pi(a) = \pi(H), \qquad \nu(a) = \nu(H), \qquad \delta(a) = 0. \tag{5}$$

PROOF. The assertion of the theorem trivially follows from Theorem 13.7.1, Proposition 1, and Sylvester's law of inertia. ■

Corollary 1. *A real polynomial $a(\lambda)$ is stable if and only if $H < 0$.*

Example 1. Let $a(\lambda) = (\lambda + 1)^2(\lambda + 2)$. Then

$$r(\lambda) = \frac{a_2(\lambda)}{a_1(\lambda)} = \frac{-2 + 4\lambda^2}{\lambda(5 - \lambda^2)} = \sum_{j=1}^{\infty} h_j \lambda^{-j},$$

where the coefficients $h_1, h_2, \ldots, h_5$ are found to be $h_1 = -4$, $h_2 = h_4 = 0$, $h_3 = -18$, $h_5 = -90$. Hence the Hankel matrix

$$H = \begin{bmatrix} -4 & 0 & -18 \\ 0 & -18 & 0 \\ -18 & 0 & -90 \end{bmatrix}$$

is negative definite (use Jacobi's method to see this), and therefore $a(\lambda)$ is stable. □

Applying Eq. (1) and Sylvester's law of inertia to the Bezout matrices B_1 and B_2, the next result follows from Theorem 13.7.2.

Theorem 2 (A. A. Markov†). *In the notation of this section, let*

$$H_1 = \begin{bmatrix} h_1 & h_2 & \cdots & h_k \\ h_2 & & \cdot^{\cdot^{\cdot}} & \vdots \\ \vdots & \cdot^{\cdot^{\cdot}} & & \\ h_k & & \cdots & h_{2k-1} \end{bmatrix}, \qquad H_2 = \begin{bmatrix} h_2 & h_3 & \cdots & h_{m+1} \\ h_3 & & \cdot^{\cdot^{\cdot}} & \vdots \\ \vdots & \cdot^{\cdot^{\cdot}} & & \\ h_{m+1} & & \cdots & h_{2m} \end{bmatrix}, \tag{6}$$

where $k = m = \frac{1}{2}l$ or $k = \frac{1}{2}(l + 1)$, $m = \frac{1}{2}(l - 1)$, according as l is even or odd, and where the Markov parameters h_j, $j = 0, 1, \ldots$, are defined by

$$\sum_{j=0}^{\infty} h_j \lambda^{-j} = \begin{cases} g(\lambda)/h(\lambda) & \text{if } l \text{ is even,} \\ h(\lambda)/\lambda g(\lambda) & \text{if } l \text{ is odd.} \end{cases} \tag{7}$$

If H_1 and H_2 are nonsingular, then $a(\lambda)$ has no pure imaginary zeros and

$$\pi(a) = \nu(H_1) + \pi(H_2),$$
$$\nu(a) = \pi(H_1) + \nu(H_2). \tag{8}$$

PROOF. Let l be even. Then $\deg g(\lambda) < \deg h(\lambda)$ and hence, in Eq. (7), $h_0 = 0$. Note that since the leading coefficient of $h(\lambda)$ is nonzero, formula (1) yields the congruence of $B_1 = B(h(\lambda), g(\lambda))$ and H_1. Furthermore, it follows that

$$\frac{\lambda g(\lambda)}{h(\lambda)} = \sum_{j=0}^{\infty} h_{j+1} \lambda^{-j}$$

† *Zap. Petersburg Akad. Nauk*, 1894. See also "The Collected Works of A. A. Markov," Moscow, 1948, pp. 78–105. (In Russian.)

for sufficiently large $|\lambda|$. Therefore by Eq. (1) the matrices $B_2 = B(h(\lambda), \lambda g(\lambda))$ and H_2 are congruent. This proves the theorem in the even case.

Now consider the case when l is odd. Here $\deg h(\lambda) \le \deg g(\lambda)$ and

$$\frac{h(\lambda)}{g(\lambda)} = \sum_{j=0}^{\infty} h_{j+1}\lambda^{-j}, \qquad \frac{h(\lambda)}{\lambda g(\lambda)} = \sum_{j=1}^{\infty} h_j \lambda^{-j}.$$

Thus H_1 (respectively, H_2) is congruent to the Bezoutian $-B_2 = B(\lambda g(\lambda), h(\lambda))$ (respectively, $-B_1 = B(g(\lambda), h(\lambda))$), and Eqs. (8) follow from Theorem 13.7.2. ■

Corollary 1. *A real polynomial $a(\lambda)$ is stable if and only if $H_1 > 0$ and $H_2 < 0$, where H_1 and H_2 are defined in Eqs. (6).*

Note that by writing the Markov parameters in the traditional form (l is even),

$$\frac{g(\lambda)}{h(\lambda)} = r_{-1} + \frac{r_0}{\lambda} - \frac{r_1}{\lambda^2} + \frac{r_2}{\lambda^3} - \frac{r_3}{\lambda^4} + \cdots, \tag{9}$$

it is found that

$$H_1 = P\tilde{H}_1 P, \qquad H_2 = -P\tilde{H}_2 P,$$

where $P = \operatorname{diag}[1, -1, \ldots, (-1)^{m-1}]$ and $m = \frac{1}{2}l$, and

$$\tilde{H}_1 = \begin{bmatrix} r_0 & r_1 & \cdots & r_{m-1} \\ r_1 & & \cdot^{\cdot^{\cdot}} & \vdots \\ \vdots & & & \\ r_{m-1} & & \cdots & r_{2m-2} \end{bmatrix}, \qquad \tilde{H}_2 = \begin{bmatrix} r_1 & r_2 & \cdots & r_m \\ r_2 & & \cdot^{\cdot^{\cdot}} & \vdots \\ \vdots & & & \\ r_m & & \cdots & r_{2m-1} \end{bmatrix}.$$

Thus, the stability criterion with respect to the form (9) is equivalent to the conditions $\tilde{H}_1 > 0$, $\tilde{H}_2 > 0$, as required.

Example 2. Let $a(\lambda)$ be as defined in Example 1. Then $h(\lambda) = 2 + 4\lambda$, $g(\lambda) = 5 + \lambda$, and

$$\frac{h(\lambda)}{\lambda g(\lambda)} = 4\lambda^{-1} - 18\lambda^{-2} + 90\lambda^{-3} + \cdots.$$

Hence

$$H_1 = \begin{bmatrix} h_1 & h_2 \\ h_2 & h_3 \end{bmatrix} = \begin{bmatrix} 4 & -18 \\ -18 & 90 \end{bmatrix}, \qquad H_2 = [h_2] = [-18],$$

and the properties $H_1 > 0$, $H_2 < 0$ yield the stability of the given polynomial. □

13.9 A Determinantal Version of the Routh–Hurwitz Theorem

A disadvantage of the previous methods for computing the inertia of a given polynomial is that this inertia is expressed in terms of the inertia of a matrix with elements that must first be calculated. In this section we determine the inertia of a *real* polynomial by computing the leading principal minors of a matrix whose non-zero elements are simply the coefficients of the polynomial.

Let $a(\lambda) = \sum_{j=0}^{l} a_j \lambda^j$ be a polynomial with real coefficients and $a_l \neq 0$, and define the $l \times l$ matrix

$$\hat{H} = \begin{bmatrix} a_{l-1} & a_{l-3} & \cdots & & \\ a_l & a_{l-2} & \cdots & & \\ 0 & a_{l-1} & a_{l-3} & \cdots & \\ 0 & a_l & a_{l-2} & \cdots & \\ 0 & 0 & a_{l-1} & \cdots & \\ 0 & 0 & a_l & \cdots & \\ \vdots & \vdots & \vdots & & \end{bmatrix}, \tag{1}$$

called the *Hurwitz matrix* associated with $a(\lambda)$. We shall show that the Hermitian matrix H from Section 13.8 and the non-Hermitian Hurwitz matrix $\hat{H}$ have the same leading principal minors. We state a more general result due to A. Hurwitz.†

Proposition 1. *Let $b(\lambda) = \sum_{j=0}^{l} b_j \lambda^j$ and $c(\lambda) = \sum_{j=0}^{l} c_j \lambda^j$ be two real polynomials with $c_l \neq 0$, and let $h_0, h_1, h_2, \ldots$ denote the Markov parameters of the rational function $r(\lambda) = b(\lambda)/c(\lambda)$. Then if we define $b_j = c_j = 0$ for $j < 0$,*

$$c_l^{2s} \det \begin{bmatrix} h_1 & h_2 & \cdots & h_s \\ h_2 & & \iddots & h_{s+1} \\ \vdots & \iddots & & \vdots \\ h_s & h_{s+1} & \cdots & h_{2s-1} \end{bmatrix} = \det \begin{bmatrix} c_l & c_{l-1} & & \cdots & c_{l-2s+1} \\ b_l & b_{l-1} & & \cdots & b_{l-2s+1} \\ 0 & c_l & c_{l-1} & \cdots & c_{l-2s+2} \\ 0 & b_l & b_{l-1} & \cdots & b_{l-2s+2} \\ \vdots & \vdots & \vdots & & \vdots \\ 0 & \cdots & c_l & \cdots & c_{l-s} \\ 0 & \cdots & b_l & \cdots & b_{l-s} \end{bmatrix}, \tag{2}$$

for $s = 1, 2, \ldots, l$.

Note that the matrices on the left and right of Eq. (2) have sizes s and $2s$, respectively.

† *Math. Ann.* **46** (1895), 273–284.

PROOF. Introduce the $1 \times 2s$ vectors

$$\boldsymbol{h}_k^{\mathrm{T}} = [0 \quad \cdots \quad 0 \quad h_0 \quad h_1 \quad \cdots \quad h_k]$$

and

$$\boldsymbol{b}_k^{\mathrm{T}} = [0 \quad \cdots \quad 0 \quad b_l \quad b_{l-1} \quad \cdots \quad b_{l-k}],$$

for $k = 0, 1, \ldots, 2s - 1$. Note that

$$\boldsymbol{h}_{2s-1}^{\mathrm{T}} = [h_0 \quad h_1 \quad \cdots \quad h_{2s-1}]$$

and

$$\boldsymbol{b}_{2s-1}^{\mathrm{T}} = [b_l \quad b_{l-1} \quad \cdots \quad b_{l-2s+1}].$$

Using Eq. (13.8.3), it is found that, for each k,

$$\boldsymbol{h}_k^{\mathrm{T}} \begin{bmatrix} c_l & c_{l-1} & \cdots & c_{l-2s+1} \\ 0 & c_l & \ddots & \vdots \\ \vdots & \ddots & \ddots & c_{l-1} \\ 0 & & 0 & c_l \end{bmatrix} = \boldsymbol{b}_k^{\mathrm{T}}. \tag{3}$$

Hence, denoting the unit vectors in $\mathbb{C}^{2s}$ by $\boldsymbol{u}_1, \boldsymbol{u}_2, \ldots, \boldsymbol{u}_{2s}$, we easily check that

$$\begin{bmatrix} \boldsymbol{u}_1^{\mathrm{T}} \\ \boldsymbol{h}_{2s-1}^{\mathrm{T}} \\ \boldsymbol{u}_2^{\mathrm{T}} \\ \boldsymbol{h}_{2s-2}^{\mathrm{T}} \\ \vdots \\ \boldsymbol{u}_s^{\mathrm{T}} \\ \boldsymbol{h}_s^{\mathrm{T}} \end{bmatrix} \begin{bmatrix} c_l & c_{l-1} & \cdots & c_{l-2s+1} \\ 0 & & \ddots & \vdots \\ \vdots & \ddots & \ddots & \\ & & & c_{l-1} \\ 0 & & 0 & c_l \end{bmatrix} = \begin{bmatrix} \boldsymbol{c}_{2s-1}^{\mathrm{T}} \\ \boldsymbol{b}_{2s-1}^{\mathrm{T}} \\ \boldsymbol{c}_{2s-2}^{\mathrm{T}} \\ \boldsymbol{b}_{2s-2}^{\mathrm{T}} \\ \vdots \\ \boldsymbol{c}_s^{\mathrm{T}} \\ \boldsymbol{b}_s^{\mathrm{T}} \end{bmatrix}, \tag{4}$$

where

$$\boldsymbol{c}_k^{\mathrm{T}} = [0 \quad \cdots \quad 0 \quad c_l \quad c_{l-1} \quad \cdots \quad c_{l-k}], \qquad k = 0, 1, \ldots, 2s - 1,$$

and $c_j = b_j = 0$ for $j > l$. Observe that the Toeplitz matrix in Eq. (4) coincides with the matrix in Eq. (3), and therefore their determinants are equal. Computing the determinant on the left side in Eq. (4), we find that it is equal to

$$(-1)^s c_l^{2s} \det \begin{bmatrix} \boldsymbol{u}_1^{\mathrm{T}} \\ \boldsymbol{u}_2^{\mathrm{T}} \\ \vdots \\ \boldsymbol{u}_s^{\mathrm{T}} \\ \boldsymbol{h}_{2s-1}^{\mathrm{T}} \\ \boldsymbol{h}_{2s-2}^{\mathrm{T}} \\ \vdots \\ \boldsymbol{h}_s^{\mathrm{T}} \end{bmatrix} = c_l^{2s} \det \begin{bmatrix} I & 0 \\ V_s & W_s \end{bmatrix},$$

where

$$V_s = \begin{bmatrix} h_0 & h_1 & \cdots & h_{s-1} \\ 0 & h_0 & \ddots & \vdots \\ \vdots & \ddots & \ddots & h_1 \\ 0 & \cdots & 0 & h_0 \end{bmatrix}, \qquad W_s = \begin{bmatrix} h_s & h_{s+1} & \cdots & h_{2s-1} \\ h_{s-1} & h_s & \cdots & h_{2s-2} \\ \vdots & \ddots & \ddots & \vdots \\ h_1 & \cdots & h_{s-1} & h_s \end{bmatrix}.$$

Thus the determinant of the matrix on the right in Eq. (2) is

$$(-1)^s c_l^{2s} \det W_s = c_l^{2s} \det \begin{bmatrix} h_1 & h_2 & \cdots & h_s \\ h_2 & & \cdot^{\cdot^{\cdot}} & \vdots \\ \vdots & & & \\ h_s & \cdots & & h_{2s-1} \end{bmatrix},$$

as required. ■

The famous Routh-Hurwitz stability result is presented in the following theorem.

Theorem 1. *If $a(\lambda)$ is a real polynomial and there are no zeros in the sequence*

$$\Delta_1, \Delta_2, \ldots, \Delta_l, \tag{5}$$

of leading principal minors of the corresponding Hurwitz matrix, then $a(\lambda)$ has no pure imaginary zeros and

$$\pi(a) = V\left(a_l, \Delta_1, \frac{\Delta_2}{\Delta_1}, \ldots, \frac{\Delta_l}{\Delta_{l-1}}\right),$$

$$\nu(a) = l - V\left(a_l, \Delta_1, \frac{\Delta_2}{\Delta_1}, \ldots, \frac{\Delta_l}{\Delta_{l-1}}\right) = P\left(a_l, \Delta_1, \frac{\Delta_2}{\Delta_1}, \ldots, \frac{\Delta_l}{\Delta_{l-1}}\right), \tag{6}$$

where V (respectively, P) denotes the number of alterations (respectively, constancies) of sign in the appropriate sequence.

PROOF. Consider the case of l even, say $l = 2k$. For $s = 1, 2, \ldots, k$, denote by $\delta_s^{(1)}$ (respectively, $\delta_s^{(2)}$) the leading principal minor of order s of the matrix H_1 (respectively H_2) defined in Eqs. (13.8.6). For $s = 1, 2, \ldots, l$, let Δ_s stand for the leading principal minor of order s of the Hurwitz matrix $\hat{H}$. Applying Eq. (2) for the polynomials $b(\lambda) = \lambda g(\lambda)$, $c(\lambda) = h(\lambda)$, where $h(\lambda)$ and $g(\lambda)$ generate the L–C splitting of $a(\lambda)$, it is found that, after s interchanges of rows in the matrix on the right in Eq. (2),

$$a_l^{2s}\delta_s^{(2)} = (-1)^s \Delta_{2s}, \qquad s = 1, 2, \ldots, k. \tag{7}$$

Now apply Eq. (2) for $b(\lambda) = g(\lambda)$, $c(\lambda) = h(\lambda)$ to obtain, for $s = 1, 2, \ldots, k$,

$$a_l^{2s}\delta_s^{(1)} = \det\begin{bmatrix} a_l & a_{l-2} & \cdots & \\ 0 & a_{l-1} & a_{l-3} & \cdots \\ 0 & a_l & a_{l-2} & \cdots \\ 0 & 0 & a_{l-1} & \cdots \\ \vdots & \vdots & \vdots & \\ 0 & 0 & \cdots & a_1 \end{bmatrix} = a_l\Delta_{2s-1}. \tag{8}$$

Using Theorem 13.8.2 and the Jacobi rule, if $\Delta_i \neq 0$ for $i = 1, 2, \ldots, l$, we obtain

$$\pi(a) = V(1, \delta_1^{(1)}, \ldots, \delta_l^{(1)}) + P(1, \delta_1^{(2)}, \ldots, \delta_l^{(2)}), \tag{9}$$

where V (respectively, P) denotes the number of alterations (respectively, constancies) of sign in the appropriate sequence. In view of Eqs. (7) and (8), we find from Eq. (9) in the case of a positive a_l that

$$\begin{aligned} \pi(a) &= V(1, \Delta_1, \Delta_3, \ldots, \Delta_{l-1}) + P(1, -\Delta_2, \Delta_4, \ldots, (-1)^k\Delta_l) \\ &= V(1, \Delta_1, \Delta_3, \ldots, \Delta_{l-1}) + V(1, \Delta_2, \Delta_4, \ldots, \Delta_l), \end{aligned} \tag{10}$$

which is equivalent (in the case $a_l > 0$) to the first equality in Eq. (6). The second equality in Eq. (6), as well as the proof of both of them for $a_l < 0$, is similar. It remains to mention that the case for l odd is established in a similar way. ■

Corollary 1. *A real polynomial $a(\lambda)$ of degree l with $a_l > 0$ is stable if and only if*

$$\Delta_1 > 0, \Delta_2 > 0, \ldots, \Delta_l > 0. \tag{11}$$

Example 1. *If $a(\lambda) = (\lambda - 1)^3(\lambda + 2)$ (see Example 13.7.3), then*

$$\hat{H} = \begin{bmatrix} -1 & 5 & 0 & 0 \\ 1 & -3 & -2 & 0 \\ 0 & -1 & 5 & 0 \\ 0 & 1 & -3 & -2 \end{bmatrix},$$

with $\Delta_1 = -1$, $\Delta_2 = -2$, $\Delta_3 = -8$, $\Delta_4 = 16$. Hence

$$\pi(a) = V(1, -1, 2, 4, -2) = 3, \qquad \nu(a) = 1,$$

as required. □

Note that the Gundelfinger–Frobenius method of Section 8.6 permits us to generalize the Routh–Hurwitz theorem in the form (10) or in the form

$$\pi(a) = V(a_l, \Delta_1, \Delta_3, \ldots, \Delta_l) + V(1, \Delta_2, \ldots, \Delta_{l-1}), \tag{12}$$

according as l is even or odd. It can be seen from the proof of Theorem 1 that this extension holds whenever Theorem 13.8.2 and the Grundelfinger–Frobenius method apply. Theorem 13.8.2 requires the nonsingularity of H_1 and H_2; that is, $\delta_k^{(1)} \neq 0$ and $\delta_m^{(2)} \neq 0$ (or, equivalently, $\Delta_{2k-1} \neq 0$ and $\Delta_{2m} \neq 0$), where k and m are as defined in that theorem.

Application of the Gundelfinger–Frobenius method to the sequences of leading principal minors of H_1 and H_2, $\delta_1^{(1)}, \delta_2^{(1)}, \ldots, \delta_k^{(1)}$ and $\delta_1^{(2)}, \delta_2^{(2)}, \ldots, \delta_m^{(2)}$, requires no two consecutive zeros in these sequences and $\delta_k^{(1)} \neq 0$, $\delta_m^{(2)} \neq 0$. In terms of the Δ's this means that there must be no two consecutive zeros in the sequences $\Delta_1, \Delta_3, \ldots, \Delta_{2k-1}$ and $\Delta_2, \Delta_4, \ldots, \Delta_{2m}$, and also $\Delta_{l-1} \neq 0$, $\Delta_l \neq 0$.

Example 2. For the polynomial $a(\lambda)$ discussed in Example 13.7.1, it is found that

$$\hat{H} = \begin{bmatrix} 0 & 2 & 0 \\ 1 & -3 & 0 \\ 0 & 0 & 2 \end{bmatrix},$$

with $\Delta_1 = 0$, $\Delta_2 = -2$, $\Delta_3 = -4$. Thus, in view of the Eq. (12),

$$\pi(a) = V(1, 0, -4) + V(1, -2) = 2,$$

as required. □

13.10 The Cauchy Index and its Applications

The notion of the Cauchy index for a rational function clarifies some properties of the polynomials involved in the quotient and, as a result, plays an important role in stability problems.

Consider a real rational function $r(\lambda) = b(\lambda)/c(\lambda)$, where $b(\lambda)$ and $c(\lambda)$ are real polynomials. The difference between the number of points at which $r(\lambda)$ suffers discontinuity from $-\infty$ to $+\infty$ as λ increases from a to b ($a, b \in \mathbb{R}$, $a < b$) and the number of points at which $r(\lambda)$ suffers discontinuity from $+\infty$ to $-\infty$ on the same interval is called the *Cauchy index* of $r(\lambda)$ on (a, b) and is written $I_a^b b(\lambda)/c(\lambda)$. Note that a and b may be assigned the values $a = -\infty$ and/or $b = +\infty$.

Exercise 1. Verify that if $r(\lambda) = \mu_0/(\lambda - \lambda_0)$, where μ_0 and λ_0 are real, then

$$I_{-\infty}^{+\infty} = \operatorname{sgn} \mu_0 = \begin{cases} 1 & \text{if } \mu_0 > 0, \\ -1 & \text{if } \mu_0 < 0, \\ 0 & \text{if } \mu_0 = 0. \end{cases}$$

Exercise 2. Verify that if

$$r(\lambda) = \sum_{i=1}^{s} \frac{\mu_i}{\lambda - \lambda_i} + r_1(\lambda),$$

where μ_i, λ_i are real ($i = 1, 2, \ldots, s$) and $r_1(\lambda)$ is a rational function without real poles, then $I_a^b r(\lambda) = \sum_i \operatorname{sgn} \mu_i$, where the summation is over all λ_i lying in (a, b).

Exercise 3. Check that

$$I_a^b \frac{b(\lambda)}{c(\lambda)} = \sum_{a<\lambda_i<b} \operatorname{sgn} \frac{b(\lambda_i)}{c'(\lambda_i)},$$

provided that all zeros λ_i of $c(\lambda)$ are simple. □

Proposition 1. *The Cauchy index $I_a^b\, c'(\lambda)/c(\lambda)$ is equal to the number of distinct real zeros in (a, b) of the real polynomial $c(\lambda)$.*

PROOF. Let $c(\lambda) = c_0(\lambda - \lambda_1)^{\mu_1}(\lambda - \lambda_2)^{\mu_2} \cdots (\lambda - \lambda_k)^{\mu_k}$, where $\lambda_i \neq \lambda_j$ for $i \neq j$, $i, j = 1, 2, \ldots, k$. Suppose that only the first p distinct zeros of $c(\lambda)$ are real. If $p = 0$, then $c'(\lambda)/c(\lambda)$ has no real poles and hence the Cauchy index is zero. If $1 \leq p \leq k$, then

$$\frac{c'(\lambda)}{c(\lambda)} = \sum_{i=1}^{k} \frac{\mu_i}{\lambda - \lambda_i} = \sum_{i=1}^{p} \frac{\mu_i}{\lambda - \lambda_i} + r_1(\lambda),$$

where $r_1(\lambda)$ has no real poles. It remains now to apply the result of Exercise 2. ■

Our first theorem links the Cauchy index of polynomials $b(\lambda)$ and $c(\lambda)$ with the inertia properties of the corresponding Bezout matrix. First recall the notion of the signature of a Hermitian matrix as introduced in Section 5.5.

Theorem 1. *Let $b(\lambda)$ and $c(\lambda)$ be two real polynomials such that*

$$\deg b(\lambda) \leq \deg c(\lambda)$$

and let $B =$ Bez(b, c) denote the corresponding Bezout matrix. Then

$$I_{-\infty}^{+\infty} \frac{b(\lambda)}{c(\lambda)} = \operatorname{sig} B. \tag{1}$$

Note that, in view of Eq. (13.8.1), the Bezout matrix B in Eq. (1) can be replaced by the Hankel matrix H of the Markov parameters of $b(\lambda)/c(\lambda)$.

PROOF. Let us establish Eq. (1) for the case when all the real zeros of the polynomial $c(\lambda)$ are simple. Then we may write the rational function $r(\lambda) = b(\lambda)/c(\lambda)$ in the form

$$r(\lambda) = \sum_{i=1}^{s} \frac{\mu_i}{\lambda - \lambda_i} + r_1(\lambda), \tag{2}$$

where $\lambda_1, \lambda_2, \ldots, \lambda_s$ are the distinct real zeros of the polynomial $c(\lambda)$ and $r_1(\lambda)$ is a real polynomial without real poles. Since

$$\frac{1}{\lambda - \lambda_i} = \sum_{k=0}^{\infty} \frac{\lambda_i^k}{\lambda^{k+1}} \tag{3}$$

for sufficiently large λ, we may substitute from Eq. (3) into Eq. (2) and find

$$r(\lambda) = \sum_{k=0}^{\infty} \left(\sum_{i=1}^{s} \mu_i \lambda_i^k \right) \Big/ \lambda^{k+1} + r_1(\lambda).$$

Therefore $h_k = \sum_{i=1}^{s} \mu_i \lambda_i^k$ are the Markov parameters of the function $r(\lambda)$, for $k = 0, 1, \ldots$.

Now observe that in view of Eq. (13.8.1) and Exercise 2, Eq. (1) will be proved for the case being considered if we establish that

$$\operatorname{sig} \begin{bmatrix} h_0 & h_1 & \cdots & h_{n-1} \\ h_1 & & & \\ \vdots & & \cdot^{\cdot^{\cdot}} & \vdots \\ h_{n-1} & & \cdots & h_{2n-1} \end{bmatrix} = \sum_{i=1}^{s} \operatorname{sgn} \mu_i, \tag{4}$$

in which $n = \deg c(\lambda)$ and $h_k = \sum_{i=1}^{n} \mu_i \lambda_i^k$, $k = 0, 1, \ldots$. But if $H_n = [h_{i+j}]_{i,j=0}^{n-1}$, a simple verification shows that

$$H_n = V_n \operatorname{diag}[\mu_1, \mu_2, \ldots, \mu_s, 0, \ldots, 0] V_n^{\mathrm{T}}, \tag{5}$$

where V_n stands for the $n \times n$ Vandermonde matrix associated with $\lambda_1, \lambda_2, \ldots, \lambda_s, x_1, x_2, \ldots, x_{n-s}$ for some distinct real numbers $x_1, x_2, \ldots, x_{n-s}$ different from $\lambda_1, \lambda_2, \ldots, \lambda_s$. Using Sylvester's law of inertia and the nonsingularity of the Vandermonde matrix V_n, the relation (4) readily follows.

In the general case of multiple real zeros of $c(\lambda)$, the reasoning becomes more complicated because the decomposition of a rational function into elementary fractions contains sums of expressions of the form $(\lambda - \lambda_i)^m$, for $m = -1, -2, \ldots, -l_i$, where l_i is the multiplicity of λ_i as a zero of $c(\lambda)$. Differentiating Eq. (3), the Markov parameters can be evaluated. The details are left as an exercise. ∎

The result of Theorem 1 has several applications. We will start with some simplifications of the Liénard–Chipart and Markov stability criteria, but we first present an interesting preliminary result.

Proposition 2. *If the real polynomial* $a(\lambda) = a_0 + a_1\lambda + \cdots + a_l\lambda^l$ *with* $a_l > 0$ *is stable, then*

$$a_0 > 0, a_1 > 0, \ldots, a_{l-1} > 0, a_l > 0. \tag{6}$$

PROOF. Any real polynomial $a(\lambda)$ of degree l can be represented

$$a(\lambda) = a_l \prod_{i=1}^{s} (\lambda + \lambda_i) \prod_{j=1}^{k} (\lambda^2 + \mu_j\lambda + \nu_j), \tag{7}$$

where $s + 2k = n$ and the numbers λ_i, μ_j, ν_j are all real. Obviously, the condition that $a(\lambda)$ has only zeros with negative real parts implies that $\lambda_i > 0$, $i = 1, 2, \ldots, s$, and also that the quadratic polynomials $\lambda^2 + \mu_j\lambda + \nu_j$, $j = 1, 2, \ldots, k$ have zeros in the open left half-plane. The last requirement yields $\mu_j > 0$, $\nu_j > 0$ for all j. Now it is obvious from Eq. (7) that since the coefficients of $a(\lambda)$ are sums of products of positive numbers, they are positive. ■

Note that the condition (6) is not sufficient for the stability of $a(\lambda)$, as the example of the polynomial $\lambda^2 + 1$ shows.

Thus, in the investigation of stability criteria we may first assume that the coefficients of $a(\lambda)$ are positive. It turns out that in this case the determinantal inequalities in (13.9.11) are not independent. Namely, the positivity of the Hurwitz determinants Δ_i of even order implies that of the Hurwitz determinants of odd order, and vice versa. This fact will follow from Theorem 4. We first establish simplifications of the Liénard–Chipart and Markov stability criteria. This is done with the help of the Cauchy index.

Theorem 2. (The Liénard–Chipart criterion). *Let* $a(\lambda) = \sum_{i=0}^{l} a_i\lambda^i$ *be a real polynomial and let*

$$a(\lambda) = h(\lambda^2) + \lambda g(\lambda^2)$$

be its L–C splitting. Then $a(\lambda)$ *is stable if and only if the coefficients of* $h(\lambda)$ *have the same sign as* a_l *and the Bezoutian* $B_1 = B(h, g)$ *is positive definite.*

PROOF. The necessity of the condition follows easily from Theorem 13.7.2 and Proposition 2. The converse statement will be proved for the even case, $l = 2k$, while the verification of the odd case is left as an exercise.

Let $B_1 = B(h, g) > O$ and assume the coefficients $a_0, a_2, \ldots, a_l$ of $h(\lambda)$ are positive. Then obviously $h(\lambda) \geq 0$ for all positive $\lambda \in \mathbb{R}$. Furthermore, if $h(\lambda_1) = 0$ for $\lambda_1 > 0$, then $g(\lambda_1) \neq 0$ since otherwise the Bezoutian B will be singular (Theorem 13.3.1). Thus, although the fractions $g(\lambda)/h(\lambda)$ and $\lambda g(\lambda)/h(\lambda)$ may suffer infinite discontinuity as λ passes from 0 to $+\infty$, there is no discontinuity in the value of the quotients from $-\infty$ to $+\infty$, or from $+\infty$ to $-\infty$. Hence

$$I_0^{+\infty}\frac{g(\lambda)}{h(\lambda)} = I_0^{+\infty}\frac{\lambda g(\lambda)}{h(\lambda)} = 0.$$

Furthermore, it is clear that

$$I_{-\infty}^{0}\frac{g(\lambda)}{h(\lambda)} = -I_{-\infty}^{0}\frac{\lambda g(\lambda)}{h(\lambda)},$$

and therefore

$$I_{-\infty}^{+\infty}\frac{g(\lambda)}{h(\lambda)} = I_{-\infty}^{0}\frac{g(\lambda)}{h(\lambda)} + I_0^{+\infty}\frac{g(\lambda)}{h(\lambda)} = I_{-\infty}^{0}\frac{g(\lambda)}{h(\lambda)} = -I_{-\infty}^{0}\frac{\lambda g(\lambda)}{h(\lambda)}. \tag{8}$$

But the Bezoutian $B_1 = \text{Bez}(h, g)$ is assumed to be positive definite, and therefore, by Theorem 1,

$$I_{-\infty}^{\infty}\frac{g(\lambda)}{h(\lambda)} = k.$$

Applying Theorem 1 again, we find from Eq. (8) that $B_2 = \text{Bez}(h, \lambda g)$ is negative definite. Theorem 13.7.2 now provides the stability of the polynomial $a(\lambda)$. ■

Note that Theorem 2 can be reformulated in an obvious way in terms of the Bezoutian $B_2 = \text{Bez}(h, g)$ and the coefficients of $h(\lambda)$ (or $g(\lambda)$).

The congruence of the matrices B_1 and H_1 defined in Theorem 13.8.2 leads to the following result.

Theorem 3 (The Markov criterion). *A real polynomial $a(\lambda)$ is stable if and only if the coefficients of $h(\lambda)$ from the L–C splitting of $a(\lambda)$ have the same sign as the leading coefficient of $a(\lambda)$ and also $H_1 > 0$.*

The remark made after Theorem 2 obviously applies to Theorem 3. This fact, as well as the relations (13.9.7) and (13.9.8), permit us to obtain a different form of the Liénard–Chipart criterion.

Theorem 4 (The Liénard–Chipart criterion). *A real polynomial $a(\lambda) = \sum_{i=0}^{l} a_i\lambda^i$, $a_l > 0$, is stable if and only if one of the following four conditions holds:*

(*a*) $a_0 > 0, a_2 > 0, \ldots;$ $\Delta_1 > 0, \Delta_3 > 0, \ldots;$
(*b*) $a_0 > 0, a_2 > 0, \ldots;$ $\Delta_2 > 0, \Delta_4 > 0, \ldots;$
(*c*) $a_0 > 0, a_1 > 0, a_3 > 0, \ldots;$ $\Delta_1 > 0, \Delta_3 > 0, \ldots;$
(*d*) $a_0 > 0, a_1 > 0, a_3 > 0, \ldots;$ $\Delta_2 > 0, \Delta_4 > 0, \ldots.$

The detailed proof is left to the reader.

We conclude this section with one more application of Theorem 1, but first we need the following statement, which is of independent interest.

Theorem 5. *The number of distinct zeros of a real polynomial $a(\lambda)$ of degree l is equal to the rank of the Hankel matrix $H = [s_{i+j}]_{i,j=0}^{l-1}$ of the Newton sums $s_0 = l$, $s_k = \lambda_1^k + \lambda_2^k + \cdots + \lambda_l^k$, $k = 1, 2, \ldots,$ where $\lambda_1, \lambda_2, \ldots, \lambda_l$ are the zeros of $a(\lambda)$.*

PROOF. We first observe that $H = VV^T$, where V denotes the $l \times l$ Vandermonde matrix associated with $\lambda_1, \lambda_2, \ldots, \lambda_l$. Hence rank $H =$ rank V and the result follows. ■

Note that the significance of this and the following theorem lies in the fact that the Newton sums $s_0, s_1, \ldots$ can be evaluated recursively via the coefficients of the polynomial. If $a(\lambda) = a_0 + a_1\lambda + \cdots + a_{l-1}\lambda^{l-1} + \lambda^l$, then

$$s_0 = l, \qquad ka_{l-k} + a_{l-k+1}s_1 + \cdots + a_{l-1}s_{k-1} + s_k = 0, \tag{9}$$

for $k = 1, 2, \ldots, l-1$, and

$$a_0 s_{k-l} + a_1 s_{k-l-1} + \cdots + a_{l-1}s_{k-1} + s_k = 0, \tag{10}$$

for $k = l, l+1, \ldots$.

Theorem 6 (C. Borhardt and G. Jacobi†). *The real polynomial $a(\lambda) = \sum_{i=0}^{l} a_i\lambda^i$ has v different pairs of complex conjugate zeros and $\pi - v$ different real zeros, where $\{\pi, v, \delta\}$ is the inertia of the Hankel matrix $H = [s_{i+j}]_{i,j=0}^{l-1}$ of the Newton sums associated with $a(\lambda)$.*

PROOF. Putting $c(\lambda) = a(\lambda)$, $b(\lambda) = a'(\lambda)$ in Theorem 1, we obtain

$$\text{Bez}(a, a') = I_{-\infty}^{+\infty} \frac{a'(\lambda)}{a(\lambda)}.$$

Since

$$a'(\lambda) = (s_0\lambda^{-1} + s_1\lambda^{-2} + \cdots)a(\lambda), \tag{11}$$

where $s_0, s_1, s_2, \ldots$ are the Newton sums associated with $a(\lambda)$, it follows that by formula (13.8.1) the matrices Bez(a, a') and the matrix H of the Newton sums are congruent. Hence the relation (1) shows, in view of Proposition 1, that the number of distinct real zeros of $a(\lambda)$ is equal to sig $H = \pi - v$. By Theorem 5, rank $H = \pi + v$ is equal to the total number of distinct zeros of $a(\lambda)$ and therefore $a(\lambda)$ has $2v$ distinct complex zeros. The result follows. ■

Exercise 4. Evaluate the number of distinct real zeros and the number of different pairs of complex conjugate zeros of the polynomial $a(\lambda) = \lambda^4 - 2\lambda^3 + 2\lambda^2 - 2\lambda + 1$.

† *J. Math. Pures Appl.* **12** (1847), 50–67, and *J. Reine Angew. Math.* **53** (1857), 281–283.

SOLUTION. To find the Newton sums of $a(\lambda)$, we may use formulas (9) and (10) or the representation (11). We have $s_0 = 4$, $s_1 = 2$, $s_2 = 0$, $s_3 = 2$, $s_4 = 4$, $s_5 = 2$, $s_6 = 0$, and hence the corresponding Hankel matrix H is

$$H = \begin{bmatrix} 4 & 2 & 0 & 2 \\ 2 & 0 & 2 & 4 \\ 0 & 2 & 4 & 2 \\ 2 & 4 & 2 & 0 \end{bmatrix}.$$

Since $\operatorname{In} H = \{2, 1, 1\}$, there is one pair of distinct complex conjugate zeros ($\nu = 1$) and one real zero ($\pi - \nu = 1$). It is easily checked that, actually, $a(\lambda) = (\lambda - 1)^2(\lambda^2 + 1)$, and the results are confirmed. □

CHAPTER 14

Matrix Polynomials

It was seen in Sections 4.13 and 9.10 that the solution of a system of constant-coefficient (or time-invariant) differential equations, $\dot{\mathbf{x}}(t) - A\mathbf{x}(t) = \mathbf{f}(t)$, where $A \in \mathbb{C}^{n \times n}$ and $\mathbf{x}(0)$ is prescribed, is completely determined by the properties of the matrix-valued function $\lambda I - A$. More specifically, the solution is determined by the zeros of $\det(\lambda I - A)$ and by the bases for the subspaces $\operatorname{Ker}(\lambda I - A)^j$ of $\mathbb{C}^n$, $j = 0, 1, 2, \ldots, n$. These spectral properties are, in turn, summarized in the Jordan form for A or, as we shall now say, in the Jordan form for $\lambda I - A$.

The analysis of this chapter† will admit the description of solutions for initial-value problems for higher-order systems:

$$L_l \mathbf{x}^{(l)}(t) + L_{l-1}\mathbf{x}^{(l-1)}(t) + \cdots + L_1 \mathbf{x}^{(1)}(t) + L_0 \mathbf{x}(t) = \mathbf{f}(t), \tag{1}$$

where $L_0, L_1, \ldots, L_l \in \mathbb{C}^{n \times n}$, $\det L_l \neq 0$, and the indices denote derivatives with respect to the independent variable t. The underlying algebraic problem now concerns the associated matrix polynomial $L(\lambda) = \sum_{j=0}^{l} \lambda^j L_j$ of degree l, as introduced in Section 7.1. We shall confine our attention to the case in which $\det L_l \neq 0$. Since the coefficient of λ^{ln} in the polynomial $\det L(\lambda)$ is just $\det L_l$, it follows that $L(\lambda)$ is regular or, in the terminology of Section 7.5, has rank n.

As for the first-order system of differential equations, the formulation of general solutions for Eq. (1) (even when the coefficients are Hermitian) requires the development of a "Jordan structure" for $L(\lambda)$, and that will be done in this chapter. Important first steps in this direction have been taken

† The systematic spectral theory of matrix polynomials presented here is due to I. Gohberg, P. Lancaster, and L. Rodman. See their 1982 monograph of Appendix 3.

in Chapter 7; we recall particularly the Smith normal form and the elementary divisors, both of which are invariant under equivalence transformations of $L(\lambda)$. In Chapter 7, these notions were used to develop the Jordan structure of $\lambda I - A$ (or of A). Now they will be used in the analysis of the spectral properties of more general matrix polynomials.

14.1 Linearization of a Matrix Polynomial

It has been seen on several occasions in earlier chapters that if $a(\lambda) = \Sigma_{j=0}^{l} a_j \lambda^j$ is a monic scalar polynomial then the related $l \times l$ companion matrix,

$$C_a = \begin{bmatrix} 0 & 1 & 0 & \cdots & 0 \\ 0 & 0 & 1 & & \vdots \\ \vdots & \vdots & \vdots & & 1 \\ -a_0 & -a_1 & -a_2 & \cdots & -a_{l-1} \end{bmatrix}, \tag{1}$$

is a useful idea. Recall the important part played in the formulation of the first natural normal form of Chapter 7 and in the formulation of stability criteria in Chapter 13 (see also Exercises 6.6.3 and 7.5.4). In particular, note that $a(\lambda) = \det(\lambda I - C_a)$, so that the zeros of $a(\lambda)$ correspond to the eigenvalues of C_a.

For a matrix polynomial of size $n \times n$,

$$L(\lambda) = \sum_{j=0}^{l} \lambda^j L_j, \qquad \det L_l \neq 0, \tag{2}$$

we formulate the generalization of the matrix in (1):

$$C_L = \begin{bmatrix} 0 & I_n & 0 & \cdots & 0 \\ 0 & 0 & I_n & & \vdots \\ \vdots & \vdots & \vdots & & I_n \\ -\hat{L}_0 & -\hat{L}_1 & -\hat{L}_2 & \cdots & -\hat{L}_{l-1} \end{bmatrix}, \tag{3}$$

where $\hat{L}_j = L_l^{-1} L_j$ for $j = 0, 1, \ldots, l-1$. Now consider the relationships between $L(\lambda)$ and $\lambda I_{ln} - C_L$. The $ln \times ln$ matrix C_L is called the (*first*) *companion matrix* for $L(\lambda)$.

The first observation is that the characteristic polynomials satisfy

$$\det L(\lambda) = \det(\lambda I_{ln} - C_L)(\det L_l). \tag{4}$$

This means that the *latent roots* of $L(\lambda)$ (as defined in Section 7.7) coincide with the *eigenvalues* of C_L. Admitting an ambiguity that should cause no difficulties, the set of latent roots of $L(\lambda)$ is called the *spectrum* of $L(\lambda)$ and is written $\sigma(L)$. From the point of view of matrix polynomials, the relation (4) says that $L(\lambda)$ and $\lambda I - C_L$ have the same invariant polynomial of highest degree. However, the connection is deeper than this, as the first theorem shows.

Theorem 1. *The $ln \times ln$ matrix polynomials*

$$\begin{bmatrix} L(\lambda) & 0 \\ 0 & I_{(l-1)n} \end{bmatrix} \quad \text{and} \quad \lambda I_{ln} - C_L$$

are equivalent.

PROOF. First define $ln \times ln$ matrix polynomials $E(\lambda)$ and $F(\lambda)$ by

$$F(\lambda) = \begin{bmatrix} I & 0 & \cdots & & 0 \\ -\lambda I & I & & & \vdots \\ 0 & -\lambda I & I & \ddots & \\ \vdots & & \ddots & I & 0 \\ 0 & \cdots & & -\lambda I & I \end{bmatrix},$$

$$E(\lambda) = \begin{bmatrix} B_{l-1}(\lambda) & B_{l-2}(\lambda) & \cdots & B_0(\lambda) \\ -I & 0 & \cdots & 0 \\ 0 & -I & \ddots & \vdots \\ \vdots & & \ddots & \\ 0 & \cdots & -I & 0 \end{bmatrix},$$

where $B_0(\lambda) = L_l$, $B_{r+1}(\lambda) = \lambda B_r(\lambda) + L_{l-r-1}$ for $r = 0, 1, \ldots, l-2$, and $I = I_n$. Clearly

$$\det E(\lambda) \equiv \pm \det L_l, \qquad \det F(\lambda) \equiv 1. \tag{5}$$

Hence $F(\lambda)^{-1}$ is also a matrix polynomial. It is easily verified that

$$E(\lambda)(\lambda I - C_L) = \begin{bmatrix} L(\lambda) & O \\ O & I_{(l-1)n} \end{bmatrix} F(\lambda), \tag{6}$$

and so

$$\begin{bmatrix} L(\lambda) & O \\ O & I_{(l-1)n} \end{bmatrix} = E(\lambda)(\lambda I - C_L)F(\lambda)^{-1}$$

determines the equivalence stated in the theorem. ■

Theorem 1 (together with Theorem 7.5.1 and its corollary) shows that *all* of the nontrivial invariant polynomials (and hence all of the elementary divisors) of $L(\lambda)$ and $\lambda I - C_L$ coincide. Here, "nontrivial" means a polynomial that is not identically equal to 1.

Exercise 1. Establish the relations (4) and (5).

Exercise 2. Show that if λ is an eigenvalue of C_L then the dimension of $\operatorname{Ker}(\lambda I - C_L)$ cannot exceed n. (Note the special case $n = 1$.)

Exercise 3. Formulate the first companion matrix and find all elementary divisors of (a) $L(\lambda) = \lambda^l I_n$ and (b) $L(\lambda) = \lambda^l I_n - I_n$. □

Observe now that there are other matrices A, as well as C_L, for which $\lambda I - A$ and $\operatorname{diag}[L(\lambda), I_{(l-1)n}]$ are equivalent. In particular, if A is a matrix similar to C_L, say $A = TC_LT^{-1}$, then $\lambda I - A = T(\lambda I - C_L)T^{-1}$, so that $\lambda I - A$ and $\lambda I - C_L$ are equivalent and, by the transitivity of the equivalence relation, $\lambda I - A$ and $\operatorname{diag}[L(\lambda), I_{(l-1)n}]$ are equivalent. It is useful to make a formal definition: If $L(\lambda)$ is an $n \times n$ matrix polynomial of degree l with nonsingular leading coefficient, then any $ln \times ln$ matrix A for which $\lambda I - A$ and $\operatorname{diag}[L(\lambda), I_{(\ell-1)n}]$ are equivalent is called a *linearization* of $L(\lambda)$.

The next theorem shows that the matrices A that are similar to C_L are, in fact, *all* of the linearizations of $L(\lambda)$.

Theorem 2. *Any matrix A is a linearization of $L(\lambda)$ if and only if A is similar to the companion matrix C_L of $L(\lambda)$.*

PROOF. It has been shown that if A is similar to C_L then it is a linearization of $L(\lambda)$. Conversely, if A is a linearization of $L(\lambda)$ then $\lambda I - A \sim L(\lambda) \dotplus I_{(l-1)n}$ (using $\sim$ as an abbreviation for "equivalent to"). But by Theorem 1, $L(\lambda) \dotplus I_{(l-1)n} \sim \lambda I - C_L$ so, using transitivity again, $\lambda I - A \sim \lambda I - C_L$. Then Theorem 7.6.1 shows that A and C_L are similar. ■

This result, combined with Theorem 6.5.1 (or 7.8.2), shows that there is a matrix in Jordan normal form that is a linearization for $L(\lambda)$ and, therefore, it includes all the information concerning the latent roots of $L(\lambda)$ and their algebraic multiplicities (i.e., the information contained in a Segre characteristic, as described in Section 6.6).

Exercise 4. Use Exercise 6.6.3 to find a matrix in Jordan normal form that is a linearization for a scalar polynomial (i.e., a 1×1 matrix polynomial $L(\lambda)$).

Exercise 5. Find matrices in Jordan normal form that are linearizations for the matrix polynomials of Exercise 3. Do the same for

$$L(\lambda) = \begin{bmatrix} \lambda^2 & -\lambda \\ 0 & \lambda^2 \end{bmatrix}.$$

SOLUTION. The solution in the latter case is either

$$\begin{bmatrix} 0 & 0 & 0 & 0 \\ 0 & 0 & 1 & 0 \\ 0 & 0 & 0 & 1 \\ 0 & 0 & 0 & 0 \end{bmatrix} \quad \text{or} \quad \begin{bmatrix} 0 & 1 & 0 & 0 \\ 0 & 0 & 1 & 0 \\ 0 & 0 & 0 & 0 \\ 0 & 0 & 0 & 0 \end{bmatrix}.$$

Exercise 6. Let $\hat{L}_j = L_j L_l^{-1}$ for $j = 0, 1, \ldots, l - 1$, and define the *second companion matrix* for $L(\lambda)$ to be the $ln \times ln$ matrix

$$C_2 = \begin{bmatrix} 0 & 0 & \cdots & 0 & -\hat{L}_0 \\ I & 0 & & \vdots & \vdots \\ 0 & I & & & \\ \vdots & & \ddots & 0 & -\hat{L}_{l-2} \\ 0 & & \cdots & I & -\hat{L}_{l-1} \end{bmatrix}.$$

Also (see Chapter 13) define the *symmetrizer* for $L(\lambda)$ to be

$$S_L = \begin{bmatrix} L_1 & L_2 & \cdots & L_{l-1} & L_l \\ L_2 & & . \cdot & L_l & 0 \\ \vdots & . \cdot & . \cdot & & \vdots \\ L_{l-1} & L_l & & & \\ L_l & 0 & \cdots & & 0 \end{bmatrix}. \tag{7}$$

Show that S_L is nonsingular, that $C_2 = S_L C_L S_L^{-1}$ and, hence, that C_2 is a linearization of $L(\lambda)$. ☐

14.2 Standard Triples and Pairs

In the spectral theory of a general matrix $A \in \mathbb{C}^{n \times n}$, the resolvent function $(\lambda I - A)^{-1}$ plays an important part on several occasions. For a matrix polynomial $L(\lambda)$, as defined in the relations (14.1.2), the matrix-valued function $L(\lambda)^{-1}$ (defined for all $\lambda \notin \sigma(L)$) plays the corresponding role and is called the *resolvent* of $L(\lambda)$. The next step in our analysis shows how the resolvent of the first companion matrix C_L can be used to determine the resolvent of $L(\lambda)$.

Theorem 1. *For every complex* $\lambda \notin \sigma(L)$,

$$L(\lambda)^{-1} = P_1(\lambda I - C_L)^{-1} R_1, \tag{1}$$

where P_1, R_1 are the matrices of size $n \times nl$ and $ln \times n$, respectively, defined by

$$P_1 = [I_n \quad 0 \quad \cdots \quad 0], \qquad R_1 = \begin{bmatrix} 0 \\ \vdots \\ 0 \\ L_l^{-1} \end{bmatrix}. \tag{2}$$

PROOF. First observe that, because of Eq. (14.1.4), the matrix $(\lambda I - C_L)^{-1}$ is defined at just those points λ where $L(\lambda)^{-1}$ exists. Now consider the proof of Theorem 14.1.1 once more. The relation (14.1.6) implies that

$$\begin{bmatrix} L(\lambda)^{-1} & 0 \\ 0 & I_{(l-1)n} \end{bmatrix} = F(\lambda)(\lambda I - C_L)^{-1}E(\lambda)^{-1}. \tag{3}$$

Using the definition of $E(\lambda)$, it is easily verified that

$$E(\lambda)\begin{bmatrix} 0 \\ \vdots \\ 0 \\ L_l^{-1} \end{bmatrix} = \begin{bmatrix} I \\ 0 \\ \vdots \\ 0 \end{bmatrix},$$

and this means that the first n columns of $E(\lambda)^{-1}$ are simply equal to R_1. Since the first n rows of $F(\lambda)$ are equal to P_1, if we equate the leading $n \times n$ submatrices on the left and right of Eq. (3), we obtain Eq. (1). ■

Three matrices (U, T, V) are said to be *admissible* for the $n \times n$ matrix polynomial $L(\lambda)$ of degree l (as defined in Eq. (14.1.2)) if they are of sizes $n \times ln$, $ln \times ln$, and $ln \times n$, respectively. Then two admissible triples for $L(\lambda)$, say (U_1, T_1, V_1) and (U_2, T_2, V_2), are said to be *similar* if there is a nonsingular matrix S such that

$$U_1 = U_2 S, \qquad T_1 = S^{-1}T_2 S, \qquad V_1 = S^{-1}V_2. \tag{4}$$

Exercise 1. Show that similarity of triples is an equivalence relation on the set of all admissible triples for $L(\lambda)$. □

Now, the triple (P_1, C_L, R_1) is admissible for $L(\lambda)$, and any triple similar to (P_1, C_L, R_1) is said to be a *standard triple* for $L(\lambda)$. Of course, with a transforming matrix $S = I$, this means that (P_1, C_L, R_1) is itself a standard triple. Note that, by Theorem 14.1.2, the second member of a standard triple for $L(\lambda)$ is always a linearization of $L(\lambda)$. Now Theorem 1 is easily generalized.

Theorem 2. *If (U, T, V) is a standard triple for $L(\lambda)$ and $\lambda \notin \sigma(L)$, then*

$$L(\lambda)^{-1} = U(\lambda I - T)^{-1}V.$$

PROOF. Since (U, T, V) is a standard triple, there is a nonsingular matrix S such that

$$P_1 = US, \qquad C_L = S^{-1}TS, \qquad R_1 = S^{-1}V.$$

Observe that $(\lambda I - C_L)^{-1} = (S^{-1}(\lambda I - T)S)^{-1} = S^{-1}(\lambda I - T)^{-1}S$, and substitute into Eq. (1) to obtain the result. ■

Thus we obtain a representation of the resolvent of $L(\lambda)$ in terms of the resolvent of any linearization for $L(\lambda)$. Our next objective is to characterize standard triples for $L(\lambda)$ in a more constructive and direct way; it is not generally reasonable to try to find the matrix S of the proof of Theorem 2 explicitly. First we will show how to "recover" the companion matrix C_L from any given standard triple.

Lemma 1. *Let (U, T, V) be a standard triple for $L(\lambda)$ and define*

$$Q = \begin{bmatrix} U \\ UT \\ \vdots \\ UT^{l-1} \end{bmatrix}. \tag{5}$$

Then Q is nonsingular and $C_L = QTQ^{-1}$.

PROOF. Recall that $P_1 = [I_n \ \ 0 \ \ \cdots \ \ 0]$ and C_L is given by Eq. (14.1.3). It is easily seen that

$$\begin{bmatrix} P_1 \\ P_1C_L \\ \vdots \\ P_1C_L^{l-1} \end{bmatrix} = I_{ln}. \tag{6}$$

Substitute for P_1 and C_L from Eqs. (4) to obtain $QS = I_{ln}$. Obviously, Q is nonsingular and $S = Q^{-1}$. In particular, $C_L = QTQ^{-1}$. ■

In fact, this lemma shows that for any standard triple (U, T, V), the primitive triple (P_1, C_L, R_1) can be recovered from Eqs. (4) on choosing $S = Q^{-1}$, where Q is given by Eq. (5).

Theorem 3. *An admissible triple (U, T, V) for $L(\lambda)$ is a standard triple if and only if the three following conditions are satisfied:*

(a) *The matrix Q of Eq. (5) is nonsingular;*
(b) $L_l UT^l + L_{l-1}UT^{l-1} + \cdots + L_1UT + L_0U = 0.$
(c) *$V = Q^{-1}R_1$, where Q and R_1 are given by Eqs. (5) and (2), respectively.*

PROOF. We first check that conditions (a), (b), and (c) are satisfied by the standard triple (P_1, C_L, R_1). Condition (a) is proved in the proof of Lemma 1. The relation (6) also gives

$$P_1 C_L^l = (P_1 C_L^{l-1}) C_L = [0 \quad \cdots \quad 0 \quad I] C_L = [-\hat{L}_0 \quad \cdots \quad -\hat{L}_{l-1}].$$

Consequently,

$$L_l P_1 C_L^l = [-L_0 \quad -L_1 \quad \cdots \quad -L_{l-1}]. \tag{7}$$

But also,

$$\sum_{j=0}^{l-1} L_j P_1 C_L^j = [L_0 \quad L_1 \quad \cdots \quad L_{l-1}] \begin{bmatrix} P_1 \\ P_1 C_L \\ \vdots \\ P_1 C_L^{l-1} \end{bmatrix} = [L_0 \quad L_1 \quad \cdots \quad L_{l-1}].$$

So, adding the last two relations, we get $\sum_{j=0}^{l} L_j P_1 C_L^j = 0$, which is just the relation (b) with (U, T, V) replaced by (P_1, C_L, R_1). In this case $Q = I$, so condition (c) is trivially satisfied.

Now let (U, T, V) be any standard triple and let

$$U = P_1 S, \qquad T = S^{-1} C_L S, \qquad V = S^{-1} R_1.$$

Then

$$Q = \begin{bmatrix} U \\ UT \\ \vdots \\ UT^{l-1} \end{bmatrix} = \begin{bmatrix} P_1 \\ P_1 C_L \\ \vdots \\ P_1 C_L^{l-1} \end{bmatrix} S = S,$$

using (6) so that (a) is satisfied. Then

$$\sum_{j=0}^{l} L_j U T^j = \left(\sum_{j=0}^{l} L_j P_1 C_L^j \right) S = 0,$$

which is condition (b). Finally, $QV = S(S^{-1} R_1) = R_1$, which is equivalent to (c).

Conversely, suppose that we are given an admissible triple for $L(\lambda)$ that satisfies conditions (a), (b), and (c). We must show that (U, T, V) is similar to (P_1, C_L, R_1). First, from the definition (5) of Q, it is obvious that $U = P_1 Q$ and, using condition (a), Q is nonsingular.

Then use condition (b) to verify that $QT = C_L Q$ by multiplying out the partitioned matrices. Thus, $T = Q^{-1} C_L Q$.

Combine these results with condition (c) to get

$$U = P_1 Q, \qquad T = Q^{-1} C_L Q, \qquad V = Q^{-1} R_1,$$

and so (U, T, V) is similar to (P_1, C_L, R_1) and hence is a standard triple. ■

This theorem makes it clear that a standard triple is completely determined by its first two members. It is frequently useful to work in terms of such a pair and so they are given a name: a *standard pair* for $L(\lambda)$ is the first two members of any standard triple for $L(\lambda)$. Thus (U, T) is a standard pair for $L(\lambda)$ if and only if conditions (a) and (b) of the theorem are satisfied.

Exercise 2. Show that, if (U, T) is a standard pair for $L(\lambda)$, then (U, T) is also a standard pair for any matrix polynomial of the form $AL(\lambda)$, where $\det A \neq 0$.

Exercise 3. Let C_2 be the second companion matrix for $L(\lambda)$, as defined in Exercise 14.1.6. Show that the admissible triple

$$[0 \quad \cdots \quad 0 \quad L_l^{-1}], \qquad C_2, \qquad \begin{bmatrix} I \\ 0 \\ \vdots \\ 0 \end{bmatrix}$$

forms a standard triple for $L(\lambda)$.

Exercise 4. Let (U, T, V) be a standard triple for $L(\lambda)$, and let Γ be a contour in the complex plane with $\sigma(L)$ inside Γ. If $f(\lambda)$ is a function that is analytic inside Γ and continuous on Γ, show that

$$\frac{1}{2\pi i} \int_\Gamma f(\lambda) L(\lambda)^{-1} \, d\lambda = U f(T) V.$$

Exercise 5. If $U, T)$ is a standard pair for $L(\lambda)$, show that in the terminology of Section 5.13, (U, T) is an observable pair of index l. (This is also a sufficient condition for (U, T) of sizes $n \times ln$ and $ln \times ln$, respectively, to be a standard pair for *some* monic matrix polynomials of degree l; see Theorem 1). □

There is a useful converse statement for Theorem 2.

Theorem 4. *Let (U, T, V) be an admissible triple for $L(\lambda)$ and assume that*

$$L(\lambda)^{-1} = U(\lambda I - T)^{-1} V. \tag{8}$$

Then (U, T, V) is a standard triple for $L(\lambda)$.

PROOF. For $|\lambda|$ sufficiently large, $L(\lambda)^{-1}$ has a Laurent expansion of the form

$$L(\lambda)^{-1} = \lambda^{-l} L_l^{-1} + \lambda^{-l-1} A_1 + \lambda^{-l-2} A_2 + \cdots, \tag{9}$$

for some matrices $A_1, A_2, \ldots$. Now let Γ be a circle in the complex plane with centre at the origin and having $\sigma(L)$ and the eigenvalues of T in its interior. Then integrating Eq. (9) term by term and using the theorem of residues,

$$\frac{1}{2\pi i}\int_L \lambda^j L(\lambda)^{-1}\,d\lambda = \begin{cases} 0 & \text{if } j = 0, 1, \ldots, l-2; \\ L_l^{-1} & \text{if } j = l-1. \end{cases}$$

But also (see Theorem 9.9.2),

$$\frac{1}{2\pi i}\int_L \lambda^j(\lambda I - T)^{-1}\,d\lambda = T^j, \qquad j = 0, 1, 2, \ldots. \tag{10}$$

It therefore follows from Eq. (8) (using $*$ to denote an $n \times n$ matrix of no immediate interest) that

$$\frac{1}{2\pi i}\int_\Gamma \begin{bmatrix} 1 & \cdots & \lambda^{l-1} \\ \lambda & \iddots & \vdots \\ \vdots & & \\ \lambda^{l-1} & \cdots & \lambda^{2l-2} \end{bmatrix} L(\lambda)^{-1}\,d\lambda = \begin{bmatrix} 0 & 0 & \cdots & 0 & L_l^{-1} \\ 0 & & & L_l^{-1} & * \\ \vdots & & \iddots & & \\ 0 & L_l^{-1} & & & \vdots \\ L_l^{-1} & * & \cdots & & * \end{bmatrix}$$

$$= \begin{bmatrix} UV & UTV & \cdots & UT^{l-1}V \\ UTV & & \iddots & \vdots \\ \vdots & & & \\ UT^{l-1}V & & \cdots & UT^{2l-2}V \end{bmatrix}$$

$$= \begin{bmatrix} U \\ UT \\ \vdots \\ UT^{l-1} \end{bmatrix} [V \quad TV \quad \cdots \quad T^{l-1}V]. \tag{11}$$

Both matrices in the last factorization are $ln \times ln$ and must be nonsingular, since the matrix above of triangular form is clearly nonsingular.

Using (10) once more we also have, for $j = 0, 1, \ldots, l-1$,

$$0 = \frac{1}{2\pi i}\int_\Gamma \lambda^j L(\lambda)L(\lambda)^{-1}\,d\lambda = \frac{1}{2\pi i}\int_\Gamma \lambda^j L(\lambda)U(\lambda I - T)^{-1}V\,d\lambda$$

$$= (L_l UT^l + \cdots + L_1 UT + L_0 U)T^j V.$$

Hence,

$$(L_l UT^l + \cdots + L_1 UT + L_0 U)[V \quad TV \quad \cdots \quad T^{l-1}V] = O,$$

and since $[V \quad TV \quad \cdots \quad T^{l-1}V]$ is nonsingular, $\sum_{j=0}^{l} L_j UT^j = O$.

It has now been shown that U and T satisfy conditions (a) and (b) of Theorem 3. But condition (c) is also contained in Eqs. (11). Hence (U, T, V) is a standard triple. ■

Corollary 1. *If (U, T, V) is a standard triple for $L(\lambda)$, then $(V^{\mathrm{T}}, T^{\mathrm{T}}, U^{\mathrm{T}})$ is a standard triple for $L^{\mathrm{T}}(\lambda) \triangleq \sum_{j=0}^{l} \lambda^j L_j^{\mathrm{T}}$, and (V^*, T^*, U^*) is a standard triple for $L^*(\lambda) \triangleq \sum_{j=0}^{l} \lambda_j L_j^*$.*

PROOF. Apply Theorem 2 to obtain Eq. (8). Then take transposes and apply Theorem 4 to get the first statement. Take conjugate transposes to get the second. ■

Exercise 6. Let (U, T, V) be a standard triple for $L(\lambda)$ and show that

$$UT^jV = \begin{cases} 0 & \text{if } j = 0, 1, \ldots, l-2; \\ L_l^{-1} & \text{if } j = l-1. \end{cases} \tag{12}$$

Hint. Examine Eqs. (11). □

The properties in the preceding example can be seen as special cases of the following useful result for standard triples.

Theorem 5. *If (U, T, V) is a standard triple for $L(\lambda)$, S_L is the symmetrizer for $L(\lambda)$ as defined in Eq.* (14.1.7), *and matrices Q, R are defined by*

$$Q = \begin{bmatrix} U \\ UT \\ \vdots \\ UT^{l-1} \end{bmatrix}, \qquad R = [V \quad TV \quad \cdots \quad T^{l-1}V], \tag{13}$$

then

$$RS_LQ = I. \tag{14}$$

PROOF. First use Exercise 14.1.6 to write $C_L = S_L^{-1}C_2S_L$ (where C_2 is the second companion matrix for $L(\lambda)$) so that, for $j = 0, 1, 2, \ldots,$

$$C_L^jR_1 = S_L^{-1}C_2^j(S_LR_1) = S_L^{-1}C_2^jP_1^{\mathrm{T}}.$$

Now it is easily verified that

$$[R_1 \quad C_LR_1 \quad \cdots \quad C_L^{l-1}R_1] = S_L^{-1}[P_1^{\mathrm{T}} \quad C_2P_1^{\mathrm{T}} \quad \cdots \quad C_2^{l-1}P_1^{\mathrm{T}}] = S_L^{-1}. \tag{15}$$

For any triple (U, T, V) there is a nonsingular S such that $U = P_1 S$, $T = S^{-1}C_L S$, and $V = S^{-1}R_1$. Thus, using Eq. (6),

$$Q = \begin{bmatrix} P_1 \\ P_1 C_L \\ \vdots \\ P_1 C_L^{l-1} \end{bmatrix} S = S$$

and, using Eq. (15),

$$R = S^{-1}[R_1 \quad C_L R_1 \quad \cdots \quad C_L^{l-1} R_1] = S^{-1} S_L^{-1}.$$

Thus, $RS_L Q = (S^{-1}S_L^{-1})S_L S = I$, as required. ■

14.3 The Structure of Jordan Triples

By definition, all standard triples for a matrix polynomial $L(\lambda)$ with nonsingular leading coefficient are similar to one another in the sense of Eqs. (14.2.4). Thus, there are always standard triples for which the second term, the linearization of $L(\lambda)$, is in Jordan normal form, J. In this case, a standard triple (X, J, Y) is called a *Jordan triple for* $L(\lambda)$. Similarly, the matrices (X, J) from a Jordan triple are called a *Jordan pair* for $L(\lambda)$. We will now show that complete spectral information is explicitly given by a Jordan triple.

Let us begin by making a simplifying assumption that will subsequently be dropped. Suppose that $L(\lambda)$ is an $n \times n$ matrix polynomial of degree l with det $L_l \neq 0$ and with all linear elementary divisors. This means that any Jordan matrix J that is a linearization for $L(\lambda)$ is diagonal. This is certainly the case if all latent roots of $L(\lambda)$ are distinct. Write $J = \text{diag}[\lambda_1, \lambda_2, \ldots, \lambda_{ln}]$.

Let (X, J) be a Jordan pair for $L(\lambda)$ and let $\mathbf{x}_j \in \mathbb{C}^n$ be the jth column of X, for $j = 1, 2, \ldots, ln$. Since a Jordan pair is also a standard pair, it follows from Theorem 14.2.3 that the $ln \times ln$ matrix

$$Q = \begin{bmatrix} X \\ XJ \\ \vdots \\ XJ^{\ell-1} \end{bmatrix} \tag{1}$$

is nonsingular. Also,

$$L_l XJ^l + \cdots + L_1 XJ + L_0 X = 0. \tag{2}$$

First observe that for each j, the vector $\boldsymbol{x}_j \neq \boldsymbol{0}$, otherwise Q has a complete column of zeros, contradicting its nonsingularity. Then note that each term in Eq. (2) is an $n \times ln$ matrix; pick out the jth column of each. Since J is assumed to be diagonal, it is found that for $j = 1, 2, \ldots, ln$,

$$L_l \lambda_j^l \boldsymbol{x}_j + \cdots + L_1 \lambda_j \boldsymbol{x}_j + L_0 \boldsymbol{x}_j = \boldsymbol{0}.$$

In other words $L(\lambda_j)\boldsymbol{x}_j = \boldsymbol{0}$ and $\boldsymbol{x}_j \neq \boldsymbol{0}$; or $\boldsymbol{x}_j \in \operatorname{Ker} L(\lambda_j)$.

Now, the nonzero vectors in Ker $L(\lambda_j)$ are defined to be the *(right) latent vectors* of $L(\lambda)$ corresponding to the latent root λ_j. Thus, *when (X, J) is a Jordan pair and J is diagonal, every column of X is a latent vector of $L(\lambda)$.*

The general situation, in which J may have blocks of any size between 1 and ln, is more complicated, but it can be described using the same kind of procedure. First it is necessary to generalize the idea of a "latent vector" of a matrix polynomial in such a way that it will include the notion of Jordan chains introduced in Section 6.3. Let $L^{(r)}(\lambda)$ be the matrix polynomial obtained by differentiating $L(\lambda)$ r times with respect to λ. Thus, when $L(\lambda)$ has degree l, $L^{(r)}(\lambda) = 0$ for $r > l$. The set of vectors $\boldsymbol{x}_0, \boldsymbol{x}_1, \ldots, \boldsymbol{x}_k$, with $\boldsymbol{x}_0 \neq \boldsymbol{0}$, is a *Jordan chain of length $k + 1$ for $L(\lambda)$ corresponding to the latent root* λ_0 if the following $k + 1$ relations hold:

$$\begin{aligned} L(\lambda_0)\boldsymbol{x}_0 &= \boldsymbol{0}; \\ L(\lambda_0)\boldsymbol{x}_1 + \frac{1}{1!} L^{(1)}(\lambda_0)\boldsymbol{x}_0 &= \boldsymbol{0}; \\ &\vdots \\ L(\lambda_0)\boldsymbol{x}_k + \frac{1}{1!} L^{(1)}(\lambda_0)\boldsymbol{x}_{k-1} + \cdots + \frac{1}{k!} L^{(k)}(\lambda_0)\boldsymbol{x}_0 &= \boldsymbol{0}. \end{aligned} \tag{3}$$

Observe that $\boldsymbol{x}_0$, the *leading vector* of the chain, is a latent vector of $L(\lambda)$ associated with λ_0, and the successive vectors of the chain each satisfy an inhomogeneous equation with the singular coefficient matrix $L(\lambda_0)$.

Exercise 1. Show that if $L(\lambda) = \lambda I - T$, then this definition of a Jordan chain is consistent with that of Eqs. (6.3.3). □

Now we will show that if (X, J) is a Jordan pair of the $n \times n$ matrix polynomial $L(\lambda)$, then the columns of X are made up of Jordan chains for $L(\lambda)$. Let $J = \operatorname{diag}[J_1, \ldots, J_s]$, where J_j is a Jordan block of size n_j, $j = 1, 2, \ldots, s$. Form the partition $X = [X_1 \ \ X_2 \ \ \cdots \ \ X_s]$, where X_j is $n \times n_j$ for $j = 1, 2, \ldots, s$, and observe that for $r = 0, 1, 2, \ldots$,

$$XJ^r = [X_1 J_1^r \quad X_2 J_2^r \quad \cdots \quad X_s J_s^r].$$

Thus, Eq. (2) implies that, for $j = 1, 2, \ldots, s$,

$$L_l X_j J_j^l + \cdots + L_1 X_j J_j + L_0 X_j = 0. \tag{4}$$

Now name the columns of X_j:

$$X_j = [\mathbf{x}_1^j \quad \mathbf{x}_2^j \quad \cdots \quad \mathbf{x}_{n_j}^j].$$

With the convention $\lambda_j^{r-p} = 0$ if $r < p$, we can write

$$J_j^r = \begin{bmatrix} \lambda_j^r & \binom{j}{1}\lambda_j^{r-1} & \cdots & & \binom{j}{n_j - 1}\lambda_j^{r-n_j+1} \\ 0 & \lambda_j^r & \ddots & & \vdots \\ \vdots & & \ddots & \lambda_j^r & \binom{j}{1}\lambda_j^{r-1} \\ 0 & \cdots & & 0 & \lambda_j^r \end{bmatrix}$$

Using the expressions in (4) and examining the columns in the order $1, 2, \ldots, n_j$, a chain of the form (3) is obtained with λ_0 replaced by λ_j, and so on. Thus, *for each j, the columns of X_j form a Jordan chain of length n_j corresponding to the latent root λ_j.*

The complete set of Jordan chains generated by $X_1, X_2, \ldots, X_s$ is said to be a *canonical set* of Jordan chains for $L(\lambda)$. The construction of a canonical set from a given polynomial $L(\lambda)$ is easy when J is a diagonal matrix: simply choose any basis of latent vectors in Ker $L(\lambda_j)$ for each distinct λ_j. The construction when J has blocks of general size is more delicate and is left for more intensive studies of the theory of matrix polynomials. However, an example is analyzed in the first exercise below, which also serves to illustrate the linear *dependencies* that can occur among the latent vectors and Jordan chains.

Exercise 2. Consider the 2×2 monic matrix polynomial of degree 2,

$$L(\lambda) = \lambda^2\begin{bmatrix} 1 & 0 \\ 0 & 1 \end{bmatrix} + \lambda\begin{bmatrix} 0 & 0 \\ 1 & -1 \end{bmatrix} + \begin{bmatrix} 0 & 0 \\ 1 & 0 \end{bmatrix} = \begin{bmatrix} \lambda^2 & 0 \\ \lambda + 1 & \lambda(\lambda - 1) \end{bmatrix},$$

and note that $\det L(\lambda) = \lambda^3(\lambda - 1)$. Thus, $L(\lambda)$ has two distinct latent roots: $\lambda_1 = 0$ and $\lambda_2 = 1$. Since

$$L(0) = \begin{bmatrix} 0 & 0 \\ 1 & 0 \end{bmatrix} \quad \text{and} \quad L(1) = \begin{bmatrix} 1 & 0 \\ 2 & 0 \end{bmatrix},$$

any vector of the form $[0 \quad \alpha]^{\mathrm{T}}$ ($\alpha \neq 0$) can be used as a latent vector for $L(\lambda)$ corresponding to either of the latent roots.

Now observe that the Jordan chain for $L(\lambda)$ corresponding to the latent root $\lambda_2 = 1$ has only one member, that is, the chain has length 1. Indeed, if $\boldsymbol{x}_0 = [0 \quad \alpha]^T$ and $\alpha \neq 0$, the equation for $\boldsymbol{x}_1$ is

$$L(1)\boldsymbol{x}_1 + L^{(1)}(1)\boldsymbol{x}_0 = \begin{bmatrix} 1 & 0 \\ 2 & 0 \end{bmatrix}\boldsymbol{x}_1 + \begin{bmatrix} 2 & 0 \\ 1 & 1 \end{bmatrix}\begin{bmatrix} 0 \\ \alpha \end{bmatrix} = \begin{bmatrix} 0 \\ 0 \end{bmatrix},$$

which has no solution.

The corresponding equation at the latent root $\lambda_1 = 0$ is

$$L(0)\boldsymbol{x}_1 + L^{(1)}(0)\boldsymbol{x}_0 = \begin{bmatrix} 0 & 0 \\ 1 & 0 \end{bmatrix}\boldsymbol{x}_1 + \begin{bmatrix} 0 & 0 \\ 1 & -1 \end{bmatrix}\begin{bmatrix} 0 \\ \alpha \end{bmatrix} = \begin{bmatrix} 0 \\ 0 \end{bmatrix},$$

which has the general solution $\boldsymbol{x}_1 = [\alpha \quad \beta]^T$, where β is an arbitrary complex number.

A third member of the chain, $\boldsymbol{x}_2$, is obtained from the equation

$$\begin{aligned} L(0)\boldsymbol{x}_2 + L^{(1)}(0)\boldsymbol{x}_1 + \tfrac{1}{2}L^{(2)}(0)\boldsymbol{x}_0 &= \begin{bmatrix} 0 & 0 \\ 1 & 0 \end{bmatrix}\boldsymbol{x}_2 + \begin{bmatrix} 0 & 0 \\ 1 & -1 \end{bmatrix}\begin{bmatrix} \alpha \\ \beta \end{bmatrix} + \begin{bmatrix} 1 & 0 \\ 0 & 1 \end{bmatrix}\begin{bmatrix} 0 \\ \alpha \end{bmatrix} \\ &= \begin{bmatrix} 0 & 0 \\ 1 & 0 \end{bmatrix}\boldsymbol{x}_2 + \begin{bmatrix} 0 \\ 2\alpha - \beta \end{bmatrix} = \begin{bmatrix} 0 \\ 0 \end{bmatrix}, \end{aligned}$$

which gives $\boldsymbol{x}_2 = [-2\alpha + \beta \quad \gamma]^T$ for arbitrary γ.

For a fourth member of the chain, we obtain

$$\begin{aligned} L(0)\boldsymbol{x}_3 + L^{(1)}(0)\boldsymbol{x}_2 + \tfrac{1}{2}L^{(2)}(0)\boldsymbol{x}_1 &= \begin{bmatrix} 0 & 0 \\ 1 & 0 \end{bmatrix}\boldsymbol{x}_3 + \begin{bmatrix} 0 & 0 \\ 1 & -1 \end{bmatrix}\begin{bmatrix} -2\alpha + \beta \\ \gamma \end{bmatrix} \\ &\quad + \begin{bmatrix} 1 & 0 \\ 0 & 1 \end{bmatrix}\begin{bmatrix} \alpha \\ \beta \end{bmatrix} \\ &= \begin{bmatrix} 0 & 0 \\ 1 & 0 \end{bmatrix}\boldsymbol{x}_3 + \begin{bmatrix} \alpha \\ -2\alpha + 2\beta - \gamma \end{bmatrix} = \begin{bmatrix} 0 \\ 0 \end{bmatrix}, \end{aligned}$$

which has no solution since $\alpha \neq 0$, whatever value of γ may be chosen. Choosing $\alpha = 1$, $\beta = \gamma = 0$, we have a chain

$$\boldsymbol{x}_0 = \begin{bmatrix} 0 \\ 1 \end{bmatrix}, \qquad \boldsymbol{x}_1 = \begin{bmatrix} 1 \\ 0 \end{bmatrix}, \qquad \boldsymbol{x}_2 = \begin{bmatrix} -2 \\ 0 \end{bmatrix},$$

corresponding to the latent root $\lambda_1 = 0$.

It is easily verified that the matrices

$$X = \begin{bmatrix} 0 & 0 & 1 & -2 \\ 1 & 1 & 0 & 0 \end{bmatrix}, \qquad J = \begin{bmatrix} 1 & 0 & 0 & 0 \\ 0 & 0 & 1 & 0 \\ 0 & 0 & 0 & 1 \\ 0 & 0 & 0 & 0 \end{bmatrix}$$

form a Jordan pair for $L(\lambda)$.

Exercise 3. Find a Jordan pair for the matrix polynomial

$$L(\lambda) = \begin{bmatrix} \lambda^2 - 2\lambda - 2 & \lambda + 2 \\ \lambda + 2 & \lambda^2 - 2\lambda - 2 \end{bmatrix}.$$

Hint. First confirm that $\sigma(L) = \{-1, 0, 1, 4\}$.

Exercise 4. As in Exercise 14.1.5, let

$$L(\lambda) = \begin{bmatrix} \lambda^2 & -\lambda \\ 0 & \lambda^2 \end{bmatrix}.$$

Verify that

$$X = \begin{bmatrix} 0 & 1 & 0 & 0 \\ 1 & 0 & 1 & 0 \end{bmatrix}, \qquad J = \begin{bmatrix} 0 & 0 & 0 & 0 \\ 0 & 0 & 1 & 0 \\ 0 & 0 & 0 & 1 \\ 0 & 0 & 0 & 0 \end{bmatrix}$$

is a Jordan pair for $L(\lambda)$.

Exercise 5. Find a Jordan pair for each of the scalar polynomials $L_1(\lambda) = (\lambda - \lambda_0)^l$ and $L_2(\lambda) = (\lambda - \lambda_1)^p(\lambda - \lambda_2)^q$, where $\lambda_1 \neq \lambda_2$ and p, q are positive integers. □

Now suppose that (X, J, Y) is a Jordan triple for $L(\lambda)$. The roles of X and J are understood, but what about the matrix Y? Its interpretation in terms of spectral data can be obtained from Corollary 14.2.1, which says that (Y^T, J^T, X^T) is a standard triple for $L^T(\lambda)$. Note that this is not a Jordan triple, because the linearization J^T is the transpose of a Jordan form and so is not, in general, in Jordan form itself. To transform it to Jordan form we introduce the matrix

$$P = \operatorname{diag}[P_1, P_2, \ldots, P_s],$$

where P_j is the $n_j \times n_j$ rotation matrix:

$$P_j = \begin{bmatrix} 0 & \cdots & 0 & 1 \\ \vdots & & 1 & 0 \\ 0 & 1 & & \vdots \\ 1 & 0 & \cdots & 0 \end{bmatrix}.$$

(Recall that $J = \operatorname{diag}[J_1, \ldots, J_s]$, where J_j has size n_j.) Note that $P^2 = I$ (so that $P^{-1} = P$) and $P^* = P$. It is easily verified that for each j, $P_j J_j^T P_j = J_j$ and hence that $PJ^TP = J$. Now the similarity transformation

$$Y^TP, \qquad J = PJ^TP, \qquad PX^T$$

of (Y^T, J^T, X^T) yields a standard triple for $L^T(\lambda)$ in which the middle term is in Jordan form. Thus, *(Y^TP, J, PX^T) is a Jordan triple for $L^T(\lambda)$*, and our discussion of Jordan pairs can be used to describe Y. Thus, let Y be partitioned in the form

$$Y = \begin{bmatrix} Y_1 \\ Y_2 \\ \vdots \\ Y_s \end{bmatrix},$$

where Y_j is $n_j \times n$ for $j = 1, 2, \ldots, s$. Then $Y^T P = [Y_1^T P_1 \quad \cdots \quad Y_s^T P_s]$, and it follows that, for each j, *the transposed rows of* Y_j taken in reverse order *form a Jordan chain of* $L^T(\lambda)$ *of length* n_j *corresponding to* λ_j.

These Jordan chains are called *left Jordan chains* of $L(\lambda)$, and *left latent vectors* of $L(\lambda)$ are the nonzero vectors in Ker $L^T(\lambda_j)$, $j = 1, 2, \ldots, s$. Thus, the leading members of left Jordan chains are left latent vectors.

Exercise 6. (a) Let $A \in \mathbb{C}^{n \times n}$ and assume A has only linear elementary divisors. Use Exercises 4.14.3 and 4.14.4 to show that right and left eigenvectors of A can be defined in such a way that the resolvent of A has the spectral representation

$$(\lambda I - A)^{-1} = \sum_{j=1}^{n} \frac{\boldsymbol{x}_j \boldsymbol{y}_j^T}{\lambda - \lambda_j}.$$

(b) Let $L(\lambda)$ be an $n \times n$ matrix polynomial of degree l with det $L_l \neq 0$ and only linear elementary divisors. Show that right and left latent vectors for $L(\lambda)$ can be defined in such a way that

$$L(\lambda)^{-1} = \sum_{j=1}^{ln} \frac{\boldsymbol{x}_j \boldsymbol{y}_j^T}{\lambda - \lambda_j}.$$

Exercise 7. Suppose that $\lambda_j \in \sigma(L)$ has just one associated Jordan block J_j, and the size of J_j is v. Show that the singular part of the Laurent expansion for $L^{-1}(\lambda)$ about λ_j can be written in the form

$$\sum_{k=1}^{v} \sum_{r=0}^{v-k} \frac{\boldsymbol{x}_{v-k-r} \boldsymbol{y}_r^T}{(\lambda - \lambda_j)^k}$$

for some right and left Jordan chains $\{\boldsymbol{x}_1, \ldots, \boldsymbol{x}_v\}$ and $\{\boldsymbol{y}_1, \ldots, \boldsymbol{y}_v\}$, respectively (see Exercise 9.5.2).

Hint. First observe that

$$(\lambda I - J_j)^{-1} = \begin{bmatrix} (\lambda - \lambda_j)^{-1} & (\lambda - \lambda_j)^{-2} & \cdots & (\lambda - \lambda_j)^{-v} \\ 0 & (\lambda - \lambda_j)^{-1} & \ddots & \vdots \\ \vdots & \ddots & \ddots & \\ & & (\lambda - \lambda_j)^{-1} & (\lambda - \lambda_j)^{-2} \\ 0 & \cdots & 0 & (\lambda - \lambda_j)^{-1} \end{bmatrix}. \qquad \square$$

14.4 Applications to Differential Equations

Consider the constant-coefficient-matrix differential equation of order l given by

$$L_l \boldsymbol{x}^{(l)}(t) + L_{l-1}\boldsymbol{x}^{(l-1)}(t) + \cdots + L_1 \boldsymbol{x}^{(1)}(t) + L_0 \boldsymbol{x}(t) = \boldsymbol{f}(t), \tag{1}$$

where $L_0, L_1, \ldots, L_l \in \mathbb{C}^{n \times n}$, $\det L_l \neq 0$, and the indices on $\boldsymbol{x}$ denote derivatives with respect to the independent variable t. The vector function $\boldsymbol{f}(t)$ has values in $\mathbb{C}^n$, is piecewise continuous, and supposed given; $\boldsymbol{x}(t)$ is to be found. The equation is *inhomogeneous* when $\boldsymbol{f}(t) \neq \boldsymbol{0}$, and the equation

$$L_l \boldsymbol{x}^{(l)}(t) + \cdots + L_1 \boldsymbol{x}^{(1)}(t) + L_0 \boldsymbol{x}(t) = \boldsymbol{0} \tag{2}$$

is said to be *homogeneous* and has already been introduced in Section 7.10. Recall the discussion in that section of the space $\mathscr{C}(\mathbb{C}^n)$ of all continuous vector-valued functions $\boldsymbol{x}(t)$ defined for real t with values in $\mathbb{C}^n$. We now discuss the relationship between Jordan chains of $L(\lambda)$ and primitive solutions from $\mathscr{C}(\mathbb{C}^n)$ of the homogeneous Eq. (2).

Observe first that if λ_0 is a latent root of $L(\lambda)$ with latent vector $\boldsymbol{x}_0$, so that $L(\lambda_0)\boldsymbol{x}_0 = \boldsymbol{0}$, then Eq. (2) has a solution $\boldsymbol{x}_0 e^{\lambda_0 t}$, and conversely. This is because

$$\left(\frac{d}{dt}\right)^j (\boldsymbol{x}_0 e^{\lambda_0 t}) = \lambda_0^j \boldsymbol{x}_0 e^{\lambda_0 t},$$

so that

$$L\left(\frac{d}{dt}\right)(\boldsymbol{x}_0 e^{\lambda_0 t}) = \sum_{j=0}^{l} L_j \left(\frac{d}{dt}\right)^j (\boldsymbol{x}_0 e^{\lambda_0 t}) = L(\lambda_0)\boldsymbol{x}_0 = \boldsymbol{0}.$$

Now suppose that $\boldsymbol{x}_0$, $\boldsymbol{x}_1$ are two leading members of a Jordan chain of $L(\lambda)$ associated with the latent root λ_0. We claim that

$$\boldsymbol{u}_0(t) = \boldsymbol{x}_0 e^{\lambda_0 t} \qquad \text{and} \qquad \boldsymbol{u}_1(t) = (t\boldsymbol{x}_0 + \boldsymbol{x}_1)e^{\lambda_0 t}$$

are linearly independent solutions of Eq. (2). We have already seen that the first function is a solution. Also, the first function obviously satisfies

$$\left(\frac{d}{dt} - \lambda_0 I\right)\boldsymbol{u}_0(t) = \boldsymbol{0}. \tag{3}$$

Now observe that, for the second function,

$$\left(\frac{d}{dt} - \lambda_0 I\right)(t\boldsymbol{x}_0 + \boldsymbol{x}_1)e^{\lambda_0 t} = \boldsymbol{x}_0 e^{\lambda_0 t}.$$

In other words,

$$\left(\frac{d}{dt} - \lambda_0 I\right)\boldsymbol{u}_1(t) = \boldsymbol{u}_0(t),$$

so that, using Eq. (3),

$$\left(\frac{d}{dt} - \lambda_0 I\right)^2 \boldsymbol{u}_1(t) = \boldsymbol{0}. \tag{4}$$

Let us represent the differential operator $L(d/dt)$ in terms of $(d/dt) - \lambda_0 I$. To do this, first write the Taylor expansion for $L(\lambda)$ about λ_0:

$$L(\lambda) = L(\lambda_0) + \frac{1}{1!}L^{(1)}(\lambda_0)(\lambda - \lambda_0) \quad + \cdots + \frac{1}{l!}L^{(l)}(\lambda_0)(\lambda - \lambda_0)^l.$$

It follows that

$$L\left(\frac{d}{dt}\right) = L(\lambda_0) + \frac{1}{1!}L^{(1)}(\lambda_0)\left(\frac{d}{dt} - \lambda_0 I\right) + \cdots + \frac{1}{l!}L^{(l)}(\lambda_0)\left(\frac{d}{dt} - \lambda_0 I\right)^l. \tag{5}$$

Now Eq. (4) implies that

$$\left(\frac{d}{dt} - \lambda_0 I\right)^j \boldsymbol{u}_1(t) = \boldsymbol{0}, \qquad j = 2, 3, \ldots.$$

Thus

$$\begin{aligned} L\left(\frac{d}{dt}\right)\boldsymbol{u}_1(t) &= L(\lambda_0)\boldsymbol{u}_1(t) + \frac{1}{1!}L^{(1)}(\lambda_0)\boldsymbol{u}_0(t) \\ &= (L(\lambda_0)\boldsymbol{x}_0)te^{\lambda_0 t} + \left(L(\lambda_0)\boldsymbol{x}_1 + \frac{1}{1!}L^{(1)}(\lambda_0)\boldsymbol{x}_0\right)e^{\lambda_0 t} \\ &= \boldsymbol{0}, \end{aligned}$$

using the first two relations of Eqs. (14.3.3), which define a Jordan chain. Thus $\boldsymbol{u}_1(t)$ is also a solution of Eq. (2).

To see that $\boldsymbol{u}_0(t)$ and $\boldsymbol{u}_1(t)$ are linearly independent functions of t, observe the fact that $\alpha\boldsymbol{u}_0(t) + \beta\boldsymbol{u}_1(t) = \boldsymbol{0}$ for all t implies

$$t(\alpha\boldsymbol{x}_0) + (\alpha\boldsymbol{x}_1 + \beta\boldsymbol{x}_0) = \boldsymbol{0}.$$

Since this is true for all t we must have $\alpha = 0$, and the equation reduces to $\beta\boldsymbol{x}_0 = \boldsymbol{0}$. Therefore $\beta = 0$ since $\boldsymbol{x}_0 \neq \boldsymbol{0}$. Hence $\boldsymbol{u}_0(t)$ and $\boldsymbol{u}_1(t)$ are linearly independent.

The argument we have just completed cries out for generalization to a sequence of functions $u_0(t), u_1(t), \ldots, u_{k-1}(t)$, where $x_0, x_1, \ldots, x_{k-1}$ is a Jordan chain and

$$u_2(t) = \left(\frac{1}{2!}t^2 x_0 + t x_1 + x_2\right)e^{\lambda_0 t},$$

and so on. It is left as an exercise to complete this argument and to prove the next result.

Theorem 1. *Let $x_0, x_1, \ldots, x_{k-1}$ be a Jordan chain for $L(\lambda)$ at λ_0. Then the k functions*

$$\begin{aligned} u_0(t) &= x_0 e^{\lambda_0 t}, \\ u_1(t) &= (t x_0 + x_1)e^{\lambda_0 t}, \\ &\vdots \\ u_{k-1}(t) &= \left(\sum_{j=0}^{k-1} \frac{t^j}{j!} x_{k-1-j}\right)e^{\lambda_0 t} \end{aligned} \tag{6}$$

are linearly independent solutions of Eq. (2).

There is also a converse statement.

Theorem 2. *Let $x_0, x_1, \ldots, x_{k-1} \in \mathbb{C}^n$ with $x_0 \neq 0$. If the vector-valued function*

$$u_{k-1}(t) = \left(\frac{t^{k-1}}{(k-1)!} x_0 + \cdots + t x_{k-2} + x_{k-1}\right)e^{\lambda_0 t}$$

is a solution of Eq. (2), *then λ_0 is a latent root of $L(\lambda)$ and $x_0, x_1, \ldots, x_{k-1}$ is a Jordan chain for $L(\lambda)$ corresponding to λ_0.*

PROOF. Let functions $u_0(t), \ldots, u_{k-2}(t)$ be defined as in Eqs. (6) and, for convenience, let $u_i(t)$ be identically zero if $i = -1, -2, \ldots$. It is found (see Eqs. (3) and (4)) that

$$\left(\frac{d}{dt} - \lambda_0 I\right)^j u_{k-1}(t) = u_{k-j-1}(t),$$

for $j = 0, 1, 2, \ldots$. Using the expansion (5),

$$\begin{aligned} 0 &= L\left(\frac{d}{dt}\right)u_{k-1}(t) \\ &= L(\lambda_0)u_{k-1}(t) + L^{(1)}(\lambda_0)u_{k-2}(t) + \cdots + \frac{1}{l!}L^{(l)}(\lambda_0)u_{k-l-1}(t). \end{aligned}$$

Now the function $e^{\lambda_0 t}$ can be cancelled from this equation. Equating the vector coefficients of powers of t to zero in the remaining equation, the relations (14.3.3) are obtained and since we assume $\boldsymbol{x}_0 \neq \boldsymbol{0}$ we see that λ_0 is a latent root of $L(\lambda)$ and $\boldsymbol{x}_0, \boldsymbol{x}_1, \ldots, \boldsymbol{x}_{k-1}$ is a corresponding Jordan chain. ■

Exercise 1. Complete the proof of Theorem 1.

Exercise 2. Use Theorem 1 and the results of Exercise 14.3.2 to find four linearly independent solutions of the system of differential equations

$$\frac{d^2x_1}{dt^2} = 0,$$

$$\frac{d^2x_2}{dt^2} + \frac{dx_1}{dt} - \frac{dx_2}{dt} + x_1 = 0.$$

SOLUTION Written in the form of Eq. (2), where $\boldsymbol{x} = [x_1 \quad x_2]^{\mathrm{T}}$, we have

$$\begin{bmatrix} 1 & 0 \\ 0 & 1 \end{bmatrix} \boldsymbol{x}^{(2)}(t) + \begin{bmatrix} 0 & 0 \\ 1 & -1 \end{bmatrix} \boldsymbol{x}^{(1)}(t) + \begin{bmatrix} 0 & 0 \\ 1 & 0 \end{bmatrix} \boldsymbol{x}(t) = \boldsymbol{0}. \tag{7}$$

So the associated matrix polynomial is $L(\lambda)$ of Exercise 14.3.2.

There is a latent root $\lambda_1 = 1$ with latent vector $[0 \quad 1]^{\mathrm{T}}$, hence one solution is

$$\boldsymbol{v}_0(t) = \begin{bmatrix} 0 \\ 1 \end{bmatrix} e^t.$$

Then there is the latent root $\lambda_2 = 0$ with Jordan chain $[0 \quad 1]^{\mathrm{T}}$, $[1 \quad 0]^{\mathrm{T}}$, $[-2 \quad 0]^{\mathrm{T}}$, and there are three corresponding linearly independent solutions:

$$\boldsymbol{u}_0(t) = \begin{bmatrix} 0 \\ 1 \end{bmatrix}, \qquad \boldsymbol{u}_1(t) = \begin{bmatrix} 1 \\ t \end{bmatrix}, \qquad \boldsymbol{u}_2(t) = \begin{bmatrix} -2 + t \\ \frac{1}{2}t^2 \end{bmatrix}.$$

It is easily verified that $\{\boldsymbol{v}_0(t), \boldsymbol{u}_0(t), \boldsymbol{u}_1(t), \boldsymbol{v}_2(t)\}$ is a linearly independent set of vector-valued functions in $\mathscr{C}(\mathbb{C}^2)$. Furthermore, it follows from Theorem 7.10.1 that they form a basis for the solution space of Eq. (7). □

14.5 General Solutions of Differential Equations

Results obtained in the preceding section are natural generalizations of Proposition 9.10.1. Indeed, this line of investigation can be pursued further to generalize Theorem 9.10.1 and show that the ln-dimensional solution space of Eq. (14.4.2) (see Theorem 7.10.1) can be spanned by linearly independent vector functions defined (as in Theorem 14.4.1) in terms of all latent roots and a canonical set of Jordan chains. The completion of this argument is indicated in Exercise 2 of this section. But first we turn our attention to the question of general solutions of initial-value problems in the spirit of Theorems 9.10.4 and 9.10.5. Our main objective is to show how a general solution of Eq. (14.4.1) can be expressed in terms of any standard triple for $L(\lambda)$.

Consider first the idea of linearization of Eq. (14.4.1), or of Eq. (14.4.2), as used in the theory of differential equations. Functions $\mathbf{x}_0(t), \ldots, \mathbf{x}_{l-1}(t)$ are defined in terms of a solution function $\mathbf{x}(t)$ and its derivatives by

$$\mathbf{x}_0(t) = \mathbf{x}(t), \qquad \mathbf{x}_1(t) = \mathbf{x}^{(1)}(t), \ldots, \mathbf{x}_{l-1}(t) = \mathbf{x}^{(l-1)}(t). \tag{1}$$

Then Eq. (14.4.1) takes the form

$$L_l \frac{d\mathbf{x}_{l-1}(t)}{dt} + L_{l-1}\mathbf{x}_{l-1}(t) + \cdots + L_0 \mathbf{x}_0(t) = \mathbf{f}(t). \tag{2}$$

If we define

$$\tilde{\mathbf{x}}(t) = \begin{bmatrix} \mathbf{x}_0(t) \\ \mathbf{x}_1(t) \\ \vdots \\ \mathbf{x}_{l-1}(t) \end{bmatrix}, \qquad \mathbf{g}(t) = \begin{bmatrix} 0 \\ \vdots \\ 0 \\ L_l^{-1}\mathbf{f}(t) \end{bmatrix}, \tag{3}$$

and let C_L be the now-familiar companion matrix for $L(\lambda)$, Eqs. (1) and (2) can be condensed to the form

$$\frac{d\tilde{\mathbf{x}}(t)}{dt} = C_L\tilde{\mathbf{x}}(t) + \mathbf{g}(t). \tag{4}$$

In fact, it is easily verified that Eqs. (1), (2), *and* (3) *determine a one-to-one correspondence between the solutions* $\mathbf{x}(t)$ *of* Eq. (14.4.1) *and* $\tilde{\mathbf{x}}(t)$ *of* Eq. (4). Thus, the problem is reduced from order l to order 1 in the derivatives, but at the expense of increasing the size of the vector variable from l to ln and, of course, at the expense of a lot of redundancies in the matrix C_L.

However, we take advantage of Theorem 9.10.5 and observe that every solution of Eq. (4) has the form

$$\tilde{\mathbf{x}}(t) = e^{C_L t}\mathbf{z} + \int_0^t e^{C_L(t-\tau)}\mathbf{g}(\tau)\,d\tau,$$

for some $\mathbf{z} \in \mathbb{C}^{ln}$. Now recall the definitions (14.2.2) of matrices P_1 and R_1 in the standard triple (P_1, C_L, R_1) for $L(\lambda)$. Noting that $P_1\tilde{\mathbf{x}}(t) = \mathbf{x}(t)$, it is found that, multiplying the last equation on the left by P_1,

$$\mathbf{x}(t) = P_1 e^{C_L t}\mathbf{z} + P_1 \int_0^t e^{C_L(t-\tau)} R_1 \mathbf{f}(\tau)\,d\tau. \tag{5}$$

We have now proved the following result.

Lemma 1. *Every solution of Eq.* (14.4.1) *has the form* (5) *for some* $z \in \mathbb{C}^{ln}$. ■

Note that this lemma achieves the preliminary objective of expressing all solutions of Eq. (14.4.1) in terms of the standard triple (P_1, C_L, R_1). Also, the general solution of Eq. (14.4.2) is obtained from Eq. (4) on setting $\boldsymbol{f}(t) \equiv \boldsymbol{0}$.

The step from the lemma to our main result is a small one.

Theorem 1. *Let* (U, T, V) *be a standard triple for* $L(\lambda)$. *Then every solution of Eq.* (14.4.1) *has the form*

$$\boldsymbol{x}(t) = Ue^{Tt}\boldsymbol{z} + U\int_0^t e^{T(t-\tau)}V\boldsymbol{f}(\tau)\,d\tau, \tag{6}$$

for some $z \in \mathbb{C}^{ln}$.

PROOF. Recall that, from the definition of a standard triple (see the proof of Theorem 14.2.2), there is a nonsingular $S \in \mathbb{C}^{ln \times ln}$ such that

$$P_1 = US, \qquad C_L = S^{-1}TS, \qquad R_1 = S^{-1}V.$$

Also, using Theorem 9.4.2, $e^{C_L\alpha} = S^{-1}e^{T\alpha}S$ for any $\alpha \in \mathbb{R}$. Consequently,

$$P_1e^{C_L\alpha} = Ue^{T\alpha}S, \qquad P_1e^{C_L\alpha}R_1 = Ue^{T\alpha}V.$$

Substitution of these expressions into Eq. (5) with $\alpha = t$ concludes the proof. ■

In practice there are two important candidates for the triple (U, T, V) of the theorem. The first is the triple (P_1, C_L, R_1), as used in the lemma that expresses the general solution in terms of the coefficients of the differential equation. The second is a Jordan triple (X, J, Y), in which case the solution is expressed in terms of the spectral properties of $L(\lambda)$. The latter choice of triple is, of course, intimately connected to the ideas of the preceding section.

Exercise 1. Suppose that, in Eq. (14.4.1),

$$\boldsymbol{f}(t) = \begin{cases} \boldsymbol{f}_0 & \text{for } t \geq 0, \\ \boldsymbol{0} & \text{for } t < 0. \end{cases}$$

Let L_0 be nonsingular and $\mathcal{Re}\ \lambda < 0$ for every $\lambda \in \sigma(L)$. Show that the solution $\boldsymbol{x}(t)$ satisfies

$$\lim_{t\to\infty} \boldsymbol{x}(t) = L_0^{-1}\boldsymbol{f}_0.$$

Exercise 2. Let (X, J) be a Jordan pair for the matrix polynomial $L(\lambda)$ associated with Eq. (14.1.2). Show that every solution of that equation has the form $x(t) = Xe^{Jt}z$ for some $z \in \mathbb{C}^{ln}$. Hence show that the solution space has a basis of ln functions of the form described in Theorem 14.4.1. □

Theorem 1 can be used to give an explicit representation for the solution of a classical initial-value problem.

Theorem 2. *Let (U, T, V) be a standard triple for $L(\lambda)$. Then there is a unique solution of the differential equation* (14.4.1) *satisfying initial conditions*

$$x^{(j)}(0) = x_j, \qquad j = 0, 1, \ldots, l-1, \tag{7}$$

for any given vectors $x_0, x_1, \ldots, x_{l-1} \in \mathbb{C}^n$. This solution is given by Eq. (6) *with*

$$z = [V \quad TV \quad \cdots \quad T^{l-1}V]S_L \begin{bmatrix} x_0 \\ \vdots \\ x_{l-1} \end{bmatrix}, \tag{8}$$

where S_L is the symmetrizer for $L(\lambda)$.

PROOF. We first verify that the function $x(t)$ given by Eqs. (6) and (8) does, indeed, satisfy the initial conditions (7). Differentiating Eq. (6), it is found that

$$\begin{bmatrix} x(0) \\ x^{(1)}(0) \\ \vdots \\ x^{(l-1)}(0) \end{bmatrix} = \begin{bmatrix} U \\ UT \\ \vdots \\ UT^{l-1} \end{bmatrix} z = Qz, \tag{9}$$

in the notation of Eq. (14.2.13). But the result of Theorem 14.2.5 implies that $QRS_L = I$, and so substitution of z from Eq. (8) into Eq. (9) shows that $x(t)$ defined in this way satisfies the initial conditions.

Finally, the uniqueness of z follows from Eq. (9) in view of the invertibility of Q. ■

14.6 Difference Equations

Consider the constant-coefficient-matrix difference equation of order l given by

$$L_l x_{j+l} + L_{l-1}x_{j+l-1} + \cdots + L_1 x_{j+1} + L_0 x_j = f_j, \qquad j = 0, 1, \ldots, \tag{1}$$

where $L_0, L_1, \ldots, L_l \in \mathbb{C}^{n\times n}$, $\det L_l \neq 0$, the sequence $\boldsymbol{f} = (\boldsymbol{f}_0, \boldsymbol{f}_1, \ldots)$ of vectors in $\mathbb{C}^n$ is given, and a sequence $\boldsymbol{x} = (\boldsymbol{x}_0, \boldsymbol{x}_1, \ldots)$, called a *solution* of Eq. (1), is to be found.

The theory of difference equations is analogous to that of differential equations developed in the previous sections, and therefore some of the proofs and technicalities are omitted from this presentation.

The following results are analogues of Theorems 14.4.1 and 14.4.2 for the *homogeneous difference equation*

$$L_l \boldsymbol{x}_{j+l} + L_{l+1}\boldsymbol{x}_{j+l-1} + \cdots + L_1 \boldsymbol{x}_{j+1} + L_0 \boldsymbol{x}_j = \boldsymbol{0}, \qquad j = 0, 1, \ldots, \tag{2}$$

associated with the matrix polynomial $L(\lambda) = \sum_{j=0}^{l} \lambda^j L_j$, where $\det L_l \neq 0$. Once again, we recall the discussion and result of Section 7.10 concerning the homogeneous equation. In particular, recall that $\mathscr{S}(\mathbb{C}^n)$ denotes the linear space of infinite sequences of vectors from $\mathbb{C}^n$. Here, it is convenient to define "binomial" coefficients $\binom{j}{r} \triangleq 0$ for integers j, r with $j < r$.

Theorem 1. *If $\boldsymbol{x}_0, \boldsymbol{x}_1, \ldots, \boldsymbol{x}_{k-1}$ is a Jordan chain for $L(\lambda)$ corresponding to λ_0, then the k sequences $(\boldsymbol{u}_0^{(s)}, \boldsymbol{u}_1^{(s)}, \ldots) \in \mathscr{S}(\mathbb{C}^n)$, $s = 0, 1, \ldots, k-1$, with*

$$\begin{aligned}
\boldsymbol{u}_j^{(0)} &= \lambda_0^j \boldsymbol{x}_0\\
\boldsymbol{u}_j^{(1)} &= \binom{j}{1} \lambda_0^{j-1}\boldsymbol{x}_0 + \lambda_0^j \boldsymbol{x}_1,\\
&\ \ \vdots\\
\boldsymbol{u}_j^{(k-1)} &= \binom{j}{k-1} \lambda_0^{j-k+1}\boldsymbol{x}_0 + \binom{j}{k-2}\lambda_0^{j-k+2}\boldsymbol{x}_1 + \cdots + \binom{j}{1}\lambda_0^{j-1}\boldsymbol{x}_{k-2}\\
&\quad + \lambda^j \boldsymbol{x}_{k-1},
\end{aligned} \tag{3}$$

for $j = 0, 1, \ldots$, are linearly independent solutions of Eq. (2).

Theorem 2. *Let $\boldsymbol{x}_0, \boldsymbol{x}_1, \ldots, \boldsymbol{x}_{k-1} \in \mathbb{C}^n$, with $\boldsymbol{x}_0 \neq \boldsymbol{0}$. If the sequence $(\boldsymbol{u}_0^{(k-1)}, \boldsymbol{u}_1^{(k-1)}, \ldots) \in \mathscr{S}(\mathbb{C}^n)$, with*

$$\boldsymbol{u}_j^{(k-1)} = \sum_{s=0}^{k-1} \binom{j}{s} \lambda_0^{j-s} \boldsymbol{x}_{k-1-s},$$

is a solution of Eq. (2), then λ_0 is a latent root of $L(\lambda)$ and $\boldsymbol{x}_0, \boldsymbol{x}_1, \ldots, \boldsymbol{x}_{k-1}$ is a Jordan chain for $L(\lambda)$ corresponding to λ_0.

Exercise 1. Solve the system of difference equations for $j = 0, 1, \ldots$:

$$\begin{aligned}
x_{j+2}^{(1)} &= 0,\\
x_{j+2}^{(2)} + x_{j+1}^{(1)} - x_{j+1}^{(1)} + x_j^{(1)} &= 0.
\end{aligned}$$

SOLUTION. Write the given system in the form (2):

$$\begin{bmatrix} 1 & 0 \\ 0 & 1 \end{bmatrix} \boldsymbol{x}_{j+2} + \begin{bmatrix} 0 & 0 \\ 1 & -1 \end{bmatrix} \boldsymbol{x}_{j+1} + \begin{bmatrix} 0 & 0 \\ 1 & 0 \end{bmatrix} \boldsymbol{x}_j = \boldsymbol{0}, \qquad j = 0, 1, \ldots,$$

where $\boldsymbol{x}_j = [x_j^{(1)} \;\; x_j^{(2)}]^{\mathrm{T}}$ for each j. Since the leading coefficient of the associated matrix polynomial $L(\lambda)$ is nonsingular ($L_2 = I$), it follows from the corollary to Theorem 7.10.2 that the given system has four linearly independent solutions. To find four such solutions, recall the result of Exercise 14.3.2 and use Theorem 1:

$$\boldsymbol{u}_j^{(0)} = [0 \;\; 1]^{\mathrm{T}}, \qquad j = 0, 1, \ldots,$$

$$\boldsymbol{v}_0^{(0)} = [0 \;\; 1]^{\mathrm{T}}, \qquad \boldsymbol{v}_j^{(0)} = \boldsymbol{0}, \quad j = 1, 2, \ldots,$$

$$\boldsymbol{v}_0^{(1)} = [1 \;\; 0]^{\mathrm{T}}, \qquad \boldsymbol{v}_1^{(1)} = [0 \;\; 1]^{\mathrm{T}}, \qquad \boldsymbol{v}_j^{(1)} = \boldsymbol{0}, \quad j = 2, 3, \ldots,$$

$$\boldsymbol{v}_0^{(2)} = [-2 \;\; 0]^{\mathrm{T}}, \qquad \boldsymbol{v}_1^{(2)} = [1 \;\; 0]^{\mathrm{T}}, \qquad \boldsymbol{v}_2^{(2)} = [0 \;\; 1]^{\mathrm{T}},$$

$$\boldsymbol{v}_j^{(2)} = \boldsymbol{0}, \quad j = 3, 4, \ldots.$$

The verification of the linear independence of the vectors $\boldsymbol{u}^{(0)}$, $\boldsymbol{v}^{(0)}$, $\boldsymbol{v}^{(1)}$, $\boldsymbol{v}^{(2)}$ in $\mathscr{S}(\mathbb{C}^2)$ is left as an exercise.

Since the solution space has dimension 4, every solution has the form

$$\alpha_1 \boldsymbol{u}^{(0)} + \alpha_2 \boldsymbol{v}^{(0)} + \alpha_3 \boldsymbol{v}^{(1)} + \alpha_4 \boldsymbol{v}^{(2)}$$

for some $\alpha_1, \alpha_2, \alpha_3, \alpha_4 \in \mathbb{C}$. □

Consider now the formulation of a general solution of the inhomogeneous difference equation (1). Observe first that the general solution $\boldsymbol{x} \in \mathscr{S}(\mathbb{C}^n)$ of (1) can be represented in the following way:

$$\boldsymbol{x} = \boldsymbol{x}^{(0)} + \boldsymbol{x}^{(1)}, \tag{4}$$

where $\boldsymbol{x}^{(0)}$ is a general solution of the homogeneous equation (2) and $\boldsymbol{x}^{(1)}$ is a fixed particular solution of Eq. (1). Indeed, it is easily checked that the sequence in Eq. (4) satisfies Eq. (1) for every solution $\boldsymbol{x}^{(0)}$ of Eq. (2) and, conversely, the difference $\boldsymbol{x} - \boldsymbol{x}^{(1)}$ is a solution of Eq. (2).

Now we take advantage of the idea of linearization to show how a general solution of Eq. (1) can be expressed in terms of any standard triple of $L(\lambda)$.

Lemma 1. *The general solution $\boldsymbol{x} = (\boldsymbol{x}_0, \boldsymbol{x}_1, \ldots)$ of the nonhomogeneous equation* (1) *is given by $\boldsymbol{x}_0 = P_1 \boldsymbol{z}$, and for $j = 1, 2, \ldots,$*

$$\boldsymbol{x}_j = P_1 C_L^j \boldsymbol{z} + P_1 \sum_{k=0}^{j-1} C_L^{j-k-1} R_1 \boldsymbol{f}_k. \tag{5}$$

where the matrices P_1 and R_1 are defined by Eq. (14.2.2) *and $\boldsymbol{z} \in \mathbb{C}^{ln}$ is arbitrary.*

PROOF. First observe that the general solution $\boldsymbol{x}^{(0)} = (\boldsymbol{x}_0^{(0)}, \boldsymbol{x}_1^{(0)}, \dots)$ of the homogeneous equation (2) is given by

$$\boldsymbol{x}_j^{(0)} = P_1 C_L^j \boldsymbol{z}, \qquad j = 0, 1, 2, \dots, \tag{6}$$

where $\boldsymbol{z} \in \mathbb{C}^{ln}$ is arbitrary. This is easily verified by substitution and by using Theorem 7.10.2 to prove that Eq. (6) gives *all* the solutions of Eq. (2).

Thus it remains to show that the sequence $(\boldsymbol{0}, \boldsymbol{x}_1^{(1)}, \boldsymbol{x}_2^{(1)}, \dots)$, with

$$\boldsymbol{x}_j^{(1)} = P_1 \sum_{k=0}^{j-1} C_L^{j-k-1} R_1 \boldsymbol{f}_k, \qquad j = 1, 2, \dots,$$

is a particular solution of Eq. (1). This is left as an exercise. We only note that the proof relies, in particular, on the use of Theorem 14.2.3 and Exercise 14.2.6 applied to the companion triple (P_1, C_L, R_1). ■

Using the similarity of the standard triples of $L(\lambda)$ and the assertion of Lemma 1, we easily obtain the following analogue of Theorem 14.5.1.

Theorem 3. *If (U, T, V) is a standard triple for $L(\lambda)$, then every solution of Eq. (2) has the form $\boldsymbol{x}_0 = U\boldsymbol{z}$, and for $j = 1, 2, \dots,$*

$$\boldsymbol{x}_j = UT^j\boldsymbol{z} + U \sum_{k=0}^{j-1} T^{j-k-1} V \boldsymbol{f}_k,$$

where $\boldsymbol{z} \in \mathbb{C}^{ln}$.

We conclude this section by giving an explicit solution of the initial-value problem for a system of difference equations (compare with Theorem 14.5.2).

Theorem 4. *Let (U, T, V) be a standard triple for $L(\lambda)$. There is a unique solution of Eq. (1) satisfying initial conditions*

$$\boldsymbol{x}_j = \boldsymbol{a}_j, \qquad j = 0, 1, \dots, l - 1, \tag{7}$$

for any given vectors $\boldsymbol{a}_0, \boldsymbol{a}_1, \dots, \boldsymbol{a}_{l-1} \in \mathbb{C}^n$. This solution is given explicitly by the results of Theorem 3 together with the choice

$$\boldsymbol{z} = [V \quad TV \quad \cdots \quad T^{l-1}V] S_L \begin{bmatrix} \boldsymbol{a}_0 \\ \boldsymbol{a}_1 \\ \vdots \\ \boldsymbol{a}_{l-1} \end{bmatrix},$$

where S_L is the symmetrizer for $L(\lambda)$.

The proof of this result is similar to that of Theorem 14.5.2.

Exercise 2. Complete the proofs of Theorems 1, 2, and 4.

Exercise 3. Suppose that $\boldsymbol{f}_j = \boldsymbol{f}$ for $j = 0, 1, 2, \ldots$ in Eq. (1), and assume that all latent roots of $L(\lambda)$ are inside the unit circle. Show that the solution of Eq. (1) satisfies

$$\lim_{j\to\infty} \boldsymbol{x}_j = (L_l + L_{l-1} + \cdots + L_0)^{-1}\boldsymbol{f}. \quad \square$$

14.7 A Representation Theorem

It has already been seen, in Theorem 14.2.2, that any standard triple for a matrix polynomial $L(\lambda)$ (with $\det L_l \neq 0$) can be used to give an explicit representation of the (rational) resolvent function $L(\lambda)^{-1}$. In this section we look for a representation of $L(\lambda)$ itself, that is, of the coefficient matrices $L_0, L_1, \ldots, L_l$, in terms of a standard triple. We will show that, in particular, the representations given here provide a generalization of the reduction of a square matrix to Jordan normal form by similarity.

Theorem 1. *Let (U, T, V) be a standard triple for a matrix polynomial $L(\lambda)$ (as defined in Eq. (14.1.2)). Then the coefficients of $L(\lambda)$ admit the following representations:*

$$L_l = (UT^{l-1}V)^{-1}, \tag{1}$$

$$[L_0 \quad L_1 \quad \cdots \quad L_{l-1}] = -L_l UT^l \begin{bmatrix} U \\ UT \\ \vdots \\ UT^{l-1} \end{bmatrix}^{-1}, \tag{2}$$

and

$$\begin{bmatrix} L_0 \\ L_1 \\ \vdots \\ L_{l-1} \end{bmatrix} = -[V \quad TV \quad \cdots \quad T^{l-1}V]^{-1}T^lVL_l. \tag{3}$$

PROOF. We first show that Eqs. (1) and (2) hold when (U, T, V) is the special standard triple (P_1, C_L, R_1). First, Eq. (1) is obtained from Eq. (14.2.12). Then Eqs. (14.2.6) and (14.2.7) give the special case of Eq. (2):

$$L_l P_1 C_L^l \begin{bmatrix} P_1 \\ P_1 C_L \\ \vdots \\ P_1 C_L^{l-1} \end{bmatrix}^{-1} = -[L_0 \quad L_1 \quad \cdots \quad L_{l-1}].$$

For any other standard triple (U, T, V), write

$$P_1 = US, \qquad C_L = S^{-1}TS, \qquad R_1 = S^{-1}V.$$

Then

$$P_1 C_L^l \begin{bmatrix} P_1 \\ \vdots \\ P_1 C_L^{l-1} \end{bmatrix}^{-1} = UT^l \begin{bmatrix} U \\ \vdots \\ UT^{l-1} \end{bmatrix}^{-1},$$

and Eqs. (1) and (2) are established.

To obtain Eq. (3), observe that Corollary 14.2.1 shows that $(V^{\mathrm{T}}, T^{\mathrm{T}}, U^{\mathrm{T}})$ is a standard triple for $L^{\mathrm{T}}(\lambda)$, so applying the already established Eq. (2),

$$-[L_0^{\mathrm{T}} \quad L_1^{\mathrm{T}} \quad \cdots \quad L_{l-1}^{\mathrm{T}}] = L_l^{\mathrm{T}} V^{\mathrm{T}} (T^{\mathrm{T}})^l \begin{bmatrix} V^{\mathrm{T}} \\ V^{\mathrm{T}} T^{\mathrm{T}} \\ \vdots \\ V^{\mathrm{T}} (T^{\mathrm{T}})^{l-1} \end{bmatrix}^{-1}.$$

Taking transposes, Eq. (3) is obtained. ■

Observe that if $L(\lambda)$ is monic, that is, if $L_l = I$, then the l coefficients $L_0, \ldots, L_{l-1}$ of $L(\lambda)$ are determined by any standard *pair* (U, T), or by a left standard pair (T, V). Note also that if the theorem is applied using a *Jordan triple* (X, J, Y) in place of (U, T, V) the coefficients of $L(\lambda)$ are completely determined in terms of the spectral properties of $L(\lambda)$.

Consider, in particular, the case in which $L(\lambda) = I\lambda - A$. Thus, $L(\lambda)$ is monic and $l = 1$. In this case Eqs. (1) and (2), written in terms of a Jordan triple (X, J, Y), yield

$$I = (XY)^{-1}, \qquad A = XJY.$$

Thus the left Jordan chains (or columns of Y) are given by $Y = X^{-1}$, and the second equation gives the expression of an arbitrary $A \in \mathbb{C}^{n \times n}$ in terms of its Jordan form and chains, namely, $A = XJX^{-1}$.

Exercise 1. Let $B, C \in \mathbb{C}^{n \times n}$ and define the matrix polynomial $L(\lambda) = \lambda^2 I + \lambda B + C$. Show that if (U, T) is a standard pair for $L(\lambda)$, then

$$[C, B] = -UT^2 \begin{bmatrix} U \\ UT \end{bmatrix}^{-1}.$$

If (U, T, V) is a standard triple for $L(\lambda)$, show also that

$$B = -UT^2V, \qquad C = -UT^3V + B^{2}\text{'} \quad \square$$

14.8 Multiples and Divisors

The idea of division of one matrix polynomial by another was introduced in Section 7.2. Recall that a matrix polynomial $L_1(\lambda)$ with *invertible leading coefficient* is said to be a right divisor of a matrix polynomial $L(\lambda)$ if $L(\lambda) = L_2(\lambda)L_1(\lambda)$ for some other matrix polynomial $L_2(\lambda)$. The reason for requiring that $L_1(\lambda)$ have invertible leading coefficient is to avoid the multiplicities of divisors that would arise otherwise. For example, a factorization like

$$\begin{bmatrix} 0 & 0 \\ 0 & 0 \end{bmatrix} = \begin{bmatrix} a\lambda^k & 0 \\ b\lambda^l & 0 \end{bmatrix} \begin{bmatrix} 0 & 0 \\ c\lambda^m & d\lambda^n \end{bmatrix},$$

where $a, b, c, d \in \mathbb{C}$ and k, l, m, n are nonnegative integers, should be excluded, not only because there are so many candidates, but mainly because such factorizations are not very informative. Here we go a little further and confine our attention to divisors that are *monic*, where the coefficient of the highest power of λ is I. In fact, it will be convenient to consider only monic divisors of monic polynomials.

It turns out that, from the point of view of spectral theory, multiples of matrix polynomials are more easily analyzed than questions concerning divisors. One reason for this concerns the question of *existence* of divisors. Thus, it is not difficult to verify that the monic matrix polynomial

$$\begin{bmatrix} \lambda^2 & 1 \\ 0 & \lambda^2 \end{bmatrix} \tag{1}$$

has no monic divisor of degree 1 in λ. A thorough analysis of the existence question is beyond the scope of this brief introduction to questions concerning spectral theory, multiples, and divisors.

Standard triples, and hence spectral properties (via Jordan triples), of products of monic matrix polynomials are relatively easy to describe if triples of the factors are known. Observe first that the following relation between monic matrix polynomials, $L(\lambda) = L_2(\lambda)L_1(\lambda)$, implies some obvious statements:

(a) $\sigma(L_1) \cup \sigma(L_2) = \sigma(L)$;

(b) Latent vectors of $L_1(\lambda)$ and $L_2^T(\lambda)$ are latent vectors of $L(\lambda)$ and $L^T(\lambda)$, respectively.

Some of the complications of this theory are associated with cases in which $\sigma(L_1) \cap \sigma(L_2) \neq \emptyset$ in condition (a). In any case, it is possible to extend statement (b) considerably. This will be done in the first corollary to the following theorem.

Theorem 1. *Let $L(\lambda)$, $L_1(\lambda)$, $L_2(\lambda)$ be monic matrix polynomials such that $L(\lambda) = L_2(\lambda)L_1(\lambda)$, and let (U_1, T_1, V_1), (U_2, T_2, V_2) be standard triples for $L_1(\lambda)$ and $L_2(\lambda)$, respectively. Then the admissible triple*

$$U = [U_1 \quad 0], \qquad T = \begin{bmatrix} T_1 & V_1 U_2 \\ 0 & T_2 \end{bmatrix}, \qquad V = \begin{bmatrix} 0 \\ V_2 \end{bmatrix} \tag{2}$$

for $L(\lambda)$ is a standard triple for $L(\lambda)$.

PROOF. With T defined in (2), it is easily seen that

$$(\lambda I - T)^{-1} = \begin{bmatrix} (\lambda I - T_1)^{-1} & (\lambda I - T_1)^{-1}V_1U_2(\lambda I - T_2)^{-1} \\ 0 & (\lambda I - T_2)^{-1} \end{bmatrix}.$$

Then the definitions of U and V give

$$\begin{aligned} U(\lambda I - T)^{-1}V &= U_1(\lambda I - T_1)^{-1}V_1U_2(\lambda I - T_2)^{-1}V_2 \\ &= L_1(\lambda)^{-1}L_2(\lambda)^{-1} = (L_2(\lambda)L_1(\lambda))^{-1} \\ &= L(\lambda)^{-1}, \end{aligned}$$

where Theorem 14.2.2 has been used. The result now follows from Theorem 14.2.4. ■

To see the significance of this result for the spectral properties of $L(\lambda)$, $L_1(\lambda)$, and $L_2(\lambda)$, replace the standard triples for $L_1(\lambda)$ and $L_2(\lambda)$ by Jordan triples (X_1, J_1, Y_1) and (X_2, J_2, Y_2), respectively. Then $L(\lambda)$ has a standard triple (not necessarily a Jordan triple)

$$[X_1 \quad 0], \qquad \begin{bmatrix} J_1 & Y_2X_1 \\ 0 & J_2 \end{bmatrix}, \qquad \begin{bmatrix} 0 \\ Y_2 \end{bmatrix}. \tag{3}$$

Although the second matrix is not in Jordan form, it does give important partial information about the spectrum of $L(\lambda)$ because of its triangular form. Property (b) of Theorem 14.2.3 implies, for example, (see also Eq. (14.3.4)) that $\sum_{j=0}^{l} L_jX_1J_1^j = 0$, where $L(\lambda) = \sum_{j=0}^{l} \lambda^j L_j$. This implies the following corollary.

Corollary 1. *Under the hypotheses of Theorem 1, $\sigma(L_1) \subset \sigma(L)$, and if $\lambda_0 \in \sigma(L_1)$, then every Jordan chain for $L_1(\lambda)$ corresponding to λ_0 is also a Jordan chain for $L(\lambda)$ corresponding to λ_0.*

Similarly, $\sigma(L_2) \subset \sigma(L)$ and every Jordan chain of $L_2^T(\lambda)$ is a Jordan chain of $L^T(\lambda)$.

Given the standard triple (3) for $L(\lambda)$, it is natural to ask when the two matrices

$$\begin{bmatrix} J_1 & Y_2X_1 \\ 0 & J_2 \end{bmatrix}, \quad \begin{bmatrix} J_1 & 0 \\ 0 & J_2 \end{bmatrix} \tag{4}$$

are similar, in which case a Jordan form for $L(\lambda)$ is just the direct sum of Jordan forms for the factors. An answer is provided immediately by Theorem 12.5.3, which states that the matrices of (4) are similar if and only if there exists a solution Z (generally rectangular) for the equation

$$J_1Z - ZJ_2 = Y_2X_1. \tag{5}$$

This allows us to draw the following conclusions.

Corollary 2. *Let (X_1, J_1, Y_1) and (X_2, J_2, Y_2) be Jordan triples for $L_1(\lambda)$ and $L_2(\lambda)$, respectively. Then* $\operatorname{diag}[J_1, J_2]$ *is a linearization for $L(\lambda) = L_2(\lambda)L_1(\lambda)$ if and only if there is a solution Z for Eq. (5); in this case*

$$\begin{bmatrix} J_1 & 0 \\ 0 & J_2 \end{bmatrix} = \begin{bmatrix} I & Z \\ 0 & I \end{bmatrix} \begin{bmatrix} J_1 & Y_2X_1 \\ 0 & J_2 \end{bmatrix} \begin{bmatrix} I & -Z \\ 0 & I \end{bmatrix}. \tag{6}$$

PROOF. Observe that

$$\begin{bmatrix} I & -Z \\ 0 & I \end{bmatrix} = \begin{bmatrix} I & Z \\ 0 & I \end{bmatrix}^{-1},$$

so that the relation (6) is, in fact, the explicit similarity between the matrices of (4). To verify Eq. (6), it is necessary only to multiply out the product on the right and use Eq. (5). ∎

An important special case of Corollary 2 arises when $\sigma(L_1) \cap \sigma(L_2) = \varnothing$, for then, by Theorem 12.3.2, Eq. (5) has a solution and it is unique. Thus, $\sigma(L_1) \cap \sigma(L_2) = \varnothing$ is a sufficient condition for $\operatorname{diag}[J_1, J_2]$ to be a linearization of $L_2(\lambda)L_1(\lambda)$.

14.9 Solvents of Monic Matrix Polynomials

As in the preceding section, we confine attention here to $n \times n$ monic matrix polynomials $L(\lambda) = \lambda^l I + \sum_{j=0}^{l-1} \lambda^j L_j$. An $n \times n$ matrix S is said to be a (*right*) *solvent* of $L(\lambda)$ if, in the notation of Section 7.2, the *right value* of $L(\lambda)$ at S is the zero-matrix, that is,

$$S^l + L_{l-1}S^{l-1} + \cdots + L_1S + L_0 = O. \tag{1}$$

This is, of course, a natural generalization of the notion of a zero of a scalar polynomial, and it immediately suggests questions concerning the factorizations of $L(\lambda)$ determined by solvents of $L(\lambda)$.

It has been shown in the Corollary to Theorem 7.2.3 that S satisfies Eq. (1) if and only if $\lambda I - S$ is a right divisor of $L(\lambda)$, that is, if and only if $L(\lambda) = L_2(\lambda)(\lambda I - S)$ for some monic matrix polynomial $L_2(\lambda)$ of degree $l - 1$. This immediately puts questions concerning solvents of $L(\lambda)$ in the context of the preceding section. In particular, recall that matrix polynomials may have *no* solvent at all (see the polynomial of Eq. (14.8.1)). Furthermore, it is deduced from Corollary 1 of Theorem 14.8.1 that when a solvent S exists, all eigenvalues of S are latent roots of $L(\lambda)$ and all Jordan chains of $\lambda I - S$ must be Jordan chains of $L(\lambda)$. Also, if $S = X_1 J_1 X_1^{-1}$, where J_1 is in Jordan normal form, then (X_1, J_1) is a Jordan pair for $\lambda I - S$. Thus, the matrix

$$R_1 = \begin{bmatrix} X_1 \\ X_1 J_1 \\ \vdots \\ X_1 J_1^{l-1} \end{bmatrix}$$

has rank n and its columns form Jordan chains for C_L. It follows that $\operatorname{Im} R_1$ is a C_L-invariant subspace of dimension n. Development of this line of argument leads to the following criterion.

Theorem 1 (H. Langer[†]). *The monic matrix polynomial $L(\lambda)$ of degree l has a right solvent S if and only if there exists an invariant subspace $\mathscr{S}$ of C_L of the form $\mathscr{S} = \operatorname{Im} Q_1$ where*

$$Q_1 = \begin{bmatrix} I \\ S \\ S^2 \\ \vdots \\ S^{l-1} \end{bmatrix}. \tag{2}$$

PROOF. If S is a right solvent of $L(\lambda)$ then, as we have seen, $\lambda I - S$ is a right divisor of $L(\lambda)$. It is easily verified that (I, S, I) is a standard triple for $\lambda I - S$, so if (U_2, T_2, V_2) is a standard triple for the quotient $L(\lambda)(\lambda I - S)^{-1}$, then Theorem 14.8.1 implies that $L(\lambda)$ has a standard pair of the form

$$X = [I \quad 0], \qquad T = \begin{bmatrix} S & U_2 \\ 0 & T_2 \end{bmatrix}.$$

[†] *Acta Sci. Math. (Szeged)* **35** (1973), 83–96.

Defining an $ln \times ln$ matrix Q by

$$Q = \begin{bmatrix} X \\ XT \\ \vdots \\ XT^{l-1} \end{bmatrix} = \begin{bmatrix} I & 0 \\ S & U_2 \\ S^2 & * \\ \vdots & \vdots \\ S^{l-1} & * \end{bmatrix}$$

(where $*$ denotes a matrix of no immediate interest), it follows, as in Lemma 14.2.1, that $C_L Q = QT$. This implies $C_L Q_1 = Q_1 S$, where Q_1 is defined by Eq. (2). Consequently, $C_L \mathscr{S} \subset \mathscr{S}$.

Conversely, let $\mathscr{S} = \operatorname{Im} Q_1$, where Q_1 is defined by Eq. (2) for some $n \times n$ matrix S, and let $\mathscr{S}$ be C_L-invariant. Then for any $\boldsymbol{y} \in \mathbb{C}^n$ there is a $\boldsymbol{z} \in \mathbb{C}^n$ such that

$$C_L Q_1 \boldsymbol{y} = Q_1 \boldsymbol{z}.$$

Using the definition (14.1.3) of C_L, this means that

$$\begin{bmatrix} S \\ S^2 \\ \vdots \\ S^{l-1} \\ -\sum_{j=0}^{l-1} L_j S^j \end{bmatrix} \boldsymbol{y} = \begin{bmatrix} I \\ S \\ \vdots \\ S^{l-2} \\ S^{l-1} \end{bmatrix} \boldsymbol{z}.$$

In particular, $\boldsymbol{z} = S\boldsymbol{y}$ and so $-\sum_{j=0}^{l-1} L_j S^j \boldsymbol{y} = S^{l-1}\boldsymbol{z}$ implies

$$(S^l + L_{l-1}S^{l-1} + \cdots + L_1 S + L_0)\boldsymbol{y} = \boldsymbol{0}.$$

Since this is true for any $\boldsymbol{y} \in \mathbb{C}^n$, Eq. (1) follows and S is a right solvent of $L(\lambda)$. ■

Note that a subspace $\mathscr{S}$ of the form $\operatorname{Im} Q_1$, with Q_1 given by Eq. (2), can be characterized as an n-dimensional subspace containing no nonzero vectors of the form

$$\boldsymbol{\varphi} = [\boldsymbol{0} \quad \boldsymbol{\varphi}_1^{\mathrm{T}} \quad \cdots \quad \boldsymbol{\varphi}_{l-1}^{\mathrm{T}}]^{\mathrm{T}},$$

where $\boldsymbol{0}, \boldsymbol{\varphi}_i \in \mathbb{C}^n$, $i = 1, 2, \ldots, l-1$. Indeed, if $\boldsymbol{\varphi} \in \mathscr{S}$ is such a vector, then

$$\boldsymbol{\varphi} = \begin{bmatrix} I \\ S \\ S^2 \\ \vdots \\ S^{l-1} \end{bmatrix} \boldsymbol{z}$$

for some $z \in \mathbb{C}^n$. Therefore, comparing the first n coordinates of $\boldsymbol{\varphi}$ and $Q_1 z$, it is found that $z = 0$ and consequently $\boldsymbol{\varphi} = 0$.

By using Theorem 1, we show in the next example that certain monic matrix polynomials have no solvents. This generalizes the example in Eq. (14.8.1).

Example 1. Consider the monic $n \times n$ $(l \geq 2)$ matrix polynomial

$$L(\lambda) = \lambda^l I_n - J_n,$$

where J_n denotes the $n \times n$ Jordan block with zeros on the main diagonal. Clearly $\sigma(L) = \{0\}$ and the kernel of $L(0)$ has dimension 1. It follows that $L(\lambda)$ has just one elementary divisor, λ^{ln}, and so the Jordan matrix associated with $L(\lambda)$ is J_{ln}, the Jordan block size ln with zeros on the main diagonal. Observe that J_{ln} has just one invariant subspace $\mathscr{S}_j$ of dimension j for $j = 1, 2, \ldots, n$, and it is given by

$$\mathscr{S}_j = \text{span}\{e_1, e_2, \ldots, e_j\} \subset \mathbb{C}^{ln}.$$

By Lemma 14.2.1, it follows that C_L also has just one n-dimensional invariant subspace and it is given by $Q\mathscr{S}_n$, where

$$Q = \begin{bmatrix} X \\ XJ_{ln} \\ \vdots \\ XJ_{ln}^{l-1} \end{bmatrix}, \qquad \mathscr{S}_n = \text{Im}\begin{bmatrix} I \\ 0 \\ \vdots \\ 0 \end{bmatrix}$$

and (X, J_{ln}) is a Jordan pair for $L(\lambda)$. If X_1 is the leading $n \times n$ block of X, then

$$Q\mathscr{S}_n = \text{Im}\begin{bmatrix} X_1 \\ X_1 J_n \\ \vdots \\ X_1 J_n^{l-1} \end{bmatrix}.$$

For this subspace to have the form described in Theorem 1 it is necessary and sufficient that X_1 be nonsingular. However, since $L^{(k)}(0) = 0$ for $k = 1, 2, \ldots, l - 1$, it is easily seen that we may choose the unit vector e_1 in $\mathbb{C}^n$ for each of the first l columns of X. Thus X_1 cannot be invertible if $l \geq 2$, and C_L has no invariant subspace of the required form. □

In contrast to Example 1, a monic matrix polynomial may have many distinct solvents, as shown in the following example.

Exercise 2. Check that the matrices

$$S_1 = \begin{bmatrix} 0 & 1 \\ 0 & 0 \end{bmatrix}, \qquad S_2 = \begin{bmatrix} 1 & 0 \\ -2 & 1 \end{bmatrix}, \qquad S_3 = \begin{bmatrix} 0 & 0 \\ 0 & 1 \end{bmatrix}$$

are solvents of the matrix polynomial

$$L(\lambda) = \lambda^2 I + \lambda\begin{bmatrix} 1 & 1 \\ -2 & -3 \end{bmatrix} + \begin{bmatrix} 0 & -1 \\ 0 & 2 \end{bmatrix}. \quad \square$$

The example of scalar polynomials suggests that we investigate the possibility of capturing spectral information about a monic polynomial of degree l in a set of n solvents. Thus, define a set of solvents $S_1, S_2, \ldots, S_l$ of $L(\lambda)$ to be a *complete set of solvents* of $L(\lambda)$ if the corresponding Vandermonde (block) matrix

$$V = \begin{bmatrix} I & I & \cdots & I \\ S_1 & S_2 & \cdots & S_l \\ \vdots & \vdots & & \vdots \\ S_1^{l-1} & S_2^{l-1} & \cdots & S_{l-1}^{l} \end{bmatrix} \in \mathbb{C}^{ln} \tag{3}$$

is nonsingular. Note that in the case $l = 2$, this condition is equivalent to the nonsingularity of the $n \times n$ matrix $S_2 - S_1$.

Clearly, in the scalar case ($n = 1$), if $L(\lambda)$ has a set of l distinct zeros then, according to this definition, they form a complete set of solvents (zeros) of the polynomial. This observation shows that not every (matrix or scalar) polynomial possesses a complete set of solvents. However, if such a set of solvents does exist, then there are remarkable parallels with the theory of scalar polynomial equations.

Theorem 2. *If $S_1, S_2, \ldots, S_l$ is a complete set of solvents of a monic $n \times n$ matrix polynomial $L(\lambda)$, then* diag$[S_1, S_2, \ldots, S_l]$ *is a linearization of $L(\lambda)$. In particular,*

$$\sigma(L) = \bigcup_{i=1}^{l} \sigma(S_i).$$

PROOF. The assertion of the theorem immediately follows from the easily verified relation

$$C_L V = V \operatorname{diag}[S_1, S_2, \ldots, S_l], \tag{4}$$

where V is given by Eq. (3) (see Exercise 7.11.14). ■

The converse problem of finding a monic matrix polynomial with a given set of solvents is readily solved; it forms the contents of the next exercise.

Exercise 3. Prove that if the Vandermonde matrix V associated with prescribed $n \times n$ matrices $S_1, S_2, \ldots, S_l$ is invertible, then there exists a monic $n \times n$ matrix polynomial $L(\lambda)$ of degree l such that $\{S_1, S_2, \ldots, S_l\}$ is a complete set of solvents for $L(\lambda)$.

Hint. Consider the $n \times ln$ matrix $[-S_1^l \quad -S_2^l \quad \cdots \quad -S_l^l]V^{-1}$. □

Theorem 3. *Let $S_1, S_2, \ldots, S_l$ form a complete set of solvents of the monic $n \times n$ matrix polynomial $L(\lambda)$ of degree l. Then every solution of the differential equation $L(d/dt)\boldsymbol{x} = \boldsymbol{0}$ is of the form*

$$\boldsymbol{x}(t) = e^{S_1 t}\boldsymbol{z}_1 + e^{S_2 t}\boldsymbol{z}_2 + \cdots + e^{S_l t}\boldsymbol{z}_l, \tag{5}$$

where $\boldsymbol{z}_1, \boldsymbol{z}_2, \ldots, \boldsymbol{z}_l \in \mathbb{C}^n$.

PROOF. Substituting Eq. (4) in the formula (14.5.5) (with $\boldsymbol{f}(t) = \boldsymbol{0}$) we obtain

$$\boldsymbol{x}(t) = P_1 e^{VDV^{-1}t}\boldsymbol{z} = P_1 V e^{Dt} V^{-1}\boldsymbol{z},$$

where $D = \operatorname{diag}[S_1, S_2, \ldots, S_l]$. Since $P_1 V = [I \quad I \quad \cdots \quad I]$ and $e^{Dt} = \operatorname{diag}[e^{S_1 t}, e^{S_2 t}, \ldots, e^{S_l t}]$, the result in Eq. (5) follows if we write $V^{-1}\boldsymbol{z} = [\boldsymbol{z}_1^{\mathrm{T}} \quad \boldsymbol{z}_2^{\mathrm{T}} \quad \cdots \quad \boldsymbol{z}_l^{\mathrm{T}}]^{\mathrm{T}}$. ■

Example 4. Consider the differential equation

$$\ddot{\boldsymbol{x}}(t) + \begin{bmatrix} 1 & 1 \\ -2 & -3 \end{bmatrix}\dot{\boldsymbol{x}}(t) + \begin{bmatrix} 0 & -1 \\ 0 & 2 \end{bmatrix}\boldsymbol{x}(t) = \boldsymbol{0}, \tag{6}$$

and the matrix equation

$$X^2 + \begin{bmatrix} 1 & 1 \\ -2 & -3 \end{bmatrix}X + \begin{bmatrix} 0 & -1 \\ 0 & 2 \end{bmatrix} = 0. \tag{7}$$

We saw in Exercise 2 that the matrices

$$S_1 = \begin{bmatrix} 0 & 1 \\ 0 & 0 \end{bmatrix}, \qquad S_2 = \begin{bmatrix} 1 & 0 \\ -2 & 1 \end{bmatrix} \tag{8}$$

satisfy Eq. (7), and since $\det(S_2 - S_1) \neq 0$, they form a complete pair of solvents for the corresponding matrix polynomial. Thus the general solution of Eq. (6) is of the form (5) with $l = 2$, where S_1 and S_2 are given by (8). To simplify this, observe that

$$e^{S_1 t} = I + S_1 t + \tfrac{1}{2}S_1^2 t^2 + \cdots = I + S_1 t = \begin{bmatrix} 1 & t \\ 0 & 1 \end{bmatrix},$$

while

$$\begin{bmatrix} 1 & 0 \\ -2 & 1 \end{bmatrix} = \begin{bmatrix} 0 & -\frac{1}{2} \\ 1 & 0 \end{bmatrix}\begin{bmatrix} 1 & 1 \\ 0 & 1 \end{bmatrix}\begin{bmatrix} 0 & -\frac{1}{2} \\ 1 & 0 \end{bmatrix}^{-1}.$$

Therefore

$$e^{S_2 t} = \begin{bmatrix} 0 & -\frac{1}{2} \\ 1 & 0 \end{bmatrix} \begin{bmatrix} e^t & te^t \\ 0 & e^t \end{bmatrix} \begin{bmatrix} 0 & 1 \\ -2 & 0 \end{bmatrix} = \begin{bmatrix} e^t & 0 \\ -2te^t & e^t \end{bmatrix}.$$

Thus, the general solution of Eq. (6) is

$$\begin{bmatrix} x_1(t) \\ x_2(t) \end{bmatrix} = \boldsymbol{x}(t) = \begin{bmatrix} 1 & t \\ 0 & 1 \end{bmatrix} \begin{bmatrix} \alpha_1 \\ \alpha_2 \end{bmatrix} + \begin{bmatrix} e^t & 0 \\ -2te^t & e^t \end{bmatrix} \begin{bmatrix} \alpha_3 \\ \alpha_4 \end{bmatrix},$$

and hence

$$x_1(t) = \alpha_1 + \alpha_2 t + \alpha_3 e^t,$$

$$x_2(t) = \alpha_2 - 2\alpha_3 te^t + \alpha_4 e^t,$$

for arbitrary $\alpha_1, \alpha_2, \alpha_3, \alpha_4 \in \mathbb{C}$. □

Note that the conditions of Theorem 3 can be relaxed by requiring only similarity between C_L and any block-diagonal matrix. This leads to a generalization of the notion of a complete set of solvents for a monic matrix polynomial and to the form $\sum_{i=1}^{l} C_i e^{X_i t} \boldsymbol{z}_i$ for the solution of the differential equation. Also, the notion of a solvent of multiplicity r of a matrix polynomial can be developed to obtain a complete analogue of the form of solutions for a linear differential equation of order l with scalar coefficients in terms of zeros of the corresponding characteristic equation.

Exercise 5. (a) Let A be a Hermitian matrix. Show that the equation

$$X^2 - X + A = 0 \tag{9}$$

has Hermitian solutions if and only if $\sigma(A) \subset (-\infty, \frac{1}{4}]$.

(b) Let A be positive semidefinite. Show that all Hermitian solutions of Eq. (9) are positive semidefinite if and only if $\sigma(A) \subset [0, \frac{1}{4}]$ and in this case $\sigma(X) \subset [0, 1]$. □

CHAPTER 15

Nonnegative Matrices

A matrix $A \in \mathbb{R}^{m \times n}$ is said to be *nonnegative* if no element of A is negative. If all the elements of A are positive, then A is referred to as a *positive* matrix. Square matrices of this kind arise in a variety of problems and it is perhaps surprising that this defining property should imply some strong results concerning their structure. The remarkable Perron–Frobenius theorem is the central result for nonnegative matrices and is established in Theorems 15.3.1 and 15.4.2.

Matrices whose *inverses* are nonnegative, or positive, also play an important part in applications, and a brief introduction to such matrices is also contained in this chapter.

If $A = [a_{ij}]$ and $B = [b_{ij}]$ are matrices from $\mathbb{R}^{m \times n}$, we shall write $A \geq B$ (respectively, $A > B$) if $a_{ij} \geq b_{ij}$ (respectively, $a_{ij} > b_{ij}$) for all pairs (i, j) such that $1 \leq i \leq m$, $1 \leq j \leq n$. With this notation, A is nonnegative if and only if $A \geq 0$. It should be noted that $A \geq 0$ and $A \neq 0$ do not imply $A > 0$. Also, if $A \geq 0$, $B \geq C$, and AB is defined, then $AB \geq AC$; and if $A \geq 0$, $B > 0$, and $AB = 0$, then $A = 0$. Note that most of these remarks apply to vectors as well as to matrices.

The duplication of notations with those for definite and semidefinite square matrices is unfortunate. However, as the latter concepts do not play an important part in this chapter, no confusion should arise.

15.1 Irreducible Matrices

The concept of reducibility of square matrices plays an important part in this chapter. This was introduced in Section 10.7; it would be wise to read the discussion there once more. Here, the combinatorial aspect of this concept will be emphasized.

It is interesting that the concept of reducibility is not connected in any way with the magnitudes or signs of the elements of a matrix but depends only on the disposition of zero and nonzero elements. This idea is developed in the concept of the directed graph associated with a matrix; we shall take only the first step in this theory to obtain a second characterization of an irreducible matrix in Theorem 1.

Let $P_1, P_2, \ldots, P_n$ be distinct points of the complex plane and let $A \in \mathbb{C}^{n \times n}$. For each nonzero element a_{ij} of A, connect P_i to P_j with a directed line $\overrightarrow{P_i P_j}$. The resulting figure in the complex plane is a *directed graph* for A. We illustrate this in Fig. 15.1 with a 3×3 matrix and is directed graph.

$$\begin{bmatrix} 1 & -2 & 0 \\ 0 & 4 & 2 \\ 1 & -1 & 0 \end{bmatrix}$$

We say that a directed graph is *strongly connected* if, for each pair of *nodes* P_i, P_j with $i \neq j$, there is a *directed path*

$$\overrightarrow{P_i P_{k_1}}, \quad \overrightarrow{P_{k_1} P_{k_2}}, \quad \ldots, \quad \overrightarrow{P_{k_{r-1}} P_j}$$

connecting P_i to P_j. Here, the path consists of r directed lines.

Observe that nodes i and j may be connected by a directed path while j and i are not.

Exercise 1. Check that the graph of Fig. 1 is strongly connected.

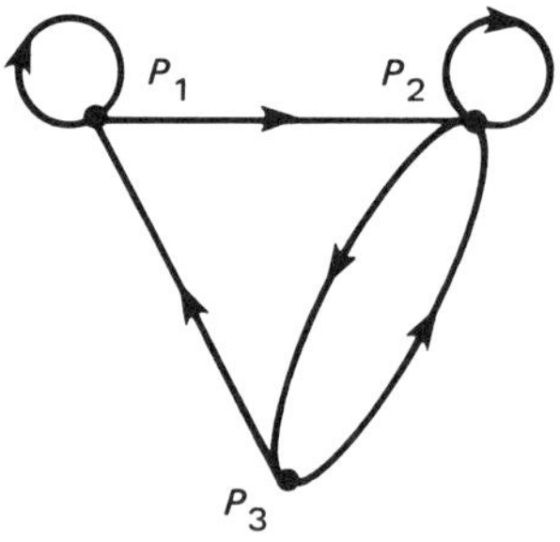

Fig. 15.1 The directed graph of a matrix.

Exercise 2. Verify that the directed graph associated with the matrix

$$\begin{bmatrix} 2 & 0 & 0 \\ 3 & 1 & 0 \\ 0 & 4 & -1 \end{bmatrix}$$

fails to be strongly connected.

Exercise 3. Show that if all the off-diagonal entries of a row A (or a column of A) are zero, then the graph of A is not strongly connected.

SOLUTION. In the case of a row of A, say row i, there are no directed lines $\overrightarrow{P_iP_j}$ starting at node i. So there is no connection *from* node i *to* any other node. So the graph is not strongly connected. □

The next two exercises form part of the proof of the next theorem.

Exercise 4. Let $A \in \mathbb{C}^{n \times n}$ have the partitioned form

$$A = \begin{bmatrix} A_{11} & A_{12} \\ 0 & A_{22} \end{bmatrix},$$

where A_{11} is $k \times k$ and $1 \le k \le n - 1$ (so that A is trivially reducible). Show that the directed graph of A is not strongly connected.

SOLUTION. Consider any directed path from a node i with $i > k$. The first segment of the path is determined by the presence of a nonzero element a_{ij} in the ith row of A. This row has zeros in the first k positions, so it is possible to make a connection from node i to node j only if $j > k$. Similarly, the path can be extended from node j only to another node greater than k. Continuing in this way, it is found that a directed path from node i with $i > k$ cannot be connected to a node less than $k + 1$. Hence the directed graph is not strongly connected.

Exercise 5. Let $A \in \mathbb{C}^{n \times n}$ and let P be an $n \times n$ permutation matrix. Show that the directed graph of A is strongly connected if and only if the directed graph of PAP^{T} is strongly connected.

SOLUTION. Observe that the graph of PAP^{T} is obtained from that of A just by renumbering the nodes, and this operation does not affect the connectedness of the graph. □

Theorem 1. *A square matrix is irreducible if and only if its directed graph is strongly connected.*

PROOF. Exercise 4 shows that if A is reducible, then the graph of PAP^{T} is not strongly connected, for some permutation matrix P. Then Exercise 5 shows that this is equivalent to the statement that A itself has a directed graph that is not strongly connected. Thus, a strongly connected directed graph implies that the matrix is irreducible. Proof of the converse statement is postponed to Exercise 15.3.3. ■

In testing particular matrices for reducibility, it is important to keep in mind the obvious fact that if A is irreducible and B is obtained from A by replacing zero entries by nonzero entries, then B is also irreducible.

Exercise 6. Check that the following matrices are irreducible. In both cases the elements a_{ij} not set equal to zero are, in fact, nonzero.

$$\begin{bmatrix} 0 & a_{12} & 0 & \cdots & 0 \\ 0 & 0 & a_{23} & & \vdots \\ \vdots & \vdots & & \ddots & 0 \\ 0 & 0 & \cdots & 0 & a_{n-1,n} \\ a_{n1} & 0 & \cdots & 0 & 0 \end{bmatrix}, \quad \begin{bmatrix} 0 & a_{12} & 0 & & \cdots & 0 \\ a_{21} & 0 & a_{23} & & \ddots & \vdots \\ 0 & a_{32} & \ddots & \ddots & & 0 \\ \vdots & \ddots & & & 0 & a_{n-1,n} \\ 0 & \cdots & 0 & & a_{n,n-1} & 0 \end{bmatrix}. \quad \square$$

15.2 Nonnegative Matrices and Nonnegative Inverses[†]

Let $A \in \mathbb{C}^{n \times n}$ with elements a_{jk} and let $|A|$ denote the nonnegative matrix in $\mathbb{R}^{n \times n}$ with elements $|a_{jk}|$. Clearly, one of the properties shared by A and $|A|$ is that of reducibility or irreducibility. In addition, many norms of A and $|A|$ take the same value (see Sections 10.3 and 10.4). However, the relationship between the spectral radii of A and $|A|$ is more delicate, and it is frequently of interest. It is plausible that $\mu_A \leq \mu_{|A|}$, and we first show that this is the case. We return in Section 15.4 to the important question of conditions under which $\mu_A < \mu_{|A|}$.

Theorem 1. *Let $A \in \mathbb{C}^{n \times n}$ and $B \in \mathbb{R}^{n \times n}$. If $|A| \leq B$, then $\mu_A \leq \mu_B$.*

PROOF. Let ε be an arbitrary positive number and define the scalar multiples

$$A_\varepsilon = (\mu_B + \varepsilon)^{-1} A, \qquad B_\varepsilon = (\mu_B + \varepsilon)^{-1} B.$$

Clearly $|A_\varepsilon| \leq B_\varepsilon$, and it is easily seen (see Exercise 1) that for $k = 1, 2, \ldots,$ $|A_\varepsilon|^k \leq B_\varepsilon^k$.

Now it is also clear that $\mu_{B_\varepsilon} < 1$, and so (see Exercise 9.8.5) $B_\varepsilon^k \to 0$ as $k \to \infty$. Since $|A_\varepsilon^k| \leq |A_\varepsilon|^k \leq B_\varepsilon^k$, it follows that $A_\varepsilon^k \to 0$ as $k \to \infty$ and so (by Exercise 9.8.5 again) $\mu_{A_\varepsilon} < 1$. But this implies $\mu_A < \mu_B + \varepsilon$, and since ε is arbitrary, $\mu_A \leq \mu_B$. ■

Note that on putting $B = |A|$, we obtain $\mu_A \leq \mu_{|A|}$ from the theorem.

Consider now the question of solving an equation $A\boldsymbol{x} = \boldsymbol{b}$, where $A \in \mathbb{R}^{n \times n}$ and $\boldsymbol{b} \in \mathbb{R}^n$ are given and A is invertible. The unique solution is, of course, $\boldsymbol{x} = A^{-1}\boldsymbol{b}$. However, if $\boldsymbol{b}$ is not known precisely and we are given only bounding vectors $\boldsymbol{b}_1, \boldsymbol{b}_2 \in \mathbb{R}^n$ in the sense that $\boldsymbol{b}_1 \leq \boldsymbol{b} \leq \boldsymbol{b}_2$, what can then be said about the solution vector $\boldsymbol{x}$? The perturbation analysis of Section 11.1 gives one answer in terms of the condition number of A. But

[†] The presentation of this section is strongly influenced by the expositions of R. S. Varga and of J. M. Ortega and W. C. Rheinboldt. See Appendix 3.

if A^{-1} is a *nonnegative* matrix, we can give precise bounds for the solution vector because, in this case,

$$A^{-1}\boldsymbol{b}_1 \le A^{-1}\boldsymbol{b} = \boldsymbol{x} \le A^{-1}\boldsymbol{b}_2.$$

A matrix $A \in \mathbb{R}^{n\times n}$ is said to be *monotone* if it is nonsingular and $A^{-1} \ge 0$. Monotone matrices frequently occur with a sign pattern among the elements; for example, main-diagonal elements positive and off-diagonal elements nonpositive. Such matrices are given their own name: a monotone matrix is called an *M-matrix* if all the off-diagonal elements are nonpositive. The matrices A and B of Exercise 2.11.17 are monotone, and B is an M-matrix.

***Exercise* 1.** Use Exercise 1.8.5 to prove that if $0 \le A \le B$, then $A^k \le B^k$ for $k = 1, 2, \ldots$.

***Exercise* 2.** Prove that if A is an M-matrix, then each main-diagonal element of A is positive.

Hint. Observe that if not, then A has a complete column that is nonpositive, say the jth column, and then $A\boldsymbol{e}_j \le 0$. □

We now discuss some criteria that ensure that a matrix $A \in \mathbb{R}^{n\times n}$ is an M-matrix. They all give sufficient conditions that are often easy to check.

Theorem 2. *Let $B \in \mathbb{R}^{n\times n}$ with $B \ge 0$. The matrix $I - B$ is an M-matrix if and only if $\mu_B < 1$.*

PROOF. If $\mu_B < 1$ then Theorems 10.3.1 and 11.1.1 imply that $I - B$ is nonsingular and

$$(I - B)^{-1} = I + B + B^2 + \cdots.$$

Since $B \ge 0$ it is clear that $(I - B)^{-1} \ge 0$. Obviously, the elements of $I - B$ off the main diagonal are nonpositive and so $I - B$ is an M-matrix.

Conversely, if $I - B$ is an M-matrix, then $(I - B)^{-1} \ge 0$. Let $\lambda \in \sigma(B)$ and $\lambda\boldsymbol{x} = B\boldsymbol{x}$ with $\boldsymbol{x} \ne \boldsymbol{0}$. Then, since $B \ge 0$, it is easily seen that

$$|\lambda||\boldsymbol{x}| \le B|\boldsymbol{x}|,$$

which implies $(I - B)|\boldsymbol{x}| \le (1 - |\lambda|)|\boldsymbol{x}|$, and hence

$$|\boldsymbol{x}| \le (1 - |\lambda|)(I - B)^{-1}|\boldsymbol{x}|.$$

Since $|\boldsymbol{x}| \ge \boldsymbol{0}$ and $(I - B)^{-1} \ge 0$, it follows that $|\lambda| < 1$. But this is the case for any $\lambda \in \sigma(B)$, so have $\mu_B < 1$. ■

Theorem 3. *Let $A = [a_{ij}]_{i,j=1}^n \in \mathbb{R}^{n\times n}$ and assume that $a_{ii} > 0$ for each i and $a_{ij} \le 0$ whenever $i \ne j$. If A is diagonally dominant, that is,*

$$a_{ii} > \sum_{j=1,\, j\ne i}^{n} |a_{ij}|, \qquad i = 1, 2, \ldots, n,$$

or, if A is irreducible and

$$a_{ii} \geq \sum_{j=1, j \neq i}^{n} |a_{ij}|, \qquad i = 1, 2, \ldots, n,$$

with strict inequality for at least one i, then A is an M-matrix.

PROOF. First consider the case in which A is diagonally dominant. Let $D = \text{diag}[a_{11}, \ldots, a_{nn}]$ and define $B = I - D^{-1}A$. Note that B has zero elements on the main diagonal and that $B \geq O$. Also, the fact that A is diagonally dominant implies that

$$\sum_{j=1}^{n} |b_{ij}| < 1, \qquad i = 1, 2, \ldots, n.$$

It follows immediately from the Geršgorin theorem (Section 10.6) that $\mu_B < 1$.

Now we have $D^{-1}A = I - B$, and it follows from Theorem 2 that $D^{-1}A$ is an M-matrix. Consequently, A is also an M-matrix.

With the second set of hypotheses we follow the same line of argument but use the corollary to Theorem 10.7.2 to ensure that $\mu_B < 1$. ■

Exercise 3. Show that, under either of the hypotheses of Theorem 3, In $A = \{n, 0, 0\}$. If, in addition, A is symmetric then A is positive definite.

Exercise 4. Let A be a lower-triangular matrix, with positive elements on the main diagonal and nonpositive elements elsewhere. Show that A is an M-matrix.

Exercise 5. Show that if A is an M-matrix and D is a nonnegative diagonal matrix, then $A + D$ is an M-matrix.

Exercise 6. Show that a symmetric M-matrix is positive definite. (Such a matrix is called a *Stieltjes matrix.*)

SOLUTION. Let A be a symmetric M-matrix and $\lambda \in \sigma(A)$. If $\lambda \leq 0$ then, by Exercise 5, $A - \lambda I$ is also an M-matrix and must therefore be nonsingular. This contradicts the assumption that $\lambda \in \sigma(A)$. Thus $\lambda \in \sigma(A)$ implies $\lambda > 0$ and hence A is positive definite. □

15.3 The Perron–Frobenius Theorem (I)

It is obvious that if $A \geq O$ then $A^p \geq O$ for any positive integer p. We might also expect that if A has a sufficiently high density of nonzero elements then, for large enough p, we would obtain $A^p > O$. Our first result is of this kind.

Proposition 1. *If the matrix $A \in \mathbb{R}^{n \times n}$ is nonnegative and irreducible, then*

$$(I + A)^{n-1} > 0.$$

PROOF. Consider a vector $\boldsymbol{y} \in \mathbb{R}^n$ such that $\boldsymbol{y} \geq \boldsymbol{0}$ and $\boldsymbol{y} \neq \boldsymbol{0}$ and write

$$\boldsymbol{z} = (I + A)\boldsymbol{y} = \boldsymbol{y} + A\boldsymbol{y}. \tag{1}$$

Since $A \geq 0$, the product $A\boldsymbol{y} \geq \boldsymbol{0}$ and so $\boldsymbol{z}$ has at least as many nonzero (and hence positive) elements as $\boldsymbol{y}$. If $\boldsymbol{y}$ is not already positive, we shall prove that $\boldsymbol{z}$ has at least one more nonzero element than $\boldsymbol{y}$. Indeed, if P is a permutation matrix such that $P\boldsymbol{y} = [\boldsymbol{u}^{\mathrm{T}} \quad \boldsymbol{0}^{\mathrm{T}}]^{\mathrm{T}}$ and $\boldsymbol{u} > \boldsymbol{0}$, then it follows from Eq. (1) and the relation $PP^{\mathrm{T}} = I$ that

$$P\boldsymbol{z} = \begin{bmatrix} \boldsymbol{u} \\ \boldsymbol{0} \end{bmatrix} + PAP^{\mathrm{T}} \begin{bmatrix} \boldsymbol{u} \\ \boldsymbol{0} \end{bmatrix}. \tag{2}$$

Hence, if we partition $\boldsymbol{z}$ and PAP^{T} consistently with the partition of $\boldsymbol{y}$,

$$\boldsymbol{z} = \begin{bmatrix} \boldsymbol{v} \\ \boldsymbol{w} \end{bmatrix}, \qquad PAP^{\mathrm{T}} = \begin{bmatrix} A_{11} & A_{12} \\ A_{21} & A_{22} \end{bmatrix},$$

then Eq. (2) implies

$$\boldsymbol{v} = \boldsymbol{u} + A_{11}\boldsymbol{u} \qquad \text{and} \qquad \boldsymbol{w} = A_{21}\boldsymbol{u}. \tag{3}$$

Now observe that the matrix PAP^{T} is nonnegative and irreducible. Hence, in particular, $A_{11} \geq 0$, $A_{21} \geq 0$, $A_{21} \neq 0$, and therefore it follows from (3) that $\boldsymbol{v} > \boldsymbol{0}$, $\boldsymbol{w} \geq \boldsymbol{0}$. But $\boldsymbol{u} > \boldsymbol{0}$ and, therefore, $\boldsymbol{w} \neq \boldsymbol{0}$. Thus, $\boldsymbol{z}$ has at least one more positive element than $\boldsymbol{y}$.

If $(I + A)\boldsymbol{y}$ is not already positive then, starting with the element $(I + A)\boldsymbol{z} = (I + A)^2\boldsymbol{y}$ and repeating the argument, it follows that it has at least two more positive elements than $\boldsymbol{y}$. Continuing this process, we find after at most $n - 1$ steps that

$$(I + A)^{n-1}\boldsymbol{y} > \boldsymbol{0}$$

for any $\boldsymbol{y} \geq \boldsymbol{0}$, $\boldsymbol{y} \neq \boldsymbol{0}$. Putting $\boldsymbol{y} = \boldsymbol{e}_j$ for $j = 1, 2, \ldots, n$, we obtain the desired result. ■

Observe that there is a simple result in the other direction. If $(I + A)^j > 0$ for any $A \in \mathbb{R}^{n \times n}$ and any positive integer j, then A must be irreducible. Otherwise, assuming that A has the partitioned form of Exercise 15.1.4, we easily obtain a contradiction with the hypothesis.

Exercise 1. If $A \in \mathbb{R}^{n \times n}$ is irreducible and D is a nonsingular, nonnegative diagonal matrix, prove that $(D + A)^{n-1} > 0$.

Exercise 2. Let $A \in \mathbb{R}^{n \times n}$ be nonnegative and irreducible, and let $a_{ij}^{(q)}$ be the i, j element of A^q. Show that there is a positive integer q such that $a_{ij}^{(q)} > 0$. Show also that if $\psi(\lambda)$ is the minimal polynomial of A, then we may choose $q \le \deg(\psi)$.

Hint. Observe that $A(I + A)^{n-1} > 0$ and use the binomial theorem. For the second part let $r(\lambda)$ be the remainder on division of $\lambda(1 + \lambda)^{n-1}$ by $\psi(\lambda)$ and use the fact that $r(A) > 0$.

Exercise 3. Let matrix A be as in Exercise 2.

(a) Show that $a_{ij}^{(q)} > 0$ if and only if there is a sequence of q edges in the directed graph of A connecting nodes P_i and P_j.

(b) Use the result of Exercise 2 to show that A has a strongly connected directed graph. □

If A is a nonnegative irreducible matrix, consider the real-valued function r defined on the nonzero vectors $\boldsymbol{x} \ge \boldsymbol{0}$ by

$$r(\boldsymbol{x}) = \min_{\substack{1 \le i \le n \\ x_i \ne 0}} \frac{(A\boldsymbol{x})_i}{x_i}, \tag{4}$$

where $(A\boldsymbol{x})_i$ denotes the ith element of the vector $A\boldsymbol{x}$. Then $r(\boldsymbol{x}) \ge 0$ and for $j = 1, 2, \ldots, n$, $r(\boldsymbol{x})x_j \le (A\boldsymbol{x})_j$, with equality for some j. Thus $r(\boldsymbol{x})\boldsymbol{x} \le A\boldsymbol{x}$ and, furthermore, $r(\boldsymbol{x})$ is the largest number ρ such that $\rho\boldsymbol{x} \le A\boldsymbol{x}$ for this $\boldsymbol{x}$.

Exercise 4. Find $r(\boldsymbol{x})$ if

$$A = \begin{bmatrix} 2 & 1 \\ 1 & 3 \end{bmatrix} \quad \text{and} \quad \boldsymbol{x} = \begin{bmatrix} 1 \\ 0 \end{bmatrix}.$$

SOLUTION. Since $A\boldsymbol{x} = [2 \quad 1]^{\mathrm{T}}$, the largest number ρ such that $\rho\boldsymbol{x} \le A\boldsymbol{x}$ is 2. Thus $r(\boldsymbol{x}) = 2$.

Exercise 5. If $A = [a_{ij}]_{i,j=1}^{n}$ and $\boldsymbol{x} = [1 \quad 1 \quad \cdots \quad 1]^{\mathrm{T}}$, show that

$$r(\boldsymbol{x}) = \min_{1 \le i \le n} \sum_{k=1}^{n} a_{ik}. \quad \square$$

Let $\mathscr{L}$ denote the domain of the function r, that is, the set of all nonzero nonnegative vectors of order n, and define the number r (not to be confused with the function r) by

$$r = \sup_{\boldsymbol{x} \in \mathscr{L}} r(\boldsymbol{x}). \tag{5}$$

From the definition of $r(\boldsymbol{x})$, we observe that r is invariant if $\boldsymbol{x}$ is replaced by $\alpha\boldsymbol{x}$ for any $\alpha > 0$. Thus, in evaluating this supremum, we need only consider the closed set $\mathscr{M}$ of vectors $\boldsymbol{x}$ such that $\boldsymbol{x} \geq \boldsymbol{0}$ and $\sum_i x_i^2 = 1$. Thus, $\mathscr{M} \subset \mathscr{L}$ and

$$r = \sup_{\boldsymbol{x} \in \mathscr{M}} r(\boldsymbol{x}).$$

If the function $r(\boldsymbol{x})$ were continuous on $\mathscr{M}$, we could equate this to max $r(\boldsymbol{x})$, $\boldsymbol{x} \in \mathscr{M}$ (see Appendix 2). However $r(\boldsymbol{x})$ may have discontinuities at points where elements of $\boldsymbol{x}$ vanish. We therefore consider the set $\mathscr{N}$ of vectors $\boldsymbol{y}$ defined by

$$\boldsymbol{y} = (I + A)^{n-1}\boldsymbol{x}, \qquad \boldsymbol{x} \in \mathscr{M}.$$

By Proposition 1 every member of $\mathscr{N}$ is a *positive* vector and so $\mathscr{N} \subset \mathscr{L}$. Now, $\mathscr{N}$ is the image of the closed and bounded set $\mathscr{M}$ under a continuous function and is therefore also closed and bounded. Furthermore, $r(\boldsymbol{y})$ is continuous on $\mathscr{N}$.

For any $\boldsymbol{x} \in \mathscr{M}$ and corresponding $\boldsymbol{y}$,

$$r(\boldsymbol{x})\boldsymbol{y} = r(\boldsymbol{x})(I + A)^{n-1}\boldsymbol{x} \leq (I + A)^{n-1}A\boldsymbol{x},$$

since $r(\boldsymbol{x})\boldsymbol{x} \leq A\boldsymbol{x}$, and hence $r(\boldsymbol{x})\boldsymbol{y} \leq A\boldsymbol{y}$ for every $\boldsymbol{y} \in \mathscr{N}$. Now, $r(\boldsymbol{y})$ is the greatest number ρ such that $\rho\boldsymbol{y} \leq A\boldsymbol{y}$ and hence $r(\boldsymbol{x}) \leq r(\boldsymbol{y})$. Thus,

$$r = \sup_{\boldsymbol{x} \in \mathscr{M}} r(\boldsymbol{x}) \leq \max_{\boldsymbol{y} \in \mathscr{N}} r(\boldsymbol{y}).$$

But since $\mathscr{N} \subset \mathscr{L}$,

$$\max_{\boldsymbol{y} \in \mathscr{N}} r(\boldsymbol{y}) \leq \sup_{\boldsymbol{x} \in \mathscr{L}} r(\boldsymbol{x}) = \sup_{\boldsymbol{x} \in \mathscr{M}} r(\boldsymbol{x}).$$

Hence

$$r = \max_{\boldsymbol{y} \in \mathscr{N}} r(\boldsymbol{y}), \tag{6}$$

and there is a $\boldsymbol{y} > \boldsymbol{0}$ such that $r = r(\boldsymbol{y})$.

There may be other vectors in $\mathscr{L}$ for which $r(\boldsymbol{x})$ attains the value r. Any such vector is called an *extremal vector* of A. Thus, a nonzero vector $\boldsymbol{z} \geq \boldsymbol{0}$ is an extremal vector of A if $r(\boldsymbol{z}) = r$ or, what is equivalent, $r\boldsymbol{z} \leq A\boldsymbol{z}$.

The reason for our interest in the number r is clarified in the next result.

Proposition 2. *If the matrix $A \in \mathbb{R}^{n \times n}$ is nonnegative and irreducible, then the number r defined by Eq. (5) is positive and is an eigenvalue of A. Furthermore, every extremal vector of A is positive and is a right eigenvector of A associated with the eigenvalue r.*

PROOF. Let $\boldsymbol{x} = [1 \quad 1 \quad \cdots \quad 1]^{\mathrm{T}}$. Then (see Exercise 5) $r(\boldsymbol{x}) = \min_i \sum_k a_{ik} > 0$. For, if any row of A consists entirely of zeros, then A is reducible. Since $r \geq r(\boldsymbol{x})$, we deduce that $r > 0$.

Let z be an extremal vector and let $w = (I + A)^{n-1}z$. Without loss of generality, we may suppose that $z \in \mathscr{M}$. Proposition 1 implies that $w > 0$, and clearly $w \in \mathscr{N}$. We also have $Az - rz \geq 0$ and, if $Az - rz \neq 0$, then

$$(I + A)^{n-1}(Az - rz) > 0.$$

Hence $Aw - rw > 0$, or $rw < Aw$, which implies that $r < r(w)$. But this contradicts the definition (5) or r, and so we must have $Az = rz$. Thus, any extremal vector z is a right eigenvector of A with associated eigenvalue r.

Finally, since $Az = rz$, we have

$$w = (I + A)^{n-1}z = (1 + r)^{n-1}z,$$

and since $w > 0$ and $r > 0$, we must have $z > 0$. ■

We can now state and prove the first part of the Perron–Frobenius theorem for irreducible matrices.

Theorem 1. *If the matrix $A \in \mathbb{R}^{n \times n}$ is nonnegative and irreducible, then*

(*a*) *The matrix A has a positive eigenvalue, r, equal to the spectral radius of A;*

(*b*) *There is a positive (right) eigenvector associated with the eigenvalue r;*

(*c*) *The eigenvalue r has algebraic multiplicity 1.*

PROOF. In Proposition 2 we have established the existence of a positive eigenvalue with an associated positive eigenvector. To complete parts (a) and (b) of the theorem it suffices to show that other eigenvalues of A cannot exceed r in absolute value.

Indeed, if we have $\alpha y = Ay$ and $y \neq 0$, then, since $A \geq 0$, it is easily seen that

$$|\alpha||y| = |Ay| \leq A|y|,$$

where if $z \in \mathbb{R}^n$, $|z|$ denotes the vector obtained from z by replacing all its components by their absolute values. Hence $|\alpha| \leq r(|y|) \leq r$, which is what we wanted to prove.

Suppose now that z is any right eigenvector associated with r. Thus, $Az = rz$ and $z \neq 0$, and, as we just showed,

$$r|z| \leq A|z|.$$

This implies that $|z|$ is an extremal vector and, by Proposition 2, $|z| > 0$. Thus, $z_i \neq 0$, $i = 1, 2, \ldots, n$. Hence the dimension of the right eigenspace of r is 1. Otherwise, we could find two linearly independent right eigenvectors z_1, z_2 and then determine numbers α, β such that $\alpha z_1 + \beta z_2$ has a zero element. This proves that the *geometric* multiplicity of r is 1.

To complete the proof,† we show that in the terminology of section 6.3, there is no generalized eigenvector of order 2 associated with r, and this will imply that the algebraic multiplicity of r is 1. To see this, let $\mathbf{x}_1 > \mathbf{0}$ and $\mathbf{y} > \mathbf{0}$ be eigenvectors of A and A^T, respectively, associated with eigenvalue r. Thus,

$$(rI - A)\mathbf{x}_1 = \mathbf{0}, \qquad (rI - A^T)\mathbf{y} = \mathbf{0}.$$

Suppose that there is a generalized eigenvector $\mathbf{x}_2 \neq \mathbf{0}$ for which $(rI - A)\mathbf{x}_2 = \mathbf{x}_1$ (see Eqs. (6.3.3)). Then, since $\mathbf{y}^T(rI - A) = \mathbf{0}^T$, we have $\mathbf{y}^T\mathbf{x}_1 = 0$. But this contradicts the positivity of the vectors $\mathbf{x}$ and $\mathbf{y}$, and r must have algebraic multiplicity 1. ■

Exercise 6. Using the above notations and taking advantage of Exercise 7.9.1 show that if $B(\lambda) = \operatorname{adj}(\lambda I - A)$, then

$$B(\lambda)(\lambda I - A) = c(\lambda)I.$$

Deduce that the derivative $c'(r) > 0$, and hence

$$B(r) = k\mathbf{x}\mathbf{y}^T > \mathbf{0}$$

for some positive k.

Exercise 7. If $A \in \mathbb{R}^{n \times n}$ is irreducible and nonnegative and $\sigma_j = \sum_{k=1}^{n} a_{jk}$, $j = 1, 2, \ldots, n$, prove that

$$\min_j \sigma_j \le r \le \max_j \sigma_j,$$

and that there is equality in either case if and only if $\sigma_1 = \sigma_2 = \cdots = \sigma_n$.

SOLUTION. Adding the first $n - 1$ columns of $rI - A$ to the last one, we obtain

$$\det(rI - A) = \det \begin{bmatrix} r - a_{11} & -a_{12} & \cdots & -a_{1,n-1} & r - \sigma_1 \\ -a_{21} & r - a_{22} & & -a_{2,n-1} & r - \sigma_2 \\ \vdots & \vdots & & \vdots & \vdots \\ & & & r - a_{n-1,n-1} & r - \sigma_{n-1} \\ -a_{n1} & -a_{n2} & \cdots & -a_{n,n-1} & r - \sigma_n \end{bmatrix} = 0.$$

Expanding by the last column and denoting the i, jth element of $B(r) = \operatorname{adj}(rI - A)$ by $b_{ij}(r)$, we obtain

$$\sum_{j=1}^{n} (r - \sigma_j) b_{nj}(r) = 0.$$

† The authors are indebted to D. Flockerzi for this part of the proof.

But we have seen in Exercise 6 that $B(r) > 0$ and so $b_{nj}(r) > 0$ for $j = 1, 2, \ldots, n$. It follows that either $\sigma_1 = \sigma_2 = \cdots = \sigma_n$ or there exist indices k, l such that $r < \sigma_k$ and $r > \sigma_l$. Hence the result.

Exercise 8. Using Exercise 7, and without examining the characteristic polynomial, prove that the following matrix has the maximal real eigenvalue 7 and that there is an eigenvalue equal to 4:

$$\begin{bmatrix} 3 & 0 & 1 & 0 \\ 1 & 6 & 1 & 1 \\ 2 & 0 & 2 & 0 \\ 1 & 3 & 1 & 4 \end{bmatrix}.$$

Exercise 9. If $A > 0$ and G is the component matrix of A associated with r, prove that $G > 0$.

Exercise 10. If A is irreducible and nonnegative and $C(\lambda)$ is the reduced adjoint of A (see Section 7.9), prove that $C(r) > 0$.

Hint. Observe that $\delta(r) = d_{n-1}(r) > 0$ (Section 7.9).

Exercise 11. If A is nonnegative and irreducible, show that A cannot have two linearly independent nonnegative eigenvectors (even associated with distinct eigenvalues).

Hint. Use Exercise 10.7.2 and Theorem 1(c) for the matrices A and A^{T} to obtain a contradiction. □

15.4 The Perron–Frobenius Theorem (II)

The main results of the preceding section generalize results established first by Perron†. Frobenius‡ published them in 1912 along with deeper results of this section, results that give more information on the structure of a nonnegative irreducible matrix A in those cases where A has more than one eigenvalue with absolute value equal to r, the spectral radius of A.

An important first step in our argument is a result due to Wielandt that is independently useful and gives a stronger result than the comparison theorem, Theorem 15.2.1, provided the bounding matrix is irreducible. Note that the proof depends heavily on the Perron–Frobenius results already established.

† *Math. Ann.* **64** (1907), 248–263.
‡ *Sitzungsber. Preuss. Akad. Wiss.* (1912), 456–477.

Theorem 1 (H. Wielandt†). *Let $A \in \mathbb{C}^{n\times n}$, $B \in \mathbb{R}^{n\times n}$ satisfy $|A| \le B$ and let B be irreducible. Then $\mu_A = r$, where μ_A and r are the spectral radii of A and B, respectively, if and only if A is of the form*

$$A = \omega DBD^{-1}, \tag{1}$$

where $|\omega| = 1$, $|D| = I$, and $\omega r \in \sigma(A)$.

Note first of all that, by Theorem 15.3.1, r is in fact an eigenvalue of B.

PROOF. If A has the form (1), it is clear that A and B have the same spectral radius.

Conversely, let $\mu_A = r$ and $\lambda \in \sigma(A)$ with $|\lambda| = \mu_A = r$. Let $\boldsymbol{a}$ be a left eigenvector of A corresponding to λ, that is, $\boldsymbol{a} \ne \boldsymbol{0}$ and $\boldsymbol{a}^*A = \lambda\boldsymbol{a}^*$. Then

$$|\lambda\boldsymbol{a}^*| = |\lambda||\boldsymbol{a}^*| = r|\boldsymbol{a}^*| \le |\boldsymbol{a}^*||A| \le |\boldsymbol{a}^*|B. \tag{2}$$

Hence for any positive vector $\boldsymbol{u}$,

$$r|\boldsymbol{a}^*|\boldsymbol{u} \le |\boldsymbol{a}^*|B\boldsymbol{u}. \tag{3}$$

By Theorem 15.3.1 there is a $\boldsymbol{u} > \boldsymbol{0}$ such that $B\boldsymbol{u} = r\boldsymbol{u}$, so for this $\boldsymbol{u}$ we have $r|\boldsymbol{a}^*|\boldsymbol{u} = |\boldsymbol{a}^*|B\boldsymbol{u}$. Thus, for this choice of $\boldsymbol{u}$ equality obtains in (3) and hence, also, throughout (2). It follows that $|\boldsymbol{a}^*|B = r|\boldsymbol{a}^*|$ and, applying Theorem 15.3.1 again, we must have $|\boldsymbol{a}^*| > \boldsymbol{0}$.

Then (2) also yields

$$|\boldsymbol{a}^*|(B - |A|) = \boldsymbol{0}^*$$

which, together with $B \ge |A|$, implies $B = |A|$.

Now define $\omega = \lambda/r$ and

$$D = \operatorname{diag}\left[\frac{a_1}{|a_1|}, \ldots, \frac{a_n}{|a_n|}\right],$$

where $\boldsymbol{a} = [a_1 \ \cdots \ a_n]^{\mathrm{T}}$. Then $|\omega| = 1$, $|D| = I$, and it is easily verified that $\boldsymbol{a}^*D = |\boldsymbol{a}^*|$. Now, $\boldsymbol{a}^*A = \lambda\boldsymbol{a}^*$ implies

$$|\boldsymbol{a}^*|D^{-1}A = \lambda|\boldsymbol{a}^*|D^{-1},$$

so that

$$|\boldsymbol{a}^*|D^{-1}AD = \omega r|\boldsymbol{a}^*|,$$

and

$$|\boldsymbol{a}^*|\omega^{-1}D^{-1}AD = r|\boldsymbol{a}^*| = |\boldsymbol{a}^*|B.$$

In other words,

$$|\boldsymbol{a}^*|(\omega^{-1}D^{-1}AD - B) = \boldsymbol{0}^*. \tag{4}$$

† *Math. Zeits.* **52** (1950), 642–648.

But $B = |A| = |\omega^{-1}D^{-1}AD|$ and, since $|\boldsymbol{a}^*| > \boldsymbol{0}^*$, it follows from Eq. (4) that $B = \omega^{-1}D^{-1}AD$, and hence conclusion (1). ■

A particular deduction from Theorem 1 will be important for the proof of Theorem 2. Suppose that, in the theorem, A is nonnegative and irreducible. Apply the theorem with $B = A$, so that $|A| \le B$ and $\mu_A = r$ are trivially satisfied. The conclusion is as follows:

Corollary 1. *Let A be nonnegative and irreducible and let $\lambda \in \sigma(A)$ with $|\lambda| = \mu_A = r$, the spectral radius of A. Then A satisfies the condition*

$$A = \frac{\lambda}{r} DAD^{-1} \tag{5}$$

for some diagonal matrix D with $|D| = I$.

Exercise 1. Let the matrix A of Theorem 1 be nonnegative. Show that $\mu_A = r$ if and only if $A = B$.

Exercise 2. Suppose that, in Theorem 1, the matrix A is obtained from B by replacing certain rows and/or columns by zero rows and/or columns, respectively. Show that $\mu_A < r$. □

Theorem 2. *Let the matrix $A \in \mathbb{R}^{n \times n}$ be nonnegative and irreducible and have eigenvalues $\lambda_1, \lambda_2, \ldots, \lambda_n$. If there are exactly k eigenvalues $\lambda_1 = r$, $\lambda_2, \ldots, \lambda_k$ of modulus r and if $\omega_1, \ldots, \omega_k$ are the distinct kth roots of unity, then $\lambda_j = \omega_j r$, $j = 1, 2, \ldots, k$.*

Moreover, the n points of the complex plane corresponding to $\lambda_1, \ldots, \lambda_n$ are invariant under rotations about the origin through $2\pi/k$. If $k > 1$, then there is a permutation matrix P such that $P^{\mathrm{T}}AP$ has the symmetrically partitioned form

$$\begin{bmatrix} 0 & A_{12} & 0 & \cdots & 0 \\ 0 & 0 & A_{23} & \ddots & 0 \\ \vdots & \vdots & \ddots & \ddots & \vdots \\ & & & & 0 \\ 0 & 0 & \cdots & 0 & A_{k-1,k} \\ A_{k1} & 0 & \cdots & 0 & 0 \end{bmatrix}.$$

PROOF. Let $\lambda_j = \omega_j r$ denote all the eigenvalues of A of modulus r, where $\omega_j = e^{i\varphi_j}$, $0 = \varphi_1 < \varphi_2 < \cdots < \varphi_k < 2\pi$. To show that $\omega_1, \omega_2, \ldots, \omega_k$ are the kth roots of unity, apply the result of Corollary 1 and write

$$A = \omega_j D_j A D_j^{-1}, \qquad j = 1, 2, \ldots, k. \tag{6}$$

By Theorem 15.3.1(b), there exists a vector $\boldsymbol{x} \in \mathbb{R}^n$, $\boldsymbol{x} > \boldsymbol{0}$, such that $A\boldsymbol{x} = r\boldsymbol{x}$. Then Eq. (6) implies that the vector $\boldsymbol{x}_j = D_j\boldsymbol{x}$ is an eigenvector of A associated with λ_j, $j = 1, 2, \ldots, k$. Moreover, since r has algebraic multiplicity 1, the algebraic multiplicity of $\lambda_j = \omega_j r$ is also 1, $j = 1, 2, \ldots, k$. Indeed, r is a simple zero of the characteristic polynomial of A and hence $\omega_j r$ is a simple zero of the characteristic polynomial of $\omega_j A$, $1 \le j \le k$. In view of Eq. (6) these polynomials coincide:

$$\det(\lambda I - A) = \det(\lambda I - \omega_j A), \qquad 1 \le j \le k,$$

and therefore the result follows.

Note that the geometric multiplicity of each of $\lambda_1, \lambda_2, \ldots, \lambda_k$ is then also 1 and hence the vectors $\boldsymbol{x}_j$, and therefore the matrices D_j, $j = 1, 2, \ldots, k$, are defined uniquely up to scalar multiples. We thus assume that $D_1 = I$ and that all ones appearing in the matrices $D_2, D_3, \ldots, D_k$ are located in the first parts of the main diagonals. With this assumption, the matrices $D_1, D_2, \ldots, D_k$ are uniquely defined.

Now observe that Eq. (6) implies the relation

$$A = \omega_j D_j A D_j^{-1} = \omega_j D_j(\omega_s D_s A D_s^{-1})D_j^{-1} = \omega_j\omega_s D_j D_s A(D_j D_s)^{-1}$$

for any pair j, s, where $1 \le j, s \le k$, and therefore the same reasoning implies that the vector $D_j D_s \boldsymbol{x}$ is an eigenvector of A associated with the eigenvalue $\omega_j\omega_s r$ of modulus r. Hence $\omega_j\omega_s = \omega_i$ for some i, $1 \le i \le k$, and, consequently, $D_j D_s = D_i$. Thus, $\{\omega_1, \omega_2, \ldots, \omega_k\}$ is a finite commutative (or Abelian) group consisting of k distinct elements.

It is well known that the kth power of any element of this group is equal to its identity element. Thus, $\omega_1, \omega_2, \ldots, \omega_k$ are all the kth roots of unity, as required.

Note that the relation (6) can now be rewritten in the form

$$A = e^{2\pi i/k} D A D^{-1} \tag{7}$$

for some diagonal D with $|D| = I$, and therefore the matrix A is similar to the matrix $e^{2\pi i/k}A$. Thus the spectrum $\sigma(A)$ is carried into itself if it is rotated bodily about the origin through $2\pi/k$.

Furthermore, the argument used above yields $D_j^k = I$, $j = 1, 2, \ldots, k$, and therefore the diagonal elements of $D = D_j$ are kth roots of unity. If $k > 1$, there exists a permutation matrix P such that

$$D_0 = P^{\mathrm{T}} D P = \operatorname{diag}[\mu_1 I_1, \mu_2 I_2, \ldots, \mu_s I_s], \tag{8}$$

where the I_l are identity matrices of appropriate sizes, $l = 1, 2, \ldots, s$, and $\mu_l = e^{i\psi_l}$, $\psi_l = n_l 2\pi/k$, $0 = n_1 < n_2 < \cdots < n_s < k$. Note that in view of Eq. (7),

$$A_0 = \varepsilon D_0 A_0 D_0^{-1}, \qquad \varepsilon = e^{2\pi i/k}, \tag{9}$$

in which $A_0 = P^T AP$. If $A_0 = [A_{ij}]_{i,j=1}^s$ is a partition of A_0 consistent with that of D_0 in (8), then Eq. (9) implies

$$A_{ij} = \varepsilon\mu_i\mu_j^{-1}A_{ij}, \qquad 1 \le i, j \le s. \tag{10}$$

In particular, this implies $A_{ii} = O$ for $i = 1, 2, \ldots, s$. Note also that no block-row in A_0 consists entirely of zeros, since otherwise we contractict the assertion of Proposition 15.3.1. Thus, starting with $i = 1$ in Eq. (10), at least one of the matrices $A_{12}, A_{13}, \ldots, A_{1s}$ is not the zero-matrix and therefore at least one of the numbers

$$\varepsilon\mu_1\mu_j^{-1} = \exp[2\pi i(1 + n_1 - n_j)/k]$$

is equal to 1. This is possible only if $n_j = 1 + n_1 = 1$, $1 \le j \le s$. Since $n_1, n_2, \ldots, n_s$ are integers and $n_1 < n_2 < \cdots < n_s$, this implies $j = 2$. Hence $A_{13} = A_{14} = \cdots = A_{1s} = O$. Putting $i = 2$ in Eq. (10), we similarly derive that $A_{2j} = O$, $j = 1, 2, \ldots, s$ but $j \ne 3$. Continuing this process, the following form of A_0 is obtained:

$$\begin{bmatrix} O & A_{12} & O & \cdots & O \\ O & O & A_{23} & \ddots & \vdots \\ \vdots & & \ddots & \ddots & O \\ O & & & O & A_{s-1,s} \\ A_{s1} & A_{s2} & \cdots & A_{s,s-1} & O \end{bmatrix}.$$

The same argument applied for the case $i = s$ in Eq. (10) yields the existence of at least one j, $1 \le j \le s - 1$, such that

$$\exp[2\pi i(1 + n_s - n_j)/k] = 1, \qquad 1 \le j \le s - 1.$$

Since $0 \le n_j < n_s < k$, $1 \le j \le s - 1$, this equality holds only if $1 + n_s - n_j = k$. But $n_1 = 0$, $n_2 = 1, \ldots, n_{s-1} = s - 2$, and therefore $n_s = k - 1$ and $n_j = 0$, hence $j = 1$. Thus $A_{s2} = A_{s3} = \cdots = A_{s,s-1} = O$. It remains to observe that $s = k$. To this end, recall that when $i = s - 1$, the relation (10) holds for $A_{ij} \ne O$ only if $j = s$. Thus $1 + n_{s-1} - n_s = 0$, and since $n_{s-1} = s - 2$ and $n_s = k - 1$, we obtain $s = k$. The proof is complete. ■

Perron's original results of 1907 can now be stated as a corollary to Theorems 2 and 15.3.1.

Corollary 2. *A positive square matrix A has a real and positive eigenvalue r that has algebraic multiplicity 1 and exceeds the moduli of all other eigenvalues of A. To the eigenvalue r corresponds a positive right eigenvector.*

PROOF. The last part of Theorem 2 implies that if $A > O$ we must have $k = 1$, and hence r exceeds the moduli of all other eigenvalues of A. ■

15.5 Reducible Matrices

We now consider the possibility of removing the assumption that A is irreducible from the hypothesis of Theorem 15.3.1. Conclusions (a) and (b) of that theorem yield results for reducible matrices because such a matrix can be thought of as the limit of a sequence of positive (and hence irreducible) matrices. Our conclusions will then follow from the continuous dependence of the dominant eigenvalue (and its normalized right eigenvector) on the matrix elements.

In contrast to this, we know from our experience with perturbation theory in Chapter 11 that the structure of a matrix (as determined by its elementary divisors) does not depend continuously on the matrix elements. For this reason, conclusion (c) of Theorem 15.3.1 yields no useful results in the reducible case.

Theorem 1. *If the matrix $A \in \mathbb{R}^{n \times n}$ is nonnegative, then*

(*a*) *The matrix A has a real eigenvalue, r, equal to the spectral radius of A;*
(*b*) *There is a nonnegative (right) eigenvector associated with the eigenvalue r.*

PROOF. Define the matrix $D(t) \in \mathbb{R}^{n \times n}$ by its elements $d_{ij}(t)$ as follows:

$$d_{ij}(t) = \begin{cases} a_{ij} & \text{if} \quad a_{ij} > 0, \\ t & \text{if} \quad a_{ij} = 0. \end{cases}$$

Then $D(t) > 0$ for $t > 0$ and $D(0) = A$. Let $\rho(t)$ be the maximal real eigenvalue of $D(t)$ for $t > 0$. Then all the eigenvalues of $D(t)$ are continuous functions of t (Section 11.3) and, since $\rho(t)$ is equal to the spectral radius of $D(t)$ for $t > 0$, $\lim_{t \to 0+} \rho(t) = r$ is an eigenvalue of $D(0) = A$. Furthermore, this eigenvalue is equal to the spectral radius of A.

We also know (see Exercise 11.7.6) that there is a right eigenvector $\boldsymbol{x}(t)$ that can be associated with $\rho(t)$ and that depends continuously on t for $t \geq 0$. By Theorem 15.3.1, $\boldsymbol{x}(t) > \boldsymbol{0}$ for $t > 0$ and hence $\boldsymbol{x}(t) \geq \boldsymbol{0}$ at $t = 0$. ■

Exercise 1. What happens if we try to extend the results of Theorem 15.4.2 to reducible matrices by means of a continuity argument?

Exercise 2. If $A \in \mathbb{R}^{n \times n}$ is nonnegative and $\sigma_j = \sum_{k=1}^{n} \sigma_{jk}$, prove that

$$\min_j \sigma_j \leq r \leq \max_j \sigma_j.$$

Hint. Note Exercise 15.3.7 and apply a continuity argument.

Exercise 3. Show that for any reducible matrix $A \in \mathbb{R}^{n \times n}$ there exists a permutation matrix P such that

$$P^{\mathrm{T}}AP = \begin{bmatrix} B_{11} & B_{12} & \cdots & B_{1k} \\ 0 & B_{22} & \cdots & B_{2k} \\ \vdots & \ddots & \ddots & \vdots \\ 0 & \cdots & 0 & B_{kk} \end{bmatrix}, \tag{1}$$

where the square matrices B_{ii}, $i = 1, 2, \ldots, k$, are either irreducible or 1×1 zero-matrices. □

The form (1) of A is often called a *normal form* of a reducible matrix.

Exercise 4. Check that the matrix

$$A = \begin{bmatrix} 0 & 0 & 1 & 0 \\ 1 & 1 & 1 & 3 \\ 1 & 0 & 2 & 0 \\ 2 & 2 & 0 & 1 \end{bmatrix}$$

is reducible and show that its normal form is

$$B = \left[\begin{array}{cc:cc} 1 & 2 & 0 & 2 \\ 3 & 1 & 1 & 1 \\ \hdashline 0 & 0 & 2 & 1 \\ 0 & 0 & 1 & 0 \end{array}\right].$$

Exercise 5. If $A \in \mathbb{R}^{n \times n}$ and $a_{jk} > 0$ for $j \neq k$, prove that the eigenvalue ρ of A with largest real part is real and has multiplicity 1. Prove also that there is a *positive* (right) eigenvector of A associated with ρ. What can be said if we have only $a_{jk} \geq 0$ for $j \neq k$?

Hint. Consider $A + \alpha I$ for large α. □

15.6 Primitive and Imprimitive Matrices

The Perron–Frobenius theorems make it clear that the structure of a nonnegative irreducible matrix A depends on the number k of eigenvalues whose moduli are equal to the spectral radius of A. In particular, it is convenient to distinguish between the cases $k = 1$ and $k > 1$. Thus, an irreducible nonnegative matrix is said to be *primitive* or *imprimitive* according as $k = 1$ or $k > 1$. The number k is called the *index of imprimitivity*.

Exercise 1. Check that any positive matrix is primitive.

Exercise 2. Check that the index of imprimitivity of the matrix

$$A = \begin{bmatrix} 0 & 1 & 0 \\ 1 & 0 & 1 \\ 0 & 1 & 0 \end{bmatrix}$$

is equal to 2. □

The index of imprimitivity is easily found (see Exercise 2) if the characteristic polynomial is known. Suppose that

$$c(\lambda) = \lambda^n + a_1 \lambda^{n_1} + a_2 \lambda^{n_2} + \cdots + a_t \lambda^{n_t} \tag{1}$$

is the characteristic polynomial of $A \in \mathbb{R}^{n \times n}$, where $a_1, a_2, \ldots, a_t$ are nonzero and

$$n > n_1 > n_2 > \cdots > n_t \geq 0.$$

If λ_0 is any eigenvalue and $\omega_1, \ldots, \omega_k$ are the kth roots of unity, then we saw in Theorem 15.3.1 that $\omega_j \lambda_0$, $j = 1, 2, \ldots, k$, are all eigenvalues of A. Thus, $c(\lambda)$ is divisible by the product

$$(\lambda - \omega_1 \lambda_0)(\lambda - \omega_2 \lambda_0) \cdots (\lambda - \omega_k \lambda_0) = \lambda^k - \lambda_0^k,$$

without remainder. Hence, for some polynomial g and an integer s, we have

$$c(\lambda) = g(\lambda^k)\lambda^s.$$

Comparing this with Eq. (1), we see that k is a common factor of the differences $n - n_j$, $j = 1, 2, \ldots, t$. Suppose that l is the greatest common divisor of these differences; then using the reverse argument we see that the spectrum of A is invariant under rotations $2\pi/l$. If $k < l$, then $2\pi/l < 2\pi/k$ and the invariance under rotations through $2\pi/l$ is not consistent with the definition of k. Hence $k = l$ and we have established the following fact.

Proposition 1. *If $A \in \mathbb{R}^{n \times n}$ is irreducible and nonnegative, and $c(\lambda)$ written in the form (1) is the characteristic polynomial of A, then k, the index of imprimitivity of A, is the greatest common divisor of the differences*

$$n - n_1, \qquad n - n_2, \ldots, n - n_t.$$

For example, if

$$c(\lambda) = \lambda^{10} + a_1 \lambda^7 + a_2 \lambda, \qquad a_1, a_2 \neq 0,$$

then $k = 3$. However, if

$$c(\lambda) = \lambda^{10} + a_1 \lambda^7 + a_2 \lambda + a_3, \qquad a_1, a_2, a_3 \neq 0,$$

then $k = 1$.

The next result provides a rather different characterization of primitive matrices.

Theorem 1. *A square nonnegative matrix A is primitive if and only if there is a positive integer p such that $A^p > 0$.*

PROOF. First suppose that $A^p > 0$ and hence that A^p is irreducible. We have seen (Exercise 10.7.1) that A reducible would imply A^p reducible, hence $A^p > 0$ implies that A is irreducible. If A has index of imprimitivity $k > 1$, then since the eigenvalues of A^p are the pth powers of those of A, the matrix A^p also has index of imprimitivity $k > 1$. But this contradicts Perron's theorem applied to the positive matrix A^p. Hence $k = 1$ and A is primitive.

Conversely, we suppose A to be primitive; hence A has a dominant real eigenvalue r of multiplicity 1 (Theorems 15.3.1 and 15.4.2). Using Theorem 9.5.1 and writing $r = \lambda_1$, we have

$$A^p = r^p Z_{10} + \sum_{l=2}^{s} \sum_{j=0}^{m_l - 1} f_{l,j} Z_{lj},$$

where $f(\lambda) = \lambda^p$ and $f_{l,j}$ is the jth derivative of f at λ_l. We easily deduce from this equation that, since r is the dominant eigenvalue,

$$\lim_{p \to \infty} \frac{A^p}{r^p} = Z_{10}. \tag{2}$$

From Exercise 9.9.7 we also have

$$Z_{10} = \frac{C(r)}{m^{(1)}(r)},$$

and the minimal polynomial m is given by

$$m(\lambda) = (\lambda - r) \prod_{l=2}^{s} (\lambda - \lambda_l)^{m_l}.$$

Since r is the largest real zero of $m(\lambda)$ and $m(\lambda)$ is monic, it follows that $m^{(1)}(r) > 0$. Then since $C(r) > 0$ (Exercise 15.3.10), we have $Z_{10} > 0$. It now follows from Eq. (2) that $A^p > 0$ from some p onward. ■

Exercise 3. If A is primitive and p is a positive integer, prove that A^p is primitive.

Exercise 4. If $A \geq 0$ is irreducible and $\varepsilon > 0$, prove by examining eigenvalues that $\varepsilon I + A$ is primitive. Compare this with Proposition 15.3.1.

Exercise 5. Check that the first matrix in Exercise 15.1.6 fails to be primitive for any $n > 1$.

Exercise 6. Check that the matrices

$$\frac{1}{10}\begin{bmatrix} 9 & 0 & 1 & 1 \\ 1 & 7 & 0 & 1 \\ 0 & 2 & 8 & 0 \\ 0 & 1 & 1 & 8 \end{bmatrix} \qquad \text{and} \qquad \begin{bmatrix} 0 & \frac{1}{2} & \frac{2}{3} \\ \frac{3}{4} & 0 & \frac{1}{3} \\ \frac{1}{4} & \frac{1}{2} & 0 \end{bmatrix}$$

are primitive. □

The number p such that $A^p > 0$ for an $n \times n$ nonnegative primitive matrix A can be estimated as follows:

$$p \leq n^2 - 2n + 2.$$

Since $p = n^2 - 2n + 2$ for the matrix

$$A = \begin{bmatrix} 0 & 1 & 0 & \cdots & 0 \\ 0 & 0 & 1 & & \\ \vdots & & & \ddots & \\ 0 & & & & 1 \\ 1 & 1 & 0 & \cdots & 0 \end{bmatrix}, \qquad A \in \mathbb{R}^{n \times n},$$

the estimation (3) is exact. We omit the proof.

15.7 Stochastic Matrices

The real matrix $P = [p_{ij}]_{i,j=1}^n$ is said to be a *stochastic* matrix if $P \geq 0$ and all its *row sums* are equal to 1:

$$\sum_{j=1}^{n} p_{ij} = 1, \qquad i = 1, 2, \ldots, n. \tag{1}$$

Matrices of this kind arise in problems of probability theory, which we shall introduce in Section 15.8. In this section we derive some of the more important properties of stochastic matrices.

Theorem 1. *A nonnegative matrix P is stochastic if and only if it has the eigenvalue 1 with (right) eigenvector given by $\boldsymbol{u} = [1 \quad 1 \quad \cdots \quad 1]^T$. Furthermore, the spectral radius of a stochastic matrix is 1.*

PROOF. If P is stochastic, then the condition (1) may obviously be written $P\boldsymbol{u} = \boldsymbol{u}$. Hence 1 is an eigenvalue with (right) eigenvector $\boldsymbol{u}$.

Conversely, $P\boldsymbol{u} = \boldsymbol{u}$ implies Eq. (1) and hence that P is stochastic.

For the last part of the theorem we use Exercise 15.3.7 to show that the *dominant* real eigenvalue of P is 1. ■

In the next theorem we see that a wide class of nonnegative matrices can be reduced to stochastic matrices by a simple transformation.

Theorem 2. *Let A be an $n \times n$ nonnegative matrix with maximal real eigenvalue r. If there is a positive (right) eigenvector $\mathbf{x} = [x_1 \quad x_2 \quad \cdots \quad x_n]^{\mathrm{T}}$ associated with r and we write $X = \operatorname{diag}[x_1, \ldots, x_n]$, then*

$$A = rXPX^{-1},$$

where P is a stochastic matrix.

Note that some nonnegative matrices are excluded from Theorem 2 by the hypothesis that $\mathbf{x}$ is *positive* (see Theorem 15.5.1).

PROOF. Let $P = r^{-1}X^{-1}AX$. We have only to prove that P is stochastic. Since $A\mathbf{x} = r\mathbf{x}$ we have

$$\sum_{j=1}^{n} a_{ij}x_j = rx_i, \qquad i = 1, 2, \ldots, n.$$

By definition, $p_{ij} = r^{-1}x_i^{-1}a_{ij}x_j$ and hence

$$\sum_{j=1}^{n} p_{ij} = r^{-1}x_i^{-1}\sum_{j=1}^{n} a_{ij}x_j = 1.$$

Thus, P is stochastic and the theorem is proved. ■

Finally, we consider the sequence of nonnegative matrices $P, P^2, P^3, \ldots,$ where P is stochastic. We are interested in conditions under which

$$P^{\infty} = \lim_{p \to \infty} P^p$$

exists. We already know that $P^{\infty} \neq 0$, for P has an eigenvalue equal to 1 and $P^r \to 0$ as $r \to \infty$ if and only if all the eigenvalues of P are less than 1 (Exercise 11.1.8).

Theorem 3. *If P is an irreducible stochastic matrix, then the matrix $P^{\infty} = \lim_{p \to \infty} P^p$ exists if and only if P is primitive.*

PROOF. We know that the maximal real eigenvalue λ_1 of P is equal to 1. Recall that, since P is irreducible, the eigenvalue 1 is unrepeated (Theorem 15.3.1(c)). So let $\lambda_2, \ldots, \lambda_k$ be the other eigenvalues of P with modulus 1 and let $\lambda_{k+1}, \ldots, \lambda_s$ be the distinct eigenvalues of P with modulus less than 1. We write $f(\lambda) = \lambda^p$ and $f_{l,j}$ for the jth derivative of f at λ_l, and using Theorem 9.5.1 again we have

$$P^p = Z_{10} + \sum_{l=2}^{k}\sum_{j=0}^{m_l-1} f_{l,j}Z_{lj} + \sum_{l=k+1}^{s}\sum_{j=0}^{m_l-1} f_{l,j}Z_{lj}. \tag{2}$$

For p sufficiently large we have

$$f_{l,j} = \frac{p!}{(p-j+1)!}\lambda_l^{p-j+1}, \tag{3}$$

and since $|\lambda_l| < 1$ for $l = k+1, \ldots, s$, it follows that the last term in Eq. (2) approaches the zero-matrix as $p \to \infty$. If we now assume that P is primitive, then the second term on the right of Eq. (2) does not appear. Hence it is clear that P^∞ exists and, furthermore, $P^\infty = Z_{10}$.

Conversely, suppose it is known that P^∞ exists; then we deduce from Eq. (2) that

$$Q = \lim_{p\to\infty} \sum_{l=2}^{k} \sum_{j=0}^{m_l-1} f_{l,j} Z_{lj}$$

exists. The $m_2 + m_3 + \cdots + m_k$ matrices Z_{lj} appearing in this summation are linearly independent members of $\mathbb{C}^{n\times n}$ (Theorem 9.5.1) and generate a subspace $\mathscr{S}$, say, of $\mathbb{C}^{n\times n}$. Since Q must also belong to $\mathscr{S}$, there are numbers a_{lj} such that

$$Q = \sum_{l=2}^{k} \sum_{j=0}^{m_l-1} a_{lj} Z_{lj}.$$

It now follows that, for each l and j,

$$a_{lj} = \lim_{p\to\infty} f_{l,j}.$$

But for $l = 2, 3, \ldots,$ we have $|\lambda_l| = 1$ $(\lambda_l \neq 1)$ and it follows from Eq. (3) that the limit on the right does not exist. Thus, we have a contradiction that can be resolved only if the existence of P^∞ implies $k = 1$. ■

It should be noted that the hypothesis that P be irreducible is not necessary. It can be proved that the eigenvalue 1 of P has only linear elementary divisors even when P is reducible. Thus, Eq. (2) is still valid and the above argument goes through. The proof of the more general result requires a deeper analysis of reducible nonnegative matrices than we have chosen to present.

Note also that, using Exercise 9.9.7, we may write

$$P^\infty = \frac{C(1)}{m^{(1)}(1)},$$

where $C(\lambda)$ and $m(\lambda)$ are the reduced adjoint and the minimal polynomial of P, respectively. If $B(\lambda) = \operatorname{adj}(I\lambda - P)$ and $c(\lambda)$ is the characteristic polynomial of P, then the polynomial identity

$$\frac{B(\lambda)}{c(\lambda)} = \frac{C(\lambda)}{m(\lambda)}$$

implies that we may also write

$$P^{\infty} = \frac{B(1)}{c^{(1)}(1)}.$$

Yet another alternative arises if we identify Z_{10} with the constituent matrix G_1 of Exercise 4.14.3. Thus, in the case in which P is irreducible, there are positive right and left eigenvectors $\boldsymbol{x}$, $\boldsymbol{y}$, respectively, of P associated with the eigenvalue 1 such that

$$\boldsymbol{y}^{\mathrm{T}}\boldsymbol{x} = 1 \qquad \text{and} \qquad P^{\infty} = \boldsymbol{x}\boldsymbol{y}^{\mathrm{T}}. \tag{4}$$

Exercise 1. A nonnegative matrix $P \in \mathbb{R}^{n \times n}$ is called *doubly stochastic* if both P and P^{T} are stochastic matrices. Check that the matrices

$$P_1 = \begin{bmatrix} \frac{2}{3} & \frac{1}{3} & 0 \\ 0 & \frac{1}{2} & \frac{1}{2} \\ \frac{1}{3} & 0 & \frac{2}{3} \end{bmatrix} \qquad \text{and} \qquad P_2 = \begin{bmatrix} \frac{1}{3} & \frac{1}{3} & \frac{1}{3} \\ \frac{2}{3} & \frac{1}{3} & 0 \\ 0 & \frac{1}{3} & \frac{2}{3} \end{bmatrix}$$

are stochastic and doubly stochastic matrices, respectively, and that P_1^{∞} and P_2^{∞} exist. □

15.8 Markov Chains

Nonnegative matrices arise in a variety of physical problems. We shall briefly present a problem of probability theory that utilizes some of the theorems developed earlier in the chapter.

For simplicity, we consider a physical system whose state, or condition, or position, can be described in n mutually exclusive and exhaustive ways. We let $s_1, s_2, \ldots, s_n$ denote these states. We suppose it possible to examine the system at times $t_0 < t_1 < t_2 < \cdots$ and to observe the state of the system at each of these times. We suppose further that there exist absolute probabilities $p_i(t_r)$ of finding the system in state s_i at time t_r for $i = 1, 2, \ldots, n$ and $r = 0, 1, 2, \ldots$. Since the system must be in one of the n states at time t_r, we have

$$\sum_{i=1}^{n} p_i(t_r) = 1, \qquad r = 0, 1, 2, \ldots. \tag{1}$$

It is assumed that we can determine the probability that state s_i is attained at time t_r after having been in state s_j at time t_{r-1}, for all i and j. Thus, we assume a knowledge of the *conditional probability*

$$q_{ij}(t_r \mid t_{r-1}), \qquad i, j = 1, 2, \ldots, n,$$

of being in state s_i at time t_r, given that the system is in state s_j at time t_{r-1}. The laws of probability then imply

$$p_i(t_r) = \sum_{j=1}^{n} q_{ij}(t_r | t_{r-1}) p_j(t_{r-1}),$$

or, in matrix–vector form,

$$\boldsymbol{p}_r = Q(t_r, t_{r-1})\boldsymbol{p}_{r-1}, \qquad r = 1, 2, \ldots, \tag{2}$$

where we define

$$\boldsymbol{p}_s = [p_1(t_s) \quad p_2(t_s) \quad \cdots \quad p_n(t_s)]^{\mathrm{T}}, \qquad s = 0, 1, 2, \ldots,$$

and $Q(t_r, t_{r-1})$ to be the matrix in $\mathbb{R}^{n \times n}$ with elements $q_{ij}(t_r | t_{r-1})$, called the *transition* matrix.

Since the system *must* go to one of the n states at time t_r after being in state j at time t_{r-1}, it follows that

$$\sum_{i=1}^{n} q_{ij}(t_r | t_{r-1}) = 1. \tag{3}$$

Finally, it is assumed that the initial state of the system is known. Thus, if the initial state is s_i then we have $\boldsymbol{p}_0 = \boldsymbol{e}_i$, the vector with 1 in the ith place and zeros elsewhere. We deduce from Eq. (2) that at time t_r we have

$$\boldsymbol{p}_r = Q(t_r, t_{r-1})Q(t_{r-1}, t_{r-2}) \cdots Q(t_1, t_0)\boldsymbol{p}_0. \tag{4}$$

The process we have just described is known as a *Markov chain.* If the conditional probabilities $q_{ij}(t_r | t_{r-1})$ are independent of t or are the same at each instant $t_0, t_1, t_2, \ldots$, then the process is described as a *homogeneous* Markov chain. We shall confine our attention to this relatively simple problem. Observe that we may now write

$$Q(t_r, t_{r-1}) = Q, \qquad r = 1, 2, \ldots.$$

Equation (4) becomes

$$\boldsymbol{p}_r = Q^r \boldsymbol{p}_0,$$

and Eq. (3) implies that the transition matrix Q associated with a homogenous Markov chain is such that Q^{T} is stochastic. For convenience, we define $P = Q^{\mathrm{T}}$.

In many problems involving Markov chains, we are interested in the limiting behavior of the sequence $\boldsymbol{p}_0, \boldsymbol{p}_1, \boldsymbol{p}_2, \ldots$. This leads us to examine the existence of $\lim_{r\to\infty} Q^r$, or $\lim_{r\to\infty} P^r$. If P (or Q) is irreducible, then according to Theorem 15.7.3, the limits exist if and only if P is primitive. Since $P = Q^{\mathrm{T}}$ it follows that Q has an eigenvalue 1, and the associated right and left eigenvectors $\boldsymbol{x}$, $\boldsymbol{y}$ of Eq. (15.7.4) are left and right eigenvectors of Q, respectively. Thus, there are positive right and left eigenvectors $\boldsymbol{\xi}$ $(=\boldsymbol{y})$ and $\boldsymbol{\eta}$ $(=\boldsymbol{x})$ associated with the eigenvalue 1 of Q for which $\boldsymbol{\eta}^{\mathrm{T}}\boldsymbol{\xi} = 1$ and

$$Q^{\infty} = \boldsymbol{\xi}\boldsymbol{\eta}^{\mathrm{T}}.$$

We then have

$$\boldsymbol{p}_{\infty} = \boldsymbol{\xi}(\boldsymbol{\eta}^{\mathrm{T}}\boldsymbol{p}_0) = (\boldsymbol{\eta}^{\mathrm{T}}\boldsymbol{p}_0)\boldsymbol{\xi},$$

since $\boldsymbol{\eta}^{\mathrm{T}}\boldsymbol{p}_0$ is a scalar. So, whatever the starting vector $\boldsymbol{p}_0$, the limiting vector is proportional to $\boldsymbol{\xi}$. However, $\boldsymbol{p}_r$ satisfies the condition (1) for every r and so $\boldsymbol{p}_0$ satisfies the same condition. Thus, if we define the positive vector $\boldsymbol{\xi} = [\xi_1 \quad \xi_2 \quad \cdots \quad \xi_n]^{\mathrm{T}}$ so that $\sum_{i=1}^{n} \xi_i = 1$, then $\boldsymbol{p}_{\infty} = \boldsymbol{\xi}$ for every possible choice of $\boldsymbol{p}_0$.

The fact that $\boldsymbol{\xi} > \boldsymbol{0}$ means that it is possible to end up in any of the n states $s_1, s_2, \ldots, s_n$. We have proved the following result.

Theorem 1. *If Q is the transition matrix of a homogeneous Markov chain and Q is irreducible, then the limiting absolute probabilities $p_i(t_{\infty})$ exist if and only if Q is primitive. Furthermore, when Q is primitive, $p_i(t_{\infty}) > 0$ for $i = 1, 2, \ldots, n$ and these probabilities are the components of the (properly normalized) right eigenvector of Q associated with the eigenvalue* 1.

APPENDIX 1

A Survey of Scalar Polynomials

The expression

$$p(\lambda) = p_0 + p_1\lambda + p_2\lambda^2 + \cdots + p_l\lambda^l, \tag{1}$$

with coefficients $p_0, p_1, \ldots, p_l$ in a field $\mathscr{F}$ and $p_l \neq 0$, is referred to as a *polynomial of degree* l *over* $\mathscr{F}$. Nonzero scalars from $\mathscr{F}$ are considered polynomials of zero degree, while $p(\lambda) \equiv 0$ is written for polynomials with all zero coefficients.

In this appendix we give a brief survey of some of the fundamental properties of such polynomials. An accessible and complete account can be found in Chapter 6 of "The Theory of Matrices" by Sam Perlis (Addison Wesley, 1958), for example.

Theorem 1. *For any two polynomials* $p(\lambda)$ *and* $q(\lambda)$ *over* $\mathscr{F}$ *(with* $q(\lambda) \not\equiv 0$*), there exist polynomials* $d(\lambda)$ *and* $r(\lambda)$ *over* $\mathscr{F}$ *such that*

$$p(\lambda) = q(\lambda)d(\lambda) + r(\lambda), \tag{2}$$

where $\deg r(\lambda) < \deg q(\lambda)$ *or* $r(\lambda) \equiv 0$. *The polynomials* $d(\lambda)$ *and* $r(\lambda)$ *satisfying Eq.* (2) *are uniquely defined.*

The polynomial $d(\lambda)$ in Eq. (2) is called the *quotient* on division of $p(\lambda)$ by $q(\lambda)$ and $r(\lambda)$ is the *remainder* of this division. If the remainder $r(\lambda)$ in Eq. (2) is the zero polynomial, then $d(\lambda)$ (or $q(\lambda)$) is referred to as a *divisor* of $p(\lambda)$. In this event, we also say that $d(\lambda)$ (or $q(\lambda)$) *divides* $p(\lambda)$, or that $p(\lambda)$ *is divisible by* $d(\lambda)$. A divisor of degree 1 is called *linear.*

Exercise 1. Find the quotient and the remainder on division of $p(\lambda) = \lambda^3 + \lambda + 2$ by $q(\lambda) = \lambda^2 + \lambda + 1$.

Answer. $d(\lambda) = \lambda - 1$, $r(\lambda) = \lambda + 3$.

Exercise 2. Show that $p(\lambda)$ divides $q(\lambda)$ and simultaneously $q(\lambda)$ divides $p(\lambda)$ if and only if $p(\lambda) = cq(\lambda)$ for some $c \in \mathscr{F}$. ☐

Given polynomials $p_1(\lambda), p_2(\lambda), \ldots, p_k(\lambda), q(\lambda)$ over $\mathscr{F}$, the polynomial $q(\lambda)$ is called a *common divisor* of $p_i(\lambda)$ $(i = 1, 2, \ldots, k)$ if it divides each of the polynomials $p_i(\lambda)$. A *greatest common divisor* $d(\lambda)$ of $p_i(\lambda)$ $(i = 1, 2, \ldots, k)$, abbreviated g.c.d.$\{p_i(\lambda)\}$, is defined to be a common divisor of these polynomials that is divisible by any other common divisor. To avoid unnecessary complications, we shall assume in the remainder of this section (if not indicated otherwise) that the *leading coefficient*, that is, the coefficient of the highest power of the polynomial being considered, is equal to 1. Such polynomials are said to be *monic*.

Theorem 2. *For any set of nonzero polynomials over* $\mathscr{F}$, *there exists a unique greatest common divisor.*

This result follows immediately from the definition of a greatest common divisor. If the greatest common divisor of a set of polynomials is 1, then they are said to be *relatively prime*.

Exercise 3. Check that the polynomials $p(\lambda) = \lambda^2 + 1$ and $q(\lambda) = \lambda^3 + 1$ are relatively prime.

Exercise 4. Construct a set of three relatively prime polynomials, no two of which are relatively prime. ☐

Theorem 3. *If* $d(\lambda)$ *is the greatest common divisor of* $p_1(\lambda)$ *and* $p_2(\lambda)$, *then there exist polynomials* $c_1(\lambda)$ *and* $c_2(\lambda)$ *such that*

$$c_1(\lambda)p_1(\lambda) + c_2(\lambda)p_2(\lambda) = d(\lambda).$$

In particular, if $p_1(\lambda)$ *and* $p_2(\lambda)$ *are relatively prime, then*

$$c_1(\lambda)p_1(\lambda) + c_2(\lambda)p_2(\lambda) = 1 \tag{3}$$

for some polynomials $c_1(\lambda)$ *and* $c_2(\lambda)$.

This important theorem requires more proof. The argument is omitted in the interest of brevity.

Let $\lambda_1 \in \mathscr{F}$. If $p(\lambda_1) = \sum_{i=0}^{l} p_i \lambda_1^i = 0$, then λ_1 is a *zero* of $p(\lambda)$ or a *root* of the equation $p(\lambda) = 0$. The fundamental theorem of algebra concerns the zeros of a polynomial over $\mathbb{C}$.

Theorem 4. *Any polynomial over $\mathbb{C}$ of nonzero degree has at least one zero in $\mathbb{C}$.*

Dividing $p(\lambda)$ by $(\lambda - \lambda_1)$ with the use of Theorem 1, the next exercise is easily established.

Exercise 5. Show that λ_1 is a zero of $p(\lambda)$ if and only if $\lambda - \lambda_1$ is a (linear) divisor of $p(\lambda)$. In other words, for some polynomial $p_1(\lambda)$ over $\mathcal{F}$,

$$p(\lambda) = p_1(\lambda)(\lambda - \lambda_1). \quad \square \tag{4}$$

Let λ_1 be a zero of $p(\lambda)$. It may happen that, along with $\lambda - \lambda_1$, higher powers, $(\lambda - \lambda_1)^2, (\lambda - \lambda_1)^3, \ldots$, are also divisors of $p(\lambda)$. The scalar $\lambda_1 \in \mathcal{F}$ is referred to as a zero of $p(\lambda)$ of *multiplicity* k if $(\lambda - \lambda_1)^k$ divides $p(\lambda)$ but $(\lambda - \lambda_1)^{k+1}$ does not. Zeros of multiplicity 1 are said to be *simple*. Other zeros are called *multiple* zeros of the polynomial.

Exercise 6. Check that λ_1 is a zero of multiplicity k of $p(\lambda)$ if and only if $p(\lambda_1) = p'(\lambda_1) = \cdots = p^{(k-1)}(\lambda_1) = 0$ but $p^{(k)}(\lambda_1) \neq 0$. $\square$

Consider again the equality (4) and suppose $p_1(\lambda)$ has nonzero degree. By virtue of Theorem 4 the polynomial $p_1(\lambda)$ has a zero λ_2. Using Exercise 5, we obtain $p(\lambda) = p_2(\lambda)(\lambda - \lambda_2)(\lambda - \lambda_1)$. Proceeding to factorize $p_2(\lambda)$ and continuing this process, the following theorem is obtained.

Theorem 5. *Any polynomial over $\mathbb{C}$ of degree l $(l \geq 1)$ has exactly l zeros $\lambda_1, \lambda_2, \ldots, \lambda_l$ in $\mathbb{C}$ (which may include repetitions). Furthermore, there is a representation*

$$p(\lambda) = p_l(\lambda - \lambda_l)(\lambda - \lambda_{l-1}) \cdots (\lambda - \lambda_1), \tag{5}$$

where $p_l = 1$ in the case of a monic polynomial $p(\lambda)$.

Collecting the equal zeros in Eq. (5) and renumbering them, we obtain

$$p(\lambda) = p_l(\lambda - \lambda_1)^{k_1}(\lambda - \lambda_2)^{k_2} \cdots (\lambda - \lambda_s)^{k_s}, \tag{6}$$

where $\lambda_i \neq \lambda_j$ $(i \neq j)$ and $k_1 + k_2 + \cdots + k_s = l$.

Given $p_1(\lambda), p_2(\lambda), \ldots, p_k(\lambda)$, the polynomial $p(\lambda)$ is said to be a *common multiple* of the given polynomials if each of them divides $p(\lambda)$:

$$p(\lambda) = q_i(\lambda)p_i(\lambda), \qquad 1 \leq i \leq k, \tag{7}$$

for some polynomials $q_i(\lambda)$ over the same field. A common multiple that is a divisor of any other common multiple is called a *least common multiple* of the polynomials.

Theorem 6. *For any set of monic polynomials, there exists a unique monic least common multiple.*

***Exercise* 7.** Show that the product $\prod_{i=1}^{k} p_i(\lambda)$ is a least common multiple of the polynomials $p_1(\lambda), p_2(\lambda), \ldots, p_k(\lambda)$ if and only if any two of the given polynomials are relatively prime. □

Exercise 7 can now be used to complete the next exercise.

***Exercise* 8.** Check that if $p(\lambda)$ is the least common multiple of $p_1(\lambda)$, $p_2(\lambda), \ldots, p_k(\lambda)$, then the polynomials $q_i(\lambda)$ $(i = 1, 2, \ldots, k)$ in Eq. (7) are relatively prime. □

Note that the condition in Exercise 7 is stronger than requiring that the set of polynomials be relatively prime.

***Exercise* 9.** Prove that if $\tilde{p}(\lambda)$ is the least common multiple of $p_1(\lambda)$, $p_2(\lambda), \ldots, p_{k-1}(\lambda)$, then the least common multiple of $p_1(\lambda), p_2(\lambda), \ldots, p_k(\lambda)$ coincides with the least common multiple of $\tilde{p}(\lambda)$ and $p_k(\lambda)$. □

A field $\mathscr{F}$ that contains the zeros of every polynomial over $\mathscr{F}$ is said to be *algebraically closed.* Thus Theorem 4 asserts, in particular, that the field $\mathbb{C}$ of complex numbers is algebraically closed. Note that the polynomial $p(\lambda) = \lambda^2 + 1$ viewed as a polynomial over the field $\mathbb{R}$ of real numbers shows that $\mathbb{R}$ fails to be algebraically closed.

If $p(\lambda)$ is a polynomial over $\mathscr{F}$ of degree l, and 1 is the only monic polynomial over $\mathscr{F}$ of degree less than l that divides $p(\lambda)$, then $p(\lambda)$ is said to be *irreducible* over $\mathscr{F}$. Otherwise, $p(\lambda)$ is *reducible* over $\mathscr{F}$. Obviously, in view of Exercise 5, every polynomial of degree greater than 1 over an algebraically closed field is reducible. In fact, if $p(\lambda)$ is a monic polynomial over an algebraically closed field $\mathscr{F}$ and $p(\lambda)$ has degree l, then $p(\lambda)$ has a factorization $p(\lambda) = \prod_{j=1}^{l} (\lambda - \lambda_j)$ into linear factors, where $\lambda_1, \lambda_2, \ldots, \lambda_l \in \mathscr{F}$ and are not necessarily distinct. If the field fails to be algebraically closed, then polynomials of degree higher than 1 may be irreducible.

***Exercise* 10.** Check that the polynomial $p(\lambda) = \lambda^2 + 1$ is irreducible over $\mathbb{R}$ and over the field of rational numbers but is reducible over $\mathbb{C}$. □

Note that all real polynomials of degree greater than 2 are reducible over $\mathbb{R}$, while for any positive integer l there exists a polynomial of degree l with rational coefficients that is irreducible over the field of rational numbers.

APPENDIX 2

Some Theorems and Notions from Analysis

In this appendix we discuss without proof some important theorems of analysis that are used several times in the main body of the book. We present the results in a setting sufficiently general for our applications, although they can be proved just as readily in a more general context. We first need some definitions.

Recall that in $\mathbb{C}^n$ any vector norm may be used as a measure of magnitude of distance (Section 10.2). Given a vector norm $\| \ \|$ on $\mathbb{C}^n$ and a real number $\varepsilon > 0$, we define a *neighbourhood* $\mathcal{N}_\varepsilon(\boldsymbol{x}_0)$ of $\boldsymbol{x}_0 \in \mathbb{C}^n$ to be the set of vectors (also called *points* in this context) $\boldsymbol{x}$ in $\mathbb{C}^n$ for which $\|\boldsymbol{x} - \boldsymbol{x}_0\| < \varepsilon$ (that is, it is the interior of a *ball* with centre $\boldsymbol{x}_0$ and radius ε; see Section 10.2). If $\mathcal{S}$ is a set of points of $\mathbb{C}^n$, then $\boldsymbol{x} \in \mathbb{C}^n$ is a *limit point* of $\mathcal{S}$ if every neighbourhood of $\boldsymbol{x}$ contains some point $\boldsymbol{y} \in \mathcal{S}$ with $\boldsymbol{y} \neq \boldsymbol{x}$. The set $\mathcal{S}$ is said to be *open* if it is a union of neighbourhoods, and *closed* if every limit point of $\mathcal{S}$ is in $\mathcal{S}$. A set $\mathcal{S}$ is *bounded* if there is a real number $K \geq 0$ and an $\boldsymbol{x} \in \mathbb{C}^n$ such that $\|\boldsymbol{y} - \boldsymbol{x}\| < K$ for all $\boldsymbol{y} \in \mathcal{S}$.

Clearly, these definitions can be applied to sets in $\mathbb{C}$ or in $\mathbb{R}$. We have only to replace the norm by the absolute value function. We can now state the first result.

Theorem 1. *Let X be a closed and bounded set of real numbers and let M and m be the least upper bound and greatest lower bound, respectively, of the numbers x in X. Then $M, m \in X$.*

The proof is a straightforward deduction from the definitions.

Now consider a function f with values in $\mathbb{C}^m$ and defined on an open set $\mathscr{S} \subseteq \mathbb{C}^n$. We say that f is *continuous* at a point $\boldsymbol{x} \in \mathscr{S}$ if, given $\varepsilon > 0$, there is a $\delta > 0$ such that $\boldsymbol{y} \in \mathscr{N}_\delta(\boldsymbol{x})$ implies

$$\|f(\boldsymbol{y}) - f(\boldsymbol{x})\| < \varepsilon.$$

We say that f is continuous on $\mathscr{S}$ if f is continuous at every point of $\mathscr{S}$. If $\mathscr{S}$ is closed, then f is continuous on $\mathscr{S}$ if it is continuous on some open set containing $\mathscr{S}$. The *range* $f(\mathscr{S})$ of the function f is a subset of $\mathbb{C}^m$ defined by saying that $\boldsymbol{a} \in f(\mathscr{S})$ if and only if there is a vector $\boldsymbol{b} \in \mathbb{C}^n$ such that $\boldsymbol{a} = f(\boldsymbol{b})$.

Theorem 2. *Let f be defined as above. If $\mathscr{S}$ is a closed and bounded set in $\mathbb{C}^n$ and f is continuous on $\mathscr{S}$, then $f(\mathscr{S})$ is a closed and bounded subset of $\mathbb{C}^m$.*

In Theorem 2 and the next theorem we may replace $\mathbb{C}^n$ by $\mathbb{R}^n$ or $\mathbb{C}^m$ by $\mathbb{R}^m$, or both. The next theorem is obtained by combining Theorems 1 and 2.

Theorem 3. *Let f be a continuous real-valued function defined on a closed and bounded subset $\mathscr{S}$ of $\mathbb{C}^n$. If*

$$M = \sup_{\boldsymbol{x} \in \mathscr{S}} f(\boldsymbol{x}), \qquad m = \inf_{\boldsymbol{x} \in \mathscr{S}} f(\boldsymbol{x}),$$

then there are vectors $\boldsymbol{y}, \boldsymbol{z} \in \mathscr{S}$ such that

$$M = f(\boldsymbol{y}), \qquad m = f(\boldsymbol{z}).$$

In this statement $\sup_{\boldsymbol{x} \in \mathscr{S}} f(\boldsymbol{x})$ is read, "the supremum, or least upper bound, of the numbers $f(\boldsymbol{x})$, where $\boldsymbol{x} \in \mathscr{S}$." Similarly, $\inf_{\boldsymbol{x} \in \mathscr{S}} f(\boldsymbol{x})$ denotes an infimum or greatest lower bound.

Theorem 2 tells us that, under the hypotheses of Theorem 3, $f(\mathscr{S})$ is a closed and bounded set of real numbers and the result then follows from Theorem 1. It may be said that, under the hypotheses of Theorem 3, f has a maximum and a minimum on $\mathscr{S}$ and we may write

$$\sup_{\boldsymbol{x} \in \mathscr{S}} f(\boldsymbol{x}) = \max_{\boldsymbol{x} \in \mathscr{S}} f(\boldsymbol{x}), \qquad \inf_{\boldsymbol{x} \in \mathscr{S}} f(\boldsymbol{x}) = \min_{\boldsymbol{x} \in \mathscr{S}} f(\boldsymbol{x}).$$

We now show in the following example that the three conditions of Theorem 3—f continuous on $\mathscr{S}$, $\mathscr{S}$ bounded, and $\mathscr{S}$ closed—are all necessary.

Exercise 1. (a) Let $f(x) = x$, and let $\mathscr{S}$ be the open interval $(0, 1)$. Here $\mathscr{S}$ is bounded but not closed and f is continuous on $\mathscr{S}$. Clearly $M = 1$, $m = 0$, and f does not attain these values on $\mathscr{S}$.

(b) With $f(x) = 1/x$ and $\mathscr{S}$ the set of all real numbers not less than 1, we have f continuous on $\mathscr{S}$ and $\mathscr{S}$ closed but unbounded. In this case $f(\mathscr{S})$ is the half-open interval $(0, 1]$ and $\inf_{x \in \mathscr{S}} f(x) = 0$ is not attained.

(c) With

$$f(x) = \begin{cases} x, & 0 \le x < 1, \\ 0, & x = 1, \end{cases}$$

we have $\mathscr{S} = [0, 1]$, which is closed and bounded, but f is not continuous on $\mathscr{S}$. Now $f(\mathscr{S}) = [0, 1]$ and $\sup_{x \in \mathscr{S}} f(x) = 1$ is not attained. □

A set $\mathscr{S}$ in the complex plane $\mathbb{C}$ is said to be *convex* if for any $\alpha_1, \alpha_2 \in \mathscr{S}$ and any $\theta \in [0, 1]$,

$$(1 - \theta)\alpha_1 + \theta\alpha_2 \in \mathscr{S}.$$

The convex set that contains the points $\alpha_1, \alpha_2, \ldots, \alpha_k$ and is included in any other convex set containing them is called the *convex hull* of $\alpha_1, \alpha_2, \ldots, \alpha_k$. In other words, the convex hull is the minimal convex set containing the given points.

Exercise 2. Prove that the convex hull of $\alpha_1, \alpha_2, \ldots, \alpha_k$ in $\mathbb{C}$ is the set of all linear combinations of the form

$$\theta_1\alpha_1 + \theta_2\alpha_2 + \cdots + \theta_k\alpha_k, \qquad \theta_i \ge 0 \quad \text{and} \quad \sum_{i=1}^{k} \theta_i = 1. \tag{1}$$

(Such sums are called *convex linear combinations.*)

Hint. First show that the set of elements given by (1) is convex, and then check its minimality (in the previous sense) by use of an induction argument. □

The notion of convexity can be extended to linear spaces in a natural way. Thus, let $\mathscr{X}$ be a linear space over $\mathscr{F}$ (where $\mathscr{F}$ is $\mathbb{R}$ or $\mathbb{C}$). The *line segment* joining members $\boldsymbol{x}$ and $\boldsymbol{y}$ of $\mathscr{X}$ is the set of elements of the form $(1 - \theta)\boldsymbol{x} + \theta\boldsymbol{y}$ where $\theta \in [0, 1]$. Then a subset $\mathscr{S}$ of $\mathscr{X}$ is said to be *convex* if, for each pair $\boldsymbol{x}$ and $\boldsymbol{y}$ in $\mathscr{S}$, the line segment joining $\boldsymbol{x}$ and $\boldsymbol{y}$ is also in $\mathscr{S}$. For example, it is easily seen that any subspace of $\mathscr{X}$ is convex.

Exercise 3. Show that any ball in $\mathbb{C}^n$ is a convex set.

Exercise 4. Let $A \in \mathbb{C}^{m \times n}$ and $\boldsymbol{b} \in \mathbb{C}^m$. Show that the solution set of $A\boldsymbol{x} = \boldsymbol{b}$ is a convex set (when it is not empty). □

APPENDIX 3

Suggestions for Further Reading

In this appendix, references to several books and papers are collected that cast more light on the subject matter of the book. They may extend the (finite-dimensional) theory further, or go more deeply into it, or contain interesting applications. Some of them have also provided sources of information and stimulation for the authors.

The list is prepared alphabetically by author, and the numbers in square brackets after each reference indicate the chapters of this book to which the reference is mainly related.

Barnett, S. *Polynomials and Linear Control Systems.* New York: Marcel Dekker, 1983. [**13**]

Barnett, S., and Storey, C. *Matrix Methods in Stability Theory.* London: Nelson, 1970. [**13**]

Baumgärtel, H. *Endlich-dimensionale Analytishe Störungstheory.* Berlin: Academie-Verlag, 1972. [**11**] To appear in English translation by Birkhäuser Verlag.

Bellman, R. *Introduction to Matrix Analysis.* New York: McGraw-Hill, 1960. [**1–15**]

Ben-Israel, A., and Greville, T. N. E. *Generalized Inverses: Theory and Applications.* New York: Wiley, 1974. [**12**]

Berman, A., and Plemmons, R. J. *Nonnegative Matrices in the Mathematical Sciences.* New York: Academic Press, 1979.[**15**]

Boullion, T., and Odell, P. *Generalized Inverse Matrices.* New York: Wiley (Interscience), 1971. [**12**]

Campbell, S. L., and Meyer, C. D. Jr., *Generalized Inverses of Linear Transformations.* London: Pitman, 1979. [**12**]

Daleckii, J. L., and Krein, M. G. *Stability of Solutions of Differential Equations in Banach Space.* Providence, R. Is.: Trans. Amer. Math. Soc. Monographs, vol. 43, 1974. [**12, 14**]

Dunford, N., and Schwartz, J. T. *Linear Operators, Part I: General Theory.* New York: Wiley (Interscience), **5, 9, 10**]

Gantmacher, F. R. *The Theory of Matrices*, vols. 1 and 2. New York: Chelsea, 1959. [**1–15**]
Glazman, I. M., and Liubich, J. I. *Finite-Dimensional Linear Analysis*. Cambridge, Mass.: M.I.T. Press, 1974. [**3–10**]
Gohberg, I., Lancaster, P., and Rodman, L. *Matrix Polynomials*. New York: Academic Press, 1982. [**14**]
Golub, G. H., and Van Loan, C. F. *Matrix Computations*. Baltimore, Md.: Johns Hopkins Univ. Press, 1983. [**3**, **4**, **10**].
Halmos, P. R. *Finite-Dimensional Vector Spaces*. New York: Van Nostrand, 1958. [**3–6**]
Hamburger, M. L., and Grimshaw, M. E. *Linear Transformations in n-Dimensional Vector Spaces*. London: Cambridge Univ., 1956. [**3–6**]
Heinig, G., and Rost, K. *Algebraic Methods for Toeplitz-like Matrices and Operators*. Berlin: Akademie Verlag, 1984. [**13**].
Householder, A. S. *The Theory of Matrices in Numerical Analysis*. Boston: Ginn (Blaisdell), 1964. [**10**, **15**]
Kato, T. *Perturbation Theory for Linear Operators*. Berlin: Springer-Verlag, 1966 (2nd ed., 1976). [**11**]
Krein, M. G., and Naimark, M. A. "The Method of Symmetric and Hermitian Forms in the Theory of the Separation of the Roots of Algebraic Equations." *Lin. and Multilin. Alg.* 10 (1981): 265–308. [**13**]
Lancaster, P. *Lambda Matrices and Vibrating Systems*. Oxford: Pergamon Press, 1966. [**14**]
Lerer, L., and Tismenetsky, M. "The Bezoutian and the Eigenvalue-Separation Problem for Matrix Polynomials." *Integral Eq. and Operator Theory* 5 (1982): 386–445. [**13**]
MacDuffee, C. C. *The Theory of Matrices*. New York: Chelsea, 1946. [**12**]
Marcus, M., and Minc, H. *A Survey of Matrix Theory and Matrix Inequalities*. Boston: Allyn and Bacon, 1964. [**1–10**, **12**, **15**]
Mirsky, L. *An Introduction to Linear Algebra*. Oxford: Oxford Univ. (Clarendon), 1955. [**1–7**]
Ortega, J. M., and Rheinboldt, W. C. *Iterative Solution of Nonlinear Equations in Several Variables*. New York: Academic Press, 1970. [**15**]
Rellich, F. *Perturbation Theory of Eigenvalue Problems*. New York: Gordon and Breach, 1969. [**11**]
Russell, D. L. *Mathematics of Finite-Dimensional Control Systems*. New York: Marcel Dekker, 1979. [**9**]
Varga, R. S. *Matrix Iterative Analysis*. Englewood Cliffs, N.J.: Prentice-Hall, 1962. [**15**]
Wilkinson, J. H. *The Algebraic Eigenvalue Problem*. Oxford: Oxford Univ. (Clarendon), 1965. [**10**]
Wonham, W. M. *Linear Multivariable Control: A Geometric Approach*. Berlin: Springer-Verlag, 1979. [**9**]

Index

A

Abel's formula, 346
Absolute norm, 367–370, 387
Adjoint matrix, 42, 274
  reduced, 272
Adjoint transformation, 169–174
Admissible pairs, 212–217
Admissible triple, for a matrix polynomial, 494, 519
Algebra, of linear transformations, 121
Algebraic multiplicity, 159, 239, 279, 349
Algebraically closed field, 556
Analytic perturbations, 391–405
Analytic function, 331
Angle, 106
Annihilating polynomial, 223–229
Appollonius's Theorem, 107
Approximate solutions of $Ax = \boldsymbol{b}$, 436–438
Augmented matrix, 57

B

Ball, 355, 356, 557
Barnett factorization, 458
Barnett, S., 560
Basis, 83
  standard, 84
  elements, 84
Bauer, F. L., 368, 387
Baumgärtel, H., 560
Bellman, R., 560
Bendixson, I., 379
Ben-Israel, A., 560
Berman, A., 560
Bessel's inequality, 109
Bezout matrix, 454–466
Bezoutian, 454–466
Bilinear, 104
Bilinear forms, 202
Binary operation, 72
  closed, 72
Binet–Cauchy formula, 39, 97
Biorthogonal systems, 113, 154, 195
Block matrix, 17, 55
Boolean sum of projectors, 199
Borhardt, C., 487
Bouillon, T., 560
Bounded set, 557

C

Campbell, S. L., 560
Canonical set of Jordan chains, 502
Cardinal polynomials, 308
Carlson, D., 448
Cartesian decomposition, 179, 219
Cauchy index, 482
  theorem, 332
Cauchy–Schwarz inequality, 42, 106, 114, 381
Cayley transform, 219
Cayley–Hamilton theorem, 165, 228, 252
Characteristic polynomial, 155–159, 228, 240, 271, 490
Chebyshev polynomial, 166
Chen, C. T., 449

Cholesky factorization, 63, 180
Chrystal's theorem, 276
Circulant matrix, 66, 146, 166
  generalized, 66
Closed set, 557
Cofactor, 33, 42
Col, 457
Column space, 80, 93
Common divisor, 554
  multiple, 555
Commuting linear transformations, 143
Commuting matrices, 166, 416–420
  normal, 420
  simple, 420
Companion matrix, 32, 36, 68, 69, 157, 262, 455
  of a matrix polynomial, 280, 490, 493
  second, 461, 490, 493, 497
Compatible norms, 360, 363, 367
Complementary subspace, 91
  orthogonal, 111
Complexification, 200
Component matrices, 314–322, 332–334, 347
  analyticity of, 393–395
Composition of transformations, 121
Condition number, 385–387
  spectral, 386
Conformable matrices, 9
Congruent matrices, 184–188, 203
Conjugate bilinear forms, 202
Consistent equations, 100
Constituent matrix, 164
Constrained eigenvalues, 291–294
Constrained eigenvectors, 291–294
Continuous function, 558
Control function, 342
Controllable pair, 216, 448–450, 470
Controllable subspace, 216, 342
Controllable system, 342–345
Convergence
  in a normed linear space, 356, 357
  in $\mathbb{C}^{m \times n}$, 325, 361
Convex hull, 559
Convex linear combination, 559
Convex set, 356
Courant–Fischer theorem, 289, 300
Courant, R., 288
Cramer's rule, 24, 60
Cyclic subspace, 231

D

Daleckii, J. L., 445, 560
Decomposable matrix, 273
Defect, of a linear transformation, 135–138, 172
Definite matrix, 179
Degree
  of a λ-matrix, 247
  of nilpotency, 12
Derivative of a matrix, 330
Derogatory matrix, 240
Determinant, 23 *et seq.*, 157, 184, 373, 379
  bordered, 65
  derivative of, 65
  of a linear transformation, 141
Diagonal, of a matrix, 25
  main, 2
  secondary, 2
Diagonal matrix, 2
Diagonally dominant matrix, 373, 531
Difference equations, 277–278, 340, 512–516
Differential equations, 161–163, 274–277, 334–340, 348, 506–612, 525
  inhomogeneous, 162, 338–340, 348, 509–512
  initial-value problem, 336–340, 510–512
  matrix, 346, 348
  steady-state solution, 340
Dimension, 86
Direct sum
  of linear transformations, 144
  of matrices, 146
Directed graph of a matrix, 528
Distance, 354
Divisor (right and left), of a matrix polynomial, 249–253, 518
Doubly stochastic matrix, 21
Dual norm, 381
Dual system, 345
Dunford, N., 560

E

Eigenbasis, 151, 173
  orthonormal, 174, 190, 193
Eigenspace, 149, 159, 175
Eigenvalue
  dominant, 361
  of a linear transformation, 148
  of a matrix, 152

Eigenvectors
left, 154
linear independence of, 150, 153
of a linear transformation, 148
of a matrix, 152
Elementary divisors, 266–271
of a function of a matrix, 313
linear and nonlinear, 266
Elementary matrices, 48, 255
Elementary operations, 47, 253
Equivalence relation, 52, 130, 184, 220, 255, 354, 494
Equivalence transformation on matrix polynomials, 255, 491
Equivalent matrices, 52
Equivalent matrix polynomials, 255, 261, 279
Equivalent norms, 353
Euclidean norm, 351, 352, 358, 378
Euclidean space, 105
Exponential function, 308, 313, 323, 329, 382
Extension
of a linear transformation, 142
of a nonnegative irreducible matrix, 535

F

Fibonacci matrix, 35
Field of values, 283–286
Fischer, E., 288
Flanders, H., 422
Flockerzi, D., 537
Fractional powers of a matrix, 323
Fredholm alternative, 174, 219
Fredholm theorem, 115
Frobenius, G., 295, 299, 415, 481, 482, 538
Fujiwara, M., 462
Full rank, 96, 111, 139, 140, 180
Function defined on the spectrum, 305
Function of a matrix, 308
composite, 324
Fundamental polynomials, 308

G

Gantmacher, F. R., 561
Gauss reduction, 49, 56
Generalized eigenspace, 229, 232–236, 239
order of, 229
Generalized eigenvector, 229
order of, 229–231
Generalized inverse, 428–438
Geometric multiplicity, 159, 239, 279
Geršgorin disks, 372
Geršgorin theorem, 371–377
Glazman, I. M., 561
Gohberg, I., 489, 561
Golub, G. H., 561
Gram matrix, 110, 218
Gramian, 110
Gram–Schmidt process, 108
Greatest common divisor, 460, 554
Greville, T. N. E., 560
Grimshaw, M. E., 561
Gundelfinger's criterion, for inertia of a Hermitian matrix, 298, 481, 482

H

Hadamard's inequality, 65
Hahn, W., 450
Halmos, P. R., 561
Hamburger, M. L., 561
Hankel matrix, 66
Heinig, G., 561
Hermite, C., 462
Hermite interpolation, 306
Hermitian forms, 203
Hermitian matrix, 3, 15, 63, 178–180, 184–188, 284–302
Hill, D. R., 452
Hirsch, A., 378
Hölder inequality, 359, 381
Hölder norms, 351, 358, 381
Homogeneous equations, 61
Householder, A. S., 387, 561
Householder transformation, 190
Hurwitz, A., 478
Hurwitz matrix, 478

I

Idempotent matrix, 11, 13, 133, 158, 164, 166, 194–199, 244, 279, 426
Image
of a linear transformation, 133
of a matrix, 77, 80
of a subspace, 86, 133
of a transformation, 116

Imaginary part of a matrix, 179
Imprimitive matrices, 544–547
Inconsistent equations, 56
Indecomposable matrix, 273
Index
  of an eigenvalue, 226, 228, 240, 269
  of imprimitivity, 544
  of stabilization, 214
Induced matrix norms, 363–367
Inertia of a matrix, 186–188, 296–300
  with respect to the real axis, 450
Inertia of a polynomial, 462–466
  with respect to the real axis, 462
  with respect to the unit circle, 466
Inf, 558
Inner product, 7, 104, 180, 186
  standard, 104
Integral of a matrix, 330
Interpolatory polynomials, 306–308
Invariant polynomials, of a matrix polynomial, 261
Invariant subspace, 143, 155, 172, 197, 201, 374
  trivial, 143
Inverse
  left, 45, 138
  of a linear transformation, 141
  of a matrix, 44, 53
  right, 45, 139
Invertible λ-matrix, 247
Invertible linear transformation, 140, 151
Invertible matrix, 44, 53, 425
Involutory matrix, 12, 44
Irreducible matrix, 374–377, 528–542
Isomorphic spaces, 113, 124, 126
Isomorphism of algebras, 126

J

Jacobi criteron, for inertia of a Hermitian matrix, 296
Jacobi, G., 487
Jacobi identity, 346
Jacobi matrix, 35
Jordan basis, 232, 234
Jordan block, 237, 244, 268, 274, 311
Jordan canonical form, 237, 239, 270, 311
  for real matrices, 242–243
Jordan chain, 230, 235
  for a matrix polynomial, 501–505, 519
  length of, 230
Jordan decomposition, 236
Jordan pair, 500–505
Jordan subspace, 230, 231
Jordan triple, 500–505

K

Kato, T., 400, 561
Kernel
  of a linear transformation, 135
  of a matrix, 77
Krein, M. G., 445, 560, 561
Kronecker delta, 79
Kronecker product, 407–413, 438, 439
  eigenvalues of, 412
  for rectangular matrices, 413, 438, 440
Kronecker sum, 412

L

λ-matrix, 246 *et seq. See also* Matrix polynomial
L–C splitting of a polynomial, 470, 485
LU decomposition, 61
Lagrange interpolating polynomial, 307
Lancaster, P., 468, 489, 561
Langer, H., 521
Laplace's theorem, 37
Latent roots, 265, 280, 281, 491
Latent vectors, 280, 501
Laurent expansion, 333
Leading principal minors, 36
Leading vector of a Jordan chain, 501
Least squares solution, 436
Left invertible linear transformation, 138
Left invertible matrices, 140, 425–427, 439
Legendre polynomials, 114
Length of a vector, 4, 105, 106
Lerer, L., 561
Lévy–Desplanques–Hadamard theorem, 373
Liénard–Chipart criterion, 470–474, 455, 486
Limit point, 557
Line segment, 559
Linear combination, 78
Linear dependence, 81
Linear hull, 79
Linear independence, 82
Linear space, 73
  finite dimensional, 84, 86
  infinite dimensional, 84, 86

Linear transformation, 117
invertible, 140
Linearization, 492 *et seq.*, 520, 524
Liouville's formula, 346
Liubich, J. I., 561
Logarithmic function, 312, 313, 317
Lower bound, 369, 370, 387
Lyapunov equation, 416, 443–450
Lyapunov stability criterion, 443

M

M-matrices, 531, 532
MacDuffee, C. C., 415, 561
Marcus, M., 561
Markov, A. A., 476
Markov chains, 550–552
Markov criterion, 475–477, 486
Markov parameters, 474, 477, 478, 483
Matrix, 1
Matrix equations, 413–424
Matrix norm, 358, 381, 382
induced, 363–367
Matrix polynomial, 246 *et seq.*, 489 *et seq.*
monic, 248
Metric, 355
Meyer, C. D., 560
Minc, H., 561
Minimal polynomial, 224, 240, 245, 271–273
Minkowski inequality, 351
Minor of a matrix, 32, 36
Mirsky, L., 561
Monotone matrices, 531
Monotonic norm, 367
Moore, E. H., 432
Moore–Penrose inverse, 432–440
Multiples of matrix polynomials, 518–520
Multiplicity
of a latent root, 279. *See also* Algebraic multiplicity, Geometric multiplicity
of a zero of a polynomial, 555

N

Naimark, M. A., 561
Natural frequencies, 210
Natural normal form (first and second), 264, 269
Negative definite matrix, 179
Negative semidefinite matrix, 179
Neighborhood, 356, 557
Newton sums, 487
Nilpotent matrix, 11, 133, 158, 244
Nonderogatory matrix, 240, 273
Nonnegative matrix, 527
Nonsingular matrix, 43, 375
Norm
euclidean, 351, 352, 358, 378
Frobenius, 358
Hölder, 351, 358, 381
infinity, 351, 352
of a linear transformation, 363
of a matrix, 358
of a vector, 105, 106, 350 *et seq.*
submultiplicative, 358
Normal coordinates, 210
Normal linear transformation, 174–177
Normal matrices, 174–177, 183, 284, 365, 390
analytic perturbation of, 404
commuting, 420
Normal modes, 210
Normalized vector, 106
Normed linear space, 351
Nullspace of a matrix, 77

O

Observable pair, 214–217, 341, 497
Observable system, 340
Odell, P., 560
Open set, 557
Ortega, J. M., 530, 561
Orthogonal complement, 111, 195
Orthogonal matrix, 46, 175, 201, 219, 346
Orthogonal projection of a vector, 112
Orthogonal projector, 195, 380, 431
Orthogonal similarity, 175
Orthogonal subspaces, 111
complementary, 111
Orthogonal system, 107
Orthogonal vectors, 107
Orthonormal basis, 108
Orthonormal system, 108
Ostrowski, A., 445
Outer product, 9

P

Parallelogram theorem, 107
Parseval's relation, 109, 112, 113

Partition of a matrix, 16
Penrose, R., 432
Perlis, S., 553
Permutation matrix, 64
Perron, O., 538, 542
Perron–Frobenius theorem, 536
Perturbation coefficients, 397–399
Perturbation of eigenvalues, 387–405
Plemmons, R. J., 560
Polar decomposition, 190, 380, 439
Polynomial in a matrix, 19
Polynomials, 553–556
Position vector, 4, 72
Positive definite matrices, 179–182, 185, 218, 219, 309, 373, 382
  Kronecker product of, 413
Positive matrix, 527
Positive semidefinite matrices, 179–182, 185, 218
Power method, 361
Preimage of a vector, 137
Primitive matrices, 544–548
Principal minors, 36
Projectors, 164, 194–199, 315, 321, 333, 426, 430
  commuting, 199
Puiseux series, 392, 401
Pythagoras's theorem, 107

Q

QR decomposition, 111, 184
Quadratic forms, 203
Quotient (right and left), 249–253

R

Range of a function, 558
Range of a matrix, 77
Rank decomposition, 97, 433
Rank
  of a linear transformation, 133–136, 172
  of a matrix, 53 *et seq.*, 93, 114, 115, 129
  of a matrix polynomial, 259–262
Rayleigh quotient, 282, 286–294, 363
Rayleigh theorem, 294
Reachable vector, 342, 345
Real part of a matrix, 179
Reduced adjoint, 272, 274, 334
Reducible matrices, 374–377, 543, 544
  normal form for, 544
Reducing subspaces, 197–198
Regular λ-matrix, 247, 259, 489
Relative perturbation, 383
Relatively prime polynomials, 554
Rellich, F., 400, 405, 561
Remainder (right and left), 249–253
Remainder theorem, 251–253
Representation
  of a linear transformation, 122, 140, 145
  of a set of vectors, 92, 95
  of a vector, 85
  standard, 123
  theorem for matrix polynomials, 516
Resolution of the identity, 315
Resolvent, 164, 315, 321, 322, 330–333, 505
  of a matrix polynomial, 493, 505
Resolvent condition, 386
Resonance, 212
Restriction of a linear transformation, 142
Resultant matrix, 460, 461
Rheinboldt, W. C., 530, 561
Right-invertible linear transformation, 139
Right-invertible matrices, 140, 425–427, 439
Rodman, L., 489, 561
Rost, K., 561
Rotation matrix, 455
Roth, W. E., 422
Routh–Hurwitz problem, 464–466
Routh–Hurwitz stability test, 480–482
Row echelon form, 50, 57
Row space, 80, 94
Russell, D. L., 561

S

Scalar matrix, 2, 133
Scalar multiplication, 6, 72
Scalar product, 7
Schneider, H., 445, 448
Schur–Cohn problem, 466–468
Schur complement, 46, 56
Schur, I., 165, 176, 377
Schwartz, J. T., 560
Schwarz matrix, 450
Segre characteristic, 241, 492
Semi-definite matrix, 179
Sequences of matrices, 325–327
Series of matrices, 327–329
Sherman–Morison formula, 64

Shift operator, 277
Signature, 187
Similar admissible triples, 494
Similar matrices, 130–133, 175, 262–271
Simple linear transformation, 146, 151, 164, 173, 177
Simple matrices, 143, 146, 147, 153, 154, 160, 239, 271, 273, 419-
  eigenvalues of, 387–390
  Kronecker product of, 413
Singular bases, 435, 437
Singular matrix, 43, 94, 155
Singular-value decomposition, 192
Singular values, of a matrix, 182–184, 192–193, 380, 386, 435
Size of a matrix, 1
Skew-Hermitian matrix, 16, 180
Skew-symmetric matrix, 14, 219, 346
Small oscillations, 208–212, 302, 303
Smith canonical form, 261, 262
Solvents of matrix polynomials, 252, 280, 520–526
  complete set of, 524
Span, 79
Spectral norm, 365, 366, 367, 381
Spectral radius, 359, 365, 377, 530–532, 539, 540
Spectral resolution, 314–320
Spectral theorem, 154, 164, 175
Spectrum, 150, 152, 178, 188, 331
  of a matrix polynomial, 281, 491
Sphere, 355
Square root of a positive (semi-) definite matrix, 180, 309, 324
Stable matrix, 414, 416, 441
  with respect to the unit circle, 451–453
Stable polynomial, 462, 465
Standard pair, 497
Standard triple, 494, 519, 520
State vector, 340
Stein equation, 451–453
Stephanos, C., 411
Stieltjes matrices, 532
Stochastic matrices, 21, 547–552
  doubly, 550
Stoer, J., 368
Storey, C., 560
Strongly connected directed graph, 528
Submatrix, 16
Subspace, 75, 105
  complementary, 91
  trivial, 76
Sum of subspaces, 87
  direct, 89, 112
  orthogonal, 112
Sup, 558
Sylvester, J. J., 450, 460
Sylvester's law of inertia, 188, 204
Symmetric matrix, 3, 14
Symmetric partition, 17
Symmetrizer
  of a matrix polynomial, 493
  of a polynomial, 455
Systems of algebraic equations, 56

T

Taussky, O., 375, 376, 445, 452
Tensor product, 407
Tismenetsky, M., 468, 561
Toeplitz matrix, 68, 69, 79
Toeplitz, O., 176
Trace of a matrix, 22, 107, 132, 157, 166
Transfer function, 70
Transformation, 116
  addition and scalar multiplication, 120
  zero, 120
Transition matrix, 98
Transposed matrix, 13
Triangle inequality, 351
Triangular matrix, 2, 44, 176
Tridiagonal matrix, 35

U

Unimodular λ-matrix, 247, 255, 268
Union of bases, 89
Unit sphere, 283
Unit vectors, 79
Unitarily similar matrices, 175
Unitary equivalence, 193
Unitary matrix, 47, 115, 174, 184, 188–190, 219, 346, 348
Unitary similarity, 175, 176, 365
Unitary space, 105, 351
Unobservable subspace, 214

V

Van Loan, C. F., 561

Vandermonde matrix, 35, 66, 69, 307, 524
  generalized, 70, 281, 307
Varga, R. S., 530, 561
Vec-function, 409, 410
Vector, 3
  position, 4
Vibrating systems, 208, 302

W

Well-conditioned equations, 383
Wielandt, H., 450, 539
Wilkinson, J. H., 561
Wimmer, H. K., 422, 452
Witzgall, C., 368
Wonham, M., 561

ISBN 0-12-435560-9

9 780124 355606 90065